PRODUCTION/OPERATIONS MANAGEMENT

From the Inside Out

PRODUCTION/OPERATIONS MANAGEMENT

From the Inside Out

Fifth Edition

ROGER W. SCHMENNER

School of Business
Indiana University

PRENTICE HALL
Englewood Cliffs, New Jersey 07632

Library of Congress Cataloging in Publication Data

Schmenner, Roger W.,
    Production/operations management: from the inside out / Roger
W. Schmenner. — 5th ed.
      p.   cm.
    Includes bibliographical references and index.
    ISBN 0-02-406871-3
    1. Production management — Case studies.   I. Title.
TS155.S322 1993
658.5 — dc20

92-6956
CIP

Editor: Charles E. Stewart
Production Supervisor: Helen Wallace
Production Manager: Su Levine
Cover Designer: Robert Vega
Cover photograph: Tim Alt/Digital Art
This book was set in Trump Medieval by Carlisle Communications.

© 1993 by Prentice-Hall, Inc.
A Simon & Schuster Company
Englewood Cliffs, New Jersey 07632

Earlier edition copyright © 1981, 1984, and 1987 by Science Research Associates,
Inc. Copyright 1990 by Macmillan Publishing Company.

ISBN   0-02-406871-3

Prentice-Hall International (UK) Limited, *London*
Prentice-Hall of Australia Pty. Limited, *Sydney*
Prentice-Hall Canada Inc., *Toronto*
Prentice-Hall Hispanoamericana, S.A., *Mexico*
Prentice-Hall of India Private Limited, *New Delhi*
Prentice-Hall of Japan, Inc., *Tokyo*
Simon & Schuster Asia Pte. Ltd., *Singapore*
Editora Prentice-Hall do Brasil, Ltda., *Rio de Janeiro*

To my father, a helluva operations manager

PREFACE

The reception that this book has received in all four of its previous editions has been very gratifying. With this fifth edition, I have once again had the opportunity to approach anew the original vision that I had for the book. I saw it as unifying the best elements of two schools of thought about production/operations management (POM): one anchored in the application of various techniques to POM problems; the other anchored in the dilemmas and decision making of the operating manager. With this edition, I am closer to that original vision.

The fifth edition has undergone a major overhaul. A recounting of the changes is instructive:

1. *Reorganization.* The plant and service tours that formerly began this text are now positioned at the back. Furthermore, the text has been rewritten so that references to the Tours do not presuppose the student having read even one of them. This restructuring of the book thus gives instructors the leeway to assign whatever topics they may want, and in whatever order that makes sense to them. Although I recommend the order I have chosen for the material, the use of the book no longer depends on that order.

   Where there used to be thirteen chapters in addition to the Tours, there are now fifteen, with some of the longer chapters split into more digestible pieces. This revamping of the material has particularly helped the exposition of some of the more traditional topics, such as production planning, inventory control, queuing, and shop floor control. Some chapters even carry different names now.

2. *Theme.* The ordering of the material and the point of view stress the learning of operations management from a bottoms up, inside out approach. This stress is so consistent that the book has been retitled *Production/Operations Management: From the Inside Out* to underscore it.

2. *Summaries.* Key ideas and tools introduced in the text are recapitulated at the end of each chapter. These recaps also highlight the distinctive features of world-class operations.

4. *Fresh Material.* Considerable amounts of fresh and up-to-date material have been added to this fifth edition. There is a new first chapter that introduces the book and that discusses key aspects of different manufacturing and service processes. Chapters 2 and 3, dealing with flows, bottlenecks, balance (including line balance), queuing, and layout, have been thoroughly revamped. Chapter 4, on quality, has likewise undergone substantial change and rewriting.

   A thorough-going revision has been applied to Chapters 6 through 10 — those dealing with the important and traditional topics of production planning, scheduling, inventory management, MRP, purchasing, logistics, shop floor control, and project management. This material is now more easily assimilated, and the coverage is more complete as well.

Additional material has been added to the chapter on just-in-time manufacturing and to the chapter on technology and design. In fact, the revisions to the latter necessitated a retitling of that chapter.

In all of the other chapters, small but helpful changes have been made throughout, making the text both more readable, and in particular places, more quantitatively rigorous.

5. *Updated Plant and Service Tours.* All of the plant and service tours except one (Thalhimers) have been brought up-to-date as of the summer and fall of 1991. There is a new tour of the General Motors plant in Oklahoma City, currently GM's best-performing domestic plant. There is also a brand new tour of a **project** (Geupel DeMars), which does a better job than ever before to describe what project management entails in practice.

ACKNOWLEDGMENTS

This edition, as with the previous ones, has benefited greatly from the cooperation I have received from a number of companies. Although I cannot acknowledge the contributions of everyone in those companies, I would like to mention a few persons who were especially helpful with this edition: James Grippe of International Paper; Joseph and Lillian Gehret of Norcen Industries;

Edward Jecelin of Jos. A. Bank Clothiers; Thomas Desser, Michael Williamson, and colleagues at General Motors (C-P-C Group—Oklahoma City); Alicia Hayes and Larry Levensky of Burger King; Al Sanfilippo of Stroh Brewery; Gregg Hutchinson of Ogle-Tucker Buick; Frank Jonas of Arthur Andersen & Co.; David Moffitt of Geupel DeMars; Kenneth Good of Black & Decker; James Krejci of Navistar International; and Frank Coyne and Victor Baldridge. My sincere thanks to them and to their colleagues.

In addition, I would like to acknowledge the helpful comments and suggestions of the following reviewers for the fifth edition: Douglas Elvers, University of North Carolina, Chapel Hill; Phillip Carter, Michigan State University; Daniel E. Shimshak, University of Massachusetts, Boston; Susan W. Engelkemeyer, Babson College; Ernest Koenigeberg, University of California, Berkeley; James Fitzsimmons, University of Texas, Austin.

Some of the problems and situations for study from the earlier editions of this book are the work of Joseph R. Biggs and Thomas F. Callarman. They reduced the burden for me at an important time in this book's early development.

Linda Baker and Cynthia Panettieri helped me prepare aspects of the manuscript for this edition, and Margaret Noble tracked down numerous bibliographic references for me.

My wife, Barbie, and my children, Will and Andrew, were again most forgiving of my attachment to this book. I apologize for all the time I spent with it that I could have spent with them. I owe them much and hope that someday they may even read what all the fuss was about.

Roger W. Schmenner

BRIEF CONTENTS

CONTENTS

PRODUCTION/OPERATIONS MANAGEMENT

From the Inside Out

CHAPTER 1

THE SPECTRUM OF MANUFACTURING AND SERVICE

Operations. For many people, mentioning the word brings on a rush of disagreeable feelings and sensations. Details. Pressure. Long hours. Inhospitable working conditions. Dull colleagues. For these people, a company's operating managers — usually typed as engineers who could not hack it in the "more creative" design aspects of the business — lead very unromantic careers. Moreover, operations is viewed as a dead-end job with no future in the upper layers of management.

This description is an exaggeration, of course, but it conveys some of the impressions that people hold about operations. This text is a crusade against that kind of stereotype of the operations manager. In fact, operations offers an exciting and dynamic management challenge that is as absorbing and rewarding as any marketing or finance can offer. A badly run operation can be a staggering drain on corporate profits and morale, as a number of even large corporations can attest; conversely, a well-run operation can be a competitive weapon every bit as potent as any in the corporate arsenal.

For many senior managers, the path to the top snaked through operations — even for those whose climb has been in so-called "marketing companies." Many others would admit that a tour of duty in operations was an important part of their general management education and background. The point is that operations is and rightly should be a key concern for most companies and one that all students of business who aspire to general management responsibilities ought to know about.

The importance of operations has been highlighted in recent years. Numerous articles have appeared in business periodicals stressing the need for young managers, especially those fresh from business schools, to seek out careers in operations jobs. The success of the Japanese has reinforced this view in many people's eyes, since the prowess of the Japanese is widely recognized to be in manufacturing.

Our standard of living today owes much to our forebears' ability to create wealth. Fortunately for us, because of their efforts, we have become better and better at transforming raw materials of all sorts into useful, high-quality products. So, too, have we become better at delivering an ever-expanding menu of services to both businesses and final consumers. We have, over the years, become increasingly efficient at these tasks — wasting fewer resources, satisfying customers more completely, requiring less physical exertion from ourselves, and frequently,

1

although not always, being kinder to the environment in the process.

These are triumphs of the operations management function within the manufacturing and service businesses of our economy. Our productivity gains, as measured by history's trace of real (inflation-adjusted) per capita income, have snowballed. We have more equipment to work with than ever before; better, more capable equipment, but also better methods and more informed ideas about what it takes to produce effectively goods and services that are valued by customers. This triumph, of course, is a shared one. Product engineering, marketing, sales, and other company functions are, in part, responsible. Nevertheless, the manufacturing or service operation itself has been in the thick of the productivity advance whose fruits we enjoy today.

THE OPERATIONS FUNCTION

Although somewhat "invisible" to the marketplace, the operations function in a typical company accounts for well over half the employment and well over half the physical assets. That, in itself, makes the operations function important. In a company's organization chart, operations often enjoys parity with the other major business functions: marketing, sales, product engineering, finance, control (accounting), and human resources (personnel, labor relations). Sometimes, the operations function is organized as a single entity which stretches out across the entire company, but more often it is embedded in the distinct, typically product-defined divisions into which most major companies are organized.

In many service businesses, the operations function is typically more visible. Service businesses are often organized into many branches, often with geographic responsibilities—field offices, retail outlets. In such tiers of the organization, operations are paramount.

The operations function itself is often divided into two major groupings of tasks: line management and support services. Line management generally refers to those managers directly concerned with the manufacture of the product or the delivery of the service. They are the ones who are typically close enough to the product or service that they can "touch" it. Line management supervises the hourly, blue-collar workforce. In a manufacturing company, line management frequently extends to the stockroom (where materials, parts, and semi-finished products—termed "work-in-process" inventory—are stored), materials handling, the tool room, maintenance, the warehouse (where finished goods are stored), and distribution, as well as the so-called "factory floor." In a service operation, what is considered line management can broaden considerably. Often, order-taking roles, in addition to order-filling roles, are supervised by service line managers.

Support services for line management's operations can be numerous. Within a manufacturing environment, support services carry titles such as quality control, production planning and scheduling, purchasing, inventory control, production control (which determines the status of jobs in the factory and what to do about jobs that may have fallen behind schedule), industrial engineering (which is work methods-oriented), manufacturing engineering (which is hardware-oriented), on-going product engineering, and field service. In a service environment, some of the same roles are played but sometimes under vastly different names.

Thus, the managers for whom operational issues are central can hold a variety of titles. In manufacturing, the titles can range from vice-president—manufacturing, works manager, plant manager, and similar titles at the top of the hierarchy, through such titles as manufacturing or production manager, general superintendent, department manager, materials manager, director of quality control, and down to general foreman or foreman. Within service businesses, "operations manager" is sometimes used but frequently the title is more general—business manager, branch manager, retail manager, and so on.

OPERATIONS: FROM THE INSIDE OUT

The title of this textbook, *Production/Operations Management: From the Inside Out*, highlights a theme that pervades the structure and philosophy of this book. Most managers, as they grow in their jobs, learn about their production processes from the bottom up. The decisions they take start with small scale, incremental improvements to existing processes. Only as those managers rise in the ranks do they then become involved in larger, more costly, more long-term, and more strategic decisions. That is, first they master their processes and then they learn how to make an ever-widening circle of decisions about them. Their appreciation for and knowledge of operations management grows *from the inside out.*

This is a natural way to learn about operations. Before one can plan, execute, or control anything about an operation, one must learn about it and how to improve it bit by bit. Similarly, if one were brought into a troubled company and asked to turn it around, one would have to start with the simple, incremental improvements before there would be the cash or vision to change the company in major, strategic ways. One starts *from the inside and works out.*

If you have seen one factory, you certainly have *not* seen them all. The same holds true for service operations. Manufacturing and service processes are wonderfully diverse. For the manufacturer, not only do machinery and product engineering differ, but also the ways in which the factory is organized and run. Similar diversity prevails in services, which can differ markedly—hospitals versus grocery stores versus bank check clearing houses versus interior decorators.

This book begins with an overview of various kinds of manufacturing and service processes. Moreover, the entire last segment of this book describes, in some detail, a collection of ten very distinct manufacturing and service operations that provide an in-depth appreciation for "process." These various tours can serve as useful groundings in the character of different operations and can be used to illustrate any number of the concepts and techniques that comprise the study of operations management. While the use of this text does not require reading these tours, and certainly not all of them, you may well want to examine at least a few of them to get an "armchair" taste of the factory or service operation to anchor your appreciation for what follows in this text.

Once the various processes are discussed, the text proceeds to examine existing operations and how to improve them. The issues dealt with are the basic building blocks of good operations decisions: analysis of materials and information flows, breaking bottlenecks, achieving balance, ameliorating quality, improving methods, empowering people, planning production, scheduling it, controlling production and inventories, and more.

Only when these basic concepts are discussed does the text proceed to the larger, more strategic issues of adding capacity, taking advantage of technology, introducing new products, restructuring supply relationships, and devising an operations strategy.

With this background on this text's structure and philosophy, let us proceed to examine the diversity of both manufacturing and service operations and the kinds of decisions that can be made concerning them.

PART ONE

MANUFACTURING
The Spectrum of
Manufacturing Processes

The factory tours in the back of this text describe a rich diversity of technologies and approaches to a variety of the issues faced by manufacturing managers. The summary chapter at the conclusion of the tours compares and contrasts the processes in considerable detail. Suffice it here to present a much more "bare bones" discussion of five major types of manufacturing processes that one can engage in. They form a spectrum (see Figure 1-1).

1. *The Project.*   The project—building a skyscraper, bringing a new product out of the R & D labs, making a movie—is sometimes excluded from consideration as a manufacturing process because, by its nature, it exists only to do a particular, unique job. This is not to say, however, that one project may not look very similar to another. Many projects, be they large buildings, new products, or movies, require much the same work to get them up and running. There are certain regularities present in a project, among them:

   • Projects typically make heavy use of certain skills and capabilities at particular

times and little or no use of those skills at other times.

   • Projects can often usefully be seen as the coordination of part-time or subcontracted skills and capabilities (people, equipment, etc.). Skills and capabilities, of course, can be many and varied.

   • The coordination of those skills and capabilities requires a lot of attention to planning and scheduling, and subsequently, attention to the control of that schedule, and any rescheduling or expediting that may be required. Great attention must be given to which tasks must be accomplished before others (a precedence diagram) and what the expected durations for those tasks are.

   • The quality of any project depends greatly on the skills and care of the members of the project team.

2. *The Continuous Flow Process.* At the other extreme of the process spectrum lies the continuous flow process. Many high volume consumer goods and commodities are made by continuous flow processes—oil refining, food processing, papermaking, lightbulb fabrication. The continuous flow process's most significant characteristic is how materials move through it—hardly ever stopping, moving constantly from one process operation to another. With a continu-

Project   Job Shop   Batch Flow   Line Flow   Continuous Flow

FIGURE 1-1   The Spectrum of Manufacturing Processes.

ous flow process, one can estimate realistically how long it takes to transform raw materials into a specific product. Work-in-process inventories exist at well-defined levels and are low relative to the value of output the continuous flow process generates. Capital investments and automation, on the other hand, are often higher than those of other processes, especially when contrasted with the workforce employed. Layouts are frequently product-specific, typically with a straight-line character to them, as the products in the making go from one operation to another.

Continuous flow processes can be very productive and very profitable, assuming normal sales levels. Only when sales levels plunge is the profitability of the continuous flow process in jeopardy.

3. *The Job Shop.* Lying next to the project in the process spectrum is the job shop. It is the most flexible process for creating a wide variety of products in significant quantities (the project, almost by definition, does not produce in quantity). Machine shops, tool and die shops, and many plastic molding operations are job shops, working to fulfill particular customer orders.

The job shop layout is often distinctly different from that of a continuous flow process; it groups similar equipment together, primarily because no single product generates enough sales volume to justify the creation of a product-specific array of equipment.[a] Often a job shop has a diverse array of equipment and capabilities to choose from.

The flow of material in a job shop can be complex and far from a straight line in character. Materials can be routed in many directions and can loop back to the same equipment later in the processing cycle. With each order (job) capable of such complexity, it is absolutely essential that information on how the order is to be routed through the factory, what is to be done to it at each step of the way, and how much time and effort is actually spent on it, follow the job. The job shop lives by its information flows. This information is vital, because job shops typically bid for work. Without good information on costs, times (run times, set up times, labor content times), routings, and process steps, a job shop would be seriously disadvantaged.

4. *The Batch Flow Process.* One step toward the continuous flow process from the job shop is the batch flow process. The job shop and the batch flow process have a good deal in common. Their layouts are similar, with equipment grouped by function rather than by product.[b] The product is regarded as moving from department to department within the factory. A batch flow operation depends on information such as routings and process steps and tracks costs and times spent. However, batch flow processes typically have a set menu of products that they produce, frequently, in set quantities (lot sizes). The batch flow operation is thus somewhat more standardized than the job shop, particularly as it relates to routings and costs.

[a] As will be described later, sometimes families of parts or products can be identified and exist in enough volume to justify the creation of a manufacturing cell (sometimes called "group technology") within the job shop. Such a manufacturing cell gathers various, different pieces of equipment together in a product family-specific configuration.

[b] Here, again, the innovation of the manufacturing cell is applicable. Defining manufacturing cells for the batch flow process rather than the job shop is likely to be much easier to do because the identification of a family of parts is likely to be easier.

While the job shop usually operates to fulfill an outside customer's order by an agreed upon due date and in whatever quantity is ordered, the batch flow operation usually produces in established lot sizes that move into an inventory from which further production proceeds or customer orders are filled. Batch flow processes are commonplace, especially when one considers all the times "fabrication" must be done. Examples of batch flow processes include much of the chemical industry, semi-conductor fabrication, apparel, much of the steel industry, and huge chunks of the metal bending, metal forming, and metal machining industries.

5. *The Line Flow Process.*   Between the batch flow and continuous flow processes, along the process spectrum, lies the line flow process. In reality it lies closer to the continuous flow process because it presents some substantive distinctions from the batch flow. The line flow process is most popularly exemplified by the moving assembly line that one finds in the auto industry, but it is also found in a host of other assembly industries such as consumer electronics and computers. In contrast to the batch flow process, the line flow process exhibits the following charactertistics:

- A product-specific layout with different pieces of equipment placed in sequence ready to perform operations on the product. There are, of course, mixed model lines that can produce distinctly different models of the basic product, but the more diverse the products made, the less satisfactory the line becomes at producing those.

- The product moves readily from one operation to another so that there is little work in process inventory, nor is there a stockroom in the product's path. This also means that there is a great need to examine the "balance" of the process so that the different tasks to be accomplished take roughly the same amount of time to perform and have the same capacities, not just over weeks of time, but over minutes of time.

- The paperwork needs of the line flow process are less demanding than the batch flow. Routings are not needed and operations sheets can frequently be simplified, if not eliminated altogether. The need for tracking labor and machine inputs to particular products/parts also fades away.

- In contrast to the continuous flow operation, the line flow is somewhat more flexible, and generally less automated and more labor intensive.

As one proceeds across the spectrum from project to continuous flow, one tends to move from a highly individualized, flexible process to one that is much more inflexible in the products it can make but, at the same time, much more productive and efficient in how it makes them.

Hybrid Processes

The five process types introduced above— project, job shop, batch flow, line flow, and continuous flow—are all "pure." Many factories are combinations of two (sometimes more) of these pure processes. Popular hybrids are the batch flow-line flow hybrid (auto engines, air conditioning, furniture) and the batch flow-continuous flow hybrid (breweries; many high-volume consumer products whose raw materials are made in batches, such as photographic film).

In these processes the first part of the flow of materials looks like a batch flow process (often, this part of the process is labelled "fabrication") while the latter part resembles a line or continuous flow process (and this part of the process is labelled "assembly" or "finishing").

Importantly, the two portions of the hybrid are separated by an inventory, typically termed a "decoupling" inventory. The batch flow process acts to fill up the inventory with parts or semi-finished product which then is drawn down by the line or continuous flow process for assembly or completion.

The reason the hybrid process is divided into two parts is that the batch flow process is not normally as nimble as the line flow or continuous flow process. The batch flow may not be as nimble because significant chunks of time may be needed to set up the existing machines for a different component of the finished product. This puts pressure on the batch flow process for longer runs than would be needed to match precisely the product mix and quantities produced by the line or continuous flow process, which normally can change over to other products more quickly. If the batch flow process tried to match the line or continuous flow process precisely (say, hourly or daily), it would lose a lot of time to setup and this downtime could rob the process of the capacity it needs to keep up with overall demand.[c] Thus, the batch flow process does not attempt a precise match of the line or continuous flow process's product mix and quantity, but rather a quantity and product mix match over a much longer period of time, say weeks or months. The batch flow process then acts to replenish the decoupling inventory, while the line or continuous flow process acts to fill particular customer orders.

MANUFACTURING CHOICES

Manufacturing managers are faced with a wide variety of choices, and how well they make

[c] Much of the push toward just-in-time manufacturing (JIT), discussed in Chapter 11, can be interpreted as a remedy for this deficiency of batch flow processes. In essence, JIT tries to make the batch flow process operate with the efficiency of the line flow process.

those choices, in large part, determines how successful they and their companies are. These operations choices define, in essential ways, the study of operations management. The remainder of this book is devoted to exploring how these choices are made and how they can be made better. Three broad categories confront the manager: (1) technology and facilities, (2) operating policies, and (3) operations organization. Let us review these categories and choices in turn by noting a number of questions about them. These questions point to some of the alternatives managers face. Do not be discouraged if you do not know exactly what these questions refer to. The remainder of this book should clarify things for you. Rather, let these questions "wash over" you to give you a feel for what operations managers face.

Technology and Facilities

These choices frequently involve large capital expenditures and long periods of time. These are the big decisions that do much to define the type of process employed.

1. *Nature of the process flow.* Is the flow of product through the plant characterized as rigid, with every product treated in the same way? At the other extreme, is the flow a jumbled one, with products routed in many different ways through the factory? Or, does the flow of product through the process fall somewhere in between?

   - Are segments of the process tightly linked to one another, or are the connections between process segments loose? How quickly can materials flow through the process?

   - Is the process a "pure" one or is it a hybrid of different process types?

2. *Vertical integration.* How much of a product's value is a direct result of factory operations? Should production involve more (or

less) integration backward toward raw materials or more (or less) integration forward toward customers? Can it involve more (or less) of either of these?

3. *Types of equipment.* Is the equipment used general purpose in design or special purpose? Can any special-purpose, seemingly inflexible equipment be linked together in innovative ways to yield production systems that are themselves more flexible than the equipment comprising them?

   • Is the equipment meant for high speeds and long runs, or not? How flexible is it for changeovers in products/models? How quickly and easily can such changes be accomplished?

   • What possibilities exist for linking one piece of equipment with another for balanced and quick throughput?

   • Is the equipment operator-controlled or automatically/computer-controlled? Is its performance monitored by operators or by computer?

   • Can the equipment be speeded up or slowed down to match production needs?

   • Does the equipment demand substantial nonoperation support (maintenance, repair, software, setup, tooling)?

   • Can the company build or modify its own machines? Does it have close ties to equipment manufacturers?

4. *Degree of capital or labor intensity.* To what degree has the equipment and/or technology permitted labor value to be driven out of the product or service? How important could reduced labor value in the product be to costs, yields, or sustained levels of "production to specifications" (good quality)?

5. *Attitude toward the process technology.* To what extent does the company pioneer advances in process technology? Is the company a leader in that regard or a follower? How closely does it track other process improvements?

   • How committed is the company to research and development? What mix is sought between the two?

   • How close is the alliance between manufacturing and engineering? To what extent are efforts made to design products so that they are easily manufactured? How are design and manufacturing teams organized for new product introductions?

   • Are engineering change orders (changes to a product's manufacturing specifications) numerous? To what can they be attributed? How disruptive to the process flow are they?

   • What investments are made in manufacturing engineering and industrial engineering relative to product design engineering?

6. *Attitude toward capacity utilization.* How close to capacity (defined as best as possible) does the company desire to operate? How willing is the company to stock out of a product because of too tight capacity?

   • Does capacity generally come in significant chunks? What can be done to keep capacity well balanced among segments of the process?

7. *Plant size.* How big does the company permit any one plant to grow? To what extent are either economies or diseconomies of scale present?

8. *Plant charters and locations.* Does it make sense to assign different product lines, product processes, or geographic markets to particular plants? How do the plant locations chosen mesh together into a multiplant operation?

Operating Policies

Once the process technology and facilities have been selected, management must still decide on a host of features concerning how the process technology is used. Three broad segments of such operating policies present themselves: loading the factory, controlling the movement of goods through the factory, and distribution.

Loading the Factory

1. *Forecasting.* To what degree is the plant's output mix known with certainty before raw materials must be gathered or equipment or personnel assigned? To what extent must forecasts be relied on to determine which raw materials ought to be ordered and which equipment/worker capacity reserved, and how much of each?

   • How reliable have past forecasts been? Does manufacturing second-guess marketing's demand forecasts? Should it?

   • What techniques work best for forecasting requirements? Is there a long product pipeline that risks pipeline momentum problems and heightens forecasting needs?

2. *Purchasing.* Given the decision on the plant's degree of vertical integration, what does the plant make for itself and what is purchased from outside suppliers? How are suppliers chosen? What kinds of contracts (e.g., long-term or spot) are sought?

   • Is purchasing formally integrated with forecasting and/or order taking? How much visibility is granted suppliers about the company's expected future needs? Are orders on suppliers made through an MRP system (a special computer-based materials management system [see Chapter 7]) or other informal means? Or are formal purchase requisitions required?

   • How important is supplier quality compared to price in determining how suppliers are chosen? How is supplier quality monitored? How well does this monitoring work? What supplier improvement programs, if needed, have been adopted?

3. *Supply logistics.* How frequently, from where, and by what mode of transportation do raw materials arrive at the plant? How sensitive are costs to changes in these factors?

   • How readily can materials be received, inspected, and stored? Is supplier quality good enough to dispense with incoming inspection? Is supplier delivery reliable enough and frequent enough to be able to feed the plant's production needs without large raw materials inventories?

   • How is materials handling accomplished within the plant? How much is automatic and how much is manual? What controls are used?

4. *Raw materials inventory system.* How much inventory of raw materials is held? What system is used (material requirements planning, reorder point, periodic reorder [see Chapters 7 and 8])? How does the inventory level vary with demand, price discounts, supplier lead time changes, supply uncertainties, or other factors? What triggers the replenishment of the raw materials inventory?

   • How are materials controlled in the stockroom? Are records accurate enough to be relied on without ceasing production for physical inventories? Is all scrap reported?

   • Are materials placed into "kits," or are they fed directly to the shop or line?

5. *Production planning.* Are goods manufactured to customer order or to forecasts, or are they manufactured to stock some finished goods inventory? Is inventory permitted to be built in advance to cover peak period

demands and thus smooth out production, or does production try to "chase" demand, with little or no buildup of inventories?

- How are workforce needs planned for? How are model or product line changes planned for? How far in advance can changes be routinely made to the general production plan? How disruptive to the process is any expediting or rescheduling?

- Are routine allowances made for such situations?

Controlling Movement Through the Factory

1. *Production scheduling and inventory control.* What triggers the specific production of goods: orders, forecasts, or reference to a finished goods inventory? How do factors like the pattern of demand and product costs influence any trigger level of finished goods inventory? How are specific departments, lines, work centers, and the like scheduled? What factors of the process, the pattern of demand, or product variations and costs affect the schedule?

   - What determines specific priorities in the jobs awaiting work in a department? How much expediting is permitted? How much rescheduling?

   - What or who determines the specific priorities of work: an MRP system, foreman discretion, specific rules, simulation results?

   - Does instituting an MRP system make sense for the plant? How much fluctuation or stability exists for bills of materials, vendor reliability on delivery, production cycle times? What levels of accuracy and integrity exist for inventory counts and records, for production/rework counts and records for scrap? What kinds of exception reports and follow-through stand to be most helpful?

2. *Pacing of production.* Is the pace of production determined by machine setting, worker effort, management pressure or discretion, or some combination? How readily can the pace be altered?

3. *Production control.* How much information, and of what type, flows within the production process, both from management to the workforce and from the workforce to management?

   - How easily can product variations, engineering change orders, product mix changes, or production volume changes be transmitted to the workforce? How soon can management react to machine breakdowns, parts shortages, or other disruptions of the normal flow of product and process on the plant floor?

   - How are machines, workers, materials, and orders monitored? What "early warning signals" are looked to? What remedies are routinely investigated?

4. *Quality control.* Does everyone in the organization truly believe that quality (performance to specification) is his or her job, not simply the function of a quality control department? What mechanisms and cooperative forums exist to assure that "work is done right the first time"? How closely linked are design engineering, manufacturing/industrial engineering, quality control, workforce training and supervision, maintenance, and production scheduling and control?

   - How is quality checked, both of process and product? How many checks are made at different steps in the process? How much authority is given to quality control· personnel?

5. *Workforce policies.* What are the skill levels required of various jobs throughout the process? How are they trained for? Is the

work content of jobs broad or narrow? Is cross-training of workers desirable? How do workers advance in the factory (by job classification, different jobs, changes of shift, into management)?

- How and how much are workers paid? Are any incentives, wage or otherwise, built into the process? How are workers' achievements and ideas encouraged and recognized?

- How does management feel about unionization? What actions concerning unions does it take?

- What is the age and sex composition of the workforce?

- What opportunities exist for job enlargement and job enrichment? Is there a place for quality of work life projects? How else can the workforce be encouraged to participate in the management of the operation?

Distribution

1. *Distribution.* What are the channels of distribution? How are they filled? What are the trade-offs between service and inventory costs?

2. *Logistics.* What are the benefits and costs of various geographical patterns of warehousing and distribution? What modes of transportation make sense? How should they be managed?

Operations Organization

As important as are the decisions about the physical composition of the factory and the systems in place to guide and control the movement of materials are decisions on how the factory is to be organized.

1. *Operations control.* Are the major operating decisions retained centrally or dispersed to individual plant units? What kinds of decisions rest primarily with the plant?

- How is the plant evaluated? What biases might that method introduce? What are the specific measures for the various levels of hierarchy in the plant and throughout the manufacturing organization?

2. *Talent.* Where within the organization are the best people put? What talents are most prized for the smooth and continued successful operation of the process?

THE PRODUCT-PROCESS MATRIX

Another useful way to visualize some of the similarities and differences among different types of production processes is to array the processes within what has been called the product-process matrix.[1] This matrix is nothing fancier than a box, one side of which describes the varieties of product mixes that are possible and another side of which describes the process patterns (the ease of flow of product through the process) that are possible. Figure 1-2 portrays this.

As we can see in Figure 1-2, the mix of products handled by the plant may range from a one-of-a-kind item (such as a work of art or a custom-designed house) to a very standard product that is produced and consumed in high volumes (such as sugar). In between, a company can produce many different products, each at relatively low volumes, or it can cut down on the number of different products offered but produce higher volumes of each.

As we might expect, management faces different sets of challenges in bringing each of these different product mixes to market. For example, a commodity item undoubtedly has to compete on price. While there may be a limited number of grades of the commodity, the

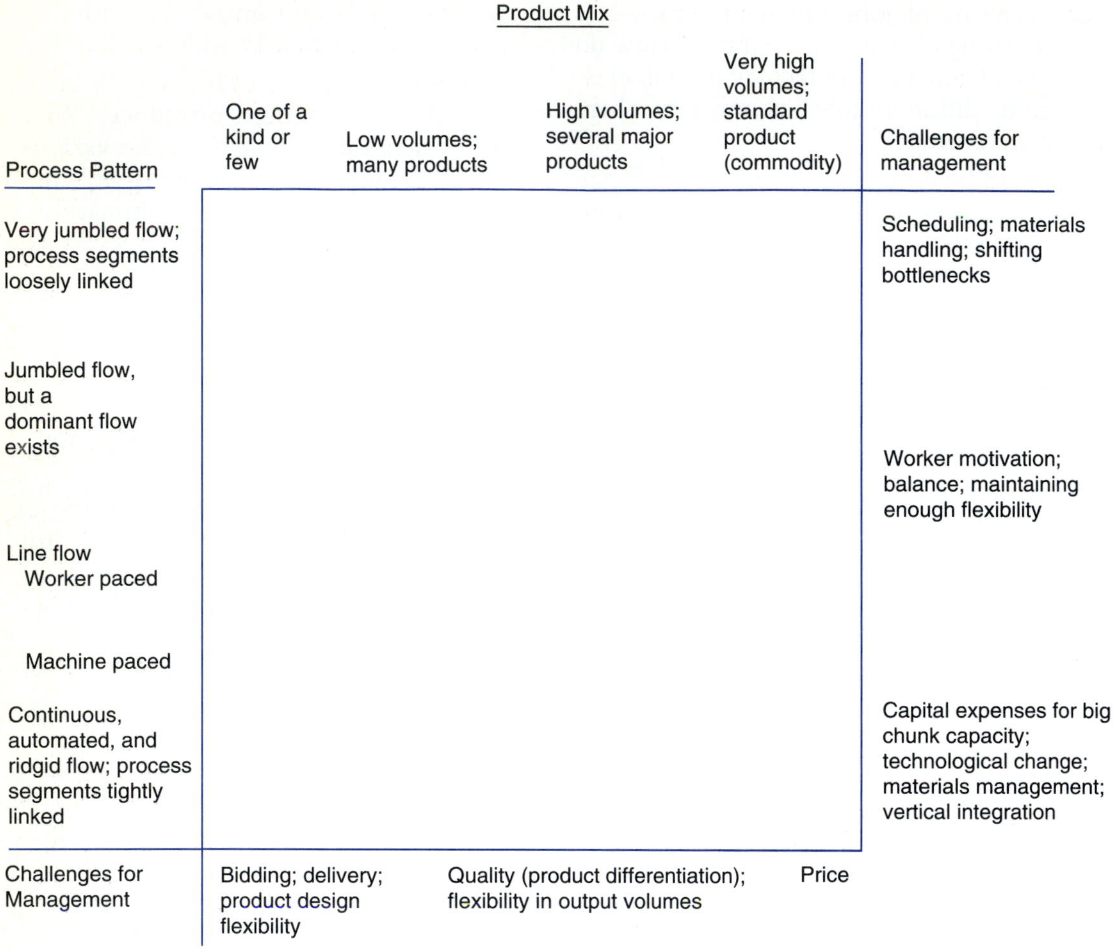

Product Mix

Process Pattern	One of a kind or few	Low volumes; many products	High volumes; several major products	Very high volumes; standard product (commodity)	Challenges for management
Very jumbled flow; process segments loosely linked					Scheduling; materials handling; shifting bottlenecks
Jumbled flow, but a dominant flow exists					Worker motivation; balance; maintaining enough flexibility

Line flow
  Worker paced
 	
  Machine paced					

Continuous, automated, and ridgid flow; process segments tightly linked					Capital expenses for big chunk capacity; technological change; materials management; vertical integration
Challenges for Management	Bidding; delivery; product design flexibility	Quality (product differentiation); flexibility in output volumes		Price	

FIGURE 1-2   A product-process matrix.

ability to offer the item at low cost is absolutely critical to company success, whether the company produces paper clips or paper.

At the other end of this product mix spectrum, the one-of-a-kind item is frequently purchased with scarcely a thought about price. Since it is a custom item, the buyer is likely to have contracted for it beforehand. Thereafter, the buyer typically becomes concerned about whether the product meets specifications and whether it can be delivered on time. These concerns place a different set of demands on the manufacturer. Instead of paying near-exclusive attention to product cost, the manufacturer must pay particular attention to product design flexibility, and delivery, since the manufacturer's success depends more on these features of the production system. Because most work is contracted for, management must be accomplished at developing bids for each, which entails thorough knowledge of costs within the factory.

In between these two extremes, management confronts somewhat different challenges. When many different products—each with low production volumes—are made or when fewer products—each with substantial production volumes—are made, product features such as quality or the ability to gear up (or down) rapidly in production become relatively more important than before. Product quality is a key element in differentiating one's product from others available, and is a chief reason why some products command price premiums. Flexibility in production volumes, while less evident to the marketplace than quality, is often a reason why some companies have been so successful—either beating their competitors to the market with seasonal, cyclical, or faddish products (for example, building supplies or fashions) or turning off production when the market for a product appears to have dried up. Depending on the situation, product design flexibility, price, and delivery can still be very important to the manufacturer; but it is likely that as product design flexibility or quality rises in importance, price falls in importance.

These then are the chief challenges that various product mixes pose for their managers. A corresponding set of management challenges derive from the process patterns depicted in Figure 1-2. As the matrix reveals, the process pattern varies from a very jumbled flow to a continuous, automated, and rigid production flow. The linking of segments of the production process varies from loose to tight. The process patterns in between exhibit varying degrees of rigidity, pacing, and process segment linking.

For the jumbled process pattern, the challenges to management include scheduling, routing materials efficiently, and coping with shifting production bottlenecks. At the other end of the process pattern spectrum, the continuous, automated, and rigid process flow demands increased attention to such items as capital expenses for large-scale capacity increases, technological change in the process, materials management, and vertical integration. Between the process poles, factors like worker motivation, balance of capacities for different process segments, and process flexibility become relatively more important.

The arrays of product mix, process patterns, and their respective challenges to management line the outside of the product-process matrix. Now to fill in the inside. As our brief survey of production processes has revealed, different process types follow different process patterns and are also better suited to particular product mixes. It is easy, then, to match process types to the appropriate position within the matrix, as Figure 1-3 does.

As is readily apparent from the chart, process types should fill out the diagonal of the product-process matrix, from projects and job shops in the upper left corner down to continuous flow processes in the lower right corner. Only by being on the diagonal do processes match properly the process pattern to the prevailing product mix.

Consider what it means to be off the diagonal. Suppose, for example, that a company tries to manufacture low volumes of many different products by using a continuous, automated, and rigid flow process that is typically capital intensive and vertically integrated. The match is clearly inappropriate since the process would have to be interrupted constantly and retooled often to permit the kind of product flexibility needed to produce low numbers of lots of products. Not only is the match inappropriate, but it is expensive. The purchase prices of the automated machines and the cost of product changeovers would be staggering, involving a great deal of out-of-pocket expense.

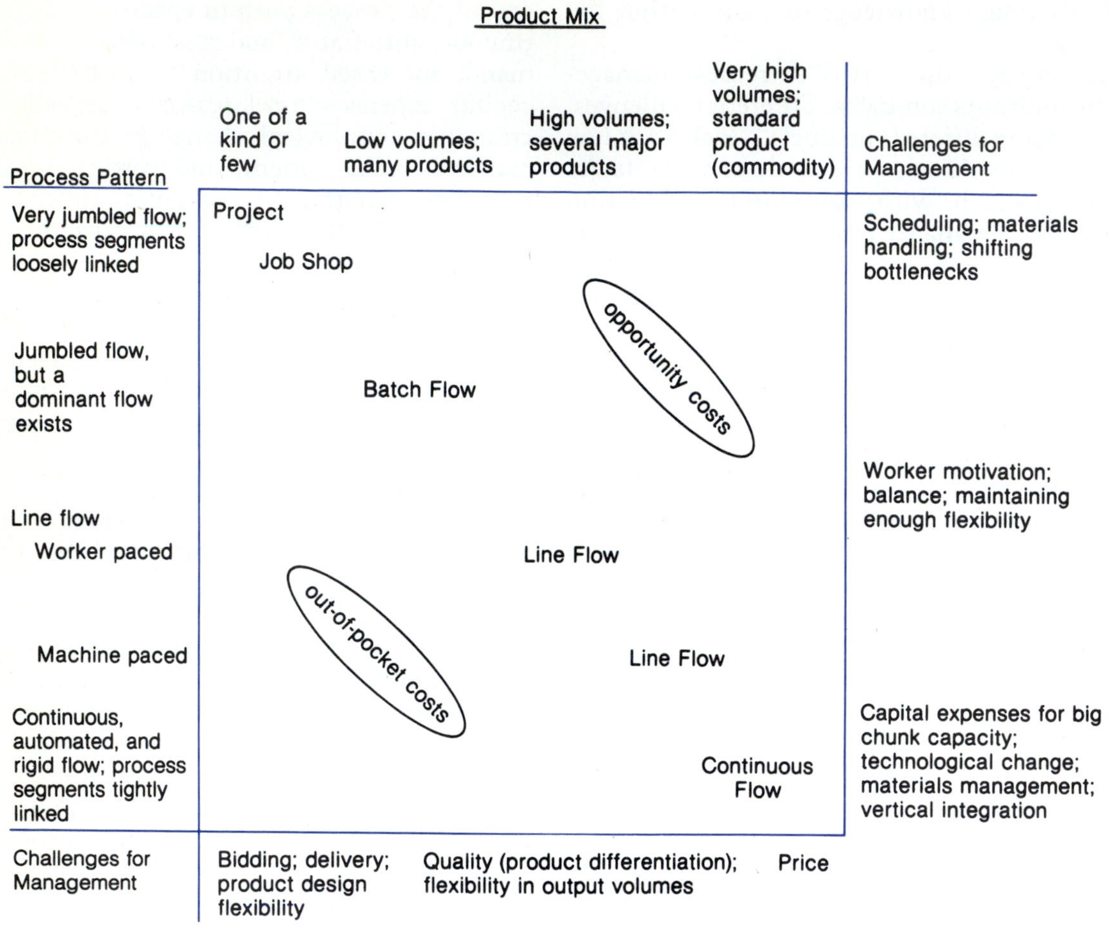

Product Mix

Process Pattern	One of a kind or few	Low volumes; many products	High volumes; several major products	Very high volumes; standard product (commodity)	Challenges for Management
Very jumbled flow; process segments loosely linked	Project Job Shop				Scheduling; materials handling; shifting bottlenecks
Jumbled flow, but a dominant flow exists		Batch Flow	*opportunity costs*		
Line flow Worker paced			Line Flow		Worker motivation; balance; maintaining enough flexibility
Machine paced		*out-of-pocket costs*		Line Flow	
Continuous, automated, and rigid flow; process segments tightly linked				Continuous Flow	Capital expenses for big chunk capacity; technological change; materials management; vertical integration
Challenges for Management	Bidding; delivery; product design flexibility	Quality (product differentiation); flexibility in output volumes		Price	

FIGURE 1-3 The product-process matrix, filled in.

Being above the diagonal brings on costs of a different kind. In this case, suppose a very standard product consumed in great quantity were manufactured in a jumbled flow process pattern. Here again the match is clearly inappropriate, though not because too many dollars go out-of-pocket to buy expensive machinery. Rather, operating costs (e.g., labor, inventories) are much higher than they could be, which means that the profit margin on each unit of the product made is much lower than it should be. By not substituting specific, special purpose machinery for high labor input and general-purpose equipment, the company forgoes profits that it would otherwise earn.

Opportunity Cost

This phenomenon goes by the name of "opportunity cost," since by not investing in more equipment and a more rigid production flow, the company is forgoing the opportunity to earn increased profits. These costs are every bit as real as the payment of dollars out-of-pocket. This notion of opportunity cost can be reinforced by a famous saying of Benjamin Franklin, no slouch himself at operations management. To make the point, however, we must make a brief excursion into logic. One truth of logic is the validity of the so-called contrapositive, which says simply that if the statement "If A, then B" is true, then it is also true that "If not B, then not A." That is, if every time A occurs B follows, then we can be sure that if B does not occur, then A did not occur as well. Enough of logic then, and back to Ben Franklin.

One of his Poor Richard sayings is that "A penny saved is a penny earned." We have all recognized the truth of that since childhood, but I assert that by this saying Ben showed us he knows everything about opportunity cost. After all, what is the contrapositive of "A penny saved is a penny earned"? A penny not earned is a penny not saved (i.e., a penny spent). All we are saying by this notion of opportunity cost is that "a penny not earned (an opportunity forgone) is a penny spent." We shall often have occasion to consider opportunity costs, as well as out-of-pocket costs, in analyzing and deciding various operations issues.

PART TWO

SERVICE OPERATIONS

Distingushing Service Operations from Manufacturing

Six of the ten tours described at the back of this textbook deal strictly with manufacturing operations. The other four, however, reveal significant distinctions between service and manufacturing. Some of the management challenges for service operations parallel those in manufacturing, but others are unique.

What is a service operation anyway? The answer to this question is surprisingly ambiguous. We may know service operations when we see them but have difficulty describing them in general. Perhaps the term can be better understood as what it isn't rather than what it is. Some have tried to define service employment as nonfarming and nonmanufacturing employment; others have also excluded government employment. This approach includes operations such as hotels, restaurants, repairs, amusements, health, education, real estate, wholesale and retail trade, transportation, and professional services like law, engineering, architecture, finance, and advertising. Although these selections may be intuitively clear, there are still some definitional problems. For example, what about public utilities? Some public utilities companies have service in their name (such as Public Service of New Hampshire), yet in many ways power generation is more manufacturing than service. And is a company like IBM a manufacturing company or a service company? As in the fable of the six blind men and the elephant, the answer may depend on what part of the beast you are touching. Surely there are aspects of IBM that are strictly manufacturing, but there are other aspects that are service (examples are field repair and the sales force). There are even aspects of IBM that are caught in-between, such a R & D.

This ambiguity is growing. As technology advances, more and more labor will be driven out of manufactured products; that is, there will be less direct labor involved in manufacturing companies and more indirect labor—the kind of labor that is oriented to support services. Even now, some high technology companies have overhead rates that are six to ten times the direct labor component, mainly because of the substantial indirect labor in these companies. The traditional management focus on direct labor is becoming less and less relevant. Service operations management affects more than just service companies per se. Furthermore, the service sector will continue to grow. About two-thirds of America's gross national product is now accounted for by services; the same fraction applies to the percentage of the workforce employed in service jobs. These fractions can only rise. Similar trends are observed in other developed countries.

Characteristics of Service Operations

Most services share certain characteristics to a greater or lesser extent. It is helpful to understand these characteristics and how they separate manufacturing from service.

1. The service provided is often something that the consumer cannot touch or feel. It may be associated with something physical (such as the food we eat, the airplane we fly in, the life insurance policy), but what is valued about the service and may be the focus of management typically involves intangibles, such as ambiance, information, or peace of mind.

2. The service is often created and delivered on the spot, in many cases with significant involvement of the customer in the service process (consider salad bars in restaurants). Because the process is often more on display in service operations than in manufacturing, whether it functions well is critical. The process cannot be saved by a quality control check at the end; it has to live with its defects. Therefore, training, process design, and employee relations are especially important to service industries.

3. Because of the visibility of the service process in many instances and the intangibility of many services, operations management and marketing are more interdependent than in manufacturing. Marketing and operations have to work together in service companies; they cannot afford to be antagonists.

4. Frequently, the consumption of a service is nearly simultaneous with its production. Services cannot be inventoried for use later on. This fact has some serious implications for capacity choice and capacity management in a service business. The site, size, and layout choices for service industries are critical. The wrong site, a size either too large or too small, or a poor layout can dramatically affect the performance of a service unit. This concern for capacity is heightened by the irregularities in the pattern of demand over time (whether a day, week, or season). Demand irregularities place a tremendous burden on a service operation to be flexible. They also force services to manage their demand via prices or the kinds of services offered.

5. Many service operations, although by no means all, require little in the way of capital investment, multiple locations, or proprietary technology. For these services, barriers to the entry of competing firms are rather low. Service companies are generally very sensitive to the real—or even potential—entry of others and must react quickly to the competition's actions and threats. Many service operations also possess rather low

barriers to exit; that is, the assets of a service organization can be sold easily (such as planes or trucks). Because firms can get in and out of the service business quickly, some service strategies can play for the "hit and run."

SERVICE OPERATIONS CHOICES

Part One of this chapter opened by reviewing a whole series of operations choices faced by manufacturing managers. Service operations managers, of course, face a similar list of choices. Table 1-1 outlines some major ones.

Analyzing Service Operations

In analyzing a service operation, it is useful to distinguish three notions: the service task, the service level, and the service delivery system. The service task refers to what a business provides the consumer; it defines the fundamental

---

TABLE 1-1   Key choices for service operations

I. Technology and Facilities

Type of facility and equipment (Used by customers themselves? How? How attractive do they have to be? How are they used during a peak versus a nonpeak situation? General-purpose versus special-purpose equipment? Anything proprietary about equipment or its use?)

Layout (Job shop-like? Fixed position? Other?)

Location (How critical to attracting customers?)

Size of the facility

Geographic spread of any multiple facilities

Degree of labor intensity

Attitude toward capacity utilization (Facility for the peak demand or not?)

II. Operating Policies

A. Planning the operation

Forecasting (Extent required? Type used?)

Logistics and inventory system used for materials employed

Manpower planning

Schedule setting (Can service provision be "leveled" in any way?)

Demand management for peak and off-peak times

B. Controlling the operation

Labor issues—hiring, training, skill levels required, job content, pay, advancement, unionism

Accounting controls used

Checklists developed

Foolproofing designed into the layout and the equipment

Quality control audits and policies

What triggers provision of the service and the pace of the operation (Customer? Forecast?)

Production control (How does the information flow within the operation? What is on track? What is not? How can anything gone amiss be fixed? How can any changes be implemented?)

III. Operations Organization

What is kept at the individual unit level, and what is centralized?

Where is talent held in the organization?

nature of the business. To get at the service task for a particular operation, we can ask, What is it about the service that the consumer values? With so much of service being intangible, the answer to this question must be clear to both management and the workforce. As was said before, in a service business, marketing and operations must cooperate. Operations cannot ignore marketing's insight into why the consumer values the service.

A service level is any measure that can be applied to the process to indicate its quality. The time standards at Burger King are examples of these measures. The service level is the guardian of quality for the service firm. Because, in most cases, a service cannot be inspected before it is provided, there must be clear checks on whether the process itself is working well.

The service delivery system is concerned with producing and delivering the service to the consumer. The service tours discussed in the back of this book are, at root, descriptions of the key elements of service delivery systems. How orders are taken, how the service is provided, how any materials are controlled are all aspects of the service delivery system. The challenge of service operations management is to devise a service delivery system that supports the service task and maintains designated service levels.

A Spectrum of Services

The following characterization of services as service factories, service shops, mass service, and professional service can be used for comparing service processes in much the same way that we compared processes across the manufacturing spectrum. This is accomplished in Table 1-2. The various features compared there are placed into various groups: service, process,

customer-orientation, labor, and management. These features undergird various challenges for management.

Much as we have thought of manufacturing enterprises in terms of different kinds of production processes, we can think of service operations in terms of distinct "processes." It is helpful to view these different service processes in terms of a matrix that contrasts the labor intensity of the process on the one hand, and the degree of interaction with and customization of the service for the consumer on the other.[2] This matrix is shown in Figure 1-4. The quadrants of the matrix roughly define four reasonably distinct service processes.

The Service Factory. Some service processes have relatively low labor intensity (and thus a greater fraction of service costs associated with the facility and its equipment) and also a low degree of customer interaction and customization. These can be characterized as service factories. Much of the transportation industry, hotels, and resorts, for example, are service factories, as are "back-of-the-house" operations for banking and financial services companies. The Burger King Restaurant in Noblesville, Indiana (Tour F) is an example of a service factory.

The Service Shop. As the degree of interaction with—or customization for—the consumer increases, the service factory gives way to the service shop, much as a line flow operation gives way to a job shop operation when customization is required in manufacturing. Hospitals and repair services of all types are prime examples of service shops. Tour G of Ogle-Tucker Buick auto repair operation is an example of a service shop.

Mass Service. Mass service processes have a high degree of labor intensity, but a rather low

TABLE 1-2   A service comparison

	SERVICE FACTORY (Example: Burger King Restaurant)	SERVICE SHOP (Example: Ogle-Tucker Buick)	MASS SERVICE (Example: Thalhimers— Cloverleaf)	PROFESSIONAL SERVICE (Example: Arthur Andersen & Co.)
Service Features				
Mix of services	Limited.	Diverse.	Limited.	Diverse.
Products compete largely on:	Price, speed, perceived "warmth" or "excitement."	Wide choice, competence.	Price, choice, perceived "warmth" or "excitement."	Competence, range of expertise.
New or unique services introduced or performed	Infrequent.	Routine.	Limited experimentation.	Routine.
Process Features				
Capital intensity	High.	High.	Low.	Low.
Pattern of process	Rigid.	Adaptable.	Rigid.	Very loose.
Ties to equipment	Integral part of process, little choice applies.	Equipment important to process, but usually several options exist for its use.	Limited ties to equipment, more tied to plant and layout.	No close ties to plant or equipment.
Importance of balance of tasks and any equipment to smooth process functioning	Balance critical.	Balance often not critical.	Balance not critical.	Balance can be critical.
Tolerance for excess capacity	Excess capacity abhorred.	Excess capacity often not a problem.	Excess capacity implies workforce adjustment that is fairly easily made.	Excess capacity abhorred.
Ease of scheduling	Sometimes tough to schedule, peak demand can be difficult.	Scheduling more easily done.	Scheduling easily done.	Sometimes tough to schedule, peak demand can be difficult.
Economies of scale	Some.	Some—permits better equipment use and thus justification.	Few, if any, except those related to any inventories.	Few, if any, although some specialization can occur.
Notion of capacity	Fairly clearcut, sometimes definable in physical terms.	Fuzzy, very dependent on mix of demands. Only definable in dollar terms.	Not as fuzzy as with service shop. Limits are often due to plant, not processing time.	Fuzzy.

continued

TABLE 1-2   A service comparison (*continued*)

	SERVICE FACTORY (Example: Burger King Restaurant)	SERVICE SHOP (Example: Ogle-Tucker Buick)	MASS SERVICE (Example: Thalhimers— Cloverleaf)	PROFESSIONAL SERVICE (Example: Arthur Andersen & Co.)
Process Features (*continued*)				
Layout	Line flow-like preferred.	Job-shop or fixed postion.	Typically fixed position although layout may change frequently, customers move through layout.	Job-shop frequently.
Additions to capacity	Can be in variable increments, requires balance of captial and labor.	Can be in variable increments, aspects of balance more murky.	Often takes big changes to plant to enact. Processing can sometimes be sped up by adding some labor.	Means adding primarily to labor in incremental fashion.
Bottlenecks	Occasionally movable, but often predictable.	Movable, frequent.	Typically well known, predictable.	Can somtimes be forecast, but otherwise are uncertain.
Nature of process change	Sometimes routine (rebalance), sometimes radical (new equipment).	Occasionally radical (new equipment and procedures).	Process change seldom occurs, although it can be radical (such as big change to plant).	Mostly incremental.
Importance of material flow to service provision	Both inventories and flow are important.	Inventories important but not so much the flow.	Inventories are often important and must be controlled.	Incidental to most services.
Customer-Oriented Features				
Importance of attractive physical surroundings to marketing of service	Can be critical.	Often insignificant.	Critical.	Often insignificant.
Interaction of customer with process	Little, brief.	Can be great.	Some.	Typically, very great.
Customization of service	Scant.	Significant.	Scant.	Significant.
East of managing demand for peaks and nonpeaks.	Can be done through price.	Some promotion of off-peak times can be done, but often difficult.	Same as for service shop.	Often very difficult to manage demand, may not be responsive to price.

TABLE 1-2   A service comparison (*continued*)

	SERVICE FACTORY (Example: Burger King Restaurant)	SERVICE SHOP (Example: Ogle-Tucker Buick)	MASS SERVICE (Example: Thalhimers— Cloverleaf)	PROFESSIONAL SERVICE (Example: Arthur Andersen & Co.)
Customer-Oriented Features (*continued*)				
Process quality control	Can be formal, amenable to standard methods (such as control charts).	Can be formal. Checkpoints can easily be established. Training can be critical to quality.	Mainly informal. Training critical to quality.	Mainly informal. Training critical to quality.
Labor-Related Features				
Pay	Typically hourly.	Varies, could include individual incentive or commision schemes.	Same as for service shop.	Salary, often with bonus of some type.
Skill levels	Generally lower skills.	High skills.	Variable, but most often lower skill.	Very high skills.
Job content	Small.	Large.	Often medium, but variable.	Very large.
Advancement	With more skills and/or seniority acquired, greater responsibility given. Seniority can lead to change in department or shift assignment.	Often, worker is an independent operator of sorts and can exert some control on what he gives and gets from job; limited hierarchical progression.	Often a hierarchy to progress upward through.	Often a pyramid, up or out. Top of pyramid exerts leverage over bottom of pyramid.
Management Features				
Staff-line needs	Large staff for process redesign, methods, forecasting, capacity planning, and scheduling. Line supervision and troubleshooting still critical.	Limited staff, mostly line operation.	Some staff, often focused on personal issues.	Limited staff, many line managers wear multiple hats.
Means of control	Variable. Can be cost or profit center.	Usually a profit center.	Usually a profit center.	Usually a profit center.

		Degree of Contact with, and Customization for, the Consumer		Challenges for Management
		Low	High	
Degree of Labor Intensity	Low	**The Service Factory** Airlines, trucking, hotels, resorts, and recreation	**The Service Shop** Hospitals, auto, and other repair services	Capital decisions, technological advances, managing demand to avoid peaks and to promote off-peaks, scheduling delivery of service
	High	**Mass Service** Retailing, wholesaling, schools	**Professional Service** Physicians, lawyers, accountants, architects	Hiring, training, methods development and control, employee welfare, scheduling workforces, control of often geographically spread locations, start-up of new units, managing growth
Challenges for Management		Marketing, making service "warm," attention to physical surroundings, managing fairly rigid hierarchy with need for standard operating procedures	Fighting cost increases, maintaining quality, reacting to customer intervention in process, managing advancement of people delivering service, managing flat hierarchy with loose subordinate-superior relationships, binding workers to the firm	

FIGURE 1-4   A matrix of service processes.

degree of interaction with or customization for the consumer. Retail operations, retail banking, schools, and wholesaling are examples of mass services. The Thalhimers department store at the Cloverleaf Mall, characterized in Tour H, is an example of mass service.

Professional Service. If the degree of interaction increases and/or customization become the watchword, mass service gives way to professional service, as provided by doctors, lawyers, consultants, architects, and the like. Arthur Andersen & Co.'s audit services in Charlotte (Tour I) provide an example of professional service.

Challenges to Management

Let us look at the challenges to management that are implied by labor intensity and interaction/customization differences.[3] The service operation that is more highly capital intensive has to look carefully at (1) its capital decisions regarding land, facilities, and equipment; and (2) technological advances that may affect them. Capital intensive processes often cannot easily augment capacity, so demand must be managed to smooth any demand peaks and to promote the offpeak times. The inflexibility of capacity also implies the scheduling service delivery is relatively more important for these process types.

For process types with high labor intensity, managing and controlling the workforce become paramount. Hiring, training, methods development and control, employee welfare, scheduling the workforce, and controlling what may be far-flung locations are critical. If new units of operations are contemplated, their startup may become a problem; managing the group of such new units can be difficult.

What about the challenges to management implied by differences in the interaction with and customization for the consumer? Those service processes that have a low degree of interaction and customization face a stiffer marketing challenge. They must try to make the service "warm" and "exciting" even though they may not give the personal attention that a customer might want. Attention to the physical surroundings and the layout become more important. With a low degree of interaction and with little customization, standard operating procedures can safely be instituted. The hierarchy of the service organization itself tends to be the classic pyramid, with fairly rigid relationships between levels.

As the service takes on a higher degree of interaction and customization, management must deal with higher costs and more-talented labor. Keeping costs down and yet maintaining quality becomes a challenge. Talented employees want attention given to how they may advance in the organization. The hierarchy of control tends to be flat, with much less rigid relationships between superiors and subordinates. Keeping workers bound to the firm rather than hopping from job to job becomes a challenge as well. Service firms with a high degree of consumer interaction also must react to frequent consumer intervention in the process.

Naturally, there are some gray areas as one passes from low to high on either dimension of this matrix. For example, while fast-food restaurants are probably best seen as service factories, the traditional restaurant is more problematic. Traditional restaurants offer a higher degree of interaction and customization for the consumer, and they are more labor intensive than fast-food restaurants. While they may be best characterized as service shops, they are fairly low in that quadrant, with at least some gourmet restaurants arguably characterized as professional services.

QUESTIONS

1. The Stroh brewery discussed in Tour E is an example of a batch/continuous flow hybrid. However, other kinds of hybrids are possible (such as batch/assembly line or job shop/batch). Give some examples of other hybrids. How would one of these other hybrids differ from the Stroh hybrid? How would they be similar?

2. Compare and contrast a "service shop" and a job shop. Compare and contrast a "service factory" and an assembly line.

3. Discuss some of the main challenges that management faces in bringing different product mixes to market. What other management challenges can you think of for a product mix that have not been included in the product–process matrix?

4. Examine some of the implications of the concept of opportunity cost for a company and its attitude toward profit. From your own experience, give some examples of the notion of opportunity cost.

5. The dimensions used for the product–process matrix were process pattern and product mix while the dimensions of the service matrix were degree of labor intensity and interaction with, and customization for, the consumer. How are these dimensions related to each other? How are they different from each other?

SITUATION FOR STUDY 1·1

DAVIS FARM—SPECIALTY VEAL OPERATION

There are three types of veal:

- Bob veal. This type of veal comes from calves less than 1 week old. The meat is pale, immature, tasteless, gelatinous, of little nutritional value, and completely unacceptable to all but the most uncultivated palates.

- Grain-fed veal. A more accurate description is puny beef. The calves are raised on a combination of milk, silage, and grass, and slaughtered at about 4 to 6 months. Though called veal, the meat is red and has a distinctive "beefy" flavor.

- Specialty veal. This type of veal is from calves raised on a strict all-milk or milk-replacer diet. It is distinguished by its light pink color, firm texture, consistent fat marbling, fitness for a wide variety of cooking techniques and recipes, and digestibility.

The Davis farm operated out of three barns: a 56-stall indoor barn, a 68-stall indoor barn, and a 116-stall outdoor enclosure (summers only). Roughly, raising veal calves is a four-step process:

1. The barns are prepared for the arrival of the calves.
2. A batch of 1- to 2-week-old bull calves is purchased and placed in pens in the barns.
3. The calves are fed, cleaned, and medicated twice daily for 12 to 16 weeks.
4. The calves are shipped to slaughter, and the process repeats.

Because young calves are highly susceptible to disease, the barns had to be as clean as possible when they arrived. Pens, floors, walls, and buckets were steam-cleaned with an antiseptic solution, and the feeding equipment was scrubbed and sterilized. To assure good ventilation, the dust filters and air screens on the vents were brushed clean and the fans were carefully checked.

Bull calves, the raw material of the veal industry, are a by-product of the dairy industry. Dairy cows must be freshened (i.e., must give birth to a calf) once a year to maintain milk production. Ideally, the cow gives birth to a heifer, which the farmer could raise to be another milk-producing cow. Half the time, however, the cow gives birth to a bull calf, which, except for the gain upon immediate sale, is useless to most dairy farmers. At an age of 1 to 2 weeks, these bull calves are taken to a regional commission sales barn and sold at auction to vealers and slaughterhouses. The ideal veal calf is healthy, and larger (weighing 90 to 120 pounds) than the average calf.

The calves purchased by Davis were trucked to the farm and penned. For the next 12 to 16 weeks, the calves would be fed, cleaned, and medicated every day at 5 A.M. and 5 P.M. The twice-daily feeding was the heart of the veal-raising process. The feed had a very high conversion ratio: on average, 1.4 pounds of the nonperishable powdered feed yielded 1 pound of growth. The feed was mostly dried milk and milk products, fortified with vitamins, antibiotics, and butterfat; the feed had a higher fat content than whole milk. It was shipped in 50-pound bags in loads of several tons, and stored on wooden pallets in the mixing room until needed. Generally, a batch of calves consumed 22 to 24 tons of the powered feed during their 12- to 16-week stay.

The milk was prepared by adding the powder to hot water and agitating the mix with an electric motor. After it had thoroughly dissolved, enough cold water was added to bring the temperature of the mix to between 100° and 105° F. The amount, type, and proportion of feed depended on the age of the batch of calves, the condition of the individual calves, and their willingness to drink. The feed was mixed in a 300-gallon bulk tank, into which water flowed from two 120-gallon drums suspended above it. A major problem with such a bulk feeding system was keeping it clean. Because calves are so susceptible to disease, it was vital that the tank, pump, and hoses be kept as clean as possible. These parts were sterilized by rinsing with hot water and hypochlorite bleach.

The normal mortality rate for bull calves (for all purposes, not just vealing) was 15 to 20 percent. This high figure was the result of both the calf's natural susceptibility and mistreatment at birth. After giving birth and before the milk comes in, a cow (or any other mammal) lactates colostrum that contains various antibodies and other enzymes that greatly improve the newborn's chances of survival. Unfortunately, many bull calves did not get colostrum because the suckling might damage the mother's teat. In addition, the calves were poorly handled at the auction house and shipped long distances to their final destinations. Usually, they got nothing to eat for several days. Davis could avoid the high mortality rate somewhat by buying only the larger, healthier calves. However, calves were always dehydrated and suffering from stress when they arrived at the barn. For most batches, 4 or 5 percent died within the first 48 hours.

The most common disease of these young calves was scours. A scouring calf has very loose feces with a distinctive odor and mottling. The cause of the disease can be either intestinal bacteria or a virus, augmented by stress. If untreated, scours lead to dehydration and death within 24 hours. Treatment consisted of massive amounts of intravenous fluids, electrolytes, and an oral antibiotic. To survive, scouring calves needed fluids three or four times daily. The most serious problem with scours, as with most other disease, was that it is highly contagious. Thus, tremendous emphasis was placed on cleanliness and observation, particularly during the first few days.

Respiratory diseases, especially pneumonia, were of particular concern at about the fourth week but could occur at any time. The animal with such a disease breathes rapidly and has a runny nose and fever. (A calf's normal temperature is 101.5° F. One gets concerned when it rises to about 103° F and panics when it goes over 104° F.) The treatment for this was an injectable antibiotic, normally a compound based on penicillin, streptomycin, or tetracycline. The veal farmer had hundreds of different drugs at his disposal and based his choice on the seriousness of the symptoms. For example, a high fever might require a tetracycline-based compound that was good for bringing down temperature. If the fever was only low grade, a penicillin compound with good long-term results could be used.

During the later days of a batch, a serious medical problem was bloat. Because of the complexity of their stomachs, calves cannot belch; any gas generated during digestion is either passed out through the intestines or absorbed. If gas generation occurs faster than the calf's natural ability to eliminate the gas, pressure builds up and slowly suffocates the animal. This condition is known as bloat. Treatment consisted of inserting a flexible tube about 5 feet down the calf's throat to permit the air to escape. If this did not work, a trocar had to be inserted into the calf's abdomen, and a tube into the rumen. Of course, the calf had to be given antibloat medication. In the case of a trocar insertion, the calf was given antibiotics as well to prevent peritonitis. A mortality rate of under 10 percent was considered good.

After feeding, all the feed buckets were hosed clean and suspended above the pens to dry. After that, the floors were washed to clean up the day's urine. Every 2 days or so, the pens were cleaned more thoroughly. Manure was scraped from under the pens and washed down a 4-inch pipe at the end of every gutter. This pipe carried the manure underground to a man-made manure lagoon (pond), where it decomposed naturally.

One hour after feeding, the calves had to be checked for bloat. At this time, any animal beginning to bloat could be easily recognized and fairly easily treated. In addition to the twice-daily bloat check, the calves were checked at lunchtime, midafternoon, and midnight. At the very beginning and very end of a batch, they were checked more often. Most of the time, these checks yielded no problems, or nothing more serious than a leaky faucet. Sometimes, though, a problem with a calf appeared—bloat, a broken stall, the beginnings of pneumonia maybe. Such a problem had to be solved immediately, or watched closely, or else it would be much more serious by the next feeding.

Calves could be shipped to market any time between the twelfth and sixteenth weeks. At this point, the cost of the feed began to exceed the benefit of resultant weight gain. This 3-week period allowed some leeway for finding the best price. A contract negotiated with a slaughterhouse specified the price (per pound of carcass) and the delivery day.

1. As described, this is a batch flow process. What would it take to transform it into a line flow process?
2. What advantages or disadvantages to such a transformation can you identify? Be specific.

REFERENCE NOTES

1.   The concept of the product-process matrix owes much to Robert H. Hayes and Steven C. Wheelwright. See the two articles by Hayes and Wheelwright in the January–February and March–April 1979 issues of the *Harvard Business Review* entitled "Link Manufacturing Process and Product Life Cycles" and "The Dynamics of Process-Product Life Cycles," respectively.

2.   See Roger W. Schmenner, "How Can Service Businesses Survive and Prosper," *Sloan Management Review*, Spring 1986, 21–32.

3.   Additional information on the management of service operations can be found in the following:

Aaker, David A., "How Will the Japanese Compete in Retail Services," *California Management Review* 33, no. 1 (Fall 1990): 54–67.

Albrecht, Karl, *At America's Service: How Corporations Can Revolutionize the Way They Treat Their Customers*, Homewood, IL: Dow Jones-Irwin, 1988.

_____ and Ron Zemke, *Service America! Doing Business in the New Economy*, Homewood, IL: Dow Jones-Irwin, 1985.

Berry, Leonard L., Valerie A. Zeithaml, and A. Parasurman, "Five Imperatives for Improving Service Quality," *Sloan Management Review* 31, no. 4 (Summer 1990): 29–38.

Bitran, Gabriel R., and Johannes Hoech, "The Humanization of Service: Respect at the Moment of Truth," *Sloan Management Review* 31, no. 2 (Winter 1990): 89–96.

Bowen, David E., Richard B. Chase, and Thomas G. Cummings, and Associates, *Service Management Effectiveness: Balancing Strategy, Organization and Human Resources, Operations and Marketing*, San Francisco: Jossey-Bass Publishers, 1990.

Brown, Steven W., Evert Gummesson, Bo Edvardsseon, and BengtOve Gustavsson, eds., *Service Quality: Multidisciplinary and Multinational Perspectives*, Lexington, MA: Lexington Books, 1991.

Chase, Richard B., "The Customer Contact Approach to Services, "*Operations Research* 29 (1981): 698–706.

Chase, Richard B., "Where Does the Customer Fit in a Service Operation?" *Harvard Business Review* 56, no. 6 (November–December 1978): 13, 42.

Collier, David A., *Service Management: The Automation of Services*. Reston, Va.: Reston, 1985.

Collier, David A., *Service Management: Operating Decisions*, Englewood Cliffs, N.J.: Prentice-Hall, 1987.

Czepiel, John A., Michael R. Solomon, and Carol F. Surprenant, eds., *The Service Encounter: Managing Employee/Customer Interaction in Service Businesses*, Lexington, MA: Lexington Books, 1985.

Davidow, William H., and Bro Uttal, "Service Companies: Focus or Falter," *Harvard Business Review* 67, no. 4 (July–August 1989): 77–85.

DiPrimio, Anthony, *Quality Assurance in Service Organizations*, Radnor, PA: Chilton Book Co., 1987.

Enderwick, Peter, "The International Competitiveness of Japanese Service Industries: A Cause for Concern?" *California Management Review* 32, no. 4 (Summer 1990): 22–37.

Fitzsimmons, James A., and Robert S. Sullivan, *Service Operations Management*. New York: McGraw-Hill, 1982.

Gronroos, Christian, "New Competition in the Service Economy: The Five Rules of Service," *International Journal of Operations & Production Management (UK)* 8, no. 3 (1899): 9–19.

———, *Service Management and Marketing: Managing the Moments of Truth in Service Competition*, Lexington, MA: Lexington Books, 1990.

Guile, Bruce R., and James Brian Quinn, eds., *Technology in Services: Policies for Growth, Trade, and Employment*, Washington, D.C.: National Academy Press, 1988.

Hackett, Gregory P., "Investment in Technology: The Service Sector Sinkhole?" *Sloan Management Review* 31, no. 2 (Winter 1990): 97–103.

Hart, Christopher W.L., James L. Heskett, and W. Earl Sasser, Jr., "Surviving a Customer's Rage," *Successful Meetings* 40, no. 5 (April 1991): 68–79.

Hart, Christopher W.L., James L. Heskett, and W. Earl Sasser, Jr., "The Profitable Art of Service Recovery," *Harvard Business Review* 68, no. 4 (July-August 1990): 148–156.

Haywood-Farmer, John, "A Conceptual Model of Service Quality," *International Journal of Operations & Production Management (UK)* 8, no. 4 (1988): 19–29.

Heskett, James L., "Lessons in the Service Sector," *Harvard Business Review* 65, no. 2 (March–April 1987): 118–126.

———, *Managing in the Service Economy*, Boston, MA: Harvard Business School Press, 1986.

Horovitz, Jacques, *Winning Ways: Achieving Zero-Defect Service*, Cambridge, MA: Productivity Press, 1990.

Japan Human Relations Association, ed., *The Service Industry Idea Book: Employee Involvement in Retail and Office Improvement*, Cambridge, MA: Productivity Press, 1990.

Kanter, Jerry, Steven Schiffman, and J. Faye Horn, "Let the Customer Do It," *Computerworld* 24, no. 35 (August 27, 1990): 75–78.

Kanter, Rosabeth M., "Service Quality: You Get What You Pay For," *Harvard Business Review* 69, no. 5 (September–October 1991): 8–9.

Levitt, Theodore, "The Industrialization of Service," *Harvard Business Review* 48, no. 5 (September–October 1970):63–74.

Lovelock, Christopher, *Managing Services: Marketing, Operations and Human Resources*. Englewood Cliffs, N.J.: Prentice-Hall, 1988.

Lovelock, Christopher, and Robert Young, "Look to Consumers to Increase Production," *Harvard Business Review* 57, no. 3 (May–June 1979): 168–79.

Lewis, Barbara R., "Customer Care in Service Organisations," *International Journal of Operations & Production Management (UK)* 8, no. 3 (1988): 67–75.

Maister, David H., "Balancing the Professional Service Firm," *Sloan Management Review* (Fall 1982): 15–29.

Maister, David H., and Christopher Lovelock, "Managing Facilitator Services," *Sloan Management Review* (Summer 1982): 19–31.

McLaughlin, Curtis P., Ronald T. Pannesi, and Narindar Kathuria, "The Different Operations Strategy Planning Process for Service Operations," *International Journal of Operations & Production Management* 11, no. 3 (1991): 63–76.

Mersha, Tigineh, "Enhancing the Customer Contact Model," *Journal of Operations Management* 9, no. 3 (August 1990): 391–405.

Mills, Peter K., *Managing Service Industries: Organizational Practices in a Post-Industrial Economy*, Cambridge, MA: Ballinger, 1986.

Murdick, Robert G., Barry Render, and Roberta Russell, *Service Operations Management*, Boston: Allyn and Bacon, 1990.

Parasuraman, A., Leonard L. Berry, and Valarie A. Zeithaml, "Understanding Customer Expectations of Service," *Sloan Management Review* 32, no. 3 (Spring 1991): 39–48.

Quinn, James Brian, Jordan J. Baruch, and Penny C. Paquette, "Exploiting the Manufacturing-Services Interface," *Sloan Management Review* 29, no. 4 (Summer 1988): 45–56.

Quinn, James Brian, Thomas L. Doorley, and Penny C. Paquette, "Beyond Products: Services-Based Strategy," *Harvard Business Review* 68, no. 2 (March–April 1990): 58–68.

Quinn, James Brian, Thomas L. Doorley, and Penny C. Paquette, "Technology in Services: Rethinking Strategic Focus," *Sloan Management Review* 31, no. 2 (Winter 1990): 79–88.

Quinn, James Brian, and Christopher E. Gagnon, "Will Services Follow Manufacturing into Decline?" *Harvard Business Review* 64, no. 6 (November–December 1986): 95–103.

Quinn, James Brian, and Penny C. Paquette, "Technology in Services: Creating Organizational Revolutions," *Sloan Management Review* 31, no. 2 (Winter 1990): 67–78.

Roach, Stephen S., "Services Under Siege—The Restructuring Imperative," *Harvard Business Review* 69, no. 5 (September–October 1991): 82–91.

Sasser, W. E., and J.L. Heskett and C. Hart, *Service Breakthroughs*, New York: The Free Press, 1990.

Sasser, W. Earl, R. Paul Olsen, and D. Daryl Wyckoff, *Management of Service Operations*. Boston: Allyn and Bacon, 1978.

Schlesinger, Leonard A., and James L. Heskett, "Breaking the Cycle of Failure in Services," *Sloan Management Review* 32, no. 3 (Spring 1991): 17–28.

Schlesinger, Leonard A., and James L. Heskett, "The Service-Driven Service Company," *Harvard Business Review* 69, no. 5 (September–October 1991): 71–81.

Shaw, John C., *The Service Focus: Developing Winning Game Plans for Service Companies*, Homewood, IL: Dow Jones-Irwin, 1990.

Sherman, H. David, "Improving the Productivity of Service Businesses," *Sloan Management Review* 25, no. 3 (Spring 1984): 11–23.

_____. *Service Organization Productivity Management*, Hamilton, Ont.: Society of Management Accountants of Canada (SCMS), 1988.

Shiffler, Ronald E., and Ray W. Coye, "Monitoring Employee Performance in Service Operations," *International Journal of Operations & Production Management (UK)* 8, no. 2 (1988): 5–13.

Shostack, G. Lynn, "Designing Services That Deliver," *Harvard Business Review 62*, no. 1 (January–February 1984): 133–39.

Teal, Thomas, "Service Comes First: An Interview with USAA's Robert F. McDermott," *Harvard Business Review* 69, no. 5 (September–October 1991): 116–127.

Thomas, Dan, "Strategy Is Different in Service Businesses," *Harvard Business Review* 56, no. 4 (July–August 1978): 158–65.

Zeithaml, Valarie A., A. Parasuraman, and Leonard L. Berry, *Delivering Quality Services, Balancing Customer Perceptions and Expectations*, New York: Free Press, 1990.

CHAPTER 2

FLOWS, BOTTLENECKS, AND BALANCE

The first tools of operations management to learn, and arguably the most important, are the process flow diagram and its sister, the information flow diagram. They are the lightning rods for analysis and change of the process. They are stepping stones, as well, in identifying and remedying bottlenecks and in balancing capacities within processes.

PROCESS AND INFORMATION FLOW DIAGRAMS

A process flow diagram is a depiction of what the process does. It is a sequential depiction, noting which operations steps are accomplished before others and which can be done in parallel fashion. Different types of operations are typically designated with different symbols. Figure 2-1 duplicates the process flow diagram from Tour A in the back of this book, the Androscoggin Mill of International Paper.

Several points about the diagram ought to be noted:

1. Actual processing operations are usually distinguished from storage points in the process. In the diagram, processing operations are indicated by rectangles and inventories by triangles. (Rectangles are used here because it is easier to fit the descriptions in them. However, typically in industrial engineering, circles are used to indicate processing operations and rectangles are used to indicate tests and inspections. [In this book, circles will represent tests and inspections.] See Figure 5-1 for the standard conventions.)

2. Several operations could be bypassed and are indicated by two arrows emanating from one operation and pointing to others. For example, all of the woodroom's output need not have gone directly to the silos but could have been placed in the woodyard for storage. Similarly, the bleachery could be bypassed by pulp destined for special uses, and the rewinders were superfluous for those orders that could be slit directly from the main winders.

3. The continuous nature of the process is evident by the very low level of inventories and by the designation of the silo chip and pulp additive inventories as temporary.

A production process is more than a series of operations performed on a collection of mate-

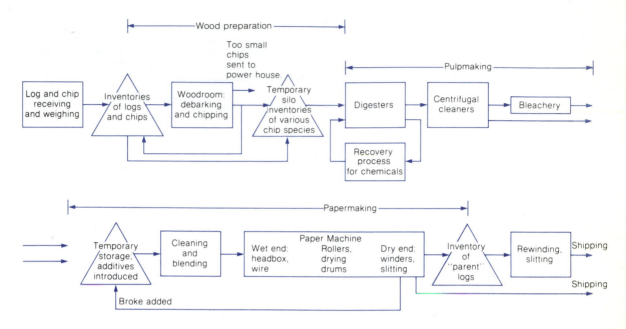

FIGURE 2-1   Process flow diagrams for International Paper Company's Androscoggin Mill.

rials. What a process flow diagram can depict — the sequencing of process steps, the choice of equipment and technology, the capacity of process steps, the tasks required of the workforce — while critical, is only part of the story. Another part of the story involves the procedures that have been put in place to direct the process flow. We can usefully think of a companion to the process flow diagram — namely, an information flow diagram. Figure 2-2 provides an example of what might be placed in such a diagram, again for the Androscoggin Mill. Note how the actions of different layers of managers and workers are distinguished in the diagram and how information is fed back up the channels of communication.

Most of the information flow in this continuous flow process is directed from the top down. Feedback is needed only to acknowledge receipt of information and to signal significant problems in the actual workings of the process. The process is designed with such care and the workforce is so carefully trained that workers do not have to be in repeated touch with management to do their jobs well. The information needs of the process simply are not great, although the thought and effort standing behind that information (such as scheduling tasks of both the corporate office and the mill itself) are considerable and sophisticated.

BOTTLENECKS

Production bottlenecks are generally considered to be temporary blockades to increased output; they can be thrown up anywhere along the course of a production process. Some are easy to identify and to remedy, while others are devilish.

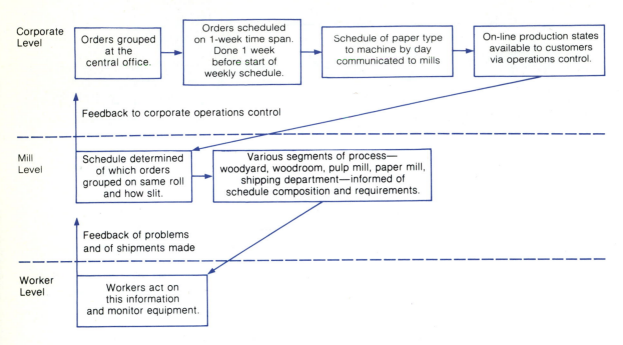

FIGURE 2-2 Information flow diagram for International Paper Company's Androscoggin Mill.

The bottleneck that is easy to cope with is stationary. Work-in-process inventory piles up quickly behind it; clearly, little is getting through. Its cause is usually also clear—a machine has broken down or key workers are absent or demand has simply outstripped the clear, rated capacity of a machine—and the remedy follows easily. Such bottlenecks often occur in service operations, causing customer waits.

More subtle are bottlenecks that shift from one part of the process to another or that have no clear cause. Inventories build up in different places and at different times. Such bottlenecks creep up on management and demand more thorough investigation. Perhaps they were detected as flaws in a product's quality caused inadvertently by one or more workers trying to keep pace with production demands that should not have been placed on them. Or, they may be caused by missing parts. They may be caused by new product startup or changes in the mix of products through the factory. In such cases the remedies are less clear-cut, and some analysis is called for.

Analyzing Bottlenecks with Process Flow Diagrams

In analyzing bottlenecks, it is always helpful to trace the production process by using a process flow diagram and to assign what capacity numbers are available to each stage of the process. Simply being systematic in this way and being as precise as possible with capacity measures can uncover primary and secondary bottlenecks straightaway. Such an analysis is naturally easier with well-delineated processes and ones where capacity is unambiguous. The process flow diagram, in that case, becomes a planning aid for breaking significant, stationary bottlenecks. It is also an aid in the design of processes themselves, as it clearly shows how the capacities of process segments can be balanced against one another.

As an example, consider the steel industry. The process is largely continuous flow; capacity, in tons, is well known and dependent primarily on equipment capabilities. A process flow diagram for a steel mill, with associated capacity measures, might well look like the one shown in Figure 2-3. The diagram makes it clear that the bottleneck at this steel mill is the blast furnace; in order to increase the mill's capacity, blast furnace capacity must be enlarged. The secondary bottleneck resides in the casting and semi-finishing operation; that is, once the blast furnace is expanded past 4.7 million tons per year, casting and semi-finishing will be the bottleneck.

The most slack element of the process flow shown in Figure 2-3 appears to be the finishing mills. In most steel mills, however, the nominal capacity of the finishing mills is purposely greater than the other elements of the process to ensure capacity under all kinds of product mixes. Thus, even in steel mills the notion of capacity becomes somewhat blurred and ambiguous. A process flow diagram must be modified somewhat and interpreted if it is to be useful for analyzing bottlenecks.

Even more modification and interpretation of process flow diagrams is needed when less rigid production processes (such as batch flow or job shop) are involved. In such processes, not only are capacity figures potentially very volatile, but the process flow itself may be indeterminate. Identifying bottlenecks in these processes becomes a formidable task. Typically, ranges of capacity must be used and different arrangements of the process flow must be tried. Still, judicious and systematic use of a process flow diagram can be a valuable tool in identifying the process elements and conditions that account for bottlenecks.

Service operations are particularly ripe for such analysis. Their process and information flows are only seldom analyzed systematically, and seemingly minor adjustments in the service delivery process can frequently yield significant improvements. It is often very helpful to indicate where the customer interacts with the process and information flows. Thus, a service process flow diagram might include a lower half that tackles the "back of the house" activities and an upper half that indicates when a customer has contact with the process.

Where Should Excess Capacity Be Kept?

The foregoing example of a steel mill raises the important issue of whether balance in a process might really mean that some excess capacity ought to be kept. We should realize that in certain circumstances, excess capacity is a very desirable thing, and that perfect balance — with no obvious bottlenecks — may not be the most appropriate state for a process.

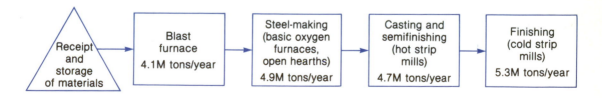

FIGURE 2-3   A process flow diagram for a steel mill.

The steel mill example pointed out that holding excess capacity at the end of a process permits the early portions of the process to continue to produce no matter what the final product mix is. Thus there are several reasons for holding excess capacity in particular locations in the process:

1. Holding excess capacity in the early stages of the process:
   - If yields drop off the further one goes in the process, having excess capacity early on is useful.
   - If changes to the product, or the product mix, are easy to make late in the process, excess capacity earlier may be desired.
2. Holding excess capacity in the late stages of the process:
   - If changes to the product, or to the product mix, are hard to make late in the process, excess capacity later is desirable.
3. Other reasons for holding excess capacity anywhere in the process:
   - Often capacity additions entail large, fixed-increment ("lumpy") investments. This can force you to hold excess capacity.
   - Sometimes capacity is cheap to build. If growth is expected, often it makes sense to hold excess capacity for what is cheap to build. This can work especially well if, over time, the bottleneck segment of the process can be broken by taking small, incremental steps.

Augmenting Process Flow Diagrams

We have discussed adding capacity figures to process flow diagrams as a means of identifying potential bottlenecks. This is a common technique, but there are other figures that can be used in conjunction with process flow diagrams, figures that augment the power of such diagrams. For example, processing times and times spent in queue and in transit can be employed with process flow diagrams to develop illustrations or estimates of throughput times for various products. (The importance of throughput times is discussed in depth in Chapter 11.) Yields at each process step, such as first-pass yields, can be used to calculate the expected yield of the entire process. In addition, value-added or costs can be assigned to specific process steps to help determine where costs are incurred and what alternative processing steps might exist. The process flow diagram is thus an important tool for any process, be it in the manufacturing or the service sector.

Analyzing Bottlenecks with Inventory Buildup Graphs

Another useful tool in analyzing bottlenecks is an inventory buildup graph, a pictorial way of accounting for the rates at which inventories are either piled up or depleted. These graphs are particularly useful when the demand on an element of the process is especially erratic, such as occurs routinely in the processing of many raw agricultural commodities but can occur in other industries. To illustrate the use of an inventory buildup graph and to provide a grounding in some classic problems of capacity and bottlenecks, consider Situation 2-1.

Discussion of Devine Nuts

In analyzing the situation Frank Coyne finds himself in at Devine Nuts, it is helpful to picture what is going on — how peanuts build up in inventory after delivery and how the inventories are drawn down by completed drying. What is of concern to Frank are the peak days, when 300 tons of soggy peanuts arrive during a 10-hour period and have to be dried for 36 hours. Figure 2-4 pictures how inventories build up at Devine Nuts during a succession of

SITUATION 2-1

Devine Nuts, Inc.

It was autumn and the peanut harvesting season was winding to a close. Frank Coyne, the proprietor of Devine Nuts, was reflecting on the past season and speculating whether he would have enough peanut drying capacity for next year, especially if the local crop increased by 10 percent, as many of the peanut farmers were saying.

Frank Coyne was a peanut broker. Devine Nuts purchased peanuts from farmers in the area (Devine, Texas, is situated about 30 miles southwest of San Antonio) and sold them to peanut processing companies and cooperatives, which used them for making peanut oil, peanut butter, candy, and a host of other items.

Devine Nuts' principal operating functions were straightforward. During the peanut harvesting season, farmers delivered loads of so-called "green" peanuts to Devine Nuts. The trucks, with their loads, were weighed and records kept by the issuance of scale tickets. After weighing, a moisture sample was taken. This inspection step was critical, since the peanut processing companies would not accept any peanuts without the government's grading certificate on them, and a grading certificate would not be issued unless the moisture content of the peanuts was 10.5 percent by weight or lower. If a delivered load of peanuts met this moisture standard and was thereby granted a certificate, another sample of peanuts would be drawn, weighed, and shelled so that the kernels could be classified into four size categories (large, regular, medium, and small). Kernel weight and size determined the price the farmer would be paid, typically about 30 cents per pound. Once the price was determined, the peanuts were loaded into railcars or trucks and shipped to a shelling plant. Sometimes immediate transportation could not be arranged, and the graded peanuts would have to be stored in Devine Nuts' bulk warehouse, which had room for 5500 tons.

More often than not, the delivered loads of peanuts would not meet the moisture standard and would have to be dried and then reweighed and graded. Devine Nuts owned a number of dryers of various sizes: 38 six-ton capacity dryers, 10 eleven-ton capacity dryers, and 14 fourteen-ton capacity dryers. How long a load of peanuts would have to remain in a dryer depended on its moisture content. Peanuts with a moisture content of only 10 to 15 percent could be dried in about 12 hours, but peanuts with a moisture content of 15 to 25 percent could take anywhere from 24 to 36 hours to dry. Devine Nuts did not have any storage for wet green peanuts; they were dumped directly from the farmers' trucks into the dryers. If dryer space was not available, trucks had no recourse but to wait until it became available.

The drying operation was critical because harvested peanuts that were wet began growing a fungus about 12 hours after harvesting. To avoid spoilage, it was essential that drying start within 12 hours after harvesting. A day's harvesting began after the morning dew had dried and continued until dusk. Deliveries to Devine Nuts began about noon and lasted until about 10 P.M. in a more or less steady stream.

Frank Coyne's success as a broker rested on his ability to meet the needs of the farmers who brought their crops to him each year. If Frank could not dry a farmer's peanuts on time, he stood an excellent chance of losing that customer forever to another broker. In addition, if Frank left too many trucks waiting too long for available dryer space, he risked souring his customer relations, since the farmers needed those trucks out in the field for the next day's harvest.

On most days, Devine Nuts' drying capacity was more than enough to handle the demand. A typical day's delivery would be about 120 tons; roughly half would have a moisture content between 10

continued

and 15 percent, and half between 15 and 25 percent. What grayed Frank Coyne's hair were the peak days that would occur right after a rainstorm. Storms usually kept the farmers from harvesting, and so loads that ordinarily might be stretched over several days would be bunched together in just a few. On such days, 300 tons of green peanuts might be delivered. What made matters much worse was that the peanuts were soaking wet from the rain and had moisture contents that required a full 36 hours of drying.

Devine Nuts had been able to handle the past season's peak days without turning away a customer, although some trucks had to wait at times to unload. Frank Coyne's immediate concern was calculating whether a 10 percent increase in the local crop (which he translated into a 10 percent increase in peak day activity) would overload his drying capacity and (1) force trucks away, (2) cause them to wait longer than dawn (6 A.M.) of the day following their delivery, or (3) cause them to wait longer than 12 hours to unload.

peak days, assuming an even distribution of truck arrivals.

Figure 2-4 assumes that at noon on day 1 all the dryers are empty. As loads of peanuts arrive from then on, at an average rate of 30 tons per hour, dryers are successively filled up. By 10 P.M. when the farmers' shipments cease for the day, 300 tons of peanuts are in the dryers. Because they are so wet, these peanuts will not be removed from the dryers until after midnight on day 2. In the meantime, another day's worth of soggy peanuts will have arrived. Most of these can be transferred directly to waiting dryers, but some will have to wait until day 1's peanuts are removed.

How many tons of peanuts will have to wait for dryers, and how many hours will they have to wait in their trucks? Devine Nuts' total drying capacity is 534 tons, figured as $[(38 \times 6) + (10 \times 11) + (14 \times 14)]$. Thus 66 tons of peanuts $(600 - 534)$ cannot be transferred immediately from trucks to dryers. Since peanuts are arriving at a rate of 30 tons per hour, dryer capacity can be expected to run out at 2.2 hours $(66/30)$ before 10 P.M. on day 2. The dryers will not be available until midnight, but at that time dryer space will be released at the rate of 30 tons per hour, the same rate that peanuts entered the dryers. The first truck that cannot dump its load immediately into a dryer will have to wait until midnight to be unloaded, a wait of 4.2

hours. Because the withdrawal rate of peanuts from the dryers exactly matches the arrival rate of peanuts to Devine Nuts, all of the trucks will wait an average of 4.2 hours. The last truck will be free to return to its farm at 2.2 hours after midnight. Thus Figure 2-4 confirms Devine Nuts' record over the past season: trucks did have to wait, but none had to be turned away or came close to spoiling the peanuts or not getting back on time.

The truck wait is represented by the shaded trapezoid in Figure 2-4. The extent of the truck wait, measured in units of ton-hours, is merely the area that is shaded. The base of the trapezoid is 6.4 hours, while the top is 2 hours. The height of the trapezoid is the quantity of peanuts in the trucks between 10 P.M. and midnight, 66 tons. Using the formula for the area of a trapezoid, $[((\text{base} + \text{top}) \times \text{height})/2]$, the extent of the truck wait is given by $[((6.4 + 2) \times 66)/2] = 277.2$ ton-hours.

How does a 10 percent increase in peak day deliveries affect operations? Figure 2-5 deals with that situation. The slightly steeper slopes pictured in this inventory buildup graph reflect the increased average arrival rate of 33 tons per hour. Since drying still must take 36 hours, Devine Nuts runs out of capacity sooner than in Figure 2-4. Specifically, Devine Nuts' capacity falls 126 tons short $(660 - 534)$ and the dryers begin overflowing at 3.82 hours $(126/33)$

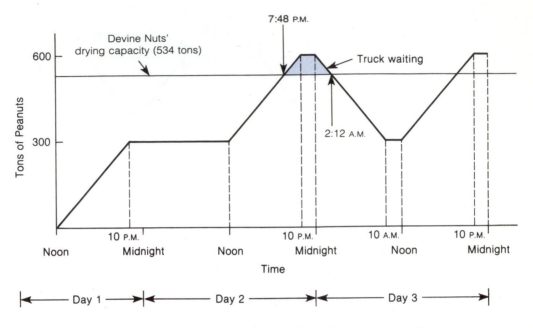

FIGURE 2-4   Inventory buildup at Devine Nuts after several peak days.

before 10 P.M. Since the dryers will not be available until midnight, each truck will have to wait an average of 5.82 hours before unloading. The latest arriving truck will still be free to return to service at 3.82 hours past midnight, well before dawn. Thus even with an increase of 10 percent in peak day tonnage, Devine Nuts can expect not to have to turn away customers for lack of capacity.

Frank Coyne may also be interested in knowing how much the current peak day demand (30 tons per hour) would have to grow before Devine Nuts' present capacity would be insufficient to free up the delivery trucks by dawn. To find the answer to this question, we need to calculate how short of the peak Devine Nuts' capacity would have to be before 6 hours (midnight to dawn) would be too little time to work down the excess.

Let x represent the factor by which we would have to multiply present demand to yield the peak day demand that strains capacity. Thus, the total tonnage of peanuts accumulating in dryers and in trucks at 10 P.M. on day 2 would be the present 600 times x. In the 6 hours after midnight, x times the present rate of 30 tons per hour would be freed from the dryers each hour. Thus we can write out the following relationship:

$$6 \times 30x = 600x - 534$$

(drying capacity freed up in 6 hours) = (wet peanuts waiting for dryers)

Solving for x, we get

$$180x = 600x - 534$$
$$534 = 420x$$
$$1.27 = x$$

Thus peak demand will have to rise by 27 percent over the present level before Devine Nuts' present capacity will be insufficient. Frank Coyne should be reassured.

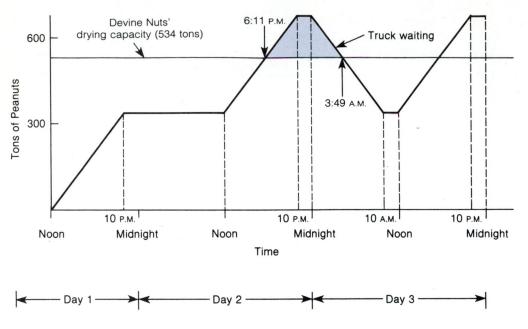

FIGURE 2-5   Inventory buildup at Devine Nuts when peak day tonnage
has increased by 10 percent.

This type of calculation, where the general question is asked "At which point does one alternative begin to dominate another?" is often called an indifference calculation. Here we asked "At what level of demand would Devine Nuts' capacity begin to be insufficient?" and the indifference point was a demand 27 percent greater than the present. As one might expect, the indifference calculation can be exceedingly useful and flexible, and we shall be using it in several guises throughout this text. It is particularly useful in gauging the sensitivity of a decision to variations in a particular variable. In fact, the indifference calculation computed here is a form of sensitivity analysis. How sensitive is Devine Nuts' capacity needs to an increase in peak demand?

With the background of these tools of analysis for finding bottlenecks—the process and information flow diagrams, the inventory buildup graph, and indifference calculations—we turn now to outlining the causes of bottlenecks of various types.

The Causes of Bottlenecks

At this point, it makes sense to discuss why bottlenecks typically occur. Such a discussion helps put several concepts in perspective. Of course, we cannot hope to cover all the remedies to the problems that cause bottlenecks in this section; indeed, avoiding and/or remedying bottlenecks of different kinds is a key concern of much of the remainder of the book. Nevertheless, outlining the most common causes of bottlenecks helps build appreciation for some of the topics to follow.

Bottlenecks can be divided into two types: episodic; and chronic, recurring ones. Of course, individual bottleneck episodes can easily develop into chronic problems. Episodic bottlenecks call on the "firefighting" skills of the workers and managers involved, while chronic bottlenecks demand more planning or design changes.[1]

Episodic Bottlenecks

Episodic bottlenecks can be classed into three major categories: (1) machine breakdowns, (2) material shortages, and (3) labor shortages. Let us consider these bottlenecks in more detail.

1. *Machine breakdowns.* Perhaps the biggest of the fires to fight in a factory involve machines, tooling, or fixtures that break. When this happens, lots of scurrying around occurs to fix the broken machine and to reroute, if possible, the job it was performing. Often the best maintenance tradespeople in a factory are assigned the firefighting role of repairing broken equipment. Plant managers can also be dragged in, if their technical skills are appropriate.

   Some machine breakdowns represent unavoidable accidents. On the other hand, many breakdowns can be avoided with some planning and preventive action. Preventive maintenance is a too frequently neglected activity; there is never enough time for it when business is booming (so it is claimed), and it is a low priority item when business is faltering. Breakdown time often exceeds the planned downtime of preventive maintenance, and planned downtime is naturally easier to cope with and does not risk the quality problems that surround breakdowns. Putting the factory's best maintenance tradespeople on the task of preventing breakdowns is increasingly being recognized as the most cost-effective policy.

   In a similar vein, attention to tooling and fixturing can eliminate unplanned downtime. Tooling or fixturing that was meant to be temporary too frequently winds up as permanent, and that can create problems. More effort and expense up front often saves frustration and cost down the road. This thinking is embodied in total preventive maintenance programs that are gaining ground in many companies.

2. *Material shortages.* If unplanned machine downtime is the biggest of the fires fought in the factory, materials shortage is the most common. The typical factory always seems to be short some part or material. Sometimes it is traceable to a vendor; sometimes it is another department within the factory that is responsible. Chapters 6–8 tackle materials management and the kinds of policies and procedures that can be used to avoid material shortages and the "work around" (i.e., work around the problem now and fix it later) that the factory is forced into to cope with them.

3. *Labor shortages.* Unexpected absences, retirements, and terminations are all examples of how an operation can be temporarily short of workers. This is more of a problem for some types of processes, of course, but it has an impact for all.

   Allied with this problem is the job bumping that is endemic to operations when there are many labor grades and when the most senior workers can bid for job openings in the next higher grade. The cascade of job openings created by workers moving up to the next grade can be many times the number of jobs initially vacated. The residual problem is that lots of workers require training for their new positions and initially are not as productive as they will become. It is as if there were a labor shortage when significant job bumping occurs. Workforce issues are treated in Chapter 5.

Chronic Bottlenecks

Chronic bottlenecks, much like episodic ones, can be categorized. Chronic bottlenecks can be thought of in terms of materials problems and process problems.

Materials Problems

1. *Ordering the wrong materials or not enough materials.*   This potentially recurring problem suggests deficiencies in production planning or in purchasing. Not all vendor shortages are caused by the vendor; some can be traced to late or incorrect purchase orders, poor forecasting of needs, incorrect and/or obsolete bills of material, inappropriate inventory policies, myopic production planning, unwillingness to release the funds, and the like. Chapters 6–8, which focus on materials management, deal with these issues.

2. *Constantly changing product mix.*   Even if the production planning and purchasing departments are on their toes, there could be bottlenecks simply because a constantly changing product mix places irregular demands on the capacity of individual departments within the factory. (Situation 2-2 is an example of this.) That is, in aggregate the factory might have enough capacity, but the character of that capacity does not fit the product mix being pushed through. This phenomenon emphasizes the importance of capacity planning, a topic discussed in Chapter 6.

Process Problems

1. *Insufficient capacity.*   Frequently, it is not merely a problem of the product mix being inappropriate in the short term for the capacity in place. Rather, the demand, in the aggregate, may well exceed capacity. In this case, if the bottleneck is to be broken, capacity must be increased with more equipment, people, and perhaps, "bricks and mortar." This is the focus of discussion in Chapter 12.

2. *Quality problems.*   Quality problems may, of course, lead to episodes such as those associated with machine breakdowns. If the fundamental causes of quality problems are not remedied, however, the problems become chronic. Quality management is a rich subject, and Chapter 4 deals with it.

3. *Poor layout.*   One of the key aspects of a process is its layout. Poor layouts—those with convoluted routings, crowded conditions, difficult and costly materials handling, and the like—can be a terrible drain on a plant's productivity. We discuss some key aspects of plant layouts in Chapter 3.

4. *Inflexible process.*   Some bottlenecks turn out to be chronic because of the process design. In these cases, a bottleneck gets designed into the process or is exposed by some shift in the pattern of demand. A classic reason for inflexibility is the purchase of a large, general-purpose piece of equipment that is supposed to do a series of automatic operations. If the equipment delivers as expected, all is well. However, often such process designs run amok because the equipment does not deliver, is not properly tooled up, takes too long to set up, or is very temperamental. Because so much is routed to it, when the equipment fails or has to be shut down for long periods of maintenance or setup, the rest of the process becomes crippled. Better, in many people's eyes, to design a process with smaller scale, easier to set up, and more special-purpose equipment that does not leave production so vulnerable. As Situation 2-2 reveals, a series of small-scale equipment can be more flexible and timely than an expensive piece of allegedly general-purpose equipment.

Bottlenecks: The Flip Side

In this chapter, we have come to understand that bottlenecks constrain a process from producing all that it can. If more capacity is required, the bottlenecks must be broken. The flip side of this insight, of course, is that non-

SITUATION 2-2

Wyoming Valley Furniture Company

It was the first part of July, time for another monthly planning meeting on the production schedule for the upcoming 16-week period. George Sowerby, owner and president of Wyoming Valley Furniture, normally relished this opportunity to grapple with what he viewed as the most important management activity in the furniture business. Today, however, he was a bit apprehensive because the newly arrived order from Pennsylvania House threatened another bottleneck in the lathe department, just like the one the company had suffered through only last month.

The Wyoming Valley Furniture Company was a small manufacturer of casegoods furniture (i.e., all-wood furniture as opposed to upholstered furniture), situated in the Wyoming Valley of Pennsylvania, near Wilkes-Barre. It was one of thousands of small firms in the furniture industry. The company did not produce its own brand; it was a subcontractor to several of the large furniture retailers, such as Pennsylvania House and Ethan Allen. A retailer like Pennsylvania House authorized Wyoming Valley Furniture to manufacture a variety of pieces (such as hutches, tables, bureaus) and each month ordered so many thousand dollars of furniture, specifying a particular mix of the items authorized. It was up to a company like Wyoming Valley Furniture to deliver the ordered pieces within its 16-week planning period and to have them pass the retailer's quality standards.

The Threatened Bottleneck

About four months ago, Pennsylvania House placed an order for 150 of a newly authorized piece, a rather elaborate dining room buffet table. Last month was the first time Wyoming Valley Furniture manufactured the buffet; the time and trouble it took to make it were greater than expected. The buffet's legs, in particular, demanded more time and care in lathe work than Wyoming Valley Furniture had allocated. (Machine setups, for ex-

ample, took 15 minutes for each lot rather than the planned 8 minutes, and run times per leg were 20 minutes rather than the planned 15 minutes. In addition, considerably more rework was required than expected.) As this became evident, the company began scheduling overtime in the lathe department. Still, the buffets were delivered late. What is more, the slowdown the order caused in the lathe department spilled over into the finish mill, and 10 workers in the finish mill were without work for about 8 percent of their time over the 2-week period the buffet order was scheduled through their shop. Management had been caught with no work suitable to give those workers, as orders got bottled up in the lathe department. This failure in production scheduling was not only embarrassing but also expensive. (Wages averaged $6 per hour, and workers generally worked 40 hours per week.)

The Furniture-Making Process and Coping with Bottlenecks

Wyoming Valley Furniture's production process was similar to that employed by most other casegoods manufacturers. It started with the purchase and storage of rough-cut lumber of various species and grades. This lumber was then dried on racks for two weeks in the predrying shed (a large enclosure in which moderately heated air was circulated) and for 2 weeks in the kiln, where higher heat was applied. From the kiln the lumber was sent to the rough mill, as it was called, where an assortment of power saws (for lengthwise ripping, mitering, and end cutting), planers, tenoners, lathes, gluing machines and presses, and other tools formed the basic parts out of which the piece of furniture was assembled. Typically, an order spent 3 weeks in the rough mill.

From the rough mill the order traveled to the finish mill, which was responsible for boring, routing, and sanding the rough milled parts and for a modest amount of subassembly. Three weeks in the finish mill

continued

was standard for an order. All of the parts needed for the assembly of a piece of casegoods furniture were gathered at the finish mill before being sent to the assembly and finishing departments. The assembly department put together each piece of furniture, and the finishing department applied the stain and lacquer to each. Inspection and packing for shipment completed the process. The typical order spent 2 weeks in assembly and finishing. Thus, the entire process took about 12 weeks. By adding a month as a time buffer, the planning period of 16 weeks was determined.

Ordinarily, Wyoming Valley Furniture was thoroughly familiar with the particular demands that the various authorized pieces of furniture it manufactured placed on machines and workers. These demands could be accommodated in the development of the production schedule so that the entire factory ran as close to perfect balance as possible. Of course, perfect balance was an ever-elusive and only momentarily achieved goal. In striving for this balance, however, the company employed a variety of rules of thumb that helped determine which orders would be grouped together in the same week and which ones shifted into other weeks. For example, only one model of dining room suite would be scheduled in one week, and only one model of hutch. In essence, bottlenecks were scheduled around whenever possible.

Most of the bottlenecks that struck the factory could not be anticipated. New products and their startups (such as the buffet) were a persistent cause of bottlenecks. New wood species also caused occasional problems (e.g., a recent switch to oak in a particular model had caused problems). Machine breakdowns, especially of one-of-a-kind machines, always created a stir in the factory. If demand was particularly high, bottlenecks might occur simply because more orders were released to the factory than the factory could realistically be expected to work through.

Most bottlenecks were overcome by throwing more labor-hours at them—usually through calling for overtime, but sometimes by hiring or transferring workers to the bottleneck task. Other bottlenecks, usually the chronic ones, were broken by buying new equipment.

Dealing with the Threatened Bottleneck

What was particularly disheartening to George Sowerby and his production scheduler, Vic Baiz, was that there seemed to be no way that the current order of 300 buffets could be scheduled through the lathe department without incurring overtime; the demand on the lathe department was unavoidably constant. Of the 400 labor-hours estimated now for the entire order (300 pieces x 4 legs per piece x 20 minutes per leg), only 250 could safely be scheduled during regular hours. The remaining 150 hours appeared headed for overtime and a pay rate of 1.5 times the regular rate.

George and Vic were eager to avoid the problem that occurred last month, when overtime in the lathe department created scheduling problems in the finish mill that could not be resolved easily by working on orders further back in the queue. To some extent, they would rely on better scheduling now to permit smoother functioning of the finish mill, but exactly how much smoother operations would be was unknown.

An alternative to this strategy of running lots of overtime during the order and hoping for the best was a strategy of working a modest amount of overtime beginning immediately and inventorying quantities of the troublesome buffet leg. This, of course, meant incurring some inventory carrying expenses. Vic Baiz estimated the inventory expense at an average of 25 cents for every buffet leg that had to be produced in overtime.

Here were two ways of coping with the upcoming bottleneck, and George Sowerby wondered which he ought to authorize. One involved some certain extra costs (inventory expenses); the other involved an uncertain, but possibly lower, cost—the disruption that significant amounts of overtime in the lathe department might have on the finish department. George wondered how little the disruption would have to be before the latter option dominated the inventory one.

bottleneck operations do not constrain the operation and can often be relegated to the back burner. Managing the bottlenecks well may demand minimizing the number of setups and changeovers and running larger lots through the bottleneck operations. The nonbottleneck operations need not be run with large lots and few setups. Rather, they are often better managed with small lots and many setups so as not to place even more burdens on the bottleneck operations. Managing with this understanding is sometimes difficult, but it pays off with more total capacity when the operation requires it.[a]

Discussion of Wyoming Valley Furniture Company Bottlenecks

Wyoming Valley Furniture knows that a bottleneck is coming, a definite advantage, but it is disadvantaged because its normal means of dealing with bottlenecks (i.e., to schedule around them) appears inadequate. If the company could routinely expect such heavy demands on the lathe department, George Sowerby might feel free to expand its equipment and workforce. However, the lathe demands made by this particular piece of furniture are unusual and, as yet, it is not a steady order. Only three or four orders for the buffet are likely in any year. Wyoming Valley Furniture seems to occupy some no-man's land between capacity expansion and the status quo.

As it stands, the company can plan for lots of overtime when the entire order is scheduled for production and hope that no disruption will occur to the finish mill. Or, the company can try to stock up early on the problem buffet leg. In this case, the same number of overtime hours will have to be scheduled, although they would be stretched over a longer period of time. In fact, stretching out production will almost

surely increase the setup times involved on the lathe, implying even more overtime.

How costly would a disruption be if it occurred again in the finish mill? The previous order of 150 pieces, half the current order, disrupted 10 workers for 8 percent of their time over 2 weeks. Thus, 64 worker hours (10 workers $\times$ 0.08 $\times$ 2 weeks $\times$ 40 hours/week) were lost at a cost of $6 each, for a total cost of $384. If twice the number of pieces causes twice the level of disruption, $768 would be sacrificed. Unfortunately, we do not know whether this level of disruption in the finish mill will be maintained or whether better production planning will leave enough slack so that the finish mill workers can work around any stalled buffet legs. This uncertainty suggests some of the same sort of sensitivity analysis that we did in the Devine Nuts situation.

How costly is the inventory building alternative? The estimated overtime is 150 hours, implying that 450 legs (150/400 $\times$ 1200 legs) will be completed using overtime. At an average inventory expense of 25 cents per leg, $112.50 would be spent for the inventory itself.

This $112.50 inventory expense is considerably below the $768 estimate for the "grin and bear it" alternative. Yet both figures are "soft" and require some interpreting. The inventory expense figure, for example, is almost surely too low, because it omits the extra setup time for the lathes to run the numerous small lots. We can make some estimate of how much setup time and expense would be required. Suppose the overtime on the legs takes 10 hours per week for 15 weeks, and that the 10 hours each week consists of one lathe operator working 1 hour longer on each weekday and 5 hours on Saturday. Thus, 6 setups of 15 minutes each would be required each week, for a total of 90 setups (6 $\times$ 15) over the next 15 weeks. At 15 minutes per setup and $6 per hour, the 90 setups would cost an additional

[a]Eli Goldratt and his theory of constraints has helped managers cope with this insight.

$135. A better estimate of the cost of the inventory building alternative would be $247.50 ($112.50 + $135.00).

The cost is still considerably below the $768 estimate for the "grin and bear it" alternative. This $768 figure, too, may need to be revised. If, for example, the order does not create any of the disruptions in the finish mill that prevailed last time, the $768 figure vanishes to nothing. The $768 figure assumes the same kind of disruption of worker time (8 percent) that occurred with the previous order. With the company on the alert now, it is likely that something less than 8 percent disruption in the finish mill will actually occur. How much less of a disruption would be needed before the "grin and bear it" option dominates the inventory buildup one? The same kind of indifference calculation that we introduced with Devine Nuts can be employed here.

Let x be the fraction of the 8 percent disruption needed to equalize the costs of the two options. Then

$$x \cdot \$768 = \$247.50$$
$$x = 0.32$$

This calculation suggests that if Wyoming Valley Furniture feels that it can lower the finish mill disruption to less than a third of its previous level, it should treat the order like any other. If, on the other hand, the company feels that more than that level of disruption would occur, the inventory buildup option should be followed. While these calculations do not make George Sowerby's decision for him, they serve to focus the issue.

Lessons from Wyoming Valley Furniture Company

Wyoming Valley Furniture, unfortunately, was caught in a very uncomfortable position. The company did not have sufficient capacity so that the rescheduling of the order could alleviate the impending bottleneck, nor was the extra demand on the lathe department steady enough to warrant an expansion of its equipment and workforce. The company was limited in its options: it could try to head off its problems by building up an inventory of parts or it could try to "gut" it out. The assessment of such nebulous bottleneck situations is one of the chronic management dilemmas of job shops and batch flow processes. In this case, a new product introduction, which changed the product mix considerably, placed excessive demands on a single department and threatened to clog up the flexibility in scheduling that keeps a job shop or batch flow process from grinding to a halt every time a problem like this is encountered. The threat to the factory's flexibility could just as easily have come from engineering changes or quality problems demanding rework or missing parts from a supplier or something similar.

Coming to a decision on how to remedy the bottleneck called for an analysis of the costs associated with each of the options. Specifically, the analysis involved comparing the extra costs incurred by each alternative, since those were the only costs that differed between the options being considered.

The importance of this type of analysis and the role of demand as well as supply considerations in confronting bottlenecks is pursued in Situation 2-3.

Discussion of Citrus Airlines Bottlenecks

Citrus Airlines is experiencing a bottleneck. Given the heavy loads it has been carrying recently, an increasing number of standbys have been left at the airport. The heavy loads and the high number of completely booked flights have no doubt caused a number of customers to switch flight times or even airlines. If such a strong demand continues, Citrus may be well advised to add equipment and offer more flights.

As a way around the current bottleneck, Dave Dove has suggested that Citrus can fly

SITUATION 2-3

Citrus Airlines

Larry Klock had to confess that he hadn't thought of it, but his operations manager, Dave Dove, had proposed an intriguing, possible solution to Citrus Airlines' present shortage of seats. Larry was the president of Citrus Airlines, a small intrastate carrier serving Miami, Jacksonville, Orlando, Fort Lauderdale, Tampa, Tallahassee, and a host of smaller Florida cities. Recently, business had been booming, largely, Larry thought, because of an upturn in the economy. Citrus Airlines was using much of its equipment to the hilt and faced the prospect of having to purchase more in order to provide more seats. Larry was leery of this option, both because of the tenuousness of the current economic recovery and the comparatively recent financial good health of the airline. Dave Dove's proposal struck an imaginative middle ground.

Dove had proposed that Citrus abolish first-class seating on all of the Boeing 727s that Citrus owned. (The 727s operated primarily between the major Florida cities, averaging 91 flights each day.) If first class were abolished, space would be provided for more seats. According to the cabin configuration of seats that Citrus had always maintained, 20 of the plane's total of 125 seats were provided for first-class passengers. If the first-class cabin were reconfigured to conform to the coach cabin, an additional eight seats could be provided. These seats could be added because coach sat five across in each row of the airplane as against four across in each first-class row and because the distance between rows in coach was slightly less (with less legroom) than in first class. Dove estimated that it would cost $15,000 to alter each of the fifteen 727s in Citrus' fleet.

Of course, abolishing first class meant abolishing first-class fares as well. At present, the average first-class fare in the Citrus system was $71 as opposed to the average coach fare of $56. On the other hand, first class was a little more costly to serve, an average of $3.28 versus $1.77, because first-class food and beverage service was more elaborate.

Larry was concerned that the abolition of first-class service would adversely affect the patronage of Citrus' first-class travelers, even though Larry knew that the availability of more seats could mean substantially more revenue. To investigate these matters, Larry had called upon Sarah Hammans, the airline's marketing manager. Hammans' report is shown in Figure 2-6.

MEMORANDUM

To: Larry Klock
From: Sarah Hammans
Subject: Abolition of First-Class Seating on 727s

You asked for analysis of two points related to the abolition of our "sun-kissed" (first-class) service. Let me address them in turn.

1. Retention of "sun-kissed" travelers. Although our retention rates will vary by city-pair depending on our competition and the prevailing flight schedules, I think we would be hard-pressed to retain more than 60 percent after abolition, if these travelers can go with another airline. If the loads continue high or if other airlines abolish first class as well, we may retain them all. It's a tough call.

2. Current loads in first class and coach. Over the past two months (61 days) on our 727s, we have averaged a first-class load factor of 61.2 percent and a coach load factor of 82.9 percent. The distribution of flights by load factor over the same period is shown in Figure 2-6.

Dave Dove's suggestion about eliminating first-class service would surely help cut down on the number of flights that leave standbys behind, but Larry Klock couldn't help but wonder whether that benefit would offset the costs of reconfiguring the first-class cabins and of losing loyal first-class passengers.

continued

"Sun-kissed"	Incidence of Load Factors (percent of 727 flights)							
	LESS THAN 50%	50–60%	60–70%	70–80%	80–90%	90–100%	1–5 STANDBYS LEFT	MORE THAN 5 STANDBYS LEFT
Service	42.3%	11.6%	12.7%	13.8%	10.9%	7.7%	1.0%	—
Coach	16.4%	10.4%	10.7%	11.3%	20.9%	18.3%	7.4%	4.6%

NOTE: The load factor (the occupied seats as a percentage of all available seats) is a useful capacity measure, not only for airlines but for many other types of service businesses..

FIGURE 2-6   The report on abolishing first class on Citrus Airlines' 727s

eight more seats on every flight if first-class service is abolished. This change has some real appeal. Not only does it solve the present over-subscription problem of some flights, but Citrus could conceivably make more money by flying an all-coach airplane.

How is this true? For every first-class passenger under the present scheme, the airline receives $71, on the average, and expends only $3.28 on services. The company thus stands to gain $67.72 for every first-class passenger. This is money it can use to pay back its debt on the planes and other facilities and the overhead it buys. This money is also the source of its profits. The figure ($67.72) is an important one; it is known as the contribution per first-class passenger.

Contribution and Variable Cost

Technically, *contribution* is the difference between a company's revenue and its so-called variable costs. Variable costs, as the name suggests, are the costs that vary directly with production activity. Drinks and dinners served on airlines are variable costs, since they are only expenses to the airline when passengers are in their seats to drink and eat. The lumber used in making furniture and the labor used to transform it are variable costs; they vary directly with the quantity of furniture produced. Similarly, the natural gas for drying peanuts and the casual labor used to run Devine Nuts are examples of variable costs.

These variable costs are easy to see, largely because they can be readily assigned to the units of output that people pay for—airplane tickets, pieces of furniture, tons of dry, graded peanuts. The "variableness" of other costs may not be quite so neat. For example, are flight attendants' salaries a variable cost? Salaries are certainly variable when it comes to devising the schedule of flights; more flights mean more flight crews. But once the flight schedule is set, the salaries paid flight attendants do not vary with the number of passengers. In terms of our situation at Citrus Airlines, the revamping of airplanes to eliminate the first-class cabin does not alter the flight schedule or the number of flight crews. Crew salaries, then, are fixed and only materials expenses are variable in this instance. This example serves to illustrate that one must be careful to think through which costs are fixed and which are variable with the particular production activity under consideration.

The $67.72 contribution per first-class passenger goes to pay crew salaries as well as to pay off airplane costs, ticketing, and a host of other costs that, for this decision, are fixed. Every additional passenger to first class who can be accommodated will contribute $67.72 to pay off these fixed charges and to secure a profit for Citrus Airlines.

Note that we can speak of contribution in a variety of forms. Total contribution is measured in dollars, as revenues less all variable costs. Contribution per unit is total contribution per unit of output and is measured in dollars per unit. In using this notion of contribution, both total contribution and contribution per unit supply the same information, but it is sometimes easier to think in terms of one rather than the other. What is essential, however, is that the costs that are considered variable be consistent with the nature of the decision for which the contribution figures are being used.

Back to Citrus Airlines

At present, the contribution per first-class passenger is $67.72. The similarly calculated contribution per coach passenger is $54.23 ($56 − $1.77). Given both an empty first-class seat and an empty coach one, Citrus Airlines would naturally prefer to have a first-class passenger. The issue, however, is not which type of passenger the airline prefers, but whether Citrus Airlines should abolish its first-class seating.

At present, a full cabin of 20 "sun-kissed" travelers contributes $1354.40 (20 seats × $67.72 contribution per passenger). If these 20 seats were eliminated and replaced by 28 coach seats, the total contribution would be $1518.44 (28 seats × $54.23 contribution per passenger). Thus, if Citrus Airlines can be assured of filling its converted first-class cabin with coach travelers, the company will make more money than it currently does−$164.04 ($1518.44 − $1354.40) per flight more.

By performing the same kind of calculation that was introduced with Devine Nuts and Wyoming Valley Furniture, we can easily see that the $15,000 cost of cabin conversion can be paid off in just 92 flights. To wit:

$$\$15,000 \text{ cost} \doteq \$164.04 \text{ contribution/flight} \cdot x \text{ flights}$$
$$x = 92 \text{ flights}$$

If each Citrus Airline plane flies one completely booked flight every four days, the cabin conversion cost would be paid off in a year.

This analysis suggests that Dave Dove's recommendation to convert the first-class cabin should be followed. Yet, on closer examination, this entire analysis hinges on the assumption that the additional eight seats placed in the converted first-class cabin will be occupied. It is not enough to state that the average load factor for coach is higher than the average load factor for first class or that, if the plane were full, more contribution would ensue if all seats were coach. The decision rests with what Citrus Airlines can expect will actually happen with the additional seats in the former first-class cabin. This mode of thinking about decisions concentrates on the incremental (often called marginal) change involved; the use of such marginal analysis is absolutely fundamental to management decision making.

Marginal Analysis and Sunk Costs

The concept of marginal (incremental) analysis for decision making is not fancy. It's just common sense that finds application in one business situation after another. In basic terms, marginal analysis states that the decision to do something should depend only on how that decision would change the situation and on nothing else. The question to ask is not whether a company will be profitable after some investment or policy change has been implemented, but whether the company will be more profitable for having made the choice than it would otherwise have been. Everything that has gone on before is irrelevant to the decision at hand.

This kind of situation is all too common: a project is almost finished when it is recognized as a turkey. Because the finish is so close, someone argues that "for just a little more money the project could be completed and the investment outlays to date will not be wasted." Marginal analysis says that what has gone before is irrelevant; the costs incurred are "sunk" and have no bearing on the decision. If the additional investment it takes to complete the project does not return additional revenues that are greater, the investment should be junked. As the old saying goes, "Don't throw good money after bad." The saying is absolutely right; the "bad" money represents sunk costs, and they have no bearing on the decision whether or not to continue with a project.

We shall have occasion many times to employ the concept of marginal analysis in this text. In fact, the analysis already completed on Wyoming Valley Furniture was couched implicitly in marginal (incremental) terms. The decision there rested on whether the additional costs from one option (build an inventory of legs) were likely to be lower than the additional costs from the other option (grin and bear it). In both cases, the additional revenues to be earned from the order were the same.

Back Again to Citrus Airlines

Can Citrus Airlines fill enough of the additional eight seats so that the total contribution of an all-coach plane exceeds that of the present airplane configuration? We have some information from the marketing report memorandum.

In the first place, we know about the standbys that have been left behind. Assuming an average of two left behind in the "one to five" category and six left behind in the "more than five" category, we can calculate the number of passengers that we could have been sure to seat on the 91 daily flights over the past 61 days. To wit:

61 days × 91 flights × 1%
    × 2 standbys/flight = 111.02
61 days × 91 flights x 7.4%
    × 2 standbys/flight = 821.55
61 days × 91 flights × 4.6%
    × 6 standbys/flight = 1532.08
111.02 + 821.55 + 1532.08 = 2464.65 people

At a contribution per coach passenger of $54.23, these stranded passengers could have

provided a total contribution of $133,660 in those two months if the first-class cabin had been remodeled.

This is not the full story, of course, since the previous first-class passengers must be accounted for. At best, Citrus Airlines will forgo the increased contribution that the first-class passengers made on all of the 727 flights. This "opportunity cost" is every bit as real a cost as the $15,000 per plane conversion change. The difference in contribution is $13.49 ($67.72 − $54.23) per first-class passenger. Over the past two months the total added contribution forgone (the opportunity cost of remodeling the cabin) would have been $916,568 (61 days × 91 flights × 20 seats × 61.2% load factor × $13.49).

This figure is naturally much larger than the gain in contribution from seating standbys in the eight additional seats provided by remodeling. The vast contribution gained by having first-class passengers on the flights that are not completely booked far outweighs the opportunity costs represented by standbys who could not get on existing flights. This comparison is all the more unfavorable to remodeling the first-class cabin if in fact customers begin deserting the airline.

Given these considerations, Dave Dove's ingenious plan for increasing capacity and breaking a bottleneck does not appear to be attractive enough to implement.

Lessons from Citrus Airlines

The Citrus Airlines situation brings together three of the basic concepts that guide good decision making at all levels of a company.

1. Opportunity cost (revenue forgone).
2. Contribution, and its companion concept, variable cost.
3. Marginal analysis, and its companion concept, sunk cost.

A useful analysis of Citrus Airlines' decision on "sun-kissed" service demands that these concepts be melded together. By the conclusion of this text these concepts should be second nature to every reader.

Let us review how each concept fits into the analysis of Citrus Airlines' situation:

1. *Opportunity cost.* By eliminating first class the airline was forgoing the difference in contribution between "sun-kissed" service and coach service. This loss would be every bit as much a cost to the airline as the cabin conversion expense of $15,000 per plane.

2. *Contribution.* For an already scheduled flight the cost of adding more passengers is small, since all the expenses of flying are fixed except for the drinks and food the airline serves. Thus, variable costs are very low, and the contribution each passenger makes to pay off the fixed investment in equipment and salaries is high. Contribution (revenue less variable costs) is the key statistic used in deciding the issue in this situation.

3. *Marginal analysis.* The decision about eliminating first-class service rested on the use of the additional eight seats put into the first-class cabin. Could the change they represented, including the opportunity cost incurred, be expected to generate enough contribution to pay off the added expense of converting that cabin?

Tools of the Trade: A Recap of Lessons Learned

This chapter, so far, has touched on a variety of issues and concepts that merit a brief recapitulation here.

Bottlenecks are vexing, that is certain, and often tricky as well. Analyzing them adequately has many advantages, and this chapter has introduced two tools for identifying and analyzing bottlenecks:

1. The process flow diagram and its sister, the information flow diagram. These diagrams can be tagged with capacity figures, times, yields, and other data that pertain to individual process steps.

2. The inventory buildup graph.

These tools help to spot the cause of a bottleneck, but other tools usually have to be deployed in order to make convincing decisions about remedying bottlenecks. The other tools that have been introduced in this chapter relate to the economics of the process and they include:

3. Indifference calculations.

4. Marginal, or incremental, analysis.

5. Opportunity cost recognition.

6. Contribution calculations.

By combining these concepts we have been able to make some sound economic decisions about the process bottlenecks that we have encountered. This kind of thinking is essential to improving processes from the inside out.

We turn now to a particular form of balance, line balance, that is a common need among line flow processes that must combine many small-scale tasks, many of which have requirements about what has to precede what else in order for the product to be made properly.

LINE BALANCE

A special case of bottlenecking is endemic to the line flow process. It is the imbalance of capacities caused not by differences in machine capacities so much as by faulty combinations of worker tasks. This is the issue of line balance.

The balance of tasks along the line is an important consideration. Indeed, when products and/or desired rates of production change, the line must often be rebalanced. The strength of a line flow process lies in the specialization of labor: a complex manufacturing assignment is split into small pieces, and a worker (any worker) is more adept at doing a few tasks repeatedly than many tasks intermittently. The observation that the *total labor content* of a product can be *reduced* by dividing up the process rather than having workers responsible for manufacturing the entire product was made famous by Adam Smith, the father of modern economics. In his *Inquiry into the Nature and Causes of the Wealth of Nations* (1776), Smith cites the fabrication of pins, noting that production was greatly increased by specializing labor.

The keys to line balance are (1) breaking down a complex product and process into its component pieces and tasks (often termed "job elements"), and (2) juggling the coordination of those pieces and tasks so that the process is smooth and no bottlenecks are built into it. Coordination is important because, after all, the output of any line is determined by the slowest workstation along it—its limiting job. No matter what else happens, if the limiting job cannot be changed, the line cannot produce at a faster rate.[2]

Needed Information and Calculations

In order to concoct a line balance, two items of information must be given beforehand:

1. How the product is made. This requires knowledge of all the job elements that comprise the product's manufacture, the time it should take to accomplish each job element, and which job elements have to precede other ones. This latter requirement is greatly aided by the construction of a precedence diagram, which portrays the individual elements that make up the fabrication and/or assembly of a product and the order they must logically or technologically follow.

2. How many units of the product are required per given time period (hour, shift, etc.)

The goal is to make all of the units required with the minimum waste of resources (idle time).

Consider an example. Our assignment for this example is to use the information in Figure 2-7 (a precedence diagram) and Table 2-1 (time estimates for each job element) to develop a line flow process capable of turning out what marketing and sales indicate the market requires, in this case, 300 units of the product in an 8-hour day.

Given this information, several questions need to be answered before a line balance is complete:

1. How big a line should be designed? What is the minimum number of workers needed in order to make the quantity desired, given the estimates of the time required to make a unit?

2. How fast must the line be run? This calculation is called the "control cycle" for each

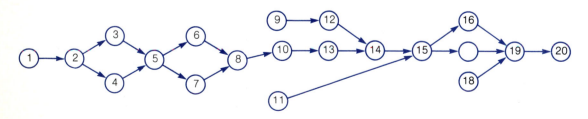

FIGURE 2-7   A sample precedence diagram for a line flow process.

TABLE 2-1   Job elements and estimated times for the process diagrammed in Figure 2-7

JOB ELEMENT	ESTIMATED TIME (minutes)	JOB ELEMENT	ESTIMATED TIME (minutes)
1	0.65	11	1.10
2	0.11	12	1.75
3	0.38	13	0.94
4	0.74	14	0.62
5	0.10	15	0.45
6	0.58	16	0.81
7	0.25	17	0.87
8	1.01	18	1.60
9	1.17	19	0.60
10	0.43	20	0.83
Total labor content			14.91

worker or workstation. It indicates how long it should take between units coming off the end of the line, or alternatively, how long a unit can afford to stay at a workstation and still have all the required number of units made.

3. Who does what on the line? How should the job elements be assigned to workers (workstations)? This is the determination of the actual line balance itself.

4. How good is the line balance? If it isn't satisfactory, it needs to be redone.

Calculating the Number on the Line and the Control Cycle

The minimum number of workers on the line can be calculated from the following formula:

$$\text{Minimum number of workers} = \frac{\text{Time needed to make the product (Units demanded} \times \text{time per unit)}}{\text{Work time available per worker}}$$

For our example, 300 units are demanded per 8-hour day, and it takes 14.91 minutes to assemble one unit. Thus,

$$\text{Minimum number of workers} = \frac{(300 \text{ Units demanded} \times 14.91 \text{ minutes per unit})}{8 \text{ hours (480 minutes) of Work time per worker}}$$

$$= 9.32 \text{ workers, or}$$

10 workers when rounded upwards. We cannot do better than assigning 10 people to the line, and quite possibly, more than 10 will have to assigned.

This answers the question of how big to make the line. The next question is how fast to make the line. What is the "control cycle" that will govern how long a unit of the product can stay with each worker? The control cycle is calculated as

$$\text{Control cycle (minutes/worker/unit)} = \frac{\text{Work time available per worker (minutes/worker/day)}}{\text{Desired output per day (units/day)}}$$

In this example, then, the control cycle can be figured as

$$\text{Control cycle (minutes/worker/unit)} = \frac{480 \text{ minutes of Work time/worker}}{300 \text{ units per day}}$$

$$= 1.60 \text{ minutes/worker/day}$$

TABLE 2-2   Worksheet for line balance assignments

WORKER (WORKSTATION)	TIME LEFT	ELIGIBLE ELEMENTS	WILL FIT	ASSIGN	TIME LEFT OVER (IDLE TIME)
1	1.6	1,9,11,18	1,9,11,18	1	0.95
	0.95	2,9,11,18	2	2	0.78
	0.78	3,4,9,11,18	3,4	4	0.04
2	1.6	3,9,11,18	3,9,11,18	3	1.22
	1.22	5,9,11,18	5,9,11	5	1.09
	1.09	6,7,9,11,18	6,7	6	0.51
	0.51	7,9,11,18	7	7	0.26
3	1.6	8,9,11,18	8,9,11,18	8	0.59
	0.59	9,10,11,18	10	10	0.16
4 & 5 alternate overcycle	3.20	9,11,13,18	9,11,13,18	9	2.03
	2.03	11,12,13,18	11,12,13,18	12	0.28
6	1.6	11,13,18	11,13,18	13	0.66
	0.66	11,14,18	14	14	0.04
7	1.6	11,18	11,18	11	0.50
	0.50	15,18	15	15	0.05
8	1.60	16,17,18	16,17,18	18	0.05
9	1.6	16,17	16,17	16	0.79
	0.79	17	17	17	0.04
10	1.6	19	19	19	1.00
	1.00	20	20	20	0.17

Assigning Job Elements to Workers (Workstations)

With the length of the control cycle established, developing the line balance involves picking 10 sets of job elements from Table 2-1 so that each of their sums totals 1.6 minutes. In this selection process, the precedence relationships portrayed in Figure 2-7 must also be adhered to.

Sometimes it is helpful to use a worksheet, such as Table 2-2 to help keep things straight as the job element choices are made.[b] The worksheet selects jobs for the ten workstations in our example. In this worksheet, the "time left" column always starts with the control cycle time and as different job elements get assigned to the workstation, the time left is reduced

by the estimated time of the assigned job element. The eligible elements are those determined by reference to the precedence diagram; elements farther along in the precedence diagram cannot be eligible until job elements before them are assigned, or unless they have no elements which precede them. The "will fit" column asks whether there is enough time left for each eligible element. The "time left over (idle time)" column looks at the time remaining after assignment. The last value in that column is the time left idle at that workstation; it is sometimes referred to as the "balance delay." Table 2-3 gives a suggested grouping of job elements into workstations. It prompts several observations:

1. We were able to develop a line balance with the minimum number of workers, 10. This, then, permitted us to have an idle time of

---

[b]This worksheet is the brainchild of Stan Stockton, Emeritus Professor of Operations Management at Indiana University.

TABLE 2-3   A suggested grouping of job elements into workstations

WORKERS (WORKSTATION)	JOB ELEMENTS	TOTAL TIME/UNIT (minutes)	IDLE TIME (minutes)	ADJUSTED IDLE TIME
1	1, 2, 4	1.56	0.04	0.00
2	3, 5, 6, 7	1.34	0.26	0.22
3	8, 10	1.44	0.16	0.12
4 & 5 alternate overcycle	9, 12	2.92	0.28	0.20
6	13, 14	1.56	0.04	0.00
7	11, 15	1.55	0.05	0.01
8	18	1.55	0.05	0.01
9	16, 17	1.56	0.04	0.00
10	19, 20	1.43	0.17	0.13
		14.91	1.09	0.69

1.09 minutes, given the control cycle of 1.60 minutes, and thus a production of 300 units per 8-hour shift. This represents a percent idle time of 6.8 percent, calculated as

$$\text{Percent idle time} = \frac{\text{Sum of idle times from all workers}}{(\text{Number of workers}) \times (\text{Cycle time})}$$

In our case this is found to be

$$\text{Percent idle time} = \frac{1.09 \text{ minutes}}{(10 \text{ workers}) \times (1.60 \text{ cycle time})}$$
$$= 6.8\% \text{ idle time}$$

This is a fine percent idle time. Indeed, industrial engineers sometimes are happy if the loss in efficiency from a trial line balance does not exceed 20 percent or so. Nevertheless, we can do better. Because all of the idle times were positive, we could contract the cycle time somewhat, to 1.56 minutes, eliminating 0.04 minutes from each worker's control cycle time. The last column of Table 2-3 shows what would happen to the idle times. Such a reduction to the cycle time would change the percent idle time to $[(0.69)/((10 \times 1.56))] = 4.4$ percent idle time. It would also increase the production available to 307 units [480/1.56].

2. Some workstations are more out-of-kilter than others. The "idle time" columns in Table 2-3 are the measures of how the total estimated time for the elements comprising each workstation (worker) differs from the control cycle, or, for the last column, the adjusted cycle time. Workers 2, 4, 5, and 10 have the most idle time, although, because the balance is so good, there is relatively little idle time anywhere.

Suppose, however, that there were significant differences in the idle times of particular workstations. What could be done about such a mix of bottleneck and slack stations? In practice, companies often try to devise work aids or other assistance to lessen the impact of any bottleneck. Both bottlenecks or significant idle times may trigger a review of the steps in manufacture and/or the product design itself. This review may lead to a new set of job elements and/or times with advantages for the balance of the line.

3. This particular assignment of workstations deliberately places two workers "overcycle." Workers 4 and 5 have to alternate work on every other unit instead of working on

every unit that passes down the line. Had not these two workers been placed on job elements 9 and 12, significant bottlenecks would have been encountered.

Determining exactly which of the eligible job elements to assign to a workstation takes judgment. To help in this task, various heuristics, or conventions, have been devised. None, however, ensure that the best grouping of job elements has been accomplished. Nothing yet devised can ensure that. Nevertheless, it is instructive to examine some of the heuristics that are sometimes used:

- *Longest processing time.* Choose the eligible job element that has the longest processing time. The rationale is to try to fit the tough ones first.

- *Most following tasks.* Choose the eligible job elements that has the most job elements following it on the precedence diagram. The rationale here is not to get caught with too much to do too late in the game.

- *Highest positional weight.* Choose the eligible job whose sum for its own processing time and the processing times for all its following tasks is the highest. This heuristic combines both processing times and following tasks, so it smacks of both of the previous heuristics.

One could also use shortest processing times or most preceding tasks as the conventions to use. The rationales are less clear with these, but their performance could be better in some circumstances. There are computer programs that can aid in the task of line balance, as well.

Line Balance in Practice

The preceding example of assembly line balance is obviously simpler and more mechanical than that faced by a company like General Motors (Tour D), but it does serve to introduce the fundamental notions of line balance. A more realistic example of a line balance problem, and one set in an automobile assembly plant, is given in Situation 2-4.

Discussion of Bumper Subassembly

We can use the data given to compute how much total time is required to assemble front and rear bumpers for car lines A, B, and C. Table 2-5 gives those computations.

These calculations show that in the aggregate, if two workers were placed on the subassembly of rear bumpers, the efficiency of those workers would fall within the 80 to 90 percent considered appropriate. The calculated required efficiency of two workers on front bumpers, however, exceeds 90 percent; that level of efficiency forces the pace too much, and management must be concerned with whether the workers can sustain the effort over long periods.

Also cause for concern is the variability in what the workers have to perform. Although, on average, the idle times do not look bad, there are some significant ranges of difference in what a worker may be required to do, depending on whether the bumpers take a certain option or not. Table 2-6 indicates the range of cycle times implied by the differences between standard and optional bumpers.

This vast range in cycle times per worker suggests that workers would be out of balance in this bumper subassembly area. When standard bumpers were called for, there would be substantial idle time on car lines A and C. However, when rub strips and bumper guards were called for, the workers would be greatly overcycled. This state of affairs, of course, is a powerful argument for the standardization of product options.

SITUATION 2-4

Bumper Subassembly

Before the front and rear bumpers are put on the automobile, several key parts must first be subassembled to these bumpers. Among these parts are mounting brackets, rub strips (rubberized impact strips that not only affect the car's appearance but also prevent minor damage to the bumpers), and bumper guards. Exactly which of these parts gets subassembled to the standard bumper depends on the style of car being assembled.

The assembly plant produces three different vehicle styles, denoted as car line A, car line B, and car line C. One of these vehicle styles is assembled every minute (i.e., 60 per hour) with the following current product mix:

Car line A	20/hour
Car line B	15/hour
Car line C	25/hour
Total	60/hour

Table 2-4 presents the elements of work and their respective task times (the time it takes to perform a designated task on a car, sometimes also referred to as cycle times) in hundredths of a minute. Note that not all cars receive bumpers subassembled with rub strips or bumper guards.

TABLE 2-4   Elements of work, front and rear bumpers: Car lines A, B, and C

FRONT BUMPER		REAR BUMPER	
Task	Task Time (hundredths of a minute)	Task	Cycle Time (hundredths of a minute)
Car Line A			
1. Transfer bumper from delivery rack to conveyor.	0.20	1. Transfer bumper from delivery rack to conveyor.	0.20
2. Install 2 rub strips, with 10 clips each, on 15%.	0.05/clip	2. Install 1 rub strip, with 20 clips, on 10%.	0.05/clip
3. Install 2 bumper guards on 40%. Install loose with required parts.	0.45/guard	3. Install 2 bumper guards on 20%. Install loose with required parts.	0.45/guard
Secure with air motor.	0.12/guard	Secure with air motor.	0.12/guard
4. Install 2 mounting brackets on all front bumpers.	0.65/bracket	4. Install 2 mounting brackets on all rear bumpers.	0.65/bracket
Car Line B			
1. Transfer bumper from delivery rack to conveyor.	0.25	Identical to front bumper.	
2. Install 1 rub strip, with 8 clips on each bumper.	0.08/clip		
3. Install 6 mounting bolts on each bumper.	0.15/bolt		
4. Car line does not require bumper guards.			

continued

Car Line C

1. Transfer bumper from delivery rack to conveyor.	0.20	1. Transfer bumper from delivery rack to conveyor.	0.20
2. Install 1 rub strip, with 12 clips, on 50%.	0.10/clip	2. Install 1 rub strip, with 12 clips, on 35%.	0.10/clip
3. Install 2 bumper guards on 5%. Each guard requires a total of 0.70, which can be divided into three elements of 0.20, 0.30, 0.20, if necessary.	0.70/guard	3. Identical to front bumper (bumper guards).	0.70/guard
4. Install 6 mounting bolts on each bumper.	0.15/bolt	4. Identical to front bumper (mounting bolts).	0.15/bolt

Because the plant is building 60 cars per hour, an operator's cycle time cannot exceed 1 minute per job; but no one can realistically be expected to work at a production pace of 60 minutes every hour. Thus, an allowance is made so that if an operator's workload falls between 80 and 90 percent of the total time available (in this case 1 minute per job), his cycle is considered "efficient." Therefore, if by combining work elements from each of the three car lines an operator is working somewhere between 48 and 54 seconds (or 80 to 90 hundredths of a minute) per cycle, he is considered efficient.

How many operators are required to subassemble these bumpers and what process considerations could assist the operators in their work?

How does a factory solve such a pernicious problem? One possibility is to put more workers on the line so that the overcycle conditions can be met, but then the factory must live with vast idle times (inefficiencies).

A more palatable solution is to improve the times (methods) required to do each task. To do this, the operators could be made more comfortable (by raising or lowering the platform, installing production aids such as air motors or lifts), or the layout of the workstation could be improved (e.g., by making materials more accessible).

Another course of action would be to subdivide the tasks required for the options work into many small component tasks and then to spread these component tasks over as many workers as is feasible. In this way, no one worker would be tremendously overcycled. Rather, several workers would be only modestly overcycled. The success of this tactic depends naturally on how neatly the major tasks can be subdivided and over how many other workers it is feasible to spread the component task.

Still another course of action would be to take as much of the options work off line as possible to decrease the variability in times faced by the worker. In this case, bumper guards could perhaps be mounted to the bumpers or rub strips attached off line, thus leaving to the assembling line merely the transfer and installation of the bumper itself with any options already mounted.

Accuracy of Time Estimates

Proper line balance depends on accurate estimates of the time it will take to perform each of the job elements of the entire assembly process. The time estimates themselves are usually derived from actual observation of the task

TABLE 2-5   Computation of total time to subassemble bumpers

FRONT BUMPER		REAR BUMPER	
TASK	TIME NEEDED (minutes)	TASK	TIME NEEDED (minutes)
Car Line A (20/hour)			
1	4.00	1	4.00
2	3.00	2	2.00
3	9.12	3	4.56
4	26.00	4	26.00
	42.12		36.56
Car Line B (15/hour)			
1	3.75	1	3.75
2	9.60	2	9.60
3	13.50	3	13.50
	26.85		26.85
Car Line C (25/hour)			
1	5.00	1	5.00
2	15.00	2	10.50
3	1.75	3	1.75
4	22.50	4	22.50
	44.25		39.75
Car Lines A, B, and C			
Total	113.22	Total	103.16
Efficiency*	94%	Efficiency*	86%

*Based on two people working and 120 minutes of work time per hour of production.

TABLE 2-6   Cycle time per worker (minutes) (assumes two workers each on front and rear bumpers)

CAR LINE	FRONT BUMPER		REAR BUMPER	
	NO OPTIONS	ALL OPTIONS	NO OPTIONS	ALL OPTIONS
A	0.75	1.82	0.75	1.82
B	0.90	0.90	0.90	0.90
C	0.55	1.85	0.55	1.85

itself or something very similar. The observation, commonly called a time and motion study, can occur in any of a variety of ways: often industrial engineers will use more than one to establish a time standard.

• *Stopwatch studies.* A traditional but less and less popular means of estimating time standards is to stand by a worker who is performing a task and take stopwatch readings of how long it takes him or her to accomplish the task. As one might expect, the presence of the industrial engineer may cause the worker to slack off somewhat, casting doubts on the estimate. Sometimes this inaccuracy is avoided by having the industrial engineer stand near one worker but actually monitor another worker at a distance. Even this bit of sneakery has not remedied all the problems associated with stopwatch studies, which remain a costly way of setting time standards.

- *Films and videotapes.* A less obtrusive way of estimating time standards is to film workers in action over an extended period of time and then take time measurements from the film itself. This has become an increasingly popular way to estimate time standards; it reduces the cost and worker-influence problems of stopwatch studies.

- *Work sampling.* Another, and less expensive, way to determine at least the relative times spent performing various tasks is to take a series of random "snapshots" of a task rather than a "full-length film." If, for example, in every 100 random looks at a particular task, 40 observe a particular action, then we can infer that 40 percent of the cycle time is spent performing that task. This technique is generally termed work sampling. It is most appropriate for nonstandard tasks.

- *Task decomposition.* Since so much information has been accumulated on the time it takes to perform so many bodily movements, sometimes time estimates can be pieced together strictly by breaking the task into its component movements and using stock time estimates for each movement. Summing these individual times is one way to estimate the time standard for a radically new job element, which can then be checked by any of the previous three observational methods.

No matter how individual job element time standards are developed, it is dangerous to ex-

pect that adding more than a few together will generate accurate time totals. Table 2-1's list of 20 job elements totals 14.91 minutes. It is foolish to expect that a single worker could actually work through all 20 steps in 15 minutes, let alone maintain that pace throughout a workday. The strength of a line flow process, as mentioned before, lies in the specialization of labor to a limited number of tasks; the corresponding time standards can be considered accurate only when relatively few job elements are strung together. The rebalance of any line flow process always entails reevaluations of new tasks and/or regroupings of elements, since so many time variations can be introduced to combinations of job elements and the materials handling and positioning requirements that accompany them.

Nontime-Related Aspects of Line Flow Process Layouts

The 20-element example of the line balancing introduced earlier does not assign any special characteristics to any of the job elements listed. In that sense the exercise was mechanical. Often, however, in many line balance situations attention has to be focused on some special features of the process that may constrain the composition of workstations in particular ways. Among these special features can be included:

- Materials handling requirements.
- Space constraints within the factory that may force the line into a particular shape.
- Special equipment needs.
- Product bulkiness or weight, requiring two or more workers to work together.
- Segregation of worker skill levels.

Any of these features can in themselves define a workstation, quite apart from the simple grouping of job elements according to time estimates alone.

Cycle Time Allowances

Some procedures for line balancing, though by no means all, factor in special allowances for fatigue. These allowances lengthen the cycle time. The time standards for individual tasks, under this system, have no slack of any kind built into them, and so it would be inappropriate to expect that they could apply for a full day's workload. Special allowances for fatigue then have to be added to make the computed cycle time realistic. Other line balance procedures try to establish sustainable time standards from the beginning, and so allowances need not be introduced.

Speed Versus Idle Time

As has been mentioned repeatedly, the appeal of a line flow process lies in its ability to reduce the labor time in a product's manufacture by breaking down the fabrication/assembly process into pieces that can be done rapidly and repeatedly by operators along the line. Generally speaking, the smaller the job content of a workstation, the more efficient labor can be. What prevents us from breaking tasks into single job elements? Apart from concern for the mind-numbing character of such jobs, the problem of idle time militates against such an atomistic approach to line balance. Individual job elements will vary greatly; the greater the variance, the more apt that some combination of job elements will have to occur to prevent high levels of idle time all down the line. Thus line flow processes with cycle times that are combinations of job elements are typically more efficient than ones with minimum cycle times.

Moreover, such longer cycle line balances are more apt to offer the kind of flexibility for multimodel lines that is increasingly valued. Cycle times that are less than about half a minute run real risks of inflexibility, precluding the development of mixed model lines.

QUESTIONS

1. Depict a hybrid process's process flow diagram. What makes the planning of production different for a hybrid process as opposed to a "pure" production process such as a batch flow or line flow process?

2. Pick a service operation and diagram its process and information flows. Be sure to diagram the contact points where customers interact with the process, say, by including those contacts as the upper part of the diagram.

3. What do you understand by the term bottleneck? Describe one method of analyzing bottlenecks.

4. What is an indifference calculation, and why is it used? Formulate your own example to show how the indifference calculation is used.

5. What is contribution? How is it useful?

6. Explain opportunity cost. Explain marginal analysis.

7. In what kinds of situations would you expect the slopes in an inventory buildup graph to be different? (In the Devine Nuts situation, the slopes in the graph were the same.)

8. Define the three important decision-making concepts embodied in the problem-solving process undertaken by Citrus Airlines. Taking an example of your own, show how these concepts can be applied.

PROBLEMS

1. In what ways can bottlenecks be remedied? Be specific.

2. What was the fundamental decision faced by Wyoming Valley Furniture? Why was the company's initial problem characterized as a bottleneck problem? Could the company have arrived at the same decision by any other management method?

3. Hope Hospital currently has 80 rooms and is able to charge $150 per day for occupancy of each room. Because Hope has had to turn away patients several times during the last 2 months, the administrator is considering expansion of the facilities to 100 rooms at a cost of $5 million. Assuming that variable costs per day average $60 per room and that all rooms will be filled, how long will it take the expansion to pay for itself?

4. If Hope Hospital expands to 120 rooms, it can do so by spending $6 million rather than the $5 million in Problem 3. In addition, however, variable costs will increase to $65 per room per day, and it is expected that the average occupancy rate will be only about 90 percent.

(a) How does this change your analysis from Problem 3?

(b) Which alternative is better?

(c) If revenue per day could be changed for the second alternative, at what revenue per day would the two alternatives break even?

5. Refer back to Situation 2-1. Calculate the amount of truck wait if 350 tons of peanuts characterized a peak day.

6. Gary Zunica manages a receiving station for various kinds of agricultural produce. Farmers typically deliver their freshly picked produce during a 12-hour period. Gary can process produce at a rate of 5 tons an hour, but during heavy days, a total of 90 tons of produce can be expected. Gary needs to finish processing within 20 hours of the day's start to leave enough time for cleaning and maintenance. The station has room to hold only 10 tons of produce in raw materials inventory prior to processing. Once this is filled, the farmers' trucks must wait to dump their contents.

(a) If produce arrives at the receiving station at an even pace all day, when is it likely that trucks will have to start waiting to unload?

(b) Draw the inventory buildup graph for this situation. How does it differ from that of Devine Nuts?

(c) What is the total extent of the truck wait experienced on a peak day?

7. The Lenz Company refurbishes old MBAs. The process takes the four sequential steps shown below.

STEP	DESCRIPTION	TIME REQUIRED PER MBA (minutes)
1	Strip off bad habits	1.5
2	Scrub and clean mind	0.8
3	Insert modern methods	1.0
4	Polish and return	1.2

One faculty member is assigned to each of these steps. They rotate jobs at the end of each 40-hour week. Mr. Lenz has been working on a contract from General Motors that requires delivery of 2000 refurbished MBAs per week. The GM Human Resources Department has just called complaining that they have not been getting the agreed-upon number of MBAs. A check of the finished goods inventory by Mr. Lenz revealed that there was no stock left. What is going on?

8. The boxing operation at a greeting card company was mostly automated. It involved placing counted cards into "packers" and then placing various packers into "shippers." The process involved a line that had a mechanical card counter that counted and placed a bundle of 12 cards on a moving belt, another machine that fed envelopes onto the belt, an Aerosonic machine that an operator used to help insert the cards into packers and two operators that placed packers into shippers. A materials handling person spent about 2 hours a day removing the shippers and filling the equipment. (This person actually helped out at two such lines and did odd jobs in the area when not working at either line.)

The equipment could operate at various speeds, but the minimum cycle times were: 0.03 minute for the card counter, 0.027 minute for the envelope feeder, 0.025 minute for the Aerosonic machine and operator, and 0.08 minute for putting packers into shippers. The company now works a full 8-hour day, and each line produces 12,000 packers per day. To meet a special contract for another company, the greeting card company would like to increase the output of each line to 16,000 cards per day without using overtime or another shift. What should they do? If overtime was permitted, how much would be required to meet their objectives?

9. A Canadian mining company struck it rich! They came into a huge deposit of ore that required no waste removal. To exploit their find, they moved in some extra-large shovels to extract the ore and load the trucks. They added a third line to the crusher so that two lines were always running, and they improved the ore concentrator's efficiency. The results were a capability to load trucks at the rate of about 10 minutes per truck for each of the three shovels used, to crush at the rate of 1000 tons per hour, and to concentrate at 1200 tons per hour. Another company had offered this mining company a premium price if they could provide 24,000 tons per day of concentrated ore.

The company was studying their truck operations. Each truck would pull into the pit, go to the next available shovel, and load. This took 10 minutes. The trip to the crusher from the pit took 15 minutes and another 10 minutes was needed to "spot" the truck and unload. The return trip took only 10 minutes. Each truck could carry 80 tons of ore. How many trucks do they need to have in operation to support the Canadian firm's proposal?

10. A new bottling line has been put into operation at the Spirit Bottling Company. The line produces a single product, Purple Burple, in 1-liter and 2-liter bottles. The demand for the two sizes is fairly constant at 16,000 cases per week for the 1-liter size and 8000 cases per week for the 2-liter size. The line speed is 1000 cases per hour for the 1-liter size or 500 cases an hour for the 2-liter size. The company works the line a full 40-hour week and currently produces each size once a week (i.e., lot sizes of 16,000 and 8000 cases). If it takes 30 minutes to change over from one size to the other, what is the minimum lot size of each bottle size that can be run, and how many changeovers does it require a week?

11. If there is a positive idle time (balance delay) at every workstation along the line, is there anything that can be done to improve the percent idle time for the entire line? What do you suggest?

12. Using the information in Figure 2-7 and Table 2-1, concoct a line balance that uses each of the following conventions to make assignments of job elements to workstations.

(a) Longest processing time.

(b) Most following tasks.

(c) Shortest processing time.

Calculate the percent idle time for each assignment technique's line balance.

13. The tasks (i.e., job elements) shown below were necessary for the assembly of a child's toy. Certain tasks, as indicated, needed to be done before others.

The factory worked a day of 450 minutes, counting all breaks. The toy production plan called for 1000 units to be assembled each day.

(a) Draw the precedence diagram for this process.

(b) Develop an assignment of tasks to workstations where the percentage idle time is less than 20 percent. Use any heuristic you want, or none at all.

(c) What is the cycle time for the line?

(d) What percentage idle time does your line balance have?

TASK	TIME ESTIMATE FOR TASK (seconds)	PREDECESSOR TASKS THAT HAD TO BE ACCOMPLISHED BEFORE THIS TASK
A	11	None
B	7	None
C	12	None
D	11	A, B
E	5	C
F	13	E
G	7	D, F
H	9	None
I	15	G, H
	90	

14. Refer back to Problem 13. Using the following heuristics, concoct a line balance.

(a) Longest processing time.

(b) Most following tasks.

(c) Highest positional weight.

What is the cycle time for each line balance? What is each line balance's percent idle time?

15. Hoosier Homes builds moderately priced houses around the state. Although some variation exists, there is a "typical" house that they build, and they use the crew inputs into this house as production standards for planned purposes. To wit:

TYPE OF CREW	WORKERS	WORKER-DAYS PER HOUSE
Carpenters	38	43
Masons	22	29.5
Plumbers	4	9
Electricians	11	15
Plasterers	19	15
Painters	20	34
Tile Setters	6	6
Cleaners	4	2

The work week is 5 days long. Comment on this process and its balance.

Problem 16.

PROCESS STEP	STANDARD TIME (seconds)	COMMENT
Stir tar, inspect viscosity	5	
Weigh-out feathers	18	Feathers are a costly material item.
Attach I.U. fan to rail:		
Lower half of body	30	Requires two people working together.
Upper half of body	20	Requires two people working together.
Pour tar over half of body	22	
Sprinkle feathers over tar	25	
Turn over fan and pour tar over remainder	20	
Sprinkle feathers over newly poured tar	22	
Spray feathers gold	45	
Touch-up	30	

16. Alfonso Wright Hudson, operations manager for Keady Tar Products, contemplated the manufacture of a new product for the company that promised significant sales, namely the tarring and feathering of I.U. basketball fans. Demand for this new product was particularly strong in the northern half of the state. Some preliminary work suggested the process steps and standard times shown in the table shown on page 62.

The Keady Tar Products Company worked a full 8-hour day. Hudson wondered how many workers he needed to process 1000 fans a day and what their specified task assignments ought to be along the assembly line he envisioned for this product.

SITUATION FOR STUDY 2-1

GREAT WESTERN TIME INC.

Adele Welch is a senior at Western Theoretical University, majoring in operations management. The university has a cooperative student work-study arrangement with the local plant of Great Western Time Inc. The students are hired for 6 months, from June to December of their senior year, as full-time first-line supervisors or the equivalent.

Adele has just started at Great Western as a supervisor over an assembly line for electronic watches. The watch assembly line is a moving belt 25 feet long at which 28 workers assemble, solder, and test watches through the complete assembly process.

Initially, an assembly line balance is developed from the order of operations, the time required for each operation, and the number of units desired. This information is also used to determine the crew size. In order to achieve the required production volume, this line balancing design usually calls for several workstations for some operations. The workstations are simple and require simple tools, so idle workstations do not represent major losses in terms of equipment.

In the actual operation the watch case is placed on the conveyor belt, and the other subassemblies are added by workers by taking the case off the line, putting it in a fixture, performing their assigned operation, and placing the watch back on the conveyor. If there are five identical workstations, each of the five workers is expected to perform that operation on every fifth watch. Since the product is so small and light, there is little difficulty in returning an unfinished watch to the beginning of the line, if necessary. However, a large number of returns indicates an imbalance in the work arrangement.

TABLE 2-7   Sequence of operations and times for producing the Slinky Slim watch (Great Western Time Inc.)

OPERATION	PRECEDING OPERATION(S)	TIME REQUIRED (minutes/unit)
1	—	0.2
2	1	0.3
3	1	0.3
4	1	0.5
5	4	0.2
6	3	0.2
7	4	0.4
8	2	0.5
9	3, 5, 7	0.6
10	8	0.6
11	9, 10	0.3
12	10, 11	0.1
13	9	0.2
14	11	0.5
15	12, 14	0.3
16	15	0.2

Adele has 23 people assigned to her; when one or more does not report for work, she has no labor pool from which to draw. Moreover, if a worker does not feel well and leaves at noon, Adele has to rebalance the assembly line quickly and adjust the belt speed to the rebalanced line. Table 2-7 shows the sequence of operations and the times involved in the manufacture of the Slinky Slim watch.

Although Great Western would like all assembly operations to go like clockwork, upper management has factored in a value for determining the "average" output requirement. The performance measure used to evaluate Adele and her colleagues is the labor content of each assembled watch compared to an engineered standard. Adele feels that her primary job is to generate a new line balance at any time. In particular, she wants to generate, ahead of time (in case absenteeism forces her to rebalance the line), a set of line balances and work assignments for the following numbers of workers on the Slinky Slim assembly line: 23, 22, 21, 20, 19, 18.

ABLE RENT-A-CAR

At the Modest Suburban Airport near Metropolis, Able Rent-A-Car shared space with two other car rental companies: Baker and Charlie. Their small, equal-sized counters were connected to each other. The rental cars were located in front of the terminal beyond the airport short-term parking and up a steep hill. To get to the rental car, the customer had to walk approximately 150 yards, including 20 steep steps.

Able divided the United States into zones, which varied in size depending on population, number of Able locations, and other practical issues. Each zone was headed by a zone manager, who made all the major administrative decisions, and had its own fleet of cars. The Modest Suburban Able was in the Metropolis zone.

Two telephone numbers were critical to the Able business. One was the main reservation center in Kansas, which served the entire United States and Canada. Kansas made reservations, distributed location business hours and phone numbers, and explained rate structures. The second key number was fleet distri-

bution; the Metropolis fleet center was responsible for ensuring that each location within the zone had enough cars in each size bracket. It was vital in the rental car business that a person who appeared at the counter with a reservation be given a suitable car ready to drive.

There were two types of employee position at Able: the service agent and the rental agent. The service agent's main responsibility was car care. At the start of each shift, the service agent used the computer, stationed behind the customer counter, to get a reservation manifest, which listed the names of customers, when and how each customer would be arriving at the airport, and what size car had been ordered. The service agent had to be sure that there were clean, rentable cars of the correct size ready for each reservation.

After reading the manifest, the service agent put into the computer the identification numbers of the cars that were in the lot and not serviced. The reason for this procedure was to ensure that each car was clear in the computer system. The computer would tell the serviceman whether certain situations preventing immediate servicing existed: (1) the car had been stolen and was now on alert; (2) the car was ready for preventive maintenance, in which case the car must be sent to the city for repair; or (3) the car had mistakenly been entered as present in a different location.

After acquiring the customer manifest and the car clearance forms from the computer, the serviceman walked to the lot to begin servicing cars. The lot was shared with Baker and Charlie; each company had its own gas pumps and parking slots. In addition, Baker and Charlie had an automatic car wash, which could wash the outside of a car in 20 seconds. Servicing cars entailed pumping gas; checking fluid levels, tire pressure, lights, mirrors, safety equipment, spare tire, and jack; and cleaning the inside windows, upholstery, outside windows, and main frame. Whereas Baker and Charlie ran their cars through the car wash, Able used a garden hose and a brush. While servicing cars, the service agent maintained a car control inventory sheet, which listed all Able cars on the lot and divided them into Metropolis zone cars, out-of-service Metropolis zone cars, and non-Metropolis-zone cars. After servicing each car, the service agent filled out a car information form, which listed the number of doors on the car, the make, model, mileage, tire type, number of passengers seated, car identification number and the parking slot number where the car would be parked.

After servicing about five cars, the service agent took the car information forms and the keys to each serviced car down to the counter in the terminal. The rental agent took the keys and the car information forms and put them on the key rack. The service agent then returned to the lot and began servicing more cars.

The responsibilities of the service agent included also inventory regulation; each week the service agent filled out an inventory need list, noting all the car care supplies and counter supplies that were running low. The agent

phoned this list in to Metropolis, where another service agent collected the needed supplies from the stockroom and drove them to Modest Suburban Airport. Other service agent duties included picking up customers from the terminal in bad weather and driving them the 150 yards to the parking lot. They also made sure the lot was clear of dirt and snow.

The other employee position at Able was the rental agent, who was stationed behind the counter in the terminal. Responsibilities of the rental agent centered around the customer: transacting rentals and returns; handling reservations that were not made through the Kansas number; answering phone inquiries about rates, car availability, and business hours; and making reservation changes. The renting process began with the customer approaching the counter and identifying himself or herself. The rental agent checked the daily customer manifest to verify the reservation and proceeded to ask the customer for a driver's license and a major credit card. The rental agent then fed the data from the two items into the computer.

Next the rental agent inquired about how long the car would be needed and where it would be returned. Using this information, the agent assigned a rental rate. There were numerous rate structures, which depended on when customers rented, when they returned, where the cars were returned, and how long customers had the cars. After assigning a rate, the rental agent estimated the cost of the rental and called a credit card toll-free number to verify the credit stability of the customer. After the credit check, the rental agent inquired about the desired car make and model, went to the key rack, and chose the appropriate car. The rental agent then entered the data on the car information form into the computer, which printed out all the data from the rental agreement. The customer signed the contract and was given all but one copy of it, with instructions to bring the contract when he or she returned the car. The rental agent gave the keys and the slot number to the customer with directions to the parking lot. The agent then filed the contract copy.

When returning a rental car, the customer left the car and the keys in the lot, noted the mileage and gas level, and returned with the contract to the terminal. The rental agent entered the present mileage and the gas level into the computer, which then calculated and printed out the bill. The customer got one copy and the rental agent took the remaining three (one was sent to Metropolis, one to billing, and one copy was kept). Almost all business was done with credit cards.

Rental agent duties also included filling out the car count inventory form and sending it through the computer to fleet distribution. This form was crucial because it informed fleet operations about Modest Suburban's car needs for the day. The car count inventory listed data from the present 8-hour period and the next two 8-hour periods. The rental agent used the car control inventory sheet made out by the service agent to list present cars available. He or she then used a forecasting book, written by the Metropolis office using historical data, to estimate how many cars would check in to Modest Suburban Airport during each 8-hour stretch. After adding present available cars and forecasted check-ins, the rental agent subtracted the reservation needs for each of the three 8-hour periods; the result showed either a surplus or a need for each period. The car count inventory form also enabled the rental agent to inform fleet of any special orders such as for station wagons or vans or distant reservations.

1. Diagram the process and information flows for this service operation.

2. How is it like a machine-paced line flow operation? How it is unlike such an operation?

3. What changes would make it more like a machine-paced line flow operation? How many of those changes are dependent on being at a large metropolitan airport to make them effective? (Consider the differences between the rental car operations at large airports and those at small airports.)

REFERENCE NOTES

1. For more on bottlenecks consult:
   Ashton, J. E., and F. X. Cook, Jr., "Time to Reform Job Shop Manufacturing," *Harvard Business Review* 67, no. 2 (March–April 1989):106–11.
   Belasco, James A., "Managing Bottlenecks," *Executive Excellence* 5, no. 3 (March 1988):15–16.
   Lambrecht, Marc R., and Lieve Decaluwe, "JIT and Constraint Theory: The Issue of Bottleneck Management," *Production and Inventory Management* 29, no.3 (Third Quarter 1988):61–66.

2. For a discussion of assembly line balance consult the following articles and their bibliographies: R. E. Gunther, G. D. Johnson, and R. S. Peterson, "Currently Practiced Formulations for the Assembly Line Balance Problem," *Journal of Operations Management* 3 (1983): 209–221; and I. Baybars, "On 'Currently Practiced Formulations for the Assembly Line Balance Problem'," *Journal of Operations Management* 5 (1985): 449–453.

CHAPTER 3

BOTTLENECKS IN SPACE AND TIME: LAYOUTS AND VARIABILITY

The previous chapter addressed itself to bottlenecks and balance in the process, and introduced some tools for analyzing those concepts. There is more to discuss, however—two more concepts, and some associated tools, that have been ignored to date. These concepts relate to space (layouts) and time (variability). These topics round out the discussion of the important, interrelated topics of flows, bottlenecks, and balance.

PLANT LAYOUTS AND BOTTLENECKS

In the last chapter we discussed the use of process flow diagrams and inventory buildup graphs as tools for identifying and analyzing bottlenecks. Those tools look essentially at process elements, their capacities, and the time involved (capacity per unit of time or inventories accumulated per unit of time). What those tools ignore is the spatial aspects of production—how materials and product move from point to point in the factory. This "geography" is too important to ignore. As noted previously, bottlenecks can be created from poor plant layouts. Poor layouts can

- Interrupt the physical flow of materials.
- Add direct labor time to the product.
- Introduce excessive handling with the risk of damage and loss.
- Cause queues of work-in-process inventory to proliferate throughout the factory.
- Keep workers and/or managers at a distance from one another when in fact they should be close.
- Increase the time it takes to manufacture the product.
- Increase setup time unduly.
- Add to overhead.
- Crowd departments into too little space.
- Contribute to poor housekeeping.
- Make any plant expansion more difficult to accomplish.
- Otherwise add costs to the operation.

Some poor layouts are born, but most are made. That is, some poor layouts are bad designs from the beginning, but most poor layouts develop bit by bit over time, as one change after another is made. New products get introduced, and the space they need is hurriedly assigned. New equipment is purchased, and a place is quickly found for it so that the rest of the process is not disturbed while it is debugged. The plant is expanded, but to avoid disruption, it is not completely re-laid out; only a few areas are. In these ways, over time, bottlenecks are built into the process and its layout.

Types of Layouts

Generally, layouts can be classed into three major categories—job shop, line flow, and fixed position—although several offshoots are possible.[1]

The Job Shop (or Process) Layout

The job shop layout (as exemplified by Norcen Industries, Tour B) groups similar machines together. Such a layout makes sense if jobs are routed all over the place, there is no clear dominant flow to the process, and tooling and fixturing need to be shared. For example, if the process sheet calls next for grinding, in a job shop layout, it is clear where the job is to be routed. The job can then enter the queue of work for the next available grinder in the group. If, on the other hand, grinding machines were scattered throughout the factory, there would be chaos. The job of production control and materials handling would be difficult; priorities and machine availabilities would be very tough to track and to execute well.

There are some other benefits to grouping like machines together, as well. For example, maintenance and setup equipment can be stored nearby. And, operators, without thorough cross-training, can run two or more pieces of equipment to enjoy the productivity gains inherent in such a scheme.

The job shop (or process) layout is attractive when routings are unique to each job, a situation that would create chaos in a more line flow layout.

The Line Flow (or Product-Specific) Layout

If there is a discernible, dominant flow to the process, then a line flow layout has tremendous advantages over a job shop layout. Materials handling can be greatly simplified, and the space necessary for production can be reduced. Production control is easier; the paperwork trail to each job in the job shop can be largely abandoned. In effect, the layout itself acts to control priorities. Work-in-process inventories can be shrunk to a fraction of what they would be in a job shop layout. Production cycle times can be similarly reduced, making the feedback of quality information that much quicker and more effective, as well.

There are a variety of line flow (or product-specific) layouts. At one extreme are the continuous flow process layouts, with their very high capital intensity (e.g., Tour A, International Paper's Androscoggin Mill). For these processes, it is very true that the process is the layout and the layout is the process. Changing one means changing the other, and any changes involve tremendous expense, and, for that reason, are accomplished only occasionally.

At the other extreme are worker-paced line flow processes that do not involve much plant and equipment (such as Tour F, Burger King). Rather, they are more labor and materials intensive. These processes are very flexible; they can be rebalanced, turned, lengthened, chopped up, and so on with comparative ease. Lying in between the extremes are machine-paced lines (as those in Tour D, General Motors-Oklahoma City) that are more flexible than continuous flow process but not as flexible as worker-paced lines.

The worker-paced lines can typically produce a number of different product "models" at the same time. This is often also true of some machine-paced lines. These mixed-model lines are equipped and their workers trained to do quick setups, where needed, and to choose (or adapt) the proper materials to the model indicated. In these mixed-model lines, similar models are typically interspersed among the other models assembled. Often, this helps balance the line by not tossing a succession of overcycle tasks at key workers. (Overcycle tasks are those that take a worker longer than the others on the assembly line; thus, more than one worker must be assigned to such tasks if the assembly line is to produce at the planned rate.)

The Fixed Position Layout

A layout in which materials are brought to a stationary product is termed a fixed position layout. It is common when the product itself is so massive and/or awkward that transporting it through the process is unreasonable. Construction and shipbuilding are readily recognizable examples of fixed position layouts. So, too, is the auto repair operation of Tour G, Ogle-Tucker Buick (Tour G).

Fixed position layouts are typically chosen by default. There are several crippling aspects of fixed position layouts that do not recommend them for a wide variety of products. First, they make materials handling more difficult because a "kit" of similar parts has to be broken up and distributed at the proper time to a variety of products-in-process rather than delivered to a single point along the line. Second, workers of different types and skills have to move from product to product, or else a single set of workers has to remain with the entire job. In the former case, scheduling worker movements becomes a chore, while in the latter case training workers to do the entire job takes time and resources. Third, quality control becomes more problematic in fixed position layouts. Inspectors often have to roam, and that may waste operators' time as they wait on

inspection. Moreover, one does not have the luxury of evaluating the process capabilities of just one machine or one station along the line; there are many stations to evaluate, as many stations as there are stalls filled with work-in-process.

In the past 20 years, there have been some attempts by automotive companies (notably Volvo) to employ a fixed position, "stall build" layout. The argument for it was enhanced responsibilities and motivation for the workers involved. Despite their best efforts, however, the stall build experiments were never as successful as the conventional assembly line. In sum, fixed position layouts are usually not the most efficient layouts, but for the products or services to which they apply, there is no good substitute.

These, then, are the three major categories of operations layouts. As we noted in Chapter 1, many processes are hybrids, and often these hybrids exhibit hybrid plant layouts. Perhaps the most common hybrid is the batch flow/assembly line hybrid process with a job shop/line flow layout.

The Power of the Line Flow Layout—Recent Trends

As described previously, the one layout that is not in some way a default is the line flow (product-specific) layout. The job shop layout works best only when there is no dominant flow to the process, and the fixed position layout applies to products and services where moving the product is just too troublesome. There is real power in the line flow layout: efficient use of labor, sure and effective materials handling, little buildup of work-in-process inventory, specialization of tooling and methods, and quick feedback of quality problems.

The spread of a concept termed "group technology" (or, often, "manufacturing cells") is testament to this power.[2] In essence, group technology is the conversion of a job shop layout into a line flow layout. Instead of grouping similar machines together, group technology may call for grouping dissimilar machines together into a line flow process all its own. In the new arrangement, a part can travel from one machine to another without waiting between operations, as would be customary in the job shop. Work-in-process queues of material are thus reduced; individual parts move more quickly through the process.

Group technology takes a hard look at the products or parts manufactured in a job shop layout and identifies those that are similar enough to share the same dominant flow. These products/parts are then grouped together and routed through a series of machines that are placed in close proximity to one another. These machine "cells," typically U-shaped, may be manned by one or by several individuals (see Figure 3-1).

Grouping the products or parts that require the same sequence of operations may demand a special, sophisticated part numbering scheme that helps to isolate the key common characteristics. Product/part characteristics that may differentiate product/part families from one another include the diameter of the part, whether the external surface is smooth, whether the internal surface is smooth, whether flat milling has to be done, and whether there are any holes to be drilled. If one develops a numbering scheme to highlight these characteristics, routings and production plans can be concocted without undue difficulty.

Many group technology layouts have been adopted by plants that have not required elaborate part numbering schemes. These plants have been clear from the start about which parts are compatible with group technology; thus, they do not need much sophistication in scheduling parts to particular cells.

Closely related to group technology is the U-line. In its most common application, the U-line is a spur off of a main assembly line that feeds the main assembly line with precisely the parts required to synchronize with the main assembly line's schedule of production (see Figure 3-2). With a U-line, parts fabrication that traditionally was accomplished elsewhere in the factory and in big batches is done adjacent to the main assembly line and in just the quantities needed by that line. The U-line is generally responsible for several different models or options of parts, so it must maintain considerable flexibility with the ability to make quick setups of the machines along its line. One of the advantages of the U-shape is that one, or just a few workers can tend the line without having to take too many steps and can even help one another out merely by turning around. There are also some materials handling advantages to the U-line layout, as well as quick feedback about quality problems. However, the most persuasive argument for U-lines is their ability to feed the main line precisely, without the buildup of any inventory and without interrupting the smooth flow of the process. U-lines were popularized in Japan, and more discussion of them, in the context of Japanese manufacturing techniques, is found in Chapter 11.

It is worth repeating: The attractiveness of both group technology and U-lines is the conversion they make possible from the job shop layout to the line flow layout. They help turn a hybrid process into a process that is line flow in character from stem to stern.

Other Trends in Plant Layouts

Although two of the most interesting trends, group technology and U-lines, have already been discussed, there are some other trends to note as well, many inspired by the Japanese. For example, in recent years, layouts are incorporating space near workcenters for any tools

A. A conventional job shop layout.

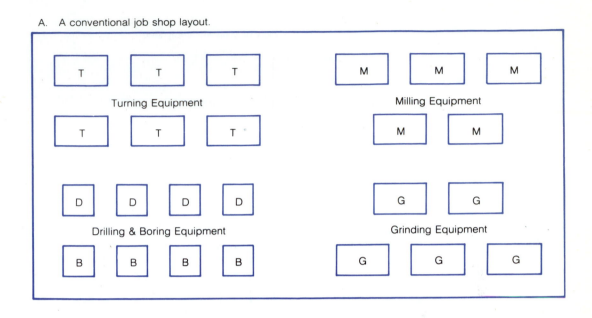

B. A group technology layout.

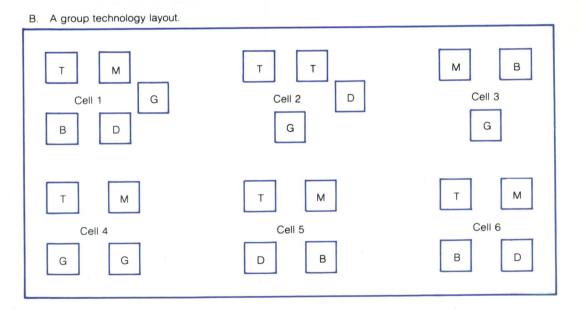

FIGURE 3-1   What group technology does to the layout.

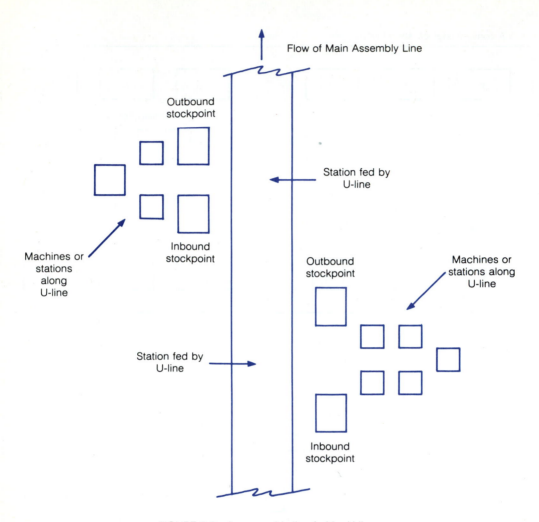

FIGURE 3-2   An assembly line fed by U-lines.

and fixtures necessary for setup, and for gages, charts, and other items needed for quality management. Often, the new layout is such that the flow of materials can progress easily in one direction through the workcenter while the flow of setup-related items can flow in a perpendicular direction through the workcenter.

Another recent trend is the plant layout in which nothing is obscured from view. In these layouts, to the extent possible, no work-in-process or raw materials inventories are stored near the process at a height greater than eye level. This protocol helps keep inventories under control and also makes any problems with the process more visible for both workers and managers. Often, the stockpoints are carefully labeled with paint or tape on the floor to indicate material storage locations. Such spatial discipline can assist production control; see Chapter 11's discussion of "kanban squares."

There has been some rethinking, as well, about materials handling. Many conventional layouts have significant investments in conveyors of various types. Besides being an expense, conveyors sometimes encourage more room between workcenters and more work-in-process inventory than is necessary. Instead, rethinking argues for simpler materials handling—gravity feeds, push-along carts—and workcenters positioned close together. Counter to this trend in one sense but consistent with it in another is the trend in some companies to use automatic guided vehicles. These riderless carts travel along guide paths either embedded or grooved into the floor. The carts carry loads of materials to specific points in the factory. They are an expense and a complication to materials handling, and some managers do not favor them for just that reason, but they do not use labor, and, in that sense, are consistent with the other moves to simplify materials handling.

Devising New Layouts

The industrial engineering of layout design is both art and science. It is beyond the scope of this book to explain in any depth the techniques used to develop new layouts or to improve existing ones. We will merely indicate some key steps one can take in devising layouts for factories or services:

1. *Examining the process flow.* Is there a dominant flow? What is it? What constraints to layout exist because of the process itself?

2. *Gaging the extent and importance of the flows.* How much material and information travels from one element of the process to another? Which flows are so important that close proximity is absolutely required?

3. *Determining area space needs.* How much space do particular operations or departments require? Are there special constraints to be aware of, such as with any existing structures? Has enough space been allocated for tooling, fixtures, and work-in-process?

4. *Arranging the layout.* What kinds of layouts satisfy the nature and extent of the flows within the process and do so within the space constraints specified? Which ones may be more flexible, or perhaps more streamlined, than others?

5. *Determining traffic flows.* What are the prevailing traffic patterns around and within the process? Do they cross or otherwise threaten congestion or safety? How will materials, tools, and fixtures be moved?

There are some computer models that use information on distance between operations, frequency of use, and special constraints that can be used to devise a trial layout. These computer models try to minimize the distance traveled by those utilizing the layout. While of interest, these models are not widely used.

VARIABILITY AND BOTTLENECKS: THE QUEUING PHENOMENON

So far we have not dealt with the impact of statistical uncertainty on bottlenecks. We have treated everything as if all were certain—certain arrival of peanuts, certain quantities delivered, certain times to dry soggy peanuts, certain setup times for wood lathes, certain run times for furniture legs, and so on. Yet life tells us all the time that we live amid statistical uncertainty, and manufacturing and service processes are no exception. The remainder of this chapter is devoted to exploring what uncertainty does to a process.

There are a variety of ways by which statistical uncertainty or variability can intrude on a production process. However, we can usually divide such intrusions into those that affect the demand on the process—that is, the timing or the quantity of items for the process to work on—and those that affect the supply of the process—that is, the speed, capacity, quality, and capability of the process itself. Any of these variabilities can cause the process to lose some of its ability to produce products at a given level of quality or quantity.

Moreover, we know that the greater the variability involved with either the demand or the supply side of the process, the greater the disruption to the process and the greater the likelihood that a significant bottleneck will occur. This unhappy state of affairs is an inevitable consequence of the phenomenon of "queuing."[3]

Perhaps the most important thing to understand about queuing phenomena is that as a process's capacity utilization nears 100 percent, waiting times grow at an increasingly rapid rate. Figure 3-3 illustrates this truism, which has been observed in countless different situations where there is uncertainty about the demand on the process and/or uncertainty about the capabilities of the process to do whatever it is supposed to do.

What is even more troubling, however, is that processes subject to particularly uncertain demand—like Devine Nuts waiting for deliveries of peanuts from the field or orders placed at a Burger King—will have waiting times increase at a faster rate than those processes whose demands are more certain and less variable. The more demand-uncertain processes will have waiting times that look more like curve A in Figure 3-4 than curve B. Theirs is a greater need for excess capacity than that for more demand-certain processes.

Similarly, those processes whose supply is more variable—whose capabilities are more erratic—will suffer the same fate, curves more like A than B. Thus job shops, which are inherently more variable than line or continuous flow processes, are more likely to experience bottlenecks that are caused by variability. Pro-

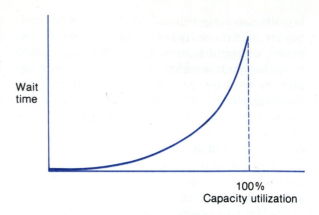

FIGURE 3-3   Wait time versus capacity utilization in queuing situations.

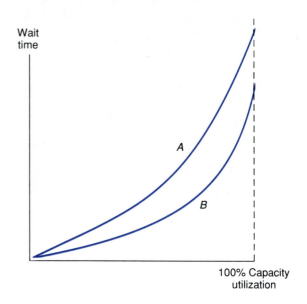

FIGURE 3-4   In some situations, wait time goes up faster than in other situations.

cesses that are well balanced and stable are less plagued by bottlenecks caused by variability than job shops or batch flow operations with irregular demands placed on them. The movement to introduce manufacturing cells can be viewed as movement to lower the variability in the process from that akin to the batch flow or job shop to more like that of the line flow process. (It should be noted, too, that production inefficiency is another explanation for the difference between curves A and B.)

Variability of all kinds saps capacity from a process and can cause shifting bottlenecks and buildups of inventory. Process capacities can thus be increased merely by figuring out ways to quell the variability in demand or supply (narrow the variance) that the process is subject to, without resort to adding more equipment or labor. There are now a number of factory and service operations simulation programs on the market that can help discover the kinds of inventory buildups and capacity breakdowns that uncertainty can bring and what can be done about it. Improvements in quality and process capabilities can lower variability as well.

Of course, having excess capacity is an important consideration for many processes and all the more important when variability is great. The job shop, given its significant variability for both demand and supply, *should* have many machines with relatively low capacity utilizations. Only then could the job shop be expected to operate well enough to keep waiting times on its deliveries to satisfactory levels.[4]

A Deeper Look at Queuing

Let's look again at Devine Nuts and some other concerns Frank Coyne has about his operation (Situation 3-1).

The General Anatomy of Waiting Line Problems

The situation at Devine Nuts is but one variant of an entire family of waiting line problems. The tremendous variety of waiting line problems is due to the many combinations of the following types of characteristics they can exhibit:

1. The arrival (or input) process
   - What kind of distribution do the arrivals follow? A popular, and useful, distribution to employ is the Poisson (discussed later), although others can be used.
   - Do the arrivals come singly, or do they sometimes also come in bulk? (Think of customers waiting to be seated at a restaurant in parties of one, two, three, four, and more.)
   - Does the length of any line serve as a deterrent to arrivals, or will the arrivals all simply take their chances?

2. Queue discipline (the order in which arrivals are served)
   - First come, first served? Last come, first served? Random?
   - Are arrivals permitted to leave the waiting line? If so, what characterizes their leaving ("reneging")?

3. The service process
   - What type of probability distribution describes how arrivals are served by the process? Is the rate of service a constant, or is it also a stochastic event, like the arrival rate? Clearly, in many types of service situations the amount of time a service employee must spend with a customer can vary markedly; we have all kicked ourselves for choosing the wrong line at the bank, or ticket counter, or supermarket.
   - Is there more than one line? How many?
   - Must arrivals go from service person/station to service person/station, or can one handle all of an arrival's needs?
   - A mixed system includes more than one line (sometimes called a channel) and more than one person/station (sometimes called a phase). What kinds of paths through any mixed systems are there?

SITUATION 3·1

Devine Nuts (Reprise)

Although Frank Coyne's major concerns were with the massive truck waits that were possible when too many wet peanuts arrived for drying, he was also concerned about the level of service that Devine Nuts provided farmers even when there was plenty of drying capacity. Even on good days, trucks generally had to wait in line for a while, especially during the early evening peak period. Frank was concerned about how a 10 to 15 percent expected increase in next year's harvest might affect this routine waiting, and naturally about whether any additional waiting might deter farmers from contracting with Devine Nuts.

During a typical day, 120 tons of peanuts were delivered between noon and 10 P.M., 40 tons of that total within the 5 to 7 P.M. time period. Trucks averaged 8 tons per delivery. Devine Nuts' unloading crew could dump only one truck's contents at a time, averaging 15 minutes per unloading.

With so many possibilities for the arrival and service processes and the queue discipline, it is easy to appreciate that most queuing problems are tackled either by (1) simplifying the problem so that mathematically tractable means can be applied, or by (2) using computer simulation techniques to ape what we think the true behavior is and draw out its implications.

Poisson Distribution

It has been discovered that many ordinarily occurring phenomena — radioactive decay, calls coming into a telephone switchboard, as well as people or trucks entering lines — come close to obeying a particular kind of probability distribution called the Poisson distribution, named for the French mathematician who first studied it in 1837.

According to the Poisson distribution, the probability that exactly k events (e.g., phone calls, people entering a line) occur in a time period of length t is equal to

$$\frac{e^{-\lambda t}(\lambda t)^k}{k!}$$

where e is the base of the natural logarithms (e ≈ 2.71828) and k! (k factorial) means $k \cdot (k - 1) \cdot (k - 2) \cdot \ldots \cdot 1$. As it turns out, this distribution has a mean (arithmetic average) of (λt) and a variance of (λt) as well.[a] We can think of λ (the Greek lowercase letter lambda) as indicating the arrival rate, the num-

---

[a]The variance of a distribution, $f(\ )$, of a random variable, x, is a measure of how dispersed the distribution is. It is obtained by squaring each deviation of x from the mean, $\bar{x}$, and weighting the squared deviations by their corresponding probabilities. Symbolically for a discrete distribution of N values, the variance (var x) can be written as

$$\text{var } x = \sum_{i=1}^{N} P_i(x_i - \bar{x})^2$$

where P_i is the probability that x_i occurs. Alternatively, it may be written as

$$\text{var } x = \sum_{i=1}^{N} P_i(x_i^2) - \bar{x}^2$$

The standard deviation is another way to express the dispersion of a distribution. It is merely

$$\text{standard deviation } s_x = \sqrt{\text{var } x}$$

The standard deviation has the advantage that it is measured in the same units as the random variable x itself.

ber of events that occur per unit of time t. This arrival rate, λ, is a key component of the various waiting line formulas that can be applied.

The Poisson distribution anchors the simplest formulas for examining waiting lines and their performance. The performance of the simplest waiting line case, where there is one line (or channel) and one station (or phase), can be described by the formulas in Table 3-1. These formulas depend on λ (the arrival rate) and μ (the service rate, which is defined in exactly analogous fashion to λ). Naturally, μ must be greater than λ; otherwise, the service could never catch up with the arrival rate, and an ever-growing waiting line would result.

Routine Wait at Devine Nuts

Let us see how these waiting line formulas can come to Frank Coyne's aid. There are four situations we can examine:

1. A routine day with no peak period
2. The peak 2-hour period within a routine day
3. A routine day with no peak period, but with 10 to 15 percent more delivered
4. The peak 2-hour period within a routine day, but with 10 to 15 percent more delivered

---

TABLE 3-1   Formulas describing the performance of a single-channel, single-phase waiting line with Poisson arrival and service rates

λ (the lower case Greek letter lambda) = arrival rate (such as units entering the line per hour)

μ (the lower case Greek letter mu) = service rate (such as units processed from the line per hour)

Average utilization = λ/μ

Mean number in the waiting line = $\lambda^2/\mu(\mu - \lambda)$

Mean number in the entire system = $\lambda/\mu - \lambda$

Mean time in the waiting line = $\lambda/\mu(\mu - \lambda)$

Mean time in the entire system = $1/(\mu - \lambda)$

Let us calculate the performance of Devine Nuts in each of these situations. If the wait time is not too great for either a situation of peak period demand within a day, or a day with more deliveries expected, then Frank Coyne can breath a sigh of relief. If, however, wait time becomes too great, then he must think of something to do.

Routine Day, No Peak Period

The arrival rate of trucks (λ) is 1.5 per hour, calculated as

$$\frac{120 \text{ tons/10-hour day}}{8 \text{ tons/truck}} = 15 \text{ trucks per 10-hour day}$$

or 1.5 trucks per hour.

The service rate (μ) is 4 trucks per hour (15 minutes for unloading each truck).

1. Average utilization is
$$\frac{\lambda}{\mu} = \frac{1.5}{4} = 37.5\%$$

2. Mean number in line is
$$\frac{\lambda^2}{\mu(\mu - \lambda)} = \frac{(1.5)^2}{4(4 - 1.5)} = 0.225 \text{ truck}$$

3. Mean number in system is
$$\frac{\lambda}{\mu - \lambda} = \frac{1.5}{4 - 1.5} = 0.60 \text{ truck}$$

4. Mean time in line is
$$\frac{\lambda}{\mu(\mu - \lambda)} = \frac{1.5}{4(4 - 1.5)} = 0.15 \text{ hour or 9 minutes}$$

5. Mean time in system is
$$\frac{1}{(u - \lambda)} = \frac{1}{4 - 1.5} = 0.40 \text{ hour or 24 minutes}$$

Note that even though the unloading crew is idle most of the time, there is a small average wait of 9 minutes per truck. Because it takes an average of 15 minutes to unload each truck, it is no surprise that the mean time in the system is 15 minutes more than the mean wait time.

Let's now examine the peak period to see how much worse it can be expected to be, and to see what to expect when the quantity delivered increases.

Peak Period, Routine Day

While the service rate, μ, is the same, the arrival rate, λ, has increased to 2.5 trucks per hour. That is calculated as

$$\frac{40 \text{ tons/peak period}}{8 \text{ tons/truck}} = 5 \text{ trucks per 2-hour}$$
$$\text{peak period}$$
$$= 2.5 \text{ trucks/hour}$$

Routine Day, No Peak Period, 10 to 15 Percent Increase in Deliveries

For convenience, assume that the number of deliveries increases from 15 per day to 17 per day, a 13.3 percent increase. Thus, $\lambda = 1.7$ per hour and μ remains the same at 4 per hour.

Peak Period, Routine Day, 10 to 15 Percent Increase in Deliveries

Again, assume 17 trucks per day, with a third of them arriving within the 5 to 7 P.M. peak period. Thus μ grows to $17/3 = 5.67$ trucks in 2 hours or 2.833 trucks per hour.

Table 3-2 summarizes the results.

Several observations can be made. Note that a 67 percent increase in the arrival rate, the peak period of a routine day — from 1.5 to 2.5 trucks per hour—leads to a 178 percent increase in waiting time (9 min. to 25 min.). And, even on a routine day, when deliveries increase, by 13.3 percent (1.5 to 1.7), the waiting time increases by 23 percent (9 min. to 11.1 min.).

And, when peak periods are compared, with deliveries increasing by 13.3 percent (2.833 trucks per hour versus 2.5 trucks per hour), the waiting time increases by 46 percent (25 min. to 36.4 min.). Observe that this peak period increase in waiting time, given the 13.3 percent increase in deliveries, is even greater than the 23 percent increase in waiting time for the no-peak-period case, given the same 13.3 percent increase in deliveries. This is an important point about waiting lines. The length of any wait can rise dramatically as capacity approaches full utilization, per Figure 3-3. This truism about waiting lines suggests that capacity choice for operations like Devine Nuts or Burger King or others is absolutely critical; a capacity utilization that is too high can have disastrous effects on customer waits and customer satisfaction.

ITEM	FORMULA	ROUTINE DAY, NO PEAK	ROUTINE DAY, PEAK	DELIVERY INCREASE, NO PEAK	DELIVERY INCREASE PEAK
λ		1.5	2.5	1.7	2.833
μ		4	4	4	4
Average utilization	$\dfrac{\lambda}{\mu}$	37.5%	62.5%	42.5%	70.8%
Mean number in line	$\dfrac{\lambda^2}{\mu(\mu - \lambda)}$	0.225 trucks	1.04 trucks	0.3 trucks	1.72 trucks
Mean number in system	$\dfrac{\lambda}{\mu - \lambda}$	0.60 trucks	1.67 trucks	0.74 trucks	2.43 trucks
Mean time in line	$\dfrac{\lambda}{\mu(\mu - \lambda)}$	9 min.	25 min	11.1 min.	36.4 min.
Mean time in system	$\dfrac{1}{\mu - \lambda}$	24 min.	40 min.	26.1 min.	51.4 min.

TABLE 3-2   Summary of Queuing Results for Devine Nuts

This phenomenon reinforces the need to manage capacity effectively. In a service operation, not only is having some spare capacity essential to avoiding lengthy waits, but that capacity must be available when needed. This typically means that labor must be scheduled to match peaks in demand and any equipment or supplies must be thoroughly prepared in advance.

There are, of course, more refined and complex formulas available for describing the performance or more complicated queuing problems. However, most of the queuing problems one finds in actual production and service operations quickly become more complicated than mathematical techniques can solve. One then confronts the choice of using mathematical techniques to approximate a solution or of investing time and effort in simulating the waiting line and thereby discovering its properties. Frequently, a queuing problem that may be complex can be broken apart into smaller problems that resemble simpler queuing phenomena. A multichannel system, such as a drive-in bank with three lanes, might be approximated as three separate single-channel systems, each with a third the demand of the full operation.[b] For many applications, however—especially if the waiting line problem is an important one to the operation—simulation is the best way to find a solution.

Simulation

The ever-increasing size and speed of computers have made simulation an ever more attractive means of obtaining some acceptable solutions to large-scale and/or complex problems, such as many queuing problems. In many ways simulation substitutes brute force (raw computer power) for the elegance (but lack of realism) of the mathematical model. Although often more realistic, simulation is often more art than science; mathematical models can often be characterized as right or wrong, whereas simulation models tend to be either good or bad. The typical simulation model is a kind of mathematical "black box" that takes some initial inputs, processes them, usually over a number of periods of time, and spits out some outputs.[5]

The first step in constructing such a model is to identify the variables of interest. Variables, quite naturally, refer to those values within the simulation that are permitted to vary over time. The choice of variables for a simulation takes judgment, since only variables that are important to the result should be included and they should be defined in units of measure that are appropriate. A simulation of truck arrivals at Devine Nuts, for example, would surely define arriving peanuts (measured in, say, tons per hour), truck wait (hours), and peanuts processed as variables. One might also include the moisture content of the peanuts as a variable as well, freeing the analysis from the assumption that all the arriving peanuts are uniformly soggy and thus need a full 36 hours in the dryer. On the other hand, the unloading time per truck may not matter much and thus not be worth including. Once identified, the variable should be set with some initial values (subject to change, of course) in different model runs.

Initial conditions apply not only to variables but also to some of the parameters of the simulation. Parameters, in contrast to variables, are those simulation values that remain fixed during any single run of a simulation. Often, successive runs of a simulation will modify one or more parameters intentionally, but during any

[b] Pooling multiple channels into the same system does however have some advantages so that the pooled multichannel system is likely to have somewhat better performance than separate, single channel systems. Thus, the bank with 3 drive-up lanes is likely to do somewhat better than 3 banks each with a single drive-up lane and one-third of the total demand.

one run all parameter values remain fixed. For example, the mean rate of arrival of peanuts to Devine Nuts and the company's total drying capacity are likely parameters for a simulation of Devine Nuts' operations.

A crucial stage in developing any simulation model is detailing the relationships among the variables and parameters and how they will vary, if at all, with the march of time. A number of these relationships will almost always be of an accounting nature, but many will serve to "drive" the model. The kinds of relationships that alter variable values in important and often systematic ways are frequently termed decision rules. They specify under which conditions variables of all kinds will be changed and how. In a simulation of Devine Nuts' situation, for example, one decision rule might check to see whether the drying capacity has been reached and then allocate arriving peanuts either to dryers or to the truck wait. Accounting relationships would then keep track of the peanuts dried and drying, and the trucks waiting.

Simulation models can be either deterministic or probabilistic. That is, the simulation can be built to yield exactly the same outputs, given the same sets of initial conditions and parameters (a deterministic model); or some elements of chance can be introduced so that the same initial conditions and parameters would yield different, although probably similar, outputs (a probabilistic model). The most useful simulation of Devine Nuts' situation would be a probabilistic one (also called a Monte Carlo simulation after the casino in Monaco), which would draw at random a value for the arriving peanuts variable from a probability distribution like the Poisson. The choice of probability distribution for such purposes is thus another important feature of building a simulation model.

Once a simulation model has been constructed—variables, parameters, initial conditions, accounting relationships, decision rules, probability distributions—it is put through its paces. Usually, a lot of tinkering goes on—more or fewer periods are run, parameters changed, initial conditions revised, relationships altered, variables added. All of this tinkering occurs to check on the simulation model's behavior. Are the results plausible? Do changes in parameters, initial conditions, or relationships yield the expected results?

After the model has been debugged and is functioning smoothly, its outputs can be used directly for decision making or (as is common for probabilistic models) can be accumulated in a data bank and investigated with statistical tools and hypothesis testing.

To understand more clearly how a simple simulation model might be developed, consider Situation 3-2.

Discussion of Burger Queen Steamer Inventory Decision

Determining the level of cooked burger pattie inventory that should be kept in the steamer is a problem ideally suited to simulation. Let us work through how such a problem could be simulated.

The parameters of this simulation—those values that stay unchanged during the simulation—include the ride time through the broiler (30 seconds), burger assembly time (20 seconds), Big One assembly time (30 seconds), waiting time interval between patties on the broiler (5 seconds), the number of patties that can be cooked per minute (12 patties), and the level of inventory in the steamer (maximum chosen for each run). Note that these service times are fixed; some simulations, of course, could have them vary, much as we assumed for the simple queuing model.

The key variable of interest is the transaction time—the time between the placement of the order and its presentation to the customer.

SITUATION 3-2

Burger Queen—Steamer Inventory Decision

The Burger Queen restaurant in Gastonville, North Carolina, operated much like a Burger King. It offered regular hamburgers and a large hamburger sandwich called "The Big One." Burger patties were broiled at a high temperature on a continuous chain broiler; they had to be placed on the chain broiler no more often than one pattie every 5 seconds. Meat patties of both sizes were completely cooked during their 30-second ride through the continuous chain broiler. After the broiler was a "burger board" where workers assembled regular burgers and Big Ones. A regular burger could be assembled, packaged, and placed in the finished goods chutes within 20 seconds; a Big One took 30 seconds.

Between the chain broiler and the burger board was a steamer that could keep cooked meat patties warm for 10 to 15 minutes without any deterioration in their taste and freshness. Burger Queen's management was interested in determining how many cooked meat patties of each type should be kept in the steamer at different times. The success of the inventory level for the steamer would be judged by whether a transaction time standard of 1 minute or less for any order could be maintained.

A typical stream of orders (a representative random sample) during a reasonably busy time of day was as follows:

ORDER NUMBER	ORDER
1	2 burgers
2	1 Big One
3	1 burger, 1 Big One
4	2 Big Ones
5	3 burgers
1 minute elapsed time	
6	2 burgers
7	3 Big Ones
8	1 burger
9	2 burgers, 1 Big One
10	1 Big One
11	4 burgers
2 minutes elapsed time	
12	2 burgers
13	1 burger
14	1 burger, 2 Big Ones
15	3 Big Ones
3 minutes elapsed time	
16	1 burger, 1 Big One
17	3 burgers
18	1 Big One
19	2 burgers, 2 Big Ones
20	1 burger
4 minutes elapsed time	

The goal is to have that time be 1 minute or less. In this somewhat simplified situation, the transaction time for any order will equal the longest time to complete any burger or Big One in the order. We are assuming that on orders of more than one burger or Big One, assembly of the multiple items is simultaneous and that board capacity is great enough that sandwich assembly is not delayed for any sandwich. We are also assuming that all Big Ones in any order are worked on first, because they have the longer assembly time. The variables that make up the computation of transaction time include the number of burgers in the order, the number of Big Ones in the order, the cooking time for each pattie, and the extent of any wait in placing patties on the chain broiler.

To calculate the longest time to complete any burger or Big One in an order, one must compare the completion times for all the sandwiches in the order. A sandwich's completion time depends on whether the pattie was drawn from the steamer or had to be broiled. If the former, the completion time equals the time it takes to assemble the sandwich. If the latter, the completion time is the sum of both the assembly time and the cooking time (which is the ride time through the broiler plus any waiting because the broiler can take only one pattie every 5 seconds). This wait can be calculated by multiplying 5 seconds times the number of patties preceding a particular one in an order through the broiler.

The simulation of transaction times is driven by the sequence of orders. One such sequence is given in the situation description, but others are possible and, indeed, desirable for computing an effective level of inventory for the steamer. Suffice it here to use the stream of orders given above.

Table 3-3 displays the completion times for each sandwich in the order (Big Ones first) and the transaction times for that order, assuming that no inventory is kept in the steamer. These simulation results show that the transaction time standard is exceeded 6 times in the 20 orders simulated, or almost one-third of the time. This finding suggests that keeping an inventory of cooked meat patties in the steamer is indeed desirable. Hence the next trial of the simulation model could attempt to keep one meat pattie of each size in the steamer to see how transaction times are affected. (The calculation of transaction times for this stream of orders, assuming steamer inventory levels of one or two patties, is part of Problem 9 at the end of the chapter.)

TABLE 3-3   Completion times and transaction times (assuming no inventory is kept in the steamer)

| ORDER NUMBER | COMPLETION TIMES (seconds) Sandwiches | | | | TRANSACTION TIME (secs) |
	No. 1	No. 2	No. 3	No. 4	
1	50	55			55
2	60				60
3	60	55			60
4	60	65			65
5	50	55	60		60
6	50	55			55
7	60	65	70		70
8	50				50
9	60	55	60		60
10	60				60
11	50	55	60	65	65
12	50*	55*			55
13	50				50
14	60	65	60		65
15	60	65	70		70
16	60	55			60
17	50	55	60		60
18	60				60
19	60	65	60	65	65
20	50				50

*Order no. 11 calls for the thirteenth and fourteenth meat patties in the previous minute's time. Thus if order no. 12 is called for within the first 10 seconds of its time period (between minutes 2 and 3), its transaction time will be later than 55 seconds because the broiler can cook only 12 patties in a minute. Its transaction time could be as high as 65 seconds. To be accurate, the simulation should keep track of capacity demands on the broiler.

How should the simulation keep track of the steamer inventory? This monitoring is clearly needed. The steamer inventory has to be constantly replenished, and it cannot be if more than 12 meat patties are ordered in a minute. (How the steamer inventory can be monitored and replenished is part of Problem 9 at the end of the chapter.)

The appendix, in broad terms, describes the sophisticated simulation model that the Burger King Corporation developed.

Managing the Queue

The raison d'etre for both the mathematical and simulation techniques for analyzing queuing problems, of course, is to provide insights for management action on the queues themselves. Typically, queues can be managed in a host of ways: adding capacity, introducing buffer inventories to the process, or investing in statistical quality control to narrow production (service) variances.

Earlier, in discussing the anatomy of waiting line problems, elements of the arrival process, queue discipline, and the service process were noted. Many of these elements can be affected by particular management actions. Indeed, in service industries, where customer happiness while waiting in line is crucial, significant management attention is focused on managing the queue properly and with techniques to make the wait feel shorter.

David Maister has provided a thought-provoking list of propositions about wait time and how to manage it.[6] They point to a variety of things that companies can do to improve the perceptions that people have about waiting.

- *Proposition 1: Unoccupied time feels longer than occupied time.* This suggests filling the wait time with something for the customer to do, preferably something that has a benefit to him.

- *Proposition 2: Pre-process waits feel longer that in-process waits.* Better to start a customer through the process than to keep him waiting at the start.

- *Proposition 3: Anxiety makes waits seem longer.* Reassure the customer that he will be served and that he is not waiting in the wrong line.

- *Proposition 4: Uncertain waits are longer than known, finite waits.* If possible, give the customer information on how long the wait is likely to be. Reservation or appointment systems are good ways to alleviate uncertain waits, as long, of course, as they are kept by the service providers.

- *Proposition 5: Unexplained waits are longer than explained waits.* If you know why there is a wait, then tell the customer.

- *Proposition 6: Unfair waits are longer than equitable waits.* Any departures from "first come, first served" systems may be seen by customers as unfair, and should thus be explained and justified.

- *Proposition 7: The more valuable the service, the longer the customer will wait.* For small scale services, even short waits can displease customers.

- *Proposition 8: Solo waits feel longer than group waits.* In waiting, as elsewhere in life, misery loves company.

Tools of the Trade: A Recap of Lessons Learned

In investigating bottlenecks, two areas of particular concern deserve our attention. Indeed, failure to devote attention to layouts and the queuing phenomenon can turn the occasional bottleneck into a persistent, chronic one.

Layouts: Layouts are more important than is sometimes believed. Sprawling layouts that force materials to snake through the factory (or office) kill productivity. Avoiding such layouts takes diligence because they creep up on the operation bit by bit. Eradicating them means returning to basics: compactness and ease of movement.

Moreover, the more line flow the layout, the greater is the chance for real efficiency. The innovations of machine cells and U-lines are really geared to converting job shop-type layouts into more line flow-type layouts.

Queuing: Almost all processes are subject to statistical variability, either in their abilities to process materials or in the demands placed on them. This statistical variability inevitably introduces wait time to the process. Knowing this fact, and being able to estimate its extent are important to the health of the process and a company's ability to wring the most out of it. Excessive wait can be avoided by adding capacity in the right places, by reducing variability in the production or service process, by adding buffers of one variety or another, or by managing demand (reducing variability in the arrival [or demand] process).

Analyzing wait times can be done either mathematically or via simulation. The mathematical models for queuing generally describe simpler situations than are faced in reality, but they, nevertheless, can provide a quick approximation that is frequently worthwhile. The chapter provides the simple queuing formulas that can often be adapted to provide the useful approximation. For the more complicated situations, computer simulations are needed, and there are a number of software packages available to model these more complex situations (e.g., XCELL+).

No matter what the wait, however, managers need to know how to manage customers' expectations so that the wait does not detract too much from the service provided.

THE BURGER KING RESTAURANT SIMULATION MODEL

The Burger King Corporation's restaurant simulation model (actually, there were several models, because each type of restaurant configuration required a different model) was one of the tools that the corporation used in the 1980s to keep it a technological leader in the fast food business. That a simulation model would be useful for helping to make many different decisions was not immediately apparent to Burger King management; the realization crept up on them gradually.

In the late 1970s Burger King experimented with a number of innovations, the most prominent of which was the introduction of several specialty sandwiches. Market research had shown an encouraging consumer response to specialty sandwiches, but it was not known how they would affect service delivery at existing restaurants. As mentioned in the Tour F discussion of Burger King, a key measure of a restaurant's operation was its speed of service; the faster people can be served a quality product, the happier they will be and the more customers a restaurant will serve, particularly during peak periods. Burger King in the late 1970s had no easy way to assess the impact of specialty sandwiches or other innovations on a restaurant's speed of service. Before something like a new specialty sandwich could be rolled out to the thousands of Burger Kings worldwide, the company had to experiment in an actual restaurant to see what its effect on operations would be. Generally speaking, several restaurants were used for each experiment. Typically, 2 weeks' worth of data were collected so that there would be enough data for different days of high and low sales volumes. A typical experiment cost between $6000 and $8000 and required a good deal of time to analyze the results.

It occurred to some farseeing Burger King managers that a mathematical description of a restaurant might be able to perform the same kind of service as a field experiment but at much lower cost. At first, a linear programming model was used to isolate particular bottlenecks in a restaurant's operations, but it was soon realized that a linear program could not adequately capture the dynamics of restaurant operations, where the most interesting problems occurred. Given this insight, it was decided to embark on building simulation models for Burger King restaurants.

THE STRUCTURE OF THE SIMULATION MODEL

As initially conceived, the simulation model, like an actual restaurant, consisted of three broad, interrelated areas: (1) customer arrivals and orders, (2) food preparation, and (3) the delivery of orders to customers. Within each of these areas, the simulation model had various modules that described particular aspects.

Customer arrivals and orders. Of course, Burger King has little or no influence on arrival and order process. Thus the simulation model simply attempted to mimic accurately what occurred in the typical restaurant on various days of the week and during various hours of the day. Particular modules isolated the behavior of in-store versus drive-thru arrivals as well as in-store versus drive-thru order generation.

These modules were based on observed facts. It had been observed, for example, that the arrival pattern of customers followed an exponential function (i.e., the plot of the cumulative percentage of arrivals versus the time between those arrivals traced out an exponential curve). It had also been observed that a relatively small number of order types account for most orders received at the restaurant (the five most common combinations of choices—such as burger, fries, and drink as opposed to specialty sandwich and shake—accounted for 45 percent of all the orders placed).

In sum, the customer arrival and order segment of the model was designed to reflect, on a day-by-day and an hour-by-hour basis, the specific orders that were typically placed at a Burger King.

Food preparation. Of the three areas, Burger King had the most control over food preparation. The company had considerable discretion in changing equipment, procedures, layout, and staffing. The production system was thus modeled in considerable detail. Elements of this modeling included descriptions of all the operating procedures for making anything that could be requested of the kitchen. There were, thus, modules that detailed how regular Whop-

pers were made, how regular chicken sandwiches were made, how french fries were prepared, and how shakes were drawn. The model's descriptions of these activities, in turn, depended on (1) the labor times for each task (such as the time it took to dress a bun crown for the Whopper Sandwich), (2) the kitchen layout that determined the distances workers had to walk between tasks, and (3) the equipment configuration, since processing times depended in part on the choice of equipment for the kitchen. Other elements of the production system included the number of workers who worked at any time and their positions and responsibilities within the kitchen, the inventory rules (e.g., 10-minute maximum hold time for all sandwiches). Typical simulation modules for the production system included such things as regular Whopper preparation, special Whopper preparation, regular burger preparation, shake preparation, broiler and steamer operation, and in-process and final inventory levels.

The order delivery system. This area of the simulation model dealt with how orders were taken, assembled, and presented to the customer. Tasks in this delivery system were divided between in-store and drive-thru tasks. The times for customers to state orders and for cashiers to punch them in, make change, and assemble orders were determined separately for different order configurations and staffing assignments of the cash registers. Representative simulation modules included in-store order assembly and delivery, and drive-thru order assembly and delivery.

The Burger King simulation models are written with a special simulation computer language called GPSS. The GPSS simulation style is to simulate activities by orienting discrete events in time. A GPSS model begins its simulation at time period 0 and inquires about which activities are to begin then or are to be continuing at that time. When the model is stepped ahead to time period 1, it keeps track of those activities that are to be continued, those that will end in that period, and those that will start then. The simulation is analogous to constructing a movie by taking a series of snapshots at numerous points in time. In 1982, a typical Burger King simulation model application required 3000 lines of GPSS code and took about 1 minute of processing time on an IBM 3033 computer to execute.

The construction of the simulation model required an enormous amount of data. Time and motion studies specific to Burger King were used along with methods time measurement (MTM) data to develop time standards for the many operations performed by workers in the kitchen or at the counter. Information was also gathered from a sampling of 40 restaurants nationwide. These restaurants offered a diversity of characteristics such as sales volumes, ownership (franchise versus company), and facilities (drive-thru versus non-drive-thru). The information gathered from these restaurants was incorporated into the model to ensure that its operation matched operations at typical Burger King restaurants.

Before the corporation was willing to use the simulation model to run low-cost "experiments" of proposed innovations, the model underwent a "live test." The company took some space in a warehouse and set up an actual restaurant kitchen and service area. Actual orders from cash register tapes, representing both high and low sales volumes, were read to the service counter people. This live test ran an entire week, during which all aspects of the operation were measured and videotaped. The same streams of orders were fed into the simulation model, and its results were compared with those from the live test. The simulation model was adjusted until the differences between the two according to measures such as service times, lengths of queues, and locations of bot-

tlenecks turned out to be very slight. This success convinced upper management of the worth of simulation for analyzing their type of operation.

The Burger King simulation model is modular in that any of the many restaurant configurations operating at any given manning level can be represented relatively easily by joining the appropriate modules together. These individual models undergo constant revision, because each innovation that the company would like to test requires reprogramming at least certain aspects of the model.

Uses of the Simulation Model

The Burger King simulation model has been used for a variety of purposes. Its first major use came in the development of the Burger King Productivity Improvement Program. For this program, manuals were developed that indicated the potential service bottlenecks and their remedies for each level of restaurant sales. Thus, a franchisee could examine a manual to discover how to break those bottlenecks and so maintain service standards as restaurant sales grew. The manual also indicated the rates of return for expenditures on remedying any of the potential bottlenecks. These studies took between 6 and 8 months to complete, with about 300 runs of the model each month.

From this success, the corporation's industrial engineering and operations research department, responsible for the development and maintenance of the simulation model, moved on to develop labor staffing standards for each restaurant configuration. About a year was spent exploring the issue and developing the standards. Two key needs were addressed:

- The minimum number of employees needed for any sales level and their duties so as to be able to satisfy corporate speed-of-service goals.

- A means for management to control the labor needs (a "labor" formula).

By repeated use of the simulation model for different store configurations, given different levels of sales and different positioning of the workforce, a set of manning tables was derived. For example, before the use of the simulation model, it was not known definitely how best the tenth worker could be positioned as sales climbed. Was that worker best added to the burger board or added as a cashier? The output of the model showed that more customers could be served and the average speed of service could be kept within the 3-minute standard by shifting that worker to the cashier position. In fact, the model's results showed that five more customers per hour (150 versus 145) could be served in this way and that the average speed of service could be maintained at 176 seconds per customer as opposed to 188 seconds. Results of this type resulted in savings, for each restaurant, of 1 to 2 percent of sales.

To address the management control needs, equations were estimated by regression analysis to predict the aggregate level of labor needed, given an optimum schedule for that labor. The equations, defined for different restaurant configurations, used real weekly sales and the number of hours open as explanatory variables.

The simulation model was instrumental in the design of the BK87 restaurant configuration, which is described in Tour F. The model permitted "experiments" that shifted equipment around the kitchen, seeing how these changes would affect service times and staffing levels. Burger King's use of a two-window drive-thru lane, rather than a multilane drive-thru as is often seen in drive-thru-banks, is another example of the use of the model. The simulation model predicted that this innovation would increase capacity by 15 percent during peak hours; this estimate has been borne out with actual increases of 14 percent.

In a drive-thru, what is the optimum stack size (number of car lengths) between the order board and the pick-up window? This was another question that Burger King wanted addressed. If that distance was too short, a car would have to wait at the pick-up window because the food was not ready; if the distance was too long, the food would be cold by the time the car got there to pick it up.

Analyzing this question through the use of the simulation model was straightforward. The model's customer arrival and order system generated cars for the drive-thru lane. The model also specified the distance between the street and the order box and between the order box and the drive-thru window. Naturally, if the drive-thru lane between the order box and the window was filled, no other car could reach the order box to place an order.

A longer stack size was simulated simply by altering the capacity of the drive-thru lane between the order box and the window. By varying this parameter from one to 11 or 12 car lengths, the company was able to see which stack sizes were most conducive to reducing the transaction time at the drive-thru window and thus to increasing the number of cars served per hour. The graph in Figure 3-5 shows that a stack size of six to eight seemed optimal.

A more challenging problem was whether to add small specialty sandwiches to the Burger King menu to complement the large specialty sandwiches that were Burger King staples. It was known, for example, that if small specialty sandwiches were introduced, they would sometimes be ordered instead of large specialty sandwiches and sometimes instead of other Burger King menu items (particularly burgers). Specialty sandwiches take somewhat longer to prepare than burgers, and small specialty sandwich patties would have to be cooked separately from

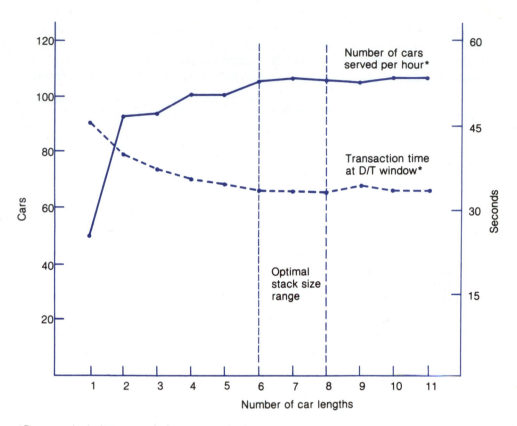

*Does not include car at window or at order box.

Based on (1) December 1979 Miami prices, (2) average company order configurations and product (3) a capacity for 20 cars from pick-up window to drive-thru entrance, and (4) K-700 kitchen layout.

FIGURE 3-5   Stack size between order box and pick-up window (in car-lengths).

large ones because the frying times are different. At issue then was how the introduction of small specialty sandwiches would affect service times, given a different pattern of orders and some additional constraints on the kitchen.

Studying such a problem involved changes to five modules in the simulation model:

1. *Order generation.* To obtain data for changing this module, a marketing test on small specialty sandwiches was performed in Wilkes-Barre, Pennsylvania.

2. *Preparation of regular specialty sandwiches.* MTM analysis was used to obtain data for adapting the module on preparation of specialty sandwiches to the smaller sandwiches.

3. *Preparation of special specialty sandwiches.* Again, MTM analysis was used to study the changes in labor standards implied by the smaller sandwiches.

4. *Fry station operations.* This module had to be changed because the patties for these small specialty sandwiches would be cooked

by the fry station operator. Moreover, at low volumes, the fry operator would also take on the responsibility of preparing the sandwich and not simply frying the pattie.

5. *Stocking rules.* The introduction of small specialty sandwiches modified some of the stocking rules for work-in-process inventory and finished good inventory in the chute. For example, the introduction of a small chicken sandwich might change the stocking rules for the large chicken sandwich, given some substitution away from the large chicken sandwich to the smaller one.

Table 3-4 reports the simulation results for small specialty sandwiches—what effect they can be expected to have on the utilization of equipment and manpower and on customer service. As can be seen in the figure, the introduction of small specialty sandwiches, even with the addition of an extra fryer, increased

average transaction times by at least 10 seconds. This deterioration of service was even more pronounced at lower sales levels. There were two reasons for this deterioration:

- Preparation times for the small specialty sandwiches remained longer than the preparation times for the burger and Whopper sandwiches that they sometimes replaced; thus they made the entire kitchen operation more labor intensive.

- The demand for small specialty sandwiches was rather low to begin with, and effective stocking was difficult, implying a high risk of waste. Consequently, an order for a small specialty sandwich was similar to an order for a special on any of the sandwiches, and these specials always took longer to prepare.

Based on these results and insights, Burger King withdrew its plans for the introduction of small specialty sandwiches.

		TABLE 3-4 Operational effects of adding small specialty sandwiches		
			UTILIZATION WITH SMALL SPECIALTY SANDWICH	
EFFECT TYPE	POSITION/EQUIPMENT/ATTRIBUTE	BEFORE SMALL SPECIALTY SANDWICH	CURRENT FRYER CAPACITY	WITH EXTRA FRYER
Equipment and manpower utilization	Fry operator (% of time busy)	52.8	54.2	57.4
	Specialty board (% of time busy)	58.5	70.2	68.7
	Multipot 1 (times down/hour)*	23	26	25
	Multipot 2 (times down/hour)†	—	—	3
Customer service	Average number of customers behind customer being served	1.14	0.95	0.93
	Average transaction time (seconds)	109	121	119

Note: The data are based on a product mix of 72 percent Whoppers and burgers, 19 percent specialty, and 9 percent small specialty.
*The frying vat that was used for cooking chicken.
† The frying vat that was added to break the potential bottleneck at multipot 1.

QUESTIONS

1. Waiting line problems have similar general characteristics. What are these? Give an example of each from the situation presented in this chapter.

PROBLEMS

1. Flowers' Gourmet Food Store has come into vogue; sales are up by 20 percent over last year. On an increasing number of occasions, however, the wait for checkout has been uncomfortably long. The store is served by only two cash registers, which are located near the exit. The store's management realizes that an additional cash register or two must be installed, but where? One option, of course, is to install the register(s) near the other two, although this would necessitate some rearrangement of the exit area. Another option, requiring less rearrangement, would be to put a cash register in a particularly busy department or two (such as deli, wine bins, butcher shop). Many customers come in for one or two items only and could use these cash registers.

(a) How should Flowers' Gourmet make such a decision?

(b) What data should be collected?

(c) What factors would tip the decision toward one option or the other?

(d) Would a simulation model (as discussed in the chapter appendix) be helpful for such a decision, and if so, what should its structure be?

2. Several hundred different parts are produced in the fabrication department of a large factory. Two of them, however, account for 30 percent of the fabrication department's total output. Both of these products follow the same general sequence of operations: turning, milling, drilling, and then grinding. The following percentages of production time are spent by the two parts on each set of machines: 20 percent turning, 40 percent milling, 10 percent drilling, and 30 percent grinding.

The inventory of machines in the department is as follows:

MACHINE TYPE	NUMBER IN DEPARTMENT	SYMBOL TO USE IN LAYOUT
Turning lathes	7	▭
Mills	8	○
Drills	3	△
Grinders	4	◇

The machine utilization has been fairly even at about 75 percent for each type of machine. Using the symbols (1 symbol = 1 machine), design a layout for the department. Explain your reasoning.

3. The Thunder Bank was a regional processor of VISA and Mastercard transactions for merchants. For transactions over a certain dollar limit, merchants were required to phone in for an authorization before the sale could be completed. The Thunder Bank handled an average of 2000 authorization calls per hour during most of the year and 4000 calls per hour during the Christmas buying season. The receptionists at the bank could handle 45 calls per hour, on average. The bank wanted to maintain a service standard of 60 seconds average wait per call.

(a) How many receptionists do you estimate the bank needs to employ during the nonpeak period of the year in order to meet its service standard? What assumptions have you made to come up with this estimate?

(b) How many receptionists does it need for the Christmas season?

4. Suppose that the Thunder Bank purchased the credit card business of another bank which had itself a substantial volume of transactions (1000 per hour average during the off-peak season and 200 calls per hour on average during the Christmas peak). How would the additional volume of calls affect the authorization process? Would you expect the number of receptionists to have to increase in the same proportion as the increase in calls (50 percent more), less than proportionately, or more than proportionately? Why?

5. Refer back to Problem 3. Suppose that the service standard was relaxed to 75 seconds average wait per call? How would that change the number of receptionists needed for both the nonpeak and the Christmas season times of the year? What would happen if the service standard were tightened to 45 seconds of wait, on average?

6. Refer back to Situation 3-1. Suppose that the unloading crew at Devine Nuts can reduce the unloading time of at truck to 10 minutes instead of 15. Recalculate the queuing performance formulas for this change.

7. Refer back to Situation 3-1. Suppose that the unloading crew at Devine Nuts can only unload a truck in 20 minutes rather than 15. Recalculate the queuing performance formulas for this change.

8. Sal's Pizza operates a delivery service that promises delivery within 40 minutes of a call-in order or the customer gets $2 off the price. Sal's operation can make a pizza every 3 minutes, on average. Pizzas take 12 minutes to bake, leaving Sal's delivery team 25 minutes to deliver the hot pizza. The travel area around the pizza shop is 10 minutes maximum. Assuming that calls for pizza come in at the rate of one every 5 minutes, and that both pizza demand and supply are Poisson distributed, what do you think of Sal's policy? Support your answer as best you can.

9. Using the order sequence provided in Situation 3-2, how would the simulation of the Burger Queen restaurant's operations change if one meat pattie of each size were kept in steamer inventory? If two meat patties of each size were kept in steamer inventory? How must you monitor and replenish the steamer inventory to get the best results you can?

10. Suppose that in Situation 3-2 we cannot assume that sandwich assembly is simultaneous and that board capacity is always great enough not to cause delays. Suppose that there were only four workers assembling sandwiches. How would that affect transaction times in the simulation? Suppose that there were five workers. What would change?

11. The Burger King restaurant description in Tour F describes a shift in the way the company managed its queuing phenomenon. Describe this shift in the language introduced in this chapter. Why would the company favor such a switch?

SITUATION FOR STUDY 3-1

BEACON GLASS WORKS

Jose Torrez, production manager at Beacon Glass Works, Marysville, West Virginia recently decided that the packaging department of his plant should package all finished goods and place the goods in stock rather than wait for customer orders.

The packaging department had four sections: finished goods, storage, package breakdown, breakdown storage, and final packaging. Figure 3-6 is a diagram of the packaging area. The finished goods storage consisted of a circular conveyor that held trays of 1000 pieces of each product. This conveyor could hold up to 5000 trays of products, depending on the mix of lengths and diameters of the products. The production process could feed up to 3000 trays of tubing per day into the storage area. The pack-

age breakdown area was necessary because different customers wanted different numbers of pieces for each order—normally, anywhere from 100 to 500 pieces per container. The breakdown area consisted of four long conveyors, each of which could handle a maximum of 400 trays per day. The foreman in charge of the breakdown area, Jesse Brown, normally scheduled breakdown by looking in the storage area to see what product had the largest quantity in storage.

Once the trays were broken down into individual containers, they were passed to a breakdown storage area with 10 gravity-feed conveyors that could store the equivalent of 5000 trays at any one time. The final packaging area had five packaging stations, fed by a conveyor

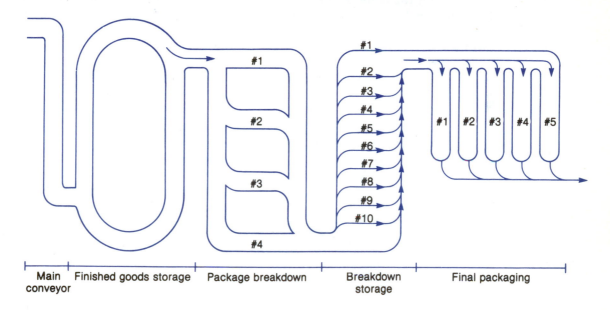

FIGURE 3-6   The flow of goods through the packaging department at Beacon Glass Works.

belt that went from the breakdown storage into the packaging area. Here packaging could be done at a rate of 500 trays per day per station. Final packaging originally had been done to customer order, but because of storage problems at the finished goods storage area, Jose had decided to package to stock.

After three months of trying the new packaging procedure, Jose noticed that (1) finished goods storage was continuing to pile up, (2) package breakdown was almost always running at full capacity, (3) breakdown storage was full, and (4) final packaging was still being done basically to customer order, with "expeditors" running around the packaging area looking for products to fill customer orders.

1. What is the information flow in the packaging area? How should information be processed in this area? Why?
2. What suggestions do you have for Jose to "break the bottlenecks" in the packaging area? How would your suggestions help?
3. What other information would you like to have to make the "best" economic decision about how to break the bottlenecks? Explain how you would use this information.

SITUATION FOR STUDY 3-2

UNIVERSITY VENDING

University Vending President Ted May had to decide soon what to do about the request from his operations chief, Dwight Bradley. One of the company's five routemen was retiring, and Bradley wanted to train a new person.

May had not approved Bradley's request for a new person yet because of his concern about the company's already high labor costs. University Vending's labor costs for the routes averaged 20 percent of total costs, while the industry average was only 16 percent. (University Vending's total costs and its average salary and benefit package of $14,500 per worker were close to the industry average.)

Ted suspected that the route structure was inefficient. The system currently had five routes, as shown in Table 3-5. Three of the routes were "product" routes (i.e., they serviced particular kinds of drink or food machines such as candy, sandwiches, and soft drinks), and the other two were "geographic" routes (i.e., they serviced all types of food and drink machines within a particular geographic area). Many locations were serviced by two product routes, and some were serviced by all three product routes. The product routes had originally been set up to minimize the costs of holding inventory; with fewer people handling

TABLE 3-5	Routes currently serviced by University Vending		
ROUTE	PRODUCT	GEOGRAPHIC COVERAGE	TIME
1	Soda, candy, snacks	Hospital	Day
2	Soda, candy, snacks	Hospital	Night
3	Candy, snacks	Both campuses	Day
4	Candy, snacks	Both campuses	Day
5	Drinks	Both campuses	Day

any one type of product, the amount of buffer stock was kept to a minimum.

The number of machines to be serviced had expanded dramatically in the last several years. As new locations and new machines were added to the system, their service requirements had simply been tacked on to the existing routes.

The routemen received a daily schedule of the machines they had to service, which they followed exactly. Because of the add-on process of scheduling new machines and the lack of strong supervision of the route schedules, the daily schedules would typically take the routemen back and forth across University Vending's service territory several times a day.

The five routes covered all the vending machines at Eusden University and Woodruff College. The vending machines were spread uniformly throughout 63 different buildings on the Eusden and Woodruff campuses. The two campuses were located four miles from each other. University Vending had the exclusive contract to service all the vending machines on both campuses, including Eusden's large, on-campus medical facility, in which University Vending's two geographic routes were located.

Each of the routes followed a 2-week cycle, during which every machine was serviced between one and six times, depending on the ma-

chine's sales volume; most machines were serviced two or four times during each 2 weeks. There were 163 machines on the two campuses, each of which was serviced an average of 3.5 times every 2 weeks. Almost all the buildings had two or more machines.

The hospital routes were unusual because they required no driving between machine locations. University Vending kept a separate vending warehouse within the hospital, from which the routemen walked to the machines they had to service.

Servicing the machines entailed not only refilling the product levels and collecting the revenues, but also cleaning the machines and making sure they were working properly. As shown in Table 3-6, time standards had been established for the service of different types of machines; these standards were found to match the actual service times closely.

Each routeman had 6 hours a day available to service the machines; the other 2 hours were spent on lunch, breaks, inventory stocking, and other nonservice activities. A recent study showed there was a wide discrepancy in the amount of time each routeman actually spent servicing the machines (see Table 3-7).

The study also indicated that even though the workers on the hospital routes had the best service ratio, some of the time they spent walking

TABLE 3-6   Time standards at University Vending

MACHINE TYPE AND NUMBER OF EACH	STANDARD SERVICE TIME/MACHINE (minutes)	NUMBER OF SERVICINGS	TOTAL STANDARD SERVICE TIME* (minutes)
Showcase (26)[†]	29	189	5481
Small candy (37)	16	88	1408
Small snack (15)	16	46	736
Soda (57)	19	156	2964
Hot drink (23)	15	61	915
Milk (5)	20	28	560

*Weighted average standard time 19 minutes per machine.
[†]Showcases (with big glass fronts) handle candy, chips, and snacks.

TABLE 3-7   Time study results for University Vending

ROUTE	PERCENTAGE OF 6 HOURS SPENT SERVICING	PERCENTAGE OF 6 HOURS SPENT DRIVING OR WALKING	MACHINES SERVICED PER DAY	LOCATIONS SERVICED PER DAY
1	79	21	15	5
2	79	21	15	5
3	55	45	9	8
4	56	44	9	8
5	55	45	9	8

to the machines was actually aimless wandering. The workers on routes 3, 4, and 5, however, were hard pressed to finish their routes in the time allotted. The study pointed out that the nonhospital routes averaged 19 minutes of driving between locations, and 4 minutes of walking to each machine. The hospital routes currently took 5 minutes to walk to each machine.

According to the study, a switch to the more efficient geographic routes would reduce the driving time on the nonhospital routes to an average of 4 minutes; it also pointed out that the walking time to any machine in the five routes should be only 4 minutes. These new estimates would result in substantial time savings. For instance, a building that had a drink and a food machine currently took 38 minutes to drive to (two routemen at 19 minutes each), while under a completely geographic system,

one person could service both machines with only 4 minutes of driving.

The current service schedule requires 351 separate stops in a 2-week period, which could be reduced to a minimum of 202 stops by getting rid of all multiple route stops, that is, by switching to geographic routes. Ted felt that the company should stop using product routes and should put in a complete system of geographic routes.

The main argument against that type of route structure was that inventory levels would have to be increased 25 to 33 percent to provide a suitable buffer stock for the additional routemen using each product. The current inventory levels are shown in Table 3-8. All inventories were replenished weekly, with ending inventories kept as low as possible. The company's cost of carrying inventory was 15 percent of the value of the inventory, and its

TABLE 3-8   Current inventory levels at University Vending

PRODUCT*	AVERAGE NUMBER OF CASES IN INVENTORY	AVERAGE COST PER CASE
Chips	100	$ 7.38
Candy bars	90	31.49
Gum	6	67.50
Soda	40	17.00
Cigarettes	10	346.80
Hot drinks	5	45.00

*Excesss milk products were picked up daily by the dairy.

opportunity cost on the money tied up in inventory was estimated at 12 percent.

Ted did not foresee any further expansion for his company in the next several months. He decided to take a detailed look at the possible savings of switching to geographic routes.

1. If University Vending switches to geographic routes, how much time would it take to service, drive to, and walk to the vending machines?

2. How many routemen would be necessary to service geographic routes?

3. How much extra cost is involved in increasing the buffer stocks of inventory?

4. Based on the answers to questions 2 and 3, what would your recommendation to Ted be?

REFERENCE NOTES

1. Francis, R.L., and J.A. White, *Facilities Layout and Location: An Analytical Approach*, Englewood Cliffs, N.J.: Prentice-Hall, 1987.
    Jacobs, F. Robert, "A Layout Planning System with Multiple Criteria and a Variable Domain Representation," *Management Science* 33, no. 8 (August 1987): 1020–34.
    Kapstein, Jonathan, "Volvo's Radical New Plant: 'The Death of the Assembly Line?'" *Business Week (Industrial/Technology Edition)* 3121 (August 28, 1989): 92–93.
    Tamashumas, Victor M., Jihan Labban, and David Sly, "Interactive Graphics Offer an Analysis of Plant Layout and Material Handling Systems," *Industrial Engineering* 22, no. 6 (June 1990): 38–43.

2. For more on group technology and manufacturing cells, consult:
    Flynn, Barbara B., and F. Robert Jacobs, "An Experimental Comparison of Cellular (Group Technology) Layout with Process Layout," *Decision Sciences* 18, no. 4 (Fall 1987): 562–581.
    Gaither, Norman, Gregory V. Frazier, and Jerry C. Wei, "From Job Shops to Manufacturing Cells," *Production & Inventory Management Journal* 31, no. 4 (Fourth Quarter 1990): 33–37.
    Gallagher, C. C., and W. A. Knight, *Group Technology Production Methods in Manufacture*, New York: Wiley, 1986.
    Greene, Timothy J., and Randall P. Saldowski, "A Review of Cellular Manufacturing Assumptions, Advantages and Design Techniques," *Journal of Operations Management* 4, no. 2 (February 1984): 85–97.
    Guerrero, Hector, H., "Group Technology: I. The Essential Concepts," *Production & Inventory Management* 28, No. 1 (First Quarter 1987): 62–70.
    _____, "Group Technology: II. The Implementation Process," *Production & Inventory Management* 28, no. 2 (Second Quarter 1987): 1–9.
    N. L. Hyer, "The Potential of Group Technology for U.S. Manufacturing," *Journal of Operations Management* 4, no. 3 (May 1984): 183–202.
    N. L. Hyer and U. Wemmerlov, "Group Technology and Productivity," *Harvard Business Review* 62, no. 4 (July–August 1984): 140–49
    Hyer, Nancy Lea, and Urban Wemmerlov, "Assessing the Merits of Group Technology," *Manufacturing Engineering* 101, no. 2, (August 1988): 107–9.
    Koelsch, James R., "Flexible Cells Take Control," *Manufacturing Engineering* 105, no. 3 (September 1990): 75–77.
    Morris, John S., and Richard J. Tersine, "A Simulation Analysis of Factors Influencing the Attractiveness of Group Technology Cellular Layouts," *Management Science* 36, no. 12 (December 1990): 1567–78.
    Vakharia, Asoo J., "Methods of Cell Formation in Group Technology: A Framework for Evaluation," *Journal of Operations Management* 6, no. 3–4 (May–August 1986): 257–271.
    Wemmerlow, Urban, and Nancy Lea Hyer, "Cellular Manufacturing Practices," *Manufacturing Engineering* 102, no. 3 (March 1989): 79–80.

3. Suggested readings for queuing phenomena include:
    James A. Fitzsimmons and Robert S. Sullivan, *Service Operations Management*, New York: McGraw-Hill, 1982.
    Gross, Donald, and Carl M. Harris, *Fundamentals of Queueing Theory*, 2nd ed. New York: Wiley, 1985.
    Howard, W. J. *A Simple Manual on Queues—The Long and Short of Waiting Lines*, Canoga Park, CA: Xyzyx Information Corporation, 1971.

Harvey M. Wagner, *Principles of Operations Research*, 2nd ed. Englewood Cliffs, N.J.: Prentice-Hall, 1975.

Winston, Wayne L., *Operations Research: Applications and Algorithms*, 2nd ed. Boston: PWS — Kent Pub. Co., 1991.

4.   See J. E. Ashton and F. X. Cook, Jr., "Time to Reform Job Shop Manufacturing", Harvard Business Review, vol. 67, no. 2 (March–April, 1989), pp. 106–111.

5.   For references on simulation see:

Carrie, Allan, *Simulation of Manufacturing Systems*, New York: Wiley, 1988.

Hurrion, R.D. *Simulation: International Trends in Manufacturing Technology*, Secaucus, N.J.: Springer-Verlag, 1986.

Law, Averill M., and W. David Kelton, *Simulation Modeling and Analysis*, 2nd ed. New York: McGraw-Hill, 1990.

Law, Averill M., and S. Wali Haider, "Selecting Simulation Software for Manufacturing Applications: Practical Guidelines & Software Survey," *Industrial Engineering* 21, no. 5 (May 1989):33–46.

6.   David H. Maister, "The Psychology of Waiting Lines" from John A. Czepiel, Michael R. Solomon, and Carol F. Surprenant, *The Service Encounter*, Lexington, MA: D. C. Heath and Company, 1985.

CHAPTER 4

MAKING QUALITY HAPPEN

During the past decade or so, the staggeringly important role of quality in both manufacturing and service operations has been hammered home by the full force of competition. This chapter approaches the topic in two parts. The first part deals with the definition of quality and the attitude to take towards it. The second part becomes more technical, explaining a number of the key tools that companies have used for making quality happen.

PART ONE

APPRECIATING QUALITY

Everybody wants high quality. The worker or the executive who proclaims the need for low quality has yet to be found. More than this, managers consistently trumpet quality's role in their marketplace successes; quality is often acknowledged as a competitive prerequisite. Yet despite the acknowledged importance of quality, it is probably the most misunderstood concept about operations.[1]

What Quality Means

When we think about a "quality" product, most of us probably think of how good the product looks, how sturdy it is, whether it performs many functions, how luxurious it may be, and how reliable it is. Judging quality for many of us is an exercise in assessing a host of subjective criteria about beauty, luxury, and performance. According to this perception of quality, Rolls-Royce and Mercedes mean quality cars, Nikon means quality cameras, Hewlett-Packard means quality calculators, Vuitton means quality leather goods, and Henredon means quality furniture.

Unfortunately, these popular perceptions of quality are not what quality professionals and enlightened managers mean by the term. Quality means only one thing to such people: conforming to the specifications for the product or service as valued by customers. If a product

99

meets the customer-valued specifications, then it is a quality product, be it a Ford Escort or a Cadillac Seville, a Kodak pocket camera or a top-of-the-line Minolta. Whereas to the popular mind, quality is a subjective judgment about the merits of one set of product or service specifications versus another, to the quality professional and to the manager, quality is simply whether or not the product or service conforms to those specifications.

Of course, a product can have specifications or design features that leave it prone to failure or quick breakdown. To the quality professional, it is not a contradiction that a quality product (in the sense that it was made to all the documented specifications) would perform shabbily due to a set of poorly conceived specifications. Making quality happen in the organization, then, means making products or delivering services according to the specifications as valued by the customer. If the specifications are found wanting by the consumer or by manufacturing, then they need to be changed, formally, so that they are more "fit" for the use to which they are put by the consumer. The means by which a product's "fitness for use" can be assessed and redefined is discussed in Part Two of this chapter.

By defining quality as conformance to specifications as valued by the customer, one removes the subjective elements about what quality may be and replaces them with objective and quantifiable ways of gauging it. Moreover, once one quantifies quality, one can then think about attaching some costs to it.

The cost of quality (or, more precisely, the cost of off-quality) becomes the cost to the company of doing things wrong, of not conforming to the specifications. When viewed this way, the cost of quality takes on an importance that goes beyond many managers' thinking about product costs. The cost of doing things wrong includes the cost of product failure to the company (often companies distinguish between external failure — failure found once the product has been sold — and internal failure — failure found while the product is still at the factory), the cost of trying to detect those failures before they are delivered, and the cost of preventing those failures in the first place. Table 4-1 lists key items in each of these cost areas. These costs frequently total 15 to 20 percent or more of a typical company's cost of sales.

If a company is to avoid most of these costs of quality, it must produce the product or deliver the service correctly the first time, and every time. To do this, however, everybody in the organization must be convinced that things have to be done right the first time. It is a cop-out to say that quality is the strict province of the quality control department or that quality problems start on the factory floor. In fact, they are more likely to start in places other than the factory — in product design, manufacturing, engineering, industrial engineering, training, purchasing, customer order processing, or elsewhere. W. Edwards Deming, one of the grand old men of quality management and the man after whom Japan's Deming Prize for quality is named, insists that management is responsible for 85 percent of the quality problems in a factory; workers are responsible for only 15 percent. Some estimate that product design is itself responsible for 50 percent or more of a product's quality problems.

As companies work diligently on quality, they often discover that not only do their total costs of quality decline but that the compositions of those costs of quality shift. Typically, the failure costs shrink in percentage importance while prevention costs expand.

IMPROVING QUALITY

Quality in an operation is improved only when everybody involved — those who handle the product and those who do not — become aware that their tasks can affect quality and that they

TABLE 4-1   Elements in the cost of quality

FAILURE COSTS	DETECTION COSTS	PREVENTION COSTS

External failure costs
  Consumer affairs (dealing
  primarily with customer
  complaints about quality)
  Field service (mostly repair of
  what should have worked)
  Product liability (insurance and
  settlements)
  Product returns, recalls, and
  corrective action costs
  Warranty costs
Internal failure costs
  Downtime due to defects
  Engineering change orders
  Purchasing change orders
  Redesign
  Retesting
  Rework in factory service
  operations
  Scrap
	Process capability measurement
	(e.g., control charts)
	Product acceptance
	Prototype inspection and tests
	Receiving inspection and tests
	Supplier surveillance
	Work-in-process and finished
	goods testing and inspection
		Design reviews
		Engineering drawing checks
		Engineering quality orientation
		program
		Engineering specifications reviews
		Preventive maintenance
		Process capability studies
		Product qualification
		Quality audits
		Quality orientation programs
		Supplier evaluations
		Supplier quality seminars
		Tool and Machines control
		Worker training and cross training

may need to redirect their procedures and habits toward preventing mistakes. The goal, of course, is to have no quality problems, to have every product be perfect. The achievement of this goal, however, requires significant changes in management and worker attitudes and a relentless pursuit of all the small things that can ruin quality by causing work to be done improperly the first time.

A controversial step toward this goal of a perfect product is to discard traditional points of view about making mistakes. There is a big jump from making perfection a "goal" to making it a "standard," but it is felt by many that such a change of attitude is essential to significant improvements in quality. According to this view, the standard is "zero defects" for everybody associated with the operation. A zero-defects standard does not say to all concerned

merely that they are to do the best they can; rather, it says that everyone is expected to produce with zero defects or change the procedure so that they will get zero defects produced. All the causes of errors or potential errors must be removed and the operation changed so that everything is done right the first time, every time. This means that workers and managers must analyze mistakes to determine why particular defects were introduced into the product or the service. Why were parts missing? Why were the wrong parts delivered? Why were operations missed or put in the wrong sequence? Why did the product fail in the field? Should it be redesigned? Should the materials be different, or should the assembly or test procedures be changed? Why was the wrong part ordered? How can the product's order blank be made foolproof? Are the tolerances in the product too tight to be held by manufacturing? Should new investment in equipment or tooling be made to permit these tolerances?

One could go on and on about the sources of errors. What is critical, though, is that each error must be removed if the product is to be produced with zero defects. This attitude change also affects the traditional roles of management and the workforce; no longer can the workers be seen merely as executing the wishes of management and engineering. Rather, management and engineering have to be viewed as resources to be applied to the solution of quality problems that are, for the most part, known first to the production workers. The workforce can solve many quality problems themselves but should not have to find and remedy all of them. Thus, if the zero defects standard is to be reached, management and engineering must be brought in to help the workforce solve particular problems. Management and engineering should serve the workforce, not the other way around.

The setting of the zero-defects standard and the importance of removing all causes of errors in the process have been debated. The debate centers on the various components of the cost of quality as noted above: failure costs, detection costs, and prevention costs. How are these costs of quality related to the number of defects in the product? As the number of defects changes, how do the total costs of quality (i.e., the sum of the costs of failure, detection, and prevention) change? Do they rise and fall? Figure 4-1 demonstrates the two opposing views on the issue. According to the traditional view (as represented in Figure 4-1a), the lowest cost is achieved at some nonzero level of defects. The other view (shown in Figure 4-1b) concludes that the zero defects standard is the lowest cost. The difference, of course, is that one side of the debate postulates that the costs of prevention and detection increase substantially as one gets closer and closer to zero defects. Supporters of this view reason that there

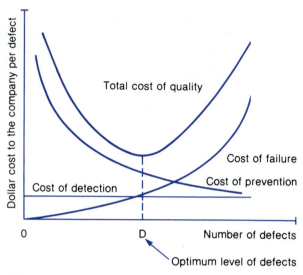

(a) The traditional view.

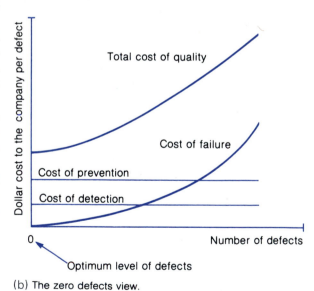

(b) The zero defects view.

FIGURE 4-1 The debate on the optimum number of defects.

have to be some diminishing returns from error removal as more and more errors are found and fewer errors persist. The last errors are thought to be the toughest to find and correct. The other school of thought states that the causes of defects in the product are so simple, although numerous, that it takes no more expenditure to remove the last cause of error than to remove the first. It may take longer to determine what that last source of error is, but, so this school of thought goes, the steps to correct it are likely to be rather simple.

This debate on whether there is, in fact, an optimal level of defects for a product is essentially an empirical debate, but as yet, no one has done a very good job of estimating what these curves look like. Are they more like Figure 4-1a or are they more like Figure 4-1b? In time, we will probably all have a better sense about whether the zero-defects standard is truly the lowest-cost way to make a product. Philip Crosby and his adherents believe that quality is "free" and that, in fact, the lowest-cost way is always to produce the product right the first time.[2] According to Crosby, the cost of quality should not be higher than about 5 percent of sales, and given the fact that in many companies the cost of quality is 15 percent or more of sales, Crosby looks to quality improvement as the chief source of profit gain, and thus, it is "free." Others are not so sure that quality, at the level of zero defects, is, in fact, free.[3] What the Japanese and others have been teaching us, however, is that the optimal level of defects is far lower than anyone had previously dared to think.

MANAGEMENT ATTITUDES TOWARD QUALITY

There is much less debate about the attitudinal and organizational changes that have to be made to promote conformity to specifications. The importance of making quality everybody's business has already been discussed; a problem in one area of the company can easily overlap into problems elsewhere, with deleterious effects on the quality of the service delivered or the product manufactured. Other attitudes about quality have to be rethought as well. In many non-quality-conscious companies, the identification of product defects too often leads to blame and recrimination, with excuses abounding and managers and workers alike seeking to avoid being singled out for them.

In a quality-conscious company, defects are not an excuse for blaming others, merely an indication that everyone needs to work together to eliminate the causes of the defects. Moreover, the notion that quality should or must be sacrificed in order to promote delivery or cost goals is replaced by the notion that only by producing a quality product can a company stay competitive and that, over time, costs will be lowered and deliveries improved when the causes of quality problems are identified and remedied. In many quality-conscious companies, the cost of quality (failure costs, detection costs, and prevention costs) is known and reported consistently, typically as a percentage of sales or of the cost of goods sold. Hunches about the costs of quality are replaced by definitive statistics kept by the accounting function of the firm. Managers in the quality-conscious firm are evaluated on their quality.

Other changes are evident as well. Suppliers become an important member of the quality team, are educated on what it takes to produce quality products or services that meet the needs of the customer company, and supplier ideas are even sought to assist product design and manufacturability. The quality department is held in high regard by everyone; it is a leader in thinking about manufacturing and changes to the product or service. It is a training and troubleshooting resource for the rest of the company. Quality control reports not to manufacturing but to a separate manager and thus is not subject to the schedule and cost measures applied to manufacturing performance.

Convincing a skeptical workforce that quality is important, when for years management has not acted as if that were truly the case, can be a long, hard pull. Often senior management's hand has to be forced to demonstrate that quality prevails over the long-held goal of meeting orders or shipment targets. When management is willing to halt shipment when there is a quality problem, and to take the heat from a customer about it, the workforce can begin to get excited about a new era at the company.

Management attitudes in many companies have been usefully altered by the adoption of formally recognized supplier–customer relationships within operations. In these progressive companies, managers and workers are imbued with an understanding that each of them is a supplier to someone else in the company and that each of them is also a customer of somebody else. They are then evaluated in part by what their customer(s) say about them, just as they help evaluate those who supply them with materials or information. Such a reorientation of operations functions that may have long felt cut off from the rest of the company has usually had a very salutary effect on people. They come to realize readily that quality, up and down the organization, depends on them.

Changes in management organization and attitude can be achieved in many ways. A number of the quality "gurus" have proposed "principles" or "plans" for quality improvement. Perhaps the most famous of these plans is Deming's "14 Points." These are the points that Deming stressed in his many educational trips to Japan, beginning in 1950, which many credit for turning around Japan's thinking about quality, and how it is achieved.

Deming's 14 Points

These points are at once philosophical and statistical in origin and reflect Deming's long years as both a statistician and a troubleshooter for industry.[4]

1. Create constancy of purpose toward improvement of product and service. This means to take the long view, to spur innovation whenever possible, and to focus on the customer.

2. Adopt the new philosophy. Take on leadership for change.

3. Cease dependence on inspection to achieve quality. Eliminate the need for inspection by building quality into the product.

4. End the practice of awarding business to the lowest bidder. Move toward long-term relationships with single suppliers for any one item.

5. Improve the system of production and service constantly and forever. Quality, productivity and cost reduction are linked.

6. Institute training on the job.

7. Institute leadership. The aim of supervision should be to help people and machines do a better job. This means removing barriers to doing the job right. It also means not treating every defect or problem as if it were a special case but to use the statistical tools available for management.

8. Drive out fear—do not let people go because they have helped to improve quality (and productivity and cost).

9. Break down barriers between departments. Promote teamwork.

10. Eliminate slogans, exhortations, and targets for the workforce asking for zero defects and new levels of productivity. Slogans do not help one to improve, and, because the vast majority of quality problems are due to the "system" and management and not the workforce, slogans can lead to resentment in the workforce.

11. Eliminate quotas on the factory floor.

12. Remove barriers that rob people (workforce, management, engineering) of their rights to pride in workmanship. This

means abolishing such things as piecerates and other production incentives that depend on quantity and not quality. Avoid numerical goal-setting for managers (e.g., increase productivity by X percent this year or lower costs by Y percent).

13. Institute a vigorous program of education and self-improvement.

14. Put everybody in the company to work to accomplish the transformation. The transformation is everybody's job.

The Deming 14 Points is not the only method, of course. Other programs exist, typically associated with other leading figures in the crusade for better quality management, such as Philip Crosby, Joseph M. Juran, and Armand Feigenbaum. Although there is much that is common across the various programs available, there are some differences in emphasis and style. For example, Crosby is sometimes criticized for being too much of a "cheerleader" and not strong enough on the techniques of quality management. Deming's is sometimes viewed as too dogmatic an approach and heavily statistical in nature, and this draws fire from some quarters. The Juran and Feigenbaum approaches are fairly eclectic in their treatment of the subject.[5]

The Malcolm Baldrige National Quality Award

In 1987, Congress instituted a new award for quality to honor a former Secretary of Commerce, Malcolm Baldrige, who had recently died in a rodeo accident. This award is analogous to the Deming Prizes in Japan, and since 1988 has been an increasingly visible and coveted award for American firms.[6] There are really three different awards that can be made in any year: a major manufacturing company award, a small company award, and a service business award. Up to two can win in any category, but there is no requirement that any award be given.

There are seven categories that are evaluated for the Baldrige award. They are each given point values that add up to 1000 points for a perfect score. Independent teams of examiners grade both written applications from the applicant companies and also make site visits to the most promising ones. The categories and point values are as follows:

1. Leadership (100 points). How the company's senior management create and sustain quality values in all areas of the company.

2. Information and Analysis (60 points). The scope of data used (internal, customers, suppliers), its validity and timeliness, statistical analyses used, root cause determinations, countermeasures, and so on. This gets to the use of "management by fact" at the company.

3. Strategic Quality Planning (90 points). How quality is integrated into the business planning of the company. The use of benchmarking against competitors and world-class companies of all types and how that benchmarking is used.

4. Human Resource Utilization (150 points). The effectiveness of the company's efforts to involve and develop all its employees. Issues are full participation, employee empowerment, continuous improvement, and organizational growth (education, training, recognition, quality of work life).

5. Quality Assurance of Products and Services (150 points). Use of total quality control in terms of product design, testing, and supplier materials/parts/services. Issues here relate to standard setting, capability studies, designation of key process characteristics to control, design of experiments, quality audits, documentation, and quality assurance outside of manufacturing itself.

6. Quality Results (150 points). What the actual results are from the adoption of quality programs based on both trends and independent assessments of customers and competitors.

7. Customer Satisfaction (300 points). The company's knowledge of its customers and its performance in serving the customer well and quickly. This includes assessment of warranties, post-sales service and contact, complaint handling, and customer feedback.

These categories are seen by many companies as a wonderful means of grading themselves and of seeking improvements to their processes, even if they do not apply for the award. The Baldrige award is certainly helping to focus management attention on quality and the mechanisms for achieving it. It is proving to be a useful way of defining, in practice, what people mean by Total Quality Management (TQM).

PART TWO

TECHNIQUES FOR MAKING QUALITY HAPPEN

As the management of quality has evolved, the inspection of products to determine which ones meet specifications has been deemphasized by many companies as the prime source of quality conformance. Instead, there has been a concerted effort to improve both product design and the design and control of the process itself. This shift of emphasis is very much in tune with the theme of "making it right the first time." Because of it, troubleshooting can more easily restore a basically stable process to a high level of conformance. If troubleshooting depended on examining the defects detected solely by product inspections rather than by process inspections, the troubleshooter would have a tougher job. This person would lose valuable information (especially that from process measurement and instrumentation) and would have to sort through more potential causes of the defects. It would take much longer to trace a symptom of poor quality back to a cause. Thus it is in everybody's interest to shift the emphasis for quality control away from the examination of poor-quality products and toward the examination of the process itself.

In keeping with the change of emphasis, in this part of the chapter we discuss primarily techniques that can be applied to the process rather than techniques for sorting good products from bad. Such product-sorting techniques (such as acceptance sampling) are called for in some circumstances, but techniques that deal directly with the process should have a greater impact. Three major questions are addressed in the remainder of this chapter, all related to keeping quality conformance high: Is the product design adequate? Is the process adequate? Is inspection adequate?

IS THE PRODUCT DESIGN ADEQUATE?

Perhaps the most basic concept of quality management is "fitness for use;" this notion largely governs all the subsequent techniques used to manage quality. Fitness for use is simply the extent to which the product, as designed, serves the real purposes of the consumer. What is it about the product that has value for the consumer? Conversely, what about the product has low value for the consumer? By carefully analyzing a product's fitness for use, managers and engineers can be sure that a product is designed expressly for its intended purpose. A product can be underdesigned or overdesigned and thus inappropriate for the use to which the consumer puts it.

Stages of Product Development

How can a new product under development meet this overarching criterion of fitness for use? Most companies have a well-defined series of steps through which their new products pass. The best companies also promote close and continuing contact among the various functions that get intimately involved with new product introductions and with suppliers, as well. The nature of the cross-function contact may vary: no formal teams, formal teams chaired by manufacturing with responsibilities upstream in the design phases of the product, formal teams chaired by engineering with responsibilities downstream in the transfer of the product to manufacturing, and special buffer managers between engineering and manufacturing. Whatever the form, it is generally acknowledged that close, sustained contact promotes better new product introductions. The typical stages of product development, briefly, include:

1. *Concept development and feasibility.* This stage in product development is devoted to thinking through the initial notions about the new product. Is it technically feasible? Is it cost-effective to develop? This stage gets engineering and marketing attention for the most part and can take a lot of time.

2. *Detailed design.* Once a new product has been deemed feasible, the detailed engineering of it must take the concept as originally developed and make it capable of being produced.

3. *Prototype.* Given the initial specifications for the product, the first units of the product need to be built and tested. In this stage, the product may perform poorly and the concept may be scrapped or the design may be reinitiated. In many companies the prototype is created by engineering; although there may be some concern for manufacturing procedures and equipment in the construction of the prototype, often the production management does not get intimately involved with the new product at this stage. Production equipment and tooling are often not employed at this stage.

4. *Manufacturing prototype.* At this stage the product is evaluated by manufacturing for its manufacturability. Typically, design changes are made at this point to simplify the manufacturing process, change materials, make use of existing manufacturing equipment or strengths, and merge the new product with any existing products. This stage may involve experimentation with new processes in pilot-plant operations. Preproduction lots of the product are made, trying as much as possible to simulate all aspects of the product's manufacture in a plant (i.e., regular equipment and materials used, production tooling) so that any potential problems can be identified early, rather than after full-scale production has begun.

5. *Ramp-up to full-scale production.* This is the transfer of the new product to manufacturing's full control and its production in existing manufacturing facilities.

6. *Post-release design changes.* Many products require design changes as suggested by customer use and manufacturing experience

once the product has been released into full-scale production. Such design changes may be minor, but there still needs to be a mechanism to accomplish them.

At each of these six stages there are means by which problems with the product under development can be identified. In the early stages there are reviews of the concept and the design. Later, there is a test of the prototype under various conditions. There are tests also of the manufacturing prototype, some done in-house and some done with consumers in trial marketing. In all cases, good records of product failure and process problems should be kept.

Product Design and Quality

As mentioned earlier, many people knowledgeable about manufacturing believe that over half of the quality problems affecting products are really flaws of the product design. Indeed, as Harmon points out, many of the "cost reductions" claimed for products are better viewed as "error corrections."[7] The issue then becomes one of how to improve the steps taken in product design so that fewer errors are introduced at the start.

Many of the techniques now being used to speed up product development (discussed in greater depth in Chapter 13) also contribute to improved product design quality. Teams of product design engineers, process engineers, quality engineers, and managers from marketing, cost accounting, and manufacturing line positions—often usefully joined by representatives of suppliers—can greatly facilitate product design speeds and quality (concurrent engineering). The entire movement for "design for manufacturability" (again, discussed at greater length in Chapter 13) also contributes mightily to improved product quality, as it argues for increased simplification and standardization of products and product components.

Keeping track of the product design and its components can frequently be a harrowing experience when all is centralized. However, decentralizing the classification and coding of product designs can help, as can establishing simple modular bills of materials and simple schemes for tracking part commonalities.[8]

Reliability, Maintainability, and Safety

Keeping good records enables companies to use some quantitative techniques to improve product development, particularly in the areas of product reliability, product maintainability, and product safety. Let us consider each of these.

Reliability. There are a number of measures of the reliability of a product. The classic one is "mean time between failures," the average time between successive failures of a product that can be repaired. Variations on this measure include the "mean time to failure," which is the average time before a failure of a nonrepairable product or the average time until the first failure of a product that can be repaired, and the "mean time between maintenance," which is the average time between specified maintenance activities. Other reliability measures include failure rates—that is, measures of the number of failures that occur for a given unit of time. Related to failure rates are "lifetime" measures, such as the "mean life" of the product, which is the average amount of time the product can be used before it wears out or is in need of major overhaul. A variation is the "median life," the time period during which half of the units of the product produced would have failed. These measures for reliability are important for setting standards for performance that the product, as designed and manufactured, must meet and also for helping to assess where the quality problems are in the product. Such analyses of reliability may lead to redesigns to make aspects of the product "redundant" or to change its materials or manufacturing procedures to enhance reliability.

Maintainability. Maintainability refers to how easily the product can be maintained by the consumer. The measures of maintainability generally refer to the "mean time to repair," the downtime needed for either scheduled or unscheduled maintenance, and the cost of maintaining the product over time. As with the measures for reliability, these measures may be used not only as goals for product design and manufacture but also as evidence in an analysis of what it takes to make maintenance simple and inexpensive.

Safety. Good product safety can scarcely be stressed enough. Good safety management means some hardheaded analysis of the product, its design, and its manufacture. Safety has to be quantified just like reliability and maintainability. Typically, various classes of hazard are identified, from catastrophic to negligible. The potential for the product to cause such safety hazards has to be realistically assessed and new product designs or methods developed.

The Voice of the Customer in Product Design: Quality Function Deployment

In the last several years, a number of companies, spearheaded by the auto industry, have adopted a Japanese concept for improving new product (and new process) designs. There is nothing particularly fancy about the technique, but it does provide a very systematic approach to the development cycle. The technique is termed *quality function deployment* because it acts to "deploy" the customer's desires throughout the organization. The technique translates customer requirements into the appropriate engineering and technical requirements for each stage of the product development process outlined above. Because of its rigor, QFD can be time-consuming to implement, but once in place it has reduced both start-up times and costs for the companies that have been diligent with it.

QFD ties together a number of features of both product and process:

1. *Attributes of the product that are recognized and valued by the customer.* No matter what the product, there are certain features or attributes which distinguish it from others. Customers of different types weigh these characteristics differently, as well. For example, suppose we were designing a new watch. Among the attributes of the watch that a customer might value could be the stylishness and the functionality of the watch. One could break down both of these concepts further. Stylishness could depend on size, shape, color, weight, type of watchband, and so on. Functionality could involve the readability of the dial, the accuracy, whether it is waterproof, whether a day/date dial was included, the life of the watch between battery changes, and so on.

2. *Customer preferences.* These attributes can be rated among competing products. One can judge the stylishness and functionality of different watches through customer surveys, focus groups, and similar means.

3. *Engineering characteristics.* Different engineering characteristics enable a product to exhibit the attributes it does. Sometimes the engineering characteristics of a product are complementary (bigger battery and readability), but often a product's design is a compromise among competing engineering characteristics. For example, small size may be valued in a watch, but it may then conflict with the watch's readability. A watch's shape

may hinder the placement of the day/date dial or the size of the battery. The engineering characteristics permitting stylishness— materials used, movement used, tolerances of the moving parts—may thus conflict with the engineering characteristics permitting functionality—battery size, waterproof casing needed, dial and hands design.

4. *Objective measures of the engineering.* Just as customers have preferences that can be measured, so too can the engineering of the watch use specific measures. One can measure the weight of the watch, the square millimeters of the dial, the relative size of the day/date to the dial's area, the length of the second and minute hands, the mean life of the battery. One can then set target values for each of them.

5. *Technical aspects of the engineering.* These include the difficulty for the design, its importance to satisfying the customer's needs, and the cost of doing it.

Quality function deployment links these attributes and characteristics in a product planning matrix (sometimes referred to as the "house of quality" because of its shape). Figure 4-2 is an example, again using our watch design project.[9]

The house of quality is built in stages, or steps:

Step 1. State the customer attributes clearly. Figure 4-2 states some of the elements of style and function along the left side. These attributes derive from marketing research, focus groups, customer surveys, and the like. They often are expressed in a hierarchy from basic points to more detailed points about the product.

Step 2. List the engineering characteristics needed to control for the customer attributes. Along the top are listed some of the corresponding engineering characteristics, such as

the area of the dial, that translate, as best as can be done, what engineering must do to achieve the desired customer attributes.

Step 3. Develop the relationship matrix between the two. The entries in Figure 4-2 indicate the relative strength of the relationships. The relative strengths help to determine whether product engineering can readily satisfy the customer attributes desired. The symbol ■ is used for a strong positive relationship, □ for a weaker positive relationship, ○ for a weaker negative relationship, and ● for a strong negative relationship. As you can see in the center of Figure 4-2, there are positive relationships between the small size desired and the area of the dial and between "not feeling it on the wrist" and the weight an engineer can measure. On the other hand, there are negative relationships between the readability of the watch and the small area of the dial and between the small size desired and the size of any numerals that can be put on the watch's face.

There are also relationships to be addressed among the engineering characteristics themselves. Some of them may conflict with each other as we can see in the "roof" of the house of quality. For example, the area of the dial conflicts with the size of the numerals that can be used.

Step 4. Enter the market evaluations of your product and the competition. This is done along the right side. This entry includes a rating for how important the customer attribute is perceived to be.

Step 5. Develop objective measures for the engineering characteristics and compare those measures of your product with the competition.

Step 6. Develop targets for each of the engineering characteristics. These may be the same as the objective measures of an existing product, or they could be different. They are based on competitive assessment of the product (Step 4) and other data, both marketing and engineering in nature.

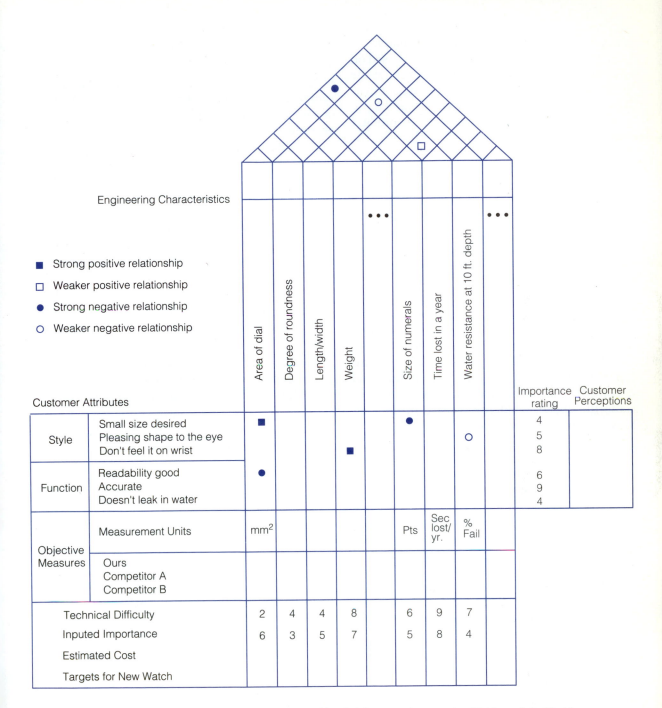

FIGURE 4-2 Product planning matrix (House of Quality) for a watch example—Matrix partially filled in.

Step 7. Select those engineering characteristics that are to be pursued (deployed) based on customer importance, the selling points of the product, and the objective engineering comparisons made and how difficult they may be.

Step 8. Carry over the results of the product planning matrix (the house of quality) to the subassemblies and components that will be needed for the product. This means creating a "component deployment chart" that lists what is critical to control in the development and manufacture of each major subassembly or component.

Step 9. This, in turn, leads to the development (deployment) of product and process checks of various types and plans for how the critical-to-control subassemblies and component parts of the product are to be manufactured. A process plan chart would determine exactly what would be tough to do and what should be checked about the process (e.g., temperature, length, weight), and how (what kind of control chart, what sampling plan, how to make the check, what the control limits are). This action completes the QFD process.

IS THE PROCESS ADEQUATE?

Improving a process, bit-by-bit in continuous fashion, demands repeated, systematic change to that process. The problem areas of the process have to be identified, investigated, and changes made. Process improvement thus requires a set procedure and a toolbox of skills for identifying and analyzing the process's problems. This portion of the chapter looks at process improvement first by outlining a "quality improvement program" and then, systematically introducing the statistical tools that can be used in a quality improvement program.

Quality Improvement Programs

The adoption of Deming's 14 Points, or of any of the other quality programs or philosophies, typically entails education in and application of a procedure for attacking quality problems of all types.

Florida Power and Light, the utility company serving most of south Florida and the first recipient of the international Deming Prize, has adopted a seven-step procedure for problem-solving that has aided it, and many other companies. It is worth examining. The seven steps, adapted from some Japanese sources, are as follows (the tools, in italics, will be discussed later):

1. *Reason for improvement.* Identify the problem area and why it is worth solving.

2. *Current situation.* Select the specific problem to work on and set a target for the problem-solving effort. *Pareto charts* may apply here as well as graphs of various sorts.

3. *Analysis.* Identify and verify the root cause(s) of the problem. This step makes use of such tools as *fishbone diagrams* and Pareto charts. It requires the gathering of data of various types as well as its analysis.

4. *Countermeasures.* These are the actions to take to counter the root causes. To many they would be known as solutions, but they may never really eliminate the problem for all time.

5. *Results.* This is the test of the countermeasures. It often requires *experimentation* on the factory floor.

6. *Standardization.* This is the identification of the actions to take to prevent the problem and its root causes from recurring. This may entail *process capability* studies and the development of *control charts.*

7. *Future plans.* Plan action for any remaining problems and evaluate the effectiveness of the problem-solving team itself.

These 7 steps are a more elaborate version of what has come to be known as the Deming cycle of Plan, Do, Check, Act (PDCA). The Deming cycle, which Deming attributes to his mentor, Walter Shewhart, the inventor of the quality control chart, goes as follows:

- *Plan.* Plan a test of available data on the problem, or plan a change to the process, an experiment.
- *Do.* Carry out the test or change, preferably on a small scale.
- *Check.* Observe the effects.
- *Act.* Study the results and ask what was learned. New knowledge can then lead to more data gathering, more testing and change, and more results to mull over. The process improves continuously by this method.

There are a variety of quality improvement programs like this one. No one has an absolute monoply on what is best. Experience with a number of them have underscored a number of points about them, however. To wit:

- Putting in a quality improvement program (QIP) is a major redirection for a company and must be led from the top. Without wholehearted support from top management, the QIP will wither and die.
- QIPs require "champions" whose role is to proselytize and encourage.
- QIPs require constant attention. They need to be consistent with the company's strategy and they need constant feedback mechanisms.
- Line management must own the QIP, with the quality department seen as a resource for training and troubleshooting.

- An extraordinarily useful concept for many companies has been the adoption of a "customer-supplier" relationship within the firm. Everyone is seen as having a customer (whoever uses the outputs of their labors) and having suppliers (those who feed them with product or information). This notion forces people to talk to their customers and suppliers and to make every effort to satisfy their requirements. It gives immediacy to the QIP process.
- What gets measured is what gets managed. This is a truism that is especially relevant for quality programs.
- Middle management, caught as they always are in the middle, will likely resist a quality improvement program unless top management provides them with the wherewithal to both understand and to be integral parts of the QIP effort.

Doubtless, other insights could be stressed here. Suffice it to say that a QIP can be an exceedingly effective management tool, but one that has to be believed in and worked through religiously if it is to give all the benefits it can to the company.

Determining What's Important: Run Charts and Pareto Charts

For almost any product (or service), we can make out a long list of specifications as to its size, shape, material content, color, functional characteristics, and so on. All of those specifications can be measured and tolerance limits for those measures established. The performance of the process in manufacturing the product to meet these specifications to within the tolerances established is a gauge as to the process's capabilities. Fortunately, it is almost always the case that only some of the specifications have to be held to truly tight tolerances for the product to be "fit for use." Often, the important specifications are known from the design phase of product development and the process itself can be designed to meet these specifications readily. Sometimes, however, problems with the product surface either in inspection or out in the field and demand special attention for their elimination.

It makes sense in these instances to make use of Pareto analysis. Pareto analysis is simply the categorization of the various problems identified and the development of frequency distributions for them. Pareto analysis tracks what percent of all the problems is due to Problem 1, what percent is due to Problem 2, and so on, so that the problems that are the most frequently occurring are readily identified. Such frequency distributions can quickly reveal what kinds of problems seem to be the most prevalent and are thus worthy of extended effort to solve. Pareto analysis can lead to the establishment of supplementary specifications for the product that, if met by manufacturing, can avoid the problems detected. Typically, data for a Pareto analysis are reasonably easy to collect, and collection is frequently done by operators and/or inspectors over the course of several days or several weeks, the time frame, of course, dependent on the frequency of problems and the output rate of the process. Pareto analysis is an extremely useful tool isolating the "vital few" problems from the "trivial many," to borrow a phrase from Juran.

Another useful tool for determining what is important is the chronological run chart, a graph of some measured characteristics plotted in the order of manufacture. Having this plot in chronological order identifies trends in the data that might be missed if just a calculation of the mean or median or standard deviation is performed. Chronological run charts differ from control charts (discussed later) in that they cannot demonstrate conclusively that a process is under statistical control, but they can offer strong evidence that something about the process has changed. Consult Figure 4-3, a run chart of the percent defective. It displays two suspicious runs, the span between time units 10 and 18, and the span between time units 21 and 30. What makes these two places suspicious is that the percent defective stays on one side of the median for at least 8 units of time in a row. That is akin to flipping a coin and having it come up eight heads in a row. It is so unlikely that it suggests a specific cause; for the section consistently below the median something very good was at work, while for the section consistently above the median something adverse was operating. These suspicious patches of time are persuasive indications that the process is not under control and that their specific causes ought to be researched.

Investigating and Correcting Causes of Quality Problems

The causes of process problems must be researched and corrective measures taken. Several things are important in this regard.

Data Gathering.   You are only as good as your data. There is nothing like good, meaningful data for tackling quality problems. Sometimes the data appropriate for investigating a problem are routinely collected. Other times the data of interest may not be readily available or subject to significant errors. If this is the case, the first step is to understand where the problem may be coming from (e.g., particular product types, particular materials, particular segments of the process) and how important it is (vital, trivial). Once these things are understood, the kinds of data appropriate to the problem's study are usually apparent. Understanding this may take protracted talks with the customer to find out precisely what the customer wants and values. It may also take the determination of how quality may best be ob-

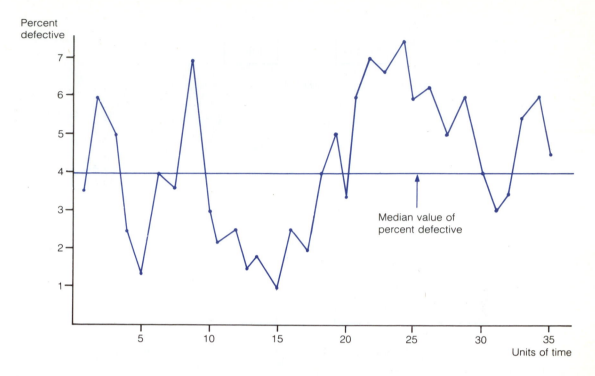

FIGURE 4-3   A chronological run chart.

tained: through product design, through materials specifications, through the production process itself. Once these things are understood, then data can be collected and analyzed.

Collecting the appropriate data may require (1) measuring aspects of the process or the product; (2) keeping track of defects and their characteristics; (3) tracing particular defects to times, lines, or machines; and (4) noting any changes or conditions in the process and its materials, equipment, and/or labor. The accumulation of these data may require a substantial investment in quality monitoring equipment and gauges and specific procedures. Some experiments may have to be conducted.

Systematic Investigation.   One of the aids to quality problems investigation that has been particularly effective is the Ishikawa or "fish-

bone" diagram, which tries to be explicit about the causes and effects in the process (see Figure 4-4). The head of the fish indicates the problem (the effect) whose causes are sought. The bones of the fish are the potential causes that are being brainstormed by this technique. The main bones are generally chosen to be some major categories of causes that one can readily examine. Often these six M's, or a subset of them, are investigated: methods, machines, manpower, material, measurements, and maintenance. Of course, other causes can be placed on the main bones of the fish. The smaller bones are more specific potential causes for the effect being investigated. They are generally decided upon by asking "why" repeatedly until one is satisfied that the root causes have been ferreted out.

Use of a fishbone diagram often leads to the collection of data or the design of an experiment

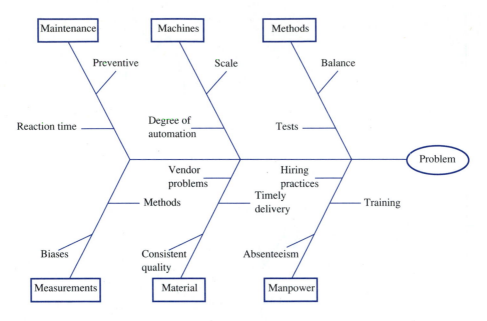

FIGURE 4-4   A "fishbone" or Ishikawa diagram with some major categories
of causes indicated.

so that causes can be checked out and effective solutions developed. Explicit diagrams of this type are reminiscent of the process flow and information flow diagrams considered earlier in this book. They, too, were means of making explicit what might otherwise be missed.

Experimentation.   Experimentation is a hallmark of science, and it is also a hallmark of scientific approaches to process improvement. Often, the experiments that factories engage in are simple ones, but that does not detract from their power. Experimentation in the factory and in product design is considered by many to be the most powerful quality weapon in the Japanese arsenal.

There are many different ways that experimentation can proceed in a factory. There are some classical experimental designs associated with the early statistical pioneers (e.g., Sir Ronald Fisher) as well as more recent ones associated with current quality gurus such as Genichi Taguchi and Dorian Shainin. Exploring the subtleties of these approaches is beyond the scope of this book. However, the debates about which approaches to use concentrate on the balancing of statistical accuracy and power with the time and expense of running the experiments. Clearly, one would like to glean as much information from an experiment as one could, but without the need to run too many experiments.

Taguchi's methods are reasonably parsimonious about the number of experiments to run, but they also risk some misleading conclusions, particularly if there are substantial interactive effects between factors that affect the product's quality. Still, they are gaining adherents in many companies.[10] Taguchi methods start with a fishbone diagram to help identify potential causes of quality problems. Once likely candidate factors are identified, the methods go on to set different levels of values for those factors, say two levels for each factor. With different settings of the factors made, a series of experiments are run. The construction of so-called signal-to-noise ratios and analysis of variance techniques are used to sort through which of the factors seems to be most important in explaining the variation in the experimental results.

Shainin argues for examination of the patterns of variation in the product that could provide clues as to the type of experiment to run or factors to focus on. For example, there may be patterns of variation that are (a) positional—same place on each unit produced, or same tooling, machine, or factory involved each time, or (b) cyclical—a group of units, perhaps consecutive, are affected, or the variation occurs from one batch to another, or (c) temporal—the variations occur from hour to hour, day to day, month to month, or something similar. With such investigation, the list of candidate factors can be pared down and the experimental design can be made simpler and more attuned to possible interactions among the factors.

No matter what technique is chosen, however, good experimentation can be exceedingly useful in testing hypotheses about why quality suffers in some instances and what can be done to remedy it. The particular techniques to employ are often less important than the raw willingness to tinker with the process in systematic ways in order to explore its weaknesses and to fix them.

Situation 4-1 is an example of a Taguchi method experiment.

Another type of often employed experimentation relates to machines in the process and whether they are the right ones for any particular task at hand. This important experiment is termed the process capability study.

Process Capability Studies, Cp, and Cp_K

Several checks should be done before a new product is released into production or before a product is assigned to a new or a different machine. One of the most useful is termed the process capability study. It is designed to assess whether the machine or process is capable of holding the tolerances that design engineering says must be held. Essentially, the process capability study is an experiment where the machine or process is put through its paces under conditions that replicate those of actual production. The equipment should be running as well as it can be (i.e., under statistical control, as discussed later), and any materials should be those that meet specifications.

With the machine or process running, the process capability study collects data on the critical specifications for all of the products or parts produced. The data are then analyzed to determine whether the machine or process can easily hold to the specification. This is done by assembling the data, usually at least 50 data points, and finding its mean and standard deviation. The process capability limits, sometimes referred to as the "natural limits" or the "natural tolerance limits," are 3 standard deviations from the mean of the sample taken.

It is important for this range of variation, the ±3 standard deviations (often called the 6-sigma range), to be well within the specifications set for the product or part by design engineering. Figure 4-6 illustrates this. What is desired is a ratio of the engineering specification

Experimentation at Navistar International, Indianapolis Engine Plant

The Navistar International engine plant in Indianapolis has embraced experimentation to improve its operations. Navistar, producer of International brand trucks, produces diesel engines for its own make trucks but also for certain varieties of Ford diesel-powered trucks.

Navistar wanted to improve the flatness of the cylinder head for the 7.3 liter engine. The specification called for a degree of flatness to the machining of the face of the cylinder head combustion chamber that the process could not hold. The prevailing Cp_k (see below) was only 0.17, considerably below the objective of 1.33.

A team of five was assigned to the task of studying the problem. Together with a group of supervisors, other operators, skilled tradesmen, and tool research people, they brainstormed 55 possible causes of the out-of-flatness condition. After additional review, 7 of the 55 variables proposed were thought worthy of additional experimental study to determine how they ought to be set for best results. The 7 variables included the following, each of which were set at 2 different levels or settings (referred to as Settings 1 and 2).

Variables	Experiment							
	1	2	3	4	5	6	7	8
Feed rate	1	1	1	1	2	2	2	2
Tool sharpness	1	1	2	2	1	1	2	2
Shaving build-up	1	1	2	2	2	2	1	1
Stock removed	1	2	1	2	1	2	1	2
Number of passes	1	2	1	2	2	1	2	1
Lubrication	1	2	2	1	1	2	2	1
Clamp pressure	1	2	2	1	2	1	1	2
Results:								
High-low average differences	37.7	38.3	44.2	38.3	15.0	28.6	21.5	41.0
Total 264.5								

1. Feed rate for the broach that shaved metal off the cylinder head (80 ft/min vs. 120 ft/min)

2. Sharpness of the tool itself (sharp vs. dull)

3. Shaving build-up behind the finish tools (loaded vs. clean)

4. The amount of stock removed (0.020" vs. 0.0028")

5. One-pass vs. two-pass operation

6. Tool lubrication (dry vs. oiled)

7. Hydraulic clamp pressure (600 lbs. vs. 500 lbs.)

It was decided to run a Taguchi method experiment using these 7 variables, 2 settings on each variable, and to run 6 pieces for each of 8 different experiments.

In a Taguchi method experiment, more than one variable is allowed to vary at a time, so as to be more economical on the number of experimental runs that have to be made. Different settings are prescribed for each experiment. The table below lists which settings were used for each variable in each of the 8 experiments.

The analysis proceeded with 6 pieces run for each of the 8 experiments. Measurements were taken at 32 points on the cylinder head surface, with the flatness of the cylinder head determined as the difference between the highest and lowest points on the surface, in ten-thousandths of an inch.

With knowledge of which settings were used in which experiments, one can construct a table showing the average high-low differences for each

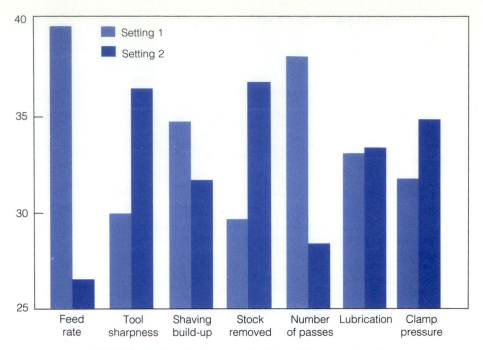

FIGURE 4-5　Graph of average high-low differences by setting.

of the settings of the seven variables. (For example, for the feed rate, Setting 1 was used for the first 4 experiments and the average of the high-low differences for the first 4 experiments is $((37.7+38.3+44.2+38.3)/4) = 39.63$.) Such a table follows and is represented in the graph in Figure 4-5:

Variables	Setting 1	Setting 2
Feed rate	39.625	26.525
Tool sharpness	29.900	36.250
Shaving build-up	34.625	31.525
Stock removed	29.600	36.550
Number of passes	37.875	28.275
Lubrication	33.000	33.150
Clamp pressure	31.525	34.625

Using these averages for each variable setting, one can perform an analysis of variance to determine which variables are the most important in explaining the total variation found in the experiments. (In the following table, the sum of squares listed for any variable works out to twice the squared difference between the averages of Setting 1 and Setting 2.) The results of such an analysis follow:

Variables	Sum of Squares/ Variance	Percent Contribution
Feed rate	343.220	46.18
Tool sharpness	80.645	10.85
Shaving build-up	19.220	2.59
Stock removed	96.605	13.00
Number of passes	184.320	24.80
Lubrication	0.045	0.01
Clamp pressure	19.220	2.59
		100.00

This analysis clearly indicates that the two most important variables affecting flatness were the feed rate and the number of passes. The string of optimum settings then becomes:

1. Feed rate for the broach—120 ft/min
2. Sharpness of the tool itself—dull
3. Shaving build-up behind the finish tools—clean
4. The degree of stock removed—0.020″
5. One-pass vs. two-pass operation—two passes
6. Tool lubrication—dry
7. Hydraulic clamp pressure—600 lbs.

This group of variable settings provided results that optimized the process.

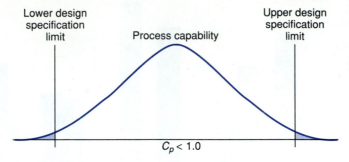

(a) Process capability unable to meet design specifications
Note: Some of the process capability falls outside
the specification limits.

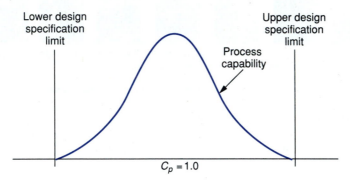

(b) Process capability just meets design specifications

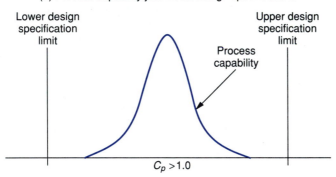

(c) Process capability exceeds design specifications

Note: The higher the C_p value, the more confident you
can be about meeting the specifications every
time.

FIGURE 4-6   Design specifications and the process
capability measure, Cp.

range to the 6-sigma process capability range that is greater than 1.25 or 1.33. This ratio is known as Cp, for "capability of the process." Such a ratio leaves enough room so that we can be confident that if the only causes of variation in the process are random ones, there will be no products or parts produced that do not meet the engineering specifications.

Naturally, the worth of any machine's capabilities are appreciated only in relation to the engineering specifications required. If a machine that is in statistical control can work to tight tolerances, with a Cp value greater than 1.25 or 1.33, all is well. The machine will likely produce almost all good items. These items produced should not undergo 100 percent inspection but should only be audited (sampled) from time to time. If, on the other hand, the machine cannot hold tolerances tighter than about three-quarters of the design tolerance (the reciprocal of a Cp value of 1.33), a different tack ought to be followed. There are four major alternatives to try in such a case:

1. Improve the machine if you can. Try different tooling, speeds, feeds, and so on.

2. Shift the item to a different machine, if that is possible.

3. Get the engineering staff to approve a looser tolerance, if they are willing.

4. If all else fails, sort good items from bad through 100 percent inspection until you can redesign the product, or purchase a new, better machine, or in some other way improve the capability of the process relative to the design tolerances.

The Cp measure is one that deals strictly with the ratio of the specification range to the 6-sigma range of process capability. It does not take into account whether the process is centered where it should be, at the precise design specification. As a means of including the position as well as the range of process capability into a meaningful measure, Cp_K has been devised. Cp_K is defined as

$$Cp_K = (1 - K) \, Cp$$

where K = (design specification − process average)/(specification range/2).

Thus if the process average equals the design specification, K = 0 and Cp_K = Cp. When the specification range is large relative to the process average, K is small and Cp_K lies close to Cp.

There are many quality experts who place great store in a process being centered well on the precise design specification. Genichi Taguchi is one who feels that the costs of quality rise dramatically when the process is not centered, even though the process may be operating within the specification limits. Taguchi argues that companies should seek the very lowest variation possible from the design specification, no matter what the range given for the specification limits. Taguchi models this belief with a quadratic "loss function":

$$L(x) = k \, (x - n)^2$$

where L(x) is the loss attributable to the measure x, k is a constant, and n is the nominal, or target, dimension. The greater x varies from n the greater the loss.

Another way that companies have highlighted the need to have process capabilities well within the design specification limits is exemplified by the Six Sigma (6σ) program of Motorola, one of the first winners of the Malcolm Baldridge National Quality Award. The Six Sigma program sets a process capability goal of 3.2 defects per million opportunities for component parts and, indeed, all that the company does. This is a very tight standard, motivated by the following two points:

1. Defects can "stack up" so that missing a specification by only a little for each of, say, 100 parts that make up the product can lead

to the product's missing specification altogether, and

2. Processes often drift from their center points by as much as 1.5 standard deviations.

If a process is to produce final products whose defects are less than 0.1 percent, but which are made up of as many as 100 components whose centering may be off by 1.5 σ, then, to be assured of meeting the 0.1 percent defect standard, the components should be made with the 6σ goal of 3.2 defects per million. Driving for such results is thus a sure way of enhancing process capability.

Getting and Keeping Control of the Process

Once the problems of the process have been identified and analyzed, so that all have some confidence that the countermeasures devised can work, then the process has to be brought into statistical control. Only with statistical control can we be assured that the process is producing the product with quality.

The Control Chart

If the process is to produce quality products, its capabilities have to be measured periodically so that management can be sure that it is performing as planned and is not drifting away from quality manufacture. In the last 50 years or so, a statistical means of assessing process capabilities has come to the fore—namely, the use of control charts to measure the adequacy of a particular process.[11]

Control charts are used to separate random causes of variation in product quality from nonrandom, "assignable" causes of poor quality (operator error, faulty setup, poor materials).

Assignable causes often demand a lot of work to remedy, but remedied they can be. Typically, such remedies may involve better training, more precise directions, improved fixtures, different materials, improved tooling, and the like. When these remedies are put in place, the process's capabilities are immediately improved; the assignable causes of poor quality have been removed. What remains is random variation in quality.

What gives the control chart power to discriminate between random causes of product nonquality and "assignable" causes is the fact that control charts depend on samples of products and not on individual units. Why does this afford power to the control chart? From statistics we know that no matter what the underlying distribution of the random causes of poor quality are, the means of samples taken from any distribution will tend to be distributed normally. This is the important Central Limit Theorem. Being able to use the normal distribution for the sample means taken permits us to say more about the process than we could by just using individual observations from the generally unknown underlying distribution of the process itself. We know lots about a normal distribution and how it behaves.

Thus, using samples in control charts is a better, more powerful way to check on a process's capabilities than merely to investigate single units as they are produced. Deming tells of a famous example at the Nashua Corporation, a producer of carbon paper. Nashua produced carbon paper by spraying carbon black on tissue paper. The carbon black was sprayed on by a nozzle, the size of whose opening could be adjusted. Naturally, spraying too little carbon black yielded poor-quality carbon paper, while spraying too much was wasteful and expensive. Prior to the use of control charts, the worker running the spraying equipment periodically measured the amount of carbon black on the finished product. If too little had been sprayed, he would open up the nozzle. If too much had been sprayed, he would close it off a bit. What was wrong with the way he was operating? Even with a perfect process, one exhibiting only random variation, half of the carbon paper examined will be above the mean thickness of carbon black and half will be below the mean thickness. A perfect process would not have to be touched at all by the worker. Yet, according to his mode of operation, he would be adjusting the nozzle all the time, even if the process were perfect. In fact, he was introducing more variation to the process than the process had naturally. Better for the worker to collect data samples, take a mean of the sample, plot it on a control chart, and take action only when the control chart indicated. In that way, the worker would not be inadvertently adding variation to the process. The control chart would have taken into account the random nature of variation in the process and would have revealed any nonrandom, assignable causes of variation. Such is the advantage of using samples for the creation of control charts.

Control charts are of two general classes: variable control charts and attribute control charts. A variable control chart plots some specific measure along the Y-axis (such as lengths, diameters, pressures) and various sample numbers along the X-axis. The most common variables charts are $\overline{X}$ and R charts. The $\overline{X}$ chart plots the average of the measurements taken on each sample, and the R chart plots the range of measurement within each sample. Attribute control charts, on the other hand, typically deal with the "defects" of whatever kind in the samples taken. They can provide useful summary data about the process, but may have to be supplemented by the use of specific variables (e.g., length, weight, pH, temperature) if particular explanations

about the causes of the defects are to be tested. Indeed, these two classes of control charts can be used at the same time, in cooperation with one another. Table 4-2 compares some of the basic variations.

Variable control charts can, in fact, be set up with an almost limitless number of quality characteristics measured on the Y-axis. In developing control charts, it is thus important to distinguish the process characteristics that are most apt to be associated with defects (the vital few) and not to worry about other aspects of the process that, by all indications, appear under control (the trivial many). The characteristics chosen for study should logically be associated with remedies that can be taken. Often, Pareto analysis can be helpful in identifying the variables to plot.

Variable Control Charts

Once it has been decided to develop a variable control chart, and the characteristics to be plotted have been picked, the next task is to contruct the chart itself. Consider the $\overline{X}$ chart first (Figure 4-7). That chart's mean line is usually the average of past data. The mean line can be determined from a process capability study or from simply a history of data, typically at least 50 data points worth. (Sometimes an engineering standard is initially used for the mean line, although actual data, of course, have the advantage.) The control limits for the chart, for example, are generally set at 3 standard deviations (of the distribution of sample means) on either side of the mean line. Such limits ensure that random causes alone would indicate false alarms in only 0.3 of 1 percent of the cases because 99.7 percent of a normal distribution lies within 3 standard deviations of the mean. This is the way a control chart "accounts" for the random variation in the process and permits the nonrandom, assignable causes to be readily identified.

TABLE 4-2   Major types of control charts

TYPE OF CHART	STATISTICAL MEASURE PLOTTED ON THE Y-AXIS	COMMENTS
$\overline{X}$	Small sample averages of particular measurements	The most common control chart; indicates change in central tendency of process
R	R = maximum-minimum values of small samples of measurements (R stands for range)	Frequently accompanies $\overline{X}$ chart; indicates when a significant loss (or gain) in uniformity has occurred
X	Individual measurements, not averages	Less sensitive than the $\overline{X}$ chart; useful only when small samples cannot be taken conveniently and one observation per lot is all that is possible
P	P = fraction defective	Most useful when "go/no go" decision is needed; can help tell when to apply pressure for quality improvements; uses "attributes" data, not measurements data
C	C = number of defects	Most useful when all samples are the same size and the number of defects possible is large but the likelihood of any single type of defect is small; avoids proliferating $\overline{X}$ or X charts in such situations; another "attributes" data control chart

With today's calculators and computers, it is easy to calculate the standard deviation of a set of sample means, and such calculations should be used. Historically, however, that calculation was more time consuming, and some formulas were developed to estimate the upper and lower control limits using the grand average (the average of the sample averages) and the average of the sample ranges. Those estimation procedures are still in use, and for the sake of completeness, they are presented here.

$$\overline{X} \text{ chart upper control limit} = \overline{\overline{x}} + 3\sigma_{\overline{x}} \approx \overline{\overline{x}} + A_2 \overline{R}$$
$$\overline{X} \text{ chart lower control limit} = \overline{\overline{x}} - 3\sigma_{\overline{x}} \approx \overline{\overline{x}} - A_2 \overline{R}$$

where $\overline{\overline{x}}$ (X double bar) = grand average (average of the sample averages), $\sigma_{\overline{x}}$ is the standard deviation of the sample means, $\overline{R}$ is the average of the sample ranges, and A_2 is the estimating fac-

tor, which varies with the size of the samples being taken. It is found in Table 4-3. Of course, the preferred means of computing the control limits now is not to use this shortcut estimation, but rather, to calculate the standard deviation of the sample means straight away.

The control limits can also be calculated from the computed process capability limits derived from a process capability study. The process capability study deals with single observations, x's. Control limits deal with samples. The mean of the process and the mean of the sample means taken are both unbiased estimates of the true process mean. However, the standard deviation of the sample means will always be smaller than the standard deviation of the individual observations. Knowing the number of observations per sample is all we need, however, to translate standard deviations of single observations into estimates of sample

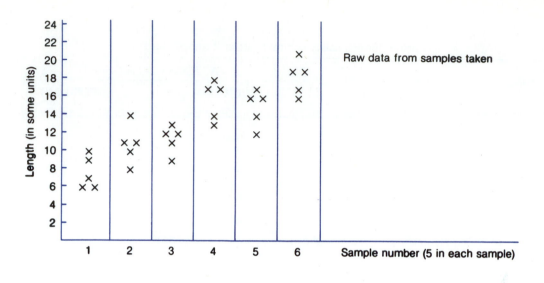

Raw data from samples taken

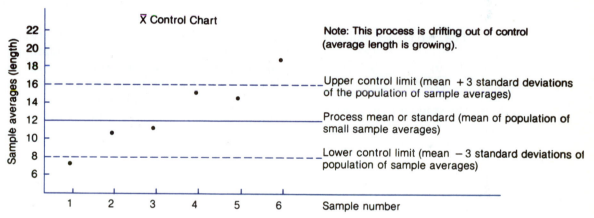

Note: This process is drifting out of control (average length is growing).

Upper control limit (mean + 3 standard deviations of the population of sample averages)

Process mean or standard (mean of population of small sample averages)

Lower control limit (mean − 3 standard deviations of population of sample averages)

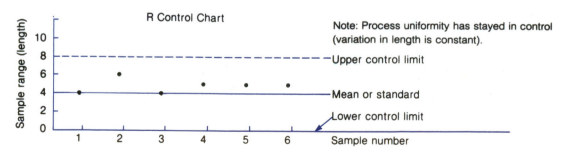

Note: Process uniformity has stayed in control (variation in length is constant).

Upper control limit

Mean or standard

Lower control limit

FIGURE 4-7   Sample $\overline{X}$ and R control charts.

TABLE 4-3   Factors for estimating control limits in $\overline{X}$ and R charts

NUMBER OF OBSERVATIONS IN SAMPLES TAKEN	FACTORS FOR $\overline{X}$ CHART FOR R, (A_2)	FACTORS FOR R CHART	
		LOWER LIMIT D_3	UPPER LIMIT D_4
2	1.880	0	3.268
3	1.023	0	2.574
4	0.729	0	2.282
5	0.577	0	2.114
6	0.483	0	2.004
7	0.419	0.076	1.924
8	0.373	0.136	1.864
9	0.337	0.184	1.816
10	0.308	0.223	1.777

SOURCE: J. M. Juran, ed., *Quality Control Handbook*. New York: McGraw-Hill, 1979, Table Y, Appendix II.

standard deviations. The two are related by the following formula:

$$\sigma_{\overline{x}} = \frac{\sigma_x}{\sqrt{n}}$$

where σ_x and $\sigma_{\overline{x}}$ are the standard deviations of the individual observations and of the samples, respectively, and n is the number of observations in each sample.

With the R chart, the mean line is precisely $\overline{R}$, the average of the sample ranges, and the control limits (1) can be calculated from the sets of samples taken, or (2) estimated from $\overline{R}$ itself.

R chart upper control limit = $\overline{R} + 3\sigma_R \approx D_4\overline{R}$

R chart lower control limit = $\overline{R} - 3\sigma_R \approx D_3\overline{R}$

where σ_R refers to the standard deviation of the sample ranges, and D_3 and D_4 are estimating factors also found in Table 4-3. Again, the preferred technique these days is to calculate the standard deviation of the sample ranges straight away.

Being in Statistical Control

A process that is operating in statistical control has no variation produced by specific, traceable, assignable causes. All its variation is at a minimum, given the process technology applied, and the variation represents merely random statistical variation, and no variation that one can label as having a particular cause. If one wants to improve a process that is in statistical control (i.e., reduce the spread of the control limits), one has to change the process technology. One implication of this truism is sometimes startling to managers: If the process is out of control, yet tighter tolerances are desired, managers should not immediately go and buy new, replacement equipment. Rather, managers should seek to uncover and remove the assignable causes of variation. Buying new equipment could be a very expensive way to tighten tolerances, and success in doing so is not ensured. Working to put the process under statistical control is the suggested prelude before any creative, and perhaps costly engineering and equipment revamping of the process.

Operationally, a process is in statistical control if it has no points outside the control limits, no suspicious-looking runs (portions where there are many points on one side of the mean line), and no definite trend that suggests the process is drifting toward the control limits and thus out of control.

As process changes are made, a control chart can be updated. A new mean and new, tighter control limits can be calculated from the ever-increasing set of data on the process. By continual improvement, variation in the process can be reduced. At some point, the process may be in such good control that managers can forgo using a control chart for the particular variable of concern. Instead of regular samples being taken, periodic quality audits can confirm whether the process continues to produce high-quality items. Sometimes, only first-and-last-piece inspections of a batch of parts/products need to be performed.

The control chart is thus a way station on the road to quality—an important way station, but nevertheless just a way station. It is of little value when the process is known to be out of control, and it may be too time consuming to keep up when the process is known to produce perfect pieces. The importance of the control chart lies in distinguishing random variation from variation caused by specific, traceable factors that can be remedied. As the process gets worked on, the control chart is a record of the process's capability.

Attribute Control Charts

So far, the discussion has centered on $\overline{X}$ and R charts, the major examples of variable control charts. Mentioned above was the other major class of control charts, attribute control charts. Attribute charts typically plot "defects" of one sort or another. These charts have to suffice when there is no clear measurement to make or where there are multitudes of defects possible, but with no single one of sufficient frequency on which to develop a variables chart. The two major types of attribute charts are the P chart, which tracks the percent defective in the samples taken, and the C chart, which tracks the number of defects in the samples

taken. The samples taken for attribute charts are usually greater than those used for $\overline{X}$ and R charts by at least an order of magnitude.

Attribute charts make the assumption that defects get produced in much the same way that "heads" turn up when coins are flipped. That kind of a process is generally termed a Bernouilli process, and there are two, similar statistical distributions that model Bernouilli processes: the binomial distribution and the Poisson distribution. These distributions make it easy to determine mean lines and control limits for attribute control charts.

The P chart assumes a binomial distribution of the percent defective created by the production process, where $\overline{p}$ is the mean percent defective and the standard deviation is

$$\sigma_p = \sqrt{\frac{\overline{p}(1 - \overline{p})}{n}}$$

where n is the size of the sample taken. Thus the control limits are defined as

P chart upper control limit $= \overline{p} + 3\sigma_p$

$$= \overline{p} + 3\frac{\sqrt{\overline{p}(1 - \overline{p})}}{n}$$

P chart lower control limit $= \overline{p} - 3\sigma_p$

$$= \overline{p} - 3\frac{\sqrt{\overline{p}(1 - \overline{p})}}{n}$$

The P chart has wide applicability and is particularly useful when you want to control a specific kind of defect that has been identified.

The C chart plots the number of defects in a sample and is applicable when the likelihood of any single type of defect is small but where there may be a large assortment of different types of defects possible. The C chart assumes that defects arrive at the process much as phone calls arrive at a telephone switchboard. The distribution that captures this phenomenon best is the Poisson, which is a particularly

---

ASIDE

A Comment on Sampling

Control charts depend on samples for their power. So far, the sizes of the samples needed for a control chart have not been discussed. Often samples of size 3, 4, or 5 are taken, but the precise sample size to use can be determined. Unfortunately, it is beyond the scope of this book to deal thoroughly with the issue of how big a sample to use in generating a control chart of a specific type. We can only hope to appreciate some of the difficulties and subtleties involved.

All sampling involves some risk. Indeed, there are two types of errors that can be made:

- *Type I error.* Rejecting a sample (or lot) when it in fact meets specifications. This is sometimes called the producer's risk, and it is usually designated by α. A 5 percent chance for this to occur is typical.

- *Type II error.* Accepting a sample (or lot) when it in fact does not meet specifications. This is sometimes called the consumer's risk and it is usually designated by B. A 10 percent chance for this to occur is typical.

These errors can be depicted on what is known as an operating characteristic curve (see Figure 4-8). There is an operating characteristic (OC) curve that can be associated with every sampling plan devised. The OC curve depicts the trade-offs inherent in a sampling plan between errors of type I and errors of type II. A type I error may cost the company money, but it is a type II error that causes ill-will with customers. For this reason, most sampling plans try especially hard to limit type II errors, even at the expense of type I errors.

Figure 4-8 is an example of an operating characteristic curve where the true proportion of defectives is 5 percent of the lot. Ideally, one would like to have an operating characteristic curve that looks like the dashed line rectangle in the figure, bounded by a 1.0 probability on the Y-axis and by 5

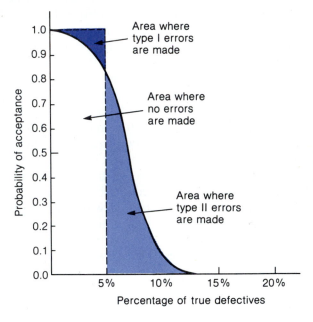

FIGURE 4-8   An operating characteristic curve.

percent defective on the X-axis. This is an unattainable goal, barring 100 percent inspection, and this reality forces us into balancing type II errors against type I errors.

The shape of an operating characteristic curve depends primarily on the sample size taken *(n)* and the acceptance number (i.e., the allowable number of defects in the sample) denoted by *C*. In general, the more critical one wishes to make the inspection—that is, the smaller the areas for both type I and type II errors—the larger the sample size and the acceptance number need to be.

friendly distribution since its standard deviation is merely the square root of its mean. The mean line on a C chart is simply $\bar{c}$, the mean number of defects. The control limits are given by

C chart upper control limit $= \bar{c} + 3\sigma_c = \bar{c} + 3\sqrt{\bar{c}}$

C chart lower control limit $= \bar{c} - 3\sigma_c = \bar{c} - 3\sqrt{\bar{c}}$

Attribute control charts, such as P and C charts, should be interpreted in the same way as variable control charts. If the process is operating in statistical control, as demonstrated by the control chart, all the variation is produced by random causes and not by assignable causes. If the percent defective is to be lowered, the process itself has to be changed.

Pre-Control: An Alternative to Control Charts

An alternative to the use of control charts, popular in some quarters, is pre-control. Pre-control can be used effectively if all that is de-

sired is a decision about whether to go on with production or to stop it, dependent on the design specification limits set.

The rules governing pre-control are simple:

1. Take half of the width of the specification range for the product and center it on the design specification. Consult Figure 4-9. These are the pre-control lines, within which we ought to expect most of the distribution of the product.

2. To determine if the process is capable, take five units in a row and see if their measurements all fall within the pre-control lines, the green zone of Figure 4-9. If they do not, examine the process more closely and reduce any variation that you can.

3. Once in production, periodically sample two units consecutively. If the following conditions prevail, act accordingly:

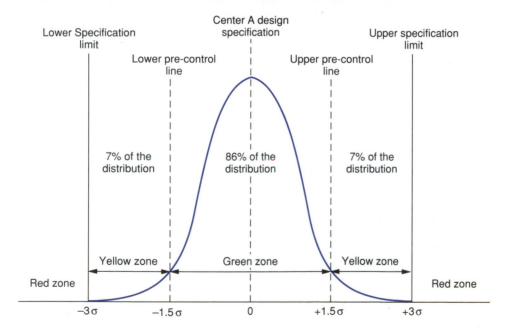

FIGURE 4-9 Pre-control chart.

a. Two units in the green
   zone   Continue
b. One in green, one in yellow
   zone   Continue
c. Two units in yellow zone   Stop
d. Any unit in red zone   Stop

4. The frequency of sampling can be determined as one-sixth of the time interval between two stoppages of production. Thus, if the process has been stopped often, the frequency of sampling will be greater than for a process that has not been stopped often. An average of 25 checks until a stoppage is considered about right.

One way to keep control of the process is to make it foolproof. Once the cause of the quality problem has been found, often the corrective measures that should be taken are designed to make the process foolproof. Foolproofing may come about through a change in the process technology, product design, or methods and procedures. Such foolproofing may call for special product design or for the use of particular foolproofing fixtures, alarms and cutoffs for machinery, multiple signals or identity cod-

ings, optical magnification or remote control viewing, formal countdown procedures, or a host of similar devices. The trick is to change the process and/or the product so that the only way to make it is the quality way.

The Notion of Dominance

Depending on the type of process involved, the production of quality usually depends on a particular aspect of the process to a much greater degree than on other aspects of the process. That is, one aspect of the operation dominates others, and if quality is to be produced, that aspect of the operation must be controlled properly. Following Juran, Table 4-4 indicates four types of dominance in the process, the kinds of production processes that typically are associated with each type, and the process control systems that can be used to keep track of process quality under each of these types of dominance. These process control systems are explained in greater detail later.

As a summary of the use of control charts and other aspects of quality management, consider Situation 4-2.

TABLE 4-4   Quality and dominance in the process

TYPE OF DOMINANCE	CHARACTER	SAMPLE PROCESSES	PROCESS CONTROL SYSTEM(S)
Setup	Process is stable and reliable; if setup is correct, product is correct	Many machine shop operations, printing	First and last piece inspection
Machine	Process subject to some drift	Papermaking, packaging (such as beer), plastics injection molding	Control charts (such as $\overline{X}$, R, X, P)
Operator	Skill of operator is critical to quality of product, foolproofing important	Apparel, fast food service, repair shops	Control charts (such as P,C), acceptance sampling
Component	Process is stable, but quality of product is dependent on quality of components	Most auto assembly and other assembly processes	Incoming inspection, acceptance sampling

SITUATION 4-2

Stroh Brewery Company Revisited, Winston-Salem Brewery, Quality Assurance

The quality assurance department at the Winston-Salem brewery of Stroh operated round the clock, 7 days a week, and employed 33 people. Quality assurance operated three separate labs at the brewery: two chemical labs (brewing and microbiology) and one physical lab (packaging). All aspects of the brewery's operations were checked systematically every day. Quality assurance wanted to make sure not only that beer shipped from the plant met the company's high standards but also that the process by which the beer was brewed and packaged was operating under control and consistently within specifications. Quality assurance viewed itself as the early warning station for the check of product and process conformance to the company's dictated specifications. There were a dozen or so manuals, several feet thick in total, that detailed the tests and specifications for product and process quality.

The Brewing Lab

The brewing lab was the larger of the two chemistry labs at the plant. The brewing lab was charged with testing each batch of beer at every stage in its brewing cycle, from the development of the wort through fermentation and aging in the ruh and government cellars. The work in the lab was divided among various technicians: alcohol technicians, government cellar technicians, water technicians, and so-called weekly technicians. The alcohol technicians were charged mainly with chemical analysis of samples taken from wort production, fermentation, and the various storage cellars. The government cellar technicians operated more in the filtering area and the finished cellars, checking air and carbon dioxide content clarity and pH every time a tank was emptied and every time the packaging lines called for a brand change. The water and weekly technicians were concerned chiefly with more specialized checks of the quality of water and other inputs (although malted barley was checked by the Detroit staff before shipment to North Carolina) and of miscellaneous items.

As an example, consider the alcohol technician's job in more detail. This person checked a number of things: specific gravity, refractive index, original gravity (a measure of the percentage of fermentables in the wort at the start); real extract (a measure of the amount of solids left in the beer), alcohol percentage, calories, VDK (a measure of the amount of an undesirable yeast by-product remaining in the beer), pH (an acidity measure), color, sodium and calcium concentrations, bitterness units (a measure of the amount of hops in the beer), foam pour (a measure of the time it takes for the foam head of the beer to collapse 1 inch when poured at a given temperature and at a specific height), foam cling (a measure of how much the foam head on a beer clings to the surface of the glass—the more it clings, the better), and haze abuse (a measure of the clarity of the beer). For all of these tests there were detailed procedures and specifications that the plant had to meet.

The test for original gravity, for example, was performed about 300 times each day, 100 or so on each shift. Whenever one of the 750 barrel tanks that were used for fermentation or aging was pumped so that the beer traveled to the next stage in the finishing process, a sample of about 250 ml was taken. This sample had to be tested within an hour or so of the pumping of the tank. This test took about 15 minutes and consisted of (1) warming the sample to 25° C, (2) passing the sample through filters, (3) reducing the temperature to 20° C, (4) placing a small portion of the filtered beer in a digital refractometer to determine its refractive index, and (5) comparing the weight of another portion of beer with the weight of water to determine its specific gravity. Given a certain refractive index and a certain specific gravity, the original gravity, alcohol percentage, and real extract was calculated by computer.

The results of such tests were promptly transmitted electronically to the brewhouse, where corrective measures could be taken, if necessary. For example, if the sampled beer was light on original gravity, so-called heavier beer from an accumulator tank could be combined with it to return the

continued

sampled beer to within specifications. Most "out of spec" conditions could be remedied by repumping and remixing the beer in question.

The brewing lab was also charged with analyzing beer that had already been distributed. Stroh employed a number of buyers who regularly bought beer in liquor stores and supermarkets and sent it back to the plant for analysis. This analysis was mainly a check on the stability of the beer over time;[a] tests were done on its clarity, foam duration and cling, and taste. The brewing lab also did an abuse test on a sample of each week's production. In this test, various cans and bottles of beer were alternately heated to 140° F and cooled to freezing over the course of a week, thus simulating all the abuse that beer could encounter before it was consumed. Clarity checks were then run.

As a check on the workings of the brewing lab, once a month the Detroit quality assurance staff sent two of the same samples to each Stroh plant for analysis. The plants submitted their results to Detroit, where the corporate quality assurance staff scrutinized the results. Any plant whose quality assurance tests were not accurate enough was so informed, and measures were taken to improve its procedures. Once a year there was a complete audit of the procedures and records of the lab. A team from Detroit came for a full week and evaluated all phases of quality assurance, not only in the brewing lab but also in the microbiology and packaging labs.

Microbiology Lab

The microbiology lab was adjacent to the brewing lab and was responsible for bacteria and yeast counts on all stages of the process. Really, there were two microbiology labs, one for draft beer and one for non-draft beer. In essence, the microbiology labs were responsible for checking that the bacteria and yeast counts in the water used in the process, in the process equipment (tanks, pipes), in the air in the plant, and in the finished beer itself

were within designated limits. Any degradation of the equipment or the materials could harm the quality of the beer, so checks were made every day on the brewery's microbiology.

The Packaging Lab

The packaging lab was located just off the packaging floor. Its mandate was to check the filling and packaging of beer into cans, bottles, and kegs. The work of the packaging lab could be divided into monitoring three major areas: (1) the activities of the packaging floor; (2) the canning and bottling processes, especially the seaming of cans and crowning of bottles; and (3) the filled cans, bottles, or kegs.

Monitoring operations on the floor meant, in large measure, keeping track of the equipment and making sure that it was working properly to produce a product that met various physical specifications. These specifications included the following:

- The proper positioning of labels and their smooth and wrinkle-free placement on bottles.
- The operation of the mechanisms that detected and removed from the line cans or bottles that were not filled to the proper levels.
- The torque on twist-off bottle caps.
- The proper temperature and water flow in the pasteurizers.
- Leaking due to improper lid placement on cans or improper crowning of bottles.
- The proper coding on cases and cans and bottles.
- Sanitation of the filling and closing machines.

Seams were examined every 4 hours, when 12 cans in a row were removed from each line. Twelve cans were removed because that was how many heads were on the seaming machine that placed the lid on the filled can of beer. For each can, a number of measurements were taken, some with the micrometer (such as seam height, seam thickness, depth of the countersink). The seam was also rated visually (tight or loose). The lids were then removed from the cans, and the seams were analyzed under a magnifying apparatus to check the overlap of the "body hook" (the flange on the can itself) with the "cover hook" (the edge of the lid crimped over). This careful examination was a check on the

---

[a] On each can and bottle were special codes that indicated exactly when and where the beer was packaged. For example, the code 06-27-91 0880607 on a can of Stroh's indicated that the beer was packaged on June 27, 1991 by the eighth plant (Winston-Salem) on line 8 at 0607 hours (military time for 6:07 AM).

strength of the seal on the can and its ability to maintain a tight, leakproof seal. Other measures were taken as well, including a test for metal exposure (an examination of the thickness and coverage of the coating on the inside of the aluminum can).

The accuracy of filling was checked every 2 hours for each line and at every tank change. At that time three containers were withdrawn and investigated for carbon dioxide content, air quantity, and the accuracy of the fill. The data derived from these samples were used to compile process control charts whose process averages and control limits were re-evaluated every 90 days. Figure 4-10 is a process control chart for air quantity and carbon dioxide content. Charts were kept on a number of characteristics: air, CO_2, labeling, seams, and many more. This chart is for bottling line 1, where 12-ounce bottles are filled, for the 4 days of October 22–25, 1985. Although data for both air quantity and CO_2 content are gathered, only the air quantity control chart graphs are shown. Those graphs are the classic $\overline{X}$ and R graphs. The key parameters for the two graphs are given in the box labeled "control data." As seen there, the overall mean is 0.42 ml of air in the bottle with upper and lower control limits of 0.56 and 0.28 ml, respectively. The mean range is 0.14 ml, with an upper control limit of 0.36 ml and a lower control limit of 0. The specifications for air quantity are well above the upper control limit. Those specifications (in the upper right corner) called for a sample average of no more than 1 ml of air and an individual sample reading of no more than 1.45 ml of air. These were the relevant data for 1985. It is interesting to note that, because of constant attention to and improvements in the process and the capabilities of the equipment, the process average for air in a 12 oz. bottle has declined from 0.42 ml to 0.30 ml.

Below the control data lies the information taken for each sample. Note that for the most part, the samples were taken at 2-hour intervals. Samples were also taken immediately after each change from one tank to another. This changeover was a source of nonrandom variation and thus was watched carefully. Note that the 2:20 P.M. changeover on October 22, involving Old Milwaukee (OM) beer from tank 24, caused a real spike in this control chart. (This chart was specifically chosen to reveal an out-of-control situation.) The notation on the chart indicates that the supervisor (Mr. Zegan)

was notified and that the beer was "held for inspection" (HFI) under HFI number 125. In fact, what happened was that 98 cases of beer were held for inspection, and upon a thorough audit, 97 cases were later released as good product. Because of the attention of the workforce to such control charts, the process was typically halted before bad product was produced. The cost of held beer at Winston-Salem was only 1 cent a barrel, a level many times lower than the industry average.

The succeeding portions of both the $\overline{X}$ and R charts document a process that is in good statistical control.

The Taste Test and Others

The Quality Assurance department was not the only one involved in quality checks. Operators along the line, for example, drew 2 cases off the line every 15 minutes and checked the aesthetics of the packaging.

Every Thusday afternoon a panel of tasters at the Winston-Salem brewery assembled to sample Stroh beers that had been sent from all the other company breweries. This test usually involved tasting three beers at a time and marking a sheet such as found in Figure 4-11. The results of this taste test were pooled with those from the taste tests done at other locations within the company. Any clearly and widely perceived deficiencies in a brewery's beer were carefully investigated and corrective measures taken. The Winston-Salem plant also tested other beer brands against Stroh to assess Stroh's quality and distinctiveness. The taste panel at the plant was qualified every year. This was done by having the Detroit staff send to the plant samples that had been "spiked" with chemicals to make the beer more hoppy, more bitter, more malty, or otherwise different than usual. A taster who did not adequately perceive these characteristics was removed from the taste panel.

The Winston-Salem brewery took pride in the quality of its beer and its packaging. Just about anything concerning the product or the process that was quantifiable was checked every day at the plant. The plant was committed to conforming precisely to the requirements established by the staff in Detroit.

continued

A. October 22 through 10 P.M. October 23.

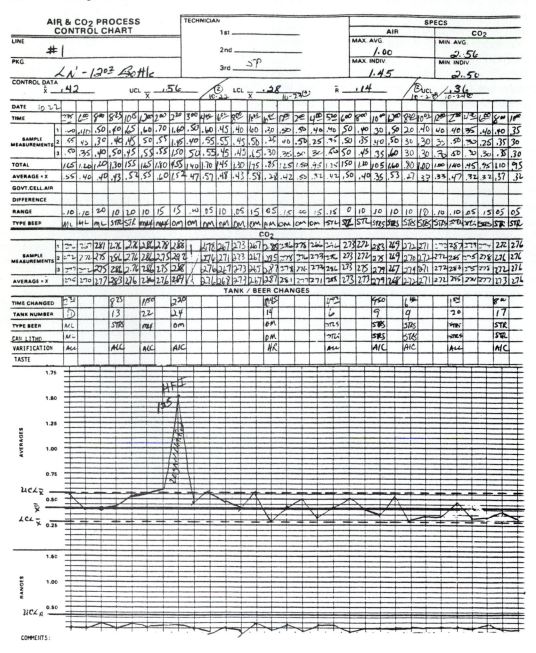

FIGURE 4-10   Air and CO_2 control charts, as used at Stroh's Winston-Salem plant.
(Courtesy of Stroh Brewery Company)

506

Stroh Brewery Company

BREWING DEPARTMENT
BEER FLAVOR PROFILE

DATE

PRODUCT ID

PANELIST ID

FLAVOR ATTRIBUTE	SCALE: IMPERCEPTIBLE		SLIGHTLY PERCEPTIBLE		MODERATELY PERCEPTIBLE		STRONGLY PERCEPTIBLE		EXTREMELY PERCEPTIBLE		SCORE
1. FRUITY	0	1	2	3	4	5	6	7	8	9	
2. HOPPY	0	1	2	3	4	5	6	7	8	9	
3. BITTER	0	1	2	3	4	5	6	7	8	9	
4. MALTY	0	1	2	3	4	5	6	7	8	9	
5. HANG	0	1	2	3	4	5	6	7	8	9	
6. TART	0	1	2	3	4	5	6	7	8	9	
7. OXIDIZED	0	1	2	3	4	5	6	7	8	9	
8. GRAINY	0	1	2	3	4	5	6	7	8	9	
9. MEDICINAL	0	1	2	3	4	5	6	7	8	9	
10. SULFUR	0	1	2	3	4	5	6	7	8	9	
11. BACTERIAL	0	1	2	3	4	5	6	7	8	9	
12. YEASTY	0	1	2	3	4	5	6	7	8	9	
13. DIACETYL	0	1	2	3	4	5	6	7	8	9	
14. MOUTHFEEL	0	1	2	3	4	5	6	7	8	9	
15. OVERALL FLAVOR ACCEPTANCE	Unaccept.		Poor		Fair		Good		Excellent		
	0	1	2	3	4	5	6	7	8	9	

NOTE: Please check the perceived attribute intensity on the scale and copy the corresponding score.

FORM 2025

FIGURE 4-11 A taste test form. (Courtesy of Stroh Brewery Company)

IS THE INSPECTION ADEQUATE?

The task of setting up and keeping control charts on the process is a type of inspection. As stated earlier, this type of inspection needs to be encouraged because it focuses on the process, not on the product. There are a host of other types of inspections, with different purposes attached to each. Table 4-5 summarizes some of these types of inspections and their purposes.

Adequate inspection, first and foremost, involves the establishment of clear operational standards for assessing an item's quality. There should be as little ambiguity as possible about what constitutes good quality. Often it is difficult to specify objective, measurable criteria, but it is an important step toward adequate in-

TABLE 4-5   Basic types of inspection

MAJOR USE	TECHNIQUE APPLIED	COMMENTS
Process control	Control charts	Can be used to examine the variability of the process along selected measured dimensions. Serves as an indication of "drift" in the process due to specific, assignable causes. Logs are often kept to help track down causes of out-of-control operations as indicated by control chart.
		Can be used with standards (tolerances) given as a base against which future operations history can be compared.
Sorting good items from bad	100% inspection	Usually done at the end of the process, but often done as well for selected vendor purchases and at selected points in the process, particularly in front of expensive processes. Often the focus for advances in automated testing.
		The types of defects detected can be analyzed as guides for improving the process.
Rating product quality	Product audits	Tests, sometimes destructive of the product are designed to determine the seriousness of various defects.
Distinguishing good lots from bad lots	Acceptance sampling	Used to classify whether a lot of material is acceptable or not. Often used for incoming inspections of vendor material.
		Measurements are made of the discrete units of lot or the specimens are taken from batch and the percentage defective is compared with the standard derived from prior data.
Inspecting inspectors	Comparison of performance versus actual	Defects detected by inspector (person or machine) are compared with actual number of defects in what was inspected.
		Can involve defects "planted" by a quality control group.

spection, and it has the added benefit of helping the workforce, as well.

Although there are a variety of issues involved in inspection, two of particular interest are how much inspection will be done and where inspection ought to be located in the process.

How Much Inspection?

At issue is whether, out of any lot, all items ought to be inspected (100 percent inspection), whether no items ought to be inspected, or whether a sample of items ought to be inspected. Let us explore the rationales for each of these alternatives, in turn.

100 Percent Inspection

The method used for 100 percent inspection is essentially a sorting operation to distinguish good items from bad ones. It can be a costly means of inspection, of course, and, in those cases, it is typically justified only if the cost of having a bad item in the product (or a bad product itself) is relatively high; 100 percent inspection is most prevalent at the finished goods stage of the process, or where product safety is a high priority.

If inspection can be made automatic, the cost of 100 percent inspection plunges. In such circumstances, 100 percent inspection of important product characteristics becomes very attractive. Many high-volume consumer goods companies such as the Stroh Brewery in Winston-Salem use automatic inspection equipment.

If there are multiple inspectors, each inspecting all the items, it is sometimes said that the inspection is 200 percent, 300 percent, and so on. Having such multiple inspection draws fire from experts such as Deming as wasteful and inefficient. They contend that 200 percent in-spection is often less effective than 100 percent inspection because the inspectors may not feel obliged to sort out every defect, relying instead on "the other guy."

No Inspection

More and more, as the movement for just-in-time production has taken hold (see Chapter 11), and as companies have become devoted to managing quality by concentrating on the process, incoming materials from trusted suppliers have been placed into production without inspection. Companies have been asking themselves why it should be necessary for items that have been inspected as finished goods in one factory to be reinspected as incoming material at the next. If one is confident in the supplier's process and inspection procedures, material ought to bypass incoming inspection. Of course, not all companies feel free enough to accept their suppliers' output without inspection, but the list is growing. In these cases, often the supplier ships along copies of the control charts relevant to the product, to demonstrate that its production process is in statistical control. In these cases, too, there are periodic audits and other checks to make sure that customer and supplier are in tune with one another, but routine inspections are dispensed with.

Acceptance Sampling

By far the most predominant current practice for inspecting incoming materials is to use a sampling procedure. (Sampling may also be used elsewhere in the process.) The idea is to sample from the entire lot of items shipped and to determine from the inspection of the sample whether to accept the shipment or to reject it. There are several, widely utilized sampling plans, including two World War II-vintage military standard procedures (Mil. Std. 105D, an attributes sampling plan, and Mil. Std. 414, a variables sampling plan) and the Dodge-Romig sampling plans (also for attributes).

Associated with any sampling plans are one or two quality indicators that serve to categorize the sampling plan as more or less stringent. Two of these quality indicators merit mention:

1. *Acceptable quality level (AQL).* The acceptable quality level is associated with the operating characteristic curve characterizing the sampling plan and its type I error (producer's risk). The AQL is a standard representing the maximum percent defective in the lot that can be considered satisfactory. It has no firm mathematical definition except that the probability of accepting the lot should be very high if its percent defective lies within the AQL set. See Figure 4-12 for a typical positioning of an AQL. The AQL

exists as a standard for accepting a lot that has been sampled. It should not be confused with the zero-defects production standard for quality, nor should it be forgotten that under any sampling plan, of no matter what AQL, some lots will be accepted with an AQL worse than standard (type II error).

2. *Rejectable quality level (RQL).* The rejectable quality level, sometimes also termed the lot tolerance percent defective (LTPD), is also associated with the operating characteristic curve characterizing the sampling plan, but it refers to the plant's type II error (consumer's risk). The RQL or LTPD is the defect level standard that defines what is unsatisfactory and should therefore be rejected. As with the AQL, there is no firm mathematical definition. Often, the RQL or LTPD is set so that if the percentage of true defects is high, the probability of accepting the lot is no more than 0.10.

Between the AQL and the RQL lies a region of ambiguity where the lot cannot be accepted straightaway but where it cannot be decisively rejected either. Sampling plans have been devised to deal with this ambiguity. The base case sampling plan is termed "single sampling." In single sampling, a single, reasonably large sample is taken and the lot is either accepted or rejected, depending on the sample's findings. Generally, a more economical means of passing judgment on a lot is first to take a small sample that can be used straightaway to accept some lots and reject others. With such a sample, however, there will be times when the sample finding will be just too borderline on which to base an accept/reject decision. In these cases, another sample is taken and added to the first. With this now larger sample, the accept/reject decision is made. This is called "double sampling." This notion of

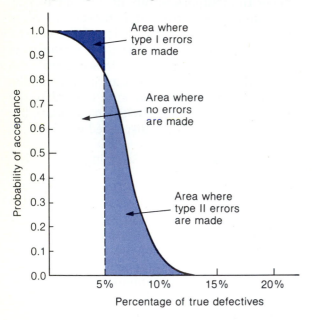

FIGURE 4-12   Positioning of the AQL and the RQL.

double sampling can be extended to include repeated sampling, embracing more than two draws from the lot. In "sequential sampling," repeated draws are made until an ambiguity about acceptance or rejection fades away with ever larger sample sizes.

Sampling plans such as Mil. Std. 105D specify how large a sample to draw (or indeed whether multiple samples ought to be drawn), how many defects will be tolerated in the sample before it is rejected, and how the sampling plan ought to be modified for different levels of performance, good or bad. These specifications naturally vary with the tightness or looseness of the AQL chosen and the size of the lots that are to be inspected.

In recent years, acceptance sampling has come under increasing criticism. Some critics object that sampling plans permit rejects at some "acceptable quality level," when the goal should be not to have any in the lot in the first place. Others note that acceptance sampling can be costly and often is not cost justified. These critics do well to call attention to the fact that acceptance sampling is sometimes misapplied.

When to Apply Acceptance Sampling and When Not To

In this regard it is useful to investigate the economics of inspection. By so doing we can determine when it is advantageous to abandon acceptance sampling altogether and either to inspect everything (100 percent inspection) or not to inspect at all.[12] The economics of inspection provide insights into those cases where acceptance sampling can still be justified, although many in the quality field acknowledge that the incidence of justifiable use of acceptance sampling is less than many previously thought. To demonstrate this, some assumptions are in order:

1. Every product made gets tested at some point in the process (such as the final test), and if any defective part/item is in it, the product will fail inspection.

2. There are no inspection errors.

Define the following costs:

k_1 = cost to inspect one part that goes into the product

k_2 = cost to dismantle the product, repair it, reassemble it, and test it again once it has failed inspection due to a defective part

K = average cost to test one or more parts in order to find a good replacement part for one that has shown up defective

p = average percent defective in the lot of parts

The cost per part of 100 percent inspection is

$$\text{Cost } 100\% = k_1 + Kp \text{ (cost to test one part and replace it if needed)}$$

and the cost per part of no inspection is:

$$\text{Cost } 0\% = p(k_2 + K) \text{ (cost of repairing the product once it was discovered bad and then of replacing the faulty part)}$$

The difference in the two costs is:

$$\text{Cost }_{100\%} - \text{Cost }_{0\%} = k_1 - pk_2$$

The break-even point between 0 and 100 percent inspection is thus when $p = k_1/k_2$. If $p > k_1/k_2$, then 100 percent inspection is merited, and if $p < k_1/k_2$, no inspection is merited.

One cannot be assured that the average percent defective in a lot will be precisely p from one lot to another. Naturally, if there is wide divergence in the percent defective from one lot to another, the decision rule of looking at k_1/k_2 becomes less of a guide. What can be said is that if the worst lot has a $p < k_1/k_2$, no inspection should be done, and if the best lot has a $p > k_1/k_2$, 100 percent inspection should be done. For situations in between, acceptance sampling makes particular sense.

It should be stressed that neither acceptance sampling—nor anything else, for that matter—is a substitute for close working ties with a supplier. To the extent that a supplier's quality can be improved to a level of known statistical control, acceptance sampling can fade away.

Placement of Inspections

Inspection stations are customarily at finished goods and in many cases upon receipt of vendor shipments. (It is important that customer and supplier test the shipments in precisely the same way, so that unnecessary misunderstandings about quality needs are avoided.) Inspection is also often routine for equipment setups. Frequently, for example, the first few pieces of any lot after a new setup are carefully inspected before the remainder of the production lot is allowed to proceed.

The positioning of other checks throughout the process depends largely on whether the marginal benefits of having the check exceed the marginal costs. This benefit/cost ratio is likely to be high in the following kinds of places within the process and so inspections are often placed there:

- Before operations that add a lot of value to the product. Thus, already defective products are detected before additional costs are incurred.

- After operations with demonstrated low yields—again, so that significant values are not added to already defective products.

- Before operations that render product repair very difficult or impossible (e.g., before beer is canned, before the body drop in an auto assembly plant, or before an automobile is painted).

- At points in the process where testing is relatively cheap. The advances made in automatic testing, particularly those making use of automated sensing equipment, are making it easier for manufacturers to introduce more inspections within the process. This fact, coupled with the use of computers and sensors to control the process better than in the past, is a heartening example of technological advance. In all cases, it is important for the feedback of the inspection to be swift and specific so that those responsible can take the steps necessary to improve the process. Sometimes companies inadvertently introduce obstacles that impede the swift flow of information that could improve the process. (Material Review Boards in some companies, charged with disposing of "discrepant material" come to mind.) These impediments, naturally, should be avoided.

A Spectrum of Inspection

While inspection is necessary, there is a gradation in which kinds of inspection are preferred. For example, using acceptance sampling for incoming materials is less desirable than employing a certification program for suppliers so that all inspection is done at the supplier's (inspection at the source). Similarly, although 100 percent final inspection is often unavoidable, within-process inspection is preferred so that value is not added to already bad product. Likewise, if the process can be controlled by statistical process control, then in-process inspections are less necessary. And, best of all is when the process can be foolproofed so that control charts are no longer needed and quality need be monitored only to assure that the foolproofing is still working properly. Thus, a spectrum from "necessary" to "desirable" could loosely be defined to run as follows: acceptance sampling of in-coming materials, 100 percent inspection of in-coming materials, in-process inspections, supplier certification programs so that all in-coming materials are inspected at the source, statistical quality control of all key facets of the process, and foolproofing of the process.

Tools of the Trade: A Recap of Lessons Learned

One way to think of making quality happen in an operation is to remove all "chaos" from it. As W. Edwards Deming has long preached, one should yearn for the statistically predictable process, where chaos has been driven out. In this chapter we have addressed quality by talking of the adequacy of design, production, and inspection. Another way to review quality management is to examine the steps needed to remove chaos. Consider the following six steps, borrowed liberally from Deming:

1. Set clear operational standards for the product. Know clearly what is and what is not good quality. This is what the tools used to improve product design have helped to do. Cross-functional design teams, quality function deployment, and recognition of the customer–supplier relationship, even within the firm have been instrumental here.

2. Analyze what may be important to correct to better the product's quality. Brainstorming, say with the use of the fishbone diagram, and data gathering for use in Pareto diagrams are typically helpful here.

3. Remove the obvious stumbling blocks to improved quality.

4. Determine process capabilities. Use control charts and process capability studies.

5. Analyze the causes of identified problems. Fishbone diagrams are helpful here for keeping track of potential problems and experimentation for pinpointing the root causes.

6. Remedy the problem. Standardize the remedy. Foolproof the process. Check that it remains so. The use of a formal quality improvement process reinforces this.

It should be expected that different products or operations will be in different stages of this six-step procedure. Control charts and fishbone diagrams should not be found in every corner of the operation, only where they are truly needed.

World Class Practices. Another useful way to recap the important aspects of quality management is to outline some of the key practices of "world-class manufacturers." Here is a sampling:

1. Everyone in the organization is knowledgeable about and committed to quality. There is no finger pointing about quality problems. The cost of quality is known.

2. There is an established quality improvement process in place. Deming's 14 Points and the PDCA cycle, or their equivalents, are pursued.

3. Quality is an important aspect of performance measurement.

4. Product design is recognized as perhaps the single most important source of potential quality problems, and steps have been taken (such as design for manufacturability and concurrent engineering) that can help the company design quality products straightaway. There are ways by which the "voice of the customer" are heard in product design.

5. Process design, using such concepts as foolproofing and the assignment of equipment and processes with capabilities well beyond the tolerances required in the product,

is recognized as tremendously helpful in assuring excellent product quality. Indeed, process design is often more powerful than the development of control charts. Control charts are seen as but a milestone on the way to perfect quality.

6. The factory (or service) is viewed as a laboratory where there is continual experimentation with product and process, all with the aim of improved quality.

7. Automation is valued because it can improve quality and not because it can save on labor.

8. Tools such as Pareto charts, run charts, fishbone diagrams, control charts, and sampling (or not sampling) are understood and used, where appropriate.

9. Suppliers are viewed as partners — in both design and manufacture. Single sourcing predominates. Feedback on and assistance with quality are given as a matter of course.

10. All areas of the company see themselves in customer-supplier relationships, be they external or internal customers or suppliers.

QUESTIONS

1. How could one resolve the debate about whether or not quality is "free?"

2. Which measures of reliability and maintainability are applicable under what conditions?

3. Which kinds of control charts are most effective for the kinds of situations faced by each of the company operations depicted in the Tours segment of this text?

4. Which kinds of inspections are most applicable to the processes described in the Tours segment of this book?

5. What are some of the main ways in which quality can be controlled and improved? Does adding another quality check always benefit an operation? What are some different kinds of quality checks?

PROBLEMS

1. Choose some popular or fashionable products or services. What does the average person say about their quality? How would you go about measuring the quality of these products or services?

2. Which of Deming's 14 steps do you think would meet with the greatest resistance at a typical company? Why?

3. What quality control practices mentioned in the chapter are evident in the operation of the Stroh Brewery in Winston-Salem?

4. The operator of a plastics extrusion machine is concerned about quality. Every so of-

ten during his shift he measures the pressure the machine is generating. If his measure is either high or low, he readjusts. What would you say about the quality practices of this operator?

5. Joe ran a small transfer line and Walt inspected the output of that line. Joe was paid an incentive wage that was tied to the output of the line. Walt was instructed to identify all the bad parts produced. Comment on this situation. How would you change things, if at all?

6. Table 4-6 presents sample measurements on a process. Previously, a process capability

TABLE 4-6	Sample measurements for problem 6			
		SAMPLE		
1	2	3	4	5
1.0050	0.9963	1.0115	0.9875	1.0085
1.0022	0.9995	1.0003	0.9930	0.9803
0.9930	1.0210	0.9915	1.0100	0.9935
0.9895	1.0005	1.0055	1.0003	1.0215

study had determined that the process could hold a mean of 1.00 with a standard deviation of 0.0035. Is this process under control? Construct an $\overline{X}$ and R control chart to see.

7. The following data were collected from a machine whose capabilities were being studied. For convenience the 50 observations collected are arranged in 10 columns of five observations each.

```
130 142 115 131 143 121 127 149 130 121
132 116 144 143 138 134 122 128 139 146
150 120 115 153 129 111 148 139 117 137
119 147 158 118 122 135 128 134 122 127
153 127 155 116 133 129 143 121 127 147
```

What is the process mean and standard deviation as determined from this capability study? What are the process capability limits? What control limits would you set for this machine for samples taken of three each?

8. Turnum and Down (T&D) have an automatic screw machine set up to run small brass shafts for a customer. The shaft diameters are supposed to be 0.375 ± 0.0025 inch. T&D have determined that the machine is capable of generating the shafts to size normally distributed with a mean of 0.375 and a standard deviation of 0.0005 inch.

(a) A sample mean has been taken with a value of 0.377 inch. What action should be taken?

(b) Three sample means of 0.3755, 0.376, and 0.3765 inch have been measured. What action should be taken?

(c) Three sample means of 0.376, 0.3745, and 0.377 inch were taken immediately after a new load of brass raw material was started. What action should be taken?

(d) Three measurements were taken of 0.3755, 0.3744, and 0.364 inch. What action should be taken?

(e) One sample mean of 0.3766 inch was measured. What action should be taken?

(f) What is the C_p for this process?

9. At the Jackson Paint Company, paint chip cards with the proper color matches were used as an important marketing tool. The data below was available on rejects of paint chip cards for poor color match. Samples of 50 were taken at any time. Is the instrumentation used to match colors of cards to standards under control?

	Rejects by sample											
Sample Number	1	2	3	4	5	6	7	8	9	10	11	12
Percent of chip cards failing color match	8	6	6	12	14	8	10	6	4	10	12	6

10. Ticketrend was a national chain of ticket vendors for sporting events, concerts, exhibitions, and the like. Offices operated out of malls and department stores and sold tickets both to walk-in customers and to those who mailed in or telephoned in orders. Management was interested in controlling the quantity of errors made in the ticketing process. A number of different errors were possible, among them: incorrect dates for the event, wrong class of ticket, incorrect number of tickets, sent to wrong address, bill incorrect, and so on. No one type of error was significantly more prevalent than the others. Management was intent on reducing all types to 1 percent or less of the volume.

From 20 samples of 200 ticket orders each, taken at random, the following number of defects were determined:

SAMPLE	NUMBER OF DEFECTS	SAMPLE	NUMBER OF DEFECTS
1	6	11	2
2	2	12	0
3	4	13	4
4	1	14	7
5	11	15	5
6	3	16	12
7	7	17	3
8	9	18	8
9	13	19	2
10	8	20	5

(a) What type of control chart is appropriate for determining the process capability of the operation? Why?

(b) Construct a control chart using these data. What observations can you make about the process?

11. The small bags of potato chips had a target weight of 0.100 kg each and were filled on a machine that had been studied for some time and that was determined to run with a process standard deviation of 0.006 kg. The table below documents the weights of 10 samples taken from the filling machine during the last shift. Is the process capable of meeting the target value? Is the process operating in control?

Sample measurements (in kilograms) of the filling machine are as follows:

SAMPLE NUMBER				
1	2	3	4	5
0.100	0.101	0.102	0.104	0.095
0.098	0.103	0.103	0.101	0.099
0.102	0.100	0.100	0.105	0.103
0.100	0.099	0.104	0.103	0.106

SAMPLE NUMBER				
6	7	8	9	10
0.107	0.104	0.112	0.097	0.106
0.102	0.110	0.108	0.106	0.110
0.100	0.101	0.105	0.100	0.101
0.097	0.098	0.101	0.108	0.105

12. H. H. Kettle's Furniture and Appliance Store on the north side of Metropolis was interested in improving its ordering process. From time to time, customers complained about errors in their orders, particularly wrong pieces ordered. The store wanted to keep order errors to less than 2 percent of the total number of orders and they had assigned Jerry Lynne to the project. Jerry looked at a number of orders, in samples of 50, and classed each order as either "OK" or "Has a problem." The results of this inspection are as follows:

SAMPLE NUMBER	ORDERS OK	PROBLEM ORDERS
1	48	2
2	50	0
3	49	1
4	47	3
5	50	0
6	48	2
7	50	0
8	46	4
9	49	1
10	48	2
11	50	0
12	50	0
13	49	1
14	48	2
15	49	1

(a) What type of control chart is appropriate for determining the process capability of the ordering operation? Why?

(b) Construct a control chart using these data. What observations can you make about the process?

13. An $\overline{X}$ and R control chart has been used to help control a manufacturing process. The sample data are consistently within the established control limits and the control limits are inside design engineering's specification limits. Some managers are confused, however, because a high percentage of product falls outside the specification limits even though the process shows itself to be within the established control limits. What is happening here? Draw this situation and clear up the confusion. How do you avoid this situation in practice?

14. The mean of the process is 2.50 and its standard deviation is 0.25. The design specification is 2.20 and the specification range has been set at between 1.00 and 2.40. Calculate both process capability indexes, Cp and Cp$_k$.

15. Should there be 100 percent inspection or no inspection of the component that evidences the following characteristics?

Average percent defective in the lots delivered	4%
Cost to inspect one part	$0.25
Cost to dismantle product, repair it, and test it again, given failure due to a defective part	$3.50

16. Piedmont Electronics manufactures some complex communications equipment (such as multiplexers) that can be regarded as the joining and testing of various subassemblies. Currently, only after final assembly is the primary functional testing done. At issue is whether some testing of individual subassemblies ought to be done before final assembly, and if so, which type of testing should be done. Here are some data to consider:

- Final assembly costs about $35 of labor and materials per unit.
- The functional test after final assembly passed two-thirds of the units. The other third went to repair. Repair cost an average of $30 per unit received. About one-fourth of the units were scrapped right away. Of the remaining units, all were final tested again and about half passed. The others were scrapped.
- The final testing and shipping cost $25 per unit.
- Each year 1000 units of this model were manufactured and sold.
- One possible test of the subassemblies would cost $20,000 in equipment and promised to identify 60 percent of the potential problems.
- The other possible test of the subassemblies would cost $12,000 in equipment but promised to identify only 40 percent of the potential problems.

Should Piedmont Electronics install the new equipment for testing the subassemblies? Which possible test should be implemented?

17. The Pork & Rind firm decided to go ahead with the production of the Puffy Cheezies (see Problem eighteen in Chapter 6), and they need to set up quality control inspection stations. The process flow is from mixer to gun to oil spray to oven and finally to packaging. The cost to put each ounce through each process (i.e., the value added for each process), the expected percentage rejected, and the costs to inspect are in Table 4-7. The tests after the mixer and the oiler require the use of the lab. The other tests are visual only. Where would you suggest that inspectors be stationed?

TABLE 4-7   Quality control considerations for problem 17

PROCESS	VALUE ADDED	EXPECTED PERCENTAGE REJECTED	COST TO INSPECT ALL ITEMS
Mixer	$0.04	10	$0.02
Gun	0.01	3	0.01
Oiler	0.04	2	0.04
Oven	0.01	8	0.01
Package	0.01	4	0.005

NOTE: Figures given per ounce of Puffy Cheezies.

SITUATION FOR STUDY 4-1

APOLLO MOTOR COMPANY, OMAHA PLANT

The Omaha Plant of the Apollo Motor Company had the worst quality rating of any of the company's plants. In addition, labor relations were exceedingly acrimonious and the plant was filthy. Attitude surveys of the workforce indicated exceedingly low morale. It was clear to Mike Garrett, the new plant manager, that without some improvement in productivity and quality the plant was sure to be closed within a year.

The plant has been making pickup trucks and Garrett knew that a new model of pickup truck was on the drawing boards at Apollo Mo-

tor's design center. The plant that would assemble this pickup truck had not yet been chosen, but Garrett thought that if the Omaha plant could somehow show significant quality improvement, it could be chosen.

The trick, of course, was how to do it. How could he turn this plant around? How could he instill quality in the work done there? How should his job change, how should the job of his supervisors change, how should the jobs of the workers change? If the Omaha plant was to show improvement, it had to do so soon.

REFERENCE NOTES

1.   For further information on quality consult the following:
Aguayo, Rafael, *Dr. Deming: The American Who Taught the Japanese About Quality*. New York: Fireside, 1990.
American Society for Quality Control Automotive Division, *Advanced Quality Planning*, Milwaukee, WI: American Society for Quality Control, 1988.
Baker, Thomas R., "Quality Engineering by Design," *Quality Progress* (December 1986): 32–42.
Box, George E. P., and Soren Bisgaard, "The Scientific Context of Quality Improvement," *Quality Progress* 20, no. 6 (June 1987): 54–61.
Burr, John T., "The Tools of Quality—Part VI: Pareto Charts," *Quality Progress* 23, no. 11 (November 1990): 59–61.

Crosby, Philip B., *Lets Talk Quality: 96 Questions You Always Wanted to Ask*, New York: Penguin Books, 1990.

Daetz, Douglas, "The Effect of Product Design On Product Quality and Product Cost," *Quality Progress* 20, no. 6 (June 1987): 63–67.

Deming, W. Edwards, *Out of the Crisis*, Cambridge, MA: Massachusetts Institute of Technology Center for Advanced Engineering Study, 1986.

Dingus, Victor R., and William A. Golomski, ed., *A Quality Revolution in Manufacturing*, Norcross, GA: Industrial Engineering and Management Press. 1988.

Dreyfuss, Joel, "Victories in the Quality Crusade," *Fortune* 118, no. 8 (October 10, 1988): 80–88.

Ebrahimpour, Maling, "An Examination of Quality Management in Japan: Implications for Management in the United States," *Journal of Operations Management* 5, no. 4 (August 1985): 419–31.

Edosomwan, Johnson Aimie, *Productivity and Quality Improvement*. Berlin: IFS (Publications) Ltd. UK, 1988.

Enrick, Norbert L., *Quality, Reliability, and Process Improvement*, 8th ed. New York: Industrial Press, 1985.

Feigenbaum, A.V., *Total Quality Control*, 3rd ed. New York: McGraw-Hill, 1991.

Garvin, David A., "Competing on the Eight Dimensions of Quality," *Harvard Business Review* 65, no. 6 (November–December 1987): 101–9.

_____, *Managing Quality: The Strategic and Competitive Edge*, New York: Free Press, 1988.

_____, "What Does "Product Quality" Really Mean?" *Sloan Management Review* 26, no. 1 (Fall 1984): 25–43.

Gitlow, Howard, and Shelly Gitlow, *The Deming Guide to Quality and Competitive Position*, Englewood Cliffs, N.J.: Prentice-Hall, 1987.

Gitlow, Howard, Shelly Gitlow, Allan Oppenheim, and Rosa Oppenheim, *Tools and Methods for the Improvement of Quality*, Homewood, IL: Irwin, 1988.

Groocock, John M., *The Chain of Quality: Market Dominance through Product Superiority*, New York: Wiley, 1986.

Handfield, Robert, "Quality Management in Japan Versus the United States: An Overview," *Production & Inventory Management* 30, no. 2 (Second Quarter 1989): 79–85.

Harrington, H. James, *Excellence —The IBM Way*. Milwaukee, Wisc.: ASQC Quality Press, 1988.

_____, *Poor-Quality Cost*. New York: Marcel Dekker, Inc. 1987.

Hauser, John R., and Don Clausing, "The House of Quality," *Harvard Business Review* 66, no. 3, (May–June 1988): 63–73.

Impoco, Jim, "Mavericks Can Win, Too," *U.S. News & World Report* 105, (November 28, 1988): 56.

Ishikawa, Kaoru, *What is Total Quality Control? The Japanese Way*. Englewood Cliffs, N.J.: Prentice-Hall, 1985.

"Japan Has a New Lesson For Detroit," *Business Week*, (October 22, 1990): 128.

Juran, J. M., *Juran on Leadership for Quality: An Executive Handbook*, New York: Free Press, 1989.

_____, *Juran on Planning for Quality*. New York: The Free Press, 1988.

Kelada, Joseph, *Integral Quality Management: The Path to Total Quality*, Dollard-des-Ormeaux, P.Q.: Quafec, Inc. 1989.

Kindlarski, Edward, "Ishikawa Diagrams for Problem Solving," (Tutorial), *Quality Progress* Vol. 17, (December 1984): 26–30.

Krantz, K. Theodor, "How Velcro Got Hooked on Quality," *Harvard Business Review* 66, no. 5 (September–October 1989): 34–40.

Kush, Thomas, "Design Manufacturing for Quality," *Automation* 35, no. 2 (February 1988): 40–44.

Landon, Wanda G., "Kanban and Deming's 14 Points," *Quality* 27, no. 9 (September 1988): 50, 52.

Lim, T.E., "Quality Improvement Using Experimental Design," *International Journal of Quality & Reliability Management (UK)* 7, no. 1 (1990): 70–76.

Maass, Richard A., and American Society for Quality Control Vendor-Vendee Technical Committee, *World Class Quality: An Innovative Prescription for Survival*. Milwaukee, WI.: ASQC Quality Press, 1988.

Main, Jeremy, "How to Win the Baldrige Award," *Fortune* 121, (April 23, 1990): 101–16.

_____, "Under the Spell of Quality Gurus," *Fortune* (August 18, 1986): 30–34.

Messina, William S., *Statistical Quality Control for Manufacturing Managers*, New York: Wiley, 1987.

Modarress, Batoul, and A. Ansari, "Quality Control Techniques in U.S. Firms: A Survey," *Production & Inventory Management* 30, no. 2 (Second Quarter 1989): 58–62.

Morse, Wayne J., Harold P. Roth, and Kay M. Poston, *Measuring, Planning, and Controlling Quality Costs*, Montvale, NJ: National Association of Accountants, 1987.

Nikkan Kogyo Shimbun, Ltd./Factory Magazine, ed., *Poka-yoke: Improving Product Quality by Preventing Defects*. Cambridge, MA: Productivity Press, 1988.

O'Neal, Kim Rogers, "Statistical Process/Quality Control Microcomputer Software Buyer's Guide," *Industrial Engineering* 19, no. 5 (May 1987): 29–41.

Ozeki, Kazuo, and Tetsuichi Asaka, eds., *Handbook of Quality Tools: The Japanese Approach*, Cambridge MA: Productivity Press, 1990.

Port, Otis, "How to Make It Right the First Time," *Business Week (Industrial/Technology Edition)*, no. 3002 (June 8, 1987): 142–143.

Port, Otis, "The Push for Quality," *Business Week (Industrial/Technology Edition)*, no. 3002 (June 8, 1987): 130–5.

Price, Frank, *Right Every Time: Using the Deming Approach*, New York: Marcel Dekker, 1990.

Propst, Annabeth L., "The Process Qualification Study," *Quality Progress* 20, no. 6 (June 1987): 70–74.

Robson, Ross E. ed., *The Quality and Productivity Equation: American Corporate Strategies for the 1990s*, Cambridge, MA: Productivity Press, 1990.

Ruiz, Oscar A., "Design of Experiments in the Rubber Industry," *Rubber World* 198, (April 1988): 26.

Schonberger, Richard J., "The Quality Concept: Still Evolving," *National Productivity Review* 6, no. 1 (Winter 1986/1987): 81–86.

Sepehri, Mehran, ed., *Quest for Quality: Managing the Total System.* Atlanta: Institute of Industrial Engineers, 1987.

Shetty, Y.K., and Vernon Buehler, eds., *Competing Through Productivity and Quality*, Cambridge, MA: Productivity Press, 1988.

Shingo, Shigeo, *Zero Quality Control: Source Inspection and the Poka-Yoke System*, Cambridge, MA: Productivity Press, 1986.

Sinha, Madhav N., and Walter O. Willborn, *The Management of Quality Assurance*, New York: Wiley, 1985.

Stratton, Q. Donald, *An Approach to Quality Improvement that Works— With an Emphasis on the White-Collar Area.* Milwaukee, WI: ASQC Quality Press, 1988.

Taylor, J. R., *Quality Control Systems: Procedures for Planning Quality Programs.* New York: McGraw-Hill, 1989.

Total Quality: An Executive's Guide for the 1990s/Ernst & Young Quality Improvement Consulting Group, Homewood, IL: Dow Jones-Irwin, 1990.

Townsend, Patrick L. with Joan Gebhardt, *Commit to Quality*. New York: Wiley, 1990.

Vaughn, Richard C., *Quality Assurance.* Ames, IA: Iowa State University Press, 1990.

Wadsworth, Harrison M., Kenneth S. Stephens, and A. Blanton Godfrey, *Modern Methods for Quality Control and Improvement*, Somerset, NJ: Wiley, 1986.

Walter, Richard M., Mark M. Higgins, and Harold P. Roth, "Applications of Control Charts," *The CPA Journal* 60, (April 1990): 90.

Walsh, L., R. Wurster, and R.J. Kimber, eds., *Quality Management Handbook*, New York: ASQC Quality Press and Marcel Dekker, 1986.

Walton, Mary, *The Deming Management Method*, New York: Dodd, Mead and Company, 1986.

Welch, William J., Tat-Kwan Yu, Sung Mo Kang, and Jerome Sacks, "Computer Experiments for Quality Control by Parameter Design," *Journal of Quality Technology* 22, no. 1 (January 1990): 15–22.

2. Crosby, Phillip B., Quality is Free. New York: McGraw-Hill, 1979.

3. Juran, Joseph M. and Frank M. Gryna, Jr., Quality Planning and Analysis, 2nd ed. New York: McGraw-Hill, 1980.

4. See Deming's book, *Out of the Crisis (MIT, CAES: 1986),* for his most detailed exposition of these 14 points, as well as his other insights into management.

5. Crosby, *Quality is Free;* Juran and Gryna, *Quality Planning and Analysis.*
    Deming, W. Edwards, *Quality, Productivity, and Competitive Position.* Cambridge, Mass.: MIT Center for Advanced Engineering Study, 1982.
    Garvin, David A., "Quality on the Line," *Harvard Business Review* (September–October 1983): 64–75.
    Garvin, David A., *Managing Quality: The Strategic and Competitive Edge.* New York: Free Press, 1988.
    Schonberger, Richard J., "Work Improvement Programmes: Quality Control Circles Compared with Traditional Western Approaches," *International Journal of Operations and Production Management* 3, no. 2. (1983): 18–32.

6. The early list of winners included: (1988) Motorola, Westinghouse (Commercial Nuclear Fuel Division), Globe Metallurgical (small company award); (1989) Milliken & Co., Xerox; (1990) Cadillac Motor Car Company, Federal Express (service award), Wallace (small company award); and (1991) Solectron, Zytek, Marlow Industries (small company award). For an article on the nature of the award, see David A. Garvin, "How the Baldrige Award Really Works", *Harvard Business Review,* (Nov.–Dec. 1991), pp. 80–93.

7. Roy L. Harmon, Reinventing the Factory II, New York: Free Press, 1992: 163.

8. Harmon, op.cit., p. 168ff.

9. For information on quality function deployment, see the following:
    Ronald M. Fortuna, "Quality Function Deployment: Taking Quality Upstream", *Target* (Winter 1987), pp. 11–16.
    John R. Hauser and Don Clausing, "The House of Quality", *Harvard Business Review*, May–June 1988, pp. 63–73.

10. For more on Taguchi methods consult:
    Bendell, A., J. Disney, and W. A. Pridmore, eds., *Taguchi Methods: Applications in World Industry*, New York: Springer-Verlag, 1989.
    Byrne, Diane M., and Shin Taguchi, "The Taguchi Approach to Parameter Design," *Quality Progress* 20, no. 12 (December 1987): 19–26.
    Ealey, Lance A., *Quality By Design: Taguchi Methods and U.S. Industry*. Dearborn MI: ASI Press, 1988.
    _____, "Taguchi Basics," *Quality* 27, no. 11 (November 1988): 30–32.
    Gunter, Berton, "A Perspective on the Taguchi Method," *Quality Progress* 20, no. 6 (June 1987): 44–52.
    Lin, Paul K.H., Lawrence P. Sullivan, and Genichi Taguchi, "Using Taguchi Methods in Quality Engineering," *Quality Progress* 23, no. 9 (September 1990): 55–59.
    Noori, Hamid, "The Taguchi Methods: Achieving Design and Output Quality," *Academy of Management Executive* 3, no. 4 (November 1989): 322–326.
    Ross, Phillip J., "The Role of Taguchi Methods and Design of Experiments in QFD," *Quality Progress* 21, no. 6 (June 1988): 41–47.
    Ryan, Nancy E. ed., *Taguchi Methods and QFD: Hows and Whys for Management*. Dearborn MI: ASI Press, 1988.
    Ryan, Thomas P., "Taguchi's Approach to Experimental Design: Some Concerns," *Quality Progress* 21, no. 5 (May 1988): 34–36.
    Sullivan, Lawrence P., "The Power of Taguchi Methods," *Quality Progress* (June 1987): 76–79.
    Turner, Joseph, "Is An Out-of-Spec Product Really Out of Spec?" *Quality Progress* 23, no. 12 (December 1990): 57–59.

11. For more on control charts, consult a source such as *Quality Control Handbook* by Joseph M. Juran, ed., 4th ed. New York: McGraw-Hill, 1988.

12. This discussion follows Deming, *Quality, Productivity, and Competitive Position*, op.cit., pp. 267 ff., although others have also dealt with this issue.

CHAPTER 5

THE WORKFORCE CONTRIBUTION

To this point, this text has stressed such concepts as the flow of the process and of information, bottlenecks, balance, and quality. These concepts have important implications for the character and effectiveness of the process. Yet, without sound management of the workforce, such concepts do not mean nearly so much. This chapter is devoted to an understanding of the contribution the workforce itself makes and of the management tools and ideas that, over the years, have proven useful. Consistent with the theme of working from the "inside out," we focus first on improving the individual worker's job and then, second, on organizing groups of workers for better contributions (employee involvement and quality of work life).

THE INDIVIDUAL WORKER'S JOB

Improving the contribution that any individual worker makes involves attention to several dimensions. Mention has already been made of such items as plant and workstation layout and the flow of materials. Later in this text items such as product design, process technology, setup, tooling, and maintenance will be dealt with in greater detail. This chapter concentrates instead on three topics:

- Worker comfort and safety
- Worker methods
- Worker attitudes and effort

Let us deal with each of these issues in turn.

WORKER COMFORT AND SAFETY

A whole study of the human body's advantages and limitations in the workplace has developed. This study, called **ergonomics,** has helped make manufacturers aware that methods changes and workplace redesign can decrease worker fatigue and improve worker satisfaction and productivity. Ergonomics studies such things as the optimal height, lighting, or positioning for doing certain jobs and how tools ought to be redesigned for less fatigue or for greater effectiveness (e.g., more leverage, greater accuracy, less eyestrain).

Nevertheless, despite its many benefits, ergonomics can go too far. Sometimes the changes considered are costly and have only marginal significance. The good methods engineer must constantly assess which improvements are necessary and cost effective and which suggestions (or demands) for workplace or tool change are not worthy of implementation. Calls by employees or unions for better workplace design are sometimes bones of contention between management and the workforce.

What is not a bone of contention between management and the workforce is worker safety. Irrespective of the efforts of OSHA (Occupational Safety and Health Act) and kindred initiatives, most businesses have long realized that it is in everyone's interests to operate as safe a working environment as possible. Worker safety is often the acknowledged first item of importance for a supervisor's performance evaluation and investments in safety are typically not subject to the company's various financial hurdles for investment decisions.

METHODS IMPROVEMENT[1]

Methods improvement and engineering are too little appreciated in many companies, especially those in the service sector. As we will see on our plant and service tours, it is frequently detailed, nitty-gritty work but it can pay off. Methods improvement basically requires an open, inquisitive mind, an eye for detail, and a passion for keeping things simple. There are no pat solutions to apply in most instances; thus methods improvements usually involve a systematic study of the current methods, an appreciation of what the job really calls for, and a disdain for any explanation that is not thoroughly persuasive.

Over time, many do's and don'ts have evolved about good methods practice, especially as regards the actions of individual workers. Good methods practice is always concerned with what tasks can be eliminated, simplified, combined with other tasks, or changed in sequence. While these kinds of change can frequently be accomplished with little or no investment in new tools, worker aids, or equipment, often methods study will lead to suggestions for capital investments of both large and small scale. Table 5-1 presents a list of suggestions for improving methods. This list provides a glimpse into the commonsense mind of the methods engineer.

Studying Methods

As with most jobs, when one attacks a task systematically, one's productivity is likely to improve. Systematic methods improvement calls for systematic methods study. Formal study of work methods—including concern for raw materials, product design, process design, tooling, plant layout, and the workers' interaction with all of these features—is sometimes referred to as motion study. While attention to methods has always existed, the formal study of work methods in operations is generally associated with the Gilbreths (Frank and Lillian), who in the late nineteenth and early twentieth centuries developed many of the charting techniques and early filming techniques that have done much to improve worker productivity. The Gilbreths had a passion for describing methods precisely through the isolation, identification, and subsequent improvement of the elements of a task.

Today, motion study of repetitive tasks can involve a number of techniques, most involving charts of one variety or another. Methods engineers have found that by charting methods and the process, they can be more effective in thinking creatively and exhaustively about potential

TABLE 5-1   Some basic principles of methods improvement for workers

A. Reduce total steps to a minimum.
B. Arrange in best order.
C. Combine steps where feasible.
D. Make each step as easy as possible.
E. Balance the work of the hands.
F. Avoid the use of the hands for holding.
G. The workplace should fit human dimensions:
  1. Can a suboperation be eliminated?
    a. As unnecessary?
    b. By a change in the order of work?
    c. By a change of tools or equipment?
    d. By a change of layout of the workplace?
    e. By combining tools?
    f. By a slight change of material?
    g. By a slight change in product?
    h. By a quick-acting clamp on jig, if jigs are used?
  2. Can a movement be eliminated?
    a. As unnecessary?
    b. By a change in the order of work?
    c. By combining tools?
    d. By a change of tools or equipment?
    e. By a drop disposal of finished material? (The less exact the release requirements, the faster the release.)
  3. Can a hold be eliminated? (Holding is extremely fatiguing.)
    a. As unnecessary?
    b. By a simple holding device or fixture?
  4. Can a delay be eliminated or shortened?
    a. As unnecessary?
    b. By a change in the work that each body member does?
    c. By balancing the work between the body members?
    d. By working simultaneously on two items? (Slightly less than double production is possible with the typical person.)
    e. By alternating the work, each hand doing the same job, but out of phase?
  5. Can a suboperation be made easier?
    a. By better tools? (Handles should allow maximum flesh contact without sharp corners for power; easy spin, small diameter for speed on light work.)
    b. By changing leverages?
    c. By changing positions of controls or tools?
    d. By better material containers? (Bins that permit a slide grasp of small parts are preferable to bins that must be dipped into.)
    e. By using inertia where possible?
    f. By lessening visual requirements?
    g. By better workplace heights? (Keep workplace height below the elbow.)
  6. Can a movement be made easier?
    a. By a change of layout, shortening distances? (Place tools and equipment as near place of use and as nearly in position of use as possible.)
    b. By changing direction of movements? (Optimum angle of workplace for light knobs, key switches, and handwheels is probably 30° and certainly between 0° to 45° to plane perpendicular to plane of front of operator's body.)
    c. By using different muscles? (Use the first muscle group in this list that is strong enough for the task.)
      (1) Finger? (Not desirable for steady load or highly repetitive motions.)
      (2) Wrist?
      (3) Forearm?
      (4) Upper arm?
      (5) Trunk? (For heavy loads, shift to large leg muscles.)
    d. By making movements continuous rather than jerky?
  7. Can a hold be made easier?
    a. By shortening its duration?
    b. By using stronger muscle groups, such as the legs, with foot-operated vises?

SOURCE: Marvin E. Mundel, *Motion and Time Study: Improving Productivity*, 5th ed. Englewood Cliffs, N.J.: Prentice-Hall, 1973, pp. 230–31.

methods changes. Four forms of these charts are worth mentioning:

- **Process flowcharts** for materials typically focus on what happens to products as they move through the process—where they get caught in bottlenecks, how materials are handled, and what has to come together before a product can proceed. A simplified form of these charts was introduced in Chapter 2, where it was called a process flow diagram. Vastly more detailed diagrams—involving maybe even floor plans—are often helpful in tracking materials movement and in documenting how long it makes to perform particular operations.

Process flowcharts such as the one in Figure 5-1 are sometimes used to document what occurs with materials used by a particular worker. Charting such a task forces the methods engineer to break down the elements of the job thoroughly and to question which elements can be improved—as by adding a worker aid, laying out the workstation differently, resequencing the operations, eliminating needless movement of workers or materials, or perhaps redesigning the product or the materials going into it. The objective of such

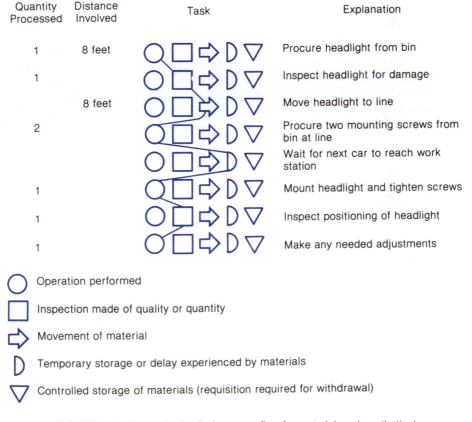

Quantity Processed	Distance Involved	Task	Explanation
1	8 feet		Procure headlight from bin
1			Inspect headlight for damage
	8 feet		Move headlight to line
2			Procure two mounting screws from bin at line
			Wait for next car to reach work station
1			Mount headlight and tighten screws
1			Inspect positioning of headlight
1			Make any needed adjustments

○ Operation performed

□ Inspection made of quality or quantity

⇨ Movement of material

D Temporary storage or delay experienced by materials

▽ Controlled storage of materials (requisition required for withdrawal)

FIGURE 5-1   A sample detailed process flow for materials: a hypothetical description of headlight mounting on an assembly line.

systematic study is to simplify the operation so that the worker can do it quickly and well.

- **Process flowcharts for workers** focus solely on what the worker does. Whereas it is usually more advantageous for methods study to track materials and what happens to them, worker-oriented flowcharts can sometimes be more effective, particularly when the worker's job forces him or her to travel around the operation, interrupting contact with the product or service. A chart like that in Figure 5-1 can be used to detail such a worker's efforts, just as it can be used to detail the flow of materials and what workers do with them. In this case, however, movement would refer to the movement of the worker and not the materials, and temporary storage or delay would refer to the delay of the worker, not the materials.

- **Worker–machine charts** are typically time-specific bar charts that track a worker together with his machine(s). This tracking can be in general terms or it can be as specific as the hand, feet, and body movements necessary to perform a given task. One popular analysis is to chart a machine's cycle time against the cycle time of the machine operator, so that any conflicts between the two can be isolated and resolved. Figure 5-2 is an example. In this analysis, the worker is seen as a "machine," with a cycle time of his own. Like the machine, the worker can be either functioning or idle. Such a worker–machine chart helps to coordinate the two so that their idle times are minimized and their outputs are maximized.

- **Worker–worker charts** investigate the potential interference and room for coordination of one worker with others. Such charts are the simple extension of the worker–machine charts in Figure 5-2. They are particularly effective in service operations, which tend to

be more labor intensive than many manufacturing operations.

The data for these charts can come from several sources. The most widely used, quite naturally, is in-person observation of the task in question by the methods engineer. This is often the quickest, easiest, and least costly means of studying the job. Given some preliminary observation to determine what the important aspects of the job are, the methods engineer can then devise the most appropriate chart(s).

For tasks that are either very complex or of long duration, methods engineers have often found that filming or videotaping is useful. Films and videotapes have a number of advantages: (1) they catch everything in the field of view of the camera; they do not inadvertently miss seeing something; (2) they can be viewed over and over again; and (3) they are accurate. For these reasons, films and videotapes are especially helpful for methods study involving great detail or many workers. And, if the film is taken at unusually slow speeds, long or complex tasks can be reviewed quickly. (This is termed *memomotion analysis.*)

The analysis of great detail—generally for small-scale tasks—for which filming or videotaping is useful is termed micromotion analysis. This technique was pioneered by the Gilbreths, who divided movements into 17 distinct categories, which they named "therbligs" (their name spelled backwards).

These therbligs (grasp, position, pre-position, use, assemble, disassemble, release load, transport empty, transport loaded, search, select, hold, unavoidable delay, avoidable delay, rest, plan, and inspect) provide a convenient and clear framework for analyzing small-scale manual tasks and suggesting ways to improve or eliminate elements of them through the use of tools, fixtures, equipment, workplace layout, or just a different sequence of motions.

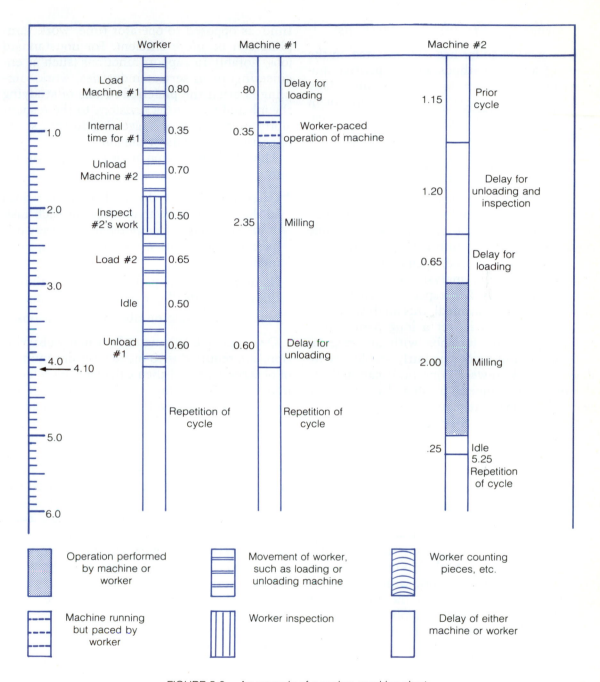

FIGURE 5-2   An example of a worker–machine chart.

Other Ways to Improve Methods

The discussion so far has concentrated on the study of worker methods when repetitive motions in standard settings are the rule. Many jobs, particularly those in support roles or in the service sector, are not repetitive, however. Memomotion analysis can be of use in such situations. Frequently, two other techniques are employed: work activity analysis and work sampling.

Work activity analysis simply calls on the worker to list chronologically the work performed, indicating the type of task, the time spent on it, and the number of actions completed (forms processed, memos written). Because work activity analysis can be time consuming for the worker (especially when many short tasks are performed), this analysis is generally not extended over a long period. It is more appropriate for jobs with an irregular daily pattern of work but fairly stable day-to-day and week-to-week content. It can indicate the need for change in a job and point up aspects of the job that require further study.

Work sampling is the random observation of a worker and the ensuing calculation of the relative amounts of time spent performing specific tasks or tied up in avoidable or unavoidable delays (e.g., 10 percent of the time making phone calls, 20 percent of the time writing reports, 5 percent of the time waiting for materials). As such, it is a good means of analyzing jobs whose pattern of work is irregular over long periods of time. While less disruptive than work activity analysis, work sampling can entail much travel throughout the operation by the analyst or supervisor involved and much time to gather enough observations. In its favor, work sampling enjoys some advantages over stopwatch or film/tape studies. For example, a single industrial engineer can study a number of tasks simultaneously; if the tasks involve a considerable amount of machine time, as opposed to operator time, work sampling can be more efficient. For nonstandard jobs, notably in support functions (such as engineering) or in service industries, work sampling is often the preferred means of studying the job and suggesting revisions to the responsibilities assigned or suggesting further study of particular aspects of the job.

The implementation of work sampling requires:

- Categorizing what the studied worker does so that the relative frequencies of each task or type of delay as computed by the study can have real application to methods improvement.

- Using an irregular pattern of observation of the worker; ideally, it should be random.

- Making a suitable number of observations.

One approach to gauging how many observations are required is to make a certain number of observations and then calculate the relative frequency of each task or type of delay. Then another set of observations is made and added to the first set. The relative frequencies for specific tasks, as calculated for the first set of observations and the enlarged second set, are compared. If the computed relative frequencies are nearly identical further sampling need not be done. However, if the relative frequencies differ by much, another set of observations should be taken and added to the other observations, relative frequencies should be computed, and comparisons should be made. When the calculated relative frequencies are judged to be stable, further sampling can cease.

Another approach to the determination of observation sample size is to use statistics. In this approach, the methods engineer states a preference for the probability that a given relative frequency is within X percent of the true relative frequency for the entire population (say a 95 percent probability that the relative frequency is within 10 percent of the true frequency). Assuming that the relative frequencies are distributed according to a binomial distribution, the sample size can be computed by the following formula:

$$N = (a/x)^2[p(1 - p)]$$

where N is the number of observations required for the study, p is the relative frequency of interest (task or type of delay), a is the number of standard deviations associated with a given probability preference (such as 2 if the preference is a 95 percent chance), and x is the accuracy desired for the frequency (such as 0.10 for an accuracy within 10 percent of the true frequency).

As is clear from Table 5-2, the higher the level of confidence that the computed relative frequency is within tolerance, or the greater the accuracy sought, the more observations that must be made. The higher the relative frequency of the task or delay studied, the more observations that are needed as well.

Balancing Methods Improvements Against Other Considerations

Often, improvements in methods can be made that are relatively easy to do, save much more money than they cost, and greatly aid product quality and/or worker safety. These are the methods improvements that are easy to adopt. Sometimes, however, a more efficient way of accomplishing a task must be balanced against other considerations—an acknowledged, or even potential, drop in the product's performance characteristics or the maintenance of a particular service standard. Consider Situation 5-1.

Discussion of Eastern Bell Telephone Company

It is clear that Charlie Harris's idea to accumulate the noncritical repair calls and bunch them up to provide better routing and thus higher productivity will reduce the company's ability to meet its established service standard on repairs. The issue is whether the service standard itself is worth keeping. The notion of having customers indicate how perturbed they are without service is probably a good one. Particularly irate customers can then be put in with the out-of-service trouble calls for immediate attention. A simple question asked of customers (such as "We can fix it the day after tomorrow. Is that all right with you?") when they phone in their complaints may placate them and yet accumulate enough calls to ease the routing problem.

Bunching orders in this way to facilitate routing may not only improve productivity in

TABLE 5-2 Representative observation sample sizes

ACCURACY AND RELATIVE FREQUENCIES	PROBABILITY PREFERENCE LEVEL OF CONFIDENCE		
	90% ($a = 1.645$)	95% ($a = 1.96$)	99% ($a = 2.576$)
Accuracy of 10% ($x = 0.10$)			
$p = 0.10$	24	35	60
$p = 0.25$	51	72	124
Accuracy of 5% ($x = 0.05$)			
$p = 0.10$	97	138	239
$p = 0.25$	203	288	498

SITUATION 5-1

Eastern Bell Telephone Company

The more Charlie Harris thought about it, the more attractive it became, but it still meant a significant change in the telephone company's repair practices. Charlie was a young district manager with responsibility for installation and repair of telephone equipment in a 200-square-mile area north of the city. In response to some higher-level pronouncements about the importance of measured installation and repair productivity, Charlie conceived the idea of changing the way telephone repair technicians were assigned to residence repair jobs.

In the prevailing system, each repair technician reporting to the garage in the morning would be assigned to the repair of a subscriber's telephone that had been reported and checked out at the central office as out-of-service. After repairing that phone, the repair technician would call in to the dispatcher, who would assign the next repair task. Over three-quarters of the repair tasks involved telephones out-of-service, but the rest involved trouble not severe enough to cut off service (e.g., an extension phone out of order, a worn cord, a sagging wire from the telephone pole to the house). The repair technician would not know, starting the day, how many jobs he or she would have to do or where any of them would be. The person was at the mercy of the dispatcher.

Because the Bell System's standards for responding to trouble calls were so demanding (respond to out-of-service calls within 24 hours), a considerable portion of the repair technician's day was spent in transit, chasing the urgent calls for service by the telephone company's customers. If the repair technician could be routed better, less

time would be spent in transit, and more service calls could be made. Better routing, however, meant bunching up more service calls geographically, which could be done only by waiting longer and collecting more trouble calls. Charlie recognized that improving work productivity almost always implied a lowering of existing service standards. Charlie knew that he could never wait to accumulate out-of-service trouble calls; they would always require immediate action. However, the less severe trouble, involving calls where the home still had telephone service on at least one phone, was a candidate for delayed service with better routing of the repair technicians.

Charlie wondered how best such a delayed response system could be implemented. Several question still plagued him:

1. How much delay would the customer accept? Should the customer be granted the right to insist upon immediate service like before?

2. How could a program of bunched orders and better routing be designed to promote worker productivity in the best way? Should the dispatching system remain the same as before, with the repair technicians calling in for their next assignment, or should they be given a full day's load from the start? Should the same person be assigned to this bunched order work, or should the work be rotated among all repair technicians? If rotated, how long should person be assigned to such work?

3. Would the benefits in increased productivity outweigh the diminishing level of service that the proposed system necessarily implied?

---

the repair of noncritical service calls, but also provide a demonstration whose effects could be broad-ranging if in fact productivity rises markedly, say from 8 calls to 12 calls per day. This improvement could make the entire repair fleet more conscious of productivity. This argument for a demonstration effect suggests that the noncritical calls be rotated among the repair technicians so that the productivity gains are experienced firsthand by everyone.

The duration of the assignment should be long enough to make an impact but short enough so that everyone can get into the act during the year. Perhaps 2 weeks to a month on this assignment would do. The likely impact of the demonstration argues further that the repair technician be given a full day's load in the morning. Counting orders then will probably make more of an impression than counting up the day's accomplishments on the way home from work.

Time Study and Time Standards[2]

Paralleling the motion study pioneering by the Gilbreths was the time study of individual worker tasks. The pioneer in this study was Frederick Taylor, who worked during the same period as the Gilbreths. Much of the work of Taylor and his followers had a similar intent to that of the Gilbreths—to improve productivity by looking closely at what workers did. However, rather than focus on the motions involved, Taylor broke down the job into elements and timed them. As in motion study, each element was studied to determine what should be kept and what should be discarded, but the stopwatch was the chief aid to the engineer rather than the movie camera or the process chart.

Use of Time Standards

The time standards developed by Taylor have at least five broad applications:

1. They can be used for planning and budgeting. In job shops, for example, jobs cannot be bid properly, the amount of labor to have available at any time (how much overtime to schedule) cannot be planned, nor the mix of operators needed decided unless the job shop knows how long a particular job should take. Standards in an operation like a job shop are critical for planning production and scheduling workforce, materials, machines, and orders. (See the Tour of Norcen Industries.)

2. They can be used to balance operations. In line flow operations, it is essential to have an accurate appraisal of how long it takes to do the various jobs along the line. Line balance requires knowledge of precedence relationships, accurate time standards, and any constraints that could influence how the line is set up. Good line balance has as little balance delay as possible. Often that means concerted work on the limiting job along the line, the job with the longest time. After all, a line can only produce as fast, and as much, as its limiting, longest time job permits it to.

   Indeed, in batch and continuous flow operations, time standards are also important for balance. Otherwise, the process may have too much capacity in some areas and not enough in others.

3. They can be used to improve performance. Industrial engineering departments are charged with improving the operation of a plant's processing or assembly operations. These goals often require time studies to be done to evaluate technique and equipment changes. (See the Tours of General Motors and Burger King.)

4. They can be used to evaluate individual workers and to serve as a basis for wage payments. In a variety of operations, time study is used as a basis for incentive wages. (See the Tour of Jos. A. Bank Clothiers.) Time standards, once they are accepted as fair by the workers, can be used to evaluate worker performance: workers who match or beat the standard can be rewarded, and workers who fail to reach the standard can be disciplined or even removed from the job. Use of time standards is discussed more later in the chapter.

5. They can be used to define a measure of labor productivity. The popular "efficiency" measure of labor productivity is conventionally defined as

$$\text{Efficiency} = \frac{\text{standard hours}}{\text{actual hours}} \times 100\%$$

6. Thus efficiencies of over 100 percent represent situations where the direct labor (i.e., those workers for whom standards have been set) has beaten the standard. This efficiency measure is sometimes combined with a measure of utilization (the fraction of time the work place was available to do work with all equipment "up" and all materials there) to yield an "effectiveness" index.

$$\text{Effectiveness} = \text{efficiency} \times \text{utilization}$$

If utilization is less than 1 (and it always is), then effectiveness is always less than efficiency, in percentage terms.

Development of Time Standards

Time standards can be developed in a variety of ways, as can be seen in the plant tours at the end of this text: history (Norcen Industries), time study (Jos. A. Bank), predetermined motion time systems (Burger King), and standard data systems (General Motors).

History

The time standards that some firms use in their planning are generally based on having worked on the same or similar parts or products before. Considering the diversity involved in a job shop, for example, using history is a very economical way by which to establish useful standards. Workers are not paid or evaluated according to these standards and thus great precision is not required. History provides enough rough but useful standards so that planning can be done effectively.

Time Study

Time study typically involves dividing a job into distinct "elements" that have unambiguous starts and stops and recording the time spent performing each of those elements. Timing is frequently accomplished by an electronic stopwatch that automatically times different elements of the job. Alternative techniques include (1) the "snap-back" method (where three different stopwatches mounted on a special clipboard are started, stopped, and reset by a lever that is punched at the beginning of every job element), and (2) the continuous reading method (where a single stopwatch is used and elements are timed by reading off the running times).

Time study recognizes that the measured actual times of workers may not be appropriate for use as standards. Two major types of corrections are usually made: performance rating and allowances.

A worker, when timed, might purposely "dog it" so that the standard is set at an easily attainable level. To counter this possibility, methods engineers typically try to "rate" the worker's efficiency in doing the job. This performance rating serves to convert the measured actual time to so-called normal time, as follows:

$$\text{Normal time} = \text{measured actual time} \times \text{performance rating}$$

A performance rating less than 1 implies that the methods engineer thought the worker was underproducing. A rating greater than 1 implies that the methods engineer did not think an average worker could sustain the pace.

Performance ratings can be made of the entire job studied, but accuracy is enhanced if individual ratings are made of each element of the job. Normal times are then computed for each element. Establishing a performance rating is a tricky endeavor. Before a job can be rated, the methods engineer must appreciate the difficulty of the job and the conditions under which it is done. The engineer must then gauge what a normal tempo would be under such circumstances and assess how the worker's efforts match up against the normal tempo. Films or videotapes are often employed to give methods engineers a sense of what the normal tempo should be; they also provide methods engineers with opportunities to practice rating workers.

Allowances, the other major correction to measured actual times, relate to delays in performing a job that are unavoidable for the worker or are irregular but that can nevertheless be expected as routine occurrences. Typically included as such delays are waits due to materials shortages, special instructions, machine adjustment or repair, quality problems, workplace housekeeping, and the like. Personal time taken to get a drink or to go to the bathroom are also generally counted as allowances. If fatigue was not taken into consideration in rating the job, it can be added as an allowance.

These allowances modify the normal time so that it can be used as a standard:

Standard time = normal time for the job
$\times$ allowance factor for the job

where

Allowance factor = $1 + \dfrac{\text{allowed minutes}}{\text{normal minutes}}$

The allowance factor is always greater than 1, often between 1.1 and 1.3. The methods engineer can arrive at the specific factor by studying the job to identify the delays inherent in it and the frequency of their occurrence. Such study can be made by continuous observation of the job for a reasonably long time (such as several hours) or by work sampling techniques.

Time study is a fairly accurate way to establish time standards for tasks for which the methods are well known. In practice, however, time study has two drawbacks: (1) it cannot be used effectively for jobs that have not been well defined and are only being contemplated, and (2) time study depends critically on the "rating" of the worker. The subjective determination of what the standard pace of the operation should be is often tricky and can become a bone of contention between industrial engineers and the workforce.

Predetermined Motion-Time Systems

The two difficulties encountered with time study techniques are to some extent remedied by predetermined motion-time systems. These systems build on the notion that any task can be decomposed into a series of basic motions. The original task, then, can be seen as the sum of its basic motions. Extensive libraries of basic motion information indicate, for example, the time it takes to move one's hands various distances to perform tasks of varying complexity, with or without tools or materials of various weights. Using this information, an analyst can develop time estimates for tasks that are contemplated; such computations can be more objective than those computed by engineers rating actual operators. Standards set by good analysts are consistently shown to be accurate, although many companies periodically use direct time studies to check their use of predetermined motion-time systems.

The most widely known of these predetermined motion-time systems is MTM (method time measurement), developed by H. B. Maynard in the late 1940s. Several versions of the MTM system (MTM-1, MTM-2, MTM-3) have been developed. The original system, MTM-1, uses very short and well-defined movements to build up a particular task. Although MTM-1 can be used successfully for even short-cycle jobs (i.e., jobs that are accomplished within a short period of time) that are repeated over long periods, its application requires great skill. The industrial engineer must be very precise in decomposing a task into all its component motions and must carefully identify which of the multitude of choices from the MTM-1 data base best characterizes each motion. Many of the problems encountered with MTM analysis are caused by engineer errors in its application rather than by inaccuracies in the time values.

Some of these applications problems are remedied in another predetermined motion-time system from the H. B. Maynard Company, which pioneered MTM. The MOST (Maynard Operation Sequence Technique) system was developed in the late 1960s and early 1970s. It was built on the observation that most worker tasks can be decomposed into three basic sequences of motions:

- A "general move" sequence for the movement of objects freely through the air. This sequence is represented by the series of letters ABGABPA where

  A = action distance — the movement through space of one part of the body or the entire body

  B = body motion — the up-and-down movement of the body or an action of the body to eliminate some obstruction to movement (e.g., opening a door)

  G = gain control — typically, the manual dexterity needed to gain control over an object

  P = place — the alignment of an object with one or more other objects before control is relinquished

- A "controlled move" sequence for the movement of objects that remain in contact with a surface or attached to another object during the movement. This sequence is represented by the series of letters ABGMXIA where A, B, and G are defined as before and

M = move controlled — the manual control of the object through its path

X = process time — the work done by machine and not by the worker

I = align — worker actions needed to align objects after the controlled move itself or after the machine's work on the object

- A "tool use" sequence for the use of common hand tools. This sequence is represented by the series of letters ABGABP ABPA, where the A, B, G, and P are defined as before and the blank space can be filled with an activity specific to the tool being used. Typical tool-related activities are fastening, loosening, cutting, treating an object's surface (painting, sanding), measuring, and recording.

Using the MOST system, the analyst follows these steps:

1. Determine the appropriate sequence from the task under study.
2. Index the movements that comprise the sequence. (Each sequence has its own index definitions. Table 5-3 presents those for the general move sequence.)
3. Add these index numbers together.
4. Multiply by 10 to compute the time standard in time measurement units (TMUs), each of which is equal to 0.00001 hour (0.036 second).

Figure 5-3 illustrates, in general terms, the MOST system for a routine task such as might occur at a fast-food restaurant. (Recall that the Burger King simulation model used MTM analysis to develop data on the making of Whoppers, drawing of drinks, and other tasks.)

By investigating sequences of motions rather than the motions themselves, the MOST system reaps some statistical benefits that help ensure the accuracy of its computed times if

TABLE 5-3 General move sequence with index numbers

	A		B	G	P
INDEX	ACTION	DISTANCE	BODY MOTION	GAIN CONTROL	PLACE
0	<2 inches				Hold or toss
1	Within reach			Move light object, or move light objects using simultaneous motions	Lay aside or loosely fit
3	1 or 2 steps		Bend and arise, 50% occurrence	Use nonsimultaneous motions; move heavy or bulky objects; move where vision is obstructed or not possible; disengage, interlock, or collect objects together	Make adjustments, use light pressure, or make two placements
6	3 or 4 steps		Bend and arise		Use care or precision; use heavy pressure; move where vision is obstructed or not possible, or make intermediate moves
10	5 to 7 steps		Sit or stand		
16	8 to 10 steps		Through door, climb on or off		

Movement Fetching a bin of sliced tomatoes from under a counter and placing the bin into its slot

General move sequence ABGABPA		Appropriate Index from Table 3-2
1. Walking a step or two	Action distance	3
2. Bending down	Body motion	6
3. Picking the bin up	Gain control	3
4. Walking back a step or two	Action distance	3
5. Dropping the bin into its slot	Body motion, Place	0,1
6. Staying in the same place to resume other work	Action distance	0

Indexed sequence $A_3B_6G_3A_3B_0P_1A_0$

Time measurement units $(3 + 6 + 3 + 3 + 0 + 1 + 0) \times 10 = 160$ TMUs, or 5.76 seconds

FIGURE 5-3 Using MOST to compute the standard time for a sample movement.

long enough sequences of movements are analyzed. Thus the MOST system can be both accurate and intuitive as well as easier and faster. For many applications the MOST system is now preferred. It can operate manually or, as with other predetermined systems, in a series of computer programs, which can expedite the updating of time standards and process descriptions.

Standard Data Systems

The MOST system represents an aggregation of the small movements that made up the original MTM system. Similarly, standard data systems can be viewed as aggregations of smaller, more specialized tasks for use by plants that have a number of routine jobs. Standard data systems are generally based on a wealth of data derived from many time studies of similar jobs. To apply them to a new or redefined job requires the breakdown of the contemplated job into not only those elements that have direct counterparts in the standard data base but also those elements that may not have precise counterparts but are related to tasks previously done and for which there is considerable data. By judiciously combining both types of elements, standards can be derived easily and with reasonable accuracy. (See the Tour of General Motors—Oklahoma City.)

Choosing Among Systems

As we have seen, each of the methods for establishing time standards has advantages and disadvantages. Some (such as time study and MTM-1) are costly; some are detailed; and some are useful for particular purposes. Although the choices are not clear-cut, direct time study and very detailed MTM analysis are generally preferred when very accurate standards are required for relatively short-cycle jobs that may be repeated often and for which worker evaluation is important. Intensive time study may also be required in processes that

pay incentive wages. For tasks that may not be repeated exactly over long periods of time, more aggregated predetermined motion time system (such as the MOST system) are useful for developing standards; such systems are suitable when there is considerable worker discretion involved in each task. Standard data systems are particularly attractive for routine work that has a rich history of time study data. Purely historical data provide a very low-cost way to establish standards if great accuracy is not required and if the tasks studied are not routine.

WORKER ATTITUDES AND EFFORT

A worker's attitude towards a job and the level of effort he or she brings to that job naturally depend on many things. In this section, we deal with three critical influences: how the worker is paid, how the job itself is designed, and how discipline is handled.

Standards, Incentives, and Wage Payments[3]

As noted previously, time standards can be used by companies in four major ways: (1) to estimate capacities and costs, (2) to improve performance, (3) to evaluate workers and to form a basis for wage payments, and (4) to measure labor productivity. The use of time standards for estimation, methods improvement and productivity measurement is widespread; in fact, almost all companies use standards for these purposes. In somewhat less than half of all manufacturing companies, time standards are used as well in the development of incentive wage schemes for at least a fraction of the workforce.[4] Thus while hourly wage payment (sometimes referred to as daywork) is more prevalent, incentive wage schemes are fairly common and widely discussed.

Incentive schemes are logically divided into two groups: individual incentives and group incentives.

Individual Incentives

The plant tour at Jos. A. Bank, found at the end of this text, introduces both key variants of individual wage incentives—the piece-rate system and the standard-hour system. The appeal of these incentive wage systems is readily understandable. The worker is paid according to how much he produces, not how long he works. It is generally accepted that workers on incentive wages work harder than those who are paid by the hour, other things being equal.

In the piece-rate system, every unit of output has its own "price," or wage reward. Workers are paid strictly according to how much they produce. Typically, there are no wage floors or ceilings, although the piece rate itself must be set so that the federally mandated minimum wage is easily met. Piece rates are generally inviolate over time except for well-recognized changes in methods, materials, equipment, or product specifications. About a third of the companies with wage incentives use a piece-rate system. At Jos. A Bank we see it used in the coat shop and the pants shop. Also, it is typical practice in a piece-rate system for most workers to earn a premium above any standard. It is hoped, however, that these standards are not so "loose" that wage costs escalate with little control.

In the standard-hour system, there are no wage "price tags" on every unit. Rather, a certain rate of output per hour is set as the standard, and pay varies according to what percentage of the standard is actually achieved. At Jos. A. Bank, the standard-hour system applies in the cutting room. There are many variations of this system. Some pay proportionally (e.g., a 115 percent average performance for the week will earn 115 percent of the standard wage); others use formulas based on a tally of "points"

earned. Many variations place a floor on earnings (a base rate, as applied in the cutting room at Jos. Bank), with the standard set so that a typical worker can expect to better it consistently (110 or 120 percent of the standard); other systems have no wage floor. These standard-hour systems are the most popular wage incentive scheme, prevailing in about three-fifths of the companies with wage incentive programs. While they function much as piece-rate systems, their advantage is that pay increases can be made simply by adjusting the base rate, which is much easier to do than adjusting the many piece rates involved so that a given percentage increase for the workforce as a whole is proportioned equitably among all piece-rate workers. (See the discussion of piece rates in the plant tour of Jos. A. Bank.)

The attractiveness of individual incentive wages is, of course, to spur worker effort. But are there unattractive features as well? Consider Situation 5-2.

Lessons from Tower Manufacturing Company

This situation uncovers a number of the problems associated with incentive wage schemes, the most important of which is the gradual erosion of the time standards on which the incentive wage is based. The erosion of standards is to be expected in many cases because workers are clever and knowledgeable about what they do and can often find ways to beat the standard. At Tower Manufacturing, for example, Bestmann found a way to beat the standards set for the new CNC machines and his pay jumped appreciably. The resulting inequities in the factory have to be dealt with.

Dealing with these inequities, however, is no easy task. Tightening up on the standards can sometimes be done, but naturally does not make management popular with the workforce. In most union contracts involving incentive pay, after a given period of time the

SITUATION 5-2

Tower Manufacturing Company

"Boy, did I just get ambushed. I haven't been dressed down like that since I forgot my wife's birthday a few years back. Those guys are steaming and we better do something about it." Bruce Hillman, foreman of the machine shop, was addressing his supervisor, Clint Oster, the manager of machining operations at Tower Manufacturing Company. Tower Manufacturing, a supplier of chassis components for heavy, customized, off-highway trucks and construction equipment, had recently added a second computer-controlled milling machine to the machine shop. It was the worker response to this second machine and its implications for worker compensation that had Bruce Hillman agitated.

Bruce continued, "I don't know how we're going to fix things. All I know is that this current arrangement simply has to go. We can't continue to pay Bestmann for piece work on both machines. His pay is way out of line. The others in the shop like him all right but pulling in almost half again the pay of the others is plain ridiculous. Everyone is clamoring for an increase in the rates or an override in the standard pay."

Clint Oster was kicking himself for not anticipating this confrontation. The machine shop at Tower Manufacturing had traditionally been a highly skilled shop, stocked with general-purpose machine tools, with workers compensated through a piece-rate incentive wage system. Times had been changing, though, and the increasing stability of Tower's major contracts had made possible the introduction of computer numerically controlled (CNC) machines into the shop. Management had agonized over the decision, since even one such machine was an investment of hundreds of thousands of dollars. But Clint was convinced that Tower had to purchase such equipment if it was going to stay competitive. Compared with what even a first-class machinist could do using conventional equipment, the computer-controlled machine was far superior. It took much longer to set up for a job, but once it was set up, its speed and quality of output were extraordinary. All the operator had to

do was feed the machine with material and remove the completed, machined output. For longer production runs, which was the situation now at Tower, the new machine was ideal.

When Tower purchased its first computer-controlled machine, Clint and Bruce assigned one of the shop's best workers, Jay Bestmann, to it. It had taken quite a while for Tower's industrial engineers and Bestmann to get the bugs out of the new machine, but, once accomplished, everyone was satisfied with the machine's capabilities. During this startup period, Bestmann had agreed to be taken off incentive pay. Instead he was paid an hourly rate equivalent to his last 2 months' average hourly pay. When the new machine was proven out, Bestmann agreed to a piece rate that was two-thirds of the rate for equivalent output from conventional machines. This rate worked out to be slightly higher than the rate of pay he was earning just prior to the introduction of the new machine, but Clint Oster had agreed to it since none of the shop's machinists wanted to work on this less-skilled job. Setups for both the new computer-controlled machine and conventional machines were paid at the same rate.

It had been planned from the beginning that one operator would be responsible for two of the new machines, since otherwise the operator would be idle much of the time, watching the machine do the job. Therefore, when the second computer-controlled machine was installed, Bestmann was assigned to it. Once again, the piece rate had to be adjusted. The engineers felt it was unlikely that both machines would always be perfectly syncopated (so that Bestmann never had to be at both machines at once); the piece rate therefore was cut in half (to one-third from two-thirds of the rate applicable to conventional machines), but rather was set at 40 percent of the rate other workers were given—also as sort of a bribe to get Bestmann to accept the assignment to the two machines.

Since this rate was agreed on, however, two things had happened to increase Bestmann's pay considerably. One was that Bestmann began setting up one machine while the other was working. This procedure entailed a lot of movement between machines so that the working one was always fed with material, but it meant that Bestmann was carrying piece-rate pay even when setting up one of the machines, something the company's industrial engineers had not planned for. The other occurrence was an increase in the lot size with which the CNC machine was loaded. This increase meant fewer setups and a lengthened time for the fast output rate of the machine. Both of these factors combined to push Bestmann's pay to levels far in excess of the pay levels for other workers.

This fact, of course, was apparent to the other workers—hence their confrontation with Bruce Hillman. If Bestmann was going to be paid so much more, they wanted a more equitable arrangement for everybody in the shop.

Clint Oster was in a quandary. He could reset the already agreed-to piece rate for the new machine and risk antagonizing Bestmann or anyone else who might be assigned to the machine. Or, he could take the machine off incentive pay and put it on hourly pay, risking not only Bestmann's displeasure but the suspicion of other workers in the shop about the future abandonment of piece work, a wage scheme that had widespread approval among them. Or, he could place a much lower-skilled worker on each of the new machines, put Bestmann back on conventional machines, and risk even more suspicion from the workforce about job trends in the shop.

standards cannot be changed without first changing the product, materials, methods, or machinery; any such change is subject to the union–management established grievance procedure.

Of course, changing a standard is far from a costless endeavor. The task must be studied, often by stopwatch, and that takes time. This fact tends to make wage incentives less popular in companies that undertake a lot of product or product mix changes or that are subject to repeated innovation in the production process itself. Wage incentives, and piece rates in particular, are more likely to flourish in processes where (1) the output of every worker is easily measured; (2) the jobs are standard, clear-cut, and unlikely to be altered much through technological changes; (3) the flow of the process is regular and thus every worker has a backlog of work to do; and (4) workmanship is either easy to judge or not critical to the product. It is no accident then that piece rates predominate in the needle trades.

The situation at Tower Manufacturing is delicate because it would be difficult, if not impossible, to change the standard now that it has been established. If the job were switched to an hourly paid one, suspicion would likely grow in the factory that all jobs in the department would be switched to hourly pay or that CNC equipment would gradually take over. These changes might mean a switch in pay schemes and a deskilling of the operation, both fearsome developments to the workforce. Some reassurances about pay policy and the direction the department is taking are definitely called for. Perhaps even the formation of a distinct new department for the CNC equipment would help allay some fears and still enable the company to continue with a program of modernization.

This predicament at Tower Manufacturing brings up another aspect of wage incentives—their effect on supervisors. Often supervisors let the wage incentives themselves manage the workforce. That is, given wages as an effective prod for production, some supervisors step back from taking an active role in discovering productivity gains or in monitoring the pace of production. Under an hourly pay scheme, the supervisor is much more directly responsible for the pace and quality of the work.

A different kind of problem with wage incentives and productivity is illustrated in Situation 5-3.

SITUATION 5-3

Baldridge Chain Saw

The numbers on the page he was staring at said so, but MacDonald still couldn't believe it. Productivity on the new lightweight chain saw assembly line had declined for the second consecutive week. There was no way that the company could meet its commitment to a large national retailing organization. That entire account, not to mention MacDonald's own job, appeared to him to be in serious jeopardy.

Dave, a recent M.B.A., was in his first year at Baldridge Chain Saw. The company was a small, family-owned manufacturer of chain saws used primarily by professional loggers. Baldridge marketed these professional chain saws under its own name. Recently, however, the growing demand for small, lightweight chain saws by what the company regulars termed "weekend woodsmen" had triggered the current project to design and manufacture a lightweight chain saw to compete with McCulloch, Homelite, and others. On the basis of its reputation, Baldridge had won a substantial trial contract from a large mass merchandiser. If quality and delivery expectations were met, Baldridge could anticipate regular orders. Dave MacDonald was the so-called product manager, responsible for shepherding the project through the company. The job was a mix of marketing, engineering, production, sales, and finance.

The new lightweight chain saw that Baldridge's engineers had developed was not very complex—at least not as complex as the company's other models. The saw had fewer parts and used many standardized parts that were readily procured either from other operations at Baldridge or from outside suppliers.

The company's industrial engineers had designed a 14-station assembly line for the new saw. This line was staffed half with assemblers pulled from other models' assembly lines and half with new employees. The new employees were paid at a fixed hourly rate for the first month (the startup) and then they went on incentive pay. During the same startup period, the experienced workers were paid at the average of their last month's wages; then they too went on incentive. The incentive, similar to the one that prevailed in Baldridge's other assembly lines, depended on the output of the entire line. If the line operated at 110 percent of standard, all of its assemblers would receive a given percentage increase in their base pay. Production at 120 percent of standard would yield a higher pay increase. The engineer-rated speed of the line was such that the average expected wage for all assembly workers was set at 110 percent of standard. Production below 100 percent of standard was paid at the startup wages. The trial contract could just be met, if the workers averaged 110 percent of standard for the 12-week life of the trial contract.

Because the workers could be expected to continue learning how to assemble more efficiently, the standard inched up higher each week after week 4 and then leveled off after week 8. The company's industrial engineers had developed this new, ever-tightening standard based on the company's past performance with other chain saws.

It was now 6 weeks into the contract's life. Productivity, relative to standard, had traced this pattern (with standards higher in weeks 5 and 6):

WEEK	PERCENT OF STANDARD
1	70
2	90
3	101
4	111
5	107
6	98

When production slipped in the fifth week, Dave MacDonald had called in the industrial engineers to check whether the line had been properly set up. Dave was concerned lest the line be unbalanced. In that situation, one or more workers might be constituting a bottleneck for the whole line. Dave's concern was heightened by a report from Vic Elgin, the line's foreman, that several of the experienced workers had been complaining about the standards. After studying the matter, however, the industrial engineers reported back that they saw no reason to alter any of the stations along the line. Dave had to wonder what was holding up production.

Lessons from Baldridge Chain Saw

The situation at Baldridge Chain Saw is suspicious, but probably not as perplexing as Dave MacDonald senses it to be. After all, the productivity drop coincides with the tightening of the group incentive wage scheme. The complaints about the standards registered by the older workers to Vic Elgin are other indications that the fall-off in productivity results from a deliberate slowdown in worker pace, as a veiled protest against what the workers would insist are too tight standards. This slowdown may also be a protest against MacDonald himself or the new lightweight saw product line. In any event, the probabilities are that the workers are "pegging their production."

What to do about it? It may already be too late to save the trial order for the mass merchandiser. If it is, the company may want to maintain the standards as they are, deliver late, accept its fate, but make a point about the company's resolve in establishing production standards. If, on the other hand, it is not too late to save the order, the tightening of the standards may have to be postponed and perhaps some overtime scheduled in order to catch up. This approach, of course, would increase costs. Neither of these alternatives is very attractive, but then the situation the company finds itself in does not leave much room for maneuvering.

Group Incentives

As the Baldridge Chain Saw situation suggests, group incentive programs try to encourage cooperation among workers by rewarding each worker according to how the group performs. Many such programs are initiated in work situations where it is difficult to calculate what each individual produces, apart from what the individuals ahead or behind produce. It is well known that the group working together can greatly advance productivity.

There are a variety of group incentive programs, but they can be divided into two major groups:

- Those group incentives that apply to small groups of workers within a single plant consider the work group as an individual. Each worker receives a rate of pay that depends on the productivity of the group as a whole. Under such a group incentive, all the group members are encouraged to work together to improve the group's productivity. Peer pressure can be great under such a system. As with individual piece rates, good time studies are essential to the smooth functioning of such group incentives. Although not widespread, small work group incentive schemes are the most likely to be found in worker-paced line flow jobs.

- Those group incentives that apply plant-wide or even companywide can be grouped under the heading "productivity gainsharing plans." These plans are devised in such a way that changes in the production process suggested by either workers or managers, as well as changes caused by increased worker effort, result in bonus payments for the workforce. Such plans are designed to increase the cooperation between all those in the plant, both direct and indirect labor, and to have everyone work smarter, not just harder.

The best established of these productivity gain-sharing plans is the so-called Scanlon plan, devised by Joseph Scanlon in the late 1930s to save the Empire Steel and Tinplate Company in Ohio. (Scanlon was the president of the steelworkers union local and had training in cost accounting.) In essence, the Scanlon plan calls for sharing, on a monthly basis, the productivity gains made that month as compared with an accepted historical standard. The "pot" of money to share is determined by multiplying a historical base ratio and the sales value of production (loosely defined as sales adjusted for any increases or decreases in inventory, with that inventory generally valued at the sales price) that month. The "base ratio," in the simplest instance, is defined as labor costs divided by the sales value of production. The labor costs are for both direct and indirect labor; they include wages, most fringe benefits, and overtime payments.

Generally excluded from labor costs are such things as social security contributions, state and federal unemployment insurance, and the pension plan. In many companies the base ratio is between 10 and 30 percent. Multiplying this base ratio by the month's sales value of production produces an "allowed labor cost" for the month. If actual monthly labor costs are less than that allowed by the formula, a bonus pool is created that is divided between the company and the employees. In a typical division of the bonus pool, half goes to all employees (including management) in proportion to their earnings for the month, a quarter to the company, and a quarter to a reserve pool for those times when there is no bonus pool because actual labor costs have exceeded those allowed.

More formally, the bonus pool in a Scanlon plan is computed as follows:

$$BP_t = ALC_t - AMLC_t$$

where BP_t is the bonus pool for the month in question, ALC_t is the allowed labor cost for the month, and $AMLC_t$ is the actual monthly labor cost. The allowed labor cost, in turn, is computed as

$$ALC_t = BR \times SV_t = (LC_b/SV_b) \times SV_t$$

Where BR is the base ratio (computed as base period or representative labor costs, LC_b, divided by base period or representative sales value of production, SV_b), and SV_t is the actual sales value of the production for 1 month.

A Scanlon plan encourages workers and managers alike to propose suggestions for the improvement of productivity. If the sales value of production increases for a given level of labor costs, the bonus pool increases, and both workers and managers benefit. The administration of a Scanlon plan thus requires a mechanism by which to evaluate and implement the suggestions offered. Some of these suggestions involve significant capital expenditures. A well-respected bipartisan group of managers and employees is needed to decide on the proposals. The implementation of a Scanlon plan also requires some sophisticated cost accounting so that the formula and any proposed variations can be straightforwardly calculated and communicated to all involved. Merely installing a Scanlon plan calls for a great deal of trust and communication; workers have to believe that they will not be manipulated by the numbers in the formula, and management must be prepared to cede some of its authority to the workforce and to communicate more thoroughly than in the past.

While there have been some notable successes with the Scanlon plan, there have been some failures as well and some criticism of its workings. Many of these criticisms relate to the base ratio. For a Scanlon plan to work well with a simple base ratio involving labor costs, the ratio of labor costs to the sales value of production must be fairly stable over time. To the extent that it is not stable, and to the extent that it is influenced by outside factors such as energy or materials price escalation or product mix changes, the size of any bonus pool will be seriously affected by forces outside the control of the workforce. When this happens, questions about the fairness of the plan arise, and often its success is threatened.

There are, of course, ways to improve the formula. For example, costs other than labor can be made part of the base ratio calculation, or individual product lines can be isolated with their own ratios to minimize the effect of product mix changes. Or, the calculation of the sales value of production can be adjusted so that it is a value-added measure, with outside purchases of material and supplies being netted out from the sales value of production. Such formula changes make the Scanlon plan more adaptable to inflation, product mix change, and other environmental considerations, although with the risk of complicating the administration of the plan and losing the understanding and thus the backing of the workforce. While some companies have been able to keep their Scanlon plans flexible and popular with the workforce, other companies have been forced to abandon them because of problems with formula calculation or plan administration.

A recently devised productivity gain-sharing plan that resembles the Scanlon plan is Improshare, meaning "improved productivity through sharing." Improshare can be simpler than the Scanlon plan. It relies less on employee involvement and administration, thus making it easier to install. Under Improshare, the bonus is calculated from the difference between the actual hours worked to produce the plant's output and some standard hours allowed, the so-called Improshare hours. The Improshare hours are calculated using "work-hour standards," based on plant history and defined as total production hours divided by the units produced. The value of the hours saved is thus greater when more units are produced with the same or fewer worker hours. As with Scanlon plans, the success of Improshare depends on the choice of a representative and equitable base period for the calculation of the work-hour standard and on a reasonable amount of trust and communication between management and the workforce.

Other Bonus and Profit-Sharing Plans

There are many other bonus and/or profit-sharing plans that various companies have instituted. Some involve only pension benefits for profit sharing while others pay cash bonuses that come from company profits.

Some of these plans encourage worker suggestions and initiative; others do not. Two bonus plans have received some comment in the press and are worth discussing briefly as interesting examples of what can be done. The Lincoln Electric Company in Cleveland uses a series of incentives and bonuses. Base rates are developed for each of the jobs, typically, the average for the industry around the Cleveland area. In addition, there are some extensive individual piece-work incentives, a large profit-sharing type of bonus, and peer review. The company also guarantees employment for a certain number of hours per week. In practice, employee bonuses are sometimes equal to annual wages. Employees own a sizable fraction of the company.

Another interesting example is that of Nucor, a steel joist and specialty steel product manufacturer that also operates many steel mini-mills (mini-mills use scrap as input to electric furnaces); these mini-mills make the angles, rounds, and other very specialized shapes that go into the products manufactured by the company. Nucor uses an ingenious series of bonuses to ensure communication and the transfer of technology across the company's geographically dispersed operating units. Every production worker is in a group and earns a given percentage bonus for every percentage increase over standard that his or her group makes. These bonuses are sometimes as large as 100 or 200 percent of the base wage. Department manager bonuses depend on whether plant goals (not department goals) are attained. Similarly, plant manager bonuses depend on whether company (not plant) goals are attained.

Job Design and Quality of Work Life[5]

Methods improvement, as discussed earlier in this chapter, could be characterized as "job dissection study," since it focuses on simplifying jobs and improving methods by breaking jobs apart and sometimes deskilling them, perhaps by adding more equipment. Many of the productivity advances since the beginning of the Industrial Revolution have involved this type of systematic analysis. But how do workers feel about doing jobs that have been simplified, deskilled? How do they feel about tending machines that do what they used to do?

Similarly, the section on wage payments earlier stressed the link between worker effort or insight and worker compensation. But do workers work only for money? Do they really consider other forms of compensation to be much less important?

Recent evidence has raised more and more questions about job dissection and money compensation. Many processes that have relied on these notions have inadvertently created situations in which workers feel that the job is not worth doing, certainly not worth doing well. In these instances, quality has declined—waste, absenteeism, and labor turnover have increased. This phenomenon has made some people wonder whether companies should shift from selecting people for particular jobs to fitting the jobs to the workers.

Many companies have tried to counter worker disaffection with innovations that make work more meaningful to people and thus more likely to receive their full attention and care. These innovations still have as a goal high company profits, but they are attentive to worker needs, especially as the character of the workforce has changed. A number of factors have emerged as important dimensions of satisfaction with one's job and workplace, and each demands some attention from managers concerned about workforce well-being and productivity. In no particular order:[6]

- *Adequate and fair compensation.*   Still an important consideration for all workers. Is the pay sufficient, and is it in line with expectations and demands of other, perhaps similar, jobs?

- *Safe and healthy working conditions.*   Is the plant structurally sound and is the machinery safe to operate? Is the workplace attractive and comfortable? Are the hours long? Is there flexibility in the time workers can start or stop work?

- *Immediate opportunity to use and develop human capacities.*   Does the job give the worker a sense of identity? Does it call for any substantial autonomy? Is the job self-contained or merely a small piece of a task with no real identity itself? Does the job call for planning as well as execution? Does it call for multiple skills? Is the job tied into the larger flow of information about the process and the control of that process?

- *Future opportunity for continued growth and security.*   What opportunities exist for advancement on the job? How secure is the employment? Will the job provide the training and development needed to tackle still larger jobs?

- *Social integration into the workplace.*   To what degree is the workplace open and supportive of each worker? To what degree is the workplace free of prejudice and status symbols? How upwardly mobile are workers?

- *Rights in the workplace.*   To what extent does the rule of law and due process characterize an individual's rights in the workplace? Are privacy, free speech, and equity fully respected?

- *Work in perspective.*   Are the demands of the job in balance with family and leisure dimensions of the worker's life?

- *Company vis-à-vis society.*   Is the company perceived as socially responsible and socially accepted?

With these kinds of considerations in mind to deal with worker dissatisfaction, a host of work restructure policies have been tested by companies like General Foods, Procter and Gamble, General Motors, TRW, Ford Motor, Goodyear, Cummins Engine, Eaton Corporation, and Mars, Inc., among others. The work restructuring that has been tested in these and other companies tends to very idiosyncratic; no two programs are alike. Goals are held to only loosely and timetables are extraordinarily flexible. The goals are dual; the intended benefits are to be reaped by both management and the workforce. The adopted programs work by altering the "culture" of the workplace.

As quality-of-work-life innovations have taken hold, there have been some clear distinctions that have emerged for workers between the old and new modes of management:[7] Jobs have become broader in scope for workers, workers have voiced more of their own concerns about the nature and direction of the workplace, labor relations have mellowed, employment security has become a higher priority, and the wage system has been altered to reward learning and/or group achievement.

There have been distinctions important for management as well: layers of management have been reduced in number, traditional work standard expectations have been replaced with more ambitious and flexible expectations, and overhead has been removed. Quality of work life (QWL) initiatives have shifted management away from exercising control over the workforce towards seeking worker commitment to the company, its goals, and its constituencies.

There are sometimes some casualties in this process, however—notably first-level supervisors. Supervisors are often caught in the middle, forced to jettison their former attitudes, strengths, and training. While some supervisors have adjusted very well to change, others have been left floundering. They no longer have the control they once enjoyed, and the skills demanded of them by QWL initiatives ("coaching," training, leadership in problem solving) are often foreign and of seemingly low esteem.

The movement to restructure work has taken several broad tacks and the variations on them have been almost as diverse as the companies involved. Broadly, however, one can divide work restructuring into job design on the one hand and QWL innovations on the other. The chief distinction here is that job design looks at altering the character of the job itself to make it inherently more interesting for the worker and thus more apt to draw his attention, care, and effort, so that both worker satisfaction and productivity are enhanced. QWL innovations, on the other hand, can be seen largely as a process that may or may not involve job design but that tries to engage workers actively in solving problems of concern in the workplace. Let us consider job design first and deal with QWL initiatives later.

Job design can be divided into two major areas: job enlargement and job enrichment. Job enlargement typically involves combining what may have been separate workplace activities, done by different people, into one enlarged job. In an enlarged job a worker may be required not only to run a machine but also to perform routine maintenance on it. To that may be added machine setup, quality checking, or a number of other tasks. While job enlargement risks performing the job more inefficiently than when individual tasks were assigned to different, specialized workers, proponents of job enlargement argue that its savings in improved quality, lower absenteeism, lower turnover, and the like more than make up for any efficiency losses implied by combining job elements.

Job enrichment takes a different approach to restructuring each individual job: it seeks to provide workers with more opportunities for planning and controlling their jobs and the surrounding environment. Job enrichment typically gives workers more autonomy and responsibility, not just more things to do. More will be said about job enrichment in the following section on employee involvement and quality of work life.

Employee Involvement and Quality of Work Life

For many, quality of work life has the broad meaning of increasing worker involvement in the workplace rather than a specific program such as job enlargement or job enrichment. Nevertheless, QWL innovations have often resulted in job enlargement and/or job enrichment programs. Thus it is often hard to disentangle the QWL process from its effects on job designs.

Although the distinction between QWL innovations and job redesign should not be pushed too strenuously, it does exist. For example, many of the real advances at numerous factories have come about not so much because of job enrichment or job enlargement changing the jobs themselves, but rather, the advances have come by involving the workforce more in the operation, with more communication and understanding on all sides. Cars are still assembled pretty much like cars elsewhere; the cycle time is still a minute, and a moving assembly line is still used. Nevertheless, at a number of plants, the quality of work life has been greatly enhanced by increasing the communication between management and the workforce and sincerely soliciting worker ideas to solve problems that are everybody's concern.

What does it take to have a QWL innovation succeed? Nobody knows for sure. These efforts are still in their youth and, in many cases, are wrought of fragile stuff. Nevertheless, some useful observations have appeared in the burgeoning literature on the subject. Successful QWL innovations, it seems, generally entail sincere commitment by both management and the workforce (and any unions involved), patience on the part of all parties, and a measure of risk taking and faith about what might be accomplished. Quality of work life should, in these instances, be regarded as an end in itself, even though other benefits, such as improved productivity, can also be realized. Successful QWL innovations have been very flexible and open to change as conditions warrant, because the means of worker involvement are just as important as the results of any problem solving. A union cannot be sidestepped; QWL innovations have to be viewed as supplements, not as replacements, to the collective bargaining process.

Most strong QWL innovations were hatched from small select groups of workers and supervisors that found themselves successfully solving some problems. They then spread, as a sort of "healthy infection," into other work groups in the plant. Typically, the QWL process cannot be forced on workers; they must embrace it in their own time. Management must make a lasting commitment to the QWL process, through rough economic times as well as smooth.

Quality of work life typically involves changing much or all of the culture of the workplace—worker attitudes toward the job, toward the product, toward fellow workers, toward management. As one might expect, new social systems sometimes step on toes. There has been some low-level management backlash to quality of work life as well as upper-level management indifference: some supervisors see the QWL process as a deskilling of their own roles that, at the same time, demands from them a lot of work and communication. Many supervisory personnel, especially in operations that have incorporated worker teams, have difficulty viewing quality of work life as a supplement to their jobs, not as a replacement for them. Supervisors and overhead function personnel (such as engineers) often have to recast the way they look at their jobs. No longer are they working for management; instead they are working for the worker teams. Rather than directing the work, they are brought in as consultants or teachers, and this change of emphasis and role is sometimes difficult for lower-level managers to accept.

Moreover, pressures on the new plant social system have to be overcome. Demands for high-volume production or for production innovations tend to lengthen hours, reduce time for team meetings, introduce rivalries, and slow personnel development. Decisions about pay and worker advancement sometimes have to be sidestepped or compromised because of the touchiness of peer evaluation or the vagaries of the business cycle and foreign competition.

On balance, however, work restructuring and the QWL process have proved to be a modest success, and much more experimentation and development of the concept can be expected. Work restructuring and quality of work life must be carefully evaluated since they have not been, nor are they likely to be, embraced wholeheartedly by all company workforces. There are still many instances where workforces resist change and do not want to assume any managerial roles. Often workers would rather have more time off than more responsibility. Any work restructuring must be carefully tuned to the situation at hand.

In this regard, it is interesting to follow one of the most recent management crazes, that of introducing quality circles (sometimes called quality control circles) into the operation. These quality circles are a much heralded aspect of Japanese management and are prevalent in many of the most successful Japanese companies.[8] Many U.S. companies have tried to emulate the Japanese. Quality circles share much with other QWL innovations in that they are problem-solving groups involving members of the same department, both workers and supervisors. (Other QWL innovations have joined workers and supervisors from different departments; quality circles deal strictly with one department's workers and supervi-

sion.) Each quality circle receives some training in problem identification, data gathering, and data analysis. After this training, the quality circle chooses a particular problem to work on. The problem may involve quality, but it may also involve production speeds, worker safety, or something else. Once the problem has been chosen, it is analyzed by gathering data to see what its causes are. Various solutions are then proposed and tested. Quality circle meetings generally take about an hour a week, typically on company time. The success of quality circles in the United States has been mixed. For some companies the time devoted to quality circle meetings is more than made up for by the savings that have resulted. In other cases, the costs of the quality circles have been greater than the savings, but worker morale is much higher, and for that reason the companies maintain them. In other situations, however, neither savings nor morale has been high enough to merit the continued use of quality circles. It is said that in Japan, quality circles are seen much more as one of the last resorts to improve quality and the flow of the process, not as one of the first things to do. It seems that quality circles, just as other QWL innovations, must be implemented with care and with real understanding of their potential problems as well as their potential advantages.

Teams and Employee Involvement[9]

In some companies, there has been a move toward autonomous, sometimes even self-managed, work groups or worker teams (of size 4 to 25, typically) that combine job enlargement and job enrichment. This is sometimes referred to as employee empowerment. The teams meet regularly and generally have charge of something that they can "own" (such as a particular subassembly or a well-defined portion of the process such as packaging). Individual team members may be trained to do more than they used to do;

the new, restructured jobs may have more elements in them or workers are rotated through different jobs more frequently. Often, such job enlargement is combined with job enrichment; a team may be charged with a number of tasks that one would not traditionally associate with blue-collar workers:

- Production scheduling
- Maintenance of all equipment within its area
- Setting up of its own equipment
- Tracking its own quality and working diligently on a quality improvement program established by the company
- Calling in engineering support for equipment modifications
- Working on inventory control with other teams elsewhere in the process
- Monitoring materials quality
- Developing budgets for their operations
- Goal setting for performance, quality, costs, delivery
- Determining which tasks the team will work on, as a group
- Meeting with customers of the team's efforts and with suppliers to the team, be they internal or external to the company
- Having the responsibility for, or at least an input to, the hiring of new members of the team, and in particularly mature teams, to have a say in team members' performance evaluations

The teams have spurred at least some companies to share information about plant performance, including profit and loss information.

From some research into the relative success of such endeavors, some managerial "tips" have sprung up:[10]

1. Have a shared view of what employee involvement is to accomplish and press issues that are consistent with that shared view.

Employee involvement programs can be associated with a wide variety of personal and organizational goals and objectives. It is best if an employee involvement program is explicitly tied to a specific purpose or orientation. This recommendation should be followed even if more than one purpose could conceivably be linked to a firm's employee involvement program. Five broad objectives or purposes of employee involvement programs are possible: (1) to focus employee efforts on continuous improvement in manufacturing processes; (2) to be used in the firm as a technique for decision making; (3) to improve communication between management and workers; (4) to improve the quality of working life for employees; and (5) to provide employees with a "voice" in the operations of the firm. That is, broadly conceived, employee involvement programs will be categorized as exhibiting either an employee (numbers 1, 2, or 3) or an organizational (numbers 4 or 5) focus. Each is legitimate and can be the foundation for an effective employee involvement program. What is necessary is that labor and management agree on the purpose of an employee involvement program. Ensuring a shared view among management and the labor force on the purpose of an employee involvement program is not a task taken lightly.

Employees report that they expect the areas of problems which they address to be tied to the explicit purpose of the employee involvement program. For example, employees are more likely to respond positively to recommendations made by managers to address problems concerning potential areas of process improvements (e.g., re-sequencing operations, reducing the length of production cycles, or quicker setups) when it is agreed that the purpose of the employee involvement program is continuous improvement rather than when the employee involvement program purpose is "employee voice." Divergent employee involvement program purposes and work issue areas are to be avoided.

2. Include staff as well as hourly employees in the program.

    Blue-collar employees report that one of the primary contributors to effectiveness in employee involvement program is to include staff employees in the employee involvement program. Employees report that staff contribute to their own efforts primarily by providing information and data to aid in their problem solving. In addition, hourly employees report that staff provide a much needed, broader perspective on manufacturing processes—"the big picture"—which is required in many efforts at process improvement.

3. Make participation mandatory.

    In nearly all firms, what is important and valued is required and what is peripheral and unimportant is voluntary. Many firms require all employees to participate to some degree without any negative reactions.

4. Meet for 1 hour each week and provide some training.

    At a minimum, hourly employees require two broad types of resources for effectiveness, time and training. Employees, it has been found, need to meet approximately one hour a week in order to make significant contributions to a firm's continuous improvement process.

    Employee involvement programs are frequently initiated without adequate employee orientation or training. Even long-time individual contributors often lack the skills necessary to identify and solve problems and work effectively in teams. Unfortunately, while firms do often provide some form of problem-solving training (e.g., fish bone techniques, SPC), programs to enhance the skills of employees in order to contribute to process improvements are rare. Thus, a well conceived training plan, and resources devoted to that end, are crucial elements of an effective employee involvement program.

5. Share information broadly, even information once thought to be too competitively sensitive.

    Employees are more effective when they operate with easy and open access to company data.

6. Financial bonuses are not needed. Rather, tie effort to the plant's existing performance evaluations. Indeed, make it a part of people's performance appraisals.

    There is always considerable interest whether it is necessary to tie financial rewards to participation of employees in an employee involvement program. However, surveys of manufacturers consistently report that only a small percentage of firms (typically 10–16 percent) link financial rewards to participation in an employee involvement program. For those firms which do provide financial rewards for participation, the most common method of payment is to compensate participants for any extra hours spent on employee involvement program on an overtime basis. For many employees, tying effort to the firm's performance evaluation system is potentially a greater and more consistent motivator than a bonus.

    What is important should be measured, evaluated, and rewarded. If in the future employee involvement programs will occupy an increasingly more central position in a firm's operational strategies, employee involvement program activities must be evaluated within the framework of a performance appraisal system.

7. Set specific goals for the employee involvement program teams.

    Goal setting is a managerial technique which has a long track record of successfully motivating employees to perform at a higher level. Goal setting can be easily adapted to the operations of an employee involvement program. Goals can focus on general group outcomes (e.g., reduce waste by 5 percent, improve yield ratios 2 percent) or on elements of group process (e.g., identify and implement at least three process improvements in the coming year) or for specific tasks (e.g., data gathering and report to be completed by a certain date).

Discipline

One of the most difficult and vexing of any supervisor's tasks is that of disciplining a worker, for whatever cause. While some employee actions are automatic causes for dismissal (theft of company property, deliberate destruction of equipment, or blatant insubordination), there are seemingly countless infractions that merit less severe action. But exactly what action should be taken is often not clear.

To help matters out, companies typically have a well-defined series of steps to follow in worker infraction cases. These steps may range from an initial reprimand to formal warning letters and on to a forced suspension and finally dismissal. Such formality serves to guard worker rights and forces the company to document, as well as it can, the instances of worker infractions of company rules and regulations.

As an example of the gray areas of management in these matters, consider Situation 5-4 and how you would handle the matter.

Some Comments on Labor Relations[11]

The trade union movement in the United States is about a hundred years old, spawned as a reaction against some sorrowful management exploitation of labor that must stand as one of the lowest points in business history. Seen against this backdrop, the accomplishments of the union movement are dramatic and heartwarming, won as they were against fierce opposition, although with too much violence from both sides.

Since World War II, however, union membership rolls have generally declined (17 million in 1989) and the percentage of the employment of nonagricultural establishments that is unionized has steadily dropped. In 1989, only 16.4 percent of the employees of nonagricultural establishments were union members, down from 33.2 percent in 1958 and from a high of about 35 percent in 1945.[12] The unionization of production workers in manufacturing is considerably higher than this, as might be expected. Still, the extent of unionism in manufacturing has declined somewhat over the last generation. Nevertheless, unions remain an important influence on managers. Wage rates in many companies are set to be comparable with union-bargained rates in other companies in the same industry or in the same geographic area. The same can be said of benefits packages. Business decisions (plant locations, for example) are often made with unions, or their avoidance, in mind.

The apoplexy with which unionism is greeted by many companies, however, is caused less by the additional wages or benefits that companies must pay because of unions. Most managers would agree that keeping unions out of a company is as expensive or even more so than having them in, at least in terms of wages and benefits. The major management complaint rests more with the inflexibility that can creep into a plant as a result of unionism. This complaint is mainly against so-called "work rules," which are largely rules and procedures governing the manning of a plant. Frequently, the work rule agreed to years ago and insignificant then comes back to haunt a manager who for a reason like technological change wants to alter the way things are done. It often takes great management foresight to avoid

SITUATION 5-4

Central Louisiana Oil

Charlie Clotfelter had been on his job as a production manager at Central Louisiana's holdings in the Lucile field for about six months. Before that time he had held a staff position in the company's headquarters. The Lucile field had been producing for decades. Central Louisiana Oil operated numerous leases in this oil field, which was owned by scores of local established families. Being an old field, it suffered from a number of disadvantages relative to new oil fields, where the most advanced recovery techniques could be used. For example, the natural gas dome that rides on the top of an oil field had been blown early in the career of the Lucile field, thus reducing the pressure that could be used to help bring the oil to the surface. Vacuum pumping had also been used at this field, which meant that pumps in the field were attracting oil to each well. This action essentially forced all of the operators in the field to use vacuum pumping to attract their fair share of oil. Central Louisiana Oil was the major operator in the field, but there were two dozen smaller operators pumping from hundreds of wells.

One result of the pumping was the unavoidable collection of a light condensate, especially in winter. This condensate was highly volatile and was typically run off in small-diameter buried pipes to an open pit near each pumping station, where it evaporated. This condensate is compatible with refined gasoline and, when mixed with leaded gasoline in a ratio of 50:50, makes a fuel that can be burned in any car. Understandably, it is termed "drip" gasoline. For years, workers in the Lucile field had tapped into the buried pipes and diverted the condensate from the pits into 55-gallon drums buried in the dense foliage that covered nearly all of the Lucile field. Each drum usually collected between 25 and 30 gallons of drip gasoline a week. In the evening, workers came along the dirt roads that crisscrossed the field to uncover their 55-gallon drums, pump out the contents, and transport it back to their homes or barns. Naturally, this drip gasoline was dangerous to handle, and the area suffered more than its share of barn fires.

The workers felt they were simply being frugal, because the drip gasoline would be vented by an operator like Central Louisiana Oil. Of course, the pipes used to divert the drip gasoline and the drums used to store it were generally taken from the operators. In the eyes of the workers, however, this was merely petty theft of company scrap pipe and drums.

Although the gathering of drip gasoline was a long-standing practice for workers of many producers, it was not known to Charlie Clotfelter when he first arrived. Nor was it known to Clotfelter's predecessor. In fact, Charlie would not have known of it at all if a disgruntled construction contractor had not tipped him off. Upon hearing of the possibility of this drip gasoline operation on Central Louisiana Oil's property, he alerted the corporation's security people to come out and check. After several weeks of sleuthing, the security people reported back to him that nearly all of Central Louisiana's 35 employees in the Lucile field had, at one time or another, been engaged in the sale of drip gasoline. Two of the three supervisors were ignorant of the practice, but one, who had risen to the supervisor level from being a field worker, knew of the operation because, as a worker, he had engaged in the practice. To make matters worse, the security people uncovered the fact that the largest dealer in drip gasoline was the president of the local union representing Central Louisiana workers, who served as the chief bargaining agent for the union in contract negotiations with the company.

This was the dilemma that Charlie Clotfelter faced. The drip gasoline operation was clearly illegal, although the workers apparently regarded it as a fringe benefit that did not harm the company since the drip gasoline would only be evaporated. What should he do about the situation? Should he fire all of his workers? Should he fire just a few? Should he make an example of the union local president, who was the biggest dealer? Should he fire the supervisor who failed to inform him of this illegal operation? If he didn't fire the workers, should he engage in some other form of discipline? What actions should he take?

future pitfalls with work rules; with a union, bad management decisions, many think, are more difficult to remedy. Moreover, the importance of work rules has gradually increased as more and more issues are bargained over by labor and management at contract time.

Despite these arguments against unionism, there are still some strong arguments in its fa-vor. Table 5-4 summarizes the major arguments for and against unionism.

Tools of the Trade: A Recap of Lessons Learned

Despite the push for ever newer and more so-phisticated technology, operations still depend heavily on people, and will so continue. Thus,

TABLE 5-4   Two views of trade unionism

	ANTI-UNION ARGUMENTS	PRO-UNION ARGUMENTS
Union Effects on Economic Efficiency	Unions raise wages above competitive levels; hence there is too little labor relative to capital in unionized firms.	Unions reduce quit rates, induce management to alter methods of production and adopt more efficient policies, and improve morale and cooperation among workers.
	Union work rules decrease productivity.	Unions collect information about the preferences of all workers; hence the firm can choose a "better" mix of employee compensation and a "better" set of personnel policies.
	Unions lower society's output through frequent strikes.	Unions improve the communication between workers and management, leading to better decision making.
Union Effects on Distribution of Income	Unions increase income inequality by raising the wages of highly skilled workers.	Unions' standard-rate policies reduce inequality among organized workers in a given company or a given industry.
	Unions create horizontal inequalities by creating differentials among comparable workers.	Union rules limit the scope for arbitrary actions concerning the promotion, layoff, recall, and the like of individuals.
		Unionism fundamentally alters the distribution of power between marginal (typically junior) and inframarginal (generally senior) employees, causing union firms to select different compensation packages and personnel practices than nonunion firms.
Social Nature of Union Organization	Unions discriminate in rationing positons.	Unions are political institutions that represent the will of their members.
	Unions (individually or collectively) fight for their own interest in the political arena.	Unions are political institutions that represent the will of their members.
	Union monopoly power breeds corrupt, and non-democratic elements.	Unions represent the political interests of lower-income and disadvantaged persons.

Source: Richard B. Freeman and James L. Medoff, "The Two Faces of Unionism," *The Public Interest*, no. 57 (Fall 1979): 75.

understanding the issues that are important for a workforce is basic to operations management.

At most factories, the first item of discussion at a staff meeting is worker safety. Great strides have been made here and with ergonomics, the study of those methods and worker aids that can help to avoid injury and fatigue on the job. These issues can only be expected to intensify with the advent of better technology.

Methods improvements are important for continued productivity increase, perhaps more so for service operations than even for manufacturing. A lot of methods improvement depends on common sense, but a variety of tools have been developed to help organize thinking about methods. The process (and information) flow chart, used in a detailed way, is an invaluable tool. Variations that track workers versus the machines, or the other workers, that they work with are also useful. Videotaping and work sampling are among other tools that have found favor with industrial engineers.

The use of time in making methods improvements is another important branch of study. Time study (using clipboards), predetermined motion-time systems (like MTM and its offshoots), and standard data systems all have a place in aiding productivity. They can also be used for planning purposes, line balance, and wage payments. This leads into the use of payment schemes of different types (individual incentives, group incentives, plant-wide incentives) to spur worker effort. A variety of such plans have been tried. Some have found happy matches to particular factory or service situations, but there is no tried-and-true formula for a contented, high effort workforce. Many operations use hourly or salaried pay as effectively as any incentive scheme.

What is clear is that job design (job enrichment, job enlargement), employee involvement, team building and similar concepts do seem to make a difference for many operations. These in-novations, especially when coupled with training in problem solving, such as was studied in Chapter 4's discussion of quality improvement programs, can offer huge rewards both to the workers involved (more job satisfaction, more "belonging" to the group or company) and to the company (better ideas and increased productivity). What guides the motivation and contentment of workers is varied. Although companies have not licked this problem yet, the decline in unionization does seem to indicate that workers no longer feel so obligated to join unions to press their concerns on management as they once did. Over the years, of course, unions have played an important role in representing the worker cause with management; that unionism is on the decline does not negate the significant contributions that unions have made to worker welfare and management attitudes about the workforce.

World-Class Practice.   Several things can be said about the workplace policies of world-class competitors, among them:

1. Employee involvement is considered critical to the company's ability to compete. Employees are trained in problem-solving and teamwork. Cross-training is valued. Job enlargement and job enrichment are pursued when they make sense. Teams are encouraged and supported.

2. Workers, indeed everyone, know how to use simple tools such as process flow charting to discuss and examine the process.

3. Management shows concern for worker safety and comfort (including ergonomics) as their tie to productivity is accepted.

4. Worker ideas are sought, appreciated, and implemented quickly. The predisposition is to act swiftly on worker advice because it is recognized that benefits will, in all likelihood, far outdistance costs. Some companies even give workers a limited budget to spend on their work areas by themselves.

5. Traditional industrial engineering is still valued, although often it is now applied more to setup reduction as opposed to direct labor operations.

6. Moves have been made to simplify the factory's organization. This means fewer levels of management between the factory floor worker and the plant manager (a frequent target is 3 levels) and fewer labor grades than has been common practice, at least in the United States.

QUESTIONS

1. Having read the chapter, how would you answer this question: "What makes a workforce supervisor truly effective?"

2. What examples of ergonomics have you observed? What benefits do you see from such ergonomics designs or practices? What costs?

3. In what ways is motion study a substitute for time study? a complement to time study?

4. Using some easily available products, consider how a company might undertake some value engineering of them. Be as specific as you can.

5. Discuss the means by which individual incentive schemes, such as piece rates, become eroded over time. Given such erosion, why are they as prevalent as they are?

6. In what situations do you think a Scanlon plan would be inappropriate or likely to fail?

7. Tower Manufacturing Company and Baldridge Chain Saw have problems. In what ways are their problems similar? In what ways are they different?

8. React to the statement "Quality of work life is a giveaway to the workers."

PROBLEMS

1. Detail how the role of the supervisor varies from one type of process to another.

2. Using a chart such as found in Figure 5-1, chart the following tasks:

(a) Cooking a mushroom omelette.

(b) Changing a flat tire.

3. A worker is charged with operating two machines simultaneously. The jobs assigned by the foreman call for the worker to engage in some loading and unloading (termed external time) of each machine and some observation and machine operation while the machines are running (termed internal time). Some machine running time requires no worker intervention. Using a device such as found in Figure 5-2, determine how many of each part the worker can make at 100 percent of standard during an 8-hour workday.

STANDARD TIMES PER PIECE	MACHINE 1 (minutes)	MACHINE 2 (minutes)
External time	0.55	0.68
Internal time	0.40	0.52
Total machine running time (including internal time)	1.90	1.65

4. The worker in Problem 3 can earn incentive pay by performing the external time for each job more quickly (neither the internal times nor the running times can be altered without re-engineering both jobs). How much better than standard (as a percentage) for each machine must the worker perform in order to earn 115 percent of the standard pay? to earn 130 percent of the standard pay?

5. The engineering department has just devised some jigs that improve the external time standards for the worker in Problem 3 from 0.55 minute to 0.40 minute for machine 1 and from 0.68 minute to 0.48 minute for machine 2. The foreman is considering adding to the worker's tasks a third machine nearby. For the job contemplated, that machine's external time is 0.35 minute, its internal time is 0.27 minute, and total machine running time is 0.95 minute. Given these additional responsibilities, how many of each part can the worker now make at 100 percent of standard during an 8-hour workday?

6. Suppose, as in Problem 4, that the worker wanted to earn 115 or 130 percent of the standard pay. At what percentages of the standard external times would he have to perform to satisfy his desires?

7. How are therbligs and the MOST system alike? How are they different?

8. An apparel firm's industrial engineer has observed a worker in the pressing and inspection department of a men's coat shop. Six ob-servations have been made, with average per-formance ratings on each element for the six observations. The times and ratings for each element and observation are shown in Table 5-5. To ascertain the extent of allowances for the job, a work sampling study was done. The job was divided into normal time working, un-avoidable delay (machine downtime, talks with foreman, out of materials), and allowance time (breaks and rests). Fifty random observations were made with the following results.

	OBSERVATIONS
Normal time working	35
Unavoidable delay	6
Allowance time	9
	50

Given this information, what should be the standard time for the job?

9. Another industrial engineer observed the same worker as found in Problem 8. His per-formance ratings for the six elements of the job, as found in Table 5-5, were somewhat dif-ferent. Respectively, they were 110, 95, 90, 90, 90, and 105 percent. Given this information, what should the standard time for the job be?

TABLE 5-5   Times and ratings (problem 8)

ELEMENT	OBSERVATION TIMES (seconds)						AVERAGE PERFORMANCE RATING (percent)
	1	2	3	4	5	6	
Examine front	7.6	7.3	8.2	7.8	8.0	8.1	115
Press front	15.5	15.9	17.0	16.5	16.2	17.5	100
Examine lapel	6.0	5.9	7.0	5.9	6.4	7.5	95
Press lapel	20.5	17.8	19.2	18.6	20.2	18.0	105
Inspect lining	8.7	7.7	7.4	8.2	6.8	9.6	90
Hang up coat and pick up new one	4.3	4.0	4.6	5.6	5.6	4.8	110

10. The task of drawing a shake from the shake machine at a fast-food restaurant for delivery to the drive-thru window was broken down as follows:

(a) Walk four steps to the machine.

(b) Pull a cup from a dispenser.

(c) Position the cup underneath the shake machine spout.

(d) Pull the lever on the shake machine to start the flow of shake into the cup. Hold the lever in the on position for the 5 seconds it takes for the shake to flow into the cup.

(e) Reach for the top to place on the cup.

(f) Put the top on the cup.

(g) Return four steps to the drive-thru window area.

Using the MOST data of Table 5-3, determine how much time this job ought to take.

11. Using the MOST system as spelled out in Figure 5-3 and some reasonable assumptions of your own, estimate a time standard for:

(a) Making a Burger King hamburger.

(b) Boxing that hamburger.

(c) Dumping fries from vat to bin and salting them.

12. A supervisor needed to calculate the efficiency and effectiveness for his operation in order to fill out a weekly production report for the plant manager. The 12 workers under his authority had worked a total of 40 hours each during the week and had produced 400 earned hours of work (i.e., standard hours). Utilization had been estimated at 80 percent.

What were the efficiency and effectiveness measures should he report?

13. Develop a piece-rate system of your own using, say, 10 different piece rates. Alter that system so that all workers receive a 10 percent wage increase. Compare what you have to do to effect such a change with a similar set of 10 jobs on a standard-hour incentive system.

14. Calculate the size of the Scanlon plan bonus percentage under the following conditions:

(a) The sales value of all production is $2 million. The allowed payroll costs are 20 percent and actual payroll runs $330,000. The company share is 25 percent; of the employee share, a reserve of 25 percent is kept. What is the average worker's bonus percentage?

(b) How does the calculation change when different allowed payroll cost percentages are introduced? Suppose that product A now accounts for 60 percent of sales (up from 50 percent) and product B accounts for 40 percent. A's allowed costs are now 25 percent, and B's are 15 percent. Actual payroll costs of $330,000 are split 50–50 between the products. What is the bonus per worker?

(c) Why might such a split between product lines be advantageous?

15. In your opinion, under what conditions does the institution of quality circles make sense, and under what conditions does their institution not make sense?

16. As Charlie Clotfelter of Central Louisiana Oil, what would you do?

SITUATION FOR STUDY 5-1

COUNTRY GELATIN COMPANY

Thomas Brewer, plant manager for Country Gelatin Company (see Tour E), noticed that every time he went into the break room, at least one of the process control technicians was drinking a cup of coffee, reading a book, or generally being nonproductive. He concluded that the technicians did not have enough work to fill their 8-hour shifts, and he asked Connie Darvin, the personnel manager, to look into the situation. Connie proceeded as follows:

1. A preliminary work sampling study was done to find out what the technicians were doing.
2. Some changes were implemented based upon the preliminary study.
3. A second work sampling study was done to determine the effects of the study.

The preliminary work sampling study was done over a 10-week period; of the 1200 working hours for the three types of technicians, about 15 percent (or 180 hours) were observed. The results of the study are shown in Table 5-6 for the three technician types. The figures represent the average percentage of an 8-hour shift that was spent performing the duties listed. The technicians spent approximately 20 to 25 percent of the time preparing for testing (such as cleaning glassware) and doing paperwork and calculations; 25 to 34 percent of the time, the technicians were idle.

Several points need to be raised about Country at the time of the study: (1) the plant was running at about three-fourths of capacity, (2) the job procedures were not clearly defined, (3) technicians were performing menial tasks and, therefore, had a low concept of job worth, and (4) technicians had serious communication problems with production foremen and others who should have been using their information.

As a result of the preliminary study, the powder technician's duties were consolidated into the duties of the line technician and solutions technician; the powder technician's job was eliminated.

After these changes, Connie observed the work of the line technicians on four different dates. The percentage of 8-hour shifts was recorded, and these results are shown in Table 5-7. Connie noticed that some changes had occurred other than just the change in work duties. For example, before the changes, the technicians would complete whatever work they were doing before taking breaks (albeit long ones); however, after the changes, the technicians took their breaks by the clock regardless of whether the particular test they were working on was completed. Also, several technicians were observed to be making the data "fit" rather than completing long tests.

TABLE 5-6 Results of preliminary work sampling (Country Gelatin Company)

DUTIES	PERCENTAGE OF SHIFT		
	SOLUTIONS TECHNICIANS	POWDER TECHNICIANS	LINE TECHNICIANS
Prepare	10	6	0
Check floor samples	18	22	26
Run tests (in lab)	17	18	15
Do paperwork and calculations	11	17	29
Consult with supervisors	2	4	2
Run extra tests and retests	8	7	3
Idle time	34	26	25

TABLE 5-7   Line technicians' duties after changes (Country Gelatin Company)

DUTIES	PERCENTAGE OF SHIFT			
	4/10	4/11	4/16	4/17
Retrain*	7	4	0	0
Take measurements and readings[†]	19	27	21	31
Obtain samples[†]	9	12	15	13
Run tests (in lab)	6	2	10	4
Do paperwork and calculations	29	38	31	36
Consult with supervisors	4	1	4	1
Run extra tests and retests	11	6	6	5
Idle time	15	10	13	10

*The line technicians had to be retrained when they took over some of the duties of the powder technicians. Preparation became a duty of the solutions technicians only.

[†]These two categories are very much the same as the category "check floor samples" in Table 5-6.

These results led Connie to question (1) whether the work sampling studies were done properly and (2) whether the changes in job were proper. What suggestions do you have for Connie?

AUTOMOTIVE TRIM PLANT

The Trim Plant of Mayflower Motors, one of Detroit's Big Three automakers, manufactured seat covers. These seat covers joined various upholstery fabrics (e.g., cloth, vinyl, velour) to polyurethane padding. Completed seat covers were shipped to Mayflower's assembly plants, where they were attached to seat frames of various types for installation in Mayflower Motors' cars.

The Trim Plant was located along an interstate highway in a major city. It was over 40 years old and employed 700 workers, mostly in sewing operations. The plant was organized by the United Auto Workers (UAW).

The plant's reputation for quality was excellent. However, given the sorry state of the automobile industry during the prevailing recession and Mayflower Motors' own perilous standing, the plant had to be cost effective as well as quality conscious in order to survive. A recent study by the corporation compared the cost of a model year's output as manufactured by the Trim Plant ($51.5 million in cost to Mayflower Motors) with that of outside vendors who used non-UAW labor ($30.8 million for the same level of output). This was a devastating comparison; other Mayflower Motors plants had been closed because of similarly poor economic comparisons. Of course, if the plant were to close, the corporation would still incur some ongoing costs (e.g., pension), estimated at $9.3 million. And if the corporation went to outside vendors, it might meet with some delivery and quality problems, the value of which was estimated at roughly $5 million per year. Still, there was an unfavorable difference of $6.4 million per year that argued powerfully for closing the plant.

The United Auto Workers' local recognized the implications of such a comparison. Moreover, the union admitted to itself that the plant was too "fat" with workers, because many sewing machine operators were finishing their daily quotas of work with an hour or more to spare. In order to keep jobs for its members, the union was willing to discuss changes in manning, work rules, quotas, and so on, if the company was willing to consider keeping the plant open. About the only thing the UAW would stand firm on was wages; no cut in wages would be considered, because that would set a bad precedent at other company plants.

This was the situation facing John Gwynne, the plant manager. The UAW was apparently willing to concede some jobs, work rules, breaks, and so on, and abide by tougher standards (i.e., quotas) in order to keep the plant running. The time was fast approaching (another two months) when the corporation had to know whether the plant could be "rehabilitated" or whether it should place orders with some outside vendors. Gwynne needed to know how much cost could realistically be saved and by which means.

1. What should he do to find out the true weaknesses in the plant?
2. What new standards, manning levels, and the like should be established?
3. What principles should be adopted for any future managing of the plant?
4. What "give" should management evidence, and how much of any job reduction should management share with the hourly workforce?

REFERENCE NOTES

1. Additional information on methods improvement can be found in the following:
    Copp, Ed, "Methods Improvement Kit Uses IE Techniques to Simplify Work," *Industrial Engineering* 20, no. 8 (August 1988): 47–50.
    Feather, John J., and Kelvin F. Cross, "Workflow Analysis, Just-In-Time Techniques Simplify Administrative Process in Paperwork Operation," *Industrial Engineering* 20, no. 1 (January 1988): 32–40.
    Halpern, David, Stephen Osofsky, and Myron Peskin, "Taylorism Revised and Revisited for the 1990s," *Industrial Management*, (January–February 1989): 20–24.
    Kurzynski, Marcia J., "Work-Flow Analysis," *Journal of Systems Management* 38, no. 1 (January 1987): 14–20.
    Mundel, Marvin E., *Motion and Time Study*, 6th ed. Englewood Cliffs, N.J.: Prentice-Hall, 1985.
    Niebel, Benjamin W., *Motion and Time Study*, 8th ed. Homewood, IL: Irwin, 1988.
    Noro, Kageyu, and Andrew Imada, ed., *Participatory Ergonomics*, New York:Taylor and Francis, 1991.
    Oborne, David J., *Ergonomics at Work*, 2nd ed. New York: Wiley, 1987.
    Ovitt, Rod, "Occupational Ergonomics on the Shop Floor," *Manufacturing Engineering* (101, no. 2 (August 1988): 97–100.
    Reilly, Bernard J., and Joseph A. DiAngelo, Jr., "From 'Hard work' to 'Smart Work': A Look at Job Redesign," *Personnel* 65, no. 2 (February 1988): 61–65.
    Salvendy, Gavriel, ed., *Handbook of Human Factors*. New York: Wiley, 1987.
    Vasilash, Gary S., "Designing Better Places to Work," *Production* 102, no. 2 (February 1990): 56–59.

2. For additional information on time study, see the following:
    Clark, Daniel O., "Step-by-Step Approach to Standard Data Development," *MTM Journal of Methods-Time Measurement* 13 (1987): 59–67.
    Karger, Delmar W., and Franklin H. Bayha, *Engineered Work Measurement*, 4th ed. New York: Industrial Press, 1987.
    Martin, John C., *Labor Productivity Control: New Approaches for Industrial Engineers and Managers*, New York: Praeger, 1991.
    May, Joseph E., and Robert Jackson, "Industrial Engineering Philosophies and Practices Can Increase Shop Floor

Productivity . . . And Our Own," *Industrial Engineering* 21, no. 6 (June 1989): 39–43.

Mayo, William, and Russell Horn, "Considerations in Developing Standards for Long Cycle Assembly," *Industrial Engineering* 22, no. 3 (March 1990): 38–42.

Mundel, Marvin E., *Motion and Time Study*, 6th ed. Englewood Cliffs, N.J.: Prentice-Hall, 1985.

Niebel, Benjamin W., *Motion and Time Study*, 8th ed. Homewood, IL: Irwin, 1988.

Polk, Edward J., et al., *Methods Analysis and Work Measurement*, New York: McGraw-Hill, 1984.

Shell, Richard L., ed., *Work Measurement: Principles and Practice*, Atlanta, GA: Industrial Engineering and Management Press, 1986.

Shinnick, Michael D., and Walter W. Erwin, "Work Measurement System Creates Shared Responsibility Among Workers at Ford," *Industrial Engineering* 21, no. 8 (August 1989): 28–30.

Zandin, Kjell B., *MOST Work Measurement Systems*, 2nd ed. New York: Marcel Dekker, 1990.

3. For additional information on incentives and wage schemes, see:

Dearden, John, "How to Make Incentive Plans Work," *Harvard Business Review* 50, no. 4 (July–August 1972): 117–24.

Dunn, J. D., and F. M. Rachel, Wage and Salary Administration. New York: McGraw-Hill, 1971.

Geare, A. J., "Productivity from Scanlon-Type Plans," *Academy of Management Review* 1, no. 3 (July 1976): 99–108.

Globerson, S., and R. Parsons, "Multi-factor Incentive Systems: Current Practices," *Operations Management Review* 3, no. 2 (Winter 1985).

Kanin-Lovers, Jill, "Motivating the New Work Force," *Journal of Compensation & Benefits* 6, no. 2 (September–October 1990): 50–52.

Kanungo, Rabindra N., and Manuel Mendoca, "Evaluating Employee Compensation," *California Management Review* 31, no. 1 (Fall 1988): 23–39.

Moore, Brian E., and Timothy L. Ross, *The Scanlon Way to Improved Productivity: A Practical Guide.* New York: Wiley, 1978.

National Commission on Productivity and Work Quality, *A Plantwide Productivity Plan in Action: Three Years of Experience with the Scanlon Plan*, May 1975.

4. See Robert S. Rice, "Survey of Work Measurement and Wage Incentives," *Industrial Engineering* (July 1977): 18–31.

5. The literature on job design, job enrichment, and quality of work life is large and growing. For some additional information, see:

Alster, Norm, "What Flexible Workers Can Do," *Fortune*, (February 13, 1989): 62.

Benson, Tracy E., "Empowerment: There's That Word Again," *Industry Week* 240, no. 9 (May 6, 1991): 44–52.

Conger, Jay A., and Rabindra N. Kanungo, "The Empowerment Process: Integrating Theory and Practice," *Academy of Management Review* 13, no. 3 (July 1988): 471–482.

Cooper, Michael R., et al., "Changing Employee Values: Deepening Discontent?" *Harvard Business Review* 57, no. 1 (January–February 1979): 117–25.

Cunningham, J. Barton, and Ted Eberle, "A Guide to Job Enrichment and Redesign," *Personnel* 67, no. 2 (February 1990): 56–61.

Davis, Louis E., and Albert B. Cherns, *Quality of Working Life: Problems, Prospects, and State of the Art.* Glencoe, Ill.: Free Press, 1975.

Dickson, Paul, *The Future of the Workplace.* New York: Weybright and Talley, 1975.

Drucker, Peter F., *Management: Tasks, Responsibilities, Practices.* New York: Harper & Row, 1973.

Drucker, Peter F., *The Practice of Management.* New York: Harper & Row, 1954.

Fein, Mitchell, "Job Enrichment: A Reevaluation," *Sloan Management Review* 15, no. 2 (Winter 1974): 69–88.

Foy, Nancy, and Herman Gadon, "Worker Participation: Contrasts in Three Countries," *Harvard Business Review* 54, no. 3 (May–June 1976): 71–83.

Goodman, P. S. "Realities of Improving the Quality of Work Life – Quality of Work Life Projects in the 1980s," *Labor Law Journal* (August 1980): 487–94.

Guest, Robert H., *Innovative Work Practices*, Work in America Institute Studies in Productivity: No. 21. Elmsford, NY: Pergamon Press, 1982.

Guest, Robert H., "Quality of Work Life – Learning from Tarrytown," *Harvard Business Review* 57, no. 4 (July–August 1979): 76–87.

Gyllenhammar, Pehr G., "How Volvo Adapts Work to People," *Harvard Business Review* 53, no. 5 (September–October 1977): 102.

Hackman, J. Richard, "Is Job Enrichment Just a Fad?" *Harvard Business Review* 53, no. 5 (September–October 1975): 129–38.

Hackman, J. Richard, and Greg R. Oldham, "Development of the Job Diagnostic Survey," *Journal of Applied Psychology* 60 (1975): 159–70.

Hackman, J. Richard, and Greg R. Oldham, *Work Redesign.* Reading, Mass.: Addison-Wesley, 1980.

Garwood, R. Dave, "Empowerment: No Longer a Luxury," *Production & Inventory Management Review* and *APICS News* 11, no. 4 (April 1991): 22, 37.

Grant, Rebecca A., Christopher A. Higgins, and Richard M. Irving, "Computerized Performance Monitors: Are They Costing You Customers?" *Sloan Management Review* 29, no. 3 (Spring 1988): 39–45.

Herrick, Neal Q., *Joint Management and Employee Participation: Labor and Management at the Crossroads,* San Francisco, CA: Jossey-Bass, 1990.

Huber, Vandra L., and Nancy Lea Hyer, "The Human Factor in Cellular Manufacturing," *Journal of Operations Management* 5, no. 2 (February 1985): 213–228.

Hulin, C. C., and M. R. Blood, "Job Enlargement, Individual Differences, and Worker Responses," *Psychological Bulletin* 69 (1968): 41–55.

Hurrell, Joseph J., Jr., and Michael J. Colligan, "Machine Pacing and Shiftwork: Evidence for Job Stress," *Journal of Organizational Behavior Management* 8, no. 2 (Fall–Winter 1986): 159–175.

Japan Human Relations Association, *The Idea Book: Improvement Through TEI (Total Employee Involvement),* Cambridge, MA: Productivity Press, 1988.

Jones, Karen, ed., *The Best of TEI: Current Perspectives on Total Employee Involvement.* Cambridge, MA: Productivity Press Inc., 1989.

Kanter, R., "Dilemmas of Managing Participation," *Organizational Dynamics,* (Summer 1982): 5–27.

Kizilos, Peter, "Crazy About Empowerment?" *Training* 27, no. 12 (December 1990): 47–56.

Klein, Janice A., "The Human Costs of Manufacturing Reform," *Harvard Business Review* 67, no. 2 (March–April 1989): 60–66.

———, "A Reexamination of Autonomy in Light of New Manufacturing Practices," *Human Relations* 44, no. 1 (January 1991): 21–38.

———, "Why Supervisors Resist Employee Involvement," *Harvard Business Review* 62, no. 5 (September–October 1984): 87–95.

Lawler, Edward E., III, and J. A. Drexler, Jr., "Dynamics of Establishing Cooperative Quality-of-Worklife Projects," *Monthly Labor Review* (March 1978): 23–28.

Lawler, Edward E., *High-Involvement Management: Participative Strategies for Improving Organizational Performance,* San Francisco, CA: Jossey-Bass, 1986.

Letize, Leta, and Michael Donovan, "The Supervisor's Changing Role in High Involvement Organizations," *Journal for Quality and Participation* (March 1990): 62–65.

Lincoln, James R., "Employee Work Attitudes and Management Practice in the U.S. and Japan: Evidence from a Large Comparative Survey," *California Management Review* 32, no. 1 (Fall 1989): 89–106.

Moore, Brian E., and Timothy L. Ross, *The Scanlon Way to Improved Productivity: A Practical Guide.* New York: Wiley, 1978.

Nadler, D. A., and E. E. Lawler, "Quality of Work Life: Perspectives and Directions," *Organizational Dynamics* (Winter 1983): 20–30.

Plunkett, Lorne C., and Robert Fournier, *Participative Management: Implementing Empowerment,* New York: Wiley, 1991.

Poole, Michael, *Towards a New Industrial Democracy: Worker's Participation in Industry.* New York: Routledge and Kegan Paul, 1986.

Richardson, Peter R., "Courting Greater Employee Involvement Through Participative Management," *Sloan Management Review* 26, no. 2 (Winter 1985): 33–44.

Robison, D., "General Motors Business Teams Advance QWL at Fisher Body Plant," *World of Work Report* (July 1977): 73, 80–81.

Rosenberg, R. D., and E. Rosenstein, "Participation and Productivity: An Empirical Study," *Industrial and Labor Relations Review* (April 1980): 355–67.

Rosow, Jerome, ed., *The Worker and the Job.* Englewood Cliffs, N.J.: Prentice-Hall, 1974.

Rubinstein, S. P., "Participation Problem Solving: How to Increase Organizational Effectiveness,"*Personnel* (January–February 1977): 30–39.

Schroeder, Dean M., and Alan G. Robinson, "America's Most Successful Export to Japan: Continuous Improvement Programs," *Sloan Management Review* 32, no. 3 (Spring 1991): 67–81.

Scobel, Donald N., "Doing Away with the Factory Blues," *Harvard Business Review* 53, no. 6 (November–December 1975): 132.

Sneen, Art, "Employee Empowerment Gives Focus to Focused Factories," *Manufacturing Systems* 9, no. 2 (February 1991): 54, 56.

Staw, Barry, M., "Organizational Psychology and the Pursuit of the Happy/Productive Worker," *California Management Review* 28, no. 4 (Summer 1986): 40–53.

Stein, B. A., and R. M. Kanter, "Building the Parallel Organization: Creating Mechanisms for Permanent Quality of Work Life," *Journal of Applied Behavioral Science* (July–August–September 1980): 371–88.

Taylor, Frederick Winslow, *Scientific Management.* New York: Harper, 1911.

Trist, E. L. et al., *Organization Choice.* London: Tavistock Institute, 1963.

Upjohn Institute for Employment Research, *Work in America.* Cambridge, Mass.: MIT Press, 1973.

Vogt, Judith F., and Kenneth L. Murrell, *Empowerment in Organizations: How to Spark Exceptional Performance,* San Diego, CA: University Associates, 1990.

Walton, Richard E., "How to Counter Alienation in the Plant," *Harvard Business Review* 50, no. 6 (November–December 1972): 70–81.

Walton, Richard E., "Improving the Quality of Work Life," *Harvard Business Review* 52, no. 3 (May–June 1974): 12 ff.

Walton, Richard E., "Successful Strategies for Diffusing Work Innovations," *Journal of Contemporary Business* (Spring 1977): 1–22.

Walton, Richard E., "Work Innovations in the United States," *Harvard Business Review* 57, no. 4 (July–August 1979): 88–98.

Walton, Richard E., and Gerald I. Susman, "People Policies for the New Machines," *Harvard Business Review* 65, no. 2 (March–April 1987): 98–106.

Yasuda, Yuzo, *40 Years, 20 Million Ideas: The Toyota Suggestion System,* Cambridge, MA: Productivity Press, 1991.

6. The following list is drawn from Richard E. Walton, "QWL Indicators—Prospects and Problems," *Studies in Personnel Psychology* 6, no. 1 (Spring 1974): 7–18.

7. For a discussion of these points, see Richard E. Walton, "From Control to Commitment: Transforming Work Force Management in the United States," Division of Research, Harvard Business School, March 1984.

8. For more on quality circles consult:

Aubrey, Charles A., and Patricia K. Felkins, *Teamwork: Involving People in Quality and Productivity Improvement,* Milwaukee, Wisc.: Quality Press, ASQC, 1988.

Berger, Roger W., and David L. Shores, eds., *Quality Circles—Selected Readings,* New York: Marcel Dekker, 1986.

Blair, John D., and Kenneth D. Ramsing, "Quality Circles and Production/Operations Management: Concerns and Caveats," *Journal of Operations Management* 4, no. 1 (November 1983): 1–10.

Crocker, Olga, Johnny Sik Leung Chiu, and Cyril Charney, *Quality Circles: A Guide to Participation and Productivity.* New York: Facts on File Publications, 1984.

Griffin, Ricky W., "Consequences of Quality Circles in an Industrial Setting: A Longitudinal Assessment," *Academy of Management Journal* 31, (June 1988): 338–58.

Hutchins, David., *Quality Circles Handbook.* New York: Nichols, 1985.

Lawler III, Edward E., and Susan A. Mohrman, "Quality Circles after the Fad," *Harvard Business Review* 63, no. 1 (January–February 1985): 65–71.

Robson, Mike, *Quality Circles: A Practical Guide,* 2nd edition, Brookfield, VT: Gower, 1988.

Shea, Gregory P., "Quality Circles: The Danger of Bottled Change," *Sloan Management Review* 27, no. 3 (Spring 1986): 33–46.

9. For more on teams, see:

Adair-Heeley, Charlene B., and R. Dave Garwood, "Helping Teams Be the Best They Can Be: The Message in the Milk Bottle," *Production & Inventory Management Review & APICS News* 9, no. 7 (July 1989): 22–25.

Blake, Robert R., Jane Srygley Mouton, and Robert L. Allen, *Spectacular Teamwork: How to Develop the Leadership Skills for Team Success.* New York: Wiley, 1987.

Buchholz, Steve, and Thomas Roth, authors. Karen Hess, ed. *Creating the High-Performance Team.* New York: Wiley, 1987.

Dyer, William G., *Team Building: Issues and Alternatives,* 2nd ed. Reading, MA: Addison-Wesley, 1987.

George, Paul S., "Team Building Without Tears," *Personnel Journal* 66, no. 11 (November 1987): 122–129.

Hardaker, Maurice, and Bryan K. Ward, "How to Make a Team Work," *Harvard Business Review* 65, no. 6 (November–December 1987): 112–120.

Huszczo, Gregory E., "Training for Team Building," *Training & Development Journal* 44, no. 2 (February 1990): 37–43.

Parker, Glenn M., *Team Players and Teamwork: The New Competitive Business Strategy.* San Francisco: Jossey-Bass Publishers, 1990.

Parker, Mike, and Jane Slaughter, "Management by Stress," *Technology Review* (October 1988): 37–44.

Pfeiffer, William, ed., *The Encyclopedia of Team-Building Activities*, San Diego, CA: University Associates, 1991.

Phillips, Steven L., and Robin L. Elledge, *The Team Building Source Book*. San Diego, CA: University Associates, Inc. 1989.

Verespej, Michael A., "When You Put the Team in Charge," *Industry Week* 239, no. 23 (December 3, 1990): 30–32.

10. The points made here are drawn from the work of my colleagues Richard Magjuka and Timothy Baldwin.

11. For more information on unions and company management, see the following:

Brett, J., "Why Do Employees Join Unions?" *Organizational Dynamics* (Spring 1980): 47–59.

Brown, Clair, and Michael Reich, "When Does Union-Management Cooperation Work? A Look at NUMMI and GM-Van Nuys," *California Management Review* 31, no. 4 (Summer 1989): 26–44.

Chaison, Gary N., and Mark S. Plovnick, "Is There a New Collective Bargaining?" *California Management Review* 28, no. 4 (Summer 1986): 54–61.

Freedman, A., and W. E. Fulmer, "Last Rites for Pattern Bargaining," *Harvard Business Review* 60, no. 2 (March–April 1982): 30–37.

Fulmer, W. E., "Step by Step Through a Union Campaign," *Harvard Business Review* 59, no. 4 (July–August 1981): 94–102.

Halley, William H., Jr., and Kenneth M. Jennings, *The Labor Relations Process*, 2nd ed. Hinsdale, IL.: Dryden Press, 1984.

Heckscher, Charles, *The New Unionism: Employee Involvement in the Changing Corporation*, New York: Basic Books, 1988.

Kochan, T. A., "How American Workers View Labor Unions," *Monthly Labor Review* (April 1979): 23–31.

Peterson, Richard B., and Lane Tracy, "Lessons from Labor-Management Cooperation," *California Management Review* 31, no. 1 (Fall 1988): 40–53.

Roth, Bill, "Quality of Worklife and Quality Improvement: A New Role for Unions," *Journal of Quality and Participation* (September 1990): 46–51.

Sandver, Marcus H., *Labor Relations: Process and Outcomes*, Boston: Little, Brown, 1987.

Sloane, Arthur A., and Fred Whitney, *Labor Relations*, 5th ed. Englewood Cliffs, N.J.: Prentice-Hall, 1985.

12. U.S. Department of Labor, Bureau of Labor Statistics, *Handbook of Labor Statistics, Statistical Abstract of the United States*, and the *Directory of U.S. Labor Associations.*

CHAPTER 6

PLANNING PRODUCTION

The goals of production planning are straightforward: figuring out which products to produce, when to produce them, and how much of each. Nevertheless, accomplishing this task effectively (at low cost and with all obligations met on time) can be devilishly difficult. There are wide variations in the means by which such decisions are made.

Factors in the Design of Production Plans

An effective production plan meets its demand obligations on time and at low cost. Being on time with production requires management to know what the process capacity is and not to promise deliveries it does not have the capacity for. Keeping costs low requires a match of resources (workforce, equipment, materials) to obligations that neither unduly stretches those resources nor leaves too much excess capacity. Short-term capacity can be augmented in several ways—adding extra workers, scheduling overtime or extra shifts, and placing auxiliary equipment into service. Sometimes, companies can subcontract with others for extra pro-

duction. All of these options, however, raise costs. Extra workers are frequently more costly to secure and less productive than regular workers. Overtime and extra shifts add costs to direct labor bills, maintenance, and overhead. Auxiliary equipment is typically less efficient and more expensive to operate and maintain. The production planner is justifiably loath to force short-term capacity increases by these means. However, the production planner also inflicts extra costs on the company if too much excess capacity is held. These costs are opportunity costs, the revenues forgone by failing to produce output that could have been sold. Striking a balance between resources and demand obligations is what production planning is all about.

Let us explore in greater detail how various kinds of processes can meet their delivery obligations at low cost.

Meeting Delivery

In all processes, there is a sincere desire to produce to exact customer specifications and to ship precisely when the customer indicates. This goal is not fully achievable in all processes because of some inherent constraints. In those that do achieve this goal, various policies are adopted:

- *Modulating capacity.* The timely delivery of a customer's precise order can be ensured by continually varying labor and materials resources and "rebalancing" the process, if need be. Scheduling overtime is the classic version of this approach. Modulating capacity is a common production planning policy in service industries because many services cannot be inventoried; thus, demands must be met primarily by adjusting the workforce and materials available on any given shift.

- *Inventory buffers.* Some processes fill their orders out of finished goods inventory. Individual products may not be generally customized, but the demands that the process can handle can be vastly different.

- *Managing demand.* In trying to ensure the timely delivery of often customized products, some companies adjust their bids for work. This is indirect and imperfect, but it can help to control the load on a process and thus its ability to deliver on time. In effect, this technique manages demand. In times of high demand on capacity, bids for business could be on the high side to help choke off the business a process would have trouble delivering on time; similarly, in times of excess capacity, lower bids can hope to draw in more business.

Sometimes processes cannot adopt these policies successfully. In such cases, the operation may have to defer production for a particular order until room appears in the schedule. The time deferred may not be particularly long, but these delivery dislocations are an inevitable consequence of the rigidity within some processes. For example, if the demand for, say, paper were particularly strong, not all deliveries would likely be made. Carrying spare pulp and paper capacity is an increasingly expensive undertaking, and so the capacity utilization of pulp and paper mills is generally high to begin with. A peak demand situation is likely, then, to imply a rationing of paper to customers.

The hybrid process (such as is described in the tour of the Stroh's Winston-Salem brewery) offers an example of another means by which deliveries can be met effectively. This type of a process is intriguing precisely because it mixes production planning systems. The first portion of many hybrid processes is triggered by either a sales forecast or the level of work-in-process inventory held; the latter portion of the process, on the other hand, is generally triggered by actual orders. What permits the company the luxury of planning production and managing materials in two different ways is the existence of the work-in-process inventory that separates the first portion of the process from the second. This inventory "decouples" one from the other. Without an inventory that functions in this manner, the latter portion of the process would probably have to be geared to the sales forecast, not to actual orders, and the company would probably have to store more in finished goods inventory, at higher cost, than it would ordinarily. Alternatively, incentives would have to be introduced to lock customers into longer lead time orders than is now the case, so that both portions of the process could both be done to order.

The key advantage of such a decoupling inventory is that it breaks up the information and materials management needs of the entire process into pieces that can be managed more easily on their own. There is a cost, of course, because the inventory must be financed, but the cost can often be substantially less than the introduction of process-wide computer systems or the financing of increased finished goods inventories.

Keeping Costs Low
Identifying the Scarce Resource

In almost all cases, the mix of resources capable of meeting the prevailing demand obligations is not perfectly balanced between the workforce and the available equipment. Either the workforce will be the scarce resource and idle equipment will exist, or equipment will be the scarce resource and idle workers will exist. What the scarce resource is has a critical impact on how production is planned. Is the production plan designed to keep the equipment loaded (long production runs), or is it designed to keep the workforce busy (perhaps through many setups and shorter production runs)?

Because more workers can generally be added to any process, low cost production plan design is essentially concerned with the economics of the process. Is equipment relatively cheaper to hold idle than labor, or is the reverse true? The question is easy to state, but in practice it may be very difficult to answer. On the one hand, it is easy to see why a paper company might devote its talent to a careful scheduling of all the paper machines that it owns, an unnecessarily idle paper machine costs the company a bundle in foregone revenues. On the other hand, it stands to reason that a job shop would house a number of machine tools that would lie idle a good deal of the time. Some machines have specialized uses and remain idle for that reason, while others can pay for themselves by being used strictly for sporadic in-tandem jobs with another machine or for addressing chronic bottlenecks. At many companies, especially job shops, holding equipment idle is cheaper than holding labor idle; production planning could call for a schedule of short runs if they would lead to on-time delivery and fewer bottlenecks.

These two examples are polar cases; in between is a sea of gray. For processes that are mostly batch and line flow, it is not always clear whether it is cheaper to hold idle some equipment or some labor, and thus whether the long production run is to be valued above a series of smaller runs. The decision depends on factors such as the costs of equipment, labor, inventory carrying, and inventory stockouts, and on production speeds, prices, and quality levels. We shall return to this point later.

Identifying the Valued Aspect of the Product or Service

If the full range of a company's product mix cannot be produced on time, which of the product mix ought to be produced? The production plan should be sensitive to those products or services that earn more contribution per unit of time or per unit of some other scarce resource. Production plans—particularly at the end of a month, quarter, or year—are often weighted heavily toward high contribution items or toward those that can easily be completed and billed. In many companies, the great end-of-month scramble to push product out the door is a scramble to push out the "winners," leaving poorer selling or poorer earning products until the next month. This kind of scheduling, of course, is more prevalent in job shop and batch flow processes; the more inflexible processes follow more rigid production plans.

An important management task is identifying how planning and scheduling can help gain or sustain high levels of contribution. The airline industry is a good case in point. Prior to the deregulation of the airlines in the late 1970s and the subsequent boom in air travel, airline load factors were not as high. The schedule of flights was used as a weapon to attract travelers; the prevailing wisdom was to schedule flights to be as convenient as possible, even if it meant cutting back on the time airplanes were scheduled to be in the air. Given the prevailing modes of competition, the valued aspect of the service was convenient time slots in the major city markets each airline served. The airlines that could provide the most convenient departure and arrival times were seen as the most successful.

With deregulation and the significant price declines that ensued, the mode of competition shifted from convenience to price. Load factors shot up, and the valued aspect of the service became airplane seats at low prices, not convenient times. There were travelers who demanded plane travel at every hour of the day, not merely at convenient hours. Airline planning and scheduling adjusted. Schedules were revised to lengthen the time airplanes were in the air, even if it meant some modestly inconvenient departure or arrival times for passengers. Many airlines adopted "hub and spoke" systems, where big city hubs (Dallas–Ft. Worth for American Airlines, Atlanta for Delta Airlines) were established and fed by numerous flights from other cities in a kind of pulsating patterns of arrivals and departures. With the scheduling adjustments, airline profits (contribution per unit of scarce resource) shot up dramatically, other things equal.

DEVELOPING A GOOD MATCH BETWEEN RESOURCES AND DEMANDS[1]

The development of a lost-cost production plan depends not only on the identification of the most valued resource and how that scarce resource should be scheduled, but also on the specific techniques that can be brought to bear on the match between resources and demands. The remainder of this part of the chapter is devoted to a discussion on those techniques that have proven helpful in matching up resources with demands. A discussion of how demands are forecast is reserved for Appendix A. Suffice it here to assume that the forecast is a given (say, from the marketing or sales department) and that the operations managers must live within its confines.

Capacity Planning And the Development of the Master Production Schedule

An effective production schedule is the heart—or at least the pacemaker—of many operations. Constructing one is an important but involved task, and the discussion of scheduling in this section is necessarily long and involved as well. With this in mind, it is appropriate to outline here the organization for this discussion of the schedule.

First we outline the evolutionary stages of production scheduling over time. This gives some perspective for the scheduling tools that are introduced throughout the remainder of the discussion. The first of the tools deals with assessing how much capacity exists for the process at any period of time (see "Capacity Planning"). Knowing capacity is a prerequisite for accepting orders and/or for scheduling production.

The discussion continues with coverage of the ways the schedule (either the master schedule, or the more general, aggregate plan) can be filled. This discussion begins in broad terms—see "Master Scheduling" and "Aggregate Planning"—but gets more precise and mathematical with the material under "Scheduling Techniques." The scheduling techniques introduced involve both flow processes, where capacities are known with some certainty, and job shops, which are more ambiguous about capacity.

Stages in the Development of a Production Schedule

The final production schedule for a plant is developed bit by bit, in stages, as more definite information on demands, materials, and capacities becomes available. Figure 6-1 depicts

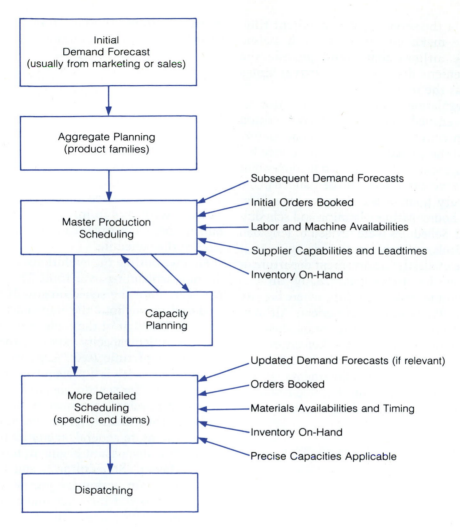

FIGURE 6-1   Stages in the development of a production schedule.

these stages. The first cut planning of production is generally concerned only with broad aggregates—entire product lines, gross materials needs, rough estimates of labor and machine availabilities—for time periods that can stretch 12 to 18 months into the future, long enough that all materials needs can be planned for but not so long that production cannot be regarded as significantly constrained by the present facility and any changes already in the works for it. Because it deals with broad aggregates of resources, this stage of production scheduling is frequently termed *aggregate planning*.

The aggregate plan serves as the first alert for marshaling resources—getting initial commitments from suppliers, checking whether more (or less) labor is needed, ensuring that inventories on hand are adequate to meet demands, determining whether machine capacity is adequate. Given this alert, the managers responsible can go about securing the resources needed to meet the plan. Decisions can be made as well about how production is going to vary to cover the expected pattern of demand. Should production remain "level," building up inventories to cover anticipated peaks of demand? Or, should production try to "chase" demand, with little or no accumulation of inventory? Or, should some mixed, in-between strategy prevail? It is in assessing the aggregate plan that decisions of this type are generally first made.

Since the aggregate plan is associated with marshaling resources, it inevitably becomes entwined with the budgeting process for the plant and the company division it reports to and, for that reason, often follows the same 12-month timetable. It also becomes a prime battleground between marketing—with its forecast of sales and any plans for product changes or new products—and manufacturing. If the planning is done well, general management will rule explicitly on the conflicts exposed by this process of schedule development. General management should resolve any differences between the desires of marketing/sales and those of manufacturing, so that both functions can be committed to meeting the schedule, as planned.

As demand firms up, the broad outlines of the production schedule, as determined by the aggregate plan, can be filled in with greater detail. Broad product line demand forecasts are replaced with more detailed forecasts of specific products and eventually, with specific product variations. Material supply capabilities and lead times are known with more certainty. More specifics are known about actual labor and machine availabilities. The aggregate plan-

ning stage gives way to successive stages of more detailed scheduling, generally termed *master production scheduling* (MPS).

The production schedule starts to reflect orders actually booked as well as those expected. The schedule begins to become firm, but it generally cannot be completely set until days or weeks before actual production. Thus, some day-to-day or week-to-week leeway is permitted in the schedule; it is the job of the production scheduler to use this leeway in devising a detailed production schedule (sometimes termed a final assembly schedule) that places as balanced a demand on the process as possible. This means that the scheduler must be thoroughly aware of the capacities of each segment of the process and where any bottlenecks are likely to show up. This kind of knowledge often leads a scheduler to concoct rules of thumb that help develop a schedule that is feasible, balanced, and reasonably low cost.[2]

The MPS must also reflect material availability. In the interest of clarity and order on the factory floor itself, it is often unwise to release an order whose component parts have not yet arrived or to release an order just to "get it moving" through the process. In both of these situations, work-in-process inventories build up; excessive work-in-process is too often disruptive and time-consuming to manage. Often companies are better off if they risk some idling workers temporarily because of no work than they would be purposely littering the floor with work-in-process. The hidden costs of work-in-process induced inefficiencies are frequently greater than the highly visible costs of idle labor. Yet, of course, many plants feel compelled to get something going in order to accommodate sales pressures, even if it means provoking inefficiencies in the process.

As the foregoing discussion has made clear, the scheduling process requires increasing refinement about capacities, demand, and material availability. The rest of this chapter examines capacities, demands, and their match. Material availability and its management are the central topics of Chapters 7 and 8; the remainder of this chapter will largely ignore problems with material availability. Instead, what follows implicitly assumes that there will be material enough when needed to avoid rescheduling. This assumption will be relaxed in Chapters 7 and 8.

Capacity Planning

The goal of capacity planning is to estimate how much process capacity will be taken up by any proposed level and mix of product demands. Effective capacity planning can indicate whether a proposed production schedule is feasible; thus, it is an important tool for the scheduling process. There are several useful capacity planning techniques. These techniques vary in their sophistication, specificity, and data requirements.[3]

Capacity Planning Using Overall Factors

The roughest cut means of assessing capacity, so called "capacity planning using overall factors," uses three readily available bits of data: (1) the output demanded for each product (or aggregation of products, such as an entire product line), (2) the labor and/or machine standards for that product (such as the hours required per unit), and (3) the historical data on average usage by each workcenter.

The first step in the calculation is to determine the expected total hours required to produce what is demanded. If the demands are as follows:

PRODUCT	TIME PERIOD		
	1	2	3
X	10	15	5
Y	20	20	25

and the labor standards for the two products are

	LABOR STANDARD (AT 100%)
Product X	0.60 hour per unit
Product Y	0.40 hour per unit

then the expected total hours required for period 1's demands are:

$$\text{Product X total hours} =$$
$$10 \text{ units} \times 0.60 \text{ hour/unit} = 16 \text{ hours}$$
$$\text{Product Y total hours} =$$
$$20 \text{ units} \times 0.40 \text{ hour/unit} = 8 \text{ hours}$$
$$\text{Total} = 14 \text{ hours}$$

The second step in the calculation uses the historical average utilizations by each workcenter (or department) to apportion the total capacity required to the various workcenters. Because these average utilizations are used to apportion the work effort across workcenters, the utilizations need to add up to 100 percent. For example, if the workcenter utilizations, determined from accounting records, were as follows:

WORKCENTER	UTILIZATION (percent)
10	50
11	30
12	20
	100

then the expected loads on each workcenter implied by the product mix demanded in each time period are:

WORKCENTER	TIME PERIOD			
	1	2	3	TOTALS
10	7.0	8.5	6.5	22.0
11	4.2	5.1	3.9	13.2
12	2.8	3.4	2.6	8.8
Total	14.0	17.0	13.0	44.0

These expected loads can be compared with actual capacity at the workcenters to suggest possible managerial adjustments (such as rescheduling some orders or working overtime). For simple, stable operations, this rough-cut technique is frequently good enough.

Capacity Bills

This technique gets more specific about the demands of each product on each of the workcenters; thus, shifts in the product mix will result, by this method, in shifts in demands on particular workcenters. Such a result is not possible in the overall factors technique because of its use of historical averages. This capacity bills technique, of course, requires more data—in particular, labor and/or machine hour standards for each product and its components, and knowledge of how the product is routed through the shop (i.e., which workcenters come into play in producing it). By using this information, a "bill of capacity" can be constructed. Such a capacity bill is an estimate of the labor and/or machine time required by each workcenter to make a unit of the product(s) in question.

Suppose that the following represents the capacity bill for products X and Y, from the previous example:

WORKCENTER	PRODUCT X (hours/unit)	PRODUCT Y (hours/unit)
10	0.25	0.10
11	0.15	0.15
12	0.20	0.15
	0.60	0.40

Note that the labor standard totals for each product (0.60 hour/unit and 0.40 hour/unit, respectively) are the same as those given in the overall factors example given previously.

The next step in the technique involves the application of each capacity bill to its respective product demand (its proposed master schedule) and summing the time estimates across products. Using the same demands as before results in expected workcenter loads as follows:

| WORK-CENTER | TIME PERIOD | | | | UTILIZATION |
	1	2	3	TOTALS	(percent)
10	4.5	5.75	3.75	14	32
11	4.5	5.25	4.5	14.25	32
12	5.0	6.0	4.75	15.75	36
	14.0	17.00	13.00	44.00	100

where workcenter 10's estimate of 4.5 hours for time period 1 is the sum of 2.5 for product X (0.25×10) and 2.0 for product Y (0.10×20).

The other capacity estimates are calculated in similar fashion.

Note that the total labor hours required in each time period match those calculated by the overall factors technique, but the expected demands on each workcenter are vastly different from those calculated via historical usage. The usage rates of the workcenters implied by the product mix proposed differ markedly from the historical averages.

Resource Profiles

In the *resource profile* technique, production lead times for each product and its components are melded with bills of capacity to compute a time pattern projection of the capacity demands for individual workcenters. For complex products whose time in the production cycle is many weeks and whose components must be worked on well in advance of final assembly, making such a time pattern adjustment is an important correction.

Capacity Requirements Planning

Capacity requirements planning (*CRP*) is the most sophisticated technique available to estimate period-by-period demand for each workcenter. This technique takes into consideration not only the time pattern of demand on the workcenter (as well as the product structure, routings, and labor and/or machine standards required to develop capacity bills) but also any lot sizing decisions for components, inventory already on-hand or expected to be received, work-in-process status (remaining work on any shop orders already opened), service parts demand, and any recent production planner changes. Thus capacity requirements planning provides a very detailed estimate of the workloads implied by the master production schedule and by the current status of individual workcenters. For complex and unstable operations, this technique can be well worth its significant computational expense and data needs. Capacity requirements planning is often an available module in computerized *materials requirements planning* (*MRP*) systems. More will be mentioned about CRP when MRP systems are described in Chapter 7.

Whichever technique is used—overall factors, capacity bills, resource profiles, or capacity requirements planning (or some other rough-cut technique)—the goal of capacity planning should be clear. That goal is namely to estimate, for any given period, how much capacity will be used up by any proposed production schedule. Without capacity planning of some sort, production planners would not know if a schedule stood a good chance of being infeasible. They would not be able to head off, in effective fashion, the resulting poor delivery performance, stranded work-in-process inventories, frustrated vendors, confused supervisors, and other ramifications of escalating chaos.

Without effective capacity planning, production planners would also not know "how much money they had left on the table" during any planning period by failing to schedule what the factory could have produced and for which they had sufficient demand in hand to justify.

Master Scheduling

Master scheduling is a process—an ongoing activity—that melds forecasts, booked orders, and knowledge of the effective capacity of the factory. It is a process that periodically and systematically "publishes" a document—the master production schedule—that indicates how many of each product the factory will produce for a specified time period or succession of time periods. The master production schedule is stated in physical units for every end item for which there is a bill of materials, or more often, for a readily recognizable aggregation of end items that could be termed a model, style, series, product line, or something similar.

If an aggregation of end items constitutes the MPS, then a more precise schedule must be developed for the specific end item variations that will actually be produced. Often, this more precise schedule is termed the *final assembly schedule* (*FAS*). The final assembly schedule paces the flow of products and materials through the process. It acts as both a planning tool and as a control document against which actual performance can be measured.

In the case where the MPS is not the most specific schedule, the MPS acts primarily as a planning tool. However, because it relates to an aggregation of specific products, it does not have a bill of materials, in conventional terms, that makes sense. A bill of materials is a precise list of all product components in exactly the quantities used. Usually, there is a different bill of materials for every model or type of the product made. In order for it to coordinate specific materials and parts effectively, then, the end item aggregation in the MPS must have its own "phantom" bill of materials that indicates, on average, how many of particular parts and how much of particular materials will be consumed in the construction of the "average" product. These bills of materials are called *planning bills* (or sometimes *super bills*). Unlike conventional bills of material, they can indicate fractional levels of usage and can change if the mix of specific end items shifts the composition of the "average" product.

The MPS is a key bridge between manufacturing and marketing/sales because it is with the MPS that manufacturing commits to making what marketing/sales has either forecast or actually booked as orders. Typically, manufacturing's commitment takes the form of a due date for the completion and shipping of the product in some agreed-upon quantity. In order for this order promising to proceed smoothly over time, manufacturing has to keep track of the production quantities that are available or could be expected to be available, and also how much of any of those quantities could actually be committed for shipment.

There are two principal techniques for such tracking: graphical and tabular. Let us explore them in turn.

A Graphical Master Production Schedule

The graphical technique typically tracks cumulative sales and production of a particular product or product line over time. It can easily show the buildup or depletion of inventory and the extent of any buffer or safety stock. Figure 6-2 in an example. Three lines have relevance in the figure: the cumulative sales forecast and two distinct cumulative MPS's that are possible, one a "level" production strategy and the other a "chase" strategy. The chase strategy matches the sales forecast week for week so that the buffer stock remains at a constant 40 units. The level strategy, on the other hand, produces a constant quantity of output in every week $(160/12 = 13.33$ units$)$—just enough so that at the end of the planning horizon, 12

weeks, the planned buffer stock of 40 units is maintained. However, during the course of the 12 weeks, sometimes the buffer stock is temporarily eroded (area A in the figure) and sometimes it is padded with excess inventory (area B in the figure).

Figure 6-3 presents a different MPS—one that incorporates batch runs of the product in lot sizes of 40. In contrast to Figure 6-2, where there was production in every period, here production is scheduled to occur only in weeks 1, 5, 8, 10, and 12. The shaded areas in Figure 6-3 represent the so-called "cycle" inventories, those created to tide the plant over until the next batch of the product is run.

Over time, one can expect that sales forecasts and orders actually booked will change. Often such changes will necessitate changes to the MPS. Thus as the schedule "rolls through time," there will be continued adjustments made to the position and shape of the MPS graph.

One can also use this graphical technique to track actual sales (or shipments) and the orders booked. Figure 6-4 depicts the same sales forecast and MPS as in Figure 6-3, but with the order backlogs for each of the first two weeks also shown. The order backlog for week 2 reflects the additional orders booked during week 1. As the figure indicates, actual sales in week 1 exceeded the forecast. As of the start of week 2, booked orders have not fully "consumed" the forecast quantity, but are very close to it. At that start of week 1, there were no booked orders past week 3, and there are no booked orders past week 4 at the beginning of week 2. The difference between the MPS and the order backlogs gives an indication of what is "available to promise" customers. For example, in week 4 there are 30 units available to be promised $(60 - 30)$, and in week 5 there are 70 units available to promise $(100 - 30)$.

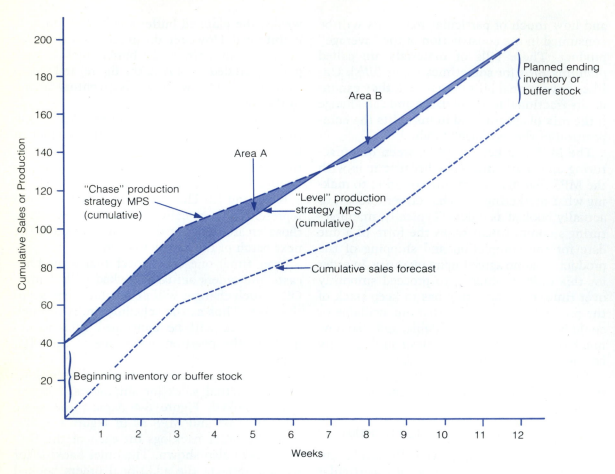

FIGURE 6-2   A graphical master production schedule, showing level and chase production strategies.

A Tabular Master Production Schedule

Another technique used to depict the MPS, and one that can easily be kept by computer, employs a tabular, time-phased record, such as that shown in Figure 6-5. The information in the figure mirrors exactly the data in the graphical presentation of Figure 6-3. While we lose the power and simplicity of the graphical approach by using a table, we do gain real precision in what is available (i.e., what the expected ending inventory is) and what is available to promise the customer. This tabular time-phased record is also easy to update from one time period to another.

Some definitions are helpful in understanding the figure:

Forecast.   The quantity forecasted to be sold during each period.

Orders.   The orders actually received for delivery in the period indicated.

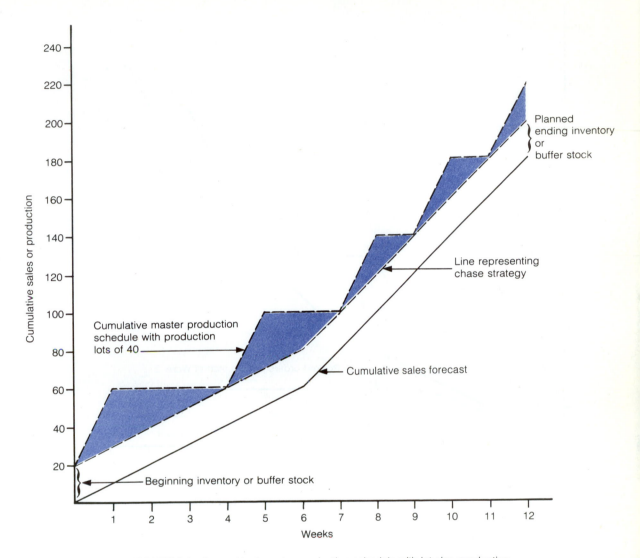

FIGURE 6-3   A graphical master production schedule with lot size production.

Available. The expected ending inventory each period computed as on-hand at the start of the period + MPS for that period − forecast that period, or, if the orders exceed the forecast, the quantity available is on-hand + MPS − orders. The quantity "available" is an impor-tant one for the production planner. When it is negative in value, it is an indication to the planner of action that needs to be taken to re-turn the value to nonnegative numbers. Nega-tive availabilities in future periods trigger pro-duction orders in preceding periods.

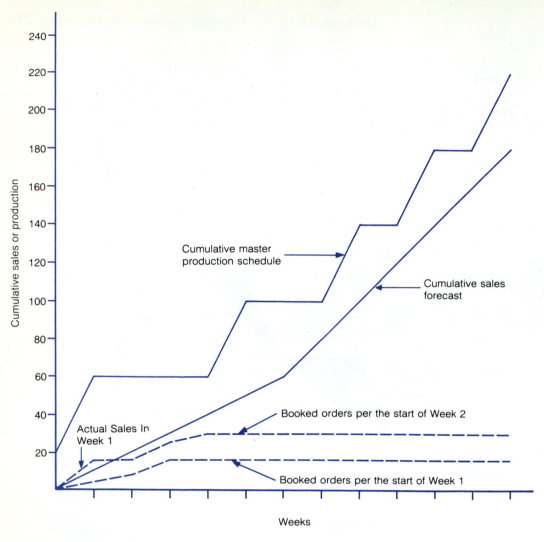

FIGURE 6-4   A master production schedule with booked orders shown.

		Weeks										
	1	2	3	4	5	6	7	8	9	10	11	12
Forecast	10	10	10	10	10	10	20	20	20	20	20	20
Orders	5	5	5									
Available	50	40	30	20	50	40	20	40	20	40	20	40
Available to Promise	45											
Master Production Schedule	40				40			40		40		40

On-hand as buffer at the start of Week 1 = 20

FIGURE 6-5   A tabular master production schedule, week 1.

Available to Promise. The quantity for which there are no pending orders that can act as claims. This quantity can be calculated as on hand + MPS − sum of orders for this period and all subsequent periods. The available-to-promise number is an important one for order taking. It indicates what the plant can commit to right away versus what must be produced later.

Master Production Schedule. The planned production of the product during each period.

As orders roll in over time, what is available to promise will change. Contrast Figure 6-5 for week 1 with Figure 6-6 for week 2. More orders have been taken, and thus the available–to-promise row has changed. (Note that we could have constructed the table with a rolling forecast of 12 weeks, but for convenience we have kept the time horizon at the twelfth week.) Of course, if orders remain strong, marketing may want to revise the forecast upward and, correspondingly, manufacturing may want to revise the MPS upward. This would be particularly true if enough orders came through to turn a forthcoming available-to-promise entry negative. Even negative available row entries could be cause for alarm, if the forecast holds.

Either a graphical or a tabular schedule can be used to track the extent of orders and forecasts versus the MPS and to indicate what is available and available to promise to custom-ers, given the prevailing master schedule. By using this technique, manufacturing management can be alert for prospective deficiencies or surpluses in production and can act to remedy the situation, either by cranking up capacity (and increasing the MPS) or by shutting down capacity (and decreasing the MPS).

The MPS lies at the heart of how the factory gets loaded. It gives the process direction and drive. A master production schedule that is feasible and low cost, but yet satisfies customer demands, can do much to remove chaos from the factory floor, enhance productivity, and flow dollars to the bottom line.

Aggregate Planning Revisited

There are typically many master production schedules for a plant, and each of them marches through time. In the best of all possible worlds, of course, they are all expected to march in step with one another. However, if actual occurrences vary too much from the expected, there simply may not be enough resources for some of the master schedules to be met. One can see, then, the usefulness of some overarching planning of resources that can govern the trade-offs that may occur between one MPS and another. Without such a planning mechanism, one can expect to be short of some products, carrying extra inventory on others, wasting capacity somewhere, or scrambling for

						Weeks						
	1	2	3	4	5	6	7	8	9	10	11	12
Forecast	10	10	10	10	10	10	20	20	20	20	20	20
Orders	14	5	7	4								
Available	46	36	26	16	46	36	16	36	16	36	16	36
Available to Promise		30										
Master Production Schedule	40				40			40		40		40

Total on-hand at the start of Week 2 = 46

FIGURE 6-6   A tabular master production schedule, week 2.

capacity somewhere else. This overarching plan is termed the *aggregate plan* or the *production plan*. As it was introduced earlier, the plan deals with aggregates of resources (materials, capacity, workers) and outputs; it extends over a time period of from 6 to 18 months. Planning that far in advance encourages discipline and enforces consistency among the various master schedules as they are concocted and refined. The plan is typically stated in dollars, labor-hours of output, or, for some commodity or near commodity products, a physical measurement such as units or tons of product. It can act as both an aid in establishing MPSs for the end products and as a control document against which the MPSs can be assessed.

A variety of aggregate plans can be chosen and followed. Just as we noted chase and level strategies with MPSs, so too can those notions be applied to aggregate production plans. A level strategy, for example, would maintain a stable workforce and build up and draw down on inventories as needed. Exactly which inventories to build up, of course, would depend on the particular MPSs adopted to be consistent with this level aggregate plan. A chase strategy, on the other hand, would try to eliminate any buildup of inventories, but permit the workforce to fluctuate. Such aggregate plans, and any modifications lying in-between these extremes, can be depicted either graphically or tabularly. Graphs and tables can be very effective ways to track (1) the conformity of specific MPSs to the aggregate plan, and (2) actual performance against the aggregate plan. In this way managers can see when and where things are going out of control, and they can then adjust schedules and production in the most fitting ways.

The Level Versus Chase Strategy Choice

Perhaps the key choice that production planners face is that of which strategy to lean toward, a chase or a level one. Those two strategies are the polar extremes that anchor production planning. Recall that the chase strategy eschews the buildup of inventory in favor of matching production in each period as closely as possible to the sales for that period. The level strategy, on the other hand, tries to avoid the ups and downs—hires and layoffs, overtime, subcontracting—that are frequently a part of a chase strategy, in favor of letting inventory be a kind of shock absorber for the factory. Rarely do companies follow either a pure chase or a pure level strategy; they usually employ a mix of the two. Still, it is always useful to understand the forces that can, and should, cause a company to lean one way or the other in devising its production plans.

Choosing between a chase and a level strategy is essentially an exercise in comparing relative costs. What kinds of costs are important? Costs of carrying inventory. Costs of (re)hiring and laying off workers. Costs of producing on overtime. Costs of subcontracting work to other producers. Costs of not making the product on time and thus of stocking out. It is the comparison of these costs relative to each other that is key in determining which strategy is more attractive to follow. Naturally, the higher are the costs of hiring, layoff, overtime, and subcontracting relative to the costs of carrying inventory, the less attractive is the chase strategy and the more attractive is the level. In addition, the chase strategy risks stockouts somewhat more than the level strategy, because it is frequently difficult to augment capacity fast enough or, alternatively, to scale back production quickly enough to match sales exactly. Thus when stockout costs are relatively high, the level strategy is preferred.

As economic conditions change, the advantages of chase versus level will also change. For example, as interest rates rise, inventory becomes more expensive to carry and the chase strategy becomes more attractive. Alternatively, company commitments to no layoff policies or new union contracts that make layoffs more difficult or expensive to institute naturally favor a level rather than a chase strategy. Production planners need to scan the economic horizon periodically for changes in the relative costs with which they have to work.

Scheduling Techniques

There are several techniques that can be used effectively to help develop both MPSs and aggregate plans. Before discussing them, however, it is important to distinguish between scheduling a flow process (continuous, line, or batch), where capacities are clearly defined, and scheduling a job shop, where capacities are muddled.

Continuous flow and machine-paced line flow processes enjoy an inherent scheduling advantage because of their rigid production cycle times. Knowing that cars can come off the assembly line at the rate of one per minute—no more, no less—or that paper of a certain weight and grade can be made at a set rate is a tremendous advantage to scheduling. The capacities are known as close to certainty as possible. Even some batch flow processes with fairly standard products can be described with clear capacity measures (225 skirts per day) that facilitate efficient scheduling. Once capacities are known, the task of matching production to demand is greatly simplified.

The task is made even simpler if the process is permitted to build up some finished goods inventory. Although there is, of course, a cost to carrying inventory, such processes can keep the flow of work balanced across products and/or time periods more easily than processes that do not enjoy such freedom.

The job shop is clearly the worst to schedule because its capacity is so ambiguous and it cannot build up finished goods inventory. We address job shop scheduling later in the chapter.

Scheduling Techniques for Flow Processes When Capacities Are Clear-Cut

To focus this discussion, consider the aggregate planning problem of Situation 6-1.

The problem posed by the situation at Ridgely Sporting Goods is a classic one of aggregate planning where one is forced to examine whether a level or a chase strategy is preferable. The tradeoff is between (1) hiring some new workers now and having them put 160 labor hours of product each into inventory, thereby incurring a carrying cost of $0.35 $\times$ 160 hours = $56 per month for every month's worth of output produced, or (2) hiring workers later so that less goes into inventory, but then having to hire more workers at $500 a crack.

Discussion of Ridgely Sporting Goods, Leather Goods Department

How the tradeoff plays out is best seen by using a spreadsheet to keep track. Tables 6-2 and 6-3 present some alternatives for consideration. Table 6-2 considers the level case strategy where the additional workers are hired early and are allowed to build up the inventory. Table 6-3, on the other hand, waits as late as possible to hire workers and still have the inventory targets met. Note that more workers have to be hired in this instance, because there are no months of inventory build-up to help tide the company over.

An examination of the two tables reveals that the lower cost alternative is to employ more of a "chase" strategy. Hiring costs are small compared with the costs of having to carry the inventory.

With this venture into aggregate planning as background, let us move to the somewhat more complicated problem where there are two products (baseball and hockey gloves) to master

SITUATION 6-1

Ridgely Sporting Goods Company, Leather Goods Department

The leather goods department at Ridgely Sporting Goods manufactured two major products: baseball gloves and hockey gloves. The company's sales department had just published its projections for the next 9 months (Table 6-1). Through some rules of thumb, the forecast had been translated into labor hours required and the Leather Goods Department faced some decisions regarding its manning over that period of time. The forecast represented an increase in labor hours relative to recent past history. Should the department hire some new workers right away and build up enough inventory to get the company by, or should the department wait a little longer, perhaps having to hire more, but not having to inventory so much for so long, thus saving costs for the entire period?

At present, there was an inventory of 6600 hours of labor effort. The cost of carrying inventory was determined to be $0.35 per month per labor hour of output produced. Hiring a new worker incurred a cost of $500, while laying off a worker cost $400.

Because of various commitments to retailers, the company had to meet two inventory target levels over the next 9 months: (1) having 6970 labor hours of product in inventory by the end of August, and (2) having 7735 labor hours of product in inventory by January. It was also considered prudent to end the 9 months with at least as much inventory as the company had now, namely 6600 labor hours of product.

TABLE 6-1   Forecast of Labor Hour Loads for the Next 9 Months for the Leather Goods Department of Ridgely Sporting Goods

LABOR HOURS FORECAST								
July	Aug.	Sept.	Oct.	Nov.	Dec.	Jan.	Feb.	March
3400	2600	3400	4000	4400	7800	4500	5600	8400

TABLE 6-2   Aggregate plan for Ridgely Sporting Goods, leather goods department, under a level production strategy

	June	July	Aug.	Sept.	Oct.	Nov.	Dec.	Jan.	Feb.	March	Total
Load hours		3400	2600	3400	4000	4400	7800	4500	5600	8400	
Employment	26	30	30	30	31	31	31	31	31	31	
Production hours		4800	4800	4800	4960	4960	4960	4960	4960	4960	
Hires		4			1						
Fires											
Overtime hours/month											
Overtime/worker/month		0	0	0	0	0	0	0	0	0	
Inventory	6600	8000	10200	11600	12560	13120	10280	10740	10100	6660	
Inventory carrying cost		2800	3570	4060	4396	4592	3598	3759	3535	2331	32641
Hire/fire cost		2000	0	0	500	0	0	0	0	0	2500
Labor cost-straight		38400	38400	38400	39680	39680	39680	39680	39680	39680	353280
Labor cost—OT		0	0	0	0	0	0	0	0	0	0
Total costs		43200	41970	42460	44576	44272	43278	43439	43215	42011	388421

TABLE 6-3   Aggregate Plan for Ridgely Sporting Goods, leather goods department, under a chase production strategy

	June	July	Aug.	Sept.	Oct.	Nov.	Dec.	Jan.	Feb.	March	Total
Load hours		3400	2600	3400	4000	4400	7800	4500	5600	8400	
Employment	26	26	26	26	26	26	36	36	37	37	
Production hours		4160	4160	4160	4160	4160	5760	5760	5920	5920	
Hires							10		1		
Fires											
Overtime hours/month											
Overtime/worker/month		0	0	0	0	0	0	0	0	0	
Inventory	6600	7360	8920	9680	9840	9600	7560	8820	9140	6660	
Inventory carrying cost		2576	3122	3388	3444	3360	2646	3087	3199	2331	27153
Hire/fire cost		0	0	0	0	0	5000	0	500	0	5500
Labor cost-straight		33280	33280	33280	33280	33280	46080	46080	47360	47360	353280
Labor cost–OT		0	0	0	0	0	0	0	0	0	0
Total costs		35856	36402	36668	36724	36640	53726	49167	51059	49691	385933

schedule, as opposed to one (labor hours). Consider Situation 6-2 which revisits Ridgely Sporting Goods.

Ridgely Sporting Goods—General Discussion

As we can tell from the situation at Ridgely Sporting Goods, developing the MPS means deciding how much capacity the factory should gear up for over the short term and how that capacity should be realized. At some point, the factory must commit itself to meeting the sales forecast or some other agreed-to pattern of shipments. But generally speaking, it can accomplish this task in several ways.

1. *Chasing demand.* The factory can raise and lower its production rate so that it produces only as much of each item as it expects to ship that month. If forecasted sales vary greatly, such a "demand chase" strategy will generally entail a good deal of (a) overtime and/or (b) layoffs and recalls or new hiring.
2. *Leveling production.* Under this alternative, no changes in the workforce size or overtime policy are planned over the planning period. All variations in shipments are absorbed by a finished goods inventory. Thus, over the planning period, inventory is built up and depleted on a regular, planned basis.

3. *A combination approach.* A pure chase strategy and a pure level strategy are extreme polar cases. In between lies a whole range of cases where some chasing of demand and some leveling of production occur.

No matter what the strategy, developing an MPS involves decisions on a number of important issues and collecting data from a variety of sources. Among them:

• *Decisions*
   a. Desired inventory levels by production period (such as a month)
   b. Desired workforce levels by production period
   c. Desired levels of overtime by production period
   d. Desired mix and pattern of production, by product and by production period

---

SITUATION 6-2

Ridgely Sporting Goods Company, Leather Goods Department Revisited

The leather goods department of the Ridgely Sporting Goods Company manufactured two major products: baseball gloves and hockey gloves. By far the larger selling item was baseball gloves. The market for them was fairly stable, tending to follow population growth trends. Recently, the modest drop in the birth rate had been compensated for by a rise in sales to girls. Sales of hockey gloves were localized predominantly in the northeast, the upper midwest, and Canada. The market for these gloves was only about a fifth the size of the baseball market but was growing at about 5 percent a year.

Sales of each type of glove paralleled the sport's season, with sales greater just before and just after the start of each season. For baseball gloves in particular, the Christmas season was also a strong sales period. The forecast for the 9 months beginning in July is given in Table 6-4. The factory had to ensure that the company's distributors were fully stocked with gloves about 2 months before the season's official start in April for baseball and in November for hockey. This timetable meant that a third of the department's forecast sales for this 9-month period had to be produced by the end of January (baseball) and August (hockey).

The baseball gloves were easier to manufacture, requiring about 48 minutes of labor per glove. The hockey gloves took about 2 hours per pair. Thus one worker working 160 hours per 4-week period (the company's "month" for planning purposes) could produce 200 baseball gloves and 80 pairs of hockey gloves. The average factory variable costs for the gloves were $17 (baseball) and $40 (hockey).

The department liked to keep the production rate steady over the course of the year because hiring and then laying off workers was disturbing and expensive. Layoffs cost the company $400 per worker in severance pay. Hiring and training costs ran to $500 per new worker. Worker pay plus fringes averaged $8 per hour. Overtime could be used, within limits, to augment capacity, with the standard time-and-a-half premium.

There were no readily identifiable costs in changing over from the production of one type of glove to another. However, given the many variations in the basic styles of each type of glove, it was considered too confusing to make both baseball and hockey gloves at the same time. Thus separate days were usually allotted to each kind of glove. A finished goods inventory was used to smooth out the demand on the factory. The inventory carrying-cost rate was determined to be 20 percent.

The leather goods department was in the process of planning production for the next 9 months to meet forecasted demand. To facilitate planning, the workweeks were grouped into fours and labeled with months in which most of the production occurred. The currently prevailing inventories of gloves were 6500 (baseball) and 700 (hockey). It was considered prudent to end the 9-months with the same level of inventory as prevailed now. At present, also, there were 26 workers in the department.

TABLE 6-4   Forecast of sales by product line, July-March
(Ridgely Sporting Goods Company, leather goods department)

PRODUCT LINE	JULY	AUG.	SEPT.	OCT.	NOV.	DEC.	JAN.	FEB.	MAR.	TOTAL
Baseball gloves (units)	4000	3000	2500	1000	500	5000	2000	4000	7000	29,000
Hockey gloves (pairs)	100	100	700	1600	2000	1900	1450	1200	1400	10,450

Note: Each month has 4 weeks of production.

• *Data*

  a. Reliability of the short-term forecast of sales in each product line
  b. Current capacity of the operation, by product line
  c. Cost and productivity of overtime
  d. Cost and productivity of second or third shift operation
  e. Cost of layoff, hiring, and recall of the workforce
  f. Cost of carrying finished goods inventory
  g. A sense of how rapidly adjustments can be made in the product mix produced, which has an impact on the size of any buffer stocks
  h. Costs and difficulties of shifting work among different products

The process of developing the production schedules that the factory follows and by which it tries to control operations takes several stages, as we have noted. The transition through these stages occurs as sales become more and more certain and as details are learned about the quantities to be manufactured of product lines, products, models, and variations such as color and extras. The scheduling process begins with rough approximations: generally, months later, firmer and firmer numbers take hold. As mentioned earlier, the initial approximation is termed the "aggregate plan" because the plant, at that stage, deals only in aggregates of workers, product models, and even products themselves. Over time, as sales forecasts become firmer, the aggregate plan can be refined and detail added to yield a "master schedule" that deals typically with the major products to be made, although perhaps not with all the models or variations on these major products. The schedule for Ridgely Sporting Goods is of this type, as it deals only with the major products and not with any specific models or variations. The master schedule is an important document because it triggers actions such as specific workforce assignments, personnel actions, and more detailed materials flows. Even more detailed production schedules, which can sequence and synchronize the work of the factor are derived (again with even firmer information) from the master schedule.

Attacking the Master Schedule at Ridgely Sporting Goods

We need to determine how many baseball and hockey gloves the company ought to produce each month and the corresponding size of the workforce, inventory levels, and overtime levels, given that the forecast of Table 6-4 must be met. A more detailed schedule would indicate the precise mix of catcher's mitts, first-base gloves, and fielding gloves in which style, model (e.g., Orel Hershiser, autograph), and variation (right- or left-handed) to be produced, and on which days. Our present task, of course, is not so detailed, but it involves the same type of thinking.

The first task is to decide whether a chase or a level strategy should guide our thinking. If holding inventory is cheaper than the costs of overtime, laying off, recalling, and/or hiring, a level production strategy is recommended. If, on the other hand, inventory cannot be held or is more costly than such personnel actions, a chase demand strategy becomes attractive. How costly, then, is inventory holding and the various personnel actions?

• Holding inventory

  1. At a carrying cost rate of 20 percent, a baseball glove costs $3.40 per year, on the average, to hold in inventory (0.20 × $17 variable cost), or about 28 cents per month.

  2. Carrying a hockey glove is more expensive, $8 per glove per year or about 67 cents per month.

TABLE 6-5 Trial master production schedule (Ridgely Sporting Goods)

	JUNE	JULY	AUG.	SEPT.	OCT.	NOV.	DEC.	JAN.	FEB.	MAR.	TOTAL
Baseball gloves											
Forecast in units		4,000	3,000	2,500	1,000	500	5,000	2,000	4,000	7,000	29,000
Ending inventory*	6,500	4,700	3,900	5,000	7,600	10,700	9,300	10,700	10,100	6,500	
Workforce devoted		11	11	18	18	18	18	17	17	17	
Production		2,200	2,200	3,600	3,600	3,600	3,600	3,400	3,400	3,400	29,000
Hockey gloves											
Forecast in units		100	100	700	1,600	2,000	1,900	1,450	1,200	1,400	10,450
Ending inventory*		700	2,120	3,540	3,800	3,240	2,280	1,420	1,090	1,010	730
Workforce devoted		19	19	12	13	13	13	14	14	14	
Production		1,520	1,520	960	1,040	1,040	1,040	1,120	1,120	1,120	10,480
Workforce	26	30	30	30	31	31	31	31	31	31	
Hires		4	0	0	1	0	0	0	0	0	
Fires		0	0	0	0	0	0	0	0	0	
Overtime hours/day		0	0	0	0	0	0	0	0	0	0
Value of inventory†		164,700	207,900	237,000	258,800	273,100	214,900	225,500	212,100	139,700	1,933,700
Inventory carrying cost		2,745	3,465	3,950	4,313.33	4,551.66	3,581.66	3,758.33	3,535	2,328.33	32,228.33
Hire/fire cost		2,000	0	0	500	0	0	0	0	0	2,500
Labor cost—straight		38,400	38,400	38,400	39,680	39,680	39,680	39,680	39,680	39,680	353,280
Labor cost—overtime		0	0	0	0	0	0	0	0	0	0
Total costs		43,145	41,865	42,350	44,493.33	44,231.66	43,261.66	43,438.33	43,215	42,008.33	388,008.3

*The inventory goals are 9667 baseball gloves by the end of January and 3483 hockey gloves by the end of August.

†The value of inventory reflects ending inventory levels for each month.

- Personnel actions
  1. Layoffs cost $400.
  2. New hires cost $500.
  3. A second shift's labor bill would cost 115 percent of the first shift.
  4. Overtime costs are 150 percent of first-shift labor costs.

With these data as background, it is useful to ask questions such as the following in gauging the extent to which a chase or a level strategy should be followed.

- How long would you be willing to keep a worker on the job (thus producing units of output that have to be carried for some time) rather than lay off the worker now and recall him or her later?
- At what point would the cost of having the existing workers work overtime give way to hiring another worker or, indeed, hiring an entire second shift?

Questions such as these are difficult to answer definitively because they often involve many hidden contingencies. For example, the perceived need for 300 more units of output 6 months from now may be met by having one worker work for 6 months, by having two workers work for 3 months each, or a similar choice. It is difficult to analyze all the situations conceivable, and each has its own cost.

Computer spreadsheet programs such as Lotus 1-2-3 and Excel are perfect vehicles for testing the consequences of chase versus level versus mixed strategies for production planning. Such programs can automatically keep track of all the related variables in the plan and the costs associated with them. What-if experiments can be tried out and the costs of each compared. Table 6-5 is a trial master schedule computed through the use of a spreadsheet.

The Trial and Error of Master Schedule Development

Having established, in rough fashion, the relative attractiveness of a level production strategy for Ridgely Sporting Goods, we need to develop a MPS that satisfies the sales forecast and the other suggested constraints, such as inventories a third of the total 9-month sales at the end of the relevant months (August for hockey, January for baseball).

If we are to produce enough to meet the expected sales totals for the next 9 months (29,000 baseball gloves and 10,450 pairs of hockey gloves), we need to be sure that we have enough manpower to do it. The average level of manpower needed is easily calculated from the information about how long it takes to make a single glove (48 minutes per baseball glove and 2 hours per pair of hockey gloves) and that there are 160 hours per worker per month of 4 weeks. The calculations reveal that 16.1 workers per month, on average, are needed to produce the 29,000 baseball gloves and that 14.5 workers per month, on average, are needed to produce the 10,450 pairs of hockey gloves. The sum for the two product lines is 30.6 workers, which means that we are short 4.6 workers, on average, for the next 9 months. We need to hire. The issue, of course, is how many to hire and when, so that total costs remain low.

Table 6-5 is a first trial at matching production to these requirements. It is based in part on the workforce needs implied by the sales forecast of Table 6-4. The trial master schedule begins by hiring four more workers immediately to bring the workforce to 30 workers. An additional worker is hired in October, so as to provide another 0.6 worker-equivalent to the process. In most months, both baseball and hockey gloves are produced and no overtime is employed. The inventory goals—9667 baseball gloves by the end of January and 3483 pairs of hockey gloves by the end of August—are met easily. The cost of this MPS is $388,000, as shown in Table 6-5. The individual components of this cost are also provided in the table's entries.

Note that no overtime was used, nor was a second shift instituted. No layoffs, only hires, were made. The convention employed is to value just the ending inventory for purposes of applying the carrying cost. Note also that we have excluded from consideration deliberately induced stockouts; positive levels of inventory are always kept. We end up with 6500 baseball gloves in inventory, exactly what we started with, as well as 730 pairs of hockey gloves, slightly more than we had at the beginning of July.

No doubt some dollars can be whittled off this cost figure by investigating alternative hiring patterns and production patterns by product and by month. Successive trial-and-error schedules, such as those handled with spreadsheets, would consider such alternatives and compare them with this first trial effort. Many actual master production schedules are derived by just such trial-and-error means. Stability of the product mix and/or sales levels over time is naturally helpful in this search for better and better MPSs.

"Optimal" and "Near-Optimal" Aggregate Plans and Master Schedules

To this point, we have used common sense and trial and error to devise aggregate plans and more detailed master schedules. As with almost any trial-and-error approach to problem solving, one always suspects that "a better way" must exist, or a way that guarantees some "optimal" solution. In the case of the Ridgely Sporting Goods Company, we cannot help but think that a lower cost means of meeting all the shipping and inventory requirements exists.

In the schedule development for Ridgely Sporting Goods, what costs should be minimized and what constraints must be satisfied?

We have already talked about them, but it is helpful to state them more formally. In essence, the costs of the plan that we want to minimize include wage costs, inventory carrying costs, and costs of hiring and laying off workers; the constraints that we want to satisfy include the inventory targets for each month, some reasonable expectations on how much overtime can be worked, and some more prosaic requirements such as how many units of which product line can be produced in any month and how inventory and the workforce can be supplemented or reduced. To be even more formal about the Ridgely Sporting Goods situation, we seek to minimize

$$1280W_1 + 1280W_2 + \ldots + 1280W_9$$

(Regular shift wage costs, since W_t is the level of the month's workforce and $1280 is monthly pay per worker)

$$+ 1920OT_1 + 1920OT_2 + \ldots 1920OT_9$$

(Overtime wage costs, since OT_t is the level of overtime worked on a monthly basis and $1920 is the cost of a month's worth of overtime)

$$+ 0.2833BI_1 + 0.2833BI_2 + \ldots + 0.2833BI_9$$

(Inventory carrying costs for baseball gloves, since 28.33 cents is the carrying cost per month for baseball gloves and BI_t is the stock of baseball gloves carried each month)

$$+ \$0.6667HI_1 + 0.6667HI_2 \text{i} + \ldots + 0.6667HI_9$$

(Inventory carrying costs for hockey gloves, since 66.67 cents is the monthly carrying cost and HI_t is the stock of hockey gloves carried each month)

+ 500 WFI_1
+ 500 WFI_2
+ ... 500 WFI_2

(Costs of increasing the workforce, since it costs \$500 for each new hire. WFI_t is the size of the workforce increase in each month)

+ 400 WFD_1
+ 400 WFD_2
+ ... + 400 WFD_9

(Costs of laying off workers since each layoff costs \$400. WFD_t is the size of the workforce decrease in each month)

subject to the following constraints:

1. The following variables must all be greater than or equal to zero for all months t, where t can range from 1 to 9:

W_t, OT_t, BI_t, HI_t, WFI_t, WFD_t, BP_t, and $HP_t \geq 0$

where

W_t = workforce in month t
OT_t = overtime in month t
BI_t = baseball glove inventory in month t
HI_t = hockey glove inventory in month t
WFI_t = workforce increase in month t
WFD_t = workforce decrease in month t
BP_t = baseball glove production in month t
HP_t = hockey glove production in month t

2. Regular workforce hours and overtime hours are composed of those devoted to baseball glove production and those devoted to hockey glove production:

$$W_t = WB_t + WH_t$$
$$OT_t = OTB_t + OTH_t$$

where

WB_t = regular time, in worker months, devoted to baseball glove production in month t

WH_t = regular time, in worker months, devoted to hockey glove production in month t

OTB_t = overtime, in worker months, devoted to baseball glove production in month t

OTH_t = overtime, in worker months, devoted to hockey glove production in month t

3. Production is governed by the following relationships:

$$BP_t = 200\ WB_t + 200\ OTB_t$$
$$HP_t = 80\ WH_t + 80\ OTH_t$$

The figure 200 is derived by noting in 4 weeks of 40 hours per week, 200 baseball gloves can be produced at an average of 48 minutes per glove; the figure 80 is derived by noting that during the same 160-hour period only 80 pairs of hockey gloves can be produced, since an average of 2 hours are required for each pair.

4. Overtime, however, has a practical ceiling, which we might peg at a quarter of the hours spent on regular shift work. Thus

$$OTB_t \leq 0.25\ WB_t$$
$$OTH_t \leq 0.25\ WH_t$$

5. Inventory for the period must be accounted for by the following:

$$BI_t = BI_{t-1} + BP_t - BS_t$$
$$HI_t = HI_{t-1} + HP_t - HS_t$$

where BP_t and HP_t represent production of baseball and hockey gloves, respectively, and BS_t and HS_t represent sales or shipments of baseball gloves and hockey gloves, respectively. BI_{t-1} and HI_{t-1}, of course, refer to inventories in the prior period.

6. The size of the workforce in any period must also be accounted for by the following:

$$W_t = W_{t-1} + WFI_t - WFD_t$$

Remember, both WFI_1 (workforce increase) and WFD_1 (workforce decrease) are defined to be nonnegative numbers.

7. Shipments must be equal to the forecasted figures of Table 6-4; inventory levels in any month must be greater than or equal to the forecasted sales in the next month; and the special inventory goals of $BI_7 \geq 9667$ and $HI_2 \geq 3483$ must also be met.

8. To be compatible with the trial master schedule of Table 6-5, some initial and closing conditions must be specified:

Initial conditions $BI_0 = 6500$, $HI_0 = 700$, $W_0 = 26$

Closing conditions $BI_9 = 6500$, $HI_9 = 730$

Fortunately, there exists a mathematical technique, linear programming, that can solve this minimization problem subject to the constraints we have specified, largely because all of the relationships specified are linear ones. Observe that all the relationships involve variables multiplied by constants and added together (geometrically, straight lines or planes) with no squared terms, trigonometric functions, inverse terms, logarithmic functions, or anything even fancier. Thus variable values rise and fall in proportion to one another and this permits the use of linear programming to aid in this scheduling.

We can state Ridgely Sporting Goods' linear program, using the symbol Σ (sigma) to indicate the summation of terms.

Objective function: Minimize

$$\sum_{t=1}^{9} (1280\ W_t + 1920\ OTB_t + 1920\ OTH_t$$
$$+ 0.2833\ BI_t + 0.6667\ HI_t$$
$$+ 500\ WFI_t + 400\ WFD_t)$$

subject to:

Nonnegativity conditions

$$W_t,\ WB_t,\ WH_t,\ OT_t,\ OTB_t,\ OTH_t,\ BI_t,\ HI_t,$$
$$WFI_t,\ WFD_t,\ BP_t,\ \text{and}\ HP_t, \geq 0$$

and structural constraints

$$W_t - WB_t - WH_t = 0$$
$$OT_t - OTB_t - OTH_t = 0$$
$$PB_t - 200WB_t - 800TH_t = 0$$
$$PH_t - 80WH_t - 80OTH_t = 0$$
$$0.25\ WB_t - OTB_t \geq 0$$
$$0.25\ WH_t - OTH_t \geq 0$$
$$BI_t - BI_{t-1}\ BP_t + BS_t = 0$$
$$HI_t - HI_{t-1} - HP_t + HS_t = 0$$
$$W_t - W_t - WFI_t + WFD_t = 0$$

$BS_1 = 4000$	$HS_1 = 100$
$BS_2 = 3000$	$HS_2 = 100$
$BS_3 = 2500$	$HS_3 = 700$
$BS_4 = 1000$	$HS_4 = 1600$
$BS_5 = 500$	$HS_5 = 2000$
$BS_6 = 5000$	$HS_6 = 1900$
$BS_7 = 2000$	$HS_7 = 1450$
$BS_8 = 4000$	$HS_8 = 1200$
$BS_9 = 7000$	$HS_9 = 1400$
$BI_0 = 6500$	$HI_0 = 700$
$BI_1 \geq 0$	$HI_1 \geq 0$
$BI_2 \geq 0$	$HI_2 \geq 3483$
$BI_3 \geq 0$	$HI_3 \geq 0$
$BI_4 \geq 0$	$HI_4 \geq 0$
$BI_5 \geq 0$	$HI_5 \geq 0$
$BI_6 \geq 0$	$HI_6 \geq 0$
$BI_7 \geq 0$	$HI_7 \geq 0$
$BI_8 \geq 0$	$HI_8 \geq 0$
$BI_9 \geq 6500$	$HI_9 \geq 730$

$$W_0 = 26$$

When this formulation of the linear program to minimize the cost of the master schedule at

Ridgely Sporting Goods is entered on a computer and run, a solution like that found in Table 6-6 results. (The entries in this table are to the nearest integer.) Several things about the optimal master schedule are worth noting:

1. The workforce is not increased right away but later on and to necessarily higher levels than the level strategy adopted in Table 6-5.

2. Inventory levels are lower than in Table 6-5. Lower inventory carrying expenses are incurred in this schedule, more than balancing off the higher hiring costs incurred.

3. Somewhat remarkably, though, the cost of the linear program's optimal solution is only a few thousand dollars less than the trial-and-error schedule developed earlier.

The last observation is important. Although linear programming is immensely attractive for developing optimal solutions to sophisticated problems, there is no guarantee that the optimal solution will be much different from a solution arrived at through common sense and trial and error. In this case, going through the pain, suffering, and expense of developing a linear program to capture Ridgely Sporting Goods' needs and of solving it seems scarcely worth the effort. Nevertheless, in other instances, such linear programming would more than pay for itself.

Furthermore, although linear programming (or its cousins like integer or mixed integer programming) can help lower the cost of a master schedule, qualitative features of production planning can override cost considerations. One instance of a huge payoff from using linear programming for schedule development, however, is International Paper Company (see Situation 6-3).

TABLE 6-6   Optimal master production schedule* as determined by linear programming (Ridgely Sporting Goods)

	JULY	AUG.	SEPT.	OCT.	NOV.	DEC.	JAN.	FEB.	MAR.[†]	TOTAL
Baseball gloves										
Forecast in units	4,000	3,000	2,500	1,000	500	5,000	2,000	4,000	7,000	29,000
Ending inventory	2,500	0	0	0	2,067	4,367	9,667	6,200	6,500	
Workforce devoted[‡,£]	0	2.5	12.5	5	2.8	36.5	36.5	2.7	36.5	
Production	0	500	2,500	1,000	2,567	7,300	7,300	533	7,300	29,000
Hockey gloves										
Forecast in units	100	100	700	1,600	2,000	1,900	1,450	1,200	1,400	10,450
Ending inventory[‡,£]	2,680	4,460	4,840	4,920	3,973	2,073	623	2,130	730	
Workforce devoted	26	23.5	13.5	21	13.2	0	0	33.8	0	
Production	2,080	1,880	1,080	1,680	1,053	0	0	2,707	0	10,480
Total workforce	26	26	26	26	26	36.5	36.5	36.5	36.5	
Hires (+)/fires (−)	0	0	0	0	0	10.5	0	0	0	

Total costs of this schedule = $385,015.54

*The optimal schedule does not have any overtime or second-shift production.

[†]To be comparable to the previous trial schedule, ending March inventories were constrained to be 6500 baseball gloves and 730 pairs of hockey gloves.

[‡]The value of inventory reflects ending inventory levels for each month.

[£]The inventory goals are 9667 baseball gloves by the end of January and 3483 hockey gloves by the end of August.

SITUATION 6-3

International Paper Company Revisited, Production Planning[a]

As Tour A of the Androscoggin Mill, found in the back of this text, discusses, the capabilities and costs associated with paper machines can vary dramatically from one machine to another. Generally speaking, new machines are wider and faster than older ones and use fewer workers and less energy per ton of output. It makes sense then to be careful about which orders for paper are assigned to which machines. The total contribution earned from one assignment of orders to machines may be very different from the total contribution earned from another assignment. Moreover, because both International Paper's mills and its customers are widely scattered and because paper companies typically absorb the costs of transporting completed orders to customer locations and because several paper machines companywide may be set up to run a particular type and grade of paper, it can be difficult to make the most cost-effective assignment of a specific order.

In response to this problem of allocating "paper machine products" destined for particular customer locations, International Paper Company (among others in the forest products industry) developed a linear programming model. The general outline of this model was straightforward, although as will be apparent, the program was very large and complex. The company wanted to assign particular orders (demands defined by the choice of paper machine product and customer location) to specific paper machines in such a way that the contribution to the company was maximized. Contribution, in turn, was defined as revenue less variable costs. Revenue was arrived at by multiplying price per ton times tons for each paper machine product made, while variable costs encompassed the costs of production as well as the costs of distribution. Variable production costs per ton included pulp and other materials, energy, direct labor, scrap, and any variable overhead expenses. The distribution cost was specific to each paper machine/customer location pair considered.

[a]This description is as of the summer of 1982.

Suppose that there are a total of L customer locations, P paper machine products, and M paper machines. The revenue from each paper machine product for each customer location had to be considered as well as the cost of making each paper machine product (PMP) on each machine and transporting it to each customer location. The linear program's objective function, the contribution to be maximized, can be expressed more formally in words as shown in Figure 6-7.

In algebraic terms, the objective function can be more easily written as total contribution to be maximized and it equals

$$\sum_{m=1}^{M}\sum_{p=1}^{P}\sum_{l=1}^{L}\{Q_{lpm} \times [P_p - (PC_{pm} + TC_{lm})]\}$$

where p indexes paper machine products, l indexes customer locations and m indexes paper machines, where P_p is the price per ton of product p, Q_{lpm} is the output in tons of product p for customer location l from machine m, PC_{pm} is the variable cost of producing a ton of product p on machine m, and TC_{lm} is the cost of transporting a ton of output from machine m to customer location l.

In the simplest case, there were two structural constraints on the objective function:

1. The sum of the specific outputs from each machine must not exceed the demands. That is, for each lp combination,

$$\sum_{m=1}^{M} Q_{lpm} \ D_{lp}$$

where D_{lp} is the demand in tons for paper machine product p for customer location l.

2. The capacity of each paper machine during the planning time period (such as one month) cannot be exceeded. That is, for each m,

$$\sum_{p=1}^{P}\sum_{l=1}^{L} (Q_{lpm} \times HPT_{pm}) \ H_m$$

where HPT_{pm} is the hours per ton for product p on machine m and H_m is the hours available during the time period for machine m.

Total contribution   = Revenue for each order — Variable cost of filling each order

or

Total contribution   = (Tons for customer location #1 of PMP #1 × Price of PMP #1 per ton)
     + (Tons for customer location #2 of PMP #1 × Price of PMP #1 per ton) + · · ·
     + (Tons for customer location L of PMP #1 × Price of PMP #1 per ton)
     + (Tons for customer location #1 of PMP #2 × Price of PMP #2 per ton) + · · ·
     + (Tons for customer location L of PMP #2 × Price of PMP #2 per ton) + · · ·
     + (Tons for customer location #1 of PMP #P × Price of PMP #P per ton) + · · ·
     + (Tons for customer location L of PMP #P × Price of PMP #P per ton)
     — (Tons for customer location #1 of PMP #1 on machine #1 × Cost of making a ton of PMP #1 on machine #1 plus transport cost from machine #1 to customer location #1) — · · ·
     — (Tons for customer location L of PMP #1 on machine #1 × Cost of making a ton of PMP #1 on machine #1 plus transport cost from machine #1 to customer location L) — · · ·
     — (Tons for customer location #1 of PMP #P on machine #1 × Cost of PMP #P on machine #1 plus transport cost from machine #1 to customer location #1) — · · ·
     — (Tons for customer location L of PMP #P on machine #1 × Cost of PMP #P on machine #1 plus transport cost from machine #1 to customer location L) — · · ·
     — (Tons for customer location #1 of PMP #1 on machine #M × Cost of PMP #1 on machine #M plus transport cost from machine #M to customer location #1) — · · ·
     — (Tons for customer location L of PMP #1 on machine #M × Cost of PMP #1 on machine #M plus transport cost from machine #M to customer location L) — · · ·
     — (Tons for customer location #1 of PMP #P on machine #M × Cost of PMP #P on machine #M plus transport cost from machine #M to customer location #1) — · · ·
     — (Tons for customer location L of PMP #P on machine #M × Cost of PMP #P on machine #M plus transport cost from machine #M to customer location L)

Figure 6-7   The linear program's objective function.

There were, of course, the standard nonnegativity constraints as well that had to be met.

The scope of this allocation problem at International Paper was staggering. In the White Papers Division, of which the Androscoggin Mill was a part, the allocation involved 600 to 700 paper machine products for between 16,000 and 20,000 customer locations assigned to 26 paper machines in seven different mills. Of course, not all machines were capable of making every product and the linear program was written to reflect this. The linear program to make the allocation was run on a large IBM 3033 computer, but it took so much time (over an hour of CPU time per run) that it could be run only at night.

The program used a time horizon for the plan of one month, although the plan generally consisted of allocating products for four "cycles" each month. Thus there was only 1 week between times when the most popular paper products were produced. The data for this linear program were continually updated. The latest demand forecasts and lists of orders received were used along with the latest price quotes from the order entry file. The program used transportation costs between mills and customers that were updated each month, and the production costs used the then prevailing standard costs for each machine. These standard costs were updated at least twice a year, and variances from actual costs were tracked carefully. If variances were too great, the standard cost was immediately updated.

The output from the linear program included its solution—the assignment of orders to machines and the contribution calculation associated with each order assignment—as well as a number of summary reports. For example, there was a report listing tons of paper in each product line produced by specific machine. Other reports listed what each machine was assigned by product, by grade, and by basis weight. Still other reports listed any orders that the program's solution had left unallocated, by grade or by region.

The discussion so far has been limited to the most basic aspects of International Paper's allocation program. There were several refinements worth describing:

1. The solution to this linear program was strictly "by the numbers," listing what maximized contribution for the company during the next month. The model ignored any longer time frame as well as the maintenance of particular customers' business.

   The linear program could be modified, however, to take into account marketing considerations of the type where customers, for one reason or another, might insist that their order come from a particular paper machine. In that case, an additional constraint specified that the output of a particular machine be directed to a certain customer location and that it be exactly X tons: that is, the constraint specified that, for specific values of l, p, and m, Q_{lpm} exactly equal X.

2. When initially constructed, the model allocated orders just to paper machines. As is mentioned in Tour A, however, the bottleneck in a paper mill can lie elsewhere, particularly in finishing, where logs of paper are rewound and slit or where sheets of paper are cut from logs. International Paper's allocation model was augmented to include the allocation of orders to finishing lines (perhaps in different mills), as well as paper machines.

3. International Paper's linear program could be used to assess what should be done if demand fluctuated. For instance, if demand far outstripped International Paper's capacity, some decisions had to be reached about which orders would be filled and which would go unfilled (unallocated). The solution to the linear program provided information on which orders yielded the lowest contributions and thus were prime candidates for not being filled. If, on the other hand, demand was weak, the linear program could be used to determine which paper machines had negative contributions and thus ought to be closed.

In regard to the third refinement, consider this chapter's earlier discussion on identifying the scarce resource. Typically, demand for International Paper's products exceeded the company's capacity. In such a case, the scarce resource was time on the mills' paper machines, and the linear program was constructed to maximize the total contribution per unit of time (the planning month). This need not have been the case. The linear program's solution during periods of weak demand could be interpreted so as to maximize contribution per ton of output rather than contribution per unit of time. In such periods, the scarce resource was not time on a paper machine but working capital for the purchase of materials, labor, and other inputs to the process. International Paper's managers recognized this feature and used their linear programming model with an eye to the maximization of either contribution per machine hour or contribution per ton of output depending on the strength or weakness of the demand for paper.

By company estimate, the use of this linear program to help plan production resulted in a 12 percent reduction in delivered costs for one of International Paper's product lines. Linear programming can be a powerful tool.

It should be noted that the linear program at International Paper typically allocated paper machine products to individual weeks within the month. This is a "rough cut" schedule. A more precise schedule had to be devised, usually by the mill itself, to assign products to particular days and shifts within the week.

Cautions About the Use of Linear Programming

In addition to this concern for the cost effectiveness of applying linear programming techniques to situations like master schedule development, there are several fundamental cautions that any manager contemplating the use of linear programming should be aware of. Among them:

1. *Meeting the linearity assumptions.* As we have observed previously, the objective function and the structural constraints must all be linear for a programming problem to be capable of solution. In particular, the production relations defined in the constraints must exhibit constant returns to scale, and all prices and costs must be fixed and invariant with quantity (no price discounts can be allowed). In many cases, of course, constant returns to scale production, especially in the short term, are a very reasonable assumption, as is the absence of any significant price or cost discounts. In fact, the applicability of these assumptions has largely facilitated the already existing use of linear programming techniques in production planning. Yet companies should be acutely sensitive to those times when such linearity assumptions stretch reality too much to be useful.

2. *The time horizon.* The solution to most linear programming problems is sensitive to the time horizon chosen. In the case of Ridgely Sporting Goods, a time horizon of 9 months was chosen, and ending inventories of certain sizes were specified. If the time horizon were extended to, say, 12 months or shortened to, say, 6 months, the specifics of the solution might vary significantly. This is, of course, a general caution about planning horizons, but it is especially relevant to techniques like linear programming since

their solutions tend to be very volatile, with no presentiments that a future beyond the problem's time horizon even exists.

3. *Problem size.* Our problem at Ridgely Sporting Goods spanned only two products and 9 months. Yet, the size of the linear program developed was substantial. For more products and more months to be included, ever larger computers must be called on, and expenses can increase dramatically.

Alternatives and Refinements to the Linear Programming Approach

It is worth mentioning, if only in passing, that there exist some alternatives and refinements to the linear programming approach to aggregate plan and master schedule development.[4] Some (like dynamic programming and the linear decision rule approach of Holt, Modigliani, and Simon) yield optimal results although at the expense of some restrictive assumptions.[5] Often, too, these techniques, for computational reasons, are simply unworkable for problems of any more than a very small size.

A refinement to the linear programming approach, the hierarchical system concept, shows some promise.[6] This approach tries to enlarge the size of the problem that linear programming can tackle by breaking it up into separate but manageable pieces (hierarchies). Naturally, something is lost by such disaggregation; but if the problem is broken apart cleverly, that "something" may be small. Different products placing demands on different equipment, for example, might be broken apart for separate analysis. Or, different levels of decisions might be separated from one another (such as the assignment of products to plants, aggregate plans within plants for broad product lines, or specific items within a product line) and analyzed separately, given the results of the next highest step in the hierarchy.

ASIDE

Other Applications of Linear Programming

As we have seen, linear programming has a natural application in the development of aggregate production plans and schedules. There are other operations problems, however, for which linear programming has been used with success.

Product Mix Decisions. Companies are constantly bumping up against capacity constraints and must make decisions about which of their products they should produce during any span of time. Some products, of course, are relatively more profitable than others. All products are likely to make different demands on a company's scarce resources. Given these scarcities, then, and the profitability of each product, what mix of products should be produced? This question is somewhat different from the one we faced in aggregate production planning. There we asked for the lowest cost way of meeting a particular schedule, given some constraints. Here we ask for the most profitable mix of products, given other constraints. This formulation of linear programming has found application in many continuous flow operations (such as oil refineries) when there is little question that all of their output can be sold on the open market with no modification to price. It is a feature of linear programming solutions that the number of products that ought to be produced will not exceed the number of structural constraints in the problem. This fact implies often that the breadth of a product line to be run at a refinery should be reduced if maximum profits are to be earned.

Product Specification Decisions. A variant of the foregoing application of linear programming is to use the technique to determine the lowest cost mix of inputs for a given product, given that certain requirements (e.g., product performance) must be met. Animal feeds, paints, gasolines, and similar products are classic examples where certain inputs can be substituted for others and where cost and performance characteristics can be balanced off against one another. Linear programming can thus be used as a purchasing tool, or even as a make-buy decision tool.

Transportation and Distribution Decisions. How should a company supply its warehouses from its plants so that its transportation costs are as low as possible? This problem is so well known that it has been termed "the transportation problem," and there are special solution algorithms that can cut down on the expense of solving problems of this type. The mathematical description of the transportation problem is:

Minimize

$$\sum_{i=1}^{m} \sum_{j=1}^{n} c_{ij} X_{ij}$$

subject to

$$\sum_{i=1}^{m} X_{ij} \leq S_i \text{ for } i = 1,2, \ldots, m$$

(a supply constraint)

$$\sum_{j=1}^{n} X_{ij} \geq D_j \text{ for } j = 1,2, \ldots, n$$

(a demand constraint)

$X_{ij} \geq$ for all i and j (the nonnegativity condition)

The X_{ij}'s are quantities shipped from plant i to warehouse j during a given time period. The cost of shipping between i and j is represented by c_{ij}. Each plant i is limited to shipping S_i of the good, and each warehouse must have at least D_j of it. There are m plants and n warehouses.

Plant Location Decisions. Some companies, particularly those in the food industry, that incur significant transportation expenses in taking their products to market use linear programming as an aid in their new plant location decisions. This application is an extension of the transportation problem. Proposed new plants are added to the existing network of plants, and the minimum costs of transport for the expected levels of sales to each geographic market are calculated. Various location trials are made and the lowest cost locations for fulfilling the expected level and pattern of demand are seriously scrutinized for new plant startups.

Other Approaches to Production Scheduling

So far we have discussed a couple of different approaches to the problem of developing increasingly detailed production schedules over time, notably the commonly used method of trial and error and the potentially cost-minimizing linear programming approach. We noted that linear programming has several offshoots that either refine its mathematical potential (integer programming, dynamic programming) or improve its applicability to broad, complex schedules (such as hierarchical planning).

Several other techniques, though infrequently used, can improve on the trial-and-error approach; these are not optimizing techniques like linear programming, but they can be used when the problem size or structure does not lend itself to linear programming or its offshoots. Of these approaches, two bear some mention: search techniques (the search decision rule of Taubert)[7] and simulation. Both of these approaches try to retain the detail of the scheduling problem as desired by the managers involved, by keeping the cost relationships true to life. They are useful because they can systematically inform management of the costs of various trial schedules.

Search techniques try to isolate good schedules by systematically probing the multidimensional surface that describes how costs differ with changes in the key elements of the production plan (such as overtime, workforce levels, inventory). If the search procedures used are effective, the techniques can yield optimal or near-optimal schedules. Simulation techniques do not search systematically for good production schedules but, rather, simply provide a means by which production planners and schedulers can test modifications to the established schedule. Simulation can be seen as a way to make trial and error easy.

Scheduling Other Than Production Schedules

We can use the logic and techniques of production scheduling to schedule not only production runs but similar items such as labor hours to be worked (especially for service operations) or vehicles to be used for distribution from a factory or warehouse. In these cases, the logic follows much the same pattern as the logic of inventory accumulation and depletion. For example, a service operation's capacity needs can be translated into labor hours and a schedule devised, say by use of a spreadsheet, to cover demand without spending too much on extraneous labor. Similarly, vehicles can be scheduled (and routed) so as to meet the expected distribution demands, but with as few extra kilometers traveled as possible. The basic accounting ($period_{n+1}$ = $period_a$ + any net change) carries over easily from application to application.

Tools of the Trade: A Recap of Lessons Learned

If an operation's existing capacity is to be well managed, then management must do a satisfactory job of planning its production. Not only should management schedule well, but it should know what it should schedule and why.

Production plans exist at several levels (aggregate plans, master schedules, final assembly schedules) and are constantly being updated. Time provides additional information that needs to be incorporated into the plan. As the plan is being developed a key question is "Is it feasible? Can we produce all that is demanded by the time requested?" Capacity planning, which can take several forms, some more sophisticated than others, is the mechanism by which feasibility is checked.

The actual schedule devised is usually a blend of two extremes: a level (constant production level) strategy and a chase (no finished goods inventories held) strategy. To which end of that spectrum of choice the plan leans depends on several things, among which are the costs of holding inventory, the costs of stocking out of the product or service, the cost of working overtime or on shifts or of subcontracting the work. How these factors combine with the scheduled or anticipated demand determines much of the character of the production plan. Techniques for devising production plans often use trial and error and rules of thumb, but with the advent of computer spreadsheets, the trade-offs involved are increasingly easy to model. For companies for which production plans are critical, investments in optimization tech-niques (e.g., linear programming) make a lot of sense.

Operations managers typically do not fore-cast the demand that drives the production planning process, but, as users of those fore-casts, knowledge of forecasting techniques can be helpful. The appendix deals with forecasting.

World-Class Practice

World-class practice tends to value the level schedule that changes only slowly from period to period. Better to make a little bit of every-thing every day than to build up inventories to "tide us over." This attitude puts an emphasis on flexible capacity and good knowledge of what that capacity is.

FORECASTING DEMAND

Of the six "factory" tours at the end of this book, only Norcen Industries — the job shop that bid on all orders — knew with certainty beforehand exactly what it was to produce. In each of the other processes, there was some uncertainty about the demands to be placed on it; forecasts had to be made to guide production. In a sense, these forecasts could be said to "drive" those processes.

In the typical company, production managers are not responsible for the forecast used to drive production schedules. The marketing department is more often the source of these forecasts. This is not to say that production managers should not be interested in the assumptions and methods used to build these forecasts. In many circles, marketers are notorious for their optimism and are resented when manufacturing gets stuck with unsold inventory. A second-guessed forecast that is actually used by manufacturing, instead of the "official" forecast, is often a sore point between marketing and manufacturing. Forecasting is one of the most important interfaces between marketing and operations, and it often helps for operations managers to be well versed in forecasting, if only for defensive purposes.

Several types of forecasts are needed to guide production. Long-term forecasting of demand is required (even in a job shop) to help decide capacity and technology choices. Medium-term (6 to 18 months) forecasts of demand for broad families of products are needed to help marshal production resources, and short-term (less than 6 months) forecasts are needed to specify demands for particular products and product variations (models, sizes, colors, and so on). Table 6-7 summarizes some of these requirements.

As one might expect, there are different kinds of forecasting techniques that can be used to satisfy different forecasting needs. Figure 6-8 divides the various forecasting techniques available into broad categories and distinguishes those that are appropriate for long-term forecasting from those that are appropriate for short-term forecasting. The next sections briefly describe each of the forecasting techniques introduced in Figure 6-8.

QUANTITATIVE TECHNIQUES USING TIME SERIES DATA

The simplest quantitative techniques use only the past history of the demand itself. These techniques try to isolate patterns that could be expected to repeat in the future. Their strength

TABLE 6-7 Forecasting requirements for different time periods

	SHORT TERM (≤ 6 months)	MEDIUM TERM (6–18 months)	LONG TERM (≥ 18 months)
Uses	Production schedules	Aggregate production plans	Capacity or process changes
	Staffing manpower plans	Levels of employment and inventory	
Needs	High accuracy	Reasonable accuracy	"Ballpark" accuracy
	Detailed, disaggregated by product, model, style	Some detail, some aggregation of products	Not detailed, high aggregation into product lines
	Forecast often by week or even day	Forecast often by month	Forecast often by either month or quarter
	Turning points less important	Turning points of modest interest	Turning points important
	Low cost, quickly done	Modest cost and speed	High cost, takes time
	Used by lower-level management	Used by middle-level management	Used by top-level management

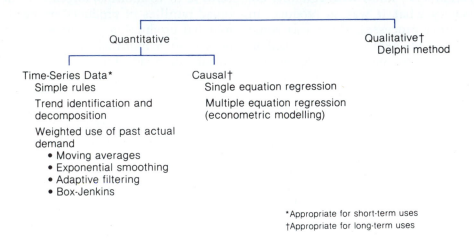

FIGURE 6-8 A classification of forecasting techniques.

lies in identifying these patterns easily and translating them into predictions.

Simple rules are the least sophisticated quantitative forecasts. They involve rules such as "this period's demand equals last period's demand," or "this period's demand equals last period's demand plus 5 percent." Although simple, such rules do not yield consistently reliable predictions.

Both trend identification and decomposition techniques are more sophisticated than simple rules because they break down the history into variations that may be cyclical or seasonal or that show a specific linear trend. These techniques, in a sense, massage the data (1) to remove seasonality or business cycle fluctuations and expose what may be general trends and (2) to isolate the extent to which variations

SITUATION 6-4

The Young Forever Boutique

The Young Forever Boutique was started five years ago in an old, renovated factory in suburban Chicago. Kathy Wilkerson, its owner, had been a buyer with a department store chain in the area, but had always wanted to satisfy an itch to run her own store. The boutique concentrated on casual and sports clothes for "baby boom generation" women, although both younger and older women felt comfortable shopping there as well.

Kathy had been pleased with the store's progress. Sales each year had grown, and the store had been turning a nice profit. In late July of the boutique's third year, Kathy had been able to relocate the store to a modern mall, not too far from the renovated factory. The rent was more expensive, but Kathy saw the opportunity to attract more customers in the mall environment.

Recently, Kathy had sensed that sales were more erratic than they had been, and she was concerned that her sales forecasts were increasingly inadequate. She also wondered about how much of a boost to sales the move to the mall had been. Kathy hoped that more sophisticated forecasting techniques would help. Monthly sales since the shop's startup are found in Table 6-8.

TABLE 6-8   Young Forever Boutique sales volumes by month

MONTH	YEAR 1	YEAR 2	YEAR 3	YEAR 4	YEAR 5
January	-	$ 6,381.07	$ 11,016.55	$ 11,965.68	$ 10,580.74
February	-	4,405.83	8,264.14	12,761.98	13,420.87
March	$ 4,566.40	9,126.95	6,843.95	18,126.70	34,476.27
April	6,941.89	10,970.45	20,109.68	19,092.90	19,108.79
May	7,047.23	10,189.28	11,148.64	15,384.01	18,109.54
June	6,784.37	7,043.85	8,365.40	12,237.00	13,743.60
July	5,814.23	6,717.37	9,469.08	13,252.53	18,339.25
August	6,906.38	9,139.72	10,221.00	13,819.77	-
September	8,075.13	11,315.19	15,161.41	21,755.64	-
October	7,913.56	10,838.40	17,713.08	24,838.10	-
November	9,054.18	11,150.94	17,508.99	15,955.60	-
December	14,114.58	17,592.20	27,566.89	19,370.70	-
Annual totals	$77,217.95 for 10 months	$114,871.25	$163,388.72	$198,560.61	$127,779.06 for 7 months

and past demands have been random. The usefulness of this data decomposition and trend identification is evident from Situation 6-4.

Discussion of the Young Forever Boutique

Kathy Wilkerson s situation at the Young Forever Boutique presents a fine opportunity for contrasting some of the more important techniques of forecasting using time series data. One simple forecasting rule that Kathy could have used rests on the premise that next year's sales will follow the same growth and pattern as this year's sales. The implications of such a rule are left as exercises at the end of the chapter.

Although simple rules can often generate perfectly acceptable forecasts, they can, just as often, be wildly incorrect. Two of the most popular alternatives to the use of simple rules are the moving average and exponential smoothing. Both of these forecasting methods rely exclusively on the historical pattern of demand, under the assumption that the future will look like the past. In contrast to the simple rule, however, moving average and exponential smoothing estimates can be designed to adapt very quickly to abrupt changes in the historical pattern.

Let us consider the *moving average* first. It is simply an average of past data and, each period, the data used in the average are updated to reflect, at least in part, what the most recent actual demand was. Moving averages can be simple in construction, with each period's value evenly weighted, or they can be more complex, with different weights applied to different time periods. When weights are used, they are typically defined so that they always sum to 1. In Kathy Wilkerson's case, for example, we could base the forecast on the past 3 months, but give the most weight (0.5) to the most recent month, less weight to the next to the last month (0.3), and the least weight to the month

furthest back (0.2). This weighted moving average can be represented mathematically as

$$F_t = 0.5D_{t-1} + 0.3D_{t-2} + 0.2D_{t-3}$$

where F_t is the forecast for this time period (month), D_{t-1} is the actual demand last period, D_{t-2} and D_{t-3} the actual demand two and three periods ago, respectively. Table 6-9 displays the forecast for year 3 for the Young Forever Boutique using this weighting scheme.

We could, of course, introduce more past data into the moving average calculation, or vary the weights applied, or specify particular time periods for use or exclusion (e.g., forecast first quarter data using only values from first quarters in the past, or exclude period $t - 1$ from the calculation). The more weight applied to recent history, the more responsive the forecast is to changes in the pattern of demand. The less weight applied to recent history—perhaps by the inclusion of many past actual demands— the more stable the forecast and the less apt it

TABLE 6-9   Young Forever Boutique moving average estimates

MONTH	FORECAST YEAR 3	ACTUAL YEAR 3
January	$ 14,309.06	$ 11,016.55
February	13,016.12	8,264.14
March	10,955.48	6,843.95
April	8,104.53	20,109.68
May	13,760.85	11,148.64
June	12,976.01	8,365.40
July	11,549.23	9,469.08
August	9,473.89	10,221.00
September	9,624.30	15,161.41
October	12,540.82	17,713.08
November	15,449.16	17,508.99
December	17,100.70	27,566.80
Annual Totals	$148,860.15	$163,388.72

Note: Because the sales in year 3 are a dramatic increase over those for year 2, the moving average estimates lag behind the year's actual sales.
   The weighting scheme ensures that the forecast does not dip as low or reach as high as actual sales do.

is to be swayed by aberrations in the data. The choice of weights and data points to use should reflect the forecaster's desires for stability versus responsiveness in the forecast.

Exponential smoothing is allied to the moving average. It represents a special weighting scheme that has the advantage of being particularly easy to compute, requiring only the most recent forecast and the most recent actual value. The forecast for the period takes the following form:

$$F_t = F_{t-1} + \alpha(A_{t-1} - F_{t-1})$$

which is built on the very logical assumption that the next period's forecast (F_{t-1}) equals the last period's forecast (F_{t-1}) plus the error in that forecast $(A_{t-1} - F_{t-1})$ adjusted by a correction factor (α).

Algebraic manipulation of this equation yields the following alternative formulation, which is often preferred for doing computations:

$$F_t = A_{t-1} + (1 - \alpha)F_{t-1}$$

where F_t is the forecast value for the time t, A_{t-1} is the actual value for the last period, and F_{t-1} is the forecasted value for the last period. The factor α can vary between 0 and 1, although in practice it is frequently limited to less than 0.5. The higher α is, the quicker the forecast follows past actual demand. If high values of α work well (i.e., lower errors than others), there is a good case that a significant trend exists in the data. That is why values greater than 0.5 are not frequently used for α.

This smoothing technique is termed exponential because by substitution with past values, the weights of prior periods decline exponentially from the most recent period on back in time as shown:

$$F_t = \alpha A_{t-1} + (1 - \alpha)\alpha A_{t-2} + (1 - \alpha)^2\alpha A_{t-3}$$
$$+ (1 - \alpha)^3\alpha A_{t-4} + \ldots$$
$$+ (1 - \alpha)^{t-2}\alpha A_1 + (1 - \alpha)^{t-1}F_1$$

The forecast for the Young Forever Boutique's year 3 using exponential smoothing (= 0.3) is found in Table 6-10.

Evaluating Forecasts

A forecast's performance can be judged after the fact by its errors. Errors, however, can be assessed in various ways. Three commonly used errors are displayed in Table 6-11: the simple difference between actual and forecast needs or sales (sometimes called *bias*), the mean absolute deviation (i.e., the mean of the absolute values of the differences between actual and forecast), and the mean squared error (using the square of the differences). Table 6-11 compares the year 3 forecasts and actuals for the moving average and exponential smoothing techniques. The table reveals how the three different measures of error react in this situation. The results indicate that

TABLE 6-10 Young Forever Boutique exponential smoothing with $\alpha = 0.3$		
MONTH	FORECAST YEAR 3	ACTUAL YEAR 3
January	$ 17,592.20	$ 11,016.55
February	15,619.51	8,264.14
March	13,412.90	6,843.95
April	11,442.21	20,109.68
May	14,642.45	11,148.64
June	13,174.31	8,365.40
July	11,731.64	9,469.08
August	11,052.87	10,221.00
September	10,803.31	15,161.41
October	12,110.74	17,713.09
November	13,791.44	17,508.99
December	14,906.71	27,566.80
Annual Totals	$160,280.29	$163,388.72

Note: The exponential smoothing technique, in this instance, yields forecasts that sum more nearly to the actual sales in year 3.

The sales extimates are really "smoothed," in the sense that they do not vary dramatically from one month to another. They are much smoother than the moving average estimates.

TABLE 6-11   Young Forever Boutique error analysis

A. Moving average estimates

MONTH	FORECAST YEAR 3	ACTUAL YEAR 3	DIFFERENCE (BIAS)	ABSOLUTE DEVIATION	SQUARED ERROR (MILLIONS)
January	$ 14,309.06	$ 11,016.55	$ −3,292.51	$ 3,292.51	$ 10.84
February	13,016.12	8,264.14	−4,751.98	4,751.98	22.58
March	10,955.48	6,843.95	−4,4,111.53	4,111.53	16.90
April	8,104.53	20,109.68	12,005.15	12,005.15	114.12
May	13,760.85	11,148.64	−2,612.21	2612.21	6.82
June	12,976.01	8,365.40	−4,610.61	4,610.61	21.26
July	11,549.23	9,469.08	−2,080.15	2,080.15	4.33
August	9,473.89	10,221.00	747.11	747.11	0.56
September	9,624.30	15,161.41	5,537.11	5,537.11	30.66
October	12,540.82	17,713.08	5,172.26	5,172.26	26.75
November	15,449.16	17,508.99	2,059.83	2,059.83	4.24
December	17,100.70	27,566.80	10,466.10	10,466.10	109.54
Annual Totals	$148,860.15	$163,388.72			
Sum of errors			$ 14,528.57	$57,466.55	$398.90
Mean difference (bias)			$ 1,210.71		
Mean Absolute deviation (MAD)				$ 4,787.21	
Mean squared error (in millions)					$ 33.24

B. Exponential smoothing with $\alpha = 0.3$

MONTH	FORECAST YEAR 3	ACTUAL YEAR 3	DIFFERENCE (BIAS)	ABSOLUTE DEVIATION	SQUARED ERROR (MILLIONS)
January	$ 17,592.20	$ 11,016.55	$ −6,575.65	$ 6,575.65	$ 43.24
February	15,619.51	8,264.14	−7,355.37	7,355.37	54.10
March	13,412.90	6,843.95	−6,568.95	6,568.95	43.15
April	11,442.21	20,109.68	8,667.47	8,667.47	75.13
May	14,642.45	11,148.64	−3,493.81	3,493.81	12.21
June	13,174.31	8,365.40	−4,808.91	4,808.91	23.13
July	11,731.64	9,469.08	−2,262.56	2,262.56	5.12
August	11,052.87	10,221.00	−831.87	831.87	0.69
September	10,803.31	15,161.41	4,358.10	4,358.10	18.99
October	12,110.74	17,713.08	5,602.35	5,602.35	31.39
November	13,791.44	17,508.99	3,717.55	3,717.55	13.82
December	14,906.71	27,566.80	12,660.09	12,660.09	160.28
Annual Totals	$160,280.29	$163,388.72			
Sum of errors			$ 3,108.43	$66,902.68	$481.25
Mean difference (bias)			$ 259.04		
Mean absolute deviation (MAD)				$ 5,575.22	
Mean squared error (in millions)					$ 40.10

while the exponential smoothing estimates offer less bias than the moving average estimates, they differ more wildly on a month-to-month basis from the actuals, having a higher mean absolute deviation and a higher mean squared error.

Note that when the mean difference (or bias) is computed, errors that are positive and negative cancel one another out to some degree. This is eliminated by taking the absolute values of the errors in calculating the mean absolute deviation. Squared errors penalize forecasts that are far off much more heavily than forecasts that are only a little off; mean squared errors are much more sensitive to outlying observations than are mean absolute errors.

The *mean absolute deviation* (*MAD*) can be a handy number, for it can be used to approximate the standard deviation of a sample of observations. If a sample is drawn from a normally distributed population, then the standard deviation (σ) is related to the MAD as follows:

$$\sigma \approx 1.25 \text{ MAD}$$

This relationship is sometimes employed in calculating safety stocks or in calculating control limits for quality control charts, topics discussed more in other chapters.

Dealing with Seasonality

The previous forecasts that use moving averages and exponential smoothing are flawed because they do not deal with the obvious seasonality of sales at the Young Forever Boutique. If adequate forecasts are to be made, this seasonality must be accounted for. There are several ways to deal with seasonality. Already

mentioned is the technique of basing, say, a quarterly forecast on past data from similar quarters. In Kathy Wilkerson's case, however, when monthly forecasts are needed and only a few years of history are available, another technique is probably preferable.

This technique employs the observation that demand can be decomposed into several components: a trend term, a seasonality term, and a randomness term.[b] Thus

$$\text{Demand} = \text{trend} \times \text{seasonality factor} \times \text{randomness component}$$

If moving averages or exponential smoothing or some other technique is to work well, then the data should be purged of their seasonality so that the trend can be identified and exploited for making the forecast. To purge seasonality, we can develop some seasonal indices (monthly, in this case) that can be used to deflate actual sales, leaving only the trend and the randomness components. These seasonal indices are typically ratios: average demand for the season (or month) divided by an average of demand for the 12 months surrounding it (or some time period not susceptible to the seasonality which you are trying to control).

There are several steps to follow to deseasonalize data in this way. Consider them for the Young Forever case.

1. Calculate a 12-month moving average of demand (where there is no seasonality, by definition) where that demand is centered on the month in question. For proper centering, we need an odd number of observations, so that developing centered moving average of demand for 12 months is itself a three-step procedure:

[b]One could also include a cyclicality term to account for business cycle fluctuations. This discussion omits consideration of cyclicality, in part because cyclicality could have been expected to play little or no role in explaining the Young Forever Boutique's steady growth. More complete discussion of forecasting demonstrates how cyclicality can be removed from an identified, underlying trend.

a. Calculate a 12-month mean, using months 1–12. This will be centered on month 6-1/2.

b. Advance the calculation by a month and calculate the new 12-month mean that is centered on month 7-1/2. This calculation uses months 2–13.

c. Average the two 12-month means to get a value centered on month 7.

Table 6-12 displays uncentered moving averages and then centered moving averages for the Young Forever data. Note that the first 6 months of data and the last 6 months are "lost" in the calculations.

2. Develop seasonal indices (or, in this case, monthly indices) by dividing the sales figures of Table 6-8 by the centered moving averages of Table 6-12 (see Table 6-13).

3. Average these seasonal indices over the years for which they were developed. That is, average all the January indices, average all the February indices, and so on (Table 6-13, panel B, column 1).

4. Sum these average indices and scale them up or down appropriately. In this case, the sum of the monthly indices is 1180.33. If sales were the same in every month, however, the monthly indices would total exactly 1200.

TABLE 6-12   Moving averages for 12 months' demand, centered and not centered

MONTH	YEAR 1	YEAR 2	YEAR 3	YEAR 4	YEAR 5
		Moving Averages (not centered, value placed into month 7)			
January	—	$ 8,332.96	$11,041.85	$15,600.72	$18,202.68
February	—	8,408.22	11,271.16	15,916.01	18,626.57
March	—	8,594.33	11,361.26	16,215.90	
April	—	8,864.34	11,681.78	16,765.42	
May	—	9,108.07	12,254.67	17,359.18	
June	—	9,282.80	12,784.51	17,229.73	
July	—	9,572.60	13,615.73	16,546.72	
August	—	9,958.89	13,694.82	16,431.31	
September	$7,333.74	10,280.42	14,069.64	16,486.21	
October	7,713.78	10,090.17	15,009.87	17,848.68	
November	8,049.50	10,851.77	14,925.14	17,850.00	
December	8,311.33	10,931.72	15,278.09	18,077.13	
		Moving Averages (centered)			
January	—	$ 8,370.59	$11,156.51	$15,758.37	$18,414.63
February	—	8,501.28	11,316.21	16,065.96	
March	—	8,729.34	11,521.52	16,490.66	
April	—	8,986.21	11,968.23	17,062.30	
May	—	9,195.44	12,519.59	17,294.46	
June	—	9,427.70	13,200.12	16,888.23	
July	—	9,765.75	13,655.28	16,489.02	
August	—	10,119.66	13,882.23	16,458.76	
September	$7,523.76	10,185.30	14,539.76	17,167.45	
October	7,881.64	10,470.97	14,967.51	17,849.34	
November	8,180.42	10,891.75	15,101.62	17,963.57	
December	8,322.15	10,986.79	15,439.41	18,139.91	

TABLE 6-13   Seasonal indices for the Young Forever Boutique

A. Data transformed to reasons indices (raw sales divided by centered moving averages)

MONTH	YEAR 1	YEAR 2	YEAR 3	YEAR 4	YEAR 5
January	—	76.96	98.75	75.93	57.46
February	—	51.23	73.03	79.43	
March	—	104.55	59.40	109.92	
April	—	122.08	168.03	111.90	
May	—	110.81	89.05	88.95	
June	—	74.71	63.37	72.46	
July	—	68.78	69.34	80.37	
August	—	90.32	73.63	83.97	
September	107.33	111.09	104.28	126.73	
October	100.40	103.51	118.34	139.15	
November	110.68	102.38	115.94	88.82	
December	169.60	160.12	178.55	106.78	
Annual Totals		1,184.41	1,211.71	1,164.41	

B. Average of seasonal indices

MONTH	UNCORRECTED	CORRECTED
January	77.09	78.37
February	68.10	69.23
March	91.29	92.81
April	134.00	136.23
May	96.27	97.87
June	70.18	71.35
July	72.83	74.04
August	82.64	84.02
September	112.36	114.23
October	115.35	117.27
November	104.46	106.20
December	155.76	158.36
Annual Totals	1,180.33	

Thus the indices as originally calculated can be scaled up by 1.0167 (1200/1180.33) (Table 6-13, panel B, column 2).

5. Divide the monthly sales figures of Table 6-8 by the seasonal (monthly) in dices to purge the seasonality from those raw sales figures. Table 6-14 displays the deseasonalized sales figures.

Identifying Trends

Once the data have been purged of seasonality, what remains are the trend and randomness factors. Normally, we choose to think about the trend as a linear function of time (sales change by $ x each time period), although, just as easily, we could think of the trend as exponential,

TABLE 6-14   Deseasonalized sales for the Young Forever Boutique

MONTH	YEAR 1	YEAR 2	YEAR 3	YEAR 4	YEAR 5
January	—	$ 8,142.34	$14,057.10	$15,268.19	$13,501.01
February	—	6,364.05	11,937.22	18,434.18	19,385.92
March	$4,920.16	9,834.02	7,374.15	19,530.98	37,147.15
April	5,095.71	8,052.89	14,761.57	14,015.19	14,026.86
May	7,200.60	10,411.04	11,391.27	15,718.82	18,503.67
June	9,508.58	9,872.25	11,724.46	17,150.67	19,262.23
July	7,852.82	9,072.62	12,789.14	17,899.15	24,769.38
August	8,219.92	10,878.03	12,164.96	16,448.19	
September	7,069.18	9,905.62	13,272.70	19,045.47	
October	6,748.15	9,242.26	15,104.93	21,180.27	
November	8,525.59	10,499.94	15,486.81	15,024.11	
December	8,912.97	11,108.99	17,407.68	12,232.07	

logarithmic, or something else. For the sake of simplicity, let us think of the trend as linear. Thus, mathematically, we view the sales data as exhibiting the following form:

$$(6\text{--}1) \qquad D_t = a + bP_t$$

where D_t is the actual sales for period t, P_t is the time period applicable, a is a constant, and b describes the trend.

The constant term a and the trend coefficient b can be estimated statistically through a technique termed linear regression. The deseasonalized sales of Table 6-14 can easily be used in a regression analysis of the suitable trend. Regression can also be used to determine whether the trend changed when Kathy Wilkerson moved the boutique from the renovated factory to the mall. This can be done by segmenting the data into before- and after-the-move groups and estimating equation 6-1 separately for each group. Table 6-15 shows the results of just such segmentation and estimation.

The results suggest that Young Forever's sales trend did increase after making the move (i.e., b increased from $223.59 per month to $307.27 per month). Such a shift in the trend is called a turning point. The differences in R^2

TABLE 6-15   Trend identification regression results for the Young Forever Boutique

Results for Before-Move Data		
Estimates of	a:	$6,007.23
	b:	$223.59 per month
		$R^2 = 0.62$
Results for After-Move Data		
Estimates of	a:	$13,783.32
	b:	$307.27 per month
		$R^2 = 0.18$

statistics also suggest that Kathy was right to sense that sales were more erratic in the mall; the trend explains less of the variation for the after-move data than it does for the before-move data.[c]

By using the estimated trend to predict a deseasonalized forecast and then by applying the relevant seasonal (monthly) index, forecasts can be made of the Young Forever Boutique's sales. Five months of forecasts are shown in Table 6-16.

[c]There are also tests to determine whether the two regressions could reasonably be expected to be truly different from one another. Refer to statistics or econometrics texts for an explanation.

TABLE 6-16 Predicting sales for the next five periods for the Young Forever Boutique		
MONTH	DESEASONALIZED ON TREND	SEASONAL ADJUSTMENT MADE
August	$21,465.07	$18,034.95
September	21,772.34	24,870.54
October	22,079.61	25,892.76
November	22,386.88	23,774.87
December	22,694.15	35,938.46

There are other, more sophisticated techniques for forecasting.[8] They are too elaborate for full explanation in a text on production/operations management, but some brief comments on them are in order:

- *Adaptive filtering.* This technique often gives better results than exponential smoothing or moving averages. It uses weighted averages of past actual demands but allows the weights to vary from period to period in such a way that the forecast error is substantially lowered. Each weight in the adaptive filtering scheme is adjusted by an equation that uses the error in the previous forecast period.

- *Box-Jenkins.* Even more sophisticated than the adaptive filtering technique is the technique devised by George Box and Gwilym Jenkins. The Box–Jenkins method is particularly appropriate for complex time periods in which one suspects that a variety of patterns exists in the data. The technique is too complex to describe briefly, and it can be expensive to run, requiring considerable past histories of observations.

Causal Models

The other major type of quantitative forecasting model uses not only time series data on the demand itself but also time series data on other variables that could be expected to influence demand in a cause and effect way. A model that can accurately capture the cause and effect relationship among the other variables and the demand can be very useful, especially in picking turning points in the demand pattern. However, these models tend to be expensive to run and the analysis required to define them properly takes more time; thus they are much more appropriate for long-term forecasting than for short-term forecasting. The two major classes of causal models are single equation regression models and multiple equation regression analysis, sometimes termed econometric modeling. In single equation regression analysis, the item forecast, or dependent variable, is postulated to vary systematically with one or more "independent variables." These independent variables can be constructed in a variety of ways, including the use of time series data (GNP, industrial production), past values of the dependent variable, and dummy variables (e.g., 1 if the month is May and 0 otherwise). Regression analysis fits the specified independent variables to the dependent variable by using historical data; this is termed estimating the coefficients of the model's specified independent variables. Predictions can then be made by assigning expected future values to the independent variables and solving the estimated equation for the dependent variable's value. Econometric modeling is a step up from simple regression analysis in that it looks at collections of regression equations where some of the dependent variables in one equation turn up as independent variables in the other equations. Econometric modeling, in particular, can be very expensive and time consuming, since the data demands are voracious and the estimation tricky.

Qualitative Techniques

The techniques previously explored make use of quantitative data, developed from history.

However, if one thinks that the future is likely not to depend much on history—because of vast disruptions in the economy or the industry, or simply because what is to be forecast is new and uncharted—one may then want to rely on the expertise of specific people. Qualitative forecasting is often applied to new technology, trying to forecast the kinds of changes to expect. The Delphi Method is perhaps the most common qualitative approach. This method uses a panel of experts who are asked particular questions. Their answers are collected and summarized (with means, ranges, standard deviations, etc.), and this information is passed out to the panel. The panel then deliberates for another round of forecasting using the information provided. Their responses are again collected, summarized, and disseminated. Numerous rounds of prediction and summary can occur. In some cases, consensus is reached after several rounds; this is particularly true if outlying responses are not held with much conviction.

On the other hand, consensus is not guaranteed if outliers are stubborn about their views.

Choosing a Forecasting Technique

As was mentioned at the start of this appendix, particular forecasting techniques are associated, in the main, with different forecasting needs. The simple time series techniques are easy to use and are generally applied to very short-term forecasts of individual product models and variations. In those cases, history is likely to be a good guide to the immediate future. The more sophisticated of the time series techniques are more useful when the historical pattern appears to be more confused and when varied trends in the data are likely to exist. The causal and qualitative techniques are more satisfactory for longer-term needs, with the qualitative techniques being at an advantage only when history is deemed to be of little significance for prediction.

QUESTIONS

1. As you understand them, what are the main differences in the key planning and control features of each of the following pairs of production processes?
(a) Job shop and continuous flow.
(b) Worker-paced line flow and machine-paced line flow.
(c) Job shop and a hybrid process.

2. What are the factors that should govern a manager's choice of forecasting technique?

3. What is the purpose of establishing a production schedule?

4. The development of a master production schedule requires serious decision making, certain types of data, and a strategy for meeting the production goals set. List the main elements in each of these that must be considered. How do these elements influence the preparation of a master schedule at Ridgely Sporting Goods?

5. The text makes a distinction between scheduling processes whose capacities are clear-cut and those whose capacities are muddled. Where do hybrid processes fit into this scheme of things? What types of problems are likely to be encountered when scheduling hybrid processes?

6. What are the advantages of a decoupled inventory to a production planner?

7. Discuss briefly how scarce resources can affect production planning.

8. What criteria should be used in determining whether long-term, medium-term, or short-term forecasting techniques are appropriate for a given production process?

9. What is capacity planning? Why is it important?

10. What are some key considerations in deciding between a "chase" strategy and a "level" strategy for a master production schedule?

11. Compare optimization and nonoptimization techniques for deriving production schedules.

12. Define and/or give the mathematical expression for the following:
(a) Objective function.
(b) Structural constraints.
(c) Nonnegativity conditions.

13. The linear programming method has numerous applications in operations problems; it also has several limitations. Discuss each of these in turn.

14. How is linear programming useful in making product specification decisions? In plant location decisions?

PROBLEMS

1. Under each of the following situations would a chase or a level production strategy be favored?
(a) Setup and changeover costs have been reduced.
(b) A strong new competitor comes into the industry, making it more likely that a good substitute could be found for your company's product.
(c) Recession strikes and your vendors and possible supplier firms are hungry for work.
(d) Your company has instituted a second shift of production.

2. As the demand on a factory falls, should its managers adopt different rules concerning the number of setups and/or the run lengths desired? Why or why not?

3. It has been observed that large-tract builders do more subcontracting than do smaller builders. This runs against the typical view that large size fosters more vertical integration. How can you explain this phenomenon in terms of production planning?

4. Owens, Inc., has been awarded a blanket 1-year contract for the requirements for the bracket in Tour B (situation for study B-1). Fred wants to start and end the year with 500 units

in inventory for safety stock. The contract is for 6000 brackets (average usage 500 per month) with the following estimated demand:

July	500
August	500
September	350
October	400
November	450
December	450
January	650
February	700
March	500
April	550
May	400
June	550

Fred has made the following estimates of production and inventory carrying costs information: startup costs to produce 1 to 500 on a single shift are $200; to run a second shift, startup costs are $250; to shut down either shift is $150. Storage will cost approximately 50 cents per unit per month, with maximum storage space for 1500 units.
(a) What is Fred's best production schedule for the year?
(b) Could you set up this problem for the linear programming technique? Support your answer.

(c) Fred has just learned that the second-shift operator will return to college in January, and Fred does not want to replace him. How will this affect your production plan?

5. The forecast for Ridgely Sporting goods has changed. The new forecast is as follows:

MONTH	BASEBALL GLOVES	HOCKEY GLOVES
July	4,500	100
August	3,000	100
September	3,000	800
October	1,200	2,000
November	300	1,900
December	5,500	2,300
January	2,000	1,500
February	3,500	1,300
March	7,500	1,200
Annual Total	30,500	11,200

In addition, the initial inventories are 5000 baseball gloves and 800 pairs of hockey gloves. The costs of hiring and layoff have been revised and now stand at $1000 for each layoff and $500 for each new hire. All other data are as found in the situation on Ridgely Sporting Goods. Concoct a production plan for baseball and hockey gloves. You may want to make use of a computer spreadsheet to help you.

6. The Nell Corporation fabricates credit card blanks for various banks and store chains. The demand forecast for the 12-month period beginning 6 months from now (i.e., months 7 to 18) is as given in the accompanying table. The seasonal peak in this growing business for the company has been in the summer (months 12 to 14), when most new residents to an area establish new accounts.

The factory currently employs 20 line workers on a single 8-hour shift. Line workers are paid $15 per hour (including all benefits).

Overtime is paid at time and a half. A maximum of 1.6 hours per day in overtime was deemed "sustainable" by management. For planning purposes, the production planning department assumes an output per worker of 5 boxes of finished cards per day, given normal lot sizes and changeovers. There were 20 work days in each month, except for month 13, when there could only be 10 workdays because the plant was shut down for vacation.

Production could be augmented by either new hires or temporary workers. Unfortunately, there was only equipment enough on the first shift for two more workers. Because of the nature of the process, subcontracting was impossible, but a second shift could be established, although for technical reasons, such a second shift would have to employ at least seven workers to run the various pieces of equipment required. Pay would be $18 per hour for second-shift work.

Temporary workers cost the company $21 per hour for first-shift work and $24 per hour for second-shift work. New hires could be attracted to the company at an average cost per hire of $200 for advertising, training, and so on. Layoffs cost the company an average of $1500 per worker (includes severance pay, counseling, retraining, and the like).

Customers whose orders were not ready one time were understandably angry. Stockout costs were estimated to be $200 per box of cards. Inventory holding costs were held to be $10 per box per month (counting carrying cost, storage, handling, damage, and obsolescence). No inventories were planned for the start of month 7.

MONTH	BOXES FORECAST	MONTH	BOXES FORECAST
7	2,000	13	2,750
8	2,150	14	3,000
9	2,300	15	2,500
10	2,250	16	2,200
11	2,300	17	1,900
12	2,700	18	2,100
			28,150

(a) Concoct a production plan for the month 7 to 18 time period. Compute its cost.

(b) Identify the strategy you used to develop your plan and state why you did it that way.

7. "Maintenance should be viewed as a machine that must be scheduled like any other." Comment on this statement.

8. Calculate the expected loads for three workcenters, given the following information. Expected total hours required to produce what is demanded:

| | TIME PERIOD | | |
PRODUCT	1	2	3
X	5	10	15
Y	15	25	30

LABOR STANDARDS:

Product X	0.75 hour/unit
Product Y	0.25 hour/unit

Historical average utilization by each workcenter:

WORKCENTER	UTILIZATION (percent)
1	60
2	25
3	15

9. Suppose that the capacity bill for the workcenters 1, 2, and 3 of Problem 9 looked as follows:

WORKCENTER	PRODUCT X (hours/unit)	PRODUCT Y (hours/unit)
1	0.35	0.10
2	0.25	0.10
3	0.15	0.05
	0.75	0.25

What would the workcenter loads be then?

10. Calculate the forecast of sales (broken down by month) for year 3, given the following information.

Sale of Textbooks		
YEAR 2	SALES/MONTH	PERCENT OF TOTAL
January	$ 71,025	7.3
February	63,429	6.5
March	82,630	8.8
April	51,345	5.1
May	74,350	7.7
June	82,964	8.4
July	94,899	9.8
August	108,465	11.2
September	97,048	10.1
October	63,748	6.7
November	83,971	8.6
December	95,437	9.8
	969,374	100.0

Calculate the forecast first using the following simple rule, which uses these three assumptions:

• Year 3's total sales will grow at the same rate at which this year's sales grew.

• The monthly breakdown of sales for year 3 will match this year's breakdown.

• Year 2's sales were 17.3 percent greater than year 1's sales.

11. Using the same data, predict monthly sales for year 3 using the moving average technique.

12. Assume that you found a trend in the data. What would be your next step in dealing with the data?

13. Using several of the techniques of forecasting described in the text (moving average, exponential smoothing, decomposition) and the data in Table 6-17, make a forecast for the next year. Do not hesitate to play with the data. How do these various techniques differ in the forecasts they make? What explains the differences?

TABLE 6-17   Actual and forecasted monthly sales (in millions of dollars) for a certain retail product (Problem 13)

MONTH	YEAR 1 ACTUAL	YEAR 1 FORECAST	YEAR 2 ACTUAL	YEAR 2 FORECAST	YEAR 3 ACTUAL	YEAR 3 FORECAST	YEAR 4 ACTUAL	YEAR 4 FORECAST
January	$ 3,934	$ 3.9	$ 2,639	$ 3.4	$ 4,016	$ 3.1	$ 3,633	$ 3.4
February	3,162	3.3	2,899	2.9	3,957	3.2	4,292	3.6
March	4,286	3.9	3,370	3.2	4,510	3.7	4,154	3.8
April	4,676	4.5	3,740	4.0	4,276	4.3	4,121	4.4
May	5,010	5.3	2,927	3.1	4,986	4.5	4,647	4.8
June	4,874	5.1	3,986	3.9	4,677	5.0	4,753	5.0
July	4,633	4.6	4,217	4.3	3,523	4.4	3,965	4.5
August	1,659	1.8	1,738	1.6	1,821	1.7	1,723	1.8
September	5,951	6.0	5,221	5.3	5,222	5.7	5,048	5.8
October	6,981	6.8	6,424	6.6	6,832	6.9	6,922	6.9
November	9,851	10.0	9,842	10.0	10,803	10.2	9,858	10.1
December	12,670	13.1	13,076	12.9	13,916	13.4	11,331	13.0

14. How good have the past forecasts in Problem 13 been? How would you measure the quality of those forecasts? Was one year's forecast better than the others?

15. The sales forecast for product number A107 is 50 units a week for the next 8 weeks. There are 65 units of A107 on hand; the master schedule calls for delivery of 75 units next week (week 2) and 60 more in week 3. The booked customer orders are 73 for week 1 (past due included), 68 for week 2, and 52 for week 3. An order of 12 has just been received. Which week is it scheduled for? Fill out the form (Figure 6-5 is an example) for this situation.

16. The Pork & Rind firm manufactures "Tater Chips" and "Cheezy Crackers." The firm has combined some old and some new techniques that have caused these two products to have a high sales volume. It can run 9000 ounces of chips or 16,000 ounces of crackers through the oil spray unit per day; 15,000 ounces of chips or 13,000 ounces of crackers can be baked in the oven daily. What's more, the basic dough is the same for both products, and the firm can mix enough for 12,000 ounces of crackers or 15,000 ounces of chips per day. Pork & Rind can purchase enough potato extract for 6000 ounces of chips and enough cheese for 14,000 ounces of crackers. Pork & Rind anticipates a profit of 14 cents per ounce of crackers and 11 cents per ounce of chips.

(a) Prepare a linear program for the daily production schedule that will generate the most profit.

(b) If Pork & Rind were to try to generate more capacity, on which constraint should it exert the most effort?

(c) The firm develops a new product, "Puffy Cheezies," which requires the same inputs as the crackers plus a gun from which to "shoot" them. This gun can produce 8000 ounces per day. The Puffy Cheezies would generate a profit of 8 cents per ounce. Generate a new production schedule that will maximize profit.

SITUATION FOR STUDY 6-1

STAINLESS TANK COMPANY, INC.

Jack Black, the president and chief salesperson of the Stainless Tank Company, Inc., has been having difficulty keeping his company "in the black." For several months, Stainless Tank had been showing a loss, even though Mr. Black and the other two salespeople were working hard on generating new sales. Mr. Black had some ideas he wanted to use to increase sales volume and thus offset this loss; but being a cautious person, he thought he would first ask an outside expert to look at his operation. Mr. Black asked Dr. Hanna Hendricks, professor of operations management at the local university, to visit Stainless Tank and provide some insight into the problem.

Mr. Black had purchased the Stainless Tank Company and moved it to its present location. At that time the firm manufactured stainless steel tanks and mounted the tanks and ancillary equipment on modified truck and trailer chassis. These tank trucks were used to transport milk for the dairy industry. This portion of the business had remained stable, both in product design and demand, for the last several years. In addition, the firm contracted with insurance companies for the repair of tanks that had been damaged in accidents.

In recent years the government environmental agency had decided that sewage plants could no longer dump their effluent into the local waterways. This regulation opened up a new market for Stainless Tank. Among the current methods of disposing of the liquid and semi-liquid output from sewage plants was to haul it to a farmer's field and spread it over or plow it under the field. The sewage tank trucks used the same basic manufacturing technology as the milk tank trucks, with only moderate differences.

The manufacturing process started with the delivery of the truck or trailer from a local dealer. First the truck was modified with extra wide fenders and tires that could be fitted before the tank was mounted. The tank fabrication entailed shearing, rolling, and welding. The rounded tank ends, the pumps, and various other parts were purchased. The tank support brackets were fabricated and mounted, after which the tank was mounted. After tank mounting, the piping was done. Some difficulty was usually encountered in getting all the components to fit as they had been designed; thus, that portion of the assembly needed some "cut-to-fit" work.

The workers in the shop were all welders; however, they were expected to do any of the other work necessary to "build" a unit. The shop foreman was the official "expert" on dealing with problems encountered in the construction of a unit. Although Mr. Black knew that this informal delegation of duties and responsibilities would not work well in a large organization, he thought it had worked well for Stainless Tank. See Figure 6-9 for a partial organization chart of the company.

After careful scrutiny of the firm and its records, Dr. Hendricks concluded that the major problem was caused by the larger volume of sales. The larger volume had created information needs that could not be handled in an informal manner. The time card for a job (Figure 6-10) recorded that Tom, Dick, or Harry had worked on a particular job for a particular length of time. These cards were the basis for calculating the cost of production both for each unit and for the plant. The total cost of production for each tank was compared with its contracted sales price to determine its profit or loss.

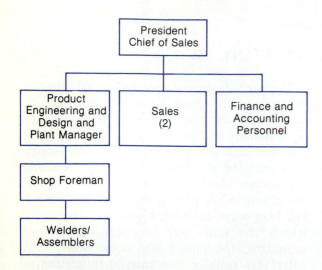

FIGURE 6-9   Partial organization chart for
Stainless Tank Company.

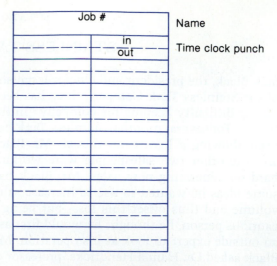

FIGURE 6-10   Time card used by
Stainless Tank Company.

Dr. Hendricks thought the first thing Mr. Black should do to get Stainless Tank back on the road to profitability was to establish a method of scheduling the production of tanks. She thought that this scheduling method would not only provide the shop foreman with the proper sequence of operations and estimated times but also provide management with a comparison of actual progress and planned progress. The analysis of the information and the reconstruction of events from this proposed method should provide insight into where and why the firm was losing money.

1. If you were Dr. Hendricks, what part of the process would you start with to establish a scheduling method? Set up a plan for your proposed investigation.

2. What information or data would you search for? What data would you want the shop to collect and to whom should this information go?

3. How should the materials be controlled in this situation?

4. What do you think of Mr. Black's desire for more volume? How would you discuss this aspect with him?

5. What method of master production scheduling would you use in this case? Support your answer with your reasons.

6. Do you think linear programming would help Mr. Black? If so, how would you set it up?

SITUATION FOR STUDY 6-2

WHITE LAWN & GARDEN EQUIPMENT

White Lawn & Garden Equipment manufactured rakes, hoes, shovels, and assorted other lawn and garden tools. Planning was done every quarter for the next 6 months, and Bill White, a member of the founding family, was charged with overseeing the production planning process.

The latest sales forecast, just for the three lines of rakes, is shown on the right.

Rake sales forecast (next 6 months in units)

PRODUCT LINES (rakes)	MONTHS					
	APR.	MAY	JUNE	JULY	AUG.	SEPT.
A models	10,000	9,000	4,000	5,000	7,500	7,000
D models	5,500	4,000	3,000	3,000	2,500	1,000
L models	1,500	2,500	2,000	2,000	3,000	4,000

Policy was to keep one-half of average monthly sales in ending inventory at the end of any quarter. The current inventory positions were: A models, 4500 units; D models, 3500 units; and L models, 1500 units.

The rake department was responsible for fabricating and assembling all three product lines. All aspects of fabrication, assembling, inspection, and packing took 1/2 hour of direct labor for the A and D models and 45 minutes of direct labor for the L models. There were 45 direct labor workers in the department who worked an average of 160 hours a month at an average wage of $9.00 per hour. Overtime was paid at the standard of time and a half. The task of making rakes was accomplished by small groups of employees working together, so change-overs were of little consequence. All three types of rakes could be made at the same time.

The standard cost of the A models was $10, that of the D models was $12, and that of the L models, $15. Inventory costs were at 25 percent of the standard cost, on a yearly basis.

1. Develop a master production schedule for the next 6 months.

2. Formulate this as a linear program.

REFERENCE NOTES

1. For additional information about the general topic of production planning, consult the following:

Avonts, Ludwig H., Ludo F. Gelders, Luke N. Van Wassenhove, "Allocating Work Between an FMS and a Conventional Jobshop: A Case Study," *European Journal of Operational Research (Netherlands)* 33, no. 3 (February 1989): 245–256.

Biemans, Frank P.M., *Manufacturing Planning and Control: A Reference Manual*, New York: Elsevier Science Publishers, 1990.

Blackstone, John H., *Capacity Management*, Cincinnati: South-Western, 1989.

Finch, Byron J., and James F. Cox, "Process-Oriented Production Planning and Control: Factors That Influence System Design," *Academy of Management Journal* 31, no. 1 (March 1988): 123–153.

Hax, Arnoldo C., and Dan Candea, *Production and Inventory Management*, Englewood Cliffs, N.J.: Prentice-Hall, 1984.

King, Barry E., and W.C. Benton, "Alternative Master Production Scheduling Techniques in an Assemble-to-Order Environment," *Journal of Operations Management* 7, no. 1–2 (October 1987): 179–201.

Leong, G. Keong, Michael D. Oliff, and Robert E. Markland, "Improved Hierarchical Production Planning," *Journal of Operations Management* 8, no. 2 (April 1989): 90–114.

Luenberger, D.G., *Linear and Nonlinear Programming*, 2nd ed. Reading, MA: Addison-Wesley, 1984.

McClelland, Marilyn K., "Order Promising and the Master Production Schedule," *Decision Sciences* 19, no. 4 (Fall 1988): 858–879.

McKay, Kenneth N., Frank R. Safayeni, and John A. Buzacott, "Job-shop Scheduling Theory: What is Relevant?" *Interfaces* 18, no. 4 (July–August 1988): 84–90.

Meal, Harlan C., "Putting Production Decisions where They Belong," *Harvard Business Review* 62, no. 2 (March–April 1984): 102–21.

Oliff, Michael D., and G. Keong Leong, "A Discrete Production Switching Rule for Aggregate Planning," *Decision Sciences* 18, no. 4 (Fall 1987): 582–597.

Parekh, Rajen, "Capacity/Inventory Planning Using a Spreadsheet," *Production and Inventory Management* 31, no. 1 (First Quarter 1990): 1–3.

Reinfeld, Nyles V., ed., *Handbook of Production and Inventory Control*, Englewood Cliffs, N.J.: Prentice-Hall, 1987.

Schroeder, Roger G., and Paul D. Larson, "A Reformulation of the Aggregate Planning Problem," *Journal of Operations Management* 6, no. 3/4 (May–August 1986): 245–256.

Tallon, William J., "A Comparative Analysis of Master Production Scheduling Techniques for Assemble-to-Order Products," *Decision Sciences* 20, no. 3 (Summer 1989): 492–506.

Vollmann, T.E., William L. Berry, and D. Clay Whybark, *Manufacturing Planning and Control Systems*, 3rd ed. Homewood, IL.: Dow Jones-Irwin, 1992.

2.   For additional information on production scheduling, see:

Baker, K. R., *Sequencing and Scheduling*, New York: Wiley, 1974.

Berry, W. L. et al., *Master Scheduling*, Washington, D.C.: American Production and Inventory Control Society, 1979.

Blackstone, John H., Jr., Don T. Phillips, and Gary L. Hogg, "A State-of-the-Art Survey of Dispatching Rules for Manufacturing Job Shop Operations," *International Journal of Production Research* 20, no. 1 (1982): 27–45.

Bowman, E. H., "Consistency and Optimality in Managerial Decision Making," *Management Science* 9, no. 2 (January 1963): 310–21.

Bowman, E. H., "Production Scheduling by the Transportation Method of Linear Programming," *Operations Research* 4, no. 1 (February 1956): 100–103.

Buffa, Elwood S., and Jeffrey G. Miller, *Production Inventory Systems: Planning and Control*, 3rd ed. Homewood, IL.: Richard D. Irwin, 1979.

Conway, R. W. et al., *Theory of Scheduling*, Reading, MA.: Addison-Wesley, 1967.

Dannenbring, D. G., "An Evaluation of Flow Shop Sequencing Heuristics," *Management Science* 22, no. 11 (July 1971): 1174–1182.

Eilon, Samuel, "Five Approaches to Aggregate Production Planning," *AIIE Transactions* 7, no. 2 (June 1975): 118–31.

Graves, S. C., "A Review of Production Scheduling," *Operations Research* 29, no. 4 (July–August 1981): 646–675.

Holt, C. C. et al., *Planning Production, Inventories, and Work Force*, Englewood Cliffs, NJ: Prentice-Hall, 1960.

Muth, John F., and Gerald L. Thompson, with collaboration from Peter R. Winters, *Industrial Scheduling*. Englewood Cliffs, NJ: Prentice-Hall, 1963.

Panwalkar, S. S., and W. Iskander, "A Survey of Scheduling Rules," *Operations Research* 25, no. 1 (January–February 1977): 45–61.

Plossl, George W., and Oliver W. Wight, *Production and Inventory Control: Principles and Techniques*. Englewood Cliffs, NJ: Prentice-Hall, 1967.

Sandman, W. E., with J. P. Hayes, *How to Win Productivity in Manufacturing*. Dresher, PA: Yellow Book of Pennsylvania, 1980.

Vollman, T. E., W. L. Berry, and D. Clay Whybark, *Manufacturing Planning and Control Systems*. 3rd ed. Homewood, IL: Dow Jones-Irwin, 1992.

3.   See William L. Berry, Thomas G. Schmitt, and Thomas E. Vollmann, "Capacity Planning Techniques for Manufacturing Control Systems: Information Requirements and Operational Features," *Journal of Operations Management* 3, no. 1 (November 1982); 13–25.

4.   For information on alternatives and refinements to the linear programming approach, see Buffa and Miller, *Production-Inventory Systems*, Chapter 6.

5.   C. C. Holt, F. Modigliani, and H. A. Simon, *Planning Production, Inventories, and Work Force*. Englewood Cliffs, NJ: Prentice-Hall, 1960. Also, Holt, Modigliani, and Simon, "A Linear Decision Rule for Production and Employment Scheduling," *Management Science* 2, no. 2 (October 1955): 10–30.

6. See particularly "Hierarchical Integration of Production Planning and Scheduling" by Arnaldo Hax and Harlan Meal in M. A. Geisler, ed., *Studies in the Management Sciences: Logistics*, Vol. 1, New York: North-Holland, 1975; and "On the Design of Hierarchical Production Planning Systems" by G. R. Bitran and Arnaldo Hax in *Decision Sciences* 8, no. 1 (January 1977): 28–55.

7. See W. H. Taubert, "A Search Decision Rule for the Aggregate Scheduling Problem," *Management Science* 14, no. 6 (February 1968): 342–59.

8. For additional information on forecasting, see:

    Barron, Michael, and David Targett, *The Manager's Guide to Business Forecasting: How To Understand and Use Forecasts for Better Business Results*, New York: Basil Blackwell Ltd., 1985.

    Bowerman, B. L., and R. T. O'Connell, *Time Series Forecasting*, Boston: Duxbury, 1986.

    Box, George E.P., and Gwilym M. Jenkins, *Time Series Analysis: Forecasting and Control*, 2nd ed. Oakland, CA.: Holden-Day, 1987.

    Cohen, Rochelle, and Fraser Dunford, "Forecasting for Inventory Control: An Example of When 'Simple' Means 'Better'," *Interfaces* 16, no. 6 (November–December 1986): 95–99.

    Gardner, Everette S., Jr., "Exponential Smoothing: The State of The Art," *Journal of Forecasting* 4, no. 1 (March 1985).

    Hanke, John E., and Arthur G. Reitsch, *Business Forecasting*, 3rd ed. Boston, MA: Allyn and Bacon, 1989.

    Makridakis, Spyros, and Steven C. Wheelwright, *The Handbook of Forecasting: A Manager's Guide*, 2nd ed. Somerset, N.Y.: Wiley, 1987.

    Makridakis, Spyros, and Steven C. Wheelwright, *Forecasting Methods for Management*, 5th ed. New York: Wiley, 1989.

    Vollmann, T.E., W.L. Berry, and D. Clay Whybark, *Manufacturing Planning and Control Systems*, 3rd ed. Homewood, IL: Dow Jones-Irwin, 1992.

    Willis, Raymond E., *A Guide to Forecasting for Planners and Managers*, Englewood Cliffs, NJ: Prentice-Hall, 1987.

    Wright, George, and Peter Ayton, eds., *Judgemental Forecasting*, New York: Wiley, 1987.

CHAPTER 7

MANAGING INVENTORIES: MATERIAL REQUIREMENTS PLANNING

Our investigation of the different production processes of Chapter 1 has already uncovered a number of places within the typical process where inventories come in handy. In broad terms, we saw inventories of raw materials, work-in-process, and finished goods.

Other things being equal, of course a company would like not to hold any inventories, because they mean tying up cash in goods that cannot improve the company's earnings the way a new piece of machinery can, or putting cash in the bank can.[a] In other words inventories represent some very real opportunity costs to the company since, by financing inventories, the company forgoes the opportunity to earn a better return by using its cash in some other, income-generating way.

As we also saw in Chapter 1, some processes are organized in certain ways expressly for the purpose of reducing inventories of one kind or another. The job shop, for example, typically produces only to order—eliminating any concern for managing a finished goods inventory.

Because many orders require very specialized raw materials either supplied by the customer or placed on special order, the job shop's need to manage raw materials inventories is very modest. It is in the realm of work-in-process inventories that the job shop is truly challenged with the control of inventories.

The inventory concerns of the continuous flow process are almost exactly the reverse of those of the job shop. The entire design of the continuous flow process revolves around speeding the manufacture of a very standard product, which implies a rock-bottom level of work-in-process inventory. Because the process's appetite for raw materials is so steady and predictable, raw materials inventories can usually be held down by insisting on long-term supply contracts with a steady flow of materials. It is only at the finished goods end where inventories can mushroom and where their control presents real management challenges.

Process types in between the poles of job shop and continuous flow generally strike

---

[a] Other things are not always equal, however. Sometimes holding large inventories can make sense; the seasonality of logging, for example, is one reason why International Paper's Androscoggin Mill (Tour A at the back of this text) has a significant raw materials inventory. Also some raw materials purchases may be unusually large for speculative reasons—to avoid an expected increase in price.

some middle ground in the importance of raw materials, work-in-process, and finished goods inventory control.

OBJECTIVES AND PHILOSOPHIES OF INVENTORY MANAGEMENT

Raw materials inventories, work-in-process inventories, and finished goods inventories all play different roles in the operation. As we have seen, the importance of those roles depends on the process. Managers should be clear about the major functions of these types of inventory for their particular processes.[1]

The major functions of raw materials inventory are:

- To protect (as a buffer) against the unreliable delivery of needed raw materials.
- To hold costs down if possible by buying large quantities or by buying at propitious times.

The major functions of work-in-process inventory are:

- To protect (as a buffer) against the unreliable delivery of materials from elsewhere in the production process.
- To permit one segment of the process to operate under a different production plan and schedule from another segment of the process (a decoupling inventory).
- To permit individual workstations or machine centers to produce parts/assemblies/materials in sizable batches, rather than individually or in smaller batches. Such "cycle" inventories act to tide the process over until the next setup.

The major functions of finished goods inventory are:

- To supply the product quickly to the consumer.

- To protect (as a buffer) against the uncertainties of consumer demand.
- To smooth (through the accumulation of finished goods inventory) the demands on the process even while demand is erratic or temporarily depressed.

The tools that production managers can wield in dealing with work-in-process inventories are frequently very different from the tools they can use to control raw materials or finished goods. As we shall soon see, the available techniques for controlling either raw materials inventories or finished goods inventories are relatively well developed and widely known. Because of this, it is often taken for granted that these inventories are managed adequately; when many production managers think about managing inventories, they think about work-in-process inventories. Here the management tools are more ill-defined, since reducing work-in-process inventory generally means grappling with the production process itself and/or with the flow of information through it.

To control work-in-process inventory levels, the manager might contemplate ways to shorten production cycle times (such as shortening process steps or coupling operations more closely) or ways to break existing production bottlenecks or ways to improve the flow of information around the process. Some situations already discussed (2-1 and 2-2) have dealt with such means of managing work-in-process inventories and more situations to come will deal with it. Managing work-in-process inventories is discussed at the end of the next chapter.

So far, we have classified inventory solely according to its place within the process (raw materials, work-in-process, finished goods). There are other classifications possible; one useful one was developed by Joseph Orlicky, who distinguishes between inventories of dependent demand items and independent demand items. Independently demanded items are typically finished goods or other items whose demands are unrelated to anything else produced or sold by the company (e.g., repair parts, production-related as opposed to product-related goods). Dependently demanded items, on the other hand, are those that can be directly linked to a specific product—typically, they are components and subassemblies that can be linked to an end item by a bill of materials. In such a case, demand for the end item automatically triggers demand for known quantities of the parts and materials that go into the product.

This distinction is useful because dependent demands fall out naturally from knowledge of their end items—there are no uncertainties—whereas independent demands are full of uncertainties as to both quantities required and timing desired. The inventory policies applicable to independently demanded items can be very different from the inventory policies applicable to dependently demanded items. For the most part, the management of finished goods inventories and, to a lesser extent, raw materials inventories is management of independently demanded items. This chapter deals with dependently demanded items, those whose use can be anticipated, and material requirements planning, which is the means by which such an approach to inventory management can be made to work. The next chapter deals with independently demanded items, time independent systems, and with managing work-in-process inventories (i.e., nonanticipatory approaches to inventory management).

A Description of MRP: An Anticipatory Approach to the Management of Inventories[2]

An anticipatory approach to inventory management has its most powerful application with dependent demand, although it can be used for end items as well as component parts and subassemblies. In this approach, commonly called material requirements planning (MRP), a company manages its inventories by anticipating their use and planning order sizes and timing accordingly to keep the stocks at some desired, low level. One ideal for such an anticipatory approach to inventory is to have no finished goods inventory over and above that expressly planned for, and little or no raw materials inventory. Work-in-process inventories then would account for by far the greatest chunk of total inventory costs. What is needed to accomplish this anticipatory approach to inventory is a firm fix on exactly what products are to be produced and when, what materials go into making each product, when in the production cycle each part or material is used, and how long it takes suppliers to deliver these parts or materials. This is a lot of information, but every bit is needed if in fact inventories are going to be matched to a company's output. In more formal phrasing, such an approach to inventory management demands the following pieces of information:

1. *Master production schedule (MPS).* When each product is scheduled to be manufactured. Usually the MPS is developed for a month or a week. The firmness by which the schedule is held to can vary markedly from company to company and from time to time. These data "drive" the entire system. Often the system also keeps track of which orders are associated with which demands in the schedule. (This tracking of parts to orders and specific customer orders to aggregated demands in the schedule is commonly called "pegging".)

2. *Bills of materials for each product.* Exactly which parts or materials are required to make each product. Indeed, each product variation needs its own bill of materials so

that the demands for all the individual component parts can be detailed.

3. *Inventory status of each component part.* How many of each part are on hand and how many are scheduled to be received prior to use.

4. *Lead times for all parts.* Parts to be purchased or fabricated so that their availability is timed exactly to meet anticipated needs.

5. *Product construction standard times and schedules.* This information details how long (in standard hours) particular tasks can be expected to take and indicates when specific parts or subassemblies will be needed in order to meet the established master schedule for the end product. This information is useful for scheduling materials as well as for assessing the capacity of the factory to produce the contemplated product mix.

6. *Routings of parts to particular workcenters.* These routings indicate which workcenters (machine tools, assembly lines, departments) are involved in the manufacture of each product or product variation. This information is critical to the capacity planning function for which an MRP system can be used. This adaptation of MRP is discussed later in this section.

7. *Standard costs (optional).* Given standard costs for particular parts and subassemblies, an MRP system can be used to calculate inventory investments and to estimate product costs for any modification to the bill of materials' product structure. This adaptation of MRP is also discussed later in this section.

These pieces of information work together to determine what should be ordered and when. The MPS, together with the bill of materials, determines what should be ordered; the MPS, the production cycle times, and the supplier lead times jointly determine when the order should be placed.

These elements form an anticipatory approach to inventory management, which lies at the heart of material requirements planning (MRP). MRP can become quite a bit more involved than this in actual practice (as we shall see), but at its roots lies this relatively simple notion of managing raw materials stocks by anticipating their use.

An MRP System in Action

To illustrate how this information gets used by, and processed through, an MRP system, let us ponder a very simplified example: a toy tractor-trailer. This toy tractor-trailer consists of (1) a yellow plastic tractor cab to which is attached two yellow plastic axles and four black plastic wheels (the wheels must first be mounted on the axles and then the axles snapped into place underneath the cab), and (2) a blue plastic, flat-bed trailer (which also has a yellow axle on which two black plastic wheels, identical to those under the tractor cab, are mounted). A "product structure" diagram of the tractor-trailer and its component parts might look like Figure 7-1.

Figure 7-1 serves as the bill of materials for the product. Notice that it can be depicted in a series of levels. These levels structure the product into various subassemblies and detail the components of each subassembly. Such a breakdown and structuring enables us to keep separate track of inventories of tractor, trailer, or axle subassemblies and their components. We will see the usefulness of this.

Figure 7-2 is a master production schedule for tractor-trailers over the next 6 weeks. The 6-week production horizon is generally somewhat arbitrary and could be extended or shortened as a matter of company policy. In any

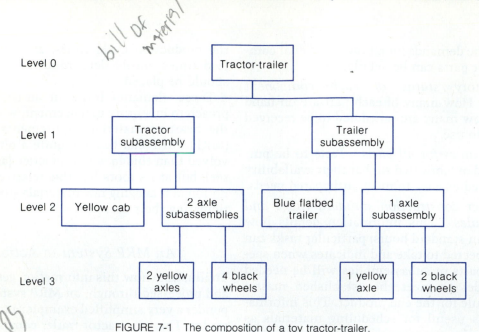

FIGURE 7-1   The composition of a toy tractor-trailer.

	Week					
	1	2	3	4	5	6
Toy Tractor-Trailers	6	10	3	12	15	8

FIGURE 7-2   Master production schedule for the toy tractor-trailer.

event, it represents a firm commitment to make so many tractor-trailer units by such-and-such a time. Exactly when during the course of the week the tractor-trailer units are desired depends on the convention used. For example, the call for 10 units in week 2 can mean that 10 units should be shipped the first day of the week or the last day of the week or even the midpoint day of the week. The choice is entirely the company's. For our example, let us adopt the convention that the week's master scheduled output is to be shipped on the final day of the week. The time period of a week is

used here since it is the most popular for scheduling purposes; but time periods of a month, a day, or something else could be used. The time period used is commonly referred to as the "time bucket" for the system. The same MRP logic applies in any case.

An inventory record for axle subassemblies is provided in Figure 7-3. Note that there are 18 axle subassemblies (black plastic wheels mounted on yellow axles) on hand that can be used to meet any immediate needs. Any other axles will take at least a week to assemble as the lead time statement in the record indicates. Similar inventory records and lead times exist for all component items and subassemblies for the product. The value of an MRP system is its ability to determine quickly how many more axle subassemblies (or whatever) to order and when those orders ought to be placed. To do this, the material requirements planning system ties the various inventory records to both the bill of materials and the master production

	Week					
Lead Time = 1 Week	1	2	3	4	5	6
Gross Requirements						
Scheduled Receipts						
Projected Available Balance	18					
Planned Order Releases						

FIGURE 7-3   Inventory record for axle subassemblies of the toy tractor-trailer.

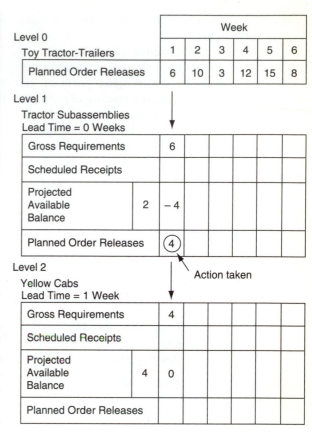

FIGURE 7-4   Linked inventory records in an MRP system and a partial explosion of requirements.

schedule. Such a tie is depicted in Figure 7-4, which shows the level-by-level link between tractor-trailers, tractor subassemblies, and yellow cabs.

It is worth devoting some attention to Figure 7-4 and what underlies it. Level 0 simply reproduces the master production schedule found in Figure 7-2. The "planned order releases" for the tractor-trailer units constitute "gross requirements" for the tractor subassemblies of level 1. That is, the need to ship, say, six toy tractor-trailers at the end of week 1 (the current week) places a demand for six tractor subassemblies for week 1. Two of the required six tractor subassemblies can be supplied out of existing inventories on hand (which we know because the bill of materials has been structured into subassemblies and thus we can keep track of them), but that leaves us four tractor subassemblies short, as indicated by the "–4" in the "on-hand" row. This shortage in turn leads us to take action—namely, to order the four yellow cabs that we will need to manufacture that subassembly. This planned order release at level 1 is translated into a gross requirement at level 2, which keeps track of the yellow cabs. Thankfully, four yellow cabs are on hand, and this order can be met out of existing inventories.

At every level, the net requirement for a part or subassembly in any period is calculated in the same way.

Net requirements for a given period
= inventory on hand from the preceding period
+ quantity on order for delivery in the given period
− gross requirements for the given period

In our diagrams, the net requirement is written in the row labeled "projected available balance." When this net requirement is negative, then some action must be taken (an order placed) if production is not to fall behind schedule

When it is zero or positive, all is well, at least for the time being.

A material requirements planning system can trace out in advance the net requirements for subassemblies or component parts at all levels. Figure 7-5 does such a tracing of requirements for both level 1 (tractor subassemblies) and level 2 (yellow cabs) under the assumption that all shortages of tractor subassemblies trigger planned order releases in the same way as traced out in Figure 7-4. On level 2 the pro-

jected available balances (net requirements) accumulate, since no orders have yet been released.

Calculating the planned order releases for the yellow cabs is not as straightforward as calculating the planned order releases for the tractor subassemblies since the yellow cabs can be fabricated only with a lead time of 1 week. For example, if the need for 10 yellow cabs in week 2 is to be met, an order for 10 cabs must be placed in week 1. Figure 7-6 follows through

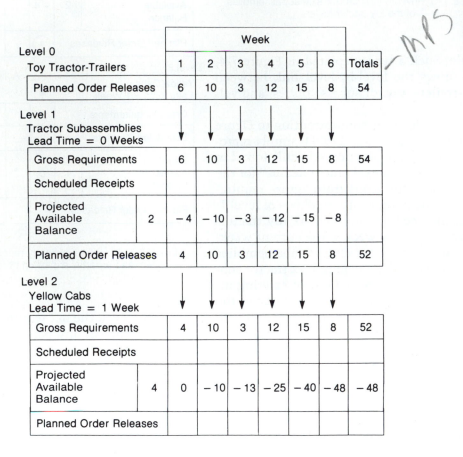

FIGURE 7-5   Net schedule requirements implied by the master production schedule (receipts not yet accounted for).

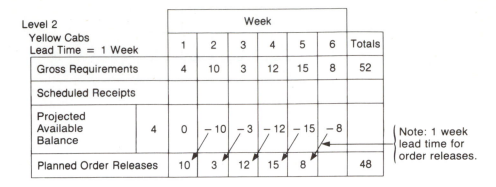

FIGURE 7-6   Planned order releases when lead time must be accounted for
(receipts not yet accounted for).

with this logic, depicting the required planned order releases for level 2.

As yet, no entries have been made in the row labeled "scheduled receipts." These receipts, however, are linked directly to planned order releases. As soon as an order is "released," it becomes a scheduled receipt. Its "receipt" position in the time phasing diagram is determined by its expected lead time. Thus the net requirements implied by the master production schedule, as indicated in Figure 7-5, can be rewritten in terms of scheduled receipts, assuming no complications with the indicated lead times. Figure 7-7 displays this. Note that the planned order release has been transferred to the scheduled receipt square for the appropriate week. The projected available balance thus reflects the emergence of a scheduled receipt.

In practice, of course, planned order releases can be expected to "mature" and be released one at a time, as in Figure 7-4, rather than in entire batteries, as Figure 7-6 may lead one to believe. That is, one cannot assume that just because an order is released that the "scheduled receipt" is a sure thing. One must wait for

confirmation or commitment to be sent before the order's receipt can be booked. Thus, scheduled receipts are not usually shown far into the future, but rather only as far as the operative planning horizon.

Before going on to introduce some "wrinkles" to this basic MRP logic (such as how to plan when the same type of part is used in more than one subassembly), let us review its fundamental steps:

1. The master production schedule is the data introduced to the MRP system that "drive" it. All the entries and actions taken must be in agreement with the master production schedule. This schedule is assumed by the MRP logic to be correct and "frozen" over some time period.

2. A detailed bill of materials for all products and product variations must also exist, and it should be structured in successive tiers to reveal subassemblies as well as individual component parts. Given the master production schedule and the relevant bills of materials, an MRP system "explodes" the bills of materials through successive tiers

		Week					
Level 0 Toy Tractor-Trailer	1	2	34	4	5	6	Totals
Planned Order Releases	6	10	3	12	15	8	54

Level 1
Tractor Subassemblies
Lead Time = 0 Weeks

Gross Requirements		6	10	3	12	15	8	54
Scheduled Receipts		4	►10					
Projected Available Balance	2	0	0	0	0	0	0	
Planned Order Releases		4	10	3	12	15	8	52

Note: Receipt by lower levels becomes receipt at next higher level, lagged by its lead time.

Level 2
Yellow Cabs
Lead Time = 1 Week

Gross Requirements		4	10	3	12	15	8	52
Scheduled Receipts		4	10					
Projected Available Balance	4	0	0	0	0	0	0	
Planned Order Releases		10	3	12	15	8		48

Note: Delivery or receipt lags order release by the lead time.

FIGURE 7-7 Scheduled receipts implied by the master production schedule and by the planned order releases.

to calculate how many parts or subassemblies of a particular type are required to satisfy the master production schedule.

MRP logic permits the computation of net requirements by adjusting the gross requirements for a particular subassembly or part by the number of units already on hand and those that are expected to be received.

3. The timing for planned order releases depends on the expected lead times for procuring/fabricating/assembling component parts or subassemblies. Orders are actually placed in anticipation of their need, as dictated by the master production schedule and the derivative explosion of the bills of materials.

The data requirements for the computer files that handle the processing for an MRP system are voluminous and, as introduced above, include the following:

• Master production schedules for the weeks indicated

- Bills of materials for all products and product variations
- Existing inventory levels for all parts and subassemblies
- Lead times for all parts and subassemblies
- Orders placed but not yet received (the scheduled receipts or often called "open orders")
- Routings of parts to particular work centers
- Standard costs for particular parts and subassemblies (optional)

The processing of an MRP system—the explosion of the bills of materials, the updating of the inventory file records, and the releasing of orders and spotting of scheduled receipts—can be an irregular event (say, once a week) or can proceed all the time. If an MRP system is run at distinctly spaced intervals, such as weekly, the system is said to be regenerative, because the computer file records on master schedules, bills of materials, inventory levels, lead times, and open orders are updated ("regenerated") in their entirety all at once. Such a regeneration conserves computer resources, but it also runs the risk of letting the information on which the entire MRP system depends deteriorate over time.

A net change MRP system, on the other hand, permits the information in any computer file on the system to be updated. More important, it also permits the consequences of that updating to be followed through, be they changes in planned order release dates or changed quantities of an item on hand or something else. A net change MRP system uses more computer resources, but it maintains the timeliness of the system. Of course, for a net change system to operate effectively, all the consequences of any transaction in any computer file must be traced to their ultimate conclusion. Records must be tied or "chained" to one another both upward and downward by

special "pointers." By following these pointers, a net change MRP system is updated by a series of partial "explosions" rather than by calling for the large, infrequent, full explosion of a regenerative MRP system.

"Wrinkles" to Add to the Basic MRP Logic

Common Parts

In our example of the toy tractor-trailer, the black plastic wheels and the yellow plastic axles are used in both the tractor and the trailer subassemblies. When it is time to review the net requirements for wheels and axles, we must be sure that the gross requirements from both tractor and trailer subassemblies are added together. The explosions of the bill of materials must be careful to keep track of all the instances where a certain part or subassembly is called for. Figure 7-8 shows the demand for black plastic wheels in various weeks emerging from both tractor and trailer assembly. Note that two wheels are needed for each axle assembly and that a tractor uses two axles and a trailer just one.

Same Parts Used at Different Levels

Although it does not happen in our toy tractor-trailer example, it is easily possible for a part to be used at both level 2 and level 3, say, of a product structure for a bill of materials. Trying to keep track of all the demands for such a part is awkward (and expensive for the computer) if some simplifying convention is not adopted. In most instances, this problem is solved by structuring the bills of materials in a way that places all of the same parts or subassemblies at the same level. In Figure 7-9, for example, part A, which could have been placed in two different levels of the bill of materials' product structure, is depicted as occurring all at the lowest

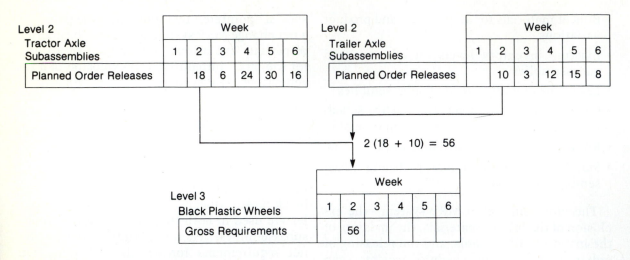

FIGURE 7-8   Dealing with common parts in an MRP system.

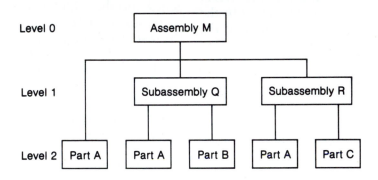

FIGURE 7-9   Bill of materials' product structure dealing with
the same part on potentially different levels.

level possible. In this way, the MRP computer program encounters it at only one level and sums up its requirements there.

Lot Sizing

Up to this point, material requirements planning has been used to place orders that represent precisely the number of units of anticipated use, no more and no less. As we have seen elsewhere in the discussion of materials

management, price discounts or some other feature may make it advisable to purchase more than the anticipated need. This possibility may make it desirable to use some economic order quantity reasoning along with the batch-for-batch (lot-for-lot) anticipatory philosophy of MRP. Figure 7-10 displays how planned order releases would look, given a specific lot size to order rather than a lot-for-lot order scheme.

Level 2 Yellow Cabs Lead Time = 1 Week Lot Size = 15		Week						
		1	2	3	4	5	6	Totals
Gross Requirements		4	10	3	12	15	8	52
Scheduled Receipts			15		15	15	15	60
Projected Available Balance	4	0	5	2	5	5	12	
Planned Order Releases		15		15	15	15		60

FIGURE 7-10   Lot sizing in an MRP system.

The MRP logic also makes it explicit that if total inventory costs are to be minimized, this must occur on a period-by-period basis, as more information is revealed about future needs, costs, and item availabilities. In practice, being so precise about minimizing a changing total inventory cost may not be worth the effort, but some lot size approximations can be built into an MRP system. Let us consider some of these lot size techniques.

The common lot size technique for MRP is to place orders for only what is anticipated—a technique known as lot-for-lot. By using lot-for-lot lot sizes, no remnants of inventory are purposely accumulated to be carried from one period to another. This is an attractive feature, and, as we shall see in later chapters, consistent with the "just-in-time" philosophy of manufacturing. Lot-for-lot is less attractive when there are significant order or setup costs that cannot be reduced or when there are frequent and significant instances where delivery is unreliable, orders get canceled or expedited, or delivery quantities are not accurate. These adverse occurrences can be eliminated (and are in "just-in-time" companies); but to the extent they exist, lot-for-lot suffers.

Economic Order Quantity (EOQ) lot size reasoning, as discussed in the next chapter, can be applied in MRP systems. There may be well-defined order or setup costs, price breaks, or other costs that can be balanced against inventory carrying costs. An EOQ formula (or computer algorithm) thus can be developed and used. One problem with using such an approach, however, is exactly what to use for D, the demand. Should the demand amount (and the concomitant carrying cost percentage) be defined for the next month? The next three months? The next 6 months? The next year? Or longer? There is no clear-cut answer, and the choice of a time horizon will alter the calculation. In general, the less uniform the demand over time, the worse any EOQ calculation performs.

There are several lot size techniques that try to skirt this time horizon problem while still balancing setup or order costs against carrying costs. These techniques allow the time horizon for which the order applies to vary from order to order. Thus, these techniques may call for one order to cover the next three periods and a subsequent order to cover the succeeding four periods. The several methods used to determine both lot sizes and the time intervals between orders differ, often trading off simplicity against optimality. We will describe two such techniques: least total cost, and part-period balancing.

Least Total Cost (LTC). This technique chooses the lot size and order interval for which the setup/order cost per unit and the carrying cost per unit are most nearly equal. This goal is similar to the rationale of the EOQ model and argues that the setup/order costs and inventory carrying costs for all lots within the time horizon will be low (if not minimized) if these costs per unit are equal.

In its calculations, the LTC technique uses what is known as an economic part-period (EPP). A part-period is one unit of inventory carried for one time period. An economic part-period is that quantity of inventory which, if carried for a single period, would result in inventory carrying costs equal to the setup or order cost. For example, if the one-period carrying cost is 10 cents per unit and setup costs are $15, then the economic part-period is 150 units ($15/$0.10).

The least total cost technique calls for a lot size and order interval whose part-period cost falls closest to the EPP. Suppose, then, that the EPP is 150 units, as above, and the net requirements over the next 12 periods are as follows:

PERIOD	NET REQUIRE-MENTS	PERIOD	NET REQUIRE-MENTS
1	20	7	
2	35	8	35
3		9	
4	25	10	50
5		11	
6	30	12	

Total net requirements: 195
On-hand inventory: 0
Lead time: 0

What is the lot size suggested by the LTC technique for period 1? The answer is 80, as shown in Table 7-1. The lot size of 80 yields a cumulative part-period sum that lies closest to

the calculated economic part-period of 150. For that reason, it is the preferred lot size.

It should be noted that LTC does not typically develop an optimal lot size, in the sense that total inventory cost is minimized. Its rationale depends on equating setup/order costs per unit with inventory carrying costs per unit, and that rationale is optimal only when the mathematical functions describing those costs are friendly. (More on this in the next chapter.)

Part-Period Balancing. This technique has the same rationale and same computations as the least total cost technique, but it adds a look ahead/look back procedure. The LTC technique chooses lot sizes one at a time, with no attention paid to what might be a proper lot size in future periods. In the previous example, the lot size of 80 would last through period 4; the LTC technique does not look ahead to see what lies beyond period 4. The total costs over the full time horizon could, in fact, be lowered by either increasing or decreasing the lot size choice for period 1, given an investigation of the next lot size choice in the future. In part-period balancing, one peeks at the next lot size decision and adjusts the initial lot size accordingly. A look ahead might indicate that the initial lot size should be increased to cover an additional period or two; a look back might indicate that the initial lot size should be decreased to cover fewer periods.

Part-period balancing can lower total inventory cost as compared with LTC. However, because it does not look ahead or back over the entire time horizon, it cannot be certain of lowering total inventory cost relative to LTC or other techniques, let alone minimizing total inventory costs.

Choosing a Lot Size Technique. The choice of a lot size technique is complicated by the fact that most MRP systems involve product struc-

	PART-PERIODS CARRIED AT END OF EACH TIME PERIOD						
PROSPECTIVE LOT SIZE	1	2	3	4	5	6	CUMULATIVE PART-PERIODS
20	0						0
55	35	0					35
80	60	25	25	0			110
110	90	55	55	30	30	0	260

TABLE 7-1 Lot sizes by part-period

tures (bills of material) that are several levels deep. The introduction of lot sizes other than lot-for-lot into the first level of the product structure can greatly complicate the lot size choices for lower levels. How complicated matters can get depends on the lot size choice itself, the complexity of the product structure, the degree of change permitted in the net requirements, and the degree of commonality of parts. Simulations of different lot size choices under various conditions have not resulted in clear recommendations for MRP users.

Having non-lot-for-lot lot size choices within a multilevel product structure also is likely to bloat inventory more than might be expected. The more complex the product structure, the more likely inventory will be bloated by most lot size techniques.

Safety Stocks

An MRP system can also accommodate safety stocks of items. In general, it is best to think of safety stocks of finished goods—"kits" of component parts—rather than individualized safety stocks for all of the component parts themselves. The accounting is easier if the safety stocks are thought of, and even actually held, in that way. The existence of safety stocks, however, goes somewhat against the grain of the material requirements planning philoso-

phy, since safety stocks cannot be thought of in terms of time phasing for anticipated use. Companies also run the constant risk that their managers, once discovering the presence of safety stocks, will adapt their own decisions to use up the safety stock early and for purposes for which they were not intended.

Safety Lead Times

A common way to cushion an MRP system against any chronic irregularities in the deliveries of either purchased or fabricated parts is to tack on some extra time to the stated lead times. Otherwise, according to this view, chronic delivery problems might result in excessive expediting or rescheduling. Of course, this situation points up the usefulness of dependable delivery schedules. One frequent means of introducing safety lead times is to add a week (the standard time bucket) to a part or subassembly as it moves from one level in the bill of materials' product structure to the next higher level. Thus a subassembly that resides in level 2 of the product structure and would otherwise be available immediately for assembly into a level 1 subassembly is given a lead time of one week for "safety's sake."

Safety stock and safety lead times are related but somewhat different concepts. Safety stock addresses uncertainty in supply (quantities) whereas safety lead time addresses uncertainty of the leadtime for a part or component.

MATERIAL REQUIREMENTS PLANNING AS A MANAGEMENT TOOL

A main function of an MRP system, of course, is to trigger purchases and factory orders so that they can meet projected production requirements. When all is working smoothly, an MRP system can do just that. However, the test of any materials management system comes not in fair weather but in foul. Events in a materials manager's life invariably go awry. It is in coping with the unexpected and in planning the future that an MRP system can often earn its own way.

A material requirements planning system can serve as a useful management tool in a variety of ways: to make schedules and revise them, to plan capacity, and to measure performance.

As a Scheduling and Schedule-Revising Device

Not only can an MRP system detail what should be ordered when, but also it can indicate how and when late items will affect other aspects of production. It can signal which orders now awaiting delivery ought to be expedited and how any tardiness will alter the existing production schedule (e.g., how much final delivery will be delayed or which demands will now be placed in subsequent weeks). Similarly, if advance supply delivery or higher yields or some other positive development occurs, an MRP system can signal which orders to de-expedite or even cancel. For these purposes, the part and order tracking mentioned earlier ("pegging") is extremely helpful.

When such positive or negative developments strike, an MRP system can automatically reschedule orders (or, more often, indicate to the materials manager the necessity for rescheduling orders), moving them either forward or backward. For example, the spoilage of some inventory or the breakdown of machinery (both of which may reduce the number available in the future) can trigger the rescheduling of an order in the proper week to avoid a stockout. Unexpected developments could even lead to a change in the master production schedule or in the lot size permitted. Typically, an MRP system identifies a range of possible actions to be taken by the materials manager.

Oliver Wight, one of the pioneers and chief advocates of MRP, referred to MRP as a "simulated shortage list."[3] Traditionally in many companies, the true schedule of production was determined by a combination of what was due out to customers and what parts were available. If parts showed up short, quickly arranged meetings had to be held and immediate decisions had to be made about expediting the parts, if possible, and rescheduling the entire production schedule to accommodate the shortage. By simulating the shortage list well ahead of time, materials managers can take action to rectify potential shortages and/or reschedule production and other materials deliveries before real trouble occurs. The factory is then supplied with what it needs when it needs it. This is not to say that materials will never again have to be expedited or the production schedule adjusted on short notice; problems will always occur and have to be dealt with. With MRP, however, the factory requires less frantic expediting and less tumultuous production rescheduling and does so with less inventory (Wight's estimate is one-fourth to one-third less) than before, because the factory will not be short of parts it needs and overstocked with parts it does not need and perhaps did not even know it had.

It has been observed that when a shortage list dictates the true production schedule, the value of any agreed-upon master production schedule diminishes greatly. The MPS cannot be used to "push" materials out into the factory; instead, the shortage list must be relied on to "pull" materials and production orders through. With a well-functioning MRP system, however, the MPS assumes its rightful place as "driving" production: materials can be "pushed" out into the factory in accordance with the master production schedule without being sidetracked by rescheduling to accommodate parts shortages.

The benefits of a smooth production schedule are numerous. Already cited are the inventory savings that can result. Labor productivity is enhanced by removing expediting, parts shortages, and the destruction of machine or line setups. Foremen are freed from expediting to work on quality, methods, worker relations, and the like. Unnecessary overtime is avoided. Vendor and supply logistics premiums are reduced. Having realistic production schedules and not having materials shortages to blame poor performance on means that workers and managers alike can be evaluated on the things they should be evaluated on (quality, meeting schedules and budgets) without having to sort through so many excuses.

As a Capacity Planning Device

An MRP system can be enlarged to cope with capacity requirements, not just materials requirements. It does this by combining several types of information: (1) the planned production schedule and the materials required to meet it, (2) the time requirements to fabricate/assemble all parts and products, and (3) the particular workcenters in the factory through which those parts and products travel. By knowing how long any part or product can be expected to be worked on (its standard hours) and where all parts or products have to be processed (their routings), an MRP system can predict the workload at any work center in the factory. This predicted workload (in hours), given the planned master production schedule, existing inventories, and any lot size conventions, can be compared with the planned (or possible) hours for the workcenter and any deviations can be noted. For example, if the predicted workload is more than the planned (or possible) hours for the workcenter, then either capacity would have to be increased (perhaps through overtime, machine adjustments, and/or subcontracting) or the MPS would have to be modified to permit a balance of production needs and capabilities. Similarly, if the predicted workload is less than the planned (or possible) hours for the workcenter, resources (such as labor time) could be freed for use elsewhere without damaging the factory's delivery performance.

By using the deviations (variances) of predicted versus planned/possible workloads as feedback on the master production schedule and material needs, an expanded MRP system can be said to "close the loop" between material needs and capacity needs. Not only can materials be planned, but work hours can be planned as well, meshing both with the MPS. The master production schedule can then be realistically deemed feasible, and the MRP system can even be used to dispatch the day's or week's production assignments to the factory's workcenters.

Using an MRP system in this "closed loop" way to calculate capacity requirements—termed *capacity requirements planning*—is the most sophisticated means of assessing whether the factory has the capacity in the short run to meet expected demands on it. There are several other procedures to accomplish the same purpose, but they are somewhat less sophisticated and use less data. These procedures were discussed in Chapter 6.

As a Performance Measuring Device

If we go one step further and assign standard costs to parts and products, we can use the closed loop MRP system to generate financial figures on production and inventories that management can live with. This further expansion of the MRP system has been dubbed manufacturing resource planning, or MRP II. Companies that use MRP systems effectively in running their operations (those that close the loop, and those that use it for financial or business planning) are often termed Class A users.

The use of MRP as a capacity planning device and as a performance measuring device highlights the possibilities of using MRP for "what if" analysis. Demands of various types can be simulated as well as capacity modifications or other changes. Such simulation can be an exceedingly useful planning aid.

The Success of MRP

The potential and promise of MRP have fired the imaginations of managers at many companies. Yet, the actual success rate of MRP to date has disappointed many observers. The failings of many companies to earn the full benefits of MRP can be traced to a lack of appreciation for the diligence that an MRP system requires. As we have seen, the data requirements of an MRP system are massive. If the system is to work well, the data must be accurate and up-to-date. Think of what this means: accurate forecasts of demand, with no last-minute changes allowed; precise bills of materials showing all variations and a formal change mechanism for any engineering change orders; regular cycle counting

of inventory so that the on-hand balances as shown are accurate; accurate lead times for vendors; accurate standard times, standard costs, and routings for each part; integrity with material usages and scrap rates, usually requiring a restricted access storage area with consistently good tracking of material flows in and out; and a realistic understanding of what production capacities are.

Accomplishing all these tasks requires management, at all levels, to understand MRP thoroughly, and to commit themselves to data integrity and all the nitty-gritty tracking and cross-checking that is required to maintain that integrity.

The success of MRP can be stirring, however. Situation 7-1 describes Black & Decker's Tarboro, North Carolina, facility and its experience with MRP.

The Cost of MRP[4]

The substantial benefits conferred upon an operation by a well-run MRP system are not reaped without some significant expenditures for hardware, software, data base development, and user training. Hardware costs are generally the easiest to estimate and the ones most subject to reduction (at least in real terms). Depending on the complexity of the system, annual lease costs for hardware for MRP can run between $50,000 and $600,000. Software costs are also subject to considerable variation. Wight estimated software lease expense at between $15,000 per year at the low end and $300,000 initial cost plus $30,000 per year maintenance at the high end. Wight also estimated the cost of programming and data base development (such as bills of material, routings, inventory records) at between 6 and 14 worker-years of effort. Users must be trained as well; that expense can vary tremendously, depending on the extent of training and whether instruction is live or via videocassettes. The

continued on p. 269

SITUATION 7-1

The Black & Decker Corporation

Materials Management at the Tarboro, N.C. Plant

The Tarboro facility was part of Black & Decker's U.S. Power Tool group. The plant manufactured products for that group and also for the Household group. It was the lead plant for injection molding of plastic parts and for the manufacture of battery packs, transformers, and chargers that were sent to a number of Black & Decker manufacturing plants worldwide. The plant specialized in a number of cordless products, about half of which were power tools and half household cleaning devices. Major products manufactured at the Tarboro facility included cordless drills, screwdrivers, ratchets, hand-held vacuum cleaners, spotlights, and mixers. About 600 different end item (finished goods) numbers were made at the plant, many being slightly different versions of the major products.

The materials manager supervised both staff people and the hourly employees (known as associates) who worked in receiving and shipping. Materials management was organized into a Central Materials Group and three Product Groups: Power Tools, Household Products, and Components, such as chargers.

The Central Materials Group handled computer systems support, shipping and receiving, and the capacity planning and loading for the plant. Each of the three product groups was responsible for materials procurement, the specific daily schedule for each line, expediting, engineering change notices coordination, bill of materials structures, and coordination with the suppliers of the 100 or so parts that were part of the "JIT schedule."

Purchasing responsibilities were not assigned to the materials manager; purchasing was operated as a separate entity under a product sourcing manager. Despite the organizational distinction between purchasing and materials management, the two departments worked closely together.

Outline of the Materials Management Systems

Black & Decker–Tarboro operated an MRP system called PACE (Planned Action and Constant Evaluation), which the company had developed in the 1960s. As a net change MRP system, PACE was updated daily. Future production requirements could be shown in PACE as far as one year in advance. Historically, forecasting these production requirements was the responsibility of each group's logistics staff, which controlled all finished goods for that group. Thus the factory's responsibility for any product ceased when it left the factory; the group's logistics function assumed the carrying costs for all finished goods. This was slated to change, however. The plant would soon "own" all its finished goods inventory and bear the responsibility for determining the master schedule for all its products.

The PACE system distinguished between "requirements" that stemmed from the 52-week forecast of the group's logistics function, and "allocations" that resulted from firm orders for the product. (See Figure 7-11 for an example of an inventory status sheet.) The PACE system had been used at Tarboro since the plant started in 1971. However, its usefulness was not fully realized until several years later, when the materials control system (MCS) was added. The MCS was a real-time computer-based system that tracked materials. Records from the receipt of all materials—whether purchased from the outside, shipped from other B & D plants, or manufactured at the Tarboro facility—were kept, as were records of the issue of all materials (both raw materials and work-in-process) from stock.

The Tarboro plant did not have a storeroom. Rather, materials (about 3 shifts' worth) were stored

05/25/91 WEEK 796 INVEN... A 50144806-98 02 I N D I V I D U A L - P A R T ORDER# NAME 21 94 DEPT NO

CATEGORY	PART NUMBER OR RAW MATERIAL CODE								
2	50144806-98 KIT BOX	50 A	970 94	19076	759 795	5 P/C 02 STD L/S	25		F/W .00

ORDER CODE	LEAD TIME S/A	COMP						STAGED	LOCATION PLANT	AREA
H		12	00	37 18	86580	63586	6779		5	

			797	798	799	800	801	802	803	804	805	806	807	WEEK 808
REQUIREMENT					5000	3300	2150	2400	5000	2200	3550		1000	1000
ALLOCATIONS	3575	4075	5000	5000				100		100				
ON ORDER	450	450		5000	5000	3300	2250	2500	5000	2300	3550		1000	
AVAILABILITY		5154	154	154	154	154	254	254	254	254	254	254	254	746-
PLANNED ORDERS		746	1750	3650	4000	2000	4400	1500	1650	1900	700	600	1200	1100

	WEEK 809	810	811	812	813	814	815	816	817	818	819	820	821	WEEK 822
REQUIREMENT	1750	3650	4000	2000	4400	1500	1650	1900	700	600	1200	1100	1800	900
ALLOCATIONS														
ON ORDER														
AVAILABILITY	2496-	6146-	10146-	12146-	16546-	18046-	19696-	21596-	22296-	22896-	24096-	25196-	26996-	27896-
PLANNED ORDERS	1600	900	3200	2000	2400									

	WEEK 823	824	825	826	827	828	829	830	831	832	833	834	835	WEEK 836
REQUIREMENT	3200	2000	2400											
ALLOCATIONS														
ON ORDER														
AVAILABILITY	31096-	33096-	35496-	35496-	35496-	35496-	35496-	35496-	35496-	35496-	35496-	35496-	35496-	35496-
PLANNED ORDERS														

	WEEK 837	838	839	840	841	842	843	844	845	846	847	TOTALS
REQUIREMENT												60450
ALLOCATIONS												14275
ON ORDER												30450
AVAILABILITY	35496-	35496-	35496-	35496-	35496-	35496-	35496-	35496-	35496-	35496-	35496-	35496-
PLANNED ORDERS												35496

ORDERS ...

STATUS

CODE	WEEK	ORDER	QUANTITY	CODE	WEEK	ORDER	QUANTITY	CODE	WEEK	ORDER	QUANTITY	CODE	WEEK	ORDER	QUANTITY
REQUIREMENTS				R	799	0014895099	2000 J	R	799	0045090099	3000 J	R	800	0014895040	400 J
R	800	0014900040	2900 J	R	801	0014700001	1500 J	R	801	0045090001	350 J	R	802	0014890002	1500 J
R	802	0014895002	600 J	R	803	0014895003	3900 J	R	803	0014900003	1100 J	R	804	0014900004	2200 J
R	805	0014700035	2100 J	R	805	0014895005	250 J	R	805	0014895005	1300 J	R	807	0014890007	500 J
R	807	0014895037	500 J	R	808	0014900008	1000 J	R	809	0014700009	1000 J	R	809	0014890009	450 J
R	809	0045090039	300 J	R	810	0014890010	900 J	R	810	0014895010	2750 J	R	811	0014895011	4000 J
R	812	0014895012	2000 J	R	813	0014900013	4400 J	R	814	0014700014	1400 J	R	814	0045090014	100 J
R	815	0014890015	1550 J	R	815	0014895016	1900 J	R	817	0014900017	700 J	R	818	0014700018	400 J
R	818	0045090018	200 J	R	819	0014890019	600 J	R	819	0045090019	600 J	R	820	0014895020	1100 J
R	821	0014895021	800 J	R	821	0014900021	1000 J	R	822	0014700022	800 J	R	822	0045090022	100 J
R	823	0014890023	1150 J	R	823	0014895023	2050 J	R	824	0014895024	2000 J	R	825	0014900025	2400 J
ALLOCATIONS				A	792	0014900092	257 JC	A	793	0014900093	128 JC	A	794	0014895026	390 JC
A	795	0000551118	2800 JC	A	795	0014890096	500 JC	A	797	0014700097	200 JC	A	797	0014890097	1200 JC
A	797	0014895097	3800 JC	A	798	0045090098	5000 JC	B	802	0L04850700	100 J	B	804	0L04855500	100 J
OPEN-ORDERS				P	794	0094595806	450 J	P	798	0094594362	5000 J	P	799	0094594535	5000 J
P	800	0095593523	3500 J	P	801	0094595095	2250 J	P	802	0094595294	2500 J	P	803	0094595585	5000 J
P	804	0094595751	2300 J	P	805	0094595976	3650 J	P	807	0094595512	1000 J				
PLND-ORDERS				J	796	0014900008	740 J	J	797	0014700009	1750 J	J	798	0014890010	3650 J
J	799	0014895011	4000 J	J	800	0014895012	2000 J	J	801	0014900013	4400 J	J	802	0045090014	1500 J
J	803	0014890015	1650 J	J	804	0014895016	1900 J	J	805	0014900017	700 J	J	806	0014700018	600 J
J	807	0014890019	1200 J	J	808	0014895020	1100 J	J	809	0014895021	1800 J	J	810	0045090022	900 J
J	811	0014890023	3200 J	J	812	0014895024	2000 J	J	813	0014900025	2400 J				
HAS WHERE USED				NO COMPONENTS				NO OP FILE							

BLACK & DECKER FORM NO 30020

FIGURE 7-11 Inventory status for kit box.

Panel A. Before review by a materials planner.

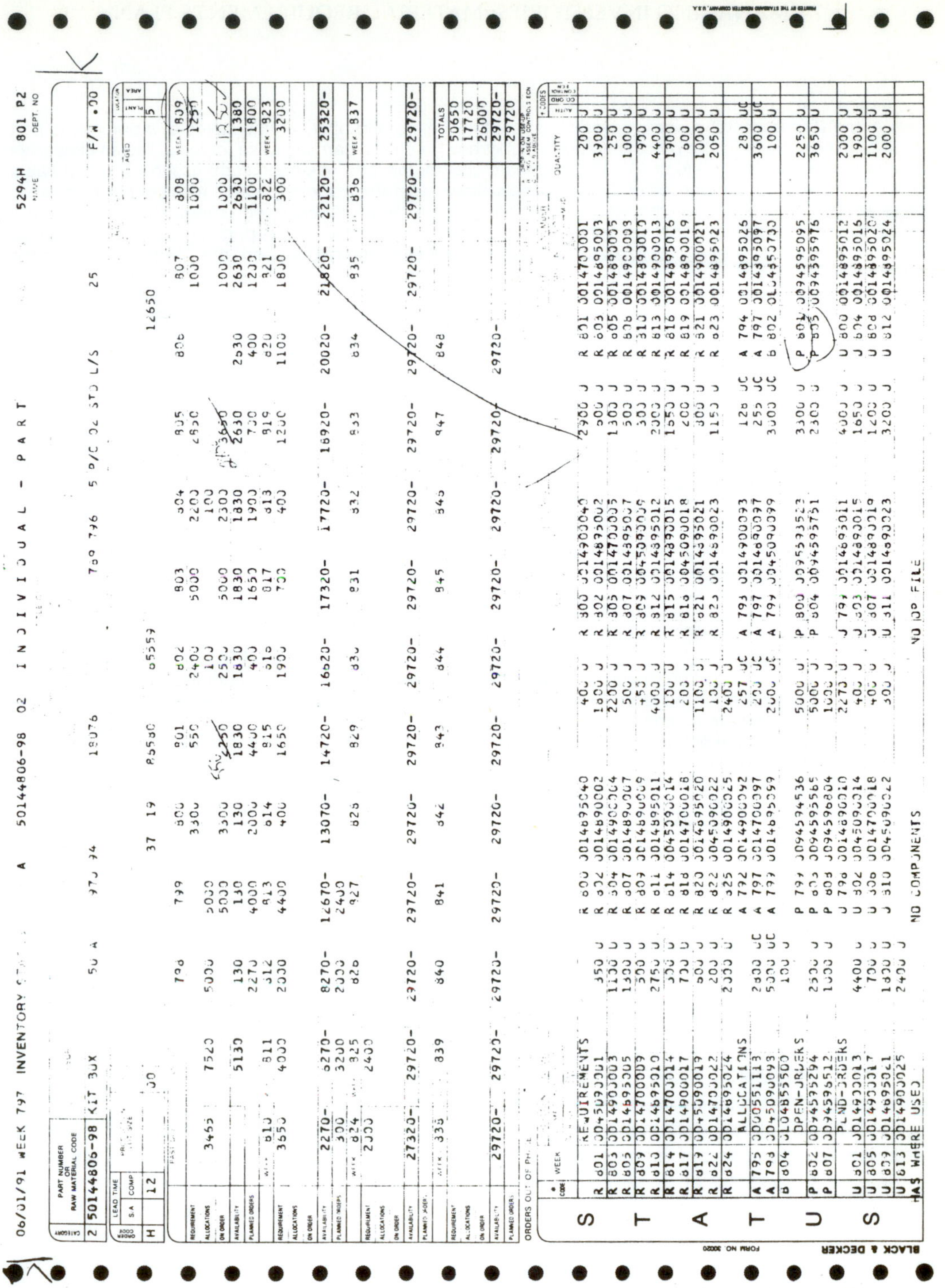

FIGURE 7-11 Panel B. After review by a materials planner.

Notes to Figure 7-11:

1. Panels A and B represent the week-to-week change, the "before and after," in the status of a purchased part [Category 2, in the upper left. The purchased part is a "kit box," a plastic carrying case for tools. It is an expensive enough part to merit an ABC code of "A". (The standard cost for this part has been suppressed.) The part has a lead time of 12 weeks.] A comparison of the two panels shows what the planner did as a result of the information in the first week, and it also shows the results of actions (such as receipts, changed orders, etc.) that occurred throughout the intervening week between these two reports. The top part of the printout displays the sum of requirements, allocations, and availabilities for weeks into the future. The bottom part, the "pegging," shows the specific order-by-order quantities that get summed for the top part.

2. The on-hand balance (in the upper right portion of the printout) is 8779 for Week 796. Against this on-hand balance is a current week allocation of 4075, which includes a past due quantity of 3575. [Examination of the pegging section's allocations (marked "A") shows that past due quantity to be composed of prior week allocations of 257, 128, 390, and 2800. The plant has thus fallen behind in its production.] The current week's additional allocation is 500. There are 450 kit boxes due in from the vendor, as well. Thus, the quantity available to be used up by additional orders (allocations) for finished tools is thus 8779 on-hand + 450 on order − 4075 allocations = 5154, which is what is listed as available in the current week.

3. Between Weeks 796 and 797, the factory caught up a bit in its production, but not completely. From Panel A we see that if the factory were to make everything "allocated" by the end of Week 797, then it would have had to make the 4075 allocation for Week 796 + the 5000 allocation for Week 797, or a total of 9075 finished tools. In Panel B, however, we still see a current allocation for 7520, which means that only 1555 finished tools were produced that used the kit box in them [9075 − 7520 = 1555].

We can see this in another way. In Panel A we had 8779 kit boxes on-hand. From Panel B we see that the "On Order" row has been changed; orders for 450 (the past due one from the vendor) and for 5000 (for Week 798) are no longer shown because they have been received. (Actually, only 426 of the 450 past due order was received.) Using this information, we can calculate the on-hand balance for week 797: 8779 on-hand in week 796 + 426 received + 5000 received − 1555 used up = 12,650 on-hand for Week 797, which is what can be seen in Panel B as the on-hand balance.

4. The availability row for Week 796 (Panel A) goes negative in Week 808, 12 weeks from the current one, which corresponds to the leadtime for this part. Consequently, the planner has placed an order for 1000, so that the availability returns to a positive number.

The inventory status for the following week (Week 797) shows that this order has indeed been placed in the system for Week 808. In addition, there are two other order changes that the planner wishes to make: Week 801's old order for 2250 has been reduced to 550, and Week 805's order for 3650 has been reduced to 2850.

5. The past due figure for Week 797 has been reduced by 110 units in response to a change in the order amount ordered for Week 794 from 390 to 280. As the "pegging" section in the lower half of the inventory sheet reveals, there are still past due allocations outstanding for 257, 128, and 2800, in addition to the revised 280 allocation.

FIGURE 7-11   (continued)

on the factory floor (in what were known as WIP bins), at the point of use. If more materials were delivered than could be deployed to the floor, they were either kept in trailers outside the plant or diverted to a warehouse elsewhere in town. (The addition of more products and sales volume to the plant necessitated this expediency.) The MCS kept track of all materials. If they were in the warehouse, MCS tracked aisle number, bin number, and height within the bin. It could assign received materials from one bin to another, or to the floor.

Backflushing and the Material Control System

Traditionally, the storeroom at the Tarboro plant had been a key point of control. Materials had been

checked into the storeroom and then checked out, as needed, to the factory floor. Although providing control over what the plant had in stock, where, and how much, the storeroom impeded the flow of materials and cost money to operate. In an effort to streamline the flow of materials, the Tarboro plant both eliminated the storeroom and modified the PACE and MCS systems to facilitate "backflushing." With backflushing, the storeroom was no longer a point of control; materials were received at the loading dock and entered onto the system but were only acknowledged as "used" when skids of finished goods were moved from the factory to shipping (when the finished goods move tag was processed) or when major subassemblies (e.g., chargers) were completed.

When materials arrived, the system searched the bills of materials to determine which product or component used them and thus where on the floor they should be located. Furthermore, the system checked how much was already on the floor and whether there was room for more. If so, the materials were granted a barcode and directed to the factory floor. If not, the materials were directed to central stores (either a trailer or the in-town warehouse).

B & D–Tarboro took considerable care with its backflushing:

1. Frequent visual checks were taken on the floor for "reasonableness" of what the system thought was being carried there.

2. Checklists were followed to reconcile suspected discrepancies (e.g., unreported scrap, mislocated parts, supplier shorted deliveries).

3. Scrap was reported daily, with documentation required so that a scrap designation could not be used indiscriminantly.

As a high-volume plant, B & D–Tarboro produced millions of units of output a year. Receiving dock workers did not count precisely either supplier receipts or the units of materials on any release of an order. Thus, at any time, an order might contain more or less than the number of pieces entered into the MCS. Nevertheless, the integrity and discipline of this backflushing system was excellent, as was evident in reports on materials usage and correct count variance. The production workers could count on receiving enough materials. The plant had not had to shut down to take a physical inventory since 1972.

The MCS system was capable of providing a wealth of data on the status of materials in the factory. Production planners could place a number of inquiries on the system to aid in their decisions and follow-up, including:

• How much of any part number (or barcode number) was on-hand and where it was located
• Which part number was located at any specific location
• What list of transactions had occurred in the "storeroom," including receiving and shipping

• Where scrapped parts/product and other "hidden location" items were held
• Which part numbers and manufacturing orders were "open" (on the floor to be worked on)
• Where any part number was being used and for which manufacturing orders
• Which purchase orders went with which part numbers

The MCS meant a great deal for the integrity of the MRP system at the Tarboro facility. Before 1974–1975, inventory turns (that is, annual sales divided by the value of average inventory held, a ratio that indicates how rapidly inventory can be viewed as turning over in a year's time) at Tarboro were only about five per year, a low number considering that the plant did not have to carry any finished goods inventory and made high-volume products. Without the MCS, it was impossible to maintain the high degree of record integrity required by the plant's high volume. The discipline provided by the MCS helped immensely to clean up the plant. In 1991, inventory turns at Tarboro were about 26 per year, and had been as high as 40. (With the change to the plant "owning" its own finished goods inventory, the calculated inventory turns would, of course, drop considerably.)

As mentioned earlier, purchasing and materials management worked closely together. In recent years the supplier base had been reduced from 400 separate suppliers to just 200. Many suppliers were now the single sources for various items, with contracts let for terms of 1 to 3 years. For 60 percent of the suppliers, accounting for 75 percent of purchased item value, there were no purchase orders typed. Instead, each week suppliers were sent an MRP system-generated document (see Figure 7-11) that displayed the plant's requirements and current forecasts. For these suppliers, purchasing controlled only the pricing and quality assessment for the items procured. The production and inventory control department actually released orders and did any expediting.

It used to be that the plant's MRP system was operated with a safety lead time of one week, as a kind of "cushion" for production. In addition, one week lead times were customarily added for each level in the bill of materials. To speed things along,

BLACK & DECKER (US) INC. 05/24/91

BUYER-NO 94		TARBORO		PAGE 00002	
VENDOR-NO 58597F		ORDER STATUS FOR XYZ METALS INC.			
PART NBR	DESCRIPTION	WEEK	DEL. DATE	PO NBR	QTY
5014480698	KIT BOX	794	05/13/91	94-593806	450 PAST DUE
		798	06/10/91	94-594362	5,000
		799	06/17/91	94-594536	5,000
		800	06/24/91	95-593523	3,300
		801	07/01/91	94-595095	2,250 CHANGE
		802	07/08/91	94-595294	2,500 CHANGE
		803	07/15/91	94-595585	5,000
		804	07/22/91	94-595751	2,300
		805	07/29/91	94-595976	3,650
		807	08/12/91	94-596512	1,000 NEW
		808	08/19/91		718 PLANNED
		809	08/26/91		1,750 PLANNED
		810	09/02/91		3,650 PLANNED
		811	09/09/91		4,000 PLANNED
		812	09/16/91		2,000 PLANNED
		813	09/23/91		4,400 PLANNED
		814	09/30/91		1,500 PLANNED
		815	10/07/91		1,650 PLANNED
		816	10/14/91		2,100 PLANNED
		817	10/21/91		700 PLANNED
		818	10/28/91		600 PLANNED
		819	11/04/91		1,200 PLANNED
		820	11/11/91		1,100 PLANNED
		821	11/18/91		1,800 PLANNED

BLACK & DECKER (US) INC. 05/31/91

BUYER-NO 94		TARBORO		PAGE 00002	
VENDOR-NO 58597F		ORDER STATUS FOR XYZ METALS INC.			
PART NBR	DESCRIPTION	WEEK	DEL. DATE	PO NBR	QTY
5014480698	KIT BOX	799	06/17/91	94-594536	5,000
		800	06/24/91	95-593523	3,300
		801	07/01/94	94-595095	2,250
		802	07/08/91	94-595294	2,500
		803	07/15/91	94-595585	5,000
		804	07/22/91	94-595751	2,300
		805	07/29/91	94-595976	3,650
		807	08/12/91	94-596512	1,000
		808	08/19/91	94–596804	1,000 NEW
		809	08/26/91		1,520 PLANNED
		810	09/02/91		3,650 PLANNED
		811	09/09/91		4,000 PLANNED
		812	09/16/91		2,000 PLANNED
		813	09/23/91		4,400 PLANNED
		814	09/30/91		1,500 PLANNED
		815	10/07/91		1,650 PLANNED
		816	10/14/91		1,900 PLANNED
		817	10/21/91		700 PLANNED
		818	10/28/91		600 PLANNED
		819	11/04/91		1,200 PLANNED
		820	11/11/91		1,100 PLANNED
		821	11/18/91		1,800 PLANNED
		822	11/25/91		800 PLANNED

Figure 7-12   Order status sheets

Notes to Figure 7-12:

1. Panels A and B are Order Status sheets for the producer of the kit box of Figure 7-11. Panel A shows that 450 kit boxes are past due for Week 794 and that 5000 are due for each of Weeks 798 and 799.
2. Panel B, a week later, shows that the 450 units are no longer past due and that Week 798's order of 5000 has been received.
3. Panel A shows that some changes had to be made to some existing orders for Weeks 801, 802, and 805. Panel B shows that a new order for 1000 has been placed for week 808, precisely what we saw in Figure 7-11 that the B&D materials planner did. Future weeks' demands are shown, but they are still not firm; rather, they are termed "planned".

however, the plant had structured its bills of material to be as "shallow" as they could be (they used as few levels as possible). More recently, the plant's materials management people had formally modified the time offsets to permit either a whole week, no offset at all, or a fractional week offset (1/2 week). This latter modification involved the creation of dummy allocations, which were in essence phony orders that were a fraction of the average weekly demand for the next quarter. By removing the traditional one week of safety lead time and replacing it with the phony order, inventory could be controlled more tightly (less than a full week's inventory was actually held).

Black & Decker's MRP system had been developed in-house, and some time ago. There were newer systems on the market with better technical capabilities: faster to operate, able to accomodate windows of display, run by accessing menus, able to offer daily "buckets" for data, as opposed to the weekly buckets that Black & Decker had to live with. These were attractive features to Black & Decker materials management, although the company's system had as much, or more, functionality to it—backflushing was easy to accomplish even with substituted materials/parts or with a need to use up certain materials before others.

Other Materials Management Tasks

Materials management was charged with several tasks in addition to the classic ones of determining what materials were to be used when, and going about their procurement. For example, materials management was charged with capacity planning, estimating inventory for financial purposes, and devising specific, final-version schedules for the plant's suppliers.

Considerable care was taken in devising the master schedule, so that it was realistic and would not necessarily lead to materials expediting or rescheduling. While orders were launched in the traditional way (i.e., the system "pushed" materials and work orders onto the factory floor), the assembly accomplished was according to well-established assembly rates and not to due dates.

typical length of time to install an MRP system is about 18 months.

Choosing MRP or the Two Non-Time-Phased Inventory Systems

Although an MRP system radiates many desirable qualities, it does so at a cost. The question still persists as to when a company should adopt an MRP system and when it should stay with a less versatile but less expensive non-time-phased system. Consider Situation 7-2.

Discussion of Hewitt-Everett Division

The lobbying effort directed at Mike Hewitt is well intentioned. An MRP system stretching

SITUATION 7-2

Hewitt-Everett Division

The Hewitt-Everett Division of Winchester International, Inc. manufactured automatic weighing and measuring equipment, which was used primarily by agricultural cooperatives for mixing feeds and fertilizers. The equipment released selected quantities of grain, nutrients, or fertilizer into a hopper for either bulk or bag loading. The heart of the product—and the toughest part to make—was the release mechanism, for it had to be both sturdy and exact. Much of the rest of the equipment was sheet metal; forming it presented no particular difficulties.

Over the past year, Hewitt-Everett had modified the release mechanism to include a new electronic sensor and control system. This electronic system was more rugged, more versatile, and quicker than the mechanical system it replaced. Its introduction had strengthened the product in the market, and sales were running strong.

Hewitt-Everett offered its weighing and measuring equipment in over 70 varieties of sizes and configurations. Since this equipment was bulky and rather expensive, no finished goods inventories were kept. Rather, production proceeded on an order-by-order basis. For the 70-odd different equipment offerings, however, there were only eight different release mechanisms required, which simplified matters considerably. A like number of electronic sensor and control systems were needed as well.

The sole supplier of the electronic sensors and controls was another division of Winchester International, which specialized in the original equipment electronics market. Hewitt-Everett was delighted with the product's quality; but recently, as sales continued their steady upward climb, deliveries from this other Winchester International division had lagged. The late deliveries were causing increasing concern among Hewitt-Everett's management because they were stretching out Hewitt-Everett's own cycle time in fabricating its weighing and measuring equipment. If the company's cycle time were stretched much further, the company would lose its ability to compete for many of the late fall orders, a prime order-taking period. That is, if the company could not deliver its equipment by late winter in time for spring planting, it was in danger of losing any order placed after the first of November.

As a remedy for this disturbing situation, a group of managers within the company had been lobbying Mike Hewitt, the company's founder and still its president, for the introduction of a material requirements planning system that might even cross division lines within Winchester International. The essence of their argument was that Hewitt-Everett needed to give its sister division more time to react to the order for sensors and controls. At present, when Hewitt-Everett received an order, only the assembly department was notified of model and quantity. That department was then responsible for placing a subsequent, derivative order with its suppliers, including the department making the release mechanism. The release mechanism department, in turn, was responsible for placing an order with the sister division. The managers lobbying Mike Hewitt thought that 2 to 3 weeks could be saved by notifying the sister division directly of the equipment orders as they were received, instead of waiting for the domino effect of department-by-department ordering. They reasoned that a material requirement planning (MRP) system could coordinate this materials ordering procedure much more quickly and efficiently than the present set of procedures could. They estimated that establishing an MRP system would cost Hewitt-Everett about $100,000 and that maintaining it would cost about $25,000 per year.

To Mike Hewitt, this sounded like a lot of money, but he also recognized that something clearly had to be done to shorten the cycle time in manufacturing. (Some figures on product cost and quantities are given here for the latest year.)

Units sold	420
Present growth rate of unit sales	9%
Average unit cost	$13,500
Average standard cost of release mechanisms	$ 3,100
Average transfer price of electronic sensors and control systems	$ 1,250

across division lines at Winchester International does promise to reduce the production cycle time of the release mechanisms that have been at the root of management's present concern. It could well reduce that cycle time by the two to three weeks its proponents claim could be shaved off. There would probably be other benefits as well, such as better control over other components of the automatic weighing and measuring systems the division sold.

Installing a material requirements planning system, however, is not the only way by which the division's production cycle time could be reduced. An alternative worthy of consideration is a simple one—namely, providing a buffer stock of the sensors and controls that have been delivered tardily. The Hewitt-Everett Division could then draw immediately on this "decoupling" inventory and at the same time place an order with the sister division that would result in the restocking of the buffer inventory. Since only eight different sensor and control systems are required anyway, the inventory would not have to be that large and diverse to be effective. Either the MRP system or the buffer inventory plan would succeed in eliminating the waiting associated with the present system of one department or division placing orders on other departments or divisions. Choosing between the two rests largely on the costs of each. The expected MRP system expenses are already provided. What must be estimated are the carrying costs of the buffer inventory.

Calculating the Carrying Cost of the Buffer Inventory

Although the particular quantities of each of the eight sensor and control systems needed would depend on the expected pattern of demand, for a back-of-the-envelope calculation we might expect to have to inventory four weeks of production, on the average. The company's inventory carrying cost percentage is not quoted in the situation, but 25 percent is not likely to be far off the mark. The expected annual inventory carrying cost would then be:

420 units/year $\times$ 4 weeks/52 weeks $\times$ \$1250 cost per system $\times$ 25 percent carrying cost
= \$10,100 per year in carrying costs

This cost is only slightly more than a third of the annual expense of maintaining an MRP system. The required weeks of inventory could be doubled and the buffer inventory costs would still be less than the MRP system's. Thus the side benefits of the MRP system would likely have to be significant for an MRP system to be preferred.

In addition, it is likely that the time needed to install a formal, computerized material requirements planning system could be significantly longer than the time it would take the sister division to build up an inventory of sensor and control systems. MRP systems typically take many months to install and debug. Although a buildup of a buffer inventory would not be quick, especially considering the recent pattern of late deliveries, at least some inventory could be built up with the next order placed on the sister division.

All things considered then, the simpler, less sophisticated, buffer inventory approach to cutting down on the division's production cycle time appears to be the preferred solution to Hewitt-Everett's present concerns.

When Implementing an MRP System Is Less Desirable[5]

As the workings of a material requirements planning system were described in this chapter, a tremendous amount of data must be accumulated and updated continually—master production schedules, bills of materials, inventory status, production cycle times, materials needs at each stage of the production cycle, supplier lead times, routings of parts, and standard costs. The more easily a company can check off the data in its master schedule, bills of materials, and the like, the more likely that it has its production well under control. Simply going through the exercise of assembling such data in preparation for the implementation of an MRP system is a tremendous learning experience for most companies and suggests a host of useful changes in operations. There is no denying that the MRP philosophy can pay huge dividends to plant management.

On the other hand, there are some circumstances under which implementing an MRP system taxes company resources too much to be worth the effort. The Hewitt-Everett Division situation is one where the benefits of an MRP system do not mesh well enough with the problem to justify its use in preference to a decoupling inventory. In other instances, an MRP system is not justified because of data accumulation and updating difficulties and costs.

For example, in situations where the master production schedule is not known with certainty well in advance, an MRP system is harder to justify. An MRP system works best when products are made to stock and not to order, when large buffer stocks of finished goods exist, and when sales are not very sensitive to abrupt seasonal or cyclical swings. A material requirements planning system is likewise tougher to implement when bills of materials for products are not stable. An MRP system is best off when the number of new product introductions over time is modest, when there are few engineering changes released, and when there is little or no tinkering on the shop floor by engineers or others.

In a similar vein, an MRP system is likely to run more smoothly and be more effective when the process itself is stable. That is, when the

process is not new or state-of-the-art itself, and when it is not likely to exhibit significant variances in product yields and quality, an MRP system stands a better chance. Also, if the time periods when various inputs are used in the process are well established, implementing an MRP system is easier. Finally, an MRP system is best off when supplier lead times are known and dependable.

MRP systems have been criticized in some quarters as inadvertently contributing to the maintenance of long production cycle times of products (that is, long time periods for the manufacture of products from initial materials acquisition through final inspection). The argument here is that because MRP systems build in excessive safety lead times and inventories, they unwittingly impede innovations that would push the process toward faster production cycles and toward running with only the slimmest stocks of inventory. Indicative of these innovations are the process advances achieved by running families of parts together, using group technology (where equipment is dedicated to a certain family of parts), and employing the just-in-time production methods of the Japanese (see the Chapter 11 discussion on production control). Although MRP systems have, no doubt, impeded some moves to more rapid production cycle times, the experience of plants like Black & Decker–Tarboro with just-in-time techniques of materials management suggests that MRP can be more flexible than might have been thought.

Tools of the Trade: A Recap of Lessons Learned

The introduction of the computer brought the power of the MRP version of anticipatory inventory management to the factory. Most factories now operate with some sort of MRP system because the management of their parts inventories can be triggered easily off of the master schedules of their finished goods. The logic of MRP is rather straightforward, although it takes a computer to keep track of the great variety of parts and uses for those parts in a factory. There are numerous "wrinkles" that can be added to the main MRP logic (e.g., lot sizing, product costing, safety lead times), but one should be careful not to adopt them simply because it is possible. The power of MRP rests with the original, simple idea of being able to manage inventories in anticipation of their use. Thus, MRP often works best when it applies merely to the ordering of parts and materials from outside the factory for delivery to the factory on a certain date, and when it does so on a lot-for-lot basis (following the master production schedule) that uses simple, shallow bills of material. Getting much fancier than that invites some problems into operations and may require more data, and more data integrity and cost, than is warranted.

World-Class Practice

The most progressive firms these days try diligently to keep their computer systems simple, knowing that managers need only a fraction of the information that could be generated by massive computer-based systems. MRP is certainly employed by these companies but typically only to help gather materials to the factory's doorstep, not to control or report on the flow of materials through the factory itself. MRP is best employed as a tool to coordinate the factory's efforts with those of its supplier base. The factory's MRP system in these progressive companies is transparent to the supplier base, giving the companies great visibility into the future and heightening their abilities to react to the changing marketplace. To be avoided is an MRP system that actually delays the manufacture of a product. What needs to continue are all the efforts to speed the flow of materials through the factory, and the MRP system is never important enough by itself to stand in the way of that all-important mission.

QUESTIONS

1. A material requirements planning system must continually keep track of numerous elements. What are these, and how are they exemplified at Black & Decker–Tarboro?

2. Review and summarize the fundamental steps in material requirements planning. Make sure that you include the data requirements for computer files for such a system. What are some of the most common complications that can arise in such a system?

3. Can a material requirements planning system be implemented successfully in a process that operates with a decoupling inventory? Support your answer.

4. Outline clearly how the inventory needs of the job shop vary from those of the continuous flow process.

PROBLEMS

1. After examining the following information on watches, derive the planned orders for watchbands, using lot-for-lot order sizing.

- Watches

  Period: 1 2 3 4 5 6 7 8 9 10
  Demand: 10 20 40 0 50 30 30 40 50 30

  On hand inventory: 50 units at the beginning of period 1
  Outstanding orders: 30 units to arrive at the beginning of period 3
  Lead time: one period

- Watchbands

  Usage: one per watch
  On-hand inventory: 0
  Outstanding order: 0
  Lead time: two periods

2. Information on product A is shown in Figure 7-13 and Tables 7-2 and 7-3. Using this information and lot-for-lot as the order sizing rule, derive the planned orders for part F.

TABLE 7-2   Factors in the production of product A (Problem 2)

ITEM	LEAD TIME (weeks)	HOW MANY USED	ON-HAND INVENTORY
A			40
B	2	2	30
C	3	2	60
D	2	1	80
E	3	1	50
F	2	2 (1 in E, 1 in C)	30
	2	(1 in C, 1 in H)	
G	1	1	20
H	1	1	10

TABLE 7-3   Master schedule for product A (Problem 2)

WEEK	REQUIREMENT	WEEK	REQUIREMENT
1	20	11	20
2	40	12	50
3	10	13	40
4	30	14	0
5	25	15	10
6	30	16	20
7	15	17	15
8	0	18	45
9	80	19	0
10	30	20	30

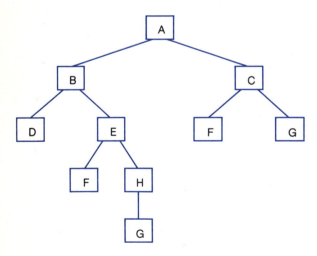

FIGURE 7-13   Components of product A.

3. Given the same information as in Problem 2, derive the planned orders for part G.

4. What is a "closed-loop" MRP system? How would you try to measure its benefits for a plant? Its costs?

5. The following list gives the parts for a particular product, indicating their level in the product structure. Higher-numbered levels fit into the next-highest-level's subassembly. Also indicated are the number of parts needed each time 1 unit of the next-highest-level's subassembly is made.

LEVEL	ITEM	QUANTITY
1	A	1
2	B	2
2	C	2
1	D	2
2	E	1
3	F	1
3	G	1
2	H	2
3	I	4
3	J	4
1	K	3

Draw the product structure tree. How many of each part do we need to make 20 of the product in question?

6. If each level in Problem 5 means a safety lead time of 1 week, what would the inventory status report look like for parts E, G, and J assuming demands for the product of 10 in week 5, 20 in week 8, and 15 in week 10?

7. Suppose that the demands for the product in Problem 6 were for 15 in week 6, 30 in week 7, and 25 in week 9. What would the inventory status report look like for parts E, G, and J?

8. How would the lot sizes be chosen by part-period balancing under the following circumstances? Setup cost is $60, an inventory carrying cost per unit per period is 40 cents, and the schedule of net requirements is as follows:

PERIOD	NET REQUIREMENTS	PERIOD	NET REQUIREMENTS
1	30	7	50
2	60	8	30
3	45	9	60
4	15	10	15
5	60	11	40
6	45	12	50

Total net requirements: 500

What would the lot sizes be like without the "look-ahead, look-back" features of part-period balancing?

9. Suppose that in Problem 8 the setup cost was reduced to $40 but that the carrying cost rose to $0.50 per unit per period. What would the lot sizes look like?

10. The master production schedule at Harris, Inc., is as follows:

MASTER PRODUCTION SCHEDULE				
	WEEK 1	WEEK 2	WEEK 3	WEEK 4
Product 101	10	15	20	15
Product 102	4	10	5	12

Component 151 is used in both products (1 unit of the component in each). Component 151 is available in lot sizes of 50 with a purchasing lead time of 1 week. Currently, there are 20 units of component 151 in inventory and an open order (scheduled receipt for 50 more (its lot size)) due at the beginning of week 2. No safety stock is held. Complete the MRP record for component 151 as of the beginning of week 1. During week 1 the following things happened:

1. The open order for 50 turned out to contain only 45 on delivery.

2. The shop scrapped 3 units of component 151 and drew out 3 replacements.

3. A customer canceled 2 units ordered for week 2.

4. The week 5 orders are for 18 product 101s and 10 product 102s.

5. A 3-unit order for 102 has been moved up from week 4 to week 2.

6. The production planner released an order for 50, due in week 2.

Complete the now-current MRP record for component 151 as of the beginning of week 2. What actions should the production planner responsible for component 151 take?

Week	1	2	3	4
Gross requirements				
Scheduled receipts				
Projected available balance				
Planned order releases				

Week	1	2	3	4
Gross requirements				
Scheduled receipts				
Projected available balance				
Planned order releases				

11. The Dunnick Tool and Die Company had the product structure shown in Figure 7-14 for one of its products, the X27.

(a) Complete the following MRP records for parts F and M given the indicated gross requirements, a lead time of 1 week for part F and 1 week for part M, lot-for-lot order quantities, and no safety stock.

Part F	Week	1	2	3	4	5
Gross requirements		12	14	10	19	22
Scheduled receipts		22				
Projected available balance	0					
Planned order releases						

Part M	Week	1	2	3	4	5
Gross requirements		42				
Scheduled receipts		40				
Projected available balance	13					
Planned order releases						

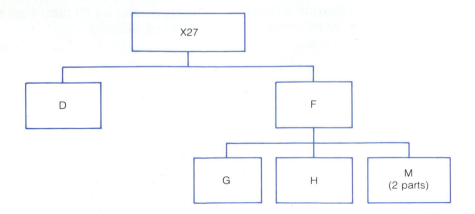

FIGURE 7-14   Product structure and MRP records for Problem 11.

(b) Suppose that lot-for-lot order quantities are forsaken in favor of lot sizes of 15 for part F and 30 for part M. How would the MRP records in part (a) change?

12. Consider the following product, a SC-121 water bath for laboratory use, one of whose components is a top/pan assembly, PN307388. The following MRP records are for these two items. Complete the inventory status records assuming that the SC-121 (a level 0 item) is produced in lots of 50 units and that the PN307388 (a level 1 item) is produced in lots of 100. The lead time for both is 2 weeks and no safety stock is required for either item. Note the inventory on hand.

13. The master schedule calls for 25 A items for delivery at the start of week 7. Using the information below, determine how many units of items E and H are needed and when any orders for them ought to be timed. Assume lot-for-lot orders. Show your work on the inventory status sheet provided.

ITEM	LEAD TIME	ON-HAND	COMPONENTS (number needed)
A	0	10	B(1), C(2)
B	1	5	D(2), E(1), F(1)
C	2	8	E(1), G(1), H(4)
D	2	12	–
E	2	5	--
F	1	5	–
G	1	3	–
H	1	15	–

14. How would the lot sizes be chosen by part-period balancing under the following circumstances? Setup cost is $60, an inventory carrying cost per unit per period is 40 cents, and the schedule of net requirements is as follows:

PERIOD	NET REQUIREMENTS	PERIOD	NET REQUIREMENTS
1	30	7	50
2	60	8	30
3	45	9	60
4	15	10	15
5	60	11	40
6	45	12	50

Total net requirements: 500

What would the lot sizes be like without the "look-ahead, look-back" features of part-period balancing?

15. Suppose that in Problem 14 the setup cost was reduced to $40 but that the carrying cost rose to $0.50 per unit per period. What would the lot sizes look like?

SC-121: Lot size = 50, lead time = 2 weeks, safety stock = 0, on hand = 47

Week		1	2	3	4	5
Gross requirements		10	10	10	10	10
Receipts						
Inventory	47					
Planned orders						
Week		6	7	8	9	10
Gross requirements		10	10	10	10	10
Receipts						
Inventory	47					
Planned orders						

PN307388; Lot size = 100, lead time = 2 weeks, safety stock = 0, on hand = 8

Week		1	2	3	4	5
Gross requirements						
Receipts						
Inventory	8					
Planned orders						
Week		6	7	8	9	10
Gross requirements						
Receipts						
Inventory	8					
Planned orders						

REFERENCE NOTES

1. See Michiel R. Leenders, Harold E. Fearon, and Wilbur B. England, *Purchasing and Materials Management*, 8th ed. Homewood, IL.: Richard D. Irwin, 1985, for a good discussion of the functions and forms of inventory.

2. Some useful sources on MRP include:
   Aggarwal, Sumar C., "MRP, JIT, OPT, FMS?" *Harvard Business Review* 63, no. 5 (September–October 1985): 8–16.
   Armstrong, David J., "Sharpening Inventory Management," *Harvard Business Review* 63, no. 6 (November–December 1985): 42–58.
   Bemelmans, Roland, *The Capacity Aspect of Inventories*, New York: Springer-Verlag, 1986.
   Duchessi, Peter, Charles Schaninger, and Don R. Hobbs, "Implementing a Manufacturing Planning and Control Information System," *California Management Review* 31, no. 3 (Spring 1989): 75–90.
   Etienne, E., "Choosing Optimal Buffering Strategies for Dealing with Uncertainty in MRP," *Journal of Operations Management* 7, no. 1 (1987): 107–120.
   Hall, Robert, "Getting the Commitment of Top Management," *Production and Inventory Management* 18, no. 1 (1977): 1–9.
   Leenders, Michiel, Harold E. Fearon, and Wilbur B. England, *Purchasing and Materials Management*, 8th ed. Homewood, IL: Irwin, 1985.
   Miller, Jeffrey G., and Linda G. Sprague, "Behind the Growth in Materials Requirements Planning," *Harvard Business Review* 53, no. 5 (September–October 1975): 83–91.
   Orlicky, Joseph A., *Material Requirements Planning: The New Way of Life in Production and Inventory Management*. New York: McGraw-Hill, 1975.
   Papesch, Roger M., "Extending Your MRP System into Your Vendor's Shop," *Production and Inventory Management* 19, no. 2 (1978): 47–52.
   Plossl, George W., *Production and Inventory Control: Principles and Techniques*, 2nd ed. Englewood Cliffs, NJ: Prentice-Hall, 1985.
   Ptak, Carol A., "MRP, MRP II, OPT, JIT, and CIM–Succession, Evolution, or Necessary Combination," *Production & Inventory Management* 32, no. 2 (Second Quarter 1991): 7–11.
   Schroeder, Roger G. et al., "A Study of MRP Benefits and Costs," *Journal of Operations Management* 2, no. 1 (October 1981):1–9.
   Silver, Edward A., and Rein Peterson, *Decision Systems for Inventory Management and Production Planning*, 2nd ed. New York: Wiley, 1985.
   Sridharan, V., and R. Lawrence LaForge, "The Impact of Safety Stock on Schedule Instability, Cost and Service," *Journal of Operations Management* 8, no. 4 (October 1989): 327–347.
   Vollmann, T.E., W.L. Berry, and D. Clay Whybark, *Manufacturing Planning and Control Systems*, 2nd ed., Homewood, IL: Dow Jones-Irwin, 1988.
   Wallace, Thomas F., *MRP II: Making it Happen: The Implementers' Guide to Success with Manufacturing Resource Planning*, 2nd ed. Essex Junction, VT: O. Wight Ltd. Publications, 1990.
   Wemmerlov, Urban, "Assemble-to-Order Manufacturing: Implications for Materials Management," *Journal of Operations Management* 4, no. 4 (1984): 347–368.
   White, Edna et al., "A Study of the MRP Implementation Process," *Journal of Operations Management* 2, no. 3 (May 1982): 145–153.
   Wight, Oliver W., MRP II: *Unlocking America's Productivity Potential*. Williston, VT.: Oliver Wight Ltd. Publications, 1981.

3. See Oliver W. Wight, *The Executive's Guide to Successful MRP II*. Englewood Cliffs, NJ: Prentice-Hall, 1982.

4. This section owes much to Wight, *The Executive's Guide to Successful MRP II*, pp. 27–35.

5. A good discussion of many of the points about implementing an MRP system is found in Robert W. Hall and Thomas E. Vollmann, "Planning Your Material Requirements," *Harvard Business Review* 56, no. 5 (September–October 1978): 105–12.

CHAPTER 8

MANAGING INVENTORIES: TIME INDEPENDENT SYSTEMS

It is not too far off the mark to visualize the problem of managing either raw materials inventory or finished goods inventory as one of managing piles of "stuff" that either the process itself or consumers in the marketplace draw down. The objective of good inventory management in this case is to offer good service to either the process or the market at reasonably low cost. This objective, in turn, means deciding how many items should be in each pile, when orders to replenish the piles ought to be placed, and how much each of those orders should contain. Managing such inventory stocks (as opposed to managing work-in-process) essentially means deciding these three questions of pile size, order time, and order size.

Because the timing of independent demands is often so uncertain, the techniques typically used for managing inventories of such demands are themselves independent of time. These techniques are sometimes called non-time-phased inventory systems, meaning simply that they do not attempt to be so precise about timing orders to expected use.[1] Their purpose is the same, however: to replenish a pile of items in timely fashion and at reasonably low cost. Retailing and distribution activities easily lend themselves to such techniques; consumer demands are frequently variable and uncertain.

The two major non-time-phased techniques have these differing philosophies:

1. Replenish the pile on a regular basis (daily, weekly, monthly) and bring it back up to the size you want (we will talk shortly about the desired size of the pile). The amount by which you replenish the pile may vary from one time to another, but you always replenish the pile. This basic strategy is often called a *periodic reorder system*.

2. Keep a constant watch over the pile. When its size dips to a predetermined level (we will talk shortly about this level), replenish the pile enough to bring it back up to the size you desire. Under this philosophy, the amount by which you replenish the pile stays the same, but the time spans between replenishments may vary. This basic strategy is frequently termed a *reorder point system*, the "reorder point" being the predetermined level that, once reached, triggers the replenishment of the pile by the same amount. More will be said about periodic reorder and reorder point systems and their application to independent demand

inventories. These same systems can also be applied to dependent demands inventories, as well, although frankly, with much less success.

Let us examine them in greater detail.

THE WORKINGS OF NON-TIME-PHASED INVENTORY SYSTEMS

Relevant Costs and Concepts

The decisions that have to be reached in non-time-phased inventory systems—those of pile size, when to replenish the pile, and how much should be in a replenishment—involve balancing a variety of costs. Several kinds of costs are relevant, as described here.

Costs of Holding the Inventory

Inventory is not costless to stock. There are a number of costs to consider, the major one being opportunity cost of capital. This is the cost incurred by having the company's capital devoted to financing the inventory rather than invested in some other, income-producing endeavor (machines, personnel, a bank account). Often the company's cost of capital is used to estimate the magnitude of this opportunity cost.

Other, lesser costs of holding inventory include:

1. *Storage and handling.* The costs of warehousing, handling, counting, and keeping track of the inventory.
2. *Insurance and taxes.* Standard, out-of-pocket costs mandated for inventories of all types.
3. *Obsolescence and shrinkage.* Costs that should be imputed to inventories because, typically, not all of the items placed into inventory can be pulled back out and used

when needed. Some items become obsolete because of engineering changes. Other items deteriorate or break or, in some other way, become unusable. This represents a cost of holding inventory.

Costs of Securing More Inventory

There are several kinds of costs incurred in securing one more replenishment for a pile of inventory. Moreover, it is important to recognize that these costs ought to be marginal (incremental) costs and not average costs. For example, consider preparing a purchase order for an outside vendor. The average costs of that order might include the time of the buyer and clerk involved, as well as the cost of the paperwork and phone calls to place and keep track of the order itself. If, as is nearly always the case, the buyer and the clerk would be employed anyway, then their costs are not relevant to the calculation of the cost of the order and thus should be excluded.

The kinds of costs incurred to secure more inventory can include:

1. *Order costs.* Costs associated with placing orders for materials from outside vendors. The marginal costs of ordering are generally very low, consisting for the most part of some telephone charges and some paperwork.
2. *Production changeover costs.* If the inventory to be secured is from within the plant, then the replenishment will likely impose some costs over and above the costs of the inventory units themselves. A new setup takes time, imposing an opportunity cost on the factory because that setup time might have been spent producing output. Sometimes the opportunity cost is zero (that is, no output would be forgone by using the time for setup), but it is a point well worth considering. The setup itself may involve

costs—fresh tooling, cleaning, new production materials, generated scrap. The replenishment may also cause the factory to increase capacity through overtime, new hires, subcontracting, or the like, and these means of enhancing capacity can be costly in incremental terms.

3. *Tracking costs.* Costs involved in determining when an order or production changeover should happen. Stashes of inventory have to be monitored, which takes people, systems, and paperwork. Frequently, inventory levels cannot be accurately determined easily, so real economies exist in tracking inventory items in bunches and in placing purchase or production orders in groups, rather than one at a time.

This is particularly true for items for which perpetual inventory records (records of the current quantities for each item, updated constantly through transactions of receipt and disbursement of the item from the stockroom) are not kept, or, at least, not kept very accurately.

4. *Volume discounts.* The considerable savings that often can be enjoyed by buying (or manufacturing) items in quantity. Many vendors publish price lists with discounts given for bulk purchases.

Costs of Not Having Enough Inventory

When a company cannot meet demand out of finished goods inventory, it can incur several (sometimes significant) costs. If the customer is willing to wait for the item—the item is then said to be backordered—there are likely to be additional costs for tracking the order: expediting it, shipping it specially, and invoicing. If the customer is not willing to wait for the item and presumably goes elsewhere to purchase it or its substitute, the company forgoes a sale and, thereby, the contribution to fixed costs and profits that sale could have made. Indeed, by stocking out, the company may forgo all future sales by that consumer, thus losing the contribution possible from future purchases. Backorders are cheap compared to this kind of loss. However, determining the cost of a stockout, or even of a backorder, is fraught with problems, because so many different assumptions can be made about the consumer's present and future actions. Sometimes a backorder is regarded as appropriate, whereas other times the penalty is stiffer: no future sales. A typical cost assigned to a stockout is the one-time loss of contribution from the item, but such an estimate could be high or low without the company knowing for sure.

Also relevant to inventory management are lead time for the item (the time it takes to receive an ordered shipment) and safety stocks held. Safety stock levels will be discussed later; for the present we will simply acknowledge their usefulness. Lead times are important to calculations of the expected demand for the item between ordering a supply of it and receiving the order.

Exactly how the costs of inventory interact among themselves and interact as well with lead times and safety stocks—to determine inventory levels, timing, and replenishment quantities—depends upon which approach is taken. The periodic reorder technique balances these costs differently from the reorder point technique. In general, though, the higher the costs of holding inventory, the less inventory we want to have on hand at any time, the more frequently we want it to be replenished, and the smaller we want any replenishment quantity to be. The higher the costs of securing more inventory or of not having enough, the more inventory we are willing to have on hand, the less frequent the replenishment, and the larger the replenishment quantity. These costs thus pull in opposing directions and must be balanced against one another. The next sections examine how those balances are made.

The Periodic Reorder System

The periodic reorder system is governed by the simple decision rule of "order enough each period (day, week, month, or whatever) to bring the pile of items inventoried back up to its desired size." This is similar to filling the car's gasoline tank every Friday. The workings of the system can be clarified by reference to Figure 8-1. The solid line traces the actual level of inventory held. The replenishments arrived at equally spaced times (1, 2, 3, 4), but their amounts differed each time. The replenishments were also ordered at equally spaced intervals (A, B, C, D), separated from the arrival times by a consistent delivery lead time. The amount ordered each time was the difference between the desired level and the actual amount on-hand at the time the order was placed. Thus, it equals the actual demand during the previous period. Note that during time period 2, the actual inventory dipped into the safety stock because of unforeseen demand.

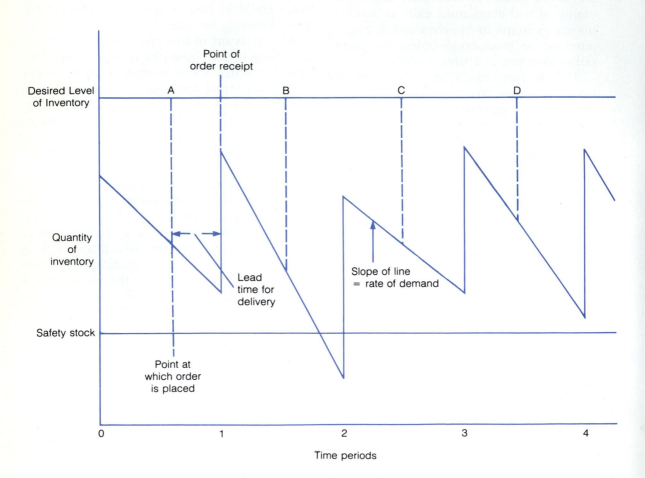

FIGURE 8-1   The periodic reorder system.

In the workings of the periodic reorder system, a key question is what the desired level of inventory is. If we know in advance exactly how much will be needed during any period, the answer is easy: order the amount that will be used and no more. So, as the truck pulls in with the supply for next period, the last unit of this period's supply is being drawn off. This technique is similar to the basic philosophy of material requirements planning — to order just what is needed. In this case of certainty of demand, Figure 8-1's desired level of inventory would change each period, along with the amounts anticipated (always correctly) to be consumed that period.

Unfortunately, a company rarely knows in advance how much of a given item will be needed. Thus the desired level of inventory cannot be determined so readily. Mere estimates of likely demand will have to be relied on, and some safety stock level will have to be set. (Safety stocks will be discussed later.)

The Reorder Point System

The reorder point inventory system follows the decision rule of "watch withdrawals from the pile of inventory until the designated reorder point is struck and then order the fixed amount needed to build the pile back up to its desired size." This is similar to filling the gasoline tank whenever the gauge reads 1/4 full. But, again, the question is, What is the desired size? Consult Figure 8-2.

The desired size in this case is not determined, as in the periodic reorder system, by the

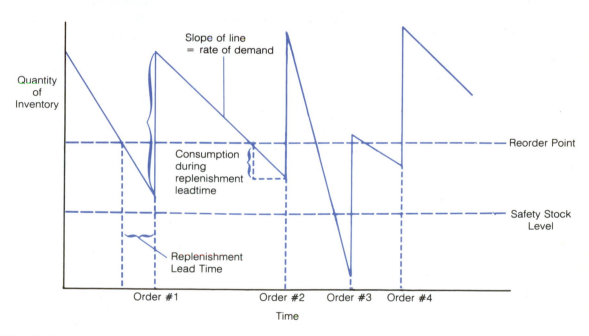

Notice that the reorders are not done at periodic times but at irregular times that are a function of the demand rate at that point.

FIGURE 8-2   The reorder point system.

expected usage over the period, since there is no regular period following which inventory needs are checked. The pile is monitored continually, not every so often. Instead, the size of the pile depends fundamentally on a quantity called the *economic order quantity* (EOQ).

The choice of reorder point depends on how soon a shipment of the EOQ can arrive. If the shipment can, once ordered, arrive instantaneously, we are safe to run the stock to zero before reordering and the reorder point is zero. If the shipment takes a while to arrive, the choice of reorder point depends on an estimate of the expected pace of demand. The order point normally should leave enough so that a stockout does not occur while waiting for delivery of the EOQ. Computing this reorder point is much like computing a safety stock for uncertain demand in a periodic reorder system. If there were no uncertainty, the reorder point would simply be the average expected rate of withdrawal from inventory per time period (x parts per day) times the number of time periods it will take to receive the order of EOQ. Given uncertainty, however, the reorder point has to be larger by the amount of a buffer or safety stock (calculating the size of the buffer stock is addressed later). No matter what the reorder point, however, the decision rule stays the same: When the reorder point is struck, order the EOQ.

How does the company determine how much the EOQ should be? The key to this order quantity and the reason it can be claimed as economic is the process by which it is found. The EOQ is chosen as the order quantity that minimizes the "total variable costs associated with changing order quantities." Every word of this phrase is meaningful. There are a variety of EOQ formulas, but what is important is the process by which the EOQ is found, not any formula that might seem to apply.

Let us consider a very simple, conventional, but suggestive case where we can discover a minimum for the total variable costs. In this case, there are only two variable costs that are in any way associated with any changing of order quantities.

1. *Inventory carrying costs.* These carrying costs vary directly with the order quantity, since the larger the order quantity, the larger is the average inventory held and thus the larger the carrying costs incurred. To illustrate this point in greater detail, let us make the common—and convenient—assumption that the demand on the inventory is absolutely steady, as is represented in Figure 8-3. Given this constant rate of draw upon the inventory, the average number of units in inventory is half of the order size. So if Q is the order size (see Figure 8-3), the average inventory is Q/2. As we shall discuss later, this is an important simplification in this computation of minimum total costs.

2. *Order costs.* The order cost for any single order is a constant. The costs of actually placing or setting up the orders for the inventory thus vary inversely with the order quantity. The larger the size of the order (for

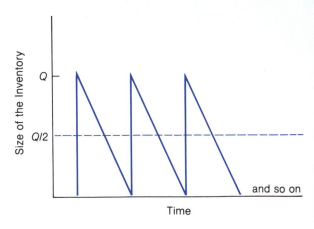

FIGURE 8-3   Graph depicting the assumption of steady demand on an inventory.

a fixed annual demand), the fewer the orders that have to be processed, and thus the lower the order costs are. These order costs can be the costs incurred in dealing with vendors or the costs incurred in setting up the order in a manufacturing department of the company. For this case, assume that (1) the time from order to receipt is a constant and (2) no stockouts are deliberately permitted. (In Figure 8-2 stockouts are deliberately permitted.)

It is important to note what is not included as a variable cost associated with changing order quantities. For example, the cost of the inventory itself is not included because, for a fixed annual demand and a fixed price per unit, the cost of the inventory does not vary with changes in the order quantity. However, if a supplier were willing to give discounts on orders above a certain size, for example, the cost of the items in inventory would become a cost that varies with different order quantities and should be included in the total cost calculation. Transportation costs are also assumed not to vary with changes in the order size, nor are any production or other related costs.

Given the assumptions spelled out above, and with the knowledge that the only variable costs we can associate with different order quantities in this case are carrying costs and order costs, we can define the total variable costs in symbols as follows:

Total variable costs = order costs + carrying costs
= (cost per order) (number of orders) + (period carrying cost as a percentage) (value of average inventory)

$$= S\left(\frac{D}{Q}\right) + i\left[C\left(\frac{Q}{2}\right)\right]$$

where

S = cost per order
D = period demand, usually annual demand
Q = order quantity
i = period carrying cost as a percentage (includes opportunity cost, physical costs of handling and storage, obsolescence)
C = *variable* cost per unit of inventory
$Q/2$ = average number of units in inventory, assuming a steady and constant demand

We want to solve for the order quantity (Q) for which the total cost is a minimum. Given the particular expressions for carrying cost and order cost, one way to solve for this minimum is to set the order cost equal to the carrying cost, since the intersection of these two lines falls directly below the low point of the graph of total cost (see Figure 8-4).*

Setting order cost equal to carrying cost and solving for Q yields

---

*In general, of course, one cannot rely on friendly mathematical functions to solve for the minimum of an expression like that for total costs. One must rely on calculus to develop an expression for the slope and solve for Q, the order quantity, at the lowest point, where the slope is zero. Differentiating total cost [TC] with respect to Q and setting the resultant expression equal to zero yields

$$TC = S\left(\frac{D}{Q}\right) + iC\left(\frac{Q}{2}\right)$$

$$\frac{\partial TC}{\partial Q} = -\frac{SD}{Q^2} + \frac{iC}{2} = 0$$

$$Q = \sqrt{\frac{2SD}{iC}}$$

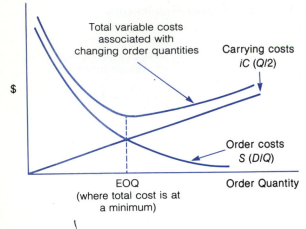

FIGURE 8-4   Costs versus order quantities for the simple case of carrying costs and order costs only.

$$s\left(\frac{D}{Q}\right) = iC\left(\frac{Q}{2}\right)$$

$$2SD = iCQ^2$$

$$Q = \sqrt{\frac{2SD}{iC}}$$

What does this solution mean for this simple example? It has the intuitively appealing properties that the cost-minimizing order quantity, the EOQ, rises with increases in period demand and cost per order and declines with increases in the inventory carrying-cost rate and in the variable cost of the inventoried item. Importantly, the EOQ varies by the square root in this simple case. That is, an increase in company sales does not require a proportional increase in inventory to service it, but merely an increase according to the square root. In other terms, a doubling of demand does not require a doubling of inventories, but merely an increase of 41 percent, all other things held constant.

To reiterate, the key thing about the determination of an EOQ within a reorder point inventory system is not the formulas themselves, but the method by which an EOQ is found— namely, by use of the framework of total variable costs associated with changing order quantities. To see this point more clearly, consider some variations on the simple case just analyzed. The EOQ calculated earlier was simple because of the numerous assumptions that were made. Some of these assumptions can be relaxed easily and a new, modified EOQ can be calculated.

Variation: Simultaneous Production and Consumption

For example, consider a case where the inventory to fill an order is not received all at once but is instead allowed to trickle in. (This is sometimes the situation when the "order" is placed on another department or plant within the company.) The units produced at rate P are withdrawn (consumed) from inventory at a rate W with P > W. Instead of resembling the sawtooth graph in Figure 8-3, the graph of simultaneous production and consumption looks more like the inventory buildup graphs in Chapter 2.

Under this revised assumption, the average level of inventory will be lower because for at least some part of the time period, both production and consumption are proceeding. In fact, the average inventory will be lower by the factor $(P - W)/P$. To see this, consult Figure 8-5. As shown in the figure, the quantity Q is produced only for the period of time T. Thus

$$Q = PT$$

where Q is the total amount of inventory produced and shipped, P is the production rate, and T is the time during which there is production. The maximum quantity ever on hand, X, can be expressed as

$$X = (P - W)T$$

At time T, the quantity of inventory held is at its peak of X and the inventory accumulates at the rate of (P - W). The average inventory held

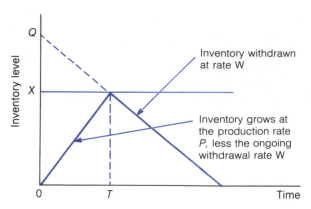

FIGURE 8-5   The case when inventory is permitted to trickle in.

is X/2. Thus average inventory in this variation is lower by

$$\frac{\left(\frac{X}{2}\right)}{\left(\frac{Q}{2}\right)} = \frac{X}{Q} = \frac{(P - W)T}{PT} = \frac{(P - W)}{P}$$

We can substitute this result for the average inventory held into the expression for total cost.

$$TC = S\left(\frac{D}{Q}\right) + iC\left(\frac{Q}{2}\right)\left(\frac{(P - W)}{P}\right)$$

When calculus is used to minimize total cost with respect to Q, we obtain a revised expression for an optimal order quantity.

$$Q = \sqrt{\left(\frac{2SD}{iC}\right)\left(\frac{P}{(P - W)}\right)}$$

Variation: Stockouts Permitted

Now let us change the stockout assumption. As mentioned earlier, the basic EOQ calculation does not permit any deliberate stockouts; it implicitly assumes that the stockout costs are infinite. Figure 8-3, however, suggests that there may be times when stockouts ought to be tolerated. Indeed, many stockouts can be handled by backordering (shipping an order late

once the EOQ replenishment has arrived) that is not too expensive. If backorders are possible without too high a penalty, it is worthwhile recalculating the economic order quantity by introducing backorder costs to the total variable costs associated with the changing of order quantities. (See Problem 21 at the end of the chapter for some hints on how the basic EOQ formula could be changed to accommodate back orders.)

Variation: Price Discounts

Another factor that the simple EOQ model does not allow for is price discounts with volume orders. It is common practice for vendors to quote reduced prices for orders of a certain size.

Thus, after certain order size thresholds, the variable cost of the unit of inventory declines. The lower cost for successive ranges of order volumes naturally changes the total variable cost and thus the computed value of the EOQ. The conventional technique for calculating the EOQ when price discounts are applicable uses the following steps:

1. Compute EOQs for each different unit cost. Compare each with the range of volumes for which it applies to see if the EOQ is feasible (i.e., check to see that the EOQ calculated with a given price lies within the quantity range for that price). Start with the lowest cost per unit range. If its EOQ is feasible, it will be the minimum-cost EOQ.

2. Compute the total costs for each price break quantity and for each feasible EOQ.

3. Select the lowest total cost from step 2. Its quantity is the minimum-cost order quantity.

Figure 8-6 illustrates this situation of price discounts for larger orders. There are three different prices applicable, C_1, C_2, and C_3, as shown

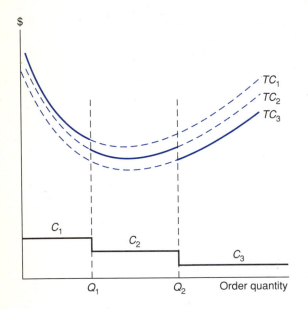

FIGURE 8-6   EOQs when price discounts exist.

in the figure. Associated with each different unit cost there is a distinct total cost curve. These cost curves are designated TC_1, TC_2, and TC_3. As the figure demonstrates, only the first part of the TC_1 curve is applicable since it is only with that first range that the unit cost C_1 prevails. After quantity Q_1, unit price C_2 applies. Thus the relevant total cost curve for quantities lying between Q_1 and Q_2 is the curve TC_2. After Q_2 what is relevant is unit cost C_3 and total cost curve TC_3. The three-step procedure outlined above is a means of investigating each section of the total cost curve systematically to see what lies at a minimum.

This three-step technique can be tedious, however, especially when there are many price break points. Kuzdrall and Britney have suggested a new method that assumes that the price discount schedule for the vendors' products is composed of fixed costs (setup costs) as well as variable costs.[2] It then calls for the use of regression analysis to estimate the fixed and

variable vendor costs. Using this information, the method then dictates several well-defined steps to develop a "critical interval" in which an EOQ of least total cost is identified. There are numerous instances where the EOQ cannot easily be calculated from a formula or even a computer algorithm. In such instances, the framework of total variable costs is essential for keeping things straight. Consider Situation 8-1.

Discussion of Gonder-Odell Manufacturing Company

We have observed that deciding how much inventory is optimal involves balancing the costs that increase with order size against the costs that decline with increases in order size. In the simple case we first considered, only carrying costs and order costs varied with changes in the order size. Deciding upon the optimal inventory in that case involved merely finding the order size that minimized the sum of carrying and order costs. The basic EOQ formula was derived in just this spirit.

The situation facing the Gonder-Odell Manufacturing Company is not nearly so simple. In particular, carrying costs and order costs are not the only ones that vary with changes in the order size; purchase costs, transport costs, and handling costs also depend on the order size. The total of all the variable costs associated with changes in the order quantity would thus involve summing up all of these costs. They are displayed and totaled in Table 8-1 for a variety of order sizes. Note that the carrying costs reflect the actual purchase price of the units and not some generalized or average figure, and that they assume a steady draw upon the inventory so that the average inventory is half the order size.

An examination of Table 8-1 reveals that an order size of 500 and average inventory holdings of 250 minimize the total variable costs. This result is not so surprising because (1) the

SITUATION 8-1

Gonder-Odell Manufacturing Company

The Gonder-Odell Manufacturing Company placed a number of orders each year with local job shops. A typical order was for some machining of small castings provided by the company. In assembling its own products, Gonder-Odell drew upon an inventory of these parts at a more or less steady rate. At issue was how large an order the company should place each time and thus how much inventory to hold. Some relevant information is as follows:

- Usage expected: 5000 units per year

- Order cost: $10. This figure constituted only costs that the company viewed as varying with each order, such as mailing and phone fees, transporting the castings to the job shops, and some necessary computer time. It did not include any clerical time, as that was regarded as a fixed cost.

- Inventory carrying cost percentage: 20 percent.

- Volume discounts on machining:
    $25 per unit if less than 100 ordered each time
    $22.50 per unit if 100 to 499 ordered each time
    $22 per unit if 500 or more ordered each time

- Transport costs: $20 for each pallet of 100 units; 30 cents for any unit not part of a 100-unit pallet. (The castings could be packaged and shipped on standard pallets in quantities of 100. This factor greatly helped transportation and unloading.)

- Handling cost: $5 for each pallet of 100 units; 10 cents for any unit not handled in 100-unit pallets.

TABLE 8-1   Order size cost calculations (Gonder-Odell Manufacturing Company)

ORDER SIZE	ORDERS PER YEAR (given 5000 unit usage)	PURCHASE COST	ORDER COST	TRANSPORT COST	HANDLING COST	CARRYING COST	TOTAL COST
50	100.00	$125,000	$1000.00	$1500	$500	$ 125	$128,125.00
100	50.00	112,500	500.00	1000	250	225	114,475.00
500	10.00	110,000	100.00	1000	250	1100	112,450.00
1000	5.00	110,000	50.00	1000	250	2200	113,500.00
400	12.50	112,500	125.00	1000	250	900	114,775.00
600	8.33	110,000	83.30	1000	250	1320	112,653.30

lowest price starts with the 500 order quantity, and (2) carrying costs rise more rapidly than order costs fall, suggesting that the company should carry as little inventory as it can. These two factors combine to yield 500, the lowest end of the lowest price category, as the optimal order quantity.

The table, except for the order size of 50, looks only at order sizes ending in even hundreds. We need not look at 501 or 499 as pos-sible optimums because of the structure of the transport and handling costs. Non-even-hundred orders involve special transport and handling cost rates for the odd pieces, and these cost increases wipe out any carrying cost or order cost savings such quantities could bring. Put another way, the graph of total variable costs against various order sizes is a discontinuous one that follows the general pattern of Figure 8-7.

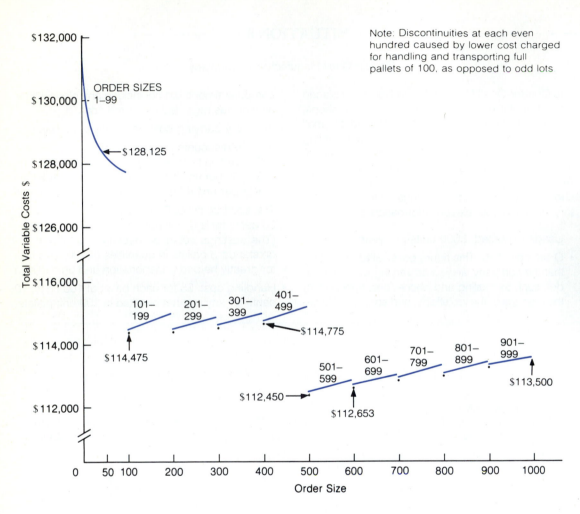

FIGURE 8-7   Total variable costs graphed against order size.

This discontinuity, coupled with the inclusion of so many variables in the calculation, makes the determination of an optimal order size something of a trial-and-error process, although in some cases a computer program can search for a solution. No neat formulas as we derived before can lead us easily to an answer. All we can do is converge on an answer in much the same manner as the table does.

Choosing Between Periodic Reorder and Reorder Point Inventory Systems

These then are the major elements of both the periodic reorder and the reorder point inventory systems. Both, remember, are non-time-phased systems. Situation 8-2 will help us consider why one might institute one system rather than the other.

SITUATION 8-2

Driscoll Lumber Company

Spurred by some informal conversation at a recent Lions Club luncheon, Aubrey Driscoll was reflecting on the way in which he stocked items for his lumber company. Together with his brother, Burney, Aubrey owned and operated the Driscoll Lumber Company in Devine, Texas. Devine, a small town southwest of San Antonio, depended largely on cattle ranching and peanut farming for its livelihood. The Driscoll Lumber Company was a major retailer of lumber, hardware, construction materials, and ranching supplies (corral gates, posts, barbed wire fencing) for the area surrounding the town.

The Lions Club discussion caused Aubrey to think of dividing his inventory of goods into distinct classes, some of which might be ordered better through a periodic reorder system rather than the standard "order when low" system that prevailed in the lumberyard now. As Aubrey saw it, his stock could be classified into several major categories:

1. *Seasonal versus nonseasonal demand.* The lumber and hardware that the Driscoll Lumber Company sold followed no seasonal pattern. Housing construction and repair in South Texas could be maintained in winter, due to the region's mild climate. Sales of barbed wire and steel fencing, however, were more likely in the late fall and winter, as ranchers and farmers shifted their attention from the raising of peanuts, hay, and other crops to the repair and construction of fields, pastures, and corrals.

2. *Items easily counted for inventory and items more difficult to count.* Most of the items the lumber company stocked were easily counted for inventory. Lumber, plywood, roofing material, cement, and the like were kept in open bins in the company's shed; low levels of any of these items were immediately obvious from just a quick glance around the shed. Some items, however, were more time consuming to count, particularly those that came in many different sizes, shapes, or grades. Some hardware items such as pipes and fittings were among these as well as special bolts stored in small compartments. Glass was another such item because it often had to be cut to order, and keeping track of the usable remains was difficult.

3. *Ordered lumber versus rolling lumber.* Most of the lumber Aubrey Driscoll sold had been ordered about 30 days in advance. What was shipped to Driscoll Lumber Company was exactly what had been ordered, usually in packaged bundles of standard lengths and a fixed count of pieces. Frequently, however, west coast sawmills specializing in Douglas fir, spruce, hemlock, and other softwoods dispensed with filling specific orders. Instead, the mill would fill railroad boxcars with just what it was sawing at the time, set the boxcars rolling on the rails toward various regions of the country, and rely on lumber brokers to sell off the boxcars before they reached the end of the line. Retail lumber companies could often enjoy significant purchasing discounts by buying lumber in this way, although it frequently meant that some lengths or grades would be overstocked for a time.

To Aubrey Driscoll's way of thinking, these were the major categories of items that he inventoried. He wondered whether he should continue to order all of them only when they were low (a reorder point system) or whether he should make periodic checks of at least some items and order variable quantities (a periodic reorder system). Which items seemed more suitable to one system or the other?

Discussion of Driscoll Lumber Company

Aubrey Driscoll's three categories of items serve as a useful starting point for a discussion of the pros and cons of either a periodic reorder system or a reorder point system.

1. *Demand seasonality.* The calculation of an order quantity for a reorder point inventory system implicitly assumes that demand is steady. That is, the minimization of total cost generally assumes that over the relevant time period the draw on the inventory is constant. Thus using an annual demand figure to calculate an order quantity is inappropriate if marked demand seasonality exists. While one could calculate different EOQs for different seasons, it is often simpler to use a periodic reorder system for those items whose demand is seasonal. Other things being equal, the Driscoll Lumber Company would be better advised to use a reorder point system for its lumber and hardware (nonseasonal) and a periodic reorder system for barbed wire and steel fencing (seasonal). If the lumber company placed barbed wire and steel fencing on a reorder point system, it might risk carrying too little inventory in fall and winter and too much the rest of the year.

2. *Ease of taking inventory.* A reorder point system requires that the inventory be monitored constantly. If inventory is difficult or expensive to take, a reorder point system is less desirable. At the Driscoll Lumber Company, lumber and other open-bin items are easily inventoried and thus can be maintained under a reorder point system. Glass and certain hardware items, however, are tough to inventory and thus argue for a periodic reorder system.

3. *Price discounts.* The possibility of price discounts on West Coast "rolling lumber" and their irregular timing disrupts both periodic reorder and reorder point inventory systems. Deciding whether or not to buy such lumber, once offered, demands a quick inventory of existing stocks and a computation of how much of the purchased lumber will be used during the ensuing time frame. In a sense, this computation lies outside the routine workings of either inventory system. The fact that an inventory of the existing stocks is required suggests that such price discounts are perhaps more compatible with a reorder point system, but they are still compatible with a periodic reorder system.

Considering all of these features then suggests that the Driscoll Lumber Company may be best off using a reorder point system for its lumber and larger hardware supplies and a periodic reorder system for its small, hard-to-inventory hardware items, its glass, its barbed wire and fencing, and similar items.

In more general terms, a periodic reorder system has a relative advantage over a reorder point system when the costs of taking a physical inventory are high or when there are savings in regard to information (such as seasonal data), production scheduling, price discounts, or transport costs from ordering items on a regular basis. If these cost savings are not significant, however, a reorder point system has the striking advantage of being able to minimize the total variable costs associated with changing order quantities. Broadly speaking, the economics of inventory control favor the reorder point system but only when the important, practical costs of inventory control and administration are not significant.

The saving grace in all of this is that failing to calculate the precisely optimal order quantity under any inventory system is not likely to be devastating. The costs of various order quantities are, in most instances, fairly near one another. In other words, a graph of inventory costs against various order quantities is likely

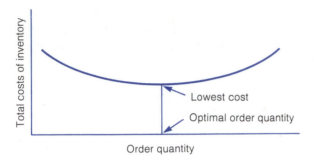

FIGURE 8-8   Inventory costs graphed against order quantities.

to be shallow (saucer-shaped, as in Figure 8-8), rather than steep (bowl- or cup-shaped). Making some sensible decisions about inventories to get oneself "into the ballpark" is thus more important than landing on precisely the best order quantity. It has been said that order timing is more important than order quantities.[3]

While prevalent in both manufacturing and nonmanufacturing businesses, non-time phased inventory systems, like the periodic reorder and reorder point systems described above, are perhaps best suited to retail and wholesale situations (Driscoll Lumber Company), where the source and precise timing of demand is seldom known with much certainty. These conditions often apply to manufacturing as well, but there are many manufacturing businesses that know far more about the source and timing of demand for their products, and hence for the parts that make up their products. It is in these instances of known or easily provided demand that the time-phased inventory system, such as material requirements planning, makes good sense.

SAFETY STOCKS

Safety stocks are useful in managing inventories precisely because companies do not know in advance exactly what the demand made on them will be. As long as replenishments to an inventory cannot occur immediately, the lead time necessary for delivery of the replenishment order is an uncertain and problematic period of time. Consult Figure 8-9. At time T, there was a delivery of inventory that replenished the stock up to the level of Q. At time T_2, it was decided to place another order. This new order could have been placed because a reorder point had been struck (level R of inventory) or because T_2 was the appropriate, periodic time to review the status of inventory (per a periodic reorder inventory system). The order is expected to arrive at time A, so that the lead time is the time between T_2 and A.

During this lead time, most anything can happen. If demand proceeds at its usual, expected rate, when the order arrives inventory should be depleted to point B. The safety stock, however, will not have been used. If demand is lighter than expected (something that one might think occurs about 50 percent of the time, if demand is a statistical creature with a distribution that is symmetrical about the mean), then at time A, inventory would stand somewhere between points B and C. If, on the other hand, demand is heavier than expected (typically, another 50 percent probability) the

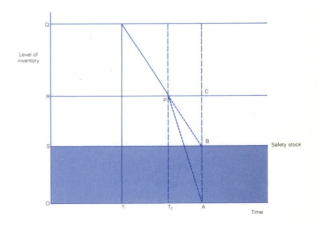

FIGURE 8-9   The usefulness of safety stock.

safety stock will have to be employed. During the lead time, the quantity AB could be used up before a stockout occurred.

Levels of Customer Service

At issue, then, is how much safety stock should a company hold? How much is enough or desirable? This is sometimes described in terms of determining the level of customer service. There are many, somewhat different measures for customer service, however. Some are more appropriate for some uses than others. Two specific measures of customer service bear mention:

1. The percentage of orders covered by the inventory, or equivalently, the probability of not stocking out. This measure indicates the probability that the company will not stock out during any given order cycle time period. A 95 percent service level, by this measure, indicates a 5 percent chance of failing to fill one or more orders during any given time period. This measure is fairly easy to work with statistically.

2. The percent of all items ordered shipped on time. This is a measure of items and not orders, and it has intuitive appeal. While similar to measure 1, this measure has some significant distinctions. Indeed, the mathematical correspondence between the measures is not a simple one, because stockouts can involve orders for any number of items. For this reason, to match a 95 percent service level by measure 1, measure 2 generally has to be greater than 95 percent. Despite the complex mathematical relationship between the two measures, for many items, rough correlations between the two measures can be devised empirically.

Let's return to the issue of determining the desired level of customer service, as measured

by either measure. The discussion of Figure 8-9 and previous comments suggest several factors that ought to influence the size of any safety stock:

1. *How frequently the order is placed.* The larger the quantity ordered at any time — and thus the less frequent the exposures to potential stockouts — the greater the level of customer service provided.

2. *Lead time.* The longer the lead time, other things being equal, the more safety stock is desired. Shorten lead times and safety stocks can be lowered. Typically, there are differences in lead time between the reorder point and the periodic reorder systems. The period of vulnerability is generally shorter with a reorder point system, only the time between the placement of the order and its delivery, than it is with a periodic reorder system. For the latter, the period of vulnerability is the length of the time period itself, not just the delivery lead time.

3. *Stability of demand.* The more erratic the demand, the greater the safety stock has to be to maintain a given level of customer service.

4. *Relative costs of stockout versus inventory carrying.* The greater the penalty to the company of not having enough inventory relative to the cost of holding inventory, the more safety stock is desirable.

For the most part, companies settle on a particular safety stock and accompanying level of customer service on a trial and error basis. A service level is established, and then, over time, performance against that service level is monitored as well as customer satisfaction with delivery. If performance is satisfactory, then the safety stock can be lowered somewhat and performance again monitored. By focusing on the service level, companies can improve their performance dramatically over rules-of-

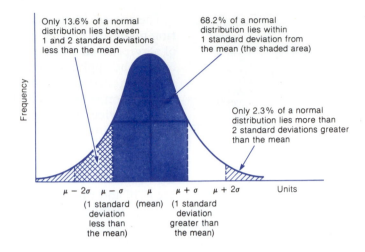

Only 13.6% of a normal distribution lies between 1 and 2 standard deviations less than the mean

68.2% of a normal distribution lies within 1 standard deviation from the mean (the shaded area)

Only 2.3% of a normal distribution lies more than 2 standard deviations greater than the mean

Frequency

$\mu - 2\sigma$ $\mu - \sigma$ μ $\mu + \sigma$ $\mu + 2\sigma$ Units

(1 standard deviation less than the mean)   (mean)   (1 standard deviation greater than the mean)

FIGURE 8-10   The normal distribution.

thumb approaches to safety stock that merely pick a given period's demand as the safety stock. It is important to keep in mind the fact that safety stocks should be used. If a safety stock is not dipped into regularly, then inventory is too high and should be reduced.

Continual trial-and-error adjustments to safety stock levels have a lot to recommend them, especially given an environment of change. There are, however, alternative schemes for setting safety stocks. One fairly simple scheme merits attention as a means to estimate what level of customer service to shoot for, namely the use of the critical fractile.

The Critical Fractile Approach

This approach to setting the level of safety stock essentially picks out a suitable level of service based on a trade-off of stockout costs and inventory carrying costs. The chosen level of service then gets translated into a "safety factor" that can be multiplied by the computed standard deviation of the distribution of demand to yield the extent of safety stock. Let us work through this process in reverse in order to understand it better.

Many distributions of demand can be approximated by the normal distribution, the familiar bell-shaped graph pictured in Figure 8-10. The two parameters needed to describe a normal distribution are its mean and its standard deviation. Naturally, a low standard deviation represents a very tight, stable demand, while a large one represents a more erratic demand. No matter what the value of the standard deviation, however, with a normal distribution, there are standard percentages of the distribution that fall within X standard deviations of the mean. This fact permits us to make the correspondence between the percentage of demand covered by a certain size of safety stock (one measure of the level of customer service discussed earlier) and the number of standard deviations above the mean the inventory represents (the safety factor, or from statistics, the z-score or z-factor which is defined as $z = (x - \mu)/\sigma$. The z-score gives the value of x for which the area under the normal curve to the left of that point equals the desired level of coverage). Some relevant safety factors to use, given a normal distribution with a 0 mean and a standard deviation of 1, are as follows:

COVERAGE OF EXPECTED DEMAND (percent)	SAFETY FACTOR (or Z-score)
99 %	2.326
97.5	1.960
95	1.645
90	1.282
85	1.038
80	0.843

Thus if the distribution of demand is normally distributed with a mean of 50 and a standard deviation of 14, a 90 percent level of coverage (i.e., a 90 percent level of customer service) would be given with an inventory of

$$x = \mu + (z \cdot \sigma) = 50 + (1.282)(14) \approx 68 \text{ units}$$

of which 18 units constitute the safety stock itself.

The critical fractile calculation helps production managers decide upon an appropriate level of service so that the corresponding safety factor can be applied to the standard deviation of the distribution of demand.

Common sense insists that the level of customer service be higher if the cost of carrying too little inventory and thus stocking out greatly exceeds the cost of carrying too much inventory. The former cost is commonly called the "cost of being under" or the cost of underage (C_u) and the latter is commonly called the "cost of being over" or the cost of overage (C_o).

The suitable percentage coverage that a company ought to seek for its inventories subject to uncertain demand is approximated by the following formula:[b]

---

[b] The ratio of $C_u/(C_o + C_u)$ is the critical fractile (or percentile). While technically it applies only to single-period holding of inventories and linear costs of underage and overage, it is a useful enough approximation to be applied more generally.

The basic argument underlying the derivation of the critical fractile is, once again, incremental in origin. Suppose we arbitrarily select q^* as the best level of inventory there could be. It must be true then that we receive an incremental gain by inventorying q^* as opposed to one unit more or less (i.e., either $q^* + 1$ or $q^* - 1$). This being the case, we can write formally the following inequality:

$$(1-1) \qquad C_u[P(d > q^*)] \le C_o[P(d \le q^*)]$$

By stocking q^*, the expected expense of being under when demand is greater than q^* (the cost of underage times the probability that demand is greater than q^*) must be less than or equal to the expected expense of being over (the cost of overage times the probability that demand is less than or equal to q^*); otherwise, it would make sense to stock another unit so that one is not caught being under. Similarly, we can write the following inequality:

$$(2-2) \qquad C_u\{P[d > (q^* - 1)]\} \ge C_o \{P[d \le (q^* - 1)]\}$$

By stocking q^*, the expected expense of being under when demand is greater then $q^* - 1$ must be greater than or equal to the expected expense of being over; otherwise, it would make sense to stock one unit less in order to save money.

Using the fact that $P(d > q^*)$ is equal to $[1 - P(d \le q^*)]$ we can rewrite inequality 1 as

$$(3-3) \qquad C_u[-P(d \le q^*)] \le C_o[P(d \le q^*)]$$

or

$$C_u \le (C_u + C_o)[P(d \le q^*)]$$

or

$$(4-4) \qquad \left(\frac{C_u}{(C_u + C_o)}\right) \le [P(d \le q^*)]$$

In a similar fashion we can rewrite inequality 2 as

$$(5-5) \qquad \left(\frac{C_u}{(C_u + C_o)}\right) \ge \{P[d \le (q^* - 1)]\}$$

Inequalities 4 and 5 suggest that $C_u/(C_u + C_o)$ is the relevant ratio for picking off the point in the cumulative probability distribution that corresponds to the optimal level of inventories, q^*.

$$\text{Critical fractile} = \frac{C_u}{(C_o + C_u)}$$

This formula meets the test of common sense in that if the cost of carrying too little and stocking out is high relative to the cost of carrying too much, then the suitable percentage coverage will be high. The formula also makes sense in that it cannot reach 1 in value nor can it fall to zero; it must range in between, just like the range for cumulative probability.

What are costs of underage and costs of overage? Let us consider the costs of overage first since they are a bit easier to think about. The costs of overage are all the costs that would be incurred by carrying one extra unit of inventory for the time period needed to avoid the stockout. These are the costs of holding inventory that were discussed previously:

- Finance charges (opportunity costs).
- Storage and handling costs.
- Insurance and taxes.
- Obsolescence and shrinkage.

As an example of the cost of overage, this cost for a year could be 30 percent of the cost of the product, calculated as 15 percent for financing costs (such as the company's prevailing borrowing rate or perhaps the rate that it expects to earn on any cash freed up by reducing inventory), 10 percent for physical costs, and 5 percent for the risk of obsolescence. In most cases, however, buffer inventory would be vulnerable to depletion for only a short time before the next delivery was due. If the period of vulnerability were only 1 to 2 weeks, the cost of overage for the relevant period for this example would be approximately 1 percent of the cost of the product.

The cost of underage is the penalty incurred by failing to have a unit of inventory on hand when demand calls for it. At worst, typically,

the cost of underage is the lost contribution to profit and overhead incurred because the unit was unavailable for sale. Again, the concept of opportunity cost intrudes into our thinking—here, in the form of contribution forgone. Forgoing all of a unit's contribution assumes that the buyer goes away empty-handed and will never again return to purchase the item. Frequently, however, stocking out of an item does not mean all of a lost sale, especially if the purchaser is willing to substitute another item for the one out of stock (another color, another model, another size) or if he or she agrees to come back when the item is received again in stock. Market research can frequently determine a reasonable expectation for this cost of underage.

As an example of the cost of underage, the contribution per unit might be 40 percent of the revenue the company receives from it; the probability that the unit will be forgone forever might be 50 percent; the probability that some other unit (with contribution of 30 percent) will be substituted might be 30 percent; and the probability that the customer may return later for it might be 20 percent. In such a case, the cost of underage could be estimated as an expected value of the contribution forgone.

$$0.5(40\%) + 0.3(40\% - 30\%) + 0.2(0\%)$$
$$= 0.5(40\%) + 0.3(10\%) + 0.2(0\%)$$
$$= 23\% \text{ of the cost of the product}$$

Costs of underage generally greatly exceed the costs of overage for most relevant time periods.

Let us then calculate the critical fractile for this example. The relevant inputs are

$C_o = 1.15\%$ (30% annual rate for 2-week period)
$C_u = 23\%$

Suppose, as well, that the demand during the period is expected to follow the normal distribution with a mean of 50 and standard deviation of 10.

Critical fractile (percentile) $= \dfrac{23}{(1.15 + 23)} = 0.95$

Safety factor (or z-score) for 95th percentile $= 1.645$

The size of the buffer or safety stock is then

$$z \cdot \sigma = (1.645)(10) = 16.45 \text{ units}$$

The size of the inventory at the start of the period should be the expected average withdrawal from inventory plus the buffer stock:

$$x = + z \cdot \sigma = 50 + 16.45 = 66.45 \text{ units}$$

To reinforce this discussion of safety stocks, consider Situation 8-3.

Discussion of the DiMarzo Costume Company

Given no discernible trend in the demand over the years for Frosty the Snowman costumes, the DiMarzo Costume Company can let the past 20 years be its guide in selecting the number of Frosty costumes it should stock.

A Historical Count Approach

The easiest, if not the most scientific, way to select an appropriate stock level so that the company can expect to cover 95 percent of the orders placed on it is to make a count of the data. The company could expect that a stock level at or above the highest number demanded over the past 20 years would be too high. Company history shows that in 95 percent of the years (19 of 20) demand was less than 512 but not more than 480. It would seem reasonable, then, to select a stock level between 480 and 512.

A Statistical Approach

A more formal way of selecting an appropriate stock level is to apply some statistics to the data.

The mean of the distribution of demand for Frosty the Snowman costumes is 398.90 and the sample's standard deviation is 50.63. These summary statistics (sample mean and sample standard deviation) are estimates of the unknown true population mean, μ, and the unknown true population standard deviation, σ, which underlie the demand for costumes at DiMarzo Costume Company. Often, we do not know how the full population of observations is distributed, and we must depend on samples from that full population to draw inferences about the population itself. Such is the case at the DiMarzo Costume Company. The summary statistics, $\bar{x}$ and s, can help us condense the Table 8-2 data into a more easily digested bar graph of the frequencies with which the data fall into various classes. Figure 8-11 displays such a bar graph. For intelligibility, the classes are set at 25, about half of the standard deviation, and the figure is centered on 400, which is just above the true mean of 398.90.

Figure 8-11's bar graph can be approximated by a normal distribution. We know that the Table 8-2 data fall roughly into a normal distribution pattern since (1) 70 percent of the 20 observations fall within one standard deviation of the mean (i.e., 14 of the 20 observations fall between 350 and 450) versus 68.2 percent in a true normal distribution, (2) 20 percent of the 20 observations fall between one and two standard deviations from the mean versus 27.2 percent in a true normal distribution, and (3) 10 percent of the 20 observations fall beyond two standard deviations from the mean versus 4.6 percent in a true normal distribution. It seems reasonable then to presume that the demand for Frosty the Snowman costumes is normally distributed with a mean of 398.90 and a standard deviation of 50.63.

In a normal frequency pattern, the top 5 percent of the distribution falls above the value of $+1.645$ standard deviations from the mean. In this case, we can presume then that 95 percent

SITUATION 8-3

DiMarzo Costume Company

The DiMarzo Costume Company rented theatrical costumes of all sorts to theater groups, colleges, schools, churches, and the like. During the Christmas season, one of the most popular costumes (aside from Santa Claus) was Frosty the Snowman. (Data on orders placed for Frosty the Snowman costumes over all 20 years of the company's history are given in Table 8-2). The company could not detect a trend in orders over the years; orders seemed to be fairly random. The company, however, wanted to be able to fill about 95 percent of all its orders for Frosty and wondered how many costumes it should stock at Christmas.

TABLE 8-2   Past demand for Frost the Snowman (DiMarzo Costume Company)

YEARS AGO	DEMANDED	YEARS AGO	DEMANDED
1	411	11	296
2	347	12	382
3	412	13	473
4	385	14	320
5	441	15	423
6	402	16	354
7	395	17	512
8	370	18	389
9	480	19	378
10	416	20	392

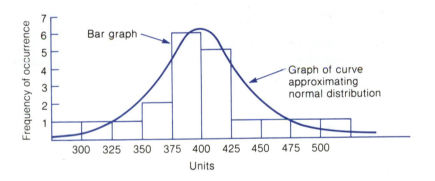

FIGURE 8-11   Frequency distribution of monthly sales for DiMarzo Costume Company.

of the demand for Frosty costumes will fall below

$$x = + z \cdot \sigma = 398.90 + (1.645)(50.63) = 482.19$$

We can take this figure, 482 or 483, as a suitable stock level to assure ourselves that 95 percent of our expected orders can be serviced.

If we were content to have only 90 percent of the expected orders serviced, then the proper stock level would be

$$x = + z \cdot \sigma = 398.90 + (1.282)(50.63) \approx 464$$

because the top 10 percent of the distribution falls above the value of + 1.282 standard deviations from the mean.

Trends and Aggregation

The DiMarzo Costume Company had to worry only about stocking Frosty the Snowman costumes for one peak time during the year. Furthermore, there were no clear trends in the demand over the years. The task of calculating an appropriate inventory level for uncertain demand is complicated further when trends are involved or when time periods are aggregated. Consider the following modifications of the procedures we have used to date for coping with demand uncertainty:

1. *Trends.* If sales have grown (or declined) over time, the distribution of demand shifts rightward (or leftward) along the x-axis of units demanded. In the simplest case, the distribution itself remains the same and merely shifts position; the mean thus changes, but the standard deviation remains the same. In more complex cases, the distribution changes along with the shift, changing both the mean and the standard deviation. When trends are involved, therefore, it is important to ascertain whether the shape of the distribution is changing along with its mean.

2. *Time period aggregation.* Suppose that we wanted to shift the horizon for a periodic reorder system from 1 month to 3 months. All the calculations about the appropriate level of inventory to hold were based on a distribution of demand for a month. How do things change if we are to define everything now in terms of 3 months? If demand follows a normal distribution, is the mean of the new distribution simply three times as large? What about the variance? If we assume a normal distribution for the demands and if those demands in any month are independent of one another (a not unreasonable assumption), then the mean of the new 3-month dis-

tribution will be three times the 1-month mean, and the variance will also be three times greater. The standard deviation, however, will not be 3 times the old standard deviation because it is the square root of the variance. Thus if the 1 month mean is 50 and the standard deviation is 10, the new 3-month distribution can be expected to have a mean of 150 and a standard deviation of

$$10 \sqrt{3} = 17.3$$

In general terms, under the assumptions of normal distribution and independence of demand, both the mean and the variance are additive.

Tools of the Trade: A Recap of Lessons Learned

Time-independent inventory systems, older than the MRP notion and often used to manage stocks of finished goods stocks or raw materials, tend to fall in one of two main categories: reorder point systems and periodic reorder systems. The reorder point system (employing the EOQ calculation) has some attractive cost-minimizing properties, but the assumptions it makes are frequently at odds with the realities of the situation. In such cases, the periodic reorder system, and the regularities it promotes, is frequently the better choice.

Safety stocks are required in both systems, and demand some knowledge of the expected time pattern of withdrawls from inventory during the replenishment period. What is clear, however, is that if the safety stocks are never called upon, then the firm is keeping too much in inventory, and the stocks ought to be reduced.

QUESTIONS

1. Compare and contrast the purpose and natures of the anticipatory (MRP) and nonanticipatory approaches (periodic reorder and reorder point systems) to inventory management.

2. Of what value is the concept and application of the critical fractile model to a situation of demand uncertainty? What else is involved in deciding inventory level under demand uncertainty?

3. In what ways do the periodic reorder system and the reorder point system differ from each other? Be specific and use graphs where relevant.

4. "The key thing about the determination of an economic order quantity within a reorder point inventory system is not the formulas themselves." What, then, is the key determinant, and how is this determinant manifested in the example of the Gonder-Odell Manufacturing Company?

5. What are some of the factors that influence the decision whether to go with an MRP or a non-time-phased system? Give an example.

6. How can work-in-process inventories build up? How can you cope with them?

PROBLEMS

1. The Wyoming Valley Furniture Company (see Situation 2-2) periodically had significant pulses in demand that were difficult to meet without delays in the delivery schedule. In particular, the finish shop would often be pinpointed as the production bottleneck. It was alleged that too much time there was being spent on rush orders and short runs and not enough time on long runs of finished parts. A number of potential solutions had been tossed out for consideration, among them (1) the introduction of a computerized material requirements planning system and (2) a decoupling inventory. How would you expect each recommendation to work in solving the bottleneck in the finishing department? What kinds of considerations would argue for one solution over the other?

2. The Digital Watch Company had demand for the year of 200,000 units. It found that its average order cost on its supplier was $30 per order and its carrying cost is $4.00 per unit per year. What is its economic order quantity?

3. Digital has found that it can order watchbands from an outside vendor on the following price schedule:

ORDER QUANTITY	PRICE
500 or less	$5.00
501 or more	$4.50

With demand information as in Problem 2 and average inventory carrying cost for a watchband of $0.75 per unit per year, what is Digital's economic order quantity?

4. Given a recent increase in the banks' prime lending rate, Digital's inventory manager now feels that she can more accurately estimate carrying cost at 30 percent of the purchase price. From the information in Problems 2 and 3, what is Digital's economic order quantity for watchbands?

5. Digital watch has found that it can produce watchbands at a rate of 5000 per week, and that demand is 4000 per week (assuming 50 weeks). Digital's setup cost is $400 per run, and its carrying cost remains $0.75 per unit per year.

(a) What is Digital's economic production quantity?

(b) For how many weeks will Digital produce at a time?

(c) How many production runs will be made in a year?

(d) What is Digital's maximum inventory?

(e) What is Digital's average inventory?

(f) If you had a choice, how would you manage this situation? Would you use an economic production quantity? Why or why not?

6. The CutCorrect Company clerk manages a retail inventory of consumer goods that are ordered from a local wholesaler. The clerk has recently read about the EOQ and wonders if it would produce a different result than their current policies. An example product has virtually constant demand of 40 units per week. It costs $20 wholesale and order costs are estimated to be $10 per order. The accountant said that the cost of carrying inventory (including cost of capital tied up, storage, and an obsolescence charge) was 23 percent of item cost. Would the use of the EOQ be different from the current policy of ordering in lots of 100 units?

7. Under what conditions would you be wary of using the critical fractile method, as outlined, to handle demand uncertainty in nonanticipatory inventory models?

8. React to the following statement. "None of these inventory systems, be they anticipatory or nonanticipatory, can divorce themselves completely from some attention to the speed at which inventories are drawn down."

9. The Spirited Bottling Company produced private label liquors for a chain of liquor stores. It purchased liquors of all types in bulk and drew off from that bulk stock when it needed to for bottling run. The bottling runs were made by a team of workers who were guaranteed 40 hours of pay a week, although they rarely had more than 35 hours of work to perform each week for the bottling runs scheduled. This team was responsible for both setups and runs of the bottling equipment. Various taxes were due. Some were paid on each bulk stock order when it was purchased, while others were due as soon as the liquor was bottled. The company scheduled production by using a reorder point system—when a set reorder point was struck, an "economic order quantity" was produced according to the following standard formula:

$$EOQ = \sqrt{\frac{2 \times \text{demand} \times \text{setup cost}}{\text{carrying cost rate} \times \text{product cost}}}$$

Some relevant information about Spirited's operations follows:

- Bottling team: 5 workers paid an average of $10 per hour for a 40-hour week.
- Setups required for bottling and leasing machines: 1/2 hour for each team member per run.
- Inventory carrying cost percentage: 20 percent
- Product-related costs, per case, for a representative product (quarts of 80-proof gin):

Materials	
Beverage	$ 3.50
Packaging	5.00
Direct labor	0.50
Taxes paid on bulk liquor	30.00
Taxes paid after bottling	25.00
Variable overhead	1.00
Fixed overhead	2.50
Full case cost	$67.50

The yearly demand was running 6000 cases of 80-proof gin at an average case price of $80. How would you advocate they decide on the length of their bottling run for gin? Feel free to question (1) which costs ought to be applied in the EOQ formula, and (2) whether the EOQ formula is appropriate to this situation.

10. Lee Plastics, a small injection-molding plastics company, did custom work for local businesses. It did not like to hold any inventory of resin but felt that it had to in order to react instantly to customer orders. (Vendor lead times were short, usually less than a week, but sometimes even that was too long.) The prices the company had to pay for its resins declined with the quantity ordered, which also argued for ordering more than needed and keeping the rest in inventory. The company used about 100,000 pounds of acetal, a typical resin, during a year. It was available at the following prices:

QUANTITY (pounds)	RESIN PRICE	TRANSPORTATION AND HANDLING COST
Under 2000	$2.69	$0.25
2001 - 4000	1.68	0.21
4001 - 6000	1.57	0.18
6001 - 13000	1.51	0.15

Placing orders more frequently was not a problem, and any extra costs were negligible. Transportation and handling varied with the size of the order, because small orders were somewhat more of a nuisance to handle. The costs ranged as in the table. The cost of carrying and storing the resin was about 20 percent of the purchase value. How much should Lee Plastics buy at a time?

11. How would your answer to Problem 10 change if Lee Plastics' use of acetal rose to 200,000 pounds per year and the carrying cost rate rose from 20 percent to 30 percent?

12. A typical value for the per pound costs of finished product for Lee Plastics (including materials, handling, labor, and variable overhead) was $4.00 per pound. Lee sold its output at an average rate of $5.75 per pound. As in the previous problem: (1) Lee's materials were typically available with a week's lead time, (2) Carrying costs were estimated at 20 percent, and (3) 100,000 pounds were sold during the year. Assuming a periodic reorder inventory system with orders made every 2 weeks, what safety stock would you recommend for Lee Plastics? How would your answer change, if at all, if Lee employed a reorder point system?

13. The weekly demand for a selected product appears to be normally distributed with a mean of 200 and a standard deviation of 30. The lead time for producing the product is generally 1 week and orders can usually be produced on demand at little or no additional cost. The inventory carrying charge per unit per month is $8. The unit's price is $300, of which $70 is the estimated contribution per unit. How many units should be stocked? How would your answer change, if at all, if the extra cost to produce the order on demand were $500? How would your answer change, if at all, if the lead time increased to 2 weeks?

14. What is the appropriate expression for total cost and the solution for the economic order quantity when back orders are possible, although with some penalty attached? The problem can be represented as in Figure 8-12 where Q is the order quantity to be optimized, H is

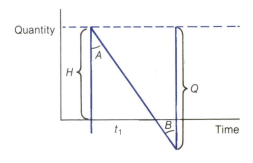

FIGURE 8-12   Order quantity when back orders are possible (Problem 14).

the maximum inventory on hand at any one time, t_1 is the time when no stockouts occur, and t_2 is the time when stockouts occur. Note that angles A and B are the same. The number of order cycles in any year is

$$\underset{\text{(order cycles)}}{N} = \frac{D \text{ (annual demand)}}{Q \text{ (order quantity)}}$$

SITUATION FOR STUDY 8-1

TIFFANY TIRE COMPANY

George Blimpo has recognized that his distribution operation is consistently carrying too much inventory. He wonders whether he can reduce that inventory by implementing some sort of inventory system.

Tiffany Tire Company is a small midwestern distributor of radial tires. It is composed of three retail outlets and one warehouse. The warehouse buys its supply of tires directly from the factory and supplies only the three retail outlets. Also, the three dealers buy only from Tiffany's warehouse.

Table 8-3 shows last year's demand for tires, by month, for the three dealers. As you can see, demand has remained fairly constant over the last three years. The variations in demand are random rather than seasonal or part of a trend.

Table 8-4 shows the results of a study done last year to determine inventory costs for the dealers and the warehouse. The differences in cost are due to the locations of the dealers and the warehouse. For example, dealer 2 is located in downtown Detroit and the cost of renting storage space is much higher than it is in Mt. Vernon, Illinois, where dealer 3 is located. Also, the dealers have higher carrying costs than the warehouse because of the transfer pricing policies of Tiffany; the dealers are paying a higher

TABLE 8-3   Demand for tires (Tiffany Tire Company)

MONTH	DEALER 1	2	3
January	400	600	500
February	300	800	400
March	200	400	900
April	600	300	600
May	800	500	200
June	600	300	600
July	300	400	800
August	200	500	800
September	400	600	500
October	350	650	500
November	250	350	900
December	600	600	300

TABLE 8-4   Inventory costs (Tiffany Tire Company)

DEALER	ORDER COST	CARRYING COST (per tire per year)	LEAD TIME (months)
1	$ 90	$1.60	1
2	45	2.20	1
3	60	0.90	1
Warehouse	150	0.75	2

price per tire than the warehouse. A similar explanation could be given for the differences in order cost.

George expects that the demand and cost estimates shown in these two tables are representative of what will happen in the future.

What would you recommend that George do to develop a new inventory system for the dealers and the warehouse, given that the existing system is just an educated guess?

CASE FOR STUDY

SPIRIT BOTTLING COMPANY

In early June 1983, Miles Seabrook, president of Spirit Bottling Company, called in his newly appointed vice-president and manager of the Soft Drink Division, Donald Whitcomb. He wanted to discuss several inventory problems.

Seabrook   Don, I am concerned about inventory levels throughout the company. Finished goods inventory levels reached $4 million in May, and May was just an average month for the company in terms of production, sales volume, and the accompanying inventory levels.

Whitcomb   I agree, Miles, our inventory levels do appear excessive. Besides, they are increasingly expensive to hold and are taking up valuable space.[c] The Rochester plant is a good example. Finished product occupies most of the storage space inside the plant, while empty returnable bottles are stored outside in an open yard.

Seabrook   I have two questions about our inventory situation. Are our finished goods inventory levels where they should be throughout the company? If not, is it possible to cut these levels, solve our problems in Rochester, and keep the 15 distribution centers supplied with production?

Whitcomb   I have already done some work to determine whether the company has a problem with excess inventory. I began an analysis of inventory and sales figures for the Mankato distribution center (See Table 8C-1). I chose this center because the number of days' supply for each product and package in Mankato was representative of companywide inventory-to-sales levels. So I was able to get a quick idea of the company's inventory situation. It appears we have been operating with excess inventory.

[c] In determining the carrying cost of inventory, the company used a carrying cost percentage of 15 percent. This percentage consisted of two elements:

Opportunity cost finance charges = 12.5%
Physical costs handling, storage, obsolescence = 2.5%

Table 8C-1   Inventory and sales analysis, May 1983 (Spirit Bottling Company, Mankato Plant)

PRODUCT (cases)	INVENTORY(cases)	SALES	INVENTORY[†]/ SALES	DAYS' SUPPLY[‡] (days)
Returnable Bottles				
Spirit 6 ½ oz	280	2454	0.11	3.5
Spirit 6 ½ oz 6-pack	65	1107	0.06	1.8
Spirit 10 oz	155	2895	0.05	1.6
Spirit 10 oz 6-pack	100	320	0.30	10.0
Spirit 16 oz	284	1418	0.20	6.2
Spirit 1 liter	1208	856	1.41	43.0
11-up 10 oz	138	294	0.47	15.3
11-up 16 oz	40	10	4.0	125
11-up 1 liter	50	19	2.63	83.0
Pep 10 oz	234	1168	0.20	6.1
Pep 16 oz	126	178	0.70	21.0
Pep 1 liter	78	68	1.15	35.4
Brite 10 oz	86	923	0.09	2.8
Brite 16 oz	130	44	2.95	92.8
Brite 1 liter	63	79	0.79	25.2
Dr. Spice 10 oz	98	448	0.22	6.8
Dr. Spice 1 liter	33	12	2.75	82.5
Dr. Spice SF 10 oz*	102	100	1.02	31.8
Jazz 10 oz	67	303	0.22	6.8
One-Way Bottles (nonreturnable)				
Spirit 10 oz	248	644	0.39	11.9
Spirit 16 oz	3206	2872	1.11	34.6
Spirit 28 oz	64	97	0.66	20.6
Spirit 2 liters plastic	2115	4205	0.50	15.59
11-up 10 oz	59	36	1.6	49.2
11-up 16 oz	75	239	0.31	9.7
11-up 28 oz	28	22	1.27	40.0
11-up 2 liters plastic	70	170	0.41	12.7
11-up SF 2 liters	101	63	1.6	50.5
plastic	48	69	0.70	21.8
Pep 10 oz	377	447	0.84	26.1
Pep 16 oz	24	11	2.18	68.5
Pep 28 oz	188	338	0.56	17.24
Pep 2 liters plastic	74	64	1.16	35.2
Brite 10 oz	315	619	0.51	15.75
Brite 16 oz	105	286	0.37	11.4
Brite 2 liters plastic	61	43	1.42	43.5
Viva GA/CS 28 oz	37	106	0.35	10.8
Viva 2 liters plastic	25	14	1.79	55.5
Viva T. Water 28 oz	51	16	3.19	102.0
Dr. Spice 10 oz	119	680	0.17	5.4

[a] SF, sugar-free; GA/CS, ginger ale/club soda; T. Water, tonic water.

[†] Total case sales for returnables, one-way bottles, and cans amount to 39,906.

[‡] Calculated as inventory divided by average daily sales. Average daily sales were determined by the previous four weeks' sales. This calculation was specific to each product and package combination.

continued

Table 8C-1 Inventory and sales analysis, May 1983 (Spirit Bottling Company, Mankato Plant)			
One-Way Bottles (nonreturnable)			
Dr. Spice 16 oz	92	53	1.74	54
Dr. Spice 2 liters plastic	127	50	2.54	79.3
Jazz 10 oz				
Cans			
Spirit	2542	7879	0.32	10.0
11-up	660	86	0.77	23.7
11-up SF	235	216	1.08	34.0
Pep	574	1803	0.32	11.2
Lean Green	85	15	5.66	117.0
Brite	862	2043	0.42	13.0
Viva	70	125	0.56	17.5
Dr. Spice	185	941	0.19	6.1
Dr. Spice SF	104	117	0.89	28.0
Jazz	807	2115	0.38	11.8
Premix (for commercial sales)			
Spirit	147	2188	0.07	2.1
11-up	27	56	0.48	15.0
Pep	84	176	0.48	15.0
Dr. Spice	13	28	0.46	14.4
Syrup (for commercial sales)			
Spirit	270	600	0.45	13.9
Pep	105	35	3.0	95.4
Dr. Spice	70	60	1.16	35.0
Others	95	150	0.63	19.7

Seabrook   I thought so. I have repeatedly asked our production plant managers to reduce inventory.

Whitcomb   Miles, in my opinion, the only way to solve the inventory problem is by changing the way we schedule production and transportation. The amount of goods the three plants produce and ship to the 15 distribution centers determines product inventory levels. I think the solution lies there.

Seabrook   Have at it then, but I want some results soon.

To answer the questions that Seabrook had raised, Whitcomb knew he had to analyze company inventory levels and scheduling methods in the production plants in Rochester, Minneapolis, and St. Cloud. If the decision were made to reduce substantially inventory levels in these plants, production management would experience the pressures of operating with low inventory levels and could no longer afford low efficiency runs or equipment breakdowns because stockouts would result.

BRIEF HISTORY OF SPIRIT BOTTLING COMPANY

Spirit Bottling Company bottled and marketed soft drinks in a franchise territory covering the southern half of Minnesota and portions of Wisconsin. In 1902 the first bottle of Spirit Cola was sold in Minnesota by a bottling plant in Rochester, which was the predecessor of Spirit's plant in Rochester. Through the years other bottling plants were built or acquired by Spirit and in May 1983, the company consisted of 15 distribution centers and three bottling plants. In 1983, the company bottled and sold 20 national brands for five companies: Spirit, Viva, Action Ade, Blink, Dr. Spice. Total sales in 1981 were 22.3 million cases, consisting of 20 different packages.

Since 1972 the company's stock had been listed on the New York Stock Exchange. During the last two years the stock price ranged from a low of 11 1/4 to a current high of 25 1/2. The financial condition of the company had remained healthy. The company was organized into three operating divisions. Individual plants reported to division managers who in turn reported to corporate.

THE PRODUCTION PLANTS

The three production plants (located in Rochester, Minneapolis, and St. Cloud) produced and shipped the company's products to 14 distribution centers, which marketed and sold them in Spirit's franchise territory. Table 8C-2 shows which plants produced and shipped the various products. Table 8C-3 gives data for the cases produced at each of the three plants.

The St. Cloud and Minneapolis plants were similar in size—112,000 square feet and 95,000 square feet, respectively—whereas the newer Rochester plant (built and equipped in 1973 for $9 million) covered 280,000 square feet.

St. Cloud had two operating lines but used only one at a time, depending on the size required. Minneapolis had only one operating line. However, each plant operated with one crew consisting of eight workers—one 10-hour shift, 4 days a week.

In contrast, the Rochester plant operated four bottling lines and two canning lines. Bottling line 1 operated two 10-hour shifts per day 4 days a week. Lines 2 through 6 operated one 10-hour shift 4 days a week. Each line was operated by an eight-worker crew. At all three plants, production runs were made Monday through Thursday; Fridays were spent on maintenance and repairs. The capacities and sizes produced by each line at the three plants are shown in Table 8C-4.

The Rochester operation, although generally well managed, had one obvious problem. There was not enough warehouse space for raw materials and finished goods. During May, for example, the plant held 452,785 cases of finished product. Lack of space within the plant cost the company in several ways:

1. The company was forced to store 86,334 cases of used returnable glass in cartons and cases outside in an open yard. The cases covered 24,500 square feet. When it rained the cardboard cartons in a case were ruined. There were four cartons in a case, each costing 5 cents.

2. During the winter (December, January, and February), approximately 20 percent of the bottles that had been stored outside would break when they were soaked in the hot caustic solution in the bottle washer. The replacement cost for returnable bottles was $4.30 per case of 24.

3. The space problem forced the Rochester plant to operate an additional 28,000-square-foot warehouse 15 miles from the production plant. The warehouse stored

continued on p. 312

TABLE 8C-2 Production assignments for each distribution center, May 1983

	MINNEAPOLIS	MANKATO	RED WING	RUSH CITY	COON RAPIDS	ST. CLOUD	NEW ULM	ARKANSAW WISCONSIN
Returnable bottles								
6 1/2 Oz. Spirit	M*	M	M	M	M	S	S	S
10 oz. Spirit	M	M	M	M	M	S	S	S
10 oz. 11 Up	M	M	M	M	M	S	S	S
10 oz. Pep	M	M	M	M	M	S	S	S
10 oz. Brite	M	M	M	M	M	S	S	S
10 oz. Action Ade		S	S			S	S	S
10 oz. Orange Orange								R
10 oz. Dr. Spice	M	M	M		M	S	S	M
10 oz. SF Orange Orange	M	S	S					M
16 oz. Spirit	M	R	M		R	S	S	M
16 oz. Brite	M	M	M	M	M	S	S	S
16 oz. 11 Up	M	M		M	M			M
16 oz. Pep	M	M	M	M	M			M
16 oz. Dr. Spice	M	M		M	M			R
1 liter all products	R	R	R	R	R	R	R	
One-way bottles								
10 oz. all products	S	S	S	S	S	S	S	S
10 oz. Viva	R	R	R	R	R	R	R	R
16 oz. all products	R	R	R	R	R	R	R	R
28 oz. all products	R	R	R	R	R	R	R	R
2 liters all products	R	R	R	R	R	R	R	R
12 oz. cans all products	R	R	R	R	R	S	R	R
Premix all products		R	R	R	R			S

TABLE 8C-2   (continued)

	ROCHESTER	ST. BONFACIUS	SAUK CENTER	PRINCETON	HUTCHINSON	LITTLE FALLS
Returnable bottles						
6 1/2 oz. Spirit	R	R	R	R	R	R
10 oz Spirit	R	R	R	R	R	R
10 oz 11-up	R	R	R	R	R	R
10 oz Pep	R	R	R	R	R	R
10 oz Brite	R	R	R	R	R	R
10 oz Action Ade	R	R	R	R	R	R
10 oz Orange Orange		M		M		R
10 oz Dr. Spice	R	R	R	R	R	R
10 oz SF Orange Orange	R	R	R	R	R	R
16 oz Spirit				M		R
16 oz Brite				R		
16 oz 11-up		M		M		
16 oz Pep		M	R	R		
16 oz Dr. Spice	R	R	R	R	R	
1 liter all products					R	
One-way bottles						
10 oz all products	R	R	R	R	R	R
10 oz Viva	S	S	S	S	S	S
16 oz all products	R	R	R	R	R	R
28 oz all products	R	R	R	R	R	R
2 liters all products	R	R	R	R	R	R
12 oz cans all products	R	R	R	R	R	R
Premix all products	R	R	R	R	R	R

* M, Minneapolis; R, Rochester; S, St. Cloud

TABLE 8C-3   Case production by facility, May 1983 (Spirit Bottling Company)

PLANT	CURRENT	BUDGET	CHANGE	PERCENT CHANGE	PRIOR YEAR	CHANGE	PERCENT CHANGE
St. Cloud							
Returnable bottles	125,353	131,500	(6,147)	(4.7)%	135,353	(10,000)	(7.4)%
One-way bottles	8,527		8,527		42,948	(34,421)	(80.1)
Total	133,880	131,500	2,380	1.8	178,301	(44,421)	(21.9)
Minneapolis							
Returnable bottles	164,764	121,300	43,464	35.8	200,226	(35,462)	(17.7)
One-way bottles	9,064		9,064			9,064	
Total	173,828	121,300	52,528	43.3	200,226	(26,398)	(13.2)
Rochester							
Returnable bottles	185,963	255,300	(69,337)	(27.2)	214,206	(28,243)	(13.2)
One-way bottles	518,707	501,000	17,707	3.5	355,782	162,925	45.8
Cans	381,871	400,400	(18,529)	(4.6)	458,053	(76,182)	(16.6)
Total	1,086,541	1,156,700	(70,159)	8.2	1,028,041	58,500	5.7
Division total	1,394,249	1,409,500	(15,251)	(1.4)	1,406,568	(12,319)	(0.9)

TABLE 8C-4   Line capacities of production plants (Spirit Bottling Company)

PLANT	LINE NUMBER	PACKAGES*	LINE CAPACITY (cases/hour)
Minneapolis	2	6 1/2 oz.R, 10 oz. R	1000
	2	10 oz. R	1000
	1	32 oz. R	600
	1	16 oz. R	670
St. Cloud	1	6 1/2 oz. R, 10 oz. R	700
	1	16 oz. R	609
	2	1 liter R	562
	2	10 oz. OWB	500
Rochester	1	2 liters plastic	960
	2	1 liter R	1149
	2	28 oz. OWB	1149
	3	10 oz. OWB	1463
	3	16 oz. OWB	1463
	4	6 1/2 oz. R, 10 oz. R	1000
	4	16 oz. R	938
	5	Cans	1875
	6	Cans	2000

*R, returnable bottles; OWB, one-way bottles.

30,000 cases of finished product in May 1983. Transport drivers were paid $10 per hour to drive a trailer load to the warehouse, unload, and drive back to the Rochester plant. The round trip took an hour and a half. During May, the Rochester transportation department made 33 trips to this warehouse.

THE PRODUCTION PROCESS

The production process began after several pallets of empty bottles were moved by forklift to the back of the bottling line. At this point, an employee loaded the inclined conveyor with cases of empty bottles, which were gravity fed to the decaser. The decaser lifted 24 bottles from a case every 2.4 seconds and placed them on an automatic steel conveyor, which fed the bottles into the bottle washer. The washer cleaned and sanitized every bottle by means of a 3.8 percent caustic solution, heated to 140°F. After the bottles rotated through the hot caustic bath, fresh water jets sprayed and rinsed each bottle. The washer then placed each row of 28 bottles onto the bottling line every 3.5 seconds. The steel conveyor carried the clean bottles through two inspection stations. Chipped or cracked bottles were removed by an empty-bottle inspector, and an electronic eye rejected bottles containing foreign particles. Before the filling process, fresh water was mixed with Spirit Cola syrup or other concentrates in the proper proportions. (A 6 1/2-ounce bottle of Spirit contained 1 ounce of syrup and 5.5 ounces of carbonated water.) The syrup and water mixture flowed through a "carbo cooler," where it was cooled to 34°F, carbonated, and released to the filler.

The washed bottles were engaged, one by one, in the rotating filler, which filled them as they moved through 360 degrees. The fillers contained 60 filling valves and were capable of speeds up to 2000 cases per hour depending on package type and size. After being filled, the bottles were sealed by the adjacent crowner, which stamped the bottle caps in place. The full bottles then went to the case packer, which dropped the bottles into an empty case every 2.8 seconds. The finished goods traveled along an overhead conveyor to the palletizer, which stacked each pallet. Forklift operators took the full pallets to transports for warehouse storage. The time required for a bottle to flow through the production process was 35 minutes.

THE PRODUCTION SCHEDULING SYSTEM

During Donald Whitcomb's first week as vice-president, he visited the three production plants and discussed production and transportation schedules with the plant managers. After asking these managers a number of questions about their scheduling procedures, Donald believed he had a good idea of what was going on. As a first step in scheduling, each production plant tabulated the previous week's sales and average daily sales by product and by package for their distribution centers. The inventory figures divided by average daily sales gave the number of days' supply "on the floor" for each product and package. Days' supply was the key factor in determining what and when to produce. A product with two days' supply would normally be produced before an item with four days' supply.

Whitcomb recalled a production scheduling meeting where he questioned the production manager from St. Cloud, John Wolfe, and the production manager from Rochester, Marcia Fox, on the issue of which product to produce first.

Donald   If you had a three-day supply of 10-ounce returnable Spirit and a 2-day supply of 10-ounce returnable 11-Up, which would you produce first?

Marcia   I would play it safe and produce 10-ounce returnable Spirit first. Spirit is a high-volume product and I would hate to lose the contribution dollars resulting from a possible stockout.

John   Well, I could stick to the days' supply calculations and produce 10-ounce returnable 11-Up. I know we have 3 days of inventory on 10-ounce Spirit. What is there to worry about?

In determining the order of production, the managers also looked at minimizing downtime due to changeovers. There were two kinds of changeovers: product changeovers (such as from Spirit to 11-Up) and package changeovers (such as from 10-ounce to 16-ounce bottles). A product changeover during producing hours cost the bottler 15 minutes of downtime. A package changeover required approximately 35 minutes. The actual costs involved in these two types of changeovers are shown in Figure 8C-1, along with data on fixed costs.

Donald recalled that the changeover issue had surfaced in his conversation with Marcia and John.

Fixed Costs	*St. Cloud*	*Minneapolis*	*Rochester*
Operating expenses	$30,583	$18,107	$191,185
Indirect labor	16,846	20,960	134,036
Payroll	2,200	2,864	21,553
Depreciation	6,429	3,538	47,640
Total fixed costs	$56,058	$45,469	$394,414

Product Changeover Cost

Washdown expense
  24 oz. syrup @ $3.92/gallon = 72 ¢
  Carbon dioxide and water = 6 ¢
  Total = 78 ¢

Labor cost for mixing concentrate
  1 man @ $8/hour
  Time required = 15 minutes
  Labor cost = $2.00
  Total cost = $2.78

Package Changeover Cost

Labor cost for resetting machine
  2 men @ $8/hour
  Time required = 35 minutes
  Total cost = $9.34

Notes: In May 1983 there were 30 product and 15 package changeovers during producing hours in the three production plants.

There was an unwritten policy of guaranteeing production workers 40 hours per week, even though actual production hours were usually fewer than 40.

A 10-hour shift in each producing plant included two 15-minute breaks and a half-hour for lunch.

FIGURE 8C-1   Production cost information for Spirit Bottling Company.

Donald   Suppose the line just finished producing 16-ounce returnable Spirit. There was a 2-day supply of 10-ounce returnable Spirit and a 3-day supply of 16-ounce returnable Brite on the floor. Which product should be produced first?

John   16-ounce returnable Brite. Package changeovers require 35 minutes of downtime. That is valuable production time. I believe in maximizing production volume because it reduces case cost.

Marcia   I disagree. The 10-ounce returnable Spirit should be produced first because (1) 10-ounce Spirit has fewer days' supply on the floor; (2) changeover costs are inexpensive; and (3) large-volume production runs can also mean large inventories.

In deciding how much of each item to produce, the managers studied the following: raw materials inventory, production time required for other products, safety stock levels, warehouse space, sales promotions, product age, and inventory costs. Sales forecasts were occasionally observed but not relied on because they were routinely off by at least 10 percent.

The production manager knew he or she should not run greater than a 90-day supply of any product and package because of quality control requirements concerning product age. The low-volume items, such as 10-ounce returnable Action Ade, were produced every 2 1/2 to 3 months. Faster selling items, like 2-liter Spirit, were produced almost daily. A safety stock of 2 to 4 days was added to each production run to allow for uncertainties in demand.

Because of present inventory levels, warehouse space was at a premium in all three pro-

duction plants. This, in addition to the demand for other products to be produced, limited the length of a production run. The actual length of a run really became a judgment call. This was evident to Donald at the scheduling meeting he attended at the Rochester plant. These meetings were held every Thursday to decide the next week's production schedule. The schedule shown in Figure 8C-2 was to be followed the next week.

In the meeting Donald questioned production manager Marcia Fox:

Donald   Why are we going to produce a 5-month supply of Lean Green cans?

Marcia   The 5-month supply requires only a 3-hour production run. A run length of less than 3 hours is not cost justified.

Donald   Long runs are great—volume is spread over the same fixed costs. But a 5-month supply is expensive to hold and product age could be a problem.

THE TRANSPORTATION SYSTEM

The company's transport fleet consisted of 17 tractors and 60 trailers. Three tractors and eight trailers were located in St. Cloud. Five tractors and 11 trailers were in Minneapolis, and the remainder were in Rochester.

During Donald Whitcomb's first visits to the production plants, he discussed transportation schedules with each transportation manager. Donald discovered that in setting up a weekly transportation schedule, each manager began by observing a day's supply figure for each product and package in each plant to which

SPIRIT BOTTLING CO.—ROCHESTER PLANT
PRODUCTION SCHEDULE

Week of _5/9 - 13/83_

Date	Line 1			Line 2			Line 3			Line 4			Canning		
	Item	Hrs.	Cases to produce	Item	Hrs.	Cases to produce	Item	Hrs.	Cases to produce	Item	Hrs.	Cases to produce	Item	Hrs.	Cases to produce
Monday 5/9	2 Liter OWB Spirit	18½	16225	1 Liter RET Spirit	9	8055	16 oz. OWB SF orange orange Spirit	5 / 3½	4408 / 3063	6½ oz. RET Spirit	9	1500 6-Pack 5250 regular carton	Spirit	9	2534
Tuesday 5/10	2 Liter OWB Pep Spirit	9 / 9	7854 / 7893	1 Liter RET Spirit	9	8055	16 oz. OWB Pep	9	7812	10 oz. RET Spirit	9	2550 6-Pack 4200 regular carton	SF Dr. Spice	8½	4469
Wednesday 5/11	2 Liter OWB Viva Spirit	9 / 9	8052 / 7893	1 Liter RET Spirit	9	8055	16 oz. OWB Brite	9	7040	10 oz. RET Spirit	9	2550 6-Pack 4200 regular carton	Lean Green Spirit	3 / 5½	8460 / 1554
Thursday 5/12	2 Liter OWB SF 11-Up Spirit	9 / 9	7875 / 7893	28 oz. OWB Viva Spirit	4 / 4½	3605 / 4095	10 oz. OWB Pep 11-Up	4½ / 4	3806 / 3264	16 oz. RET Spirit	9	7155	Brite	9	2538
Friday 5/13															

Note: OWB = one-way bottle; RET = returnable.

FIGURE 8C-2   Production schedule for the Rochester plant of Spirit Bottling Company.

they shipped. Products with the lowest days' supply were shipped first.

Each plant had warehouse space assigned for every product and package. For example, in the St. Cloud warehouse, there were three rows of 2-liter plastic Spirit capable of holding 90 pallets of product. The actual amount shipped to a distribution plant is sufficient to bring the plant's inventory level to the maximum level for the shipped products. For example, if there were 1000 cases of Pep cans on inventory in a distribution plant and the maximum inventory level for Pep cans was 2500 cases, then 1500 cases would be shipped to the distribution plant.

CONCLUSION

Donald Whitcomb knew that Miles Seabrook would soon hold a follow-up meeting to the June meeting. Seabrook expected Whitcomb to

TABLE 8-C5   Sales and inventories held by supplying plant (cases)

	SALES BY SUPPLYING PLANT (CASES)			INVENTORY TOTALS BY SUPPLYING PLANT		
	MINNEAPOLIS	ST. CLOUD	ROCHESTER	MINNEAPOLIS	ST. CLOUD	ROCHESTER
Returnable bottles						
6 1/2 oz Spirit	25,695	24,081	31,690	11,162	7,188	11,930
10 oz Spirit	18,996	24,614	30,439	5,686	6,497	8,835
10 oz 11-Up	1,735	2,021	3,399	665	538	1,342
10 oz Pep	6,439	6,522	8,289	1,355	3,980	3,278
10 oz Brite	5,570	5,484	7,797	891	5,181	2,760
10 oz Action Ade	0	0	481	677	841	589
10 oz Orange Orange	0	7,480	2,669	0	3,278	1,165
10 oz Dr. Spice	1,946	0	0	827	0	0
10 oz SF Orange Orange	0	3,362	0	1,677	11,826	
16 oz Spirit	53,708	14,200	63,031	6,348	4,655	980
16 oz Brite	7,559	2,033	2,606	2,275	1,513	0
16 oz 11-Up	3,909	0	0	858	0	0
16 oz Pep	12,114	0	0	2,263	0	0
16 oz Dr. Spice	1,774	0	0	1,069	0	0
1 liter all products	0	0	99,497	0	0	43,777
One-way bottles						
10 oz all products	0	0	26,953	0	0	14,725
10 oz Viva	0	2,249	0	0	2,570	0
16 oz all products	0	0	118,942	0	0	58,140
28 oz all products	0	0	18,159	0	0	13,296
2 liters all products	0	0	119,316	0	0	90,386
12 oz cans all products	0	0	406,687	0	0	155,304
Premix all products	0	6,980	58,984	0	2,178	3,633

analyze company inventory levels. If there were problems, Whitcomb would be expected to correct them. Possibly a new scheduling method, which would reduce inventory levels and overall costs, would have to be developed. Donald had compiled company sales and inventory figures for May 1983 (see Table 8C-5). This information, along with the other information he had collected, would enable him to construct a sample production schedule for each production plant for the first week of June (see Figure 8C-2).

It appeared that the Rochester plant had outgrown itself. Marcia Fox was convinced the plant should be expanded, at a cost of $21 per square foot. Donald wondered what he should recommend.

REFERENCE NOTES

1. Good sources of additional information on nonanticipatory approaches include:
    Backes, Robert W., "Cycle Counting—A Better Way for Achieving Accurate Inventory Records," *Production and Inventory Management* 21, no. 2 (Second Quarter 1980): 36–44.
    Buffa, Elwood S., and Jeffrey G. Miller, *Production-Inventory Systems: Planning and Control*, 3rd ed. Homewood, IL.: Richard D. Irwin, 1979.
    Flores, Benito E., and D. Clay Whybark, "Implementing Multiple Criteria ABC Analysis," *Journal of Operations Management* 7, no. 1–2 (October 1987): 79–85.
    Hadley, George, and T. M. Whitin, *Analysis of Inventory Systems*, Englewood Cliffs, NJ: Prentice-Hall, 1963.
    Magee, J. F. and D. M. Boodman, *Production Planning and Inventory Control*, 4th ed. New York: McGraw-Hill, 1980.
    Peterson, Rein, and Edward A. Silver, *Decision Systems for Inventory Management and Production Planning*, New York: Wiley, 1979.
    Plossl, George W., and W. Evert Welch, *The Role of Top Management in the Control of Inventory*, Reston, VA.: Reston, 1978.
    Plossl, George W., and Oliver W. Wight, *Production and Inventory Control: Principles and Techniques.* Englewood Cliffs, NJ: Prentice-Hall, 1967.
    Reisman, A., B. Dean, and M. Oral, *Industrial Inventory Control*, New York: Gordon and Breach, 1972.
    Starr, Martin K., and D. W. Miller, *Inventory Control: Theory and Practice*, Englewood Cliffs, NJ: Prentice-Hall, 1962.
    Vollmann, Thomas E., *Operations Management: A Systems Model-Building Approach*, Reading, MA.: Addison-Wesley, 1973.
    Vollmann, Thomas E., William Berry, and D. Clay Whybark, *Manufacturing Planning and Control Systems*, 3rd ed., Homewood, IL.: Dow Jones-Irwin, 1992.

2. See Paul J. Kuzdrall and Robert R. Britney, "Total Setup Lot Sizing with Quantity Discounts," *Decision Sciences* 13, no. 1 (January 1982): 101–12, and R. R. Britney, P. J. Kuzdrall, and N. Fartuch, "Note on Total Setup Lot Sizing with Quantity discounts," *Decision Sciences* 14, no. 2 (April 1983): 283–91.

3. Plossl, George W., *Production and Inventory Control: Principles and Techniques*, 2nd ed., Englewood Cliffs, NJ: Prentice-Hall, 1985, p. 97.

CHAPTER 9

PROCUREMENT, INVENTORY CONTROL, AND LOGISTICS

Considerable management attention is focused on ensuring that the proper materials are directed to the right places in timely fashion and at low cost. Managing materials well is not easy. A surprisingly large fraction of many operations' overhead personnel are devoted to the task, often a quarter or more. Chapter 6 addressed the important area of production planning and control, which has much to say about how materials get managed. Chapters 7 and 8 dealt with the different styles of inventory management. What remains to discuss are issues of procurement, inventory control, and logistics.

In the first part of this chapter we tackle procurement, including the important make–buy decision. Inventory control is discussed in the second part. With all of these elements of materials management thus introduced, the third portion treats selected issues in logistics. A fourth segment discusses the organization of the materials function, as it embraces procurement, production planning and control, and logistics.

PROCUREMENT[1]

The old stereotype of the procurement function (commonly termed purchasing) runs something like this. The purchasing department is staffed with buyers, each concentrating on a different kind of part or material. The objective of the buyer is to line up many different suppliers (typically termed vendors) to bid on the work that the company has (work that the engineering department has thrown "over the wall" to purchasing). The lowest bid is accepted, but several suppliers are given the business, just in case something unexpected happens to them. A good buyer is one who can negotiate well, thus reducing the costs of procurement to the company. Fortunately, much of this old stereotype is falling by the wayside. The work of the purchasing department has greatly changed in recent years.

The common duties of a modern purchasing department include all of the following.

1. Understanding Material Needs for Engineering and Manufacturing

One of the purchasing group's tasks, too frequently overlooked, is to question the materials specifications established by design engineering (perhaps in consultation with manufacturing engineering). Sometimes purchasing can suggest substitute materials or product redesigns that use standard materials, thus lowering the costs implied by engineering's initial specifications.

Purchasing should be engaged in any value engineering that occurs. (Recall the discussion of value engineering in Chapter 5.)

How can purchasing perform this role effectively? Often, in the course of meeting with a supplier's salespeople and in keeping abreast of developments in the industry, purchasing departments can uncover materials and products that may have been overlooked by design engineering. More and more, too, purchasing departments are asking their suppliers to be more innovative in what they do for the company. Often this means not defining the specifications so rigidly for the product and actively seeking the supplier's ideas for improving quality or costs.

2. Selecting and Evaluating Suppliers

Many production and operations people believe that purchasing's chief obligation is to obtain materials at the lowest cost. This, however, is a narrow view of purchasing. Low cost is only one of the aspects of a supplier's performance that a purchasing department needs to monitor and foster. Exactly what constitutes a good supplier varies, of course, with the situation, but among the characteristics that purchasing departments should seek in addition to low cost are

- Quality (conformance to the established specifications).
- Delivery reliability.
- Ability to deliver frequently, and/or speed of delivery.
- Trustworthy and timely communication on any problem encountered.
- Initiative in product design or materials; suggestions.

Taking a lesson from Japanese companies, many U.S. companies (such as the auto companies) have recently undergone a massive reorientation of purchasing, shifting the emphasis from price to quality, special performance characteristics, and delivery. (In some people's eyes this is a move away from "purchasing" and toward "procurement.") Many companies like to have at least two suppliers supplying each raw material: they feel that competition for their business encourages lower prices and better service and that reliance on a single supplier can be risky in case of strikes, natural disasters, and other uncommon causes of delay. Other companies feel that dealing with a single supplier can be more advantageous, especially if the customer can exercise clout with that supplier by being a significant customer. By employing a sole supplier, a company can often foster the quality and delivery performance that meshes best with its own operations; in a very real sense the supplier can act as an extension of the company's operations. Such a relationship can often stimulate some spillover benefits—improved productivity at the customer company, less costs incurred in inspecting and returning nonconforming in-coming materials, teamwork that can tackle special problems or new product initiatives. For reasons such as these, the current trend among Western companies is to reduce the supplier base. In some cases, this reduction has been dramatic.

No matter what the company decides about single versus multiple suppliers, the purchasing department has to act as the company's eyes and ears in the raw materials marketplace. Suppliers have to be sought for new materials or products, and the quality level of suppliers has to be assessed before orders can be placed. Thereafter, purchasing must continue to evaluate the suppliers it uses and help them to improve their performance.

Increasingly, companies are establishing formal supplier qualification or certification programs. This has been inspired mainly by increased devotion to quality conformance and the desire to move the supplier's product right into production (or storage) without having to inspect it. As one might expect, the stringency of the various programs varies markedly from company to company. In some companies a supplier can be "qualified" or "certified" merely by shipping two consecutive orders that have no detected defects. In other companies, the suppliers must surmount a number of hurdles that may take years. For companies serious about quality and willing to use a single source for their purchases, such stringent programs can lead to situations in which both customer and supplier win.

Companies sometimes distinguish between product certification and supplier certification.

Product certification. Product certification can include the following steps. Exactly which ones, however, vary from company to company.

- Inspection of their processes, via an extensive visit.
- Formal supplier rating done, using extensive checklists.
- Process quality documentation—process capability studies (Cp_k development), inspection capability studies, control charts for identified key characteristics.
- Development of control plans—inspection, maintenance, and product review plans to ensure continued adherence to customer quality requirements.

Supplier certification extends beyond product certification to include such things as on-time performance, technical support for manufacturing and design work, and a competitive cost position.

3. Negotiating Price, Delivery, and Quality

Purchasing typically must decide whether it wants a long-term (multiyear) contract with the supplier or a short-term or even spot (single-transaction) contract. Frequently, a long term contract is required even to interest a supplier in providing a specific item, particularly if the item is nonstandard. Otherwise, price expectations and views of supply risk usually play a major role in choosing between long- and short-term contracts. On many long-term contracts the specific price may not be settled ahead of time but instead may be negotiated at regular intervals.

Many companies are using fewer suppliers and longer-term contracts. Many of these contracts are "blanket orders," which cover a variety of items, or "open-end orders," which permit items to be added and/or time to be extended. Customer companies using such tactics are flexible about prices and often are willing to share their demand forecast and order backlog information, but they demand consistent quality and frequent deliveries in return. The plant tours at the end of this book demonstrate how the huge demands of operations like GM–Oklahoma City and Stroh's–Winston-Salem permit those companies to seek—and be accorded—regular deliveries, thereby cutting down on inventory storage and carrying costs.

In large multiplant companies, purchasing can proceed at different levels. The corporate staff, for example, may negotiate and monitor the large supply contracts that feed a number of plants. By centralizing this activity, the company often can exert more clout over price and delivery terms than could plants each acting on their own. In addition, centralizing such purchasing may aid plant-to-plant coordination of materials. Smaller supply contracts may be left to individual plant negotiations. In some other companies, individual plants' purchasing departments may take the lead in keeping abreast of new developments in particular industries and negotiating companywide contracts with suppliers in those industries.

Quality checking varies from contract to contract. Many companies inspect incoming supplies (100 percent or a sampling). Some supply contracts allow the posting of teams within the supplier's plant to monitor quality and delivery and to aid that plant in meeting the requirements specified by the customer; both supplier and customer can reduce costs and overhead if incoming inspection in the customer plant can be avoided through a joint quality effort. Supplier prices, even for single-source suppliers, are often monitored, and benchmarked against other desirable suppliers, so that both the supplier and customer companies can stay abreast of the best technology and ideas.

4. Placing Purchase Orders with Suppliers

Traditionally, suppliers did not deliver orders unless they had received purchase orders from the customers. This traditional mode of operation is still prevalent. However, with longer contracts and use of fewer suppliers, the actual release of the supplier's products to the customer is becoming more flexible. With closer customer/supplier ties and innovations like blanket or open-end long-term contracts, supplier deliveries can be released more informally, through telephone calls, computer-to-computer transactions, or plant paperwork that is less formal. In many instances, the contracts have been negotiated by a centralized purchasing function in the company that may have a more talented and savvy staff than any plant could hope to have. Nevertheless, the delivery of needed parts may be triggered by the plant itself and not by the centralized purchasing group.

5. Working Out Supplier Problems and Monitoring Orders

Here is where the procurement function, particularly at the plant level, wins its reputation, for good or for ill. Rarely does all go smoothly with orders and deliveries. The typical purchasing department must spend much time dealing with supplier quality or delivery problems—going to the supplier's plant, arranging meetings between engineering staffs, or tracing and expediting orders that are needed before the supplier's quoted lead time or for which shipment has been delayed. The purchasing department can thus act as an early warning alarm system for materials shortage or quality problems that can affect the organization's schedule and quality performance.

World-Class Suppliers and World-Class Customers

Some of the elements of the latest thinking about manufacturing (often captured by the general rubric of just-in-time production) have been discussed, and more will be addressed in subsequent chapters, particularly Chapter 11. Several elements of this thinking relate to the purchasing function. Among them are the following:

- Quality conformance viewed as more important than price; indeed, it is viewed as the basis of the relationship. World-class customers do not talk quality and then buy strictly according to price.

- Long-term relationships with suppliers, and relationships with suppliers that extend past the purchasing department itself to include customer–supplier contacts in production, engineering, and other functions within both customer and supplier companies. Both suppliers and customers work hard to understand one another's processes and systems and to fulfill promises made.

- Many single-source contracts, although several suppliers may be qualified to supply a particular item.

- Frequent deliveries of small quantities with the aim of being able to deliver directly to the factory floor, with no incoming inspection.

- Pursuit of very capable, typically computer-based systems to aid in the speedy transmittal of information (e.g., Electronic Data Interchange [EDI], quick response programs), or to aid in the development of quality-imbued products and processes (e.g., SQC).

- More stable production schedules for suppliers to meet; sharing of schedule information, giving visibility into the future of the stream of orders from the customer.

- Encouragement of supplier ideas for improving product design and manufacturability. Often, early on in the design cycle with new products, the customer company provides only the sketchiest ideas about what is required of the supplier. This calls on the supplier to be as innovative as possible, and can lead to both better designs for the customer and lower costs as well, because poor designs for manufacturing or assembly can be averted.

- Open and frequent communications between supplier and customer companies. More and more this communication is via purchasing or commodity teams.

World-class companies, acting as either suppliers or customers, work at the relationship over long periods of time. As more and more companies adopt aspects of just-in-time manufacturing, they are finding out how effective it and these nontraditional ideas about procurement can be. For some companies, however, just-in-time production does not mean implementing fundamental change on the factory floor. Rather, it constitutes an excuse to take advantage of suppliers, forcing them to hold the inventory the customer does not want to hold, under the guise that the customer should receive it "just in time." This attitude does not really reflect the just-in-time concept. Companies adopting true just-in-time production typically work first on cleaning up their own operations before they try to work with suppliers on theirs. The idea is to lower inventories throughout the pipeline and not just at the cus-

tomer's plant. World-class customers simply do not try to take reckless advantage of their suppliers, but try to make the relationship a profitable one for all concerned. Companies with this mentality talk of building trust between customer and supplier so that the products produced are of high quality and low cost.

The ABC Classification

Many companies have organized their purchasing staffs to mesh with the ABC classification of parts and materials. It has long been observed that in most companies a small fraction of the number of parts accounts for a disproportionately large fraction of purchasing expenditures. This small group of high-value items is typically classified as A items. The A items might account for 5 to 10 percent of the number of parts but 75 percent of the value of all parts. At the other extreme are the C items: 80 percent of the number of parts that account for only 10 percent or so of their value. The B group falls between the extremes—perhaps 10 to 15 percent of the number of parts and 10 to 15 percent of their value. Such an ABC classification provides a means of breaking down the tasks of dealing with suppliers and tracking the materials needed by the operation. Specific staff members can be assigned to deal with A items only, B items only, or C items only; in this way, purchasing department resources can be matched well to the importance particular parts have to purchasing expenditures. The ABC classification can be used by the production and inventory control groups to assign responsibilities for tracking materials and orders.

THE MAKE–BUY DECISION

Although typically not the responsibility of the purchasing department, the decision as to what the factory should make for itself and what it should purchase on the outside is one for which the purchasing department is a key source of information. The make–buy decision is a common one, especially in companies that have purposely split their supply needs between in-house operations and outside suppliers.

What does a make–buy decision entail? Consider Situation 9-1.

SITUATION 9-1

Dulaney Toy Company

He knew that the business was sometimes boom or bust, but that still did not mean that he liked the "bust" part. Rick Jerauld, the manager of the injection molding department at the Dulaney Toy Company, was on the horns of a dilemma. After several years of record sales and growth, Dulaney Toy appeared to be in for a slow year. Sales for Dulaney's two latest entries in the fickle toy market—the Incredible Smiling Monster and the Cynical Santa—simply did not appear to be taking off as the company's marketing people had claimed they would. Given the sales declines of some of Dulaney's older toys, which was characteristic of product life cycles in the industry, many of Dulaney's manufacturing departments had spare capacity. The injection molding department, which was responsible for all the plastic injection-molded parts of Dulaney's toys, was among them.

Rick could, but was loath to lay off any of the workers in his department, particularly given the company's goal of lifetime employment for its loyal and productive workforce. What this left was for Rick to start pulling back to his department the manufacture of plastic parts that Dulaney had subcontracted to other firms. Fortunately, Dulaney had a battery of such plastic injection-molded parts, but this abundance did not lessen Rick Jerauld's concern about which of the parts now produced by suppliers (vendors) he should pull back to his own department. As Jerauld thought about it, several criteria appeared to be appropriate to deciding the issue:

1. Whether the part demanded many labor-hours because of its complexity or because it was produced in large quantities.

2. Whether the part could be manufactured as cheaply at Dulaney Toy as at the supplier's. At present, Dulaney Toy's formula for such cost comparisons called for comparing the supplier's price for the part (adjusted upward for the expected costs Dulaney Toy would incur in purchasing the part and in having it transported and inspected) against the calculated, internal factory cost for the part, which included all materials, labor, and overhead.

3. What effect pulling the part back would have on the supplier and on Dulaney Toy's expected future use of the supplier in boom times when Dulaney Toy was short of capacity. Rick was naturally leery of doing anything to antagonize his supplier base to the point where many vendors would be reluctant to bid for Dulaney Toy's business in the future.

4. What risk the part carried for becoming obsolete in the near future.

5. What impact the part would make on shop operations, given the prevailing conditions on the shop. That is, would the part need special tooling or require setup times that would make its coordination with other parts running through the shop difficult?

One of Rick's assistants had developed a sheet of candidate parts for pulling back to the factory (Table 9-1). Some were made exclusively by suppliers, while others were made both by the injection molding department and by suppliers. The candidate parts identified to date derived from two of the company's more successful

TABLE 9-1   Candidate parts for pulling back to the factory (Dulaney Toy Company)

PART	AVERAGE VOLUME PER MONTH	CURRENT STATUS	VENDOR PRICE (includes adjustment)	INTERNAL COST—ACTUAL OR ESTIMATE				COMMENTS
				MATERIALS	LABOR	OVERHEAD	TOTAL	
Caboose under-carriage	6000	Long-term contract; made in shop before	11.97	3.20	1.37	5.20	9.77	Vendor problems with quality and delivery
Caboose roof	6000	Long-term contract	9.60	2.77	3.93	12.57	19.27	
Caboose railings	12,000	Short-term contract	3.79	0.88	3.03	5.75	9.66	Same vendor as caboose roof
Crib gym bar	2000	Long-term contract; made in shop before	22.39	5.91	6.70	9.53	22.14	
Crib gym flower	8000	Long-term contract; always made outside	18.26	3.58	4.43	11.66	19.67	$1500 in tooling needed

toys, Walter the Caboose and the Little Monkey Crib Gym.

Rick needed to decide which of the parts should be pulled back inside his department and which should be left with the suppliers. Also, Rick wanted to develop a system whereby he could put priorities on parts that should be brought back first and ones that could be allowed to wait.

Discussion of Dulaney Toy Company

The decision Rick Jerauld faces, that of pulling in some subcontracted parts, is one that companies in cyclical industries face regularly. The five criteria that Jerauld consider appropriate are all issues that a manager should consider for a make–buy decision. A part should not be pulled back to the injection molding department unless (1) it would make a significant dent in the excess labor the company is experiencing, (2) it costs less to do it inside (more comment on this later), (3) the supplier would not be so adversely affected that it could not be counted on for future contracts, (4) the part is not in danger of becoming obsolete in the near future, and (5) the part could be fitted into the shop fairly smoothly and manufactured with good quality.

Judgment is required in assessing many of these. The cost comparison is more mechanical, but it should be noted that Dulaney Toy's present method for comparing costs is plainly incorrect. The present cost comparison ignores the fact that the make–buy decision is a marginal one, just like the Citrus Airlines situation examined in Chapter 2. That is to say, although the supplier's price, as adjusted, is a variable cost to Dulaney Toy, not all of the internal factory costs vary with the factory's level of production. Overhead expenses, for example, would have to be covered no matter what the level of factory output. And if Dulaney Toy is serious about lifetime employment for its workers, it can be argued that, for the short run at least, the factory's labor costs are fixed as well. Let's assume labor is a variable cost, however.

Thus, by bringing in-house any product for which the materials-plus-labor factory cost is less than the supplier quote-plus-adjustment, Dulaney Toy's contribution (see Chapter 2) would be enhanced. Table 9-2 computes the contribution for each of the candidate parts listed in Table 9-2.

Table 9-2 suggests that under no circumstances should the caboose railings be pulled back to the factory. In fact, the supplier producing both the caboose railings and the caboose roof looks like it should be encouraged to produce even more than it is, since its prices look very good indeed. Perhaps the short-term contract for railings should be extended to long-term. The possibility remains, however, that the supplier price is not low, but that the factory's estimated cost is high. This should be investigated.

Dulaney Toy would earn additional contribution by pulling in any of the parts except the caboose railings. Even the $1500 tooling expense for the crib gym flower would be paid for out of added contribution within two months. The issue remains, however, in what order the various parts should be pulled back to the factory, at least according to the prevailing economics. In a sense, what we are looking for is to pull back the parts with the most "bang for the buck." Here this translates into the most contribution per some unit of resource, which might be something like factory time or labor cost. In fact, time and labor costs are likely to be highly correlated. This means that an index constructed as *(contribution per unit)/(labor cost per unit)* should give us a fair picture of which parts can add the most contribution quickly. Table 9-3 displays the values for this index of contribution per penny of labor costs.

It is clear from Table 9-3 that the company is best served by pulling in-house the caboose undercarriage first, then the crib gym flower, followed by the crib gym bar, and the caboose roof. From the comments in Table 9-1 this

TABLE 9-2	Contribution figures for candidate parts (Dulaney Toy Company)			
PART	VENDOR PRICE PER UNIT (includes adjustment)	INTERNAL VARIABLE COST (materials and labor)	CONTRIBUTION PER UNIT (if pulled back to factory)	CONTRIBUTION PER MONTH
Caboose undercarriage	11.97	4.57	7.40	$ 444.00
Caboose roof	9.60	6.70	2.90	174.00
Caboose railings	3.79	3.91	−0.12	−14.40
Crib gym bar	22.39	12.61	9.78	195.60
Crib gym flower	18.26	8.01	10.25	820.00

TABLE 9-3   Index of contribution per penny of labor cost (Dulaney Toy Company)

PART	CONTRIBUTION PER UNIT (if pulled back to factory)	LABOR COST PER UNIT	CONTRIBUTION PER PENNY OF LABOR COST
Caboose undercarriage	7.40	1.37	5.40
Caboose roof	2.90	3.93	0.74
Caboose railings	−0.12	3.03	−0.04
Crib gym bar	9.78	6.70	1.46
Crib gym flower	10.250	4.43	2.31

priority appears consistent with some of the other criteria mentioned for the make-buy decision. We must be sure, however, to include in the evaluation whether the crib gym flower, with the tooling required, will be pulled back for a long enough time to justify pulling it back.

A sensitivity analysis can help here. Let x be months. We want to calculate the number of months it would take before the crib gym flower would contribute as much as the crib gym bar, knowing that we would rather pull in the flower because its contribution per penny of labor cost is higher. Drawing from the contribution per month figures in Table 9-2,

Contribution from bar = contribution from flower
$$\$195.60x = \$820.00x - \$1500$$
$$x = 2.40 \text{ months}$$

If we plan to produce crib gym flowers for 2.4 months or more, we can expect to gain more contribution than by producing the crib gym bar, in addition to a higher contribution per penny of labor cost.

Perils of the Make–Buy Decision

For many companies, the make–buy decision is a tricky one, and one that many do not approach correctly. Too many companies try to compare the outside, quoted supplier price to the internal, fully burdened (that is, all fixed cost or period allocations assigned to it). The problem is that the fixed, or period, expenses are generally those that in the short term, at least, would still be incurred by the company. Thus, they should be removed from the make–buy price comparison so that the outside supplier price is compared only against the variable costs that would actually be incurred by the company if it decided to make the item rather than buy it. When companies do not make the proper comparison, the outside price, almost invariably, will be lower, and the company decides to outsource. If the company then engages in a series of such comparisons and decisions, more and more of its fixed charges will be borne by fewer units of its own production, and the tendency will be to shift more and more away to outside suppliers. This improper examination of costs thus tends to "hollow out" the company, pulling it away from the industrial competence that could have been its traditional strength. One must be diligent about such make–buy comparisons.

INVENTORY CONTROL

Once materials are in the factory, they need to be properly controlled. The factory needs to know how much it has of which parts and materials, and where in the factory they are. Accomplishing this requires discipline in receiving the raw materials and discipline in controlling them once received.

The functions of receiving and inspecting incoming materials and then storing them are sometimes overlooked. Nonetheless, they can be very important for the smooth functioning of the operation. Processing incoming orders is becoming increasingly automated and linked better to the rest of the process. Several issues surface when we examine these functions, and these issues are helpful in suggesting some of the alternatives available.

Receiving and Inspection

1. Did the supplier ship the proper items and in the quantities requested? Was the order received on time?

2. Is the receipt properly reported in the plant's materials system so that its status can be tracked accurately?

   For most plants, now, this information is computerized. Terminals in the receiving area log in this information and reconcile it with purchasing and production control information.

3. Are the incoming materials packaged in convenient ways for use by the factory floor? Less popular today are the old practices of shipping bulk quantities of material that a factory's receiving operation has to break down into more convenient quantities. Instead, shipments are not only being made in smaller quantities, as already noted, but also in containers and packaging that facilitates their use in the process.

4. What kind of inspection is appropriate for incoming materials?

   What orders require 100 percent inspection, where every piece is inspected and sorted into good or bad categories?

   What orders receive acceptance sampling, where a sample from every lot of material shipped is inspected and inferences are then drawn about the disposition of the entire lot?

   When can purchases receive no inspection at all on a routine basis; perhaps, only some occasional auditing done at random?

There has already been considerable discussion of quality matters in Chapter 4. Remember that although acceptance sampling is currently the most popular means of doing incoming inspection, doing no inspection at all may make more sense in some situations. Current trends are moving toward reduced inspection of incoming materials.

5. What is done with any rejected material?

The discipline with which plants return rejected material to suppliers varies considerably. Some plants are exacting and prompt with returns, and eager to help their suppliers learn what it takes to satisfy them. Others hang on to rejected material just in case it might turn out to be needed. Happily, the former attitude is prevailing in more and more companies.

1. How many storage areas are there, and is access to them controlled?

2. How automated are the storeroom's operations?

3. How is control kept?

Storage

Many manufacturing plants have a single, restricted access storeroom and a stringent routine for checking material in or out. Other plants have multiple store rooms. Traditionally, storage operations have used standard racks, bins, boxes, and pallets to store and to stage materials either for the factory floor or for shipment. Such systems are fairly labor-intensive in both materials handling (forklift truck operators, workers to break down bulk lots) and materials control functions. There are two conventional types of layouts for storerooms. The first keeps all the material of a particular kind together in a unique location in the storeroom. When an order has to be filled (either for the factory floor or for shipment), storeroom personnel travel the aisles and pick the required materials from their many locations. A second alternative saves labor by diminishing the travel from bin to bin; however, it requires more space and racks. This alternative adds some staging area storage so that some racks are used for bulk storage and others strictly as intermediate storage. Thus, when orders are filled, they are filled from the intermediate, staging area racks only. From time to time, as the material in the staging area dwindles, it is replenished from the bulk racks. The appropriateness of one alternative versus the other depends, naturally, on the balance of storage space and rack costs versus personnel costs. The higher a company's personnel costs are relative to space and rack costs, the more attractive is the intermediate, staging area layout. Conventional storage systems have some endemic problems, however. They often use space ineffectively, suffer from inaccurate counts of material, and misdirect materials.

There have been some recent developments that have improved on the traditional systems, however. Bar coding and automated storage and retrieval systems merit mention.

Bar coding involves the use of codes, like the universal product codes now found on grocery items, to identify particular lots of material. Storeroom workers can pass light scanning wands over these codes easily to ensure that accurate information on the status of material is placed into material management's computerized records. Bar codes are also used to direct boxes, trays, and pallets along a plant's (or distribution center's) conveyor system. Stations along the conveyor automatically read the bar coding and route the material to the proper area.

Automated storage and retrieval systems (AS/RS) use specially constructed aisles, stacks, and bins with computer controlled cranes to store and retrieve automatically what conventional systems do with operators riding forklift trucks or manually picking items. The AS/RS system's computer keeps track of where all the material is, when it was stored, and the remaining quantity. The computer can also route the crane's movement along the aisle and up and down the stacks most efficiently, so that the crane does not become a bottleneck. AS/RS systems can be very expensive, but they offer labor and space savings, increased accuracy, and the capability of always using the oldest material first.

There are, of course, some less-automated options that still improve on the conventional stacks. These semi-automated options may involve man-aboard cranes and routings that are not controlled by computer. They can still offer substantial improvements for the storeroom, however.

Control of What's in Stock

With the advent of computer systems, control over what is in the storeroom has become easier. The computer can easily track the transactions (in, out, inspected or not, returned, scrapped) entered by the storeroom's personnel.

What remains critical is the discipline of the storeroom's personnel and systems. Are discrepancies identified quickly and reconciled immediately and completely? If the personnel are permitted to be sloppy about handling transactions, the control of the storeroom will suffer. Although most plant auditors still require annual *physical inventories* to count parts (where production comes to a halt for a day or two while workers log what inventories are), more and more plants are maintaining control effectively without the hassles and lost production time involved with physical inventories. Rather, companies have been able to use *cycle counting* — the counting of one item, or family of items, at a time, at periodic intervals. Cycle counts often make use of the ABC classification scheme whereby high-valued items are cycle counted more regularly (say, monthly or quarterly) than lower-valued items (annually or semi-annually). Sometimes cycle counts are used prior to the launch of particular orders, so as to insure that the materials will be on hand for the manufacture of those orders.

The results of any physical inventories or cycle counts need to be entered into the computerized system that keeps track of all of the factory's part numbers. That is, the records kept of the parts in question are updated (reconciled) with the data from the factory floor. These records of what is on hand are crucial inputs into any MRP system that the factory may run.[a]

As just-in-time production philosophies and techniques proliferate, more and more factories are shrinking, and even eliminating completely, their former stockrooms. Instead, the parts and material received are directed out to the factory floor. There is where the materials have a "home," either on the line, or by a department, or perhaps in smaller, more open storerooms close by the production. The formal transactions involved in the movement of materials in and out of the often restricted-access stockrooms are going by the boards in favor of a system of more visual controls that

are monitored by the production workforce itself. Thus, the vast recordkeeping that existed in many factories is being transcended by a system where there are known numbers of standard containers in which materials are kept. Thus, there are only so many parts or materials quantities available, and they are kept in particular locations, easy to keep visually tracked. Discipline is required in this system; one has to report scrap religiously and monitor the various parts or materials stored. Nevertheless, such a system can be much more effective, because the production floor is very conscious of what its needs are and can easily monitor those needs and call for action when impending shortages show up.

ISSUES IN LOGISTICS[2]

Logistics is chiefly concerned with the movement of products to markets and of materials to manufacturing facilities. Entwined with the actual movement of material, of course, is its storage and handling. Logistics has been a frequently neglected aspect of operations, but one whose potential impact on costs and profits is greater than most realize, as evidenced by the figures in Table 9-4. Total distribution costs, as shown there, are roughly as high a percentage of sales as direct labor costs, and, for the most part, more amenable to cost cutting.

In managing logistics, several issues loom of central concern:

1. What should be the structure of the distribution system? Where should various inventories, especially finished goods inventories, be held within the system?

2. Which geographic locations make sense as stocking points?

3. How much inventory should be held at each location?

[a] When inventory records are updated at set intervals they are said to be "periodic" in nature. When records are adjusted at each addition or withdrawl from inventory, they are said to be "perpetual."

TABLE 9-4 Distribution costs as a percentage of sales dollar

	OUTBOUND TRANSPORTATION	INVENTORY CARRYING	WAREHOUSING	ADMINISTRATION	RECEIVING AND SHIPPING	PACKAGING	ORDER PROCESSING	TOTAL
All manufacturing companies	6.2	1.3	3.6	0.5	0.8	0.7	0.5	13.6
Chemicals and plastics	6.3	1.6	3.3	0.3	0.6	1.4	0.6	14.1
Food manufacturing	8.1	0.3	3.5	0.4	0.9	—	0.2	13.4
Electronics	3.2	2.5	3.2	1.2	0.9	1.1	1.2	13.3
Paper	5.8	0.1	4.6	0.2	0.3	—	0.2	11.2
Machinery and tools	4.5	1.0	2.0	0.5	0.5	1.0	0.5	10.0
Pharmaceutical	1.4	—	1.2	0.7	0.5	0.1	0.5	4.4
All other	6.8	1.0	2.9	1.2	1.4	0.4	0.4	14.1
All merchandising companies	7.4	10.3	4.2	1.2	0.6	1.2	0.7	25.6
Industrial goods	5.9	13.7	2.9	0.7	0.2	2.0	1.0	26.4
Consumer goods	8.1	8.5	4.0	1.3	0.9	0.9	0.5	24.2

SOURCE: Data from B. J. LaLonde and P. H. Zinszer, *Customer Service: Meaning and Measurement.* Chicago: National Council of Physical Distribution Management, 1976.

4. How should the flow of goods throughout the distribution system be coordinated? Which products should be treated in which ways?

5. Which modes of transportation should be used?

General Motors' service parts network, as mentioned in the service tour of Ogle–Tucker Buick, is a good example of these key issues in logistics. GM's distribution system, as described there, has three tiers to it: national distribution controlled from Flint; various regional distribution centers, such as Lansing; and various local distribution centers, such as Chicago. One can expect that the most commonly used parts were abundantly stocked in the local distribution centers, but that uncommon, rarely used parts are available routinely only from the regional or national distribution centers. As that service tour relates, there are several ways orders could be placed with the parts network, some calling for wider searches and faster deliveries than others.

To discuss these logistics issues in greater depth, they can be organized into three broad categories: (1) distribution system structure and operation, (2) distribution center location, and (3) transportation mode choice.

Distribution System Structure and Operation

In deciding how the distribution system should be structured, a critical decision turns on how centralized or decentralized the system ought to be. Should the company keep its inventory in just a few locations, or should it disperse its goods into a number of regional or metropolitan areas? This question can be a subtle one, and an answer to it depends on balancing the costs and benefits of centralization versus decentralization.

The Argument for Decentralization

The argument in favor of dispersing inventories in a number of locations is clear and compelling, namely that customer service can be greatly enhanced. The time it takes to fill an order can be cut markedly if sales and service people can call on a nearby facility. Fewer errors are likely to be made as well, because field people can be in close, informed touch with the distribution center. Sales may even be stimulated if customers react favorably to knowledge of a facility nearby dedicated to serving them.

Moreover, delivery time can be reduced, not only because the order may be filled more quickly, but also because there is less distance to travel and swift modes of transportation can be used. This contraction in delivery time means that safety stocks can be reduced because fewer days of uncertainty need to be covered.

The Argument Against Decentralization

Despite all the aid decentralized distribution can promise field sales and service people, it can be costly. Transportation and handling expenses typically increase with decentralization. Decentralization can also lead to an expansion of inventory to be carried and a loss of control over that inventory. These points against decentralization are argued in greater detail in the following discussions.

Transportation and Handling Expenses

How do the costs of transporting and handling a product differ between a centralized strategy, where the customer is served directly, and a decentralized strategy, where the customer is served through various tiers? The answer is not clear-cut, because there are costs that cut both ways. Naturally, the fewer the number of intervening locations, the lower the transportation expense, suggesting that centralization and direct customer shipments are more economical. On the other hand, shipment of small quantities is invariably more expensive than shipment in bulk, and often, the replenishment of decentralized distribution centers can be done cheaply through bulk shipments. This point argues for at least some decentralization. How these conflicting costs total up naturally depends on the situation—what the full truckload versus less-than-truckload rates are, where the customer is, where the distribution centers are, and how much the order is for.

Often, however, the telling expense is in the handling. If the product has to be handled from one tier of the distribution system to another, additional costs are incurred. These can swamp any cost savings attributable to freight consolidation.

Administering Decentralized Distribution

Inventory spread among a number of facilities is more difficult to keep track of than when it is in one or just a few facilities. Record keeping is more voluminous and thus subject to error. Swaps between facilities may be so informal that records are not even entered. Moreover, the sales and service people may become so possessive of "their" facility that a hoarding mentality takes over, especially with items that are scarce and that a central authority may want to call in for allocation somewhere else. Such items may purposely become "lost" in the distribution system.

Irregular Demand and Buffer Stocks: The Law of Large Numbers

The demand in any one region of a company's market area is likely to follow a more irregular pattern than the company's overall demand. Greater-than-average demand for a particular product in one region may well be balanced by below-average demand for the same product in another region. Thus, the buffer stock that a centralized facility would have to carry is likely to be considerably less than the sum of the buffer stocks of decentralized facilities, for the same level of protection against stockouts. This result is a variation of the "law of large numbers." The variance of demand for a single, large service center will almost always be proportionally less than the variance of demand faced by smaller, dispersed service centers offering the same coverage against stockouts.

Scattered Warehousing and the Optimal Inventory Calculation

The logic of our EOQ calculation can be applied to the case of many, scattered warehouses versus a single, large warehouse. Much as we derived earlier, let

D = expected period demand over the entire company
i = inventory carrying cost percentage
S = cost per order
C = variable cost per unit of inventory
N = number of scattered warehouses, all of the same size

From our calculations before for the simple case where carrying cost and order cost were the only costs that varied with changes in the order quantity, the optimal order quantity, Q, was found to be

$$(1) \qquad Q = \sqrt{\frac{2SD}{iC}}$$

Now let us consider a small warehouse, one of N identical warehouses that the company operates. The demand that each small warehouse will face during the period is D/N. We can substitute this into our formula to find the optimal order quantity, Q^*, for the small warehouse.

$$(2) \qquad Q^* = \sqrt{\frac{2S\dfrac{D}{N}}{iC}}$$

But there are N small warehouses and so the total of all the order quantities would be

$$(3) \quad NQ^* = N\sqrt{\frac{2S(D/N)}{iC}} = \sqrt{N}\ \sqrt{\frac{2SD}{iC}}$$

We can substitute equation 1 in to find a relationship between Q^* and Q, namely,

$$(4) \quad NQ^* = \sqrt{N}\ \sqrt{\frac{2SD}{iC}} = \sqrt{NQ}$$

Thus, merely by dispersing inventory in N facilities and allowing each to optimize its order quantities under a reorder point inventory system, we find that the order quantities demanded of the factory—and by implication inventories since in this case average inventory is $Q/2$ or $Q^*/2$—increase by the square root of the number of warehouses. If we establish four small, identical warehouses instead of one, and if we follow a reorder point system for each one, then inventories will be double what they would be if we established a single warehouse.

This argument is meant to be more suggestive than exact. The regional warehouses of most companies are not organized so that individual reorder point systems prevail. But they are organized differently in many cases precisely to avoid the heavy inventory burden that the line of reasoning sketched above leads to.

Tiered Inventories and Pipeline Momentum

Many distribution systems are tiered. They have a centralized inventory that sits above a number of regional inventories; these, in turn, sit above local, metropolitan inventories. In some of these systems, the central inventory feeds products to the regional inventories, and the regional inventories feed the local ones. This kind of division can work quite well. However, if the decision rules for holding inventory are disjointed rather than coordinated, such a tiered system can lead to substantial swings in demand and somewhat higher levels of inventories than a completely coordinated system.

To illustrate what is meant, think of an inventory system of three tiers and each tier is mandated to hold 2 weeks' worth of demand. For a weekly demand of 20, then, the corresponding levels of inventory and orders on the next tier would look like Table 9-5. Note that in each tier the desired inventory is twice the weekly sales, and orders on the next tier proceed accordingly. Observe, however, what happens when demand falls by a single unit (5 percent) at the level of tier A (the local level) and stabilizes there (at 19). The period-by-period adjustments implied by the myopic decision rule of "hold two weeks' worth of sales in inventory" are traced in Table 9-6.

As the table shows, this tiered inventory system with its myopic replenishment rule takes four periods to settle into a steady state. What is more, the orders placed on successive tiers become more and more erratic and vary wildly from period to period. When the initial demand reduction occurs, it is covered by the inventory already in the pipeline and no orders are placed on the factory. However, the pipeline's inventory is soon depleted and the next period's factory order is huge. This is an example of the "pipeline momentum" that can afflict a tiered inventory system or tiers of suppliers in a multilevel supply network. Not only does it cause great variability in period-by-period demands on each tier, but it can sometimes cause more inventory to be held, over time, than necessary.

Avoiding the whiplash of this pipeline momentum effect calls either for more coordination among the tiers (so that tier C can adjust immediately to the problems tier A is experiencing) or more foresight in the lower tiers (so that these tiers can anticipate the erratic demands that will be placed on them). In either case, avoiding the pipeline momentum effect means junking the myopia that is implied by a seemingly innocuous decision rule like "hold 2 weeks' worth of sales in inventory."

TABLE 9-5   Tiered inventory system and disjointed decision rules—the steady state

TIER	ITEM	AMOUNT
A (Local)*	Beginning inventory	40
	Sales in week	20
	Ending inventory	20
	Desired inventory	40[†]
	Order on the next tier	20
B (Regional)	Beginnning inventory	40
	Sales in week	20
	Ending inventory	20
	Desired inventory	40[†]
	Order on the next tier	20
C (Companywide)	Beginning inventory	40
	Sales in week	20
	Ending inventory	20
	Desired inventory	40[†]
	Order on factory or to be absorbed by buffer stocks	20

*Weekly sales demand at level of tier A is 20.
[†]Twice the demand of 20.

TABLE 9-6   Tiered inventory system and disjointed decision rules—reaction over time to 5 percent demand reduction

TIER	ITEM	PERIOD 0	PERIOD 1	PERIOD 2	PERIOD 3	PERIOD 4
A	Beginning inventory	40	38	38	38	38
	Sales	20	19	19	19	19
	Ending inventory	20	21	19	19	19
	Desired inventory	40	38	38	38	38
	Order on the next tier	20	17	19	19	19
B	Beginning inventory	40	40	34	38	38
	Sales	20	17	19	19	19
	Ending inventory	20	23	15	19	19
	Desired inventory	40	34	38	38	38
	Order on the next tier	20	11	23	19	19
C	Beginning inventory	40	40	29	46	38
	Sales	20	11	23	19	19
	Ending inventory	20	29	6	27	19
	Desired inventory	40	22	46	38	38
	Order on factory or to be absorbed by buffer stocks	20	0	40	11	19

In any event, this pipeline momentum effect can be one more argument against a decentralization of distribution centers.

Distribution System Operation

No matter how the network of distribution centers is configured, there are several ways to link the operations of the distribution centers to customer demand. Three major techniques are used to operate distribution systems, all variations of the familiar reorder point, periodic reorder, and MRP techniques already discussed. Let us examine their application to distribution systems.

1. *Reorder point.* Using a reorder point technique is a straightforward, very decentralized way to operate a distribution system. Each distribution center has its established reorder point levels for the products it handles. When they are replenished by the next level up in the distribution network, which could be another distribution center or a factory, the replenishment is an economic order quantity's worth, typically using either past demand or a future projection as its basis.

This technique is decentralized because the inventory records are maintained strictly at the distribution centers and not by some central authority. Often, the setting of reorder points and economic order quantities is left to the individual distribution centers. This decentralization of information and decision making tends to bloat the level of inventory in the system and to make any allocations of temporarily scarce inventories difficult to achieve. It is sometimes said that the reorder point technique "pulls" inventory out into the field.

2. *Base stock.* The base stock technique is essentially a periodic reorder technique. With this technique, the distribution center is replenished on a regular, periodic basis. The various inventories held are brought up to levels that are expected to hold up, with a given level of customer service, until the next replenishment arrives. The replenishment is thus of variable quantity.

The base stock technique is a more centralized one because the resupply point requires information enough to assemble periodically what each distribution center needs to tide itself over until the next replenishment. The information processing required of the base stock technique is thus greater than that of the reorder point technique, but the field inventories are likely to be lower and better coordinated. This technique "pushes" inventory out into the field, rather than "pulls" it.[b]

It is interesting to note that the typical just-in-time manufacturing relationships with suppliers call for frequent, periodic replenishments of inventory to bring the factory inventory back up to some base stock level (the level dictated by the number and size of the standard containers used in the process).

3. *Distribution requirements planning* (DRP). DRP is analogous to MRP in that the levels of inventory in the field are tied to projected, time-phased demand. The projected demand works like the gross requirements line of an MRP inventory status record, and the conventional MRP logic can be applied to determine how much inventory should be supplied to each distribution center and when it should be shipped. This approach is often also termed the "time-phased order point."

[b] Note that the push or pull associated with distribution networks is different from the push and pull associated with production control within a factory. The latter is discussed first in the next chapter and then again in Chapter 11 dealing with just-in-time manufacturing.

Much as with MRP, various lot sizing techniques can be applied, often to group differently timed needs together to ship replenishments economically. And, like MRP, rescheduling can be accomplished easily if forecasts or actual demands change.

Of course, DRP is more information-intensive than either the reorder point or base stock techniques. It can also be linked to MRP in the following way: The demands forecast for individual distribution centers can be summed and used as a basis for the development of a master production schedule, the schedule that drives an MRP system.

Distribution Center Location

The location decision for distribution centers is dominated by consideration of transportation costs, and this makes it a much easier decision than plant location, which is dealt with in Chapter 12. The distribution center decision lends itself to some mathematical analysis, and numerous studies for companies have been done using mathematical techniques to zero in on low-cost locations.

The heart of the decision is finding the location (or set of locations) that minimizes the cost of transporting goods to customers. Doing this analysis requires information on the quantities demanded by all customers, the quantities that can be supplied from each distribution center, where those customers are located, and the freight rates applicable from the contemplated locations to each customer. Different locations will offer different distances and hence transportation costs to customers, and a systematic way to examine these costs is called for. Simulations can be performed and are often most appropriate for complex, many-factor decisions. For simpler cases, however, linear programming or one of its offshoots is often the most effective tool.

Variations on the simple transportation problem permit the distribution center to be subject to fixed as well as variable costs. Consult a book on operations research topics for a discussion of the sophisticated methods employed to solve these more true-to-life problems.

It should be stressed, however, that for most companies the solution given by these mathematical techniques is frequently modified by some qualitative considerations that could not be incorporated into the formulation.

Transportation Mode Choice

Transportation mode choice tends to be a very specialized field of inquiry. Mode choice decisions require reams of very detailed information on transport costs and delivery times by mode for different classes of shipments characterized by weight, volume, and specialized shipping requirements (such as refrigerated trucks, open flatcars, pressurized compartments). Less than truckload or less than carload rates tend to be very different from full truckload or full carload rates. Mode choice is further complicated by a proliferation of rates. A company's traffic department cannot merely think of truck, rail, water, or air but must think in terms of common truck carrier, owner-operator trucks, own fleet trucking; regular rail service, piggyback (truck-rail) service; ocean, lake, or river barges; commercial jet shipping or private jet shipping; and a number of other variations. Rates and deliveries vary, as does the impact of federal and state regulation, which can for no good reason dictate substantially different shipping rates from seemingly equivalent shipping points or different rates between two points depending on which direction shipment is made.

Other than to note some of the many peculiarities that afflict the nation's transportation system and thus the mode choice decisions of companies, this text will leave to more specialized and detailed treatments this issue of transport mode choice.[3]

ORGANIZING THE MATERIALS FUNCTION

This chapter began by calling attention to the three major divisions of materials management: (1) procurement, (2) materials control (which includes both inventory control and production planning and control), and (3) logistics. In traditionally organized companies, these three functions often report to three different department heads—procurement to a general manager, materials control to a manufacturing manager, and logistics to a marketing manager. In the last decade, however, considerable organizational restructuring of the materials function has occurred, elevating the importance and visibility of the "materials manager" within the corporate organization.

Four organizational variants have emerged in many companies, and they seem to be about equally popular. Borrowing from Miller and Gilmour, these variants follow the organization charts depicted in Figure 9-1 and are labeled, respectively:[4]

- *Integrated approach.* All materials functions report to a single materials manager.
- *Supply-oriented approach.* The materials manager assumes authority over procurement and materials control, leaving logistics to report elsewhere.
- *Distribution-oriented approach.* The materials manager assumes authority over materials control and logistics, leaving procurement to report elsewhere.
- *Manufacturing-oriented approach.* The materials manager assumes authority over the

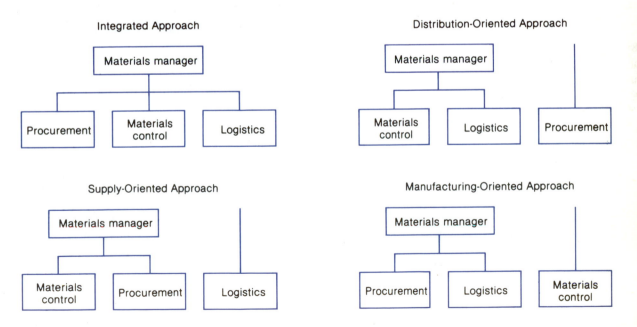

FIGURE 9-1 Four possible organization charts for materials management.

procurement and logistics, leaving materials control to report directly to a manufacturing manager.

The choice of organization appears very much to depend on the types of materials problems that the company is experiencing. If the problems straddle the traditional boundaries of procurement, materials control, and logistics, then it is much more likely that a materials manager will be assigned to oversee the functions in question. For example, if the company is having difficulty deciding how to balance inventory carrying costs against volume discounts from purchasing in quantity, a materials manager who oversees both procurement and materials control may help to resolve the issue. Similar types of problem overlaps across the traditional functional lines explain many of the recent materials management reorganizations that have been sweeping across the industrial landscape.

Tools of the Trade:
A Recap of Lessons Learned

World-Class Practice. As summarized earlier in the discussion of world-class suppliers and world-class customers, procurement has, in recent years, changed dramatically. The move to just-in-time manufacturing (Chapter 11) has forced a major rethinking about procurement. Gone is the old mentality about multiple vendors kept at arm's length, competing almost exclusively on price, and shipping nothing without a formal purchase order. In is a new mentality that stresses cooperation, long-term, single-supplier relationships, quality rather than price, frequent deliveries triggered by factory need, and early involvement in new product planning and development. The new mentality is altering everything that went on in the old "purchasing department."

Other changes are evident as well in the control of in-coming materials and their storage and handling. Gone are full-scale physical inventories; these are replaced by cycle counting and much more discipline in the handling of materials.

The classic make–buy decision still remains, although many companies are becoming smarter about how they apply costs to that comparison. Clearly, the internal comparison should not involve fully burdened (all overhead assigned) costs, but only the incremental and variable costs that would truly change if the product were brought in-house from outside (or vice versa).

Logistics has also changed in recent years, in part because of changes in transportation regulation and in part because of changes in how inventories are managed. Central to the logistics function is the decision on where to carry field inventories, centrally or dispersed regionally. The argument is one that pits customer service considerations against the costs incurred by tiers of inventory in the field. Centralized information systems, fortunately, help out in any case. Thus the tendency has been to go for more centralized warehousing and distribution than has been the case in years past.

These changes are also affecting the organization of the firm's materials function. More integration of all facets of the materials management of the company is becoming prevalent.

QUESTIONS

1. How does the situation at Dulaney Toy Company illustrate the main principles of procurement and the principles of a make–buy decision?

2. What is the rationale behind the ABC classification system?

3. How are purchasing agents' jobs changing?

4. What are some of the advantages and disadvantages of a decentralized distribution strategy? How would you quantify them?

5. What are the most popular variants for organizing the materials function?

PROBLEMS

1. Under what conditions and process types would you make certain that your director of purchasing is one of the two or three best talents you have on your staff? Under what conditions do you not care so fervently?

2. Joe Little had been assigned the task of designing and acquiring 12 engine oil containers for the Major Motors Corporation. The containers were to have a capacity of 110 gallons plus; they had to be able to be moved by forklift trucks and to be stored on existing racks. The containers were needed to make the handling of the oil easier and the engine oil-fill operation more efficient. A preliminary analysis indicated that the estimated cost savings would generate a payback within 6 months at most.

Joe's design required only basic "metal bending" and welding processes, and these processes were part of the plant capacity of Major Motors. Joe received bids from three external shops and summarized them as shown at the top of the second column.

Joe's next step was to make an analysis of having the maintenance and machine shop departments fabricate the containers in-house. The resulting figures are on the right.

His analysis also indicated that the maintenance workers could be allowed to work on the containers only in their spare time because their primary function was to keep the plant

VENDOR	COST PER CONTAINIER	DELIVERY
1	$565	In 6 months
2	578	Maybe in 2 months, but in 4 months for sure
3	535	In 1 month, but this vendor had a reputation of not meeting delivery dates and of taking shortcuts

Material	$294
Labor at $7 per hour wages + $7 per hour fringes = $14 per hour to shear, bend, assemble, and weld; 3.5 hours for maintenance; and 6.5 hours for machine shop	140
Factory overhead at 2.5 x direct labor = $17.50 per hour 175	175
Total costs	$609

operating. The machine shop had a backlog of orders, but the necessary work could be squeezed in within the next month.

Analyze the various alternatives for Joe and discuss the various implications of the non-quantified information.

3. Woodfield Sound Systems fabricated various specialty products for the professional sound reproduction market. In the current economic downturn, the company, which practiced a policy of no layoffs, was seeking to bring in some parts that usually were made by its vendors. The company's current policy was to compare (1) the vendor's price adjusted upward by the freight, order, and inspection costs Woodfield incurred itself in procuring the part, with (2) the plant's fully burdened costs. Some representative parts' data follows:

PART	VENDOR PRICE INCLUDING UPWARD ADJUSTMENT
B	$ 24.64
D	$ 49.50
E	$140.50
F	$ 1.87
G	$ 10.72

INTERNAL FACTORY COST			
MATERIALS	LABOR	OVERHEAD	TOTAL
$ 9.56	$ 2.75	$10.44	$ 22.75
$24.95	12.66	48.58	86.19
$28.65	17.90	68.03	114.58
$ 1.28	0.20	0.65	2.13
$ 5.52	1.02	3.88	10.42

The plant needed to decide which parts to bring in-house, and in which order.

4. Refer to Problem 3. How would your answer change if you knew that economic conditions were such that the material Woodfield Sound Systems relied on was being rationed, effectively constraining the quantity of what the company could produce? Which parts would you then bring in-house, and in which order?

5. Show that the buffer stock needed by a centralized warehouse is likely to be less than the sum of buffer stocks needed by decentralized warehouses. Assume independent demand in two geographic areas described by the following:

UNITS DEMANDED DURING PERIOD	DEMAND PROBABILITY
0	0.2
1	0.6
2	0.2

What would be the expected loss in sales if only 2 units were held in inventory, either as 2 units in a centralized warehouse or 1 unit each in the two decentralized warehouses?

REFERENCE NOTES

1.   For additional information on procurement consult the following:
Bailey, P. J. H., *Purchasing and Supply Management*, 5th ed. Methuen, N.Y.: Chapman and Hall, 1987.
Burt, David R. "Managing Product Quality through Strategic Purchasing," *Sloan Management Review* 30, no. 3 (Spring 1989): 39–48.
_____, "Managing Suppliers Up to Speed," *Harvard Business Review* 67, no. 4 (July–August 1989): 127–35.
Dumond, Ellen, and William Newman, "Closing the Gap Between Buyer and Vendor," *Production & Inventory Management* 31, no. 4 (Fourth Quarter 1990): 13–17.
Freeland, James R., "A Survey of Just-in-Time Purchasing Practices in the United States," *Production & Inventory Management Journal* 32, no. 2 (Second Quarter 1991): 43–50.
Heinritz, Stuart F., Paul V. Farrell, and Clifton L. Smith, *Purchasing: Principles and Applications*, 7th ed. Englewood Cliffs, N. J.: Prentice-Hall, 1986.
Ho, Chrwan-jyh, and Phillip L. Carter, "Using Vendor Capacity Planning in Supplier Evaluation," *Journal of Purchasing & Materials Management* 24, no. 1 (Spring 1988): 23–30.
Lee, Sang M., and A. Ansari, "Comparative Analysis of Japanese Just-in-Time Purchasing and Traditional U.S. Purchasing Systems," *International Journal of Operations and Production Management* 5, no. 4 (1985): 5–14.

Leenders, Michiel R., and David L. Blenkhorn, *Reverse Marketing: The New Buyer-Supplier Relationship*, New York: Free Press, 1988.

Leenders, Michiel R., Harold E. Fearon, and Wilbur B. England, *Purchasing and Materials Management*, 8th ed. Homewood, IL: Irwin, 1985.

McMillan, John, "Managing Suppliers: Incentive Systems in Japanese and the United States Industry," *California Management Review* 32, no. 4 (Summer 1990): 38–55.

Newman, Richard G., "Insuring Quality: Purchasing's Role," *Journal of Purchasing & Materials Management* 24, no. 3 (Fall 1988): 14–21.

Newman, Richard G., "The Buyer–Supplier Relationship Under Just-In-Time," *Production & Inventory Management* 29, no. 3 (Third Quarter 1988): 45–50.

Pettit, R. E., "Vendor Evaluation Made Simple," *Quality Progress* Vol. 17 (March 1984): 19–22.

2.   For more on logistics consult:
Ballou, Ronald H., *Basic Business Logistics: Transportation, Materials Management, Physical Distribution*, 2nd ed. Physical Distribution Series, Englewood Cliffs, NJ: Prentice-Hall, 1987.

_____, *Business Logistics Management Planning and Control*, 2nd ed. Englewood Cliffs, NJ: Prentice-Hall, 1985.

Magee, John F., William C. Copacino, and Donald B. Rosenfield, *Modern Logistics Management and Physical Distribution* (Wiley Series on Marketing and Management), New York: Wiley, 1985.

Robeson, James F., and Robert G. House, eds., *The Distribution Handbook*, New York: Free Press, 1985.

Shapiro, Roy D., and James L. Heskett, *Logistics Strategy: Cases and Concepts*, St. Paul, Minnesota: West, 1985.

Stock, James R., and Douglas M. Lambert, *Strategic Logistics Management*, 2nd ed. Homewood, IL: Irwin, 1987.

Surdell, Gregg J., "Capacity Planning Applies to Warehouses Too!" *Production & Inventory Management and APICS News* 11, no. 2 (February 1991): 42–47.

Van Rijn, C. F. H., ed., *Logistics: Where Ends Have to Meet*, Oxford, England: Pergamon Press, 1989.

3.   An issue related to mode choice is the determination of when deliveries ought to be expedited, perhaps by using a faster, more expensive mode of transportation. See D. P. Herron, "Managing Physical Distribution for Profit," *Harvard Business Review* 57, no. 3 (May–June 1979): 121–32.

4.   Miller, Jeffrey G. and Peter Gilmour, "The Emerging Materials Manager," working paper 79–17 of the Graduate School of Business Administration, Harvard University (January 1979).
_____ "Materials Managers: Who Needs Them?" *Harvard Business Review* 57, no. 4 (July–August 1979): 143–53.

CHAPTER 10

THE QUESTION OF CONTROL

Prior chapters have addressed the planning of production (how much of what to make, when) and the gathering of enough resources (manpower, materials) to make it happen. Planning is an essential aspect of operations management, but planning, by itself, is not enough. One has to execute the plan. That is where "controls" come in. One needs controls (often termed production controls) in place to assure that the product is made (or the service delivered) in accordance with the plan.

Controls come in all varieties. We have already in this text discussed quality controls, the set of procedures that help to assure the product is made according to the specifications. These controls included such things as control charts, inspection reports, and checklists. Inventory control was discussed in the previous chapter. Still, there are many more types of controls: controls on the schedule (delivery), and controls on materials (including cash controls).[a] These controls are the province of this chapter.

The chapter is divided into three parts. The first part deals with controls for manufacturing, where actions are repetitive. This topic is commonly known as shop floor control. The second part addresses controls when actions are unique, such as occurs with projects, a topic known as project management. The last part deals with service operations, and some of the unique aspects of control that they present.

---

[a] Cost controls are still another kind of control. This text does not address itself to cost controls specifically, although systems that track orders and labor and materials usage for an order (See Tour B and the discussion of job shops) are systems that provide the wherewithal to track costs and thus be tagged as cost controls.

PART ONE

SHOP FLOOR CONTROL[b]

More than any other aspect of manufacturing management, shop floor control (also known as production control) gets misunderstood. It means different things to different people. In reality, it means many things because many things are required to execute a manufacturing plan well.

A manufacturing plan devises a series of orders for the factory floor. Exactly how these orders get carried out, however, is the domain of shop floor control. The activities that constitute shop floor control can be broken out into five headings:

- Order review/release
- Detailed assignment of the order
- Data collection/monitoring
- Feedback/corrective action
- Order disposition

Let us examine these headings in more detail.

Order review/release. The orders devised by the plan must be reviewed for feasibility before they can be "released" to the factory floor with the expectation that the workers will work on and complete them by the time desired. Such a review demands attention to:

- *Order documentation.* Can the order be tracked effectively? Does it have a number? Does it have a routing? Are any necessary production standards with it (time or cost)? Are the materials specified? Is any or all tooling specified?
- *Material checking.* Are the materials required available?
- *Capacity evaluation.* Is enough capacity available?

Most of this order review and release can be done by reference to a computer that keeps track of orders, routings, standards, inventories, and past order status, among other things. Modules of the typical computerized MRP system attend to these matters.

Detailed assignment of the order. Typically, the manufacturing order itself does not specify exactly when or where on the factory floor the product is to be made. It leaves those decisions to shop floor control. A detailed assignment states the precise when and where for each order. A system to do this formally requires that the detailed assignment indicate: (1) order sequencing, (2) any scheduled maintenance, and (3) any scheduled down time. Order sequencing, also commonly called dispatching, specifies which machines are to take on the order and in which sequence. If a machine is not scheduled for production, then it may be scheduled for maintenance or for forced idleness.

There are a number of ways such dispatching may occur and they are the subject of a subsequent section.

Data collection/monitoring. Orders, once on the factory floor, need to be tracked so that management knows where they are, what has been done to them, and thus how much is left

[b] This section owes much to the fine book by Melnyk, Carter, Dilts, and Lyth, *Shop Floor Control*, Dow Jones-Irwin, Homewood, Illinois, 1985.

to do. How such orders can be monitored is discussed further in a subsequent section.

Feedback/corrective action.   The data on the status of the order can be compared with either expectations or desires and, if need be, corrective action can be taken. These actions vary. They can include: scrapping the order or some part of it, or changing the pace of work on the order, usually to speed it up. This latter action may involve overtime, subcontracting, rerouting, or order splitting (i.e., part of the order is allowed to go ahead of another part). Such actions may require the expediting of materials as well, changes in the quantity ordered, and/or changes in the date due.

Order disposition.   After the work is completed, the order needs to be disposed of, as either good work, work to be reworked, or work to be scrapped. It is here, too, that the necessary data are collected for determining how well the factory has performed (labor hours, material yields, quality, time, tooling).

These are the tasks of shop floor control. Several are elaborated on later.

All of these functions, of course, are related. One cannot devise an appropriate dispatch list without monitoring the schedule, and one cannot expedite orders without constant attention to the dispatch list. Nevertheless, it is instructive to focus on these individual functions. Before we do, however, it is helpful to be grounded again on the information requirements of various processes.

What is sometimes forgotten in the rush to satisfy orders and bird-dog materials is that some of the wealth of information that can help guide and control the process relates to the workforce. Workers need to know several things before they can plunge comfortably into doing their work: what they should work on, how they should do it, how long it should

take, and where their output ought to be directed. Sometimes this information must be transmitted to the workers repeatedly; other times they already know what is required of them. Table 10-1 summarizes the worker information requirements that are typical of various processes.

MORE ON DISPATCHING

The specific, sequential assignment of jobs to workcenters—the dispatching task—is, quite naturally, more of a problem for the job shop or batch flow operation. Line flow and continuous flow processes are designed to handle some product variations, but those variations have to be carefully planned into the line balance of the process itself. The freedom to schedule almost anything through the job shop or batch flow factory makes its dispatching task a much more difficult one.

Scheduling Techniques for the Job Shop or Batch Flow Operation

By all accounts the most difficult scheduling task is to schedule a job shop, be it a machine shop, a print shop, a plastics fabricator, or whatever. Fine minds have spent years trying to devise priority rules and scheduling procedures that job shops could successfully adopt.

The job shop and, to a lesser extent, the batch flow process are devilish to schedule because the workcenters or departments within them are so disjointed and independent and because the products manufactured there place such diverse demands on machine and/or worker capacity. What determines how quickly a job can proceed through a job shop depends fundamentally on the priorities assigned to it. Jobs with rush priorities can sail through the shop, but at the risk of stalling those jobs with

TABLE 10-1   Worker information requirements in different processes

PROCESS	WHAT SHOULD BE WORKED ON	HOW IT SHOULD BE DONE	HOW LONG IT SHOULD TAKE	WHERE PRODUCT IS TO BE DIRECTED NEXT
Job shop	Told by foreman	Written on process sheet attached to order	Written on job sheet attached to order	Written on a routing slip attached to order
Batch flow	Told by foreman or implied by a priority numbering scheme for batch	Either known from training or written on process sheet attached to batch	Implied by piece rate or written on process sheet accompanying batch	Written on a routing slip
Line flow	Determined by sequencing of line	Basically known from training, though options may be indicated on a process sheet	Implied by known worker standards or mechanical pacing of line	Known from the layout of the line
Continuous	Determined by process layout	Known from training; few options available indicated by independent information from managment	Determined by process layout	Determined by process layout

low priorities. Thus, scheduling a job shop is very much an exercise in sequencing the many jobs awaiting work by the various workcenters. This task of sequencing particular jobs is sometimes termed dispatching and, depending on one's scheme of things, may be classified as one element of what is called shop floor control.

Dozens of priority setting rules for determining job sequence have been tested over the years, both in actual shops and in computer simulations.[1] Consider the following additional priority rules:

- Work first on the job with the closest due date.
- Work first on the oldest job (first come, first served).

- Work first on the job with the shortest processing time to completion.
- Work first on the job with the longest processing time to completion.
- Work first on the job with the most successor operations.
- Work first on the job with the least "slack," that is, the least difference between the time remaining until the job is due and the time it is expected to take to do the job.

The shortest processing time has a desirable characteristic. In the simple case of scheduling jobs through a single machine or work center, the shortest processing time rule gives the lowest average wait across all jobs. This stands to reason, of course. When the small jobs go first, they avoid the wait that would be imposed by having to follow the lengthy, "big" jobs, and thus the average wait time for all of the jobs taken together is lowered. However, if more than one machine or work center has to be scheduled, this desirable characteristic, unfortunately, no longer necessarily holds.

These rules could be termed "pure," or simple, priority rules. There are other "mixed" or compound, rules that have been considered as well:

- Work first on the job with the lowest critical ratio as defined by

$$\text{Critical ratio} = \frac{\text{due date} - \text{current date}}{\substack{\text{standard processing time} \\ \text{plus standard queue time} \\ \text{remaining}}}$$

(The rationale is that jobs with low critical ratios are most likely to be late and thus deserve the most attention.)

- When there is slack in the shop, work first on the job with the closest due date; but when there is no slack in the shop, work first on the job with the shortest processing time to completion.

What is apparent from all the testing done on these various rules is simply the intuitively appealing notion that no one priority setting rule—simple or compound—can meet all order due dates and simultaneously spend the least amount of shop time working on those orders.

To see some of the difficulties inherent in job shop scheduling by using priority rules, consider Situation 10-1.

Discussion of Morris Machine Shop

Tracing the implications of each priority rule is easy enough using a chart of operations by hours and filling in the boxes with the appropriate job number. For the traditional, closest

due-date priority rule, the priority sequence is jobs 1, 5, 3, 2, 4. As shown in Figure 10-1, under this rule both jobs 2 and 3 take more than eight hours to perform, with job 2 not finishing until the thirteenth hour. Job 2's lathe setup was broken by job 1's higher priority, as was job 4's grinding setup. The machine utilization implied by this priority rule is not good.

For the priority rule of "first come, first served," the sequence of jobs is 1, 2, 3, 4, 5. As shown in Figure 10-2, the performance of this priority rule is as bad, or worse, than the closest due-date rule. Three of the five jobs exceed 8 hours, and the grinding setup for job 4 is interrupted by job 1.

For the priority rule of "work on the job with the shortest processing time to completion," the sequence of jobs is 4, 1, 5, 2, 3, with jobs 1 and 5 both having 6 hours of processing time required. As shown in Figure 10-3, two jobs are slow, with job 3 completed only in hour 13. Several setups are broken as well.

The final priority rule to be tried, "work first on the job with the longest processing time to completion," yields a schedule of jobs 3, 2, 1, 5, 4. Its performance is shown in Figure 10-4. Of the priority rules for this test of five jobs, this one performs the best. Only one job is late. Still, it is not a perfect schedule, and in another test of five jobs, its performance could be as poor as the others tested here.

Approaching the Job Shop Scheduling Problem from a Different Angle

As we have seen, traditional priority rules for scheduling job shops do not seem to work. No one rule seems to dominate any other. Most managers, when confronted with the task of scheduling jobs for expensive machine tools, try to make sure that those machines are fully loaded with work. This typically means that these managers will tolerate substantial

---

SITUATION 10–1

Morris Machine Shop

The Morris Machine Shop in Bristol, Connecticut, specialized in machining brass parts as a subcontractor to larger firms in the brass industry of western Connecticut. David Lee, the shop's foreman, had always tried to schedule jobs by assigning highest priorities to those jobs with the closest due dates. Glenn Morris, the shop's owner and founder had recently visited some other shops that used different priority rules. Curious, Glenn asked David to work out the implications of three other priority rules on the last five jobs that had been released to the shop: (1) first come, first served; (2) work first on the job with the shortest processing time to com-

pletion; and (3) work first on the job with the longest processing time to completion.

David was curious himself. The sequences and times for the last five jobs released were as shown in Table 10-2. The total time for each machine (drill, lathe, grinder, and mill) was 8 hours. A perfect schedule would use all four machines and finish all five jobs in 8 hours. The jobs' due dates were such that job 1 had the closest due date followed in order by jobs 5, 3, 2, and 4. David wondered which priority rule would work best and whether any of them would yield an absolutely perfect schedule.

TABLE 10-2   Sequence and times (in hours) for last five jobs
(Morris Machine Shop)

	JOB 1		JOB 2		JOB 3		JOB 4		JOB 5	
	Drill	1	Lathe	2	Grind	2	Drill	1	Mill	2
	Lathe	2	Mill	2	Drill	2	Grind	2	Drill	2
	Grind	1	Grind	1	Mill	2	Lathe	2	Lathe	1
	Mill	2	Drill	2	Grind	1	Grind	1	Lathe	1
Total		6		7		8		5		6

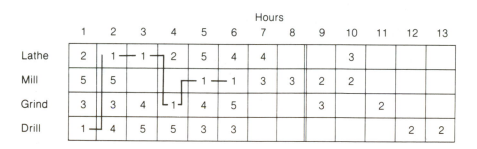

Figure 10-1   Chart of tasks versus hours—closest due date. (1, 5, 3, 2, 4)
Path for job 1 is traced.

Hours

	1	2	3	4	5	6	7	8	9	10	11	12
Lathe	2	1	1	2		4	4	5				3
Mill	5	5			1	1	2	2	3	3		
Grind	3	3	4	1	4				2	5	3	
Drill	1	4	3	3	5	5				2	2	

Figure 10-2   Chart of tasks versus hours—first come, first served.
(1, 2, 3, 4, 5)

Hours

	1	2	3	4	5	6	7	8	9	10	11	12	13
Lathe	2	2	1	4	4	1	5						3
Mill	5	5	2	2				1	1	3	3		
Grind	3	4	4	3	2		1	5				3	
Drill	4	1	5	5	3	2	2	3					

Figure 10-3   Chart of tasks versus hours—shortest processing time.
(4, 1, 5, 2, 3)

Hours

	1	2	3	4	5	6	7	8	9	10	11	12
Lathe	2	2	1	1	4	4		3	5			
Mill	5	5	2	2	3	3	1	1				
Grind	3	3	4	4	2	1	3			5		
Drill	1	4	3	3	5	2	2	5				

Figure 10-4   Chart of tasks versus hours—longest processing time.
(3, 2, 1, 5, 4)

queues of jobs awaiting work by those machines. In many job shops, 95 percent of a job's time on the shop floor is spent in one queue or another; only 5 percent of the time is it actually being worked on.

This fact has led some observers to concentrate on reducing queues. The potential advantages of reduced job queues are seductive. For one thing, massive queues represent substantial investments in work-in-process inventory.

Reduce queues and capital is freed for other uses. The many choices to work on in any queue invite expediting and rescheduling, with the result that machine setups are broken and total processing time increases. With reduced queues there is less time devoted to placing work in queues and retrieving it later. Reduced queues also mean that bottleneck operations in the shop are more visible and thus more likely to receive management attention—more rerouting around the bottleneck, more switching of labor from underutilized operations to bottleneck operations, more of the most effective overtime, more helpful subcontracting.

But how can one reduce queues in a job shop? Two major ways have proved effective: (1) using simulation techniques to plan the life of each job in the shop as well as to plan the life of each workcenter, and (2) turning the job shop into a line flow process. A third way, the repetitive lots concept, shows promise.

Simulation Techniques

Priority rules act only to plan the life of each machine or workcenter in the shop, with the result that queues themselves go unmanaged. Through simulation techniques[2] it is now possible to plan the position of every job in the shop so that jobs can be assigned to successive workcenters with little or no buildup of queues. These techniques can control, as opposed to merely keep track of, the jobs in the shop. They can regulate the amount and sequencing of work so that workcenters neither get starved for work (through underutilization) nor swamped with work (through excessive queuing). Although it is not ensured, a simulation approach may well have happened upon the perfect schedule for the Morris Machine Shop: namely, the sequence of jobs 3, 5, 2, 4, 1 (see Figure 10-5).

One of the key insights of these simulation models is that if the shop is to produce more, it has to schedule jobs efficiently through the

Hours

	1	2	3	4	5	6	7	8
Lathe	2	2	1	1	4	4	5	3
Mill	5	5	2	2	3	3	1	1
Grind	3	3	4	4	2	1	3	5
Drill	4	1	3	3	5	5	2	2

Figure 10–5 Chart of tasks versus hours—perfect schedule. (3, 5, 2, 4, 1)

bottleneck operations. The known bottlenecks constrain capacity so it is important to schedule them for maximum output, namely to run large lots through them and have few setups. For the nonbottleneck operations, on the other hand, it does not matter whether there are few or many setups; all that matters is that the bottleneck operations are fed work in time. Simulation techniques thus often split lots into different sizes and keep some buffer stocks in front of the bottleneck operations. One vision for these simulation techniques is to devise the schedule so that the job shop looks like a line flow process; the goal is to have jobs planned so that they never have to stop for long.

Simulation techniques require nightly computer runs to yield a schedule for the next day that indicates the exact sequence of jobs for each workcenter. The data needed to run the simulation are considerable. Accurate data on setups, run times, and routings for all products and capacities for each workcenter are required. Neither technique can guarantee perfect schedules, and, of course, machine breakdowns, faulty materials, and the like can ruin any schedule. Because of these data demands and the prevalence of changes visited on the process, simulation techniques have frequently fallen on account of their weight.

These techniques, however, can point out the bottlenecks and plan the flows, and this information can lead to capacity adjustments (overtime of the right kind, switching of labor to the bottleneck operation), workload adjustments (subcontracting), or job mix adjustments (split orders, reroutings) that can increase throughput in the shop. What these techniques do not permit is expediting.

Rearrangement

The attractiveness of simulation scheduling is that conceptually it plans out what has to happen for the job shop to act like a line flow process. Line flow processes use minimal work-in-process inventories because they are designed that way; simulation scheduling, in conceptual terms, rearranges the job shop into a line flow process to achieve the same result.

Suppose that instead of using a schedule to yield the "line flow effect" in a job shop, one physically rearranged the job shop into a line or series of lines. The physical layout of the factory would be altered so that queues of work could not build up. Workcenter capacities along each "line" would be balanced. Extensive work-in-process inventories would be designed out of the process. Some of the flexibility of the job shop would be lost but the increases in factory throughput could be significant. The notion of manufacturing cells and group technology, discussed in Chapter 3, seeks just this kind of change from job shop to line flow.

Of course, not all job shops could be rearranged. Tour B (Norcen Industries) certainly could not; its jobs are too varied. There are, however, job shops—many of which are part of a hybrid process—that need not be managed as job shops. Demands, in these instances, are not so variable that with some modest investments in equipment, a re-layout of the plant, and attention to easy machine changeovers, a "line flow" process could not be devised.

This insistence on defining the process to limit the buildup of work-in-process inventory smacks of the "just-in-time" (JIT) approach to manufacturing. The process, and the production control within it, are geared to the "just-in-time arrival" of materials at workstations. The process is defined to remove queuing problems; with attention to quick changeovers, remarkable flexibility can be maintained and scheduling can be simplified. (More on JIT will be presented later in this chapter and in Chapter 11.)

The Repetitive Lots Concept

Recently, Jacobs and Bragg have introduced a concept they call "repetitive lots;" it shows good promise for improved job shop scheduling.[3] The repetitive lots method combines the advantages of small batches (lower inventories, shorter queues, quicker throughput times through the process) with the advantages of limited set ups (less downtime devoted to setup, greater workcenter capacity per unit of time). Jacobs and Bragg distinguish among three different kinds of batches: (1) "order batch" (the batch released to the shop that often matches the customer order), (2) "transfer batch" (the batch, usually smaller than the work order, that is actually moved from one workcenter to another), and (3) "operation batch" (the batch that is actually worked on at a given time at a workcenter). Splitting an order into several batches is an old trick for expediting things in a job shop (sometimes known as overlapping production; it is effective because it permits work to be done on units of the order in parallel rather than series. What repetitive lots do is to provide some discipline to this traditional practice.

With repetitive lots, the order batch is split into an integer number of smaller, independent "transfer" batches. At any workcenter, the next job to be worked on is determined by:

1. A traditional priority rule such as first-come, first-served or earliest due date worked on first, unless

2. In the queue of jobs awaiting work there is a transfer batch that matches what is already set up for the workcenter. In this case, that transfer batch is pulled in, and thus the "operation" batch size can be greater than the transfer batch size. The logic here is that, by so doing, setup time is reduced.

Although there are still a number of questions to be asked of this approach (such as what the size of transfer batches is and which priority rules to use), its initial results are encouraging. While repetitive lots may not improve on simulation, it certainly is easier to implement.

Different Types of "Loading"

The prior section on job shop scheduling already discussed some aspects of dispatching—that element of shop floor control devoted to the precise sequencing of jobs to particular workcenters. We saw two major ways job shops or batch flow processes) can be scheduled: (1) with priority rules to guide the selection of which job to pull into the workcenter, or (2) with simulation to plan how each job should move from workcenter to workcenter.

Those are two characteristically different approaches to dispatching, and they carry different general labels: *vertical loading* and *horizontal loading*. Vertical loading views the problem from the standpoint of the workcenter and asks which job (or succession of jobs) should be assigned to it. Horizontal loading, on the other hand, views the problem from the standpoint of the job itself and asks how and when the job will be sequenced through all the workcenters. Vertical loading leads one to consider priority rules for the selection process, an essentially static procedure. Horizontal loading is more dynamic because it forces the dispatcher to consider what happens to orders and workcenters over time. As noted in the section on job shop scheduling, while it is an appealing notion, horizontal loading is difficult to accomplish well, even with computers. Nevertheless, many think it holds more promise than the static, simpler approach represented by vertical loading.

The oldest, and least sophisticated, form of horizontal loading technique is the *Gantt chart*, which is essentially a bar chart visual aid to the scheduler named after Henry Gantt, who popularized them in the early twentieth century. Despite their simplicity, Gantt charts can be very useful tools for loading a workcenter (or for scheduling any variety of activities).

To construct a Gantt chart we need to know when the item is due and how long each required operation (including setup time, run time, and move time to the next operation) can be expected to take. Starting with the due date, one can then follow an operation backwards, figuring out when the start date has to be. One has the choice of including queue time in these calculations or not. Queue time can often be 75 to 95 percent of the time an item spends on the shop floor. Naturally, the better the schedule, the lower the amount of unproductive queue time.

The charts used in the section on job shop scheduling (Figures 10-1 to 10-5) are a type of Gantt chart. They show the sequence of work at various workcenters as well as the quantity of work expected (i.e., expected utilization levels) at each. Their use forces the scheduler to think of the dynamics of the situation, the germ idea for sophisticated simulations. In fact, the simulation of a workcenter over time so that jobs are chosen not to exceed its capacity has been given the special term of *finite loading*. The contrast, of course, is called *infinite loading*, which permits an assignment of jobs past rated capacity, necessarily triggering action such as overtime to take up the excess. (Priority rules are compatible with infinite loading because their use does not recognize explicitly the capacity of the workcenter). Figure 10-6 shows a load profile that could exist when infinite loading of a workcenter is done.

Finite loading essentially shaves off any more-than-capacity loads on workcenters and distributes that excess into later time periods, invariably delaying the expected completion of some jobs. While often more realistic than infinite loading, finite loading can confuse workers and foremen, because they may not be so easily accepting of the logic that may keep a workcenter idle in anticipation of a job to come. In fact, if they are measured by standard efficiency measures (standard hours/actual hours), they may strenuously resist finite loading for the equipment under their control.

This discussion of dispatching focuses on job shops and batch flow processes. Line flow and continuous flow processes, of course, require the same functions of sequential assignment of jobs for the process and of monitoring the performance to the schedule. However, with the rigidness of routing, the relative certainty of capacity, and the attention to balance in the flow, what gets started in a line or continuous flow process is generally what gets finished, and in the same order. And unless there are malfunctions in the process, monitoring the schedule is a relatively easy matter. In such processes, the final assembly schedule is the dispatch list, and priorities do not have to be assigned to jobs, workcenters merely work on the jobs that come to them, one at a time. The final assembly schedule paces production, and control for the process is largely a matter of tracking that schedule and acting to fix any irregularities.

MORE ON MONITORING A PRODUCTION SCHEDULE

One of the key tasks of shop floor control, that of monitoring the status of job orders, is information intensive. Vast amounts of paperwork must be consumed by an effective production control staff. Consider the following kinds of information and paperwork that production control must keep abreast of:

- Orders opened, closed, or worked on, by workcenter or department, with quantities tracked
- Routings for each order. (Sometimes routing information can be combined with order status to show the status of orders at workcenters or departments immediately preced-

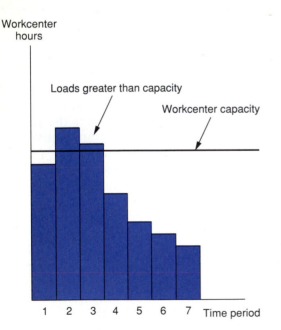

Figure 10-6   A workcenter load profile given infinite loading.

ing or following a job's current location. This helps give visibility to foremen and workers, so that they will more often work on exactly what is needed.)

- Stock status lists for inputs, including material shortage lists
- Bills of material (the link between inputs and outputs)
- Workcenter capacities, and efficiencies
- Standards for run time, setup time, queue time, and move time, by part or item
- Master production schedules and/or final assembly schedules
- Dispatch lists (that production control may generate itself) indicating the most recent assignment of jobs to workcenters, and thus the availability of workcenters for additional work.

All these pieces of information are requested by the production control function in order to know:

1. What the schedule is.
2. Which orders are where in the process and thus, how they stack up to the schedule (whether they are on time or behind schedule).
3. Where orders are due next in the process.
4. Which of those orders have material enough to be completed and which do not.
5. How much of each type of operation remains to be done to the order.
6. Whether the workcenters stand a chance of completing the order per schedule, given the other work assigned to them.

Integrating this information in a way that makes sense for the process is an important undertaking. Usually, monitoring schedules is helped greatly by using some visual aids.

Schedule-monitoring aids can be laid out in many ways. One common way is a column tracking scheduling chart.

Figure 10-7 is an example of a schedule-monitoring device that tracks progress down a column indicating the present day or week of production. The figure illustrates job/batches that were scheduled to begin on different production days and whose expected durations varied. Expected durations are indicated by the box outline itself while completed production is indicated by the shaded portion of the box outline. Those job/batches whose shaded-in portions fall left of the panel indicating the present production day are behind schedule; those whose shaded-in portions are right of the panel are ahead of schedule.

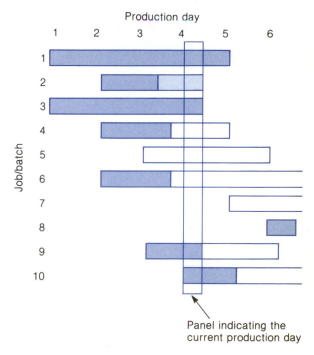

FIGURE 10-7   A chart for monitoring progress in production.

Tasks with scheduled starts and stops that differ can be readily portrayed. Also, job/batches that take up unscheduled time can be depicted (e.g., job/batch 2 and the lighter shaded portion of the portion of the line.) Because the chart is laid out for production days rather than tasks, different tasks can be represented by the same production day. Of course, the capacities and capacity utilizations of various tasks or machines must be accounted for separately so that too much is not asked of any task or machine.

Such visual aids lend themselves to a variety of graphic conventions. For example, different symbols or colors can distinguish wait time or maintenance time from actual production. Different symbols or colors can indicate the task or factory department in which the job/batch is, was, or will be. This choice of convention depends solely on what information is useful and what is available: and various charts like

these can be bought commercially for scheduling up to a year ahead.

A somewhat different method for monitoring production is illustrated in Figure 10-8. The Gantt chart, originated by Henry Gantt in the early 1900s, tracks time (along the bottom) and particular activities along the side). Different jobs are indicated by the letters A and B in the chart. The brackets display the planned time for each job at each activity (here designated as workcenters) and any space between the brackets, such as found for workcenter 2, is the scheduled downtime for that activity.

The lines running through each bracket show how much has been accomplished. The V at the top indicates what day it is. Any lines that do not extend as far as the V are behind schedule, whereas those extending past the V are ahead of schedule. In Figure 10-8, we see that workcenters 1 and 2 are on schedule, while workcenter 3 is ahead and workcenter 4

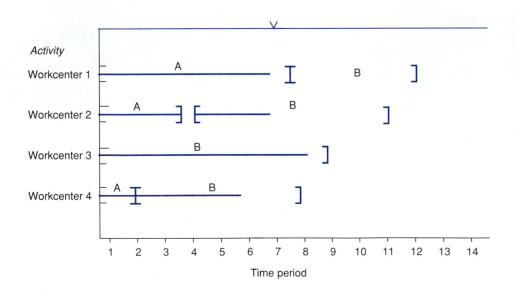

FIGURE 10-8   Gantt chart used to monitor production.

is behind. Gantt charts are useful for both planning a production schedule and for monitoring actual adherence to that schedule by each workcenter.

MANAGING THE WORK-IN-PROCESS

As should be abundantly evident by now, inventory systems (especially those of an anticipatory nature) are inextricably interwoven with planning and controlling production itself. This is nowhere more evident than in managing work-in-process inventories, for it is here that managers must keep alert to the interactions of the process and inventory.

How can work-in-process inventories build up, and how can they be reduced, and controlled? The process descriptions of Chapter 1 help to isolate some general reasons. Among them are production bottlenecks, line changeovers, parts shortages, and uncoordinated production control.

1. Production Bottlenecks

As we have already observed, the production bottleneck is a classic explanation for the buildup of work-in-process inventories and the bane of the job shop in particular. When bottlenecks occur, good inventory management means taking a close look at the balance of capacities in the process. In particular, if demand has increased or if the mix of products going through the shop or the production methods or equipment has changed, a careful process flow diagram-type analysis can work wonders for diagnosing the problem and suggesting possible solutions.

2. Line Changeovers

Again, as noted in Chapter 1, the typical line flow process is a rather inflexible agent for producing a broad mix of products. Often, different products must be run separately and serially on the line. Thus, if the company is to offer a complete product mix, inventories must be built up for each product in turn. Naturally, the more inflexible the line flow and the more dependent the company is on a small number of lines, the greater the inventory buildup is apt to be. In this case, the impact of the process on inventories can be dramatic. Frequently, a company's flexibility is enhanced by establishing production lines that are smaller but more numerous rather than production lines that are fewer in number and larger in size.

3. Parts Shortage

A classic and all-too-common explanation for the buildup of work-in-process inventories is the lack of one or more parts for completing a unit. In many continuous and line flow processes, the lack of a part would shut down operations altogether; but in other processes parts shortages can frequently be worked around, and are, especially when pressure for delivery is exerted. Unfortunately, although working around a parts shortage may be the quickest way to manufacture a given unit, an entire operation that works in this way reduces productivity considerably. Invariably, extra time is spent figuring what to work around, how to do it, placing work-in-process in limbo somewhere on the factory floor, fetching it for rework, figuring out how to proceed again, actually reworking the unit, and interrupting ongoing operations all along the way. Often, by having to cope with parts shortages and the factory floor difficulties they pose, the company suffers the consequences of poor raw materials inventory management. For many operations, the stockout penalty for a part in raw materials inventory does not include the many petty but cumulatively significant problems that a parts shortage can introduce to the factory floor.

More frequently still than a faulty raw materials safety stock decision, a parts shortage develops because of delivery problems with a supplier. The purchasing department may have ordered enough and may have given the supplier sufficient notice considering past delivery performance, but the supplier simply falls down on the job. There may not be a lot the manufacturer can do in such a situation. We have already discussed (1) the make–buy option, which is sometimes a way out of chronic delivery problems, (2) working with a supplier on managerial and technical means to improve its production capabilities and quality performance, and (3) the development of systems (such as MRP) that can give suppliers as much visibility into the future as the customer company enjoys.

A fourth means of coping better with such a situation may be the division of attention within the purchasing department to supplies of various kinds. One popular way of dividing up the responsibilities for ordering, monitoring, and expediting supply deliveries is the ABC system described earlier. Different people can be assigned to each group. The A group people, for example, might keep close check on the delivery status of their items, mainly the ones where delivery is likely to be a problem. The C group people are probably dealing with commodity items where their concerns are less with delivery and more with price differences by supplier, price discounts for large volumes, and EOQ-type calculations. (With an MRP system in effect, however, an ABC system is somewhat redundant; the speed and comprehensiveness of an MRP system makes it fairly easy to treat a C item like an A item.)

4. Uncoordinated Production Control

As we have seen elsewhere, there can be a mismatch in the product mix processed in different segments of the operation, especially when broad, changeable, and/or variably demanded product mixes are involved. This mismatch in production signals can be purposely sought (The "decoupling" inventory in the Hewitt-Everett situation in Chapter 7) or it can be inadvertent. In either case, a work-in-process inventory builds up. Note that this reason for an expansion of work-in-process results not from a production bottleneck per se, because the aggregate capacities of different segments of the process may match up very well. Rather, this work-in-process inventory develops because one segment's output, on a specific product-by-product basis, does not match in the short run a succeeding segment's output needs. In the longer run, of course, unless there is obsolescence, the two segments will produce matching product mixes. The challenge here lies in understanding when production and inventory control is better served and more effective with a work-in-process inventory that buffers segments of the full production process and their information needs from one another, and when such an inventory is merely taking up space, swallowing financing dollars, and confusing and disrupting operations.

One common way that inadvertent work-in-process accumulates in this way is when one portion of the process demands a substantially different setup expenditure than another portion. The typical reaction to a high setup cost is to make the run long. While such a strategy often makes good sense, it is not always the best course of action. Long runs may build up too much inventory for too long a period and risk stockouts of items with more pressing demands. The balance of setup costs against inventory expenses and stockout penalties ought to be done in every case.

Another common way that uncoordinated production control can stack up undesired work-in-process inventory is when expediting within the factory rages out of control. As more and more rush orders are placed on the factory floor, the regular flow of the process becomes increasingly disrupted. This raises work-in-process inventory and lowers productivity (labor time per unit increases). Existing machine setups are likely to be broken repeatedly, necessitating time-consuming rework of the setups themselves. As Chapter 1's process descriptions revealed, rush orders are incompatible with line flow and continuous flow operations. In job shop and batch-flow operations, production control policies become critical. This is not to imply that rush orders should not be permitted. Rather, it is an admonition that a company whose work-in-process inventories seem to be ballooning ought to look carefully at the extent of rush orders within the process and whether a restructuring or expansion of the process, a different means of production scheduling, or a segregation of rush orders might be a more effective way to maintain both productivity and service levels.

The Japanese Contribution to Production Control[4]

Much has been written about the major Japanese manufacturers, those companies making autos, cameras, consumer electronics, and other items that have been able to compete to tremendous effect in all of the world's marketplaces. Much more will be said about the Japanese in Chapter 11, particularly about how various manufacturing techniques, many inspired by the Japanese, can be integrated into a manufacturing philosophy that merits careful study by all students of production. Nevertheless, it is worthwhile introducing here the key concept about Japanese-style production control, for it contrasts sharply with the traditional practice of many Western companies.

Coordination has been a key topic of discussion in this and earlier chapters. Materials ought to be managed so that they are coordinated effectively with the production schedule, no matter how that production schedule is derived. To avoid idle worker and machinery time in such an endeavor, however, the traditional Western outlook is to make sure that there are queues of work ahead of each machine and that there are banks of parts awaiting work or assembly by the workforce. This attitude toward production scheduling and control could be characterized as a "push," where materials are to be pushed out of material stockrooms or out of particular manufacturing departments according to a set schedule that is influenced heavily by demand and due dates. In such a system, considerable time and effort are spent making sure that particular problems can be expedited efficiently if changes to the schedule have to be made on short notice. Extensive coordination is needed to be certain that the materials are made available according to the dictates of the often changing production schedule. Buffer stocks, safety lead times, and other tactics are often used to make sure that the materials are there "just in case" they may be needed.

Some of the finest Japanese companies have a very different perspective on production scheduling and control. As Chapter 11 relates, the foundations for this perspective lay more with company attitudes toward quality improvement than with schedule flexibility. Suffice it, at this point, to note that the Japanese view is to have production be "just in time" rather than "just in case." The just-in-time notion is to have as little inventory in the process as possible, yet for the process to flow smoothly. Raw materials, fabricated parts, subassemblies, and the like are to be delivered to the respective segments of the process just before they are required by the workers there, much as in the continuous flow process. Indeed, much of Japanese manufacturing can be seen as a movement to try to change batch flow operations into as close a facsimile of a continuous flow process as possible. In contrast, the traditional

Western view is to develop information-intensive schemes such as MRP to salvage the flexibility of batch flow processes.

If this kind of batch-to-continuous flow transformation is to occur, one cannot expect to "push" materials out into the factory without risking a buildup of inventory. After all, if the process cannot handle the production as scheduled, inventory will accumulate. For this reason, the Japanese do not value as highly the philosophy of coordination underlying an information-intensive MRP system. Using an MRP system to release orders to the factory floor risks increasing inventories and abandoning the concept of just-in-time production.

The implications of just-in-time production are far-reaching: low inventory levels, better product quality, smoother production flow, and increased worker involvement. Just-in time production is indicative of a point of view dif-ferent from that implicit in MRP systems as they are commonly adopted. If materials are not to be "pushed" out of the materials stockroom or out of one manufacturing department and into another, what kind of production control is consistent, then, with the just-in-time production perspective? The answer is to adopt a system that "pulls" materials through the factory. The authority to draw materials out of the stockroom or to begin production in a department is given only when it is called for by an operation downstream. Production is thus triggered not so much by the production schedule as by the needs of operations that follow. The most famous of the Japanese "pull" production control systems is the kanban system developed over many years by the Toyota Motor Company. It, and its offshoots, are discussed in more detail in Chapter 11 on just-in-time production.

PART TWO

PROJECT MANAGEMENT

A project—constructing the World Trade Center or sending a man to the moon—shares many characteristics with other types of production processes. We can picture a project with the same kind of precedence relationship with which we pictured the line flow process. Certain activities must be completed before others begin, and each activity can be expected to take a given period of time, just as in other process types. But our concern for balance in the factory, so prominent in our thinking about production processes, does not trouble us in thinking about a project since a project is, by definition, a one-time endeavor and workers can generally expect to work only so long on it and then move on to something else. It is typical for a project to employ wildly fluctuating numbers of people, many with different skills. We do not worry about idleness—construction workers move on to the next project as do space scientists and engineers.

What does trouble us in a project is getting it done on time, for projects often have important deadlines to meet. Scheduling is thus absolutely critical to the management of a project. Several techniques have been developed to highlight the project activities that must be accomplished on time (or else risk delaying the entire project) and the activities that can be delayed somewhat. One such scheduling technique is called the critical path method; what follows is a brief description of its rationale and workings.

The Critical Path Concept

As an illustration, let us consider the simplified example of a project many companies face — the startup of a new plant. The major tasks involved in a plant startup are arrayed and described in Table 10-3. The table makes clear that some tasks cannot be started until other tasks (predecessors) have been completed.

The precedence relationships of the startup can be depicted in a graph or network, as shown in Figure 10-9. This network follows a particular convention known as *activity-on-node*. This convention is intuitively appealing and somewhat less confusing than the alternative convention, known as *activity-on-arrow*.

The key principle of the critical path concept is that a project cannot be completed any faster than the longest time path between the project's start and its finish. In Figure 10-9 the longest path from the start (project approval) to the finish (plant startup) involves the following activities — B, E, H, and K; they take a total of 17 months. The longest path between project approval and plant startup is termed the *critical path*, primarily because any delay along this path of activities sets back the entire project. It is this critical path that merits the most management attention.

All the other paths from start to finish enjoy at least a month's slack time, and so they can experience delays of varying lengths and still not harm the 17-month completion schedule. For example, the next two longest paths after the critical path involve the following activities:

1. B-E-H-L   16 months
2. A-D-G-K   16 months

JOB NAME	JOB DESCRIPTION	IMMEDIATE PREDECESSORS	TIME (months)
	TABLE 10-3 A plant startup project		
A	Selection of plant manager and management personnel	—	3
B	Site survey soil test	—	1
C	Extension of roads, water, utilities, sewer	—	6
D	Selection and purchase of equipment	A	2
E	Final engineering plans for construction	B	3
F	Employment interviews and hiring	A	3
G	Equipment delivery	D	9
H	Construction of facility	E	11
I	Precise layout of plant	D, E	1
J	Institution of management systems for control of production, inventory, purchasing, accounting	A	4
K	Worker training	F, G, H	2
L	Equipment and system installation	G,H,I,J	1

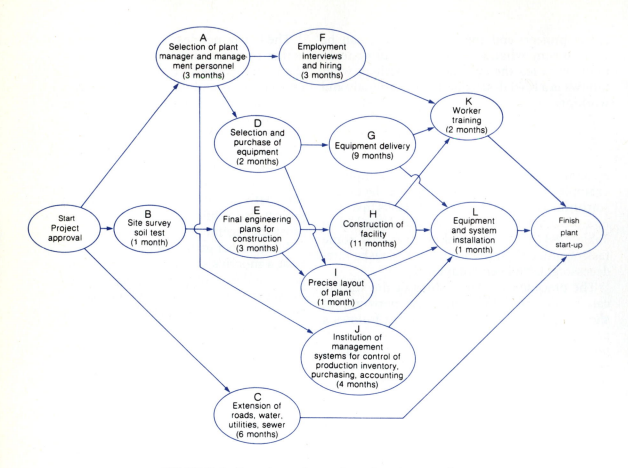

FIGURE 10-9   Precedence diagram for a plant start-up project.

There can be delays of up to a month on these two paths without setting back the entire project. However, note that the first path is very much like the critical path itself; it differs only in that activity L is substituted for activity K. The slack in this path then can be taken only on activity L (equipment and system installation). If a delay occurred on any other activity, say H (construction of the facility), the critical path would also be affected, and the entire project would be delayed.

The second path is much more flexible about where its slack can be used up. Only one of its activities, K (worker training), is shared with the critical path. Thus its month of slack can be used up on activities A, D or G without affecting the critical path in any way.

Other paths, of course, have considerably more slack and can be accomplished at a more leisurely pace, if need be. For example, the uppermost path in the diagram (A-F-K) is expected to take 8 months, with only activity K on the critical path. Activity F could be delayed as much as 9 months without making the project late. However, activity A could not be delayed that long, since it is on one of the second longest paths and can be delayed only a month before it would affect the project's completion time.

This example raises an important point about critical paths and project scheduling. As time wears on in a project's life, delays and even some speed-ups can be expected. What may look like the critical path at a project's start (such as B-E-H-K) may not remain the critical path if many delays (or speed-ups) strike the project. This means that the project manager should periodically recalculate the critical path to check that he or she is focusing on the activities that really matter to finishing a project in the least time.

Gantt Charts and Critical Paths

Gantt charts can capture some of what the critical path gives us, but not all. A Gantt chart of the table above could be constructed. (See Tour J for both a Gantt chart and a precedence diagram of the same project steps.) Whereas the precedence characteristics can be depicted, and some feel for the slack in the system can be shown, the links between activities are missing and the critical path itself is not evident from the diagram itself; it must be imposed on it from an outside calculation.

Calculating the Critical Path[5]

To this point we have calculated the critical path solely by inspecting all the paths and choosing the one with the longest duration. In simple cases, such as this generalized plant startup, inspection is a perfectly feasible and reasonable way of selecting the critical path. When projects get more complicated, with many more activities, inspection as a means of computing the critical path bogs down. Happily there is an alternative procedure or algorithm, which can cut down the speed of solution markedly.

In essence, the critical path algorithm is a procedure to identify which activities have

some slack time and which do not. Those with no slack time, of course, make up the critical path or paths.

Before describing the algorithm, however, it is helpful to define some terms:

Early Start (ES)—The earliest time a job can begin, which is the latest time that any of the job's predecessors are finished.

Early Finish (EF)—The earliest time a job can finish, which is the early start time plus the time it takes to do the job.

For any job i, the early start and early finish can be represented symbolically.

$$ES(i) = \max [EF \text{ (all of } i\text{'s predecessors)}]$$

or the start time of the project if we are considering the beginning jobs.

$$EF(i) = ES(i) + t(i)$$

where t(i) is the time to perform job i.

Late Start (LS)—The latest a job can start without causing the entire project to be completed any later, which is the late finish time less the time for the job.

$$LS(i) = LF(i) - t(i)$$

Late Finish (LF)—The latest time a job can finish, which is the earliest time that any of the job's successors have to be started. These can be represented symbolically, as well, for any job i.

$$LF(i) = \min [LS(\text{all of } i\text{'s successors)}]$$

or the earliest finish time for the project if we are considering the ending jobs.

Total Slack (TS)—The difference between the early and late starts for a job, or the early or late finishes. Total slack is the time a particular job could be delayed without delaying the completion of the project. Symbolically,

$$TS(i) = LS(i) - ES(i) = LF(i) - EF(i)$$

The critical path algorithm involves calculating *ES, EF, LS,* and *LF* for all the jobs in the network and then comparing them to discover which nodes have zero total slack.[c] These nodes are the ones comprising the critical

[c] Or a minimum total slack if there is some discretion about the completion time possible for the project. Note the later discussion of windows of time.

path(s). The calculation and comparison involve three steps:

1. *Forward pass.* The forward pass through the network calculates the early start (ES) and early finish (EF) times. It does so by moving from start to finish, figuring first the early start time and then adding the job's duration to it to figure the early finish. The early start for any job is calculated by scan-

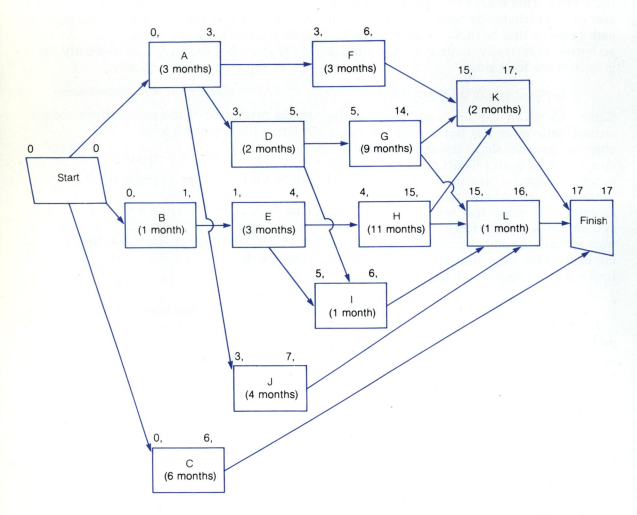

Figure 10-10   Early start (number at upper left of box) and early finish (number at upper right of box) months for the plant startup project depicted in Figure 10-9.

ning all of its immediate predecessor activities (those with arrows pointing into it in the path) and choosing as the job's early start the latest of all the predecessors' early finish times. The ES and EF times for our plant startup example are noted in Figure 10-10.

2. *Backward pass.* The backward pass through the network calculates the late start (LS) and late finish (LF) times. It does this by moving from the finish back to the start, figuring first the late finish times and then deriving the late start times by subtracting the job's duration from the late finish time. The late finish for any job is calculated by scanning all of its immediate successor activities (those with arrows leading away from it on the path) and choosing as the job's late finish the earliest of all the successors' late start times. The LF and LS times for our plant startup example are noted in Figure 10-11.

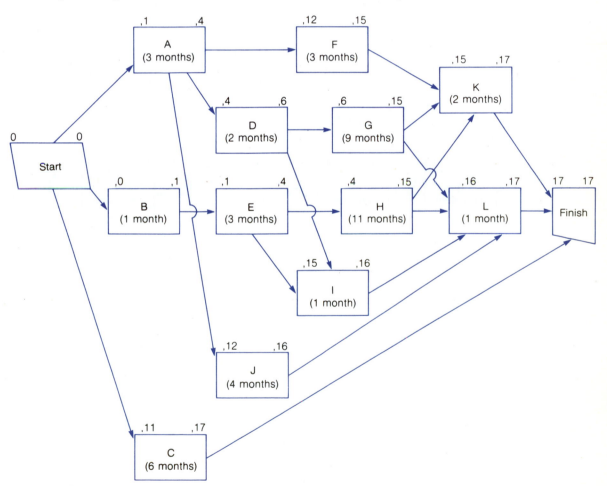

FIGURE 10-11   Late start (number at upper left of box) and late finish (number at upper right of box) months for the plant startup project depicted in Figure 10-9.

3. *Comparison.* The comparison of either early start with early finish or late start with late finish is the way to figure whether the job has any total slack (TS) associated with it. If the LS(i) − ES(i) or LF(i) − EF(i) calculations are zero, the job has no total slack and is on the critical path. Figure 10-12 combines all the times in one diagram and displays the critical path.

For very large projects, of course, and ones that go into significant detail, even following the critical path method can become tedious. To aid in the analysis, there are computer programs that calculate the critical path (See Tour J). These can be used with ease to determine how differences in expected completion times for various activities (and even the introduction of variations in the network of activities itself) af-

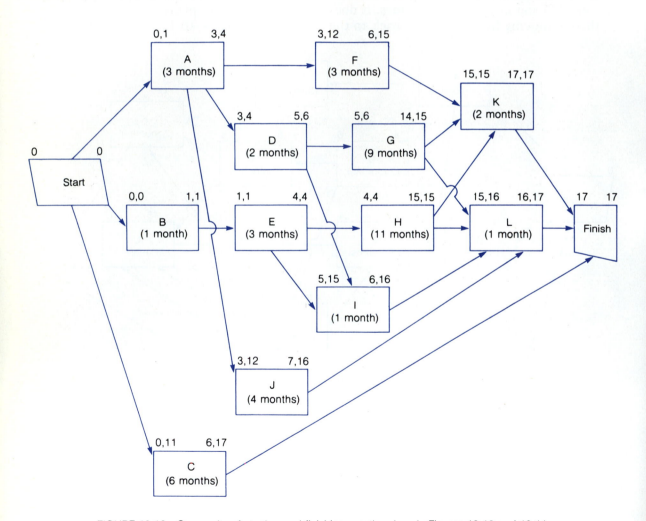

FIGURE 10-12   Composite of starting and finishing months given in Figures 10-10 and 10-11 for the plant startup project.

fect the critical path(s). Because so many things can go awry in a project, it is advantageous to compute the critical path on a regular basis so as not to be caught napping.

Project Costs and Timing and the Critical Path

Frequently, actual projects (such as the new plant startup outlined in Figure 10-9) need not be finished in the least time possible but, rather, can be finished during any portion of a "window" of time. When during the acceptable window of time the project's finish ought to be targeted depends largely on the project's managers. Often they are influenced by cost considerations, because it is not uncommon for a project's contract to stipulate rewards for an early finish and penalties for a late finish. These rewards and penalties must be weighted against the costs that hurrying up any of the project's activities could impose on the project's budget.

Windows of Time and the Critical Path

Windows of time can be easily incorporated into the calculation of the critical path; in fact, the terminology is already in place to handle them. The acceptable window of time can be denoted as the difference between the early finish and the late finish (or between the early start and the late start). To return to our example, if the plant startup could be completed between 17 and 21 months from the start, the values for Figure 10-12 would be recast as in Figure 10-13.

Building a window of time in the critical path computation makes it important to distinguish between two different kinds of slack time during the project: total slack and free slack. Total slack, as discussed previously, is the difference between the early and late finishes (or early and late starts) for any activity.

An activity's total slack time is the maximum time that activity may be delayed beyond its early start without forcing the delay of the entire project. As previously noted, this notion of total slack can be used to define the critical path; the critical path is composed of those activities that have the lowest total slack (the smallest differences — possibly zero — between early and late starts or finishes).

Free slack, on the other hand, is the time by which an activity can be delayed without delaying the early start of any other activity. Free slack is defined as the difference between an activity's early finish time and the earliest of the early start times for all its immediate successor activities. Symbolically, the free slack (FS) for any job i can be defined as

$$FS(i) = min \ [ES \text{ all of } i\text{'s immediate successors}] - EF(i)$$

Consider the path A-J-L in Figure 10-13. Activity L can be delayed at most a month without risking the delay of the entire project; as it is, its early finish time (16 months) is but 1 month shy of the project completion's early start month (17 months). Its free slack is 1 month. Activity J, on the other hand, has more free slack. Its early finish time of 7 months is 8 months shy of its only successor activity's (L's) early start time of 15 months; its free slack is thus 8 months. Activity A, the predecessor to activity, has no free slack because its early finish time (3 months) is equal to the early start time of activities D, F, and J.

This lack of any free slack does not mean, however, that if activity A were inadvertently delayed, the entire project would be delayed. Activity A, after all, does not lie on the critical path. It has a positive total slack and thus could be delayed by one month without necessarily delaying the project. It is often useful to employ a network-based Gantt chart as an aid in tracking and scheduling noncritical activities

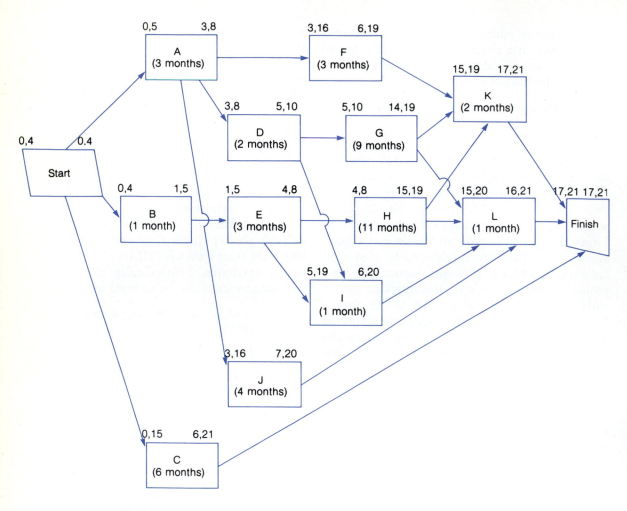

FIGURE 10-13   The addition of four months (a window of time) to the plant startup project.

(see Figure 10-14). Such charts usually display two panels, one depicting all activities at their earliest starts and the other depicting all activities at their latest starts. Network-based Gantt charts can also include data on manpower schedules per time period and cost incurred per time period. Figure 10-14 refers back to Figure 10-12's network where there is no window of time for the completion of the project. Note how the activities on the critical path coordinate with one another on the chart.

Trading Off Time Against Cost

Suppose up to two months could be shaved off both the time it took to construct the new plant and the time it took to deliver the necessary equipment, although at a cost. The costs that would have to be incurred to "crash" each of these activities are shown in Table 10-4. Suppose also that having the new plant on stream a month or two early was estimated to be worth $70,000 per month, since the new plant

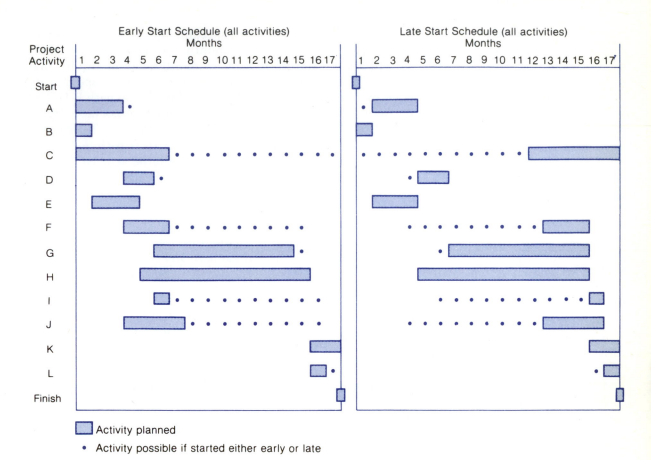

FIGURE 10-14   A network-based Gantt chart for the plant startup project.

Early Start Schedule (all activities)
Months
Late Start Schedule (all activities)
Months
Project
Activity
Activity planned
• Activity possible if started either early or late

TABLE 10-4 Costs to "crash" three activities in the plant startup project		
ACTIVITY	COST TO CRASH FIRST MONTH	COST TO CRASH SECOND MONTH
Construction of facility (H)	$50,000	$75,000
Equipment delivery (G)	20,000	40,000
Selection of plant manager and management personnel (A)	5,000	Impossible

promised capacity that was unobtainable any other way. Should these activities be "crashed" by 1 or 2 months? How can the critical path aid in the analysis?

As should be increasingly evident by this point in the text, this decision is one for which marginal analysis is appropriate. That is, it pays to speed up the project if the additional benefits enjoyed are greater than the additional costs incurred. If the costs outweigh the benefits, it is best to keep the project on its original schedule.

This use of marginal analysis to decide which activities are to be "crashed", and how, is a central element of the *critical path method*/(CPM), which is one of the key tools for project management. It is particularly popular with construction companies, although it has a wide following elsewhere.

The use of the critical path is important for this kind of analysis, for it indicates, for any project length chosen, exactly which activities have to be speeded up and which can be left on the original schedule. Ideally, of course, we would like to leave slack all that can be left slack and speed up only what must be speeded up (and if a choice exists, to speed up the least costly activities). Put another way, to the extent possible, we would like every path to be a critical path because in so doing, the least expenditure for project speed-up will be incurred. What does this mean for our example? To complete the plant startup in 16 months rather than 17, activity H (plant construction) must be crashed since it (not activity G, equipment delivery) lies on the critical path. Activity G can be left as originally scheduled. The costs incurred for speeding up the project to 16 months are thus only $50,000 and the expected benefit from the early completion is $70,000. It is clear that the project ought to be speeded up by at least 1 month. Note that there are now two critical paths: B-E-H-K and A-D-G-K.

What about speeding up the project from 16 months to 15 months? To complete the plant startup in only 15 months would require that plant construction (activity H) be speeded up an additional month (for critical path B-E-H-K) and that either equipment delivery (activity G) or personnel selection (activity A) be speeded up by one month (for critical path A-D-G-K). The additional cost would be at least $80,000 ($75,000 for crashing activity H again and $5000 for crashing activity A), which exceeds the benefits of $70,000 for being another month early. Speeding up the project from 16 months to 15 months is thus not advised. Note that the critical path and marginal analysis are both called for in analyzing trade-offs of time and cost in project scheduling.

An Alternative Program: Evaluation and Review Technique (PERT)

At about the same time as the CPM was developed in the late 1950s, a similar technique called PERT (program evaluation and review technique) was also developed. Both CPM and PERT require the construction of a network diagram, and they both share the concept of the critical path as determining the least time for project completion. In contrast to our discussion so far, PERT offers a different graphical technique and an added wrinkle.

Graphing PERT

Where the networks we prepared previously place activities to be performed at the nodes of the network diagram and use arrows only to indicate which activities follow others, PERT uses a different convention. Activities are placed on arrows and the nodes of the network serve as events — the starts and completions of various activities. In order to preserve the proper precedence relationships when using this graphical convention, dummy activities and events must sometimes be introduced. Figure 10-15 illustrates this with a portion of the precedence relationships from the plant startup example. As Figure 10-15 implies, PERT is more cumbersome to graph than the activity-on-node networks.

Time Estimate Variations

PERT can be used to provide some insight into likely variations in the completion time of the project. Individual estimates (an optimistic one, a pessimistic one, and a most likely one) of the variance in performing each activity are made. This is accomplished by assuming a certain distribution for these variances — the *beta* *distribution* (a skewed distribution that permits the possibility of occasional very late events, but no very early events — and by insisting on the independence of each activity from others. An estimate of the time variance in accomplishing the entire project (and thus the probability for completing it by various dates) can be made by summing the variances along the critical paths. The assumptions ensure that the distribution of the summed variances will be normal.

The expected value of the time for each activity is

$$T = \frac{oe + 4ml + pe}{6}$$

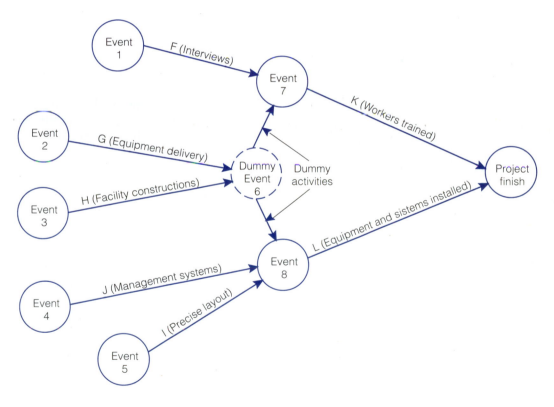

FIGURE 10-15   A sample PERT network using a portion of the plant start up example. The dummy event and activities are used only to clarify that both activity G and activity H must precede activities K and L.

where *oe* is the optimistic estimate, *pe* is the pessimistic estimate, and *ml* is the most likely estimate. The variance of the time for each activity is

$$o^2_T = \left(\frac{(pe - oe)}{6} \right)^2$$

It is the sum of these variances along any path in the network that is normally distributed. By using the summed variance for the path (e.g., the critical path) and the fact that the summed variance is distributed normally, a probability of completion of any event can be computed.

As may be surmised, using PERT in this way to calculate a project completion time variance and probabilities of completion by specified dates can be a bit concocted. The variance for the entire project can be only as good as the variance estimates for individual activities, and often there is little information on which to base such variance estimates. Using PERT in this way can be more trouble than it is worth. Nevertheless, PERT does make explicit a concern for time variations and the fact that the critical path itself cannot be known with certainty. These insights are useful reminders to any project manager.

Some Principles for Project Managers

Although critical paths, the critical path method, and PERT have been exceedingly useful to managers in scheduling projects, effective project management means more than the adept use of these techniques. Drawing on his own vast experience in project management, Herbert Spirer has arrived at a dozen principles that serve as a useful checklist for project managers:[6]

1. A project has to have a clear objective or set of objectives, and these objectives ought to be stated in terms of specific items (tangible or intangible) to be delivered to the project's sponsor.

2. The project manager ought to be able to structure the project clearly, detailing which bundles of major activities (work packages) are required and how they comprise the deliverable items that meet the project's objectives. Spirer calls this a "work breakdown structure," and he sees it as a hierarchical representation of the project.

3. The work packages making up the work breakdown structure are composed of individual activities. These activities have to be listed and organized into groups that can be overseen by specific people. The specific activities for a major project can number in the thousands. Spirer recommends, however, that no one manager have direct responsibility for more than 50 specific activities. Top-level management itself should not be responsible for more than 30 to 50 aggregations of these activities, a precept that highlights the need for an effective work breakdown structure.

4. These activities should then be placed in a precedence diagram or network, and a critical path developed with the concomitant early and late start and finish times and slack calculations. The development of the precedence diagram or network is an important task in itself; analyzing such a network can lead to various suggestions or to resequencing activities. Calculations of the critical path need to be done periodically as the project proceeds.

5. The calendar of activities and times that comes out of the network creation and the critical path analysis must be combined with some knowledge of resource constraints to produce a schedule that tells

management when activities will be done, not simply when they must be done.

6. Managing the resources needed to meet the schedule is eased by making use of Gantt charts and plots of cumulative resource use such as are applied in manufacturing situations.

7. Activity times can be estimated often by analyzing similar activities and modifying their times by using common sense. Sometimes more analytic methods (regressions, cross tabulation) can also be used. The estimation of activity time is often helped by referring explicitly to the network and the critical or near-critical paths that are implied by the first-round estimates.

8. Use pessimistic, optimistic, and most likely estimates for as many activities as possible. It is always helpful to recognize the uncertainty inherent in projects.

9. Tracking the progress of a project is helped by establishing milestones, which are easily grouped events. Such milestones can help serve as motivation for the project as well as checkpoints for the project's progress.

10. Assign one and only one person to be accountable for every activity.

11. Use the plan to control the project. This is done by keeping the plan current and monitoring the actual performance in timely fashion. Determine how the actual performance measures up to the plan and what variances are implied by that contrast. Take any corrective action that is necessary, assessing the likely consequences and adjusting the plan accordingly.

12. Assess the status of the project by using various "earned value concepts" such as the budgeted cost of work scheduled, the budgeted cost of work performed, and the actual cost of work performed. By comparing the budgeted cost of work scheduled against the budgeted cost of work performed, one can come up with a general measure of the on-schedule performance of the project. Similarly, by comparing the budgeted cost of the work performed with the actual cost of the work performed, a measure of the cost variance of the project can be ascertained.

PART THREE

CONTROL IN SERVICE BUSINESSES

Control for service businesses is a recurrent issue and a constant fight. Control may be even more important for services than for manufacturing because it is harder to do. A number of things contribute to this difficulty:

• Demand for services often cannot be levelled as one can level the demand for manufacturing, and thus the erratic demand induces queues that can be sometimes dramatic in services. This stretches the control system.

- Customization is often rampant in service businesses, and this hinders control.

- Often, customers can themselves intrude on the service process, and this naturally impedes control.

- With services often there are fewer variables that can be measured as the dimensions of a manufactured product can. Thus, quality control tends to center around attributes control charts rather than variables control charts. And, attributes control charts do not tell you as much.

- For many services, cash must be handled. When cash gets transferred, cash controls are needed and accounting taken care of.

These features make control more difficult for the service business, and thus a real managerial issue. With respect to the service matrix introduced in Chapter 1, controls are easiest in the service factory. Customization is lowest. The service factory may be at a distance from the customer. The process is more capital-intensive. Standard operating procedures are the norm. All of these things make control easier to accomplish. When one gets to the other extremes of the service matrix, however, control issues push into the limelight.

There are a variety of different types of controls that prevail in service businesses. Some of them have direct analogies to manufacturing. For example, many services deal with materials that need to be inventoried, and the kinds of inventory controls that were discussed for manufacturing apply to service (see Tour F, Burger King). Often the inventory is managed with non-time phased systems like the reorder point and periodic reorder systems. Service businesses often need the same kinds of safeguards on inventory: locked storerooms, sealed trucks, separate accounts kept for inventory items, checklists, and crosschecks on what has been handled, as is discussed in the Thalhimer's tour at the end of this book.

Quality controls are another common feature between service and manufacturing businesses. As one might expect, the service factory, with its more standard operating procedures, can have quality controls that are more conventional than those services away from the service factory. As you get away from the service factory, training becomes more critical and more long-term. Service standards tend to be tougher to define. Reviews and audits may be more common.

There are a number of other classic controls that apply to service delivery. Orders can be tracked in services like auto repair (Ogle-Tucker Buick in the back of this book) just like orders get tracked on the factory floor. Services can get dispatched in just the way that manufacturing is, say, with the use of Gantt charts. Simulations are often helpful to services, especially if they involve geography. Telephone service, electric service, airlines, and other transportation services use this kind of process control.

Just as manufacturing orders can be split, so that they move through the factory faster, services can be staged so that the bottlenecks can be smoothed out. Delivery in this case occurs bit by bit and not all at once.

More likely to be seen in service than in manufacturing, is triage. If one can identify the customer's particular needs early in the delivery process, and assign that customer to a specific service path through the process, often customer satisfaction, time, and productivity are enhanced. Hospitals use triage in their emergency rooms. Often now triage is done when you phone in to find out information about your bank account or mutual fund, and it is done in automated fashion to smooth out the delivery.

Services often also have controls on demand. One wants to shift the demand off the service peak times and influence exactly how much needs to be processed at any one time. One can do this through price sensitivity or by offering specific services (early bird specials). One could also have marketing initiatives such as promotions that can shift demand. Almost unique to services, one can use reservations to schedule the customer into the service process (airlines, restaurants, doctors).

Service businesses may also require financial controls. The service workers also often handle cash and cash is yet another kind of inventory that must be controlled.

Tools of the Trade: A Recap of Lessons Learned

The question of how a process is best controlled is a universal one, but the precise controls differ somewhat, depending on the environment — manufacturing, projects, or service. Budgeting and delivery (schedule) controls are common to all three environments. The documentation needs are typically fairly similar:

- Orders must be released and that usually means that capacity must be judged to be adequate.
- Orders must then be tracked through the process (monitoring through the use of column-tracking and Gantt charts).
- Materials and accessories (e.g., tooling, change-over equipment) must also be tracked so that all is ready for the order.

More and more, world-class practice uses "pull" (as opposed to "push") production control systems (for example, kanban cards or kanban squares) to control the movement of materials through an operation. Computer-generated peperwork is on the way out as a means of determining a work center's priorities; more physical, real-time signals are being adopted.

The techniques used to control delivery vary more. In manufacturing, Gantt charts are common. With projects, Gantt charts are typically not enough, especially if the project is complex. There, techniques such as the critical path method and PERT are extremely useful. With service, one typically can use Gantt charts for the service delivery process, but other scheduling devices are frequently appropriate that can actually manage the demand (given that inventories cannot be used). Reservations can be taken and various price and non-price incentives can be employed to shift the timing of the demand.

Other types of control affect how work is assigned specifically to the process. Such assignment becomes an issue most when routings are not standard, such as occurs in the job shop. Nothing is a sure-fire success with a job shop; various rules have been used and techniques such as simulation have been tried. In neither case are sure-fire, perfect schedules possible routinely.

Projects build the assignment of work right into their statements of work. Problems occur only when snags are hit and resources committed to one use may be needed elsewhere at the same time.

Corrective actions, such as expediting, are similar to all three environments.

QUESTIONS

1. Compare and contrast the worker information requirements in the batch flow and line flow processes.

2. What is meant by the term "production control"?

3. Why are job shops difficult to schedule? How are job priorities determined?

4. What is an alternative to scheduling by priority rules?

5. Discuss the concepts of vertical and horizontal loading. How do they differ from one another?

6. What is the "just-in-time" philosophy? How is it different from the "just-in-case" philosophy? What determines the flow of production in each philosophy?

7. Briefly describe the critical path method and discuss its relation to project costs and timing. How does PERT differ from CPM?

8. According to the critical path concept, what is the shortest amount of time in which a product may be completed?

9. Why is it important for managers to continually recalculate the critical path?

10. How is marginal analysis used in the critical path concept?

11. Compare shop floor controls to controls on services.

PROBLEMS

1. The following five jobs are being routed through a single machine. What is the average wait time per job under the following priority rules: first come, first served; earliest due date, shortest processing time, and longest processing time?

	Job 1	Job 2	Job 3	Job 4	Job 5
Due	3/22	3/21	3/26	3/24	3/25
Time	5	7	6	4	8

2. Suppose production through a second operation is required for the 5 jobs of Problem 1. What is the average wait time per job under the following priority rules: first come, first served; earliest due date, shortest processing time, and longest processing time?

	Job 1	Job 2	Job 3	Job 4	Job 5
Due	3/22	3/21	3/26	3/24	3/25
Operation 1	5	7	6	4	8
Operation 2	7	3	5	5	5
Total time	12	10	11	9	13

3. Return to the Morris Machine Shop (Situation 10-1) and consider another set of jobs to be released to the shop as shown in Table 10-5. The total times for each machine are: lathe (8 hours), mill (10 hours), drill (3 hours), grinder (9 hours).

What is the best schedule for the jobs? How do the various priority rules perform: first-come, first-served; work first on shortest processing time; work first on longest processing time; work on earliest due date (the order of which is 1, 3, 5, 4, 2)?

TABLE 10-5 Sequences and times (in hours) for five jobs (Morris Machine Shop—Problem 1)									
	JOB 1		JOB 2		JOB 3		JOB 4		JOB 5	
	Lathe	2	Drill	1	Drill	1	Mill	2	Lathe	2
	Mill	3	Lathe	1	Mill	2	Drill	1	Mill	2
	Grind	2	Mill	1	Grind	2	Lathe	3	Grind	2
			Grind	1			Grind	2		
Total		7		4		5		8		6

4. Using some Gantt charts, schedule the jobs listed using the following priority rules: first come, first served (jobs are numbered consecutively when they are released to the shop floor), earliest due date, and shortest processing time. Which one does the best? Can you devise a perfect schedule?

DUE	JOB 1 11/30		JOB 2 11/27		JOB 3 11/28		JOB 4 11/29		JOB 5 11/26	
	Mill	1	Drill	2	Lathe	1	Grind	1	Lathe	2
	Grind	2	Mill	1	Grind	2	Mill	1	Drill	2
	Mill	1	Lathe	2	Drill	1	Drill	1	Mill	2
	Drill	2	Grind	1	Mill	2	Lathe	2	Grind	1
			Lathe	1			Grind	1		
	Hours	6		7		6		6		7

5. Christopher Carson has investigated a future project and has determined the sequence of activities, their duration (both normal and if speeded up), and the cost of overtime (Table 10-6). Mr. Carson has found that for each week over 25 that it takes to finish the project, his firm will have to pay a penalty cost of $325.

(a) What is the critical path, and how long is it?

(b) Where should Mr. Carson use the overtime, if at all?

(c) After the project had been started, Mr. Carson found that a vital piece of equipment had broken down, causing activity F to be delayed an extra 2 weeks. What, if any, action should he take?

6. Refer back to Problem 3. Christopher Carson has run into some difficulties with some of the activities in his project. In particular, the time to complete several of the activities has lengthened by a week each. The affected activities are B, D, E, H, I, and K. How does this affect the critical path of the project? Can Mr. Carson achieve his desired schedule in any way?

TABLE 10-6   Time and money considerations (Problem 3)

ACTIVITY	FOLLOWS	TIME TO COMPLETE (weeks)	LENGTH ACTIVITY CAN BE SHORTENED (weeks)	EXTRA COST FOR OVERTIME (per week)
A		2	1	$300
B		4	1	250
C	A	1	None	
D	A	4	1	275
E	B	1	None	
F	C	10	3	250
G	D	7	1	400
H	F,G	4	1	300
I	E	6	2	325
J	H	3	None	
K	I	3	None	
L	J, K	6	2	350

REFERENCE NOTES

1. For a taste of the literature on job shop scheduling and other aspects of shop floor control, see:
    Adams, Joseph, Egon Balas, and Daniel Zawack, "The Shifting Bottleneck Procedure for Job Shop Scheduling," *Management Science* 34, no. 3 (March 1988): 391–401.
    Baker, Kenneth R., and Gary D. Scudder, "Sequencing with Earliness and Tardiness Penalties: A Review," *Operations Research* 38, no. 1 (January/February 1990): 22–36.
    Barrett, Robert T., and Sukran N. Kadipasaoglu, "Dispatching Rules for a Dynamic Flow Shop," *Production & Inventory Management* 31, no. 1 (First Quarter 1990): 54–58.
    Duchessi, Petter, Charles M. Schaninger, and Don R. Hobbs, "Implementing a Manufacturing Planning and Control System," *California Management Review* 31, no. 3, (Spring 1989): 75÷90.
    Flynn, Barbara B., "Repetitive Lots: The Use of A Sequence-Dependent Set-Up Time Scheduling Procedure in Group Technology and Traditional Shops," *Journal of Operations Management* 7, no. 1–2 (October 1987): 203–16.
    Fry, Timothy D., and Patrick R. Philipoom, "A Dispatching Rule to Allow Trade-Offs Between Inventory and Customer Satisfaction," *International Journal of Operations & Production Management (UK)* 9, no. 7 (1989): 72–78.
    Fry, Timothy D., Patrick R. Philipoom, and John H. Blackstone, "A Simulation Study of Processing Time Dispatching Rules," *Journal of Operations Management* 7, no. 3, 4 (December 1988): 77–92.
    Funk, Paul N., "Throughput Planning Instead of Capacity Planning Is Next Logical Step After MRP II," *Industrial Engineering* 21, no. 1 (January 1989): 40–44.
    Goldratt, Eliyahu, "Computerized Shop Floor Scheduling," *International Journal of Production Research* 26, no. 3 (1988): 443–55.
    Goldratt, Eliyahu, and Jeff Cox, *The Goal*, rev. ed. New York: North River Press, 1986.
    Hancock, Terence M., "Effects of Lot-Splitting under Various Routing Strategies," *International Journal of Operations & Production Management (UK)* 11, no. 1 (1991): 68–74.
    Jones, Marilyn S., and Roberta A. Russell, "Multiple Performance Measures in the Selection of a Sequencing Rule," *International Journal of Operations & Production Management (UK)* 10, no. 8 (1990): 29–41.
    McLeavey, Dennis W., and Seetharama L. Narasimhan, *Production Planning and Inventory Control*, Boston: Allyn & Bacon, 1985.
    Narasimhan, Seetharama L., and Paul M. Mangiameli, "A Comparison of Sequencing Rules for a Two-Stage Hybrid Flow Shop," *Decision Sciences* 18, no. 2 (Spring 1987): 250–265.
    Philipoom, Patrick R., Robert E. Markland, and Timothy D. Fry, "Sequencing Rules, Progress Milestones and Product Structure in a Multistage Job Shop," *Journal of Operations Management* 8, no. 3 (August 1989): 209–229.
    Putnam, Arnold O., "Shop Floor Control and Priority Scheduling," *Production & Inventory Management* 28, no. 2 (2nd Quarter 1987): 50–58.
    Ragatz, Gary L., and Vincent A. Mabert, "An Evaluation of Order Release Mechanisms in a Job-Shop Environment," *Decision Sciences* 19, no. 1 (Winter 1988): 167–189.
    Ritzman, Larry P., Barry E. King, and Lee J. Krajewski, "Manufacturing Performance—Pulling the Right Levers," *Harvard Business Review* 62, no. 2 (March-April 1984): 143–52.
    Sandman, William E., and John P. Hayes, *How to Win Productivity in Manufacturing*, New York: AMACOM, 1982.
    Sridharan, V., and William L. Berry, "Freezing the Master Production Schedule Under Demand Uncertainty," *Decision Sciences* 21, no. 1 (Winter 1990): 97–120.
    Sridharan, Sri V., William L. Berry, and V. Udayabhanu, "Measuring Master Production Schedule Stability Under Rolling Planning Horizons," *Decision Sciences* 19, no. 1 (Winter 1988): 147–166.

2. The two most common simulation techniques are OPT and Q-Control. Both OPT and Q-Control are proprietary software packages, so not much is known about their inner workings. OPT, which stands for "optimized production technology," has received more media and business attention, and somewhat more is known about how it operates. For Q-Control, see William E. Sandman and John P. Hayes, *How to Win Productivity in Manufacturing*, Dresher, PA: Yellow Book of Pennsylvania, 1980; for OPT, see Gene Bylinsky, "An Efficiency Guru with a Brown Box," *Fortune* (September 5, 1983): 120–32. See also the book, *The Goal*, by Eliyahu M. Goldratt and Jeff Cox (North River Press: 1984) for a discussion of OPT principles.

3. Jacobs, F. Robert, and Daniel J. Bragg, "Repetitive Lots: Flow-Time Reductions Through Sequencing and Dynamic Batch Sizing," *Decision Sciences* 9, no. 2 (Spring 1988): 281–294.

4. There are many readings to draw on now for information on Japanese production control techniques. Chapter 11 lists them in earnest. The first two readings in English have both had significant impacts on Western manufac-

turers. They are worthy of a special mention here. See Richard Schonberger, *Japanese Manufacturing Techniques: Nine Hidden Lessons in Simplicity*. New York: Free Press, 1982; and Robert W. Hall, *Zero Inventories*. Homewood, IL: Dow Jones-Irwin, 1983.

5.   For information on critical paths and PERT, consult:

Davis, E. W., "Project Scheduling Under Resource Constraints—Historical Review and Categorization of Procedures," *AIIE Transactions* 5, no. 4 (December 1973): 297–313.

Kelley, J. R., "Critical Path Planning and Scheduling: Mathematical Basis," *Operations Research* 9, no. 3 (1961).

Levy, F. K. et al., "The ABCs of the Critical Path Method," *Harvard Business Review* 41, no. 5 (September–October 1963): 98–108.

Malcolm, D. G. et al., "Application of a Technique for Research and Development Program Evaluation," *Operations Research* 7 (September-October 1959): 646–69. Reprinted as Chapter 29 in *Readings in Production and Operations Management*, E. S. Buffa, ed. New York: Wiley, 1966.

Meredith, J. R., and S. J. Mantel, *Project Management: A Managerial Approach*. New York: Wiley, 1985.

Smith, Larry A., and Joan Mills, "Projection Management Network Programs," *Project Management Quarterly* (June 1982): 18–29.

Wiest, Jerome D., and Ferdinand K. Levy, *A Management Guide to PERT/CPM: With Gert-PDM, DCPM, and Other Networks*, 2nd ed. Englewood Cliffs, NJ: Prentice-Hall, 1977.

Wiest, Jerome D., "Precedence Diagramming Method: Some Unusual Characteristics and Their Implications for Project Managers," *Journal of Operations Management* 1, no. 3 (February 1981): 121–30.

6.   See Herbert F. Spirer, "The Basic Principles of Project Management," *Operations Management Review* 1, no. 1 (Fall 1982): 7–8ff.

CHAPTER 11

JUST-IN-TIME MANUFACTURING

During the last decade, a revolution has been sweeping over much of American manufacturing, and indeed, manufacturing in other areas of the world. It is a revolution that owes much to the Japanese because it has been Japanese companies such as Toyota Motor that have led the way. Western companies have seen the power that can come from following some of the practices and thinking of the Japanese, and they are emulating those with increasing fervor and effect. As the Western success stories have piled up, more and more managers have been convinced that the prowess of the Japanese companies in the 1970s and 1980s has been due less to cultural differences (e.g., Japanese work attitudes, cultural homogeneity in Japan) and more to management practice that can be applied anywhere. For many manufacturers, these have been trying but exciting times.

This revolution in management thinking about manufacturing now goes under the name of "just-in-time manufacturing" (JIT). Unfortunately, it is not a particularly apt appellation. Most accounts of JIT in the business press stress the arrival of parts from vendors just as they are needed by the customer company as being what JIT is all about. As we discuss in this chapter, however, managing vendor arrivals is one of the least important aspects of JIT. The topic is much richer than that.[1]

This chapter begins with a discussion of what I term the JIT manufacturing philosophy because I believe that JIT is best viewed as a philosophy rather than as a grab bag of tricks or techniques. Following this discussion, some practical aspects of JIT implementation are addressed.

JIT MANUFACTURING PHILOSOPHY

The resounding worldwide success of many Japanese businesses over the last decade or so has forced companies of every type to take notice. A host of industries—from steel and automobiles to cameras and consumer electronics—have been affected dramatically by Japanese manufacturers. The impact has been all the more striking when one considers that as recently as the 1950s, the words "Made in Japan" were synonymous with shoddy merchandise. Ironically, the world's reception of Japan's shoddy merchandise immediately after World War II laid the groundwork for today's broad acceptance of the quality and reasonable price of Japanese

378

goods. In essence, Japanese industry diagnosed its problems and embarked on far-reaching changes that lifted the competitiveness of many Japanese companies to new highs. With improved quality as their goal (quality being defined, as in Chapter 4, as "conformance to the requirements"), major companies in Japan have reworked their manufacturing choices so that quality in manufacturing is supported at every turn.

To understand how the Japanese, and increasingly a number of Western companies, have realigned their manufacturing choices to produce quality merchandise and to become effective worldwide competitors in manufacturing, it is useful to contrast the JIT manufacturing philosophy with what can best be described as the "received tradition"—the precepts under which at least some companies in the United States and in other countries have implicitly operated. This received tradition is not a carefully constructed "philosophy" of manufacturing. Rather, it is a series of policies that have sprung up willy-nilly in response to the barrage of demands routinely made on Western manufacturing—demands particularly for quick delivery and for customization of the product. Also important to the development of this tradition has been the ascendancy of short-run financial measures by which to gauge the performance of business units in Western companies. I hope that the received tradition's precepts smack, at least a little bit, of what you have heard or thought before, so that the innovation in thinking that is represented by JIT can stand in sharper relief.

The distinctions, then, between the received tradition and the JIT manufacturing philosophy are best seen as contrasts in the policies derived from distinctly different competitive strategies: on the one hand, preoccupation with delivery, customization, and the short run; on the other hand, preoccupation with

quality. That so many manufacturers worldwide are studying the Japanese is testament to the effect quality has had in the competitive arena. If this philosophy had not been so effective, manufacturers here and elsewhere would not be studying it so diligently.

This analysis contrasts the JIT manufacturing philosophy with the "received tradition" via 17 contrasting pairs of precepts. Each pair of precepts will be presented first with a discussion about the received tradition (designated RT) and then with a discussion about the JIT manufacturing philosophy (designated JIT). By placing the precepts together here, I hope to encourage your understanding of how the pieces of JIT manufacturing practice fit together into an overreaching "philosophy."

1. Quality

RT: It costs money to make quality products.

This precept states that conforming to the requirements is often difficult and thus costly. According to the received tradition, there is a real, acknowledged trade-off between cost and quality; one cannot expect to be both the low-cost producer and the producer with the highest quality ratings. This precept contrasts starkly with the JIT philosophy.

JIT: Quality is free.

In their campaign to improve quality, Japanese manufacturers have shown the world that, in the long run at least, working to improve quality can actually lead to reduced costs as well. Doing things right the first time ensures not only quality products but also low-cost ones, as discussed in Chapter 4. Before the United States began its own march to improvement, the differences in the cost of comparable U.S. and Japanese goods were significant and were not fully explained by differences in wage rates or governmental policies. For example, detailed study of Japanese and U.S. automakers has put Japan's cost advantage, per comparable car, at about $ 2000 (early 1980s), although this cost difference has narrowed in recent years.[2]

This notion of quality being free rests at the heart of the JIT manufacturing philosophy. Improving quality, eliminating waste, and reaping the profit rewards of so doing, are much of what JIT is all about. Many of the precepts that follow build upon, and support, what perfect quality can mean to a company, its operation, and its customers. The fundamental importance of quality is what makes this set of precepts the first one.

2. Expertise

RT: Engineers and managers are experts. Workers serve their desires.

This received tradition acknowledges that charting the direction for the company is the responsibility of its management and engineering staff. These people are highly educated and are paid to be take-charge types. Workers are the means by which the directions set by these managers and engineers are fulfilled.

JIT: Workers are experts. Managers and engineers serve them.

According to the JIT philosophy, if the company is to do things right the first time, the workers have to do it. It is the workers who have to build quality in — workers on the line actually fabricating or assembling parts, workers testing the items, workers handling orders, workers handling materials. Under this philosophy, it is the workers who know the problems that are encountered in getting things done right the first time. They know what the problems are although they may not know exactly how to solve them. Managers and engineers under this philosophy provide the wherewithal and skills to work on the quality problems identified by the workforce. Many of the initiatives taken in the quality-of-work-life and quality circle movements are in this spirit. These initiatives put the worker at the center

of change in the organization, and view management and engineering as aids for the eyes, ears, and minds of the workforce — not simply as the authorities from whom orders come for directing the movements of the hands, feet, and backs of the workforce.

3. Mistakes

RT: Mistakes are inevitable and have to be inspected out.

The received tradition recognizes the truism that all of us are imperfect and that we are liable to make mistakes in our work. If these mistakes are not to reach consumers, we have to inspect products to rout out any that contain defects. The process should be designed, therefore, to make it fairly easy to identify products that have defects and to have those defects remedied (in a rework station) if possible and, if not, to have the defective products scrapped.

JIT: Mistakes are treasures, the study of which leads to improvement.

Under the JIT philosophy, mistakes are not inevitable. Recognizing that a defect exists is not enough, nor is purging defects by final inspection. This philosophy goes beyond the received tradition by holding to the view that every mistake is an opportunity to understand why the process is not foolproof. By tenaciously investigating every defect and its cause, one can gradually, bit by bit and project by project, improve the process so that it does not create defects. In this philosophy, zero defects is not just a goal but a standard obtainable by diligent investigation of the errors created in the process.

To adhere to the standard of zero defects, the JIT company tries to sustain a "habit of improvement," where all personnel of the company are engaged in continual efforts to improve product designs and process performances. Here, again, the workers are viewed as experts. Only they, once trained and equipped, can monitor all aspects of the process. To do such monitoring well, however, clear and compelling measures of product and process quality are required that are well communicated to all those associated with the process. (Charts and displays tracking quality measures are everywhere in JIT factories.) JIT workers are responsible for correcting their own errors, which they are often also required to uncover through 100 percent inspections of completed parts. To make it easy to identify defects, considerable time, thought, and expense are devoted to developing foolproof devices and automatic defect checks (termed *bakayoke* or *pokayoke* in Japanese) that monitor product and/or process characteristics continuously.

Not only are workers expected to monitor quality and to correct their mistakes (no rework stations on which to discard faulty work), they are also frequently part of any study of process improvements. Workers are schooled in the use of statistical techniques and cause-and-effect analysis (such as in the "fishbone" chart). Quality circles are one forum for the use of these skills in the workforce.

JIT companies typically reject certain notions about the acceptance sampling of lots. Why sample lots of completed parts or products when you would like to be sure that every piece is perfect? Only 100 percent inspection is desirable in that case. Inspection of every piece is particularly appropriate if the process is relatively "unstable" (i.e., not highly automated or lacking a long history of operation). With "stable" processes, sampling may occur, but it usually involves sampling the first and last pieces, to be assured that the process has not inadvertently wandered off course.

4. Inventory

RT: Inventory is useful—it keeps production rolling along.

The received tradition recognizes that inventory can buffer a manufacturing process from defects and other problems (parts shortages, late vendor delivery) that could otherwise cause the operation to grind to a halt. Inventory is like grease for the process, enabling it to keep functioning despite adversities.

JIT: Inventory is evil—it hides problems that should be allowed to surface.

According to the JIT philosophy, not only does inventory take up space and cost you money to carry, but it also lets you trade off that which is not perfect. If the process is to be refined continually so that it turns out only perfect products, all the problems in the process must be identified and worked on. If excess inventory exists in the process, there is a temptation to avoid working to perfect the process, and thus problems that should be uncovered are not. The Japanese term *jidoka*, meaning "making things visible," is at odds with the notion of inventory. What better way to focus everyone's attention on the need for quality than to have the process rigged so that there is no place to hide? Banish inventory and you expose nonquality work.

Japanese managers often refer to a flowing stream as a parable for production (see Figure 11-1). Water does not always flow evenly and at the same pace everywhere along a stream. Water sometimes gets trapped in deep pools, blocked by rocks, and other obstacles that sometimes remain hidden under the surface. These rocks and obstacles impede the smooth, swift flow of the stream. In this parable, the flowing stream is the flow of materials in the process. The pools of water are inventories, and the rocks in the stream are the imperfections in the process (such as quality problems; problems fostered by distrust such as double orders; extra safety stock; poor maintenance). If the stream is to run fast and clear, the rocks and obstacles must be removed. To accomplish this, however, the water in the stream has to be lowered so that even the hidden rocks are exposed. The meaning of

FIGURE 11-1   Making things visible: A parable.

the parable is inescapable. To uncover the imperfections in the process, inventories should be lowered. Otherwise, those imperfections are likely to remain hidden from view and the process will never be smooth flowing.

JIT companies recognize, of course, that some inventory usually is necessary and that no process can run with no work-in process. But they would argue that the quantity of inventory that is minimally required is generally much smaller than anyone thinks. Running a process with little work-in-process inventory is no small task; keeping product moving through the system often is done by applying considerable pressure to the worker to perform at the set rate so that other workers are steadily supplied with product on which to work.

This pressure is sometimes deliberately applied either by removing inventory from the system or by shifting selected workers off the line and into other work. Such "experiments" to see how the operation copes are done to uncover the bottlenecks and quality problems in the process. In some operations, there are lights above each worker that can be lit when a worker falls behind or when quality is jeopardized (a yellow light signals that a worker cannot keep up the pace without making errors; a red light signals that something fundamental is wrong and the process should be stopped). An operation with few yellow lights lit may indicate that too much labor or inventory is available, and management may react by removing workers from the line or tightening up on inventory until more yellow lights are lit. Red lights are valued because under them treasures of mistakes are found.

This important concept can be illuminated in still another way. Product design engineers often push their designs to the limits, so that the product fails. By understanding why the product failed, the engineers can begin to create another version that does not have that problem. The situation with the production process itself is analogous. You want to push it to its limits and discover what makes it fail. But how can you break a process? By removing its inventory. That is a way to expose the weaknesses of the process and indicate what needs to be improved.

5. Lot Sizes

RT: Lot sizes should be economic.

As noted in Chapter 8, good inventory management principles involve balancing the cost of carrying inventory against other costs involved in (1) setting up production, (2) granting price breaks for volume purchases, or (3) dealing with other variable costs associated with changing the order or production quantity. The simplest example of this balancing of costs is characterized by the celebrated economic order quantity formula:

$$EOQ = \sqrt{\frac{2DS}{ic}}$$

In the received tradition, manufacturers can fill in the variables in the equation (i.e., demand for the period, setup cost, prevailing interest rate, and variable cost of the product) and solve for the economic order quantity. This is straightforward, impeccable logic.

JIT: Lot sizes should be small, preferably 1.

In contrast, according to the JIT manufacturing philosophy, the EOQ equation, although thoroughly logical, is not seen as an expression to solve for some economic order quantity.

Rather, the formula is a mandate to lower setup costs, because if EOQ is to be low, the setup costs should decline. JIT companies have made tremendous investments in reducing the time required for each changeover in the factory and making sure that machine setups can be done quickly and routinely. The difference between the changeovers in many JIT factories and those in conventional U.S. factories is like the difference between a typical motorist changing a flat tire and the pit crews operating at the Indianapolis 500. Whereas the typical motorist may take 15 minutes to do the job, the pit crews at Indianapolis take 15 seconds.

There are stories told of tire-curing molds now changed in 10 minutes (as opposed to the 3 or 4 hours that manufacturers typically took) and of car panel stamping presses changed in under 10 minutes (as opposed to 6 hours historically). JIT companies strive for "single setup" (i.e., single-digit or less than 10 minutes setup) or even "one-touch setup" (less than 1 minute).

How do they accomplish this? Not too differently from the pit crews at Indianapolis, and usually, nothing fancy has to be done. Some precepts apply to setup reduction. Among them:

1. Approach setup reduction with the same techniques of industrial engineering and methods improvement that are routinely applied to operator tasks. This means documenting how the setup is done now (videotaping is popular) and then looking for ways to eliminate steps, improve the times of other steps, and make any adjustments either trivial or automatic. It means establishing a procedure and looking to improve it all the time.

2. Do as much of the setup as you can while the machine is still running. Have all materials

ready for the setup so that the machine is down for as little time as possible. To foster this "external" setup reduction, work stations are often designed so that setup materials can be stored close by. Attention may also have to be paid to preparing for the next step carefully (cleaning, sharpening, adjusting) well before the next setup is required.

3. Modify the operating equipment to permit easy setup and little adjustment. This can mean designing cartridge-type connections, color coding, multiple connection pins or plugs, special clamps, and similar measures.

4. Know what you want the machine to be set up for. Do not make the machine any more general purpose than you need to. This means understanding what kinds of parts are going to be assigned to it and what kinds of tools and fixtures are to be used.

5. Let one person do as much of the setup as possible. This may mean designing special carts for storing tools/fixtures/attachments, perhaps at the same height as the machine, so that physical effort is reduced.

6. Practice doing the setup. Practice is as important for reducing setup times as it is for reducing operator process times.

With quick setups and small lot production, the level of inventory can be kept lower for two reasons: (1) inventory need not accumulate at work stations into large lots to be transported later to other work stations, and (2) with quick changeovers, the process becomes more flexible. Production can be scheduled to match, often precisely, the current mix of demand for product variations. No longer must orders be filled out of inventories that are kept solely because expensive setups dictate long production runs. Orders can be filled by a very responsive production process.

6. Queues

RT: Queues of work in process are needed to be sure that machine utilization stays high.

The received tradition recognizes that machines represent a substantial investment and that they can best earn their return if they are kept busy. Thus having a queue of work in process in front of each machine is a way to make sure that the machine will stay busy, earning its keep.

JIT: Once in motion, always in motion. Production should be just in time; there should be no queues of work in process.

Under the JIT philosophy, if inventory is evil and if lot sizes are to be small, all materials have to keep moving through the process continuously. For that to happen the process needs to be tightly coupled, with work done just in time, as it is needed. This may make it vulnerable to disruption, but only by tight coupling can the performance of the process be made visible to all and the success or failure of segments of the process be easily determined and traced.

In the JIT philosophy, quality and just-in-time (tightly coupled) production are closely linked. Just-in-time production uncovers problems quickly, permits feedback to the workforce that is timely and meaningful, enhances the interdependence of workers and process segments, raises worker motivation and pride, and gives new meaning to cross-training. When these things happen, quality is worked on and improved; rework and scrap are reduced. The factory's systems created to cope with low quality or high inventories are no longer required and can be phased out. Productivity measures show great increases.

With improved quality, inventories can be reduced even more, space saved, and the process made even more tightly coupled. Thus quality consciousness and just-in-time production often feed one another, lowering costs even as quality conformance improves.

For some managers, unfortunately, just-in-time production is simply an excuse for berating suppliers. For these managers, going just-in-time means forcing the supplier to hold all the inventory and blaming the supplier for any quality problems encountered. In reality, just-in-time production is less a supplier policy than it is a prescription for linking together elements of the process so that materials are seldom required to sit idle. To the extent that a factory can make its future materials needs highly visible to suppliers and that the suppliers themselves can improve their own process capabilities, just-in-time production can be applied to the entire manufacturing pipeline, and inventories for all firms in the pipeline can be reduced. Some managers inappropriately make procurement policy step one in a shift toward just-in-time production. In fact, procurement changes should follow a plant's conversion to just-in-time operation, with all that such a conversion means for process redesign; production control; quality management; and leveled, steady, and highly visible production plans.

7. The Value of Automation

RT: Automation is valued because it drives labor out of the product.

In the received tradition, most appropriations for new capital are justified primarily by the labor savings that they generate. Automation is thus seen primarily as a capital-for-labor substitution.

JIT: Automation is valued because it facilitates consistent quality.

Under the JIT philosophy, machines are seen as a means to perform tasks the same way every time. This view assigns a value to automation that is over and above any labor savings that may occur; labor savings are secondary to the ability of automation to make the process perfect.

This point of view also implies that automation does not have to be sophisticated to be valued. Many JIT machine tools are simple, in-house creations of the company that are designed for a single purpose. Robots, mainly of an unsophisticated type, are widespread. This last point has received much press. While generally impressed with the numbers of robots, some observers have downplayed the Japanese commitment to robotics because the robots in use there are less sophisticated than some had thought. These people argue that Japan is not as much in the forefront of technology in robotics as some would have it. To me, this criticism misses the point. The Japanese use lots of robots; if one values producing quality, the extent of Japanese robotics should be praised, because those robots help improve the quality of the products being produced.

8. Sources of Cost Reduction

RT: Cost reduction comes by driving labor out of the product and by having high machine utilization. High rates of production are valued.

In the received tradition, the two most analyzed elements of manufacturing cost are labor wages and capital equipment expense. Considerable engineering and management effort is devoted to assessing how the direct labor component of a product can be lowered or whether machines of various types will be utilized enough to justify their capital expenditure. In line flow processes, line balancing is frequently management's chief concern. The dynamics of process change are seldom considered explicitly.

The lead in cost reduction comes from an examination of the factory's cost structure, not the dynamics of process change. Indeed, many companies fall victim to the following scenario: (1) cost pressure, from whatever source, strikes; (2) management examines the elements of the product's cost; and (3) questions are asked about how, and when, different components of cost can be reduced. Make versus buy is investigated, with the lure of shipping production offshore looming large. A cost-reduction investigation that starts with the cost structure can often lead to chasing low-cost labor across the globe.

The production process is evaluated highly if it can run large quantities of product through all the time whether or not the output is needed by the marketplace. The more that can be manufactured, given the assets and people in place, the more productive is the factory and the more highly it is held in esteem.

> **JIT: Cost reduction comes by speeding the product through the factory. Quick throughput times are valued.**

In the JIT philosophy, there is no particular focus on wages or capital expenditures or other aspects of the cost structure. The focus is on collapsing the time it takes for materials to run through the factory. Why is it better to concentrate on collapsing throughput times? The old fable of the tortoise and the hare can help our understanding. As we all recall, the tortoise won the race because it never stopped moving. Each step was itself slow, but the tortoise was diligent and went about its business single-mindedly. The hare, on the other hand, was all raw speed, but with a lot of stopping and starting and a propensity to be diverted from the race.

A lot of factories run like hares (see Figure 11-2). They operate equipment that is fast and fancy—when it is set up and actually running. However, materials in those factories spend a lot of time dozing, typically in a work-in-process inventory somewhere, waiting for the hare to get off its haunches to add value to them.

Fast equipment, but not well-maintained

Materials wait a lot to be worked on

Lots of WIP   Lengthy setups

Scattered layout; materials move a lot, often into and out of the storeroom

Jerky, stop & start schedules, often interrupted

No help sought of others

(a)

Slow equipment, always able to run

Materials do not wait long to be worked on

Quick setups   Little WIP

Compact, rational layout; materials move economically between operations

Smooth schedules, few interruptions

Help asked of all

(b)

FIGURE 11-2   The tortoise's factory versus the hare's.

Setup times are long and equipment breakdowns frequent. The tortoise-type factory, on the other hand, can actually make things faster, not because its equipment runs at higher speeds, but because materials keep moving slowly but surely through every process step. Value is added for a greater fraction of time spent in the process. Making things fast, then, does not mean operating the fastest machines or having the most automation. It does mean designing and organizing the factory so that materials are always moving forward through well-maintained equipment that is easy to changeover from one product to another.

ASIDE

The Case for Throughput Time Reduction

Consider some of the actions that focusing on throughput time reduction encourages:

1. *Good quality.* Throughput time is best defined as the weighted average time through the process of all the units of a representative batch or order. Thus, if throughput time is to be lessened, managers cannot afford to let an order languish in a rework station or sit idly by waiting for scrapped output to be replaced. All of the order needs to be made right the first time if the throughput time is to reach its lowest value. Feedbacks on problems must be quick and solutions timely. Problem-solving skills take on new importance and urgency, as does the desirability of involving everybody in the factory in the task. Throughput time reduction thus complements the push to improve quality, a campaign that many managers cite as greatly aiding their competitiveness.

2. *Low inventories.* Inventories are perhaps the chief culprit in hindering the speedy travel of materials through the process. Throughput time reduction encourages a variety of means for lowering inventories:

   - Make a bit of everything demanded every day. Rather than build large inventories to tide the factory over until the product is next produced, build in small volumes in just the quantities sold in the marketplace.

   - Rational layouts make storing work-in-process inventory inconvenient and make moving materials between workstations foolproof and effortless.

   - Setup time reduction facilitates small batches.

   - Supplier relationships emphasize good quality and reliable deliveries.

   - Production controls employ relatively few points of control in the factory. The fewer the "artificial" barriers to the flow of materials, the swifter that flow can be.

3. *Process rationalization.* Throughput time reduction can focus attention on which steps of the process truly add value and which do not. This can then lead to the removal of unnecessary steps and/or the modification of others.

4. *Attention to bottlenecks.* With throughput time reduction as an objective, bottlenecks must be addressed without delay. As it is, bottlenecks are vivid indications of problems with quality, maintenance, layouts, flows, operator training, and the like. One simply cannot dismiss the hard decisions involved with bottleneck breaking by observing that the squeeze on capacity will evaporate when the schedule is changed or when the "heat" is off. These problems must be analyzed and solved, preferably permanently. Here again, the involvement of everybody in the factory assists this process.

5. *Diminished chaos and confusion.* Chaos and confusion have always been recognized as detrimental to productivity gain. Fluctuations in the product mix and in volumes, expediting of shop orders and vendor deliveries, engineering and purchasing change orders, and materials shortages have long been decried by factory management. Concentrating on throughput time reduction, however, helps managers avoid the all-too-compelling sense of urgency these conditions present. Better to work toward and to be measured on a lower average throughput time for representative orders than to rush around trying to expedite the "hot" one.

Next consider four key spillover benefits:

6. *Overhead elimination.* In many factories, the cost item that is growing most rapidly is overhead. And much of that overhead can be attributed to the factory's numerous transactions, many on the materials side of the business—purchasing, materials receipt, inspection, production planning, inventory control, materials handling, production control and

the accounting for it all. If the need for these transactions and the accounting that tags along with each of them can be eliminated, the overhead personnel associated with these transactions can be eliminated as well. I know of no better way to do this than by focusing on throughput time reduction. The popular Pareto "80–20" rule may apply—80 percent of the factory's headaches, especially for the overhead functions, coming from 20 percent of the products, the 20 percent that for one reason or another linger in the factory longer. With a quick flow of materials through the factory, many fewer resources have to be spent planning them, tracking them, adjusting their status, and costing them. Simple, quick flows within the process can breed simple, quick management systems to plan and control them.

7. *Quick response to the market.*   Quick delivery of exactly what the customer wants has always been an effective selling tool. Concentration on throughput time helps to align capacity and output rates to the quantities actually demanded in the marketplace. Manufacturing can then support the sales effort better, and everybody wins.

8. *Improved capital appropriations.*   The conventional means of justifying new equipment focuses on direct labor cost reduction, machine utilization increases, and sometimes, quality improvements. Lately, the narrowness of this conventional practice has come under attack. The change in throughput time that any new investment can make should also be considered. With lower throughput times, flexibility can be enhanced and quicker customer service can be achieved. Space and inventories decline and quality improves. Other things equal, a capital expenditure that promises to reduce throughput time should be preferred to one that does not, even if it means holding "excess capacity" (which is better than "excess inventory"). Indeed, the potential of any new investment or technology is fully realized only when it can speed up the flow of materials through the process.

9. *More times to practice.*   When the time to do anything is diminished, just as when setup times are diminished, one can do more of it during the same calendar period. And, given that practice does make perfect, more opportunities to practice anything is a real advantage. More learning can take place and more experimentation. Throughput time reduction can thus lead to a more rapidly paced process of improvement than ordinarily occurs.

How does a factory cease being a hare and become more like a tortoise? Throughput time reduction does not improve productivity by itself. Rather its attractiveness is as a broad-gauge objective—an umbrella concept—that stimulates a host of complementary actions and tactics within the factory that, in turn, improve productivity.

The Power of Linking

The attractiveness of throughput time reduction as a principle to guide thinking about productivity improvement is powerfully illustrated by what many plants are now doing with manufacturing cell (group technology) concepts, U-lines, and multimodel lines for particular classes of products. Where some years ago many plants' processes could be described as hybrids—part batch flow and part line flow—now significant numbers of them have linked their former batch operations together directly with their line flow operations. Doing so usually means altering the way the factory is organized and measured, but results in greatly improved productivity, flexibility, space utilization, and quality.

Figure 11-3 illustrates the point. The upper part of the figure depicts a common hybrid operation, where the first part of the process is fabrication, typically done in sizable batches

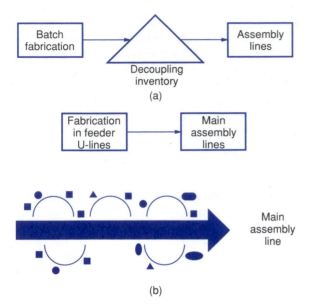

Batch fabrication → Decoupling inventory → Assembly lines

(a)

Fabrication in feeder U-lines → Main assembly lines

Main assembly line

(b)

FIGURE 11-3   The hybrid process.

on a series of machines that are laid out like the usual job shop, with similar machines located together. The second part of the process is a standard assembly line that tries to match its production to customer orders. The fabrication portion of the factory is frequently viewed as too slow and clumsy to match its production rate to that of the assembly line, and thus the production plans and schedules for the two halves of the factory are different. A "decoupling" inventory has to be kept to buffer the requirements of the assembly line from the slower reacting batch flow operation.

The genius of the new thinking — as captured by the goal of consistently reducing throughput time — is the rejection of the need for a decoupling inventory whose purpose is to keep labor efficiencies and machine utilizations high. Instead, the emphasis turns to setup time reduction and process simplification to improve the reaction time of the fabrication operation and to link it directly — both physically and through

production planning — to the assembly line. This is where the innovation of the manufacturing cell or the U-line plays a part. These innovations greatly aid the flexibility of fabrication, in part by limiting their role to a family of parts. The hybrid process thus gets converted to a more powerfully productive line flow operation (the bottom part of the figure). When the character of a process can get transformed in this way, the productivity and flexibility of the entire process soars, even though particular departments may suffer by conventional measures. Better to suffer more "idle," but flexible labor that produces just what the market is buying than to keep labor busy producing what is not really needed.

9. Material Flow

RT: Materials should be coordinated and pushed out into the factory.

Most conventional materials management systems try to coordinate the procurement and handling of materials to the established master production schedule, so that the factory departments involved will have their materials just before they need them. This is the concept behind MRP systems. This notion of production scheduling and control could be characterized as a "push," where materials are to be pushed out of material stockrooms or out of particular manufacturing departments according to a set schedule that is influenced heavily by demand. In such a system, considerable time and effort are spent making sure that particular problems can be expedited efficiently if changes to the schedule have to be made on short notice. Extensive coordination is needed to be certain that the materials are made available according to the dictates of the often-changing production schedule. Buffer stocks, safety lead times, and other tactics are often used to make sure that the materials are there "just in case" they may be needed.

In practice, this often involves the accumulation of materials for a set period of time, be it a day or a week or so, and giving the supervisors on the factory floor the ability to select which jobs they will perform first from among the bundled materials available. Much of the shop floor control function revolves around the precise scheduling of work to be done by different areas of the factory and the determination of the status of all orders. In such a conventional system, the materials function can be seen as driving the pace of production in the factory. To avoid idle worker and machinery time, however, the traditional outlook is to make sure that there are queues of work ahead of each machine and that there are banks of parts awaiting work or assembly by the workforce.

JIT: Materials should be pulled through the factory.

Under the JIT philosophy, to ensure that no work-in-process queues exist, the production schedule and the shop floor control system must have the "downstream" operations triggering the initiation of work in the "upstream" operations. Otherwise, there is a greater risk of accumulating work-in-process inventory ahead of any slow operations in the process. By this philosophy, providing a unit more than is demanded is as bad as providing a unit less than demanded. Under such a philosophy, there can still be an MRP system that explodes requirements through the bill of materials to ensure that materials arrive at the factory in time to be used, but the MRP system must not be used to authorize the release of an order to the factory floor; only processes downstream can act to release such orders. Stockrooms are a nemesis in such a philosophy, as they can interrupt smooth-flowing signals between process steps. The ideal is for all the inventory to be on the factory floor, visible to all.

The most famous example of this type of "pull" system is that operated by Toyota Motor Company, the *kanban* system. It should be remembered that the kanban system works well only when demands are fairly stable, when the production schedule is frozen for fairly long periods of time, and when few, if any, irregularities in product models have to be coped with.

ASIDE

Kanban, a Japanese Contribution to Production Control.

The most celebrated of the Japanese "pull" production control systems is the kanban system developed over many years by the Toyota Motor Company. Kanban is Japanese for card, and the Toyota system uses a plasticized card for indicating what upstream operations should be doing to keep downstream operations functioning. The Toyota system uses two kanban cards: (1) a "production" kanban, which triggers production in the manufacturing department upstream, and (2) a "move" kanban, which indicates use by the downstream department. Figure 11-4 displays the contents of a move and a production kanban card.

The move card includes the following information: part number, capacity of the standard container the part is held in, the card number (such as 2 of 3), the supplying workcenter (its outbound stockpoint number), and the using workcenter (its inbound stockpoint number). The production card includes this information: the part number to be produced, the standard container's capacity, the supplying workcenter number, the card number (such as 2 of 3), the materials required by the workcenter and where these materials can be found, and often useful information such as tooling or fixturing needed.

A. Typical move card

FROM SUPPLYING WORK CENTER #52 PAINT	PART NO. A575 GAS TANK MOUNT	TO USING WORK CENTER #2 ASSEMBLY
OUTBOUND STOCKPOINT NO. 52-6	CONTAINER: TYPE 2 (RED) NO. IN EACH CONTAINER: 20 CARD NO. 3 NO. CARDS ISSUED: 5	INBOUND STOCKPOINT NO. 2-1

B. Typical production card

WORK CENTER #52 PAINT
PART NO. A575 GAS TANK MOUNT

PLACE AT STOCKPOINT NO. 52-6

MATERIAL REQUIRED: PAINT #5 ST. BLACK
PY 372 STAMPING

FOUND AT: WORK CTR. 31 PRESS SHOP
STOCKPOINT NO. 31-18

C. Move card used by Hino Motors Ltd. to pick up parts from a supplier

Supplier and pick up location   Customer and delivery location

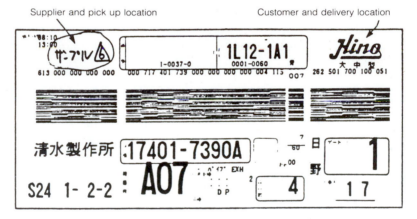

Note the bar code. Only move cards sent to suppliers are imprinted with the bar. Upon receipt, the bar code is read, and a cumulative total is kept on-line by computer. Hino uses it only for development of receipts to pay the supplier's invoice, an accounting function not related to control of the part for production. However, the cumulative counts can be used by production control to keep a cumulative count of parts received.

The times posted at the upper left refer to delivery times, twice daily for this part. Sometimes the times posted refer, more precisely, to the times the move cards must be picked up to give to the truck driver bringing a new delivery of parts.

Source: Robert W. Hall, *Zero Inventories*, Dow Jones-Irwin, 1983, pp. 44–45.

FIGURE 11-4   Kanban cards.

continued

The use of these kanban cards is illustrated by reference to Figure 11-5 and the following discussion. Note that workcenter 1 feeds workcenter 2.

1. A container of parts is selected for use at workcenter 2's inbound stockpoint. The container's move card is removed and stored in a hold box.

2. Later, the move card is picked up from its hold box and taken to workcenter 1's outbound stockpoint, where several full containers of parts are located. The production card in a container there is re-

moved and replaced by the move card, and the container is transported from workcenter 1's outbound stockpoint to workcenter 2's inbound stockpoint.

3. The production card that was removed at workcenter 1's outbound stockpoint is stored in a hold box. From time to time, when the workcenter is ready to perform more work, the production hold box is checked. Each production card is authorized to produce a standard container of parts. A worker merely goes over the workcenter's inbound stockpoint, removes the move card from a con-

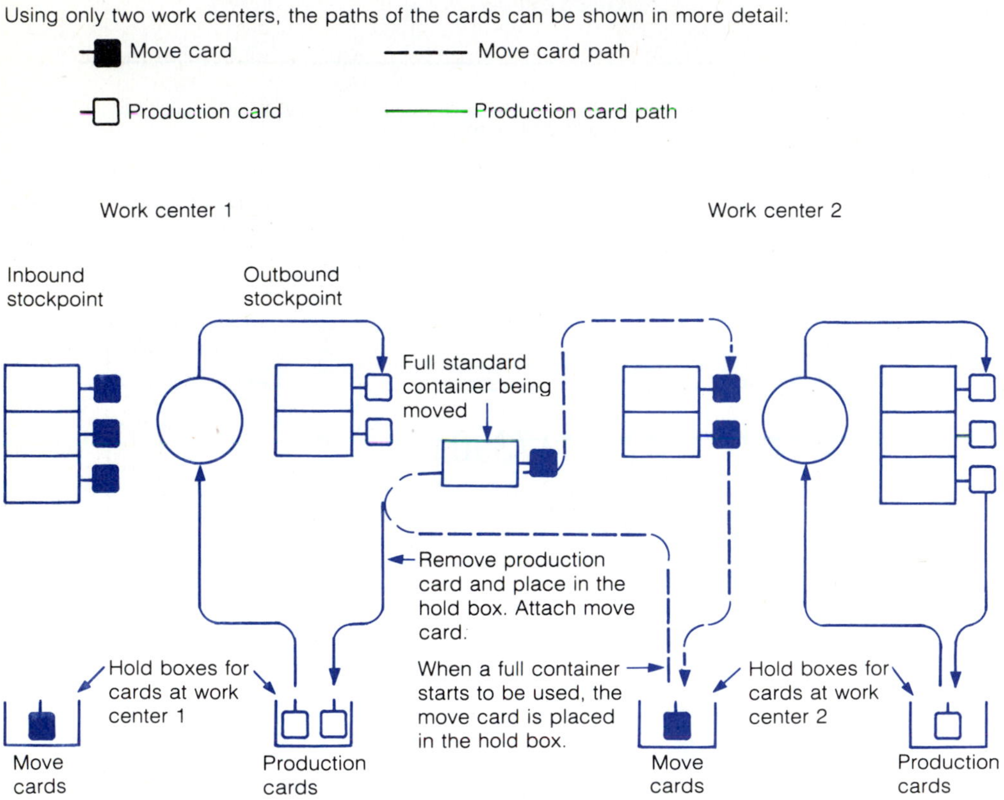

Using only two work centers, the paths of the cards can be shown in more detail:

Source: Robert W. Hall, *Zero Inventories,* Dow Jones-Irwin, 1983, p. 44.

FIGURE 11-5   The operation of a two-card kanban system.

tainer (as in point 1), substitutes the production card, and proceeds with the work.

There are several rules to follow as workers use the kanban cards:

1. Either a move card or a production card must be attached to any container full of parts.

2. Containers of parts are never moved from a supplying workcenter's outbound stockpoint to a using workcenter's inbound stockpoint without a move card attached.

3. Standard containers are always used and filled only to their indicated capacity.

4. Work is accomplished at a workcenter only when there is an unattached production card to authorize it.

This Toyota-type kanban system is appropriate for those just-in-time production processes that make relatively high-volume parts under demand conditions that do not fluctuate severely. Such a kanban system, therefore, is not advised for all types of production.

There is another type of kanban system that is also used in Japanese companies: a single-card kanban system. This system uses only the move card that serves to regulate deliveries to user workcenters. Parts are produced or bought according to a schedule, but delivery is controlled by a kanban. This single-card kanban system has much in common with the well-known two-bin system that is employed in many companies to control the stock of parts flowing to the process. In a two-bin system, orders of parts are triggered when the last container (the "second bin") is opened and parts are withdrawn from it.

The single-card kanban system works much like the two-bin system in that it triggers replenishment only when available parts have been depleted to a fixed level. The single-card kanban system, however, is somewhat more disciplined than the typical two-bin system. Only standard containers are used, and the quantity per container is kept exact. The number of full containers is severely limited, and the quantity in any container is generally small, so that at least one container is used up daily.

The single-card kanban system does not exercise as much control over production as does the Toyota two-card kanban system, but it can still act to keep inventories lower than they might otherwise be. They help orient production control more toward pull rather than push.

A variation on the single-card kanban system that has been popularized by Hewlett-Packard and others is the kanban square. In this variation, colored tape is placed on the floor at and between various workcenters. Workers are instructed not to work on anything unless there is free space in the kanban square between their workcenter and the next, and not to bring in any work that cannot fit into their work area.

The kanban squares thus work like a limited number of standard containers to restrict the amount of work-in-process permitted on the factory floor. By making the squares either smaller or larger, management can adjust the quantity of work-in-process on the floor.

The success of these just-in-time production control techniques has been, at times, spectacular. Work-in-process inventory levels and space requirements typically drop sharply (often 50 percent or more) and throughput typically increases, sometimes substantially. The use of kanban cards, of course, is not wholly responsible; there are the other elements of manufacturing in this environment. Nevertheless, kanban cards are an indication of the potential that good production control has for improving operations.

Production planning is one of the toughest factors in a pull system. Planners must take care not to introduce too much change nor to press production so much that there is no slack in capacity. Yet, on the other hand, the process needs to be kept reasonably flexible and free of sizable pockets of excess capacity. A master production schedule (and final assembly schedule) that meets these criteria is often a delicate thing to concoct.

10. Flexibility

RT: You get flexibility at a cost — excess capacity, general-purpose equipment, inventories, overhead, and so on.

Under the received tradition, flexibility, like quality, is attained by sacrificing other things. It can be achieved, but only by specifically planning for it and building up a stock of assets — space, machinery, people, materials — so that changes in the product mix or in product volumes or in new product introductions do not cause havoc in the factory.

JIT: Flexibility comes from contracting all lead times — factory throughput times, vendor lead times, new product development cycles, order entry and production planning cycles, engineering change order lead times, and other lead-times-to-change.

With the JIT philosophy, flexibility, while difficult to achieve, does not necessarily mean the imposition of additional costs. Rather, the strategy is to shrink all the lead times that affect the factory and its operations. If suppliers can react quickly, if orders can be handled quickly, if production can be planned and scheduled quickly, if engineering changes can be made quickly, if new products can be brought into production quickly, if all that the company does can be done without any wasted motion, then flexibility can be achieved along with low cost and high quality. Indeed, the more flexible the process can become, the fewer the restrictions there are to impede production planning. One can thus get closer and closer to the ideal of being able to schedule through the factory today what was ordered yesterday.

Lead times outside the factory — in the office, in the drafting room, and so on — are slashed in exactly the same ways that factory throughput times are slashed, by purging waste from them and by diligently investigating where value is not added and how operations can be subsequently modified to permit more added value. No element of the job is too small to ignore. Removing waste for all the small operations/workstations of the company will remove waste from the big ones as well. This takes discipline to do effectively, but it can be done.

Here, again, the notion of reducing throughput time is a useful way to identify where the waste is — be it the entry of an order, or the handling of payables/receivables, or the month-end accounting close, or the processing of inquiries from the sales force. When delays occur, there lies waste in the process. Diagramming the flow of the process and of information, and attaching times to it, is a proven technique for aiding the efficiency of the overhead activities of a company.

11. The Role of Overhead

RT: Overhead functions are essential.

At the same time that the received tradition pushes for lower direct labor cost and higher machine utilization, it often implicitly acknowledges the requirements for overhead functions (such as purchasing, industrial engineering, production scheduling, inventory control, quality control, materials handling). Much of this overhead is dedicated to either coordinating aspects of the process or studying it. Although manufacturers operating under the received tradition may grumble about overhead rates that in recent years have soared to 500 percent or more of direct labor, typically they do not take positive action to arrest such growth in indirect labor.

Not only does the received tradition promote the growth of "staff" versus "line" management, but also the people picked for such staff assignment are apt to be trained for their "specialty" and desire little, if any, rotation into other staff or line assignments. Their careers are often wholly within their discipline. Their allegiance as well may be more to the discipline itself than to the company.

JIT: Any labor that does not directly add value to the product is waste.

JIT manufacturers have a fervor about ridding the process of extraneous labor, both direct labor and indirect labor. This may be done, for example, by some simple materials handling schemes (one worker positioned close enough to another to hand over the product). More important, the whole push for getting the process clean and functioning perfectly results in fewer inspectors, fewer rework stations, less scrap to be handled, and simpler production schedules.

Moreover, JIT companies are much more committed to limiting the size of any staff and to rotating managers through many different kinds of jobs, rather than promoting specialization. Through just-in-time production and the tight coupling of the process, many of the coordination tasks needed in job shop/batch flow process are eliminated. Overhead can actually shrink as its waste and the process-related waste it supports are purged from the operation. JIT companies also require more thought and study from their line managers and workers than happens with many other companies. JIT companies can get away with adding such burdens to their line people, because the process itself is simplified and smoothed so that managers and line workers do not have to fight fires continually. Managerial time is viewed as better spent improving the process permanently rather than trying to get a "hot" order through.

12. The Cost of Labor

RT: Labor is a variable cost.

Under the received tradition, labor is one cost that can be forgone when demand drops. In many industries, layoffs are a hard fact of life; although Western companies do not like to lay off their people, losses are expected to occur when times are tough.

JIT: Labor is a fixed cost.

The notion of lifetime employment in Japan has received much attention in the popular press. In fact, lifetime employment is a reality for only a quarter or so of the Japanese workforce, and only then in the very strongest companies, those which in the last decade or so have become household words in the United States. In such companies, during downturns in demand, workers may be put on special maintenance assignments or a considerable degree of subcontracting may be brought in-house. Supplier firms that are less strong are made to suffer proportionately more than the larger companies that guarantee employment for their workers and managers. Despite this caveat, lifetime employment remains a goal for even the weaker companies in Japan. The benefits of lifetime employment are clear both for society as a whole and for the companies involved. The dedication of the workforce is generally high, although this dedication is becoming less strong in today's Japan. For JIT companies everywhere, a no-layoff policy is a typical goal. The chief implication of this view of fixed labor costs pertains to the attitude that the company holds toward personnel. When workers are lifetime employees, the company is forced to take seriously its hiring, training, cross-training, management development, job rotation, and career path counseling. Personnel departments in JIT companies are strong; their work is not ignored, nor are end runs made around them, as happens in so many other companies.

13. Machine Speed

RT: Machines are sprinters, and pulled hamstrings are to be expected.

According to the received tradition, machines are expensive and if they are to return that expense (and more), they must run at capacity. Breakdowns may then be expected. Moreover, in the received tradition, large, multipurpose machines are valued more than specialized, less flexible machines.

JIT: Machines are marathon runners, slow but steady and always able to run.

According to the JIT philosophy, machines should be run only as fast as they can turn out perfect pieces consistently. Machine speeds in JIT companies are often not set as high as they could be, and more important, preventive maintenance is performed religiously. Too often, the practice is to perform maintenance "when we can get to it," rather than at definite times that may be inconvenient for the production delivery goals set for the process. The best maintenance people are too frequently assigned to the false heroics of fixing equipment that has broken, rather than devising and accomplishing better preventive maintenance procedures.

The freedom to perform religiously systematic preventive maintenance and daily machine checks in JIT factories comes in large part from the practice of deliberately planning for less production than is theoretically possible. Rather than put pressure on the process to produce at capacity, JIT companies provide for the slack required for maintenance, process improvement, and the like. What JIT companies sacrifice in lost output, they more than make up for in reduced machine downtime, higher yields, less rework, and the like.

JIT companies also do not fear using specialized equipment, often of their own devising. Such equipment is frequently lower cost, slower, and more inflexible than is typically contemplated by other companies. Yet, for JIT companies, such equipment is valued because it may help make the process more tightly coupled; its use, along with that of other inflexible equipment, can result in a production system

that is at once more flexible, quicker to respond, and less subject to breakdown than a production system that is built around a large, multipurpose, and expensive piece of equipment. As noted earlier in this book, many small production centers or production lines are typically easier to schedule than one large one, and inventories can be kept lower at the same time.

14. Procurement

RT: Procure from multiple vendors.

Vendors in the received tradition should be managed with an eye toward the benefits of competition. Vendor should be set against vendor so that the resulting competition keeps prices low and service attractive.

JIT: Procure from a single supplier.

JIT companies view their suppliers as extensions of their companies. Indeed, they prefer to use the word "supplier" rather than the word "vendor," which has a somewhat pejorative cast. They are part of the company, to be worked with to improve their quality and to guarantee their delivery to be "just in time." Price is not nearly as important as quality or delivery dependability. Why risk compromising quality and threatening just-in-time production for a minor break in price?

To JIT companies, purchasing is a critical function. Much of management in Japan, for example, consists not so much of organizing people within a company as organizing other companies to do the work. There is less vertical integration in Japan (e.g., materials as a percentage of the cost of goods sold is about 80 percent for Toyota versus about 50 percent for General Motors), and this leaves the management of suppliers and subcontractors as a very important endeavor.

The activities of a JIT company purchasing department have much less to do with the price negotiations and vendor qualifications engaged in by purchasing departments operating within the received tradition. Rather, the purchasing department's mandate is to support product quality and just-in-time production. This implies a number of changes:

- Developing long-term contracts with suppliers for whom the company's business is significant.
- Encouraging suppliers to produce and ship in small lots of exact quantities, demanding frequent deliveries.
- Enabling suppliers to react to "emergencies" (e.g., tool breakage, surprise demands) better.
- Practicing "family-of-parts" sourcing so that the supplier provides a series of related parts so that product mix changes do not cause vast volume fluctuations for the supplier.
- Encouraging supplier process quality so that incoming inspections are eliminated.
- Providing minimal part specifications to suppliers so that supplier talent and value engineering can be devoted to improving one's own product performance and cost even with new product introductions.
- Insisting on reliable transportation. Frequent deliveries are increasingly made by trucks modified for lots of less-than-truckload shipments. These trucks then enter a fixed-route cycle of pickups and deliveries, much like a bus route.

In essence, the time freed from letting out quotes for work is devoted instead to working with suppliers so that they can truly meet the full needs of the customer company. If all goes well, inventories throughout the entire pipeline are reduced, not just those held by the customer company.

There is always concern that if a company goes to a single supplier, it will be too vulnerable to strikes and Acts of God. Such concerns are typically overdrawn. First, even though one supplier may supply the part or material, other suppliers may also be qualified to supply. This helps to diversify the risk, which in the case of a strike, is known with considerable lead time anyway. It is useful to distinguish between single sourcing and sole sourcing. The latter is what occurs when only one source can even be qualified. The former can be managed fairly easily.

15. Expediting

RT: Expediting and "work around" are ways of life.

If, as in the received tradition, delivery and customization are viewed as essential to one's competitive success, one must then be able to cope with the whims of consumers. Many of the factory's systems are developed with expediting in mind, and much of the time of the factory's managers is spent expediting material from suppliers, expediting orders through the factory, and scrambling to devise ways to work around the materials that are not yet there.

JIT: Expediting and "work around" are sins.

The surest way to destroy just-in-time production and quality is to change continually the production scheduling or the engineering documentation on the product. In the JIT philosophy, rush tags and engineering change orders are inventions of the devil. Instead, production schedules are set well in advance. They are frozen for longer periods than is the case in other companies, and production is run with less chasing of demand and more of a level production strategy. If some product variety and customization have to be forgone, so be it. To do otherwise compromises quality too much.

This is not to say that JIT companies cannot provide a wide range of products. Their concern for quick changeovers and small production lots permits greater product variety and flexibility than one might think. The configuration of production lines (U-shaped rather than straight line), the cross-training of the work-force, and concern for keeping the entire process tightly coupled argue for considerable changes in worker job assignments and enhanced flexibility in JIT factories. Yet marketing whims are less apt to disrupt JIT companies than a company that accepts the received tradition.

16. Housekeeping

RT: Work means getting your hands dirty.

Just as factories under the received tradition accept mistakes as inevitable, they also often accept dirt and untidiness as the price one pays to get the product out on time.

JIT: Buddha and Confucius like things clean.

This statement is a variation on "Cleanliness is next to godliness." Cleanliness, according to the JIT philosophy, goes hand in hand with the notion of making things visible. The cleaner and the more organized the factory floor, the better able the workers will be to identify problems in the process and call everyone's attention to them.

17. Horizons

RT: Quick and dirty often has to be "good enough."

In the received tradition, results are often expected, even demanded, in a relatively short time. Much of the corporation's attention is spent on making sure that the quarterly income goals are met. This concern naturally laps over into manufacturing philosophy and the pressures on plant managers. Short-term goals often carry more weight than long-term goals, and short-term results that deviate somewhat from long-run plans may set off a premature redefinition of the entire long-term strategy.

JIT: Patience is more than its own reward.

In the JIT company, short-run goals are kept in perspective. This attitude implies that patience will prevail as the process is purged of its problems and the workforce becomes accustomed to performing to the standard of zero defects. Similarly, capital appropriation request procedures may not be viewed as nearly as high a hurdle as is sometimes the case in many companies. Japanese companies are generally tied to longer-term goals, such as market share. Short-run financial concerns, largely because of Japan's system of industrial organization, do not have much of an impact on corporate policy.

In Conclusion

These precepts from the received tradition and the JIT manufacturing philosophy are listed for comparison in Table 11-1. What should be evident from examining them is that the different motivations for these contrasting philosophies have yielded some points of view about manufacturing that often radically differ from one another. I hope that by now you are convinced that JIT is a philosophy, a set of principles that hang together and that support the improvement of quality and a culture of never-ending improvement.

A caveat is in order. Although the JIT manufacturing philosophy has proven to be very effective for a number of Japanese companies, mainly in high-volume, repetitive industries, this philosophy may not be applicable in all its particulars to diverse sets of businesses, even in Japan. Yet for many companies it does portend a better way. Many U.S. companies have recently adopted this type of philosophy and have been very satisfied with the changes that it has brought to their operations. The precepts listed in the table can be applied in any country or culture.

The power of JIT can be seen by a comparison of two automobile factories (Table 11-2), due to a research study by Robert Hall and Jin-

TABLE 11-1   JIT manufacturing philosophy versus the "received tradition": 17 precepts

THE RECEIVED TRADITION	JIT MANUFACTURING PHILOSOPHY
1. It costs money to make quality products.	1. Quality is free.
2. Engineers and managers are experts. Workers serve their desires.	2. Workers are experts. Managers and engineers serve them.
3. Mistakes are inevitable and have to be inspected out.	3. Mistakes are treasures, the study of which leads to process improvement.
4. Inventory is useful—it keeps production rolling along.	4. Inventory is evil—it hides problems that should be allowed to surface.
5. Lot sizes should be economic.	5. Lot sizes should be small, preferably one.
6. Queues of work-in-process are needed to be sure that machine utilization stays high.	6. Once in motion, always in motion. Production should be just in time; there should be no queues of work-in-process.
7. Automation is valued because it drives labor out of the product.	7. Automation is valued because it facilitates consistent quality.
8. Cost reduction comes by driving labor out of the product and by having high machine utilization. High rates of production are valued.	8. Cost reduction comes by speeding the product through the factory. Quick throughput times are valued.
9. Materials should be coordinated and pushed out into the factory.	9. Materials should be pulled through the factory.
10. You get flexibility at a cost—excess capacity, general purpose equipment, inventories, overhead, etc.	10. Flexibility comes from contracting all lead times—factory throughput times, vendor lead times, new product development cycles, order entry and production planning cycles, engineering change order lead times, and other lead-times-to-change.
11. Overhead functions are essential.	11. Any labor that does not directly add value to the product is waste.
12. Labor is a variable cost.	12. Labor is a fixed cost.
13. Machines are sprinters, and pulled hamstrings are to be expected.	13. Machines are marathon runners, slow but steady and always able to run.
14. Procure from multiple suppliers.	14. Procure from a single supplier.
15. Expediting and work around are ways of life.	15. Expediting and work around are sins.
16. Work means getting your hands dirty.	16. Buddha and Confucius like things clean.
17. Quick and dirty often has to be "good enough."	17. Patience is more than its own reward.

ichiro Nakane.[3] This is a comparison of a Japanese and an American plant from 1986. Needless to say, since that time, and even before, the American auto companies have made giant strides in process improvement that have narrowed the gaps shown here. (The description of the General Motors assembly plant in Oklahoma City, at the end of this text, is a good indication.) What this comparison highlights, however, is the power and pervasiveness of JIT, and how it spills over benefits to all aspects of company operations.

Another indication of the spillover benefits of JIT is due to Kim Clark and Takahiro Fujimoto.[4] They studied the new product development times of all the world's major automakers and

TABLE 11-2   Comparison of Japanese and American Auto Production, 1986
(Based on two specific auto assembly plants)

Item for Comparison	Japan	United States
Number of body styles (models) at plant	8	6
Floor space index	100	600
Number of employees index (2 shifts)	100	460
Average leadtime to domestic customer	12 days	5-6 weeks
Hours of cars on line	10.8	24.7
Operators per line station	3.5 (teams)	1.3
No. of line stations (index)	100	1000
Parts per operator at lineside	23	3.3
Material on hand (in assembly hours)	4	128
Container sizes	Totebox	Larger wire baskets
Container positioning	Easy reach	Long walk
Line speed	48/hr.	75/hr.
Line balance idle time	6% idle time	22% idle time
Line speed variation	15% each mo.	Rarely changed
Jidoka	3–4% stop time allowance	Only supervisors stop line
Daily production volume changes possible	10%	1%
No. of supplier shipping points	100	500
Parts per supplier shipping point	60	13
Normal engineering changes	Weeks	Months
New tooling built at plant	20%	<1%
Equipment modified at plant	Almost all	Very little
Relations with supplier personnel	Close	Distant

Source: R. W. Hall and J. Nakane, *Flexibility: Manufacturing Battleground of the 90s*, AME Research Report, 1990.

noted that the Japanese are able, as of the late 1980s, to bring new products to market about a year and a half quicker that either American or European producers. Much of the Japanese advantage stems from their prowess in manufacturing; Japanese tool builders are able to provide tooling to the auto companies (e.g., for stamping out parts) in so much quicker time that entire projects for the auto companies benefit substantially.

Adopting JIT does require a change in a company's culture. One can envision a pyramid (See Figure 11-6) of capabilities that are due to a company's investment in its people and the process of continuous improvement that is at the core of the JIT manufacturing philosophy.[5]

IMPLEMENTING JUST-IN-TIME MANUFACTURING

Given this JIT manufacturing philosophy, how can managers best put it to use in transforming existing operations? This question is the focus of the remainder of this chapter. To concentrate our efforts, let's consider Situation 11-1.

Discussion of Longstone Metal Products Company

Management is right to think that the principles of JIT can be applied in this case. In fact, there is hardly a factory that cannot find some application of JIT somewhere within its walls.

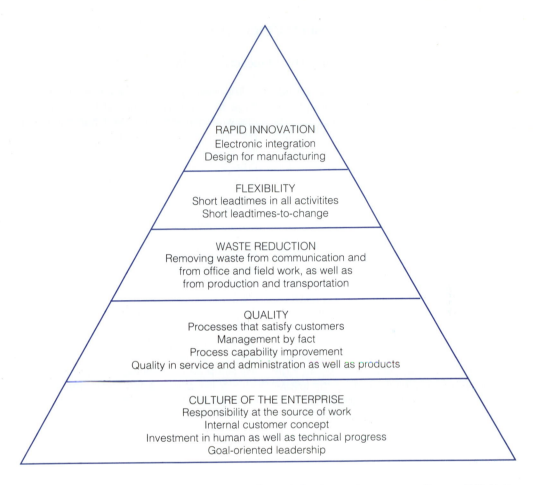

FIGURE 11-6   Cumulative Human Development Through Continuous Improvement. Source: R.W. Hall.

Even plants that consider themselves "process industries" can often find JIT very useful because they are frequently, at core, batch operations in chemicals or foods or something else.

This situation can be a useful one for stepping through some of the key aspects of JIT implementation.[6]

1. Workplace Organization.   One of the first things that veterans of JIT implementation stress is the organization of individual workplaces. This means attention to things like housekeeping and the elimination of anything at the workstation that is not needed. The flow of work through the workstation then has to be investigated—using flow charts and even physical models of the workstation itself—to see exactly what may be required and how the workstation might be altered to serve the purpose even better. It may mean, for example, that some or all of the equipment for the workstation be moved into a different configuration and a re-layout of the workstation accomplished. The situation at Longstone

---

SITUATION 11-1

Longstone Metal Products Company

The Longstone Metal Products Company, an 80-year-old manufacturer based in the midwest, fabricated a wide assortment of metal housings for motors and other electrical equipment. While the products required basically the same kinds of production processing, they differed in size, in the gauge of metal used, and in the type of metal used (carbon steel, aluminum, stainless steel, brass). Longstone worked off long-term contracts with its various customers. Its customers insisted that Longstone keep finished goods inventories of the housings they used, with the call-off on those orders on those inventories coming every week. Demand was typically stable from week to week as to quantity shipped, although not to model. Housings were fabricated in batches, usually of at least 500 at a time. Materials for each batch were assembled in the controlled, fenced-off storeroom and then released to the factory floor, along with the appropriate paperwork. The major process steps through which a typical housing went included the following: (1) metal cut to length from coiled stock, (2) stamping of key parts on variously sized stamping presses, (3) welding of selected parts, (4) drilling and riveting of other parts, (5) deburring, (6) painting, if that was called for by the customer, and (7) labeling the finished pieces, usually with the customer's logo.

The metal pieces for each batch were kept in large bins that were moved by forklift truck from one work center to another. Changing the setup of particular machines for each new batch took differing amounts of time. For example, the stamping presses took several hours for most changeovers, while drilling and welding took only about a half an hour each to alter their fixtures and tooling.

The factory was organized conventionally, with the different manufacturing functions each in their own department (i.e., a cut-to-length department, a stamping department, a welding department, a paint department). Each department had its own supervisor, with the supervisors reporting to a manufacturing superintendent, who in turn reported to the plant manager. The supervisors were evaluated on their delivery performance, their labor efficiency (earned hours versus total available hours), and machine utilization (percent of time available actually setup and producing parts).

Over the previous few years, management had begun to worry that inventories, especially work-in-process, had been growing out of control, that quality had been slipping somewhat, and that productivity had not budged much at all. Management wondered whether just-in-time manufacturing could be applied in this instance, and just how it might work.

---

is not described in sufficient detail for us to investigate workplace organization at that factory, but one can easily see how such an investigation might reveal some problems or offer some suggestions for tasks like welding or press work.

2. *Study the Flow of the Whole Process.* Once the flow of materials through individual work stations is understood and evaluated, one can go on to examine the flow of materials through all the work stations of the process. The process flow diagram, used at whatever level of detail seems most appropriate, is the simple but powerful tool to use for this. Analyzing the process flow means understanding more than the various routings used, but also the volumes associated with different machines and workstations. Importantly, it also means examining the flows of work for product families that can be used to group like-processed products together. The idea here is to be able to design self-contained units of the process to handle

each particular product family. Thus, attention must be paid to issues like capacity, materials handling, maintenance, and setup.

At Longstone Metal Products, there is a case to be made for product families of various types. It may be possible to segregate production of different size and gauge housings and certainly of different metals. Such segregation can mean the isolation of particular machines to particular products and the creation of simpler routings and flows as a result.

3. *Make the Process Visible.*   That is, create a process that can be managed visually, without recourse to computer printouts to check the status of either orders or parts. You want to be able to walk on the factory floor and know immediately how the process is faring. Accomplishing this step feeds off the analysis of materials flows for the entire process. It argues for a number of important items:

a. Emphasis on quality production—process capabilities known and in control.

b. The creation of manufacturing cells to cope with the various product families identified.

c. The abandonment of process (job shop) layouts wherever possible, and their replacement by product-specific layouts.

d. The renunciation of buffer inventories between segments of the process in favor of coupling process segments together to facilitate more rapid production cycles, lower inventories, and other attractive goals. Stockrooms should be contracted as more and more inventories are found, and managed, on the factory floor itself in the form of stockpoints of limited capacity but constant location.

e. Avoidance of "super-machines" that may end up as bottlenecks and which often lack flexibility and can disrupt the flow of materials.

There are a number of things that can be said about making the process at Longstone more visible and its management more visual. Attention to quality is certainly one critical aspect. When quality is emphasized, management can push for quality training for everyone, and for the use of SQC techniques, cause-and-effect analysis, experimentation, and feedback to product design on issues of quality. All the progressive things discussed in Chapter 4 can come to life in a JIT environment. Indeed, to make JIT work, quality must be high so that good products can be reliably produced without risking sustained shutdown of the process. JIT manufacturing philosophy calls for surfacing problems in the process so that they can be corrected, but if paralysis is not to take over, quality must be high before a shift to JIT manufacturing can be seriously entertained. Thus, working on quality in the process is an important aspect of preparation for JIT production.

Promoting a process that can be managed visually has other implications for the situation at Longstone. Stemming from the identification of sensible product families can come the creation of manufacturing cells for the production of key parts or subassemblies of the products. These cells can then be linked together for the production of entire products, and stockrooms avoided, if at all possible.

One way to spur this is to reorganize the plant out of its old style job shop layout and into a product-specific layout. The departments, as currently designated, can be abolished in favor of product line managers whose tasks would be to shepherd their grouping(s) of products through the entire process, using machines that for the most part are dedicated to particular groupings and not to all types of products, as before. Product line-oriented manufacturing managers would also be less likely to invest in large-scale equipment, just to have it. Rather, smaller and more versatile equipment would be prized.

By reorganizing in this way, management could also change the incentives given the supervisors. Currently, all the incentives are to force production out of the factory: missed deliveries are penalties, lots of earned hours are encouraged, and keeping both labor and machinery busy receives high marks. This orientation leads to some problems, however. Quality, and preventive maintenance, are not rewarded, just quantity. Inventories are allowed to grow large because a concentration on labor efficiency and machine utilization puts the limelight on long production runs. The factory is permitted to become sluggish and inflexible in pursuit of such incentives.

With reorganization, however, labor efficiency and machine utilization measures can be scrapped in favor of new measures that could target quality, throughput times, and levels of WIP inventory. Quick setups would be encouraged as well as shorter production runs that are better suited to producing what the marketplace is buying during any one week. While the old departmental goals were at odds with items such as flexibility and quick-response production, the new reorganization would have no built-in biases that way.

4. *Foolproof the Process.* When the process is made visible and weaknesses with the process are easily uncovered, management and the workforce can begin to alter the process in clever ways so that it produces only good product. They can foolproof the process by adding special fixtures, limit switches, counters, timers, sensors, vision systems, or trouble lights to the existing equipment. They may have to change product designs as well so that assembly or testing or adjustment is made in only one way, the quality way. (More on design for manufacturing in Chapters 4 and 13.)

Exactly what should be done at Longstone Metal Products to foolproof its process is, of course, not immediately obvious. It would take greater familiarity with the process to suggest something, but no doubt some things could be done easily and inexpensively.

5. *Setup Time Reduction.* The JIT manufacturing philosophy argues for low levels of inventory, small lot sizes, and quick setups. At Longstone Metal Products, there appears to be some opportunity for reducing setups, as the presses are taking several hours to change over and the welding and drilling equipment is taking half an hour.

Setup time reduction is mostly the application of long-understood principles of industrial engineering to the neglected aspects of setup. Successful reduction programs have sought to simplify or even to eliminate any adjustments, using standard fixtures and parts, trying to do as much as you can while the machine is running so that the machine is down the least amount of time it can be, videotaping the setup and then discussing what was taped to identify areas for improvement, and putting a team on the job to do the setup, and even to practice it! Many who have achieved significant reductions in setup, say, on the order of 10 percent or less of their old times, state that new setups that are half as long as old ones are rather easily accomplished.

Such ideas are readily adaptable to the situation at Longstone Metal Products. The press changeovers might be aided by changing all the dies so that they operate at the same height in the press and have the same simple clamping systems to them. Special rolling tables can be used to bring new dies to each press at the same height that they will be used, so that old and new dies can be quickly swapped without much lifting or adjustment. Similar techniques could be used at the drilling and welding work stations. Here the location of special magazines to hold the needed tools could be designed, and lazy-susan-type tables might be created to hold the fixtures utilized so that they can be accessed quickly. Special clamps that can be set and released quickly could also apply here. There is every reason to believe that lot sizes can be slashed dramatically from the 500 levels prevailing now. Exactly how low they could go is something to experiment with, and will depend on things such as setup reduction, excess capacity, and the pattern of demand.

6. Regularly Timed Cycles. If the process is to lose the pressing, but disruptive, urgency of firefighting, it must become a routine activity for all concerned. Thus materials handling should strive for standard places to put all parts, standard containers for those parts, and standard moves of those parts throughout the process. In addition, the production schedules should be fairly level ones, with no expediting and only modest changes from one day to the next. You want to be able to produce a little bit of everything every day. Checks of quality, tooling, and preventive maintenance all need to be made at regular intervals.

Clearly, Longstone Metal Products could adopt a layout that dispensed with controlled storage and dispersed inventory to specific areas of the layout instead. In such a layout, parts destined for particular families of parts would have just a single area on the factory floor that would be theirs exclusively. The control of those parts would thus be simplified; any worker looking at the special containers of parts could easily see how many were left and how many should be there. The standard movement of those parts in the process could also be readily identified, and any irregularities immediately apparent. Companies that have used such techniques for their own "visible" production processes have found that their inventory control is always much better for their visual systems than for their computer-based systems and the parts managed by those more vulnerable and less visual systems.

Most of Longstone's products were kept in inventory, being on long-term contract to their customers. Thus production in long runs was not required by the customer. The production schedule could be made more level, but it would require many setups during any week.

7. "Pull" Production Control. Moving to a pull system, using kanban cards, or kanban squares, or simple signals of one type or another, is one of the last things to accomplish in shifting to JIT production. At Longstone Metal Products, given the creation of manufacturing cells and special, known places for parts of all types, it would not be difficult to attach special cards to the bins of parts. When a container of parts was pulled off the shelf for use, the card could be detached and routed to the area in the process which fabricated (or ordered) that part. Thus the downstream operations could trigger the upstream operations in simple fashion. The lot size to produce would always be a standard container's worth.

8. A Technological Admonition. Shifting to JIT production typically also means going slow on new automation. The goal is to simplify the process, and not necessarily to automate. Certainly, you want to simplify the process before you automate it. Considerations of flexibility in the product mix often argue as well for smaller-scale automation. Large-scale equipment can frequently lead to bottlenecked operations and materials flows that must be altered to accommodate the equipment rather than the marketplace.

9. Organizational Aspects. As one can well imagine, a shift to JIT production is not done casually. It can take 1 to 2 years to really get under way, and in a very real sense, the shift is never fully implemented. There is always room for more improvement.

Adopting JIT implies a number of things for an organization:

1. Commitment to educate workers and management alike in the tenets of JIT.
2. Creation of teams to study the process, SQC, problem solving, and experimentation.
3. Shoving staff onto the factory floor so that the workforce can truly be supported by management and engineering.
4. The development of new measures of performance for the factory, specifically measures of trends in quality, throughput times, inventory levels, production to schedule, setup times, skill levels in the workforce, and costs. The measures to discard are the conventional ones on labor efficiencies and machine utilizations.

The adoption of JIT will mean changes in all aspects of the factory. There is not a job that will remain untouched by the new philosophy. For some managers this has meant a reluctance to dive into the change, for fear that poor planning will dilute the gains. For the pros who have been through it, the advice is clear, however. Do not worry about it, simply jump in, and you will soon realize what the next steps ought to be. JIT is a revolution, but it is a revolution that can be won by most any factory.

Tools of the Trade: A Recap of Lessons Learned

One way to view JIT is as a pyramid or a layer cake with the following levels:

Level 1 (top): Enhanced profitability, the end result. This enhanced profitability is due to:

Level 2: Quality improvement and the elimination of waste of all kinds. The well-known Japanese engineer and author, Shigeo Shingo, has developed a list of the different kinds of waste that one can have: defects, inventories, overproduction, transportation, waiting, unnecessary motion, unnecessary processing. Removing any of these wastes from the process helps.

The policies and actions that, in turn, lead to the improvement of quality and the elimination of waste are the following:

Level 3: Throughput time reduction, the development of analysis skills in the workforce, and the unleashing of the workforce on the processes of the company. The development of the workforce takes training and the development of teams, and an organization of manufacturing that puts management and engineering at the service of the first line people who actually touch the product and add value to it.

How does one reduce throughput time? What is required?

Level 4 (base): Implementing JIT and reducing throughput time takes the study of the following items:

- Workplace (workstation) organization
- Flows of material and of information — Where is value added, and where is the waste?
- Process visibility: — Creation of cells, lowering of inventories, use of SQC and other aids to quality
- Foolproofing, including both process changes (which may take experimentation to determine) and product changes (design for manufacturing)
- Setup time reduction
- "Pull" production control
- Careful, small-scale automation; no super-machines

- Regularly timed cycles for the scheduling of production, maintenance, quality, and so on.
- Development of measures of physically meaningful things such as time, defects (first time yields, et al.), performance to schedule, actual costs, etc.

It takes quite some doing to "eat" one's way through such a layer cake, but, increasingly, this "cake" makes good eating.

World-Class Practice. The implementation of the JIT philosophy is the epitome of world-class practice. Achieving implementation broadly within a company takes time and effort, often several years' worth. Many companies have smaller scale JIT implementations in progress; rather fewer have implemented JIT broadly. These latter are the companies others look to as "world-class."

QUESTIONS

1. Why is just-in-time manufacturing described as a "philosophy" in this chapter?

2. The popular business press often describes JIT as a program that emphasizes the delivery of supplies to a factory just before they are used. How is this description of JIT flawed?

3. How relevant is JIT for service operations as well as manufacturing operations?

4. "JIT is mainly a program for discrete parts manufacture, not process industries." What is your reaction to this statement?

PROBLEMS

1. Just-in-time manufacturing is sometimes seen as primarily an inventory reduction program. How would you react to such a characterization?

2. How would you determine how many containers to have for a particular portion of the operation?

3. Assume that you want to eliminate the stockroom as much as possible and that you want to put as much directly on the factory floor as you can. How would you decide on the location of a part that is used in two locations within the factory, and how would you replenish that part when it fell to low levels?

4. In choosing families of products for JIT production, what criteria do you think would be important to employ?

5. E. Dilly and Company is a midwestern pharmaceutical producer. The production process

at Dilly is divided into two major parts, biochemical manufacturing (biochem for short) and a pharmaceutical manufacturing (pharma). Biochem takes standard commodity chemicals and transforms them into the specific active ingredients of Dilly's patented pharmaceuticals. This transformation often involves complex multistaged chemical reactions that yield several chemical intermediates before the active ingredient itself is developed. The process, which is repeated for each intermediate, could involve such steps as fermentation, extraction, isolation, crystallization, and drying. The operations are done in "rigs" of various sizes composed of tanks, centrifuges, dryers, and so on, all connected with pipes and capable of precisely controlling the chemical reactions contained within them. These rigs are erected in multifloor buildings so that gravity could flow the product from top to bottom.

The pharma operation took the active ingredients together with other inert chemicals and created and packaged drugs that patients actually used. The drug could be in many forms: tablets (coated or uncoated), capsules, and sterile liquid injectables, in different dosages and various sizes and types of package. The pharma operation typically involved such processes as granulating, compressing, coating, sorting, bottling, labeling, and packaging.

What arguments do you see, both pro and con, for scheduling this process with significant lot sizes? Significant lot sizes means lot sizes that build up finished goods inventories to tide the company over for several months before manufacture has to be resumed. If lot sizes were to be reduced, what kinds of actions would be required?

6. Several hundred different parts are produced in the fabrication department of a large factory. Two of them, however, account for 30 percent of the fabrication department's total output.

Both of these products follow the same general sequence of operations: turning, milling, drilling, and then grinding. The following percentages of production time are spent by the two parts on each set of machines: 20 percent turning, 40 percent milling, 10 percent drilling, and 30 percent grinding.

The inventory of machines in the department is as follows:

MACHINE TYPE	NUMBER IN DEPARTMENT	SYMBOL TO USE IN LAYOUT
Turning lathes	7	□
Mills	8	○
Drills	3	△
Grinders	4	◇

The machine utilization has been fairly even at about 75 percent for each type of machine. Using the symbol (1 symbol = 1 machine), design a layout for the department. Explain your reasoning.

SITUATION FOR STUDY 11-1

COLLINS MUFFLER COMPANY

Collins Muffler produced mufflers and exhaust systems for the automobile after-market in Europe. [It did not produce for original equipment manufacturers (OEMs).] It did so in two factories, one across the street from the other. The older factory specialized in component manufacture, while the newer one housed the assembly operation and the distribution center. The muffler business in Europe was somewhat different from the one in the United States in that mufflers were seldom sold by themselves; most often they were sold, together with the exhaust pipes, in one large assembly that could be easily and quickly placed on a car. As a result, however, the exhaust systems were fairly specialized to particular car models.

Component manufacture entailed the cutting to length of coiled steel, stamping of that steel in the shapes and sizes required, and other forming operations on the metal, such as bending and curling. Once the basic components were created, they were shipped across the street, where they were combined with sound absorbing material and assembled into the muffler portion of the product. Pipes of various sizes and diameters were then formed into the specific shape required and the muffler and the pipes were all welded together.

The separate operations that produced an exhaust system were performed in their own portions of the factory to batches of materials that were routed their way. Batch sizes were typically 500, although daily sales for only the best-selling models exceeded 250. There were about 150 different models that had to be produced and kept in stock for shipment from the warehouse. Capacity was about 5000 exhaust systems per day.

All orders from the company's wholesalers were filled from the distribution center. Production was planned to replenish the distribution center. Stockouts generally implied lost sales, so that Collins was careful to maintain significant inventories of finished goods in the distribution center.

The workforce was generally assigned to particular tasks, be they steel cutting, forming, assembly of mufflers, welding of mufflers to pipes, or bending of pipes prior to welding. All workers were union members.

Given this situation, what could or should be done to implement just-in-time manufacturing at Collins Muffler?

REFERENCE NOTES

1. Some useful readings on JIT and Japanese manufacturing techniques and philosophies include:
    Abegglen, James C., and George Stalk, Jr., *Kaisha*. New York: Harper & Row, 1985.
    Clark, Rodney, *The Japanese Company*. New Haven, CT.: Yale University Press, 1979.
    Drucker, Peter F., "What We Can Learn from Japanese Management," *Harvard Business Review* 49, no. 2 (March–April 1971): 110–22.
    Drucker, Peter F., "Behind Japan's Success," *Harvard Business Review* 59, no. 1 (January–February 1981): 83–90.
    Hall, Robert W., *Zero Inventories*. Homewood, IL.: Dow Jones-Irwin, 1983.
    Hall, Robert W., *Attaining Manufacturing Excellence*. Homewood, IL.: Dow Jones-Irwin, 1987.
    Harmon, Roy L., *Reinventing the Factory II*. New York: Free Press, 1992.
    Hayes, Robert H., "Why Japanese Factories Work," *Harvard Business Review* 59, no. 4 (July–August 1981): 56–66.
    Hout, Thomas M., and George Stalk, Jr. *Competing Against Time: How Time-based Competition Is Reshaping Global Markets*. New York: Free Press, 1990.
    Imai, Masaaki, *Kaizen*. New York: Random House, 1986.
    Juran, Joseph M., "Japanese and Western Quality: A Contrast in Methods and Results," *Management Review* (November 1978): 27–45.
    Lu, David J. (translator), edited by the Japanese Management Association, *Kanban, Just-in-Time at Toyota: Management Begins at the Workplace*. Stamford, CT.: Productivity Press, 1986.
    Schonberger, Richard J., *Japanese Manufacturing Techniques: Nine Hidden Lessons in Simplicity*. New York: Free Press, 1982.
    Schonberger, Richard J., "Applications of Single-Card and Dual Card Kanban," Interfaces 13, no. 4 (August 1983): 56–67.
    Schonberger, Richard J., "The Vital Elements of World-Class Manufacturing," *International* Management 41, no. 5, (May 1985): 76–78.
    Schonberger, Richard J., *World Class Manufacturing: The Lessons of Simplicity Applied*. New York: Free Press, 1986.
    Shingo, Shigeo, *A Study of the Toyota Production System from an Industrial Engineering Viewpoint*, Norwalk, CT: Productivity Press, 1989.
    Suzaki, Kiyoshi, *The New Manufacturing Challenge*. New York: Free Press, 1987.
    Voss, C.A., ed., *Just-In-Time Manufacture: International Trends in Manufacturing Technology*, New York: Springer-Verlog, 1987.
    Wheelwright, Steven C., "Japan—Where Operations Really Are Strategic," *Harvard Business Review* 59, no. 4 (July-August 1981): 67–74.
    Yoshino, Michael J., *Japan's Managerial System: Tradition and Innovation*. Cambridge, MA.: MIT Press, 1968.
    Yoshino, Michael J., *Japan's Multinational Enterprises*. Cambridge, MA.: Harvard University Press, 1976.

2. Cole, Robert E. and Yakushiji, Taizo; *The American and Japanese Auto Industries in Transition*, Center for Japanese Studies, University of Michigan and Technova., Inc., Tokyo, 1984.

3. Robert W. Hall and Jinichiro Nakane, *Flexibility: Manufacturing Battlefield of the 90s*, Association for Manufacturing Excellence Research Report (Executive Summary), 1990.

4. Kim B. Clark and Takahiro Fujimoto, *Product Development Performance*, Cambridge, Harvard University Press, 1991.

5. Again, this insight is due to my colleague, Robert Hall.

6. This discussion draws heavily on the work of my colleague Robert W. Hall, particularly his excellent book, *Attaining Manufacturing Excellence*, Dow Jones-Irwin, 1987.

CHAPTER 12

DEALING WITH CAPACITY CHANGE

So far in this book, we have dealt with improving existing operations. To this end, we have discussed short-term measures to increase an operation's capacity. Among such measures were:

1. Overtime.
2. Second or third shifts.
3. A more level production rate over time for the purpose of building up finished goods inventory for peak periods of demand.
4. Altering the product mix or the production schedule to reduce setups of machines or work groups and thereby to coax more production out of the operation.
5. Adding more labor to an existing stock of capital — for example, by rebalancing a line flow process.
6. Improving information or materials movement within a process — for example, by introducing material requirements planning systems or decoupling inventories.
7. Modest investments, product redesigns, process modifications, or management innovations that decrease manufacturing cycle times, freeing resources for additional capacity. Many of these measures, with possible long-term implications as well, will be discussed more thoroughly in the subsequent chapter on technological change.
8. Subcontracting products or pieces of the production process. This measure is frequently not an immediately implemented change; it can remain a long-term policy of the company as well as a short-term one.

These measures, either separately or in combination, can have dramatic effects on a plant's capacity. Most of them, however, are short-term substitutes for securing additional space and equipment for production. Overtime, shiftwork, revamped production scheduling, and some others can often meet peaks in demand, but they usually cannot be sustained over a long period of time. Improvements to the process, product design, information, or materials handling systems, although desirable long-term policies in themselves, are often incapable of boosting production enough to eliminate a plant's need for more space in the face of sustained increases in demand. Although a company normally looks first to such a list of short-term means of adding capacity, more substantial space additions are generally considered at the same time.

411

In this chapter we concentrate on the three "bricks and mortar" choices for adding capacity: (1) on-site expansion of an existing facility, (2) the establishment of a new branch operation, and (3) the relocation of an existing operation to larger quarters. Among the issues that surround these choices are:

- How much should any capacity expansion be?
- When should it be timed?
- Where should it be located?

The discussion of these issues and choices comprises the agenda for this chapter. After discussing the basic elements behind the size and timing of capacity expansion in a manufacturing enterprise, we examine similar issues in a service industry. Finally, we consider, in more depth, the decision whether to add capacity on site or to establish a new branch or to relocate to larger quarters, and the decision of where to place a new facility.

The decision about contracting capacity is every bit as involved as the decision about increasing capacity—and markedly more difficult for most managers to make. While most of the chapter deals, implicitly if not explicitly, with capacity expansion, some portions treat plant closing and contraction as well.[1]

AN OVERVIEW OF CAPACITY MANAGEMENT

In most companies, most short-term measures to increase capacity (1) can be planned for in a matter of weeks, (2) involve only limited sums of corporate capital, (3) are studied at the plant (rather than the corporate) level, and (4) originate and are carried through in an informal way. Although they are concerns of top management, such short-term capacity measures seldom absorb much of the energies of the top rank of company managers. These short-term measures are apt to be too routine and too devoid of strategic importance to demand the careful study of the layers of management that sit above the particular concerns of individual plants.

Longer-term, bricks-and-mortar capacity expansions occupy top management time to a much greater degree. This is because they involve considerable expense, far above the spending authority of even the chief executive officer, and they often have strategic and competitive implications for the company. Moreover, the initiation of capacity expansions is apt to be the result of a formal companywide planning and review procedure rather than the result of a plant's reaction to day-to-day demands.

A typical planning and review procedure is an annual exercise that looks 5 years into the future. (Five years is the typical number. Highly capital-intensive companies like those in the chemical and steel industries may choose 10 years; more consumer-sensitive industries like apparel or fast growing industries like electronics may choose 3 years.) This exercise is driven by a marketing forecast for each year (and quarter, perhaps) of the planning horizon and for each product produced. Available production capacities are estimated as well and the two are meshed together to determine to what extent capacity is in surplus or in deficit. Naturally, both marketing and production estimates are much more nebulous the further into the future the company looks.

More than simply identifying capacity shortfalls and their likely timing over a multiyear horizon, the planning exercise concentrates on alternatives for dealing with such shortfalls. The on-site expansion of certain plants, the character, size, and region for possible new branch plants, or the relocation of some plants may be considered as viable alternatives for coping with capacity shortfalls. The review of the submitted plans normally comes to some tentative decisions about which avenues for remedying capacity shortfalls ought to be shelved, at least for the time being. It is unlikely that firm commitments will be made on any specific proposal but the waters will be tested on many.

The planning and review procedure generally works from the bottom up. (See the organization chart in Figure 12-1.) While the particular individuals involved vary from one organization to another, generally plant managers and their staffs work out the lowest tier of corporate

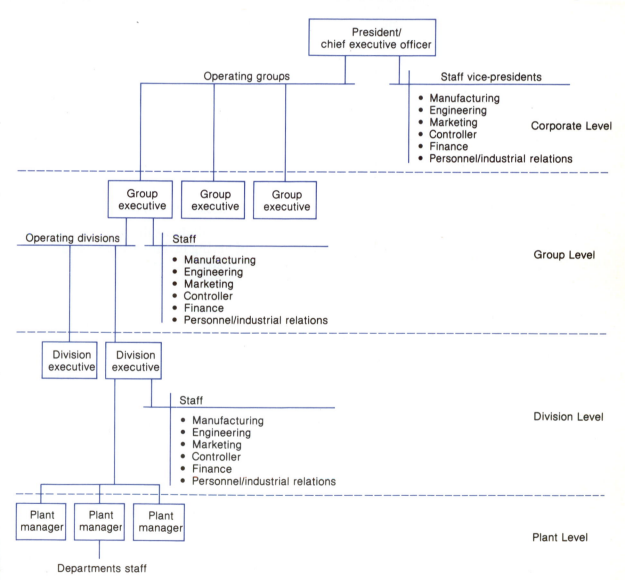

FIGURE 12-1   Simplified organization chart for a large corporation.

needs and priorities. The divisions under which the plants operate review and expand on the plant proposals and needs and address more far-reaching issues. The same kind of review and expansion occurs at the group level (groupings of divisions), if such a level exists, before the corporate level operations committee (or some similar body) reviews the refined proposals. For our purposes, on-site expansion is probably initiated at the plant level and approved at successively higher levels in the corporation, although it could also be initiated at the division or group level. New branch plants and relocations, on the other hand, usually originate not at the plant level (understandably, since the plant itself is not responsible for such an action) but at the division, group, or corporate levels.

As a result of the planning and review procedure, the divisions (and groups) are given a mandate to develop in-depth analyses of the on-site expansion, new branch plant, or plant relocation proposals that surfaced during the planning process. These proposals must be translated from ideas and plans into specific projects to be manned and funded by the corporation. Most corporations have a formal "capital appropriation request procedure" that roughly parallels the hierarchy through which capacity plans are reviewed. The capital appropriation request often merely makes formal and explicit what has been discussed during the company's business planning. This request typically consists of a justification for the project and some in-depth analyses of it. These analyses include engineering specifications and cost/timing estimates; specific size, product, equipment, worker and manager requirements, and location decisions; and an economic analysis (sometimes called a capital budgeting analysis). The appropriation request, after it has been completed and approved at the group and division levels, makes its way to the company's board of directors for final approval and the authorization to spend the necessary funds.

In broad outline, this planning and review procedure is fairly typical for capacity expansion or contraction.[2] It may sound straightforward, and in many cases it is, but there are many other cases where the choices and issues are complex and require clear and systematic analysis.

DECIDING HOW MUCH CAPACITY TO ADD AND WHEN

How can managers decide how much capacity to add and when to have that capacity in place? To ground a discussion of these uncertainty-riddled decisions, consider Situation 12-1.

Discussion of Kemper Games

Kemper Games finds itself in the joyous predicament of having invented a successful game that now threatens to overtake the company's ability to produce it. What sales will be is rather uncertain, and the marketing department has spun out two scenarios. Both foresee a dramatic increase in sales in the short run. After the first year, however, the two forecasts steadily diverge; one assumes that the game peaks as a fad while the other sees it as a long-lasting success. The marketing people have underestimated sales to date; one might well wonder whether the sales forecasts are still on the low side. It would be reasonable to expect any capacity plan for the future to be robust enough to cope with another sizable increase in forecasted sales.

The first order of business is to assess whether all three courses of action being considered (expansion, third shift, and subcontracting) are feasible, given the prevailing market forecast. Subcontracting seems feasible since no mention was made of any constraints

SITUATION 12-1

Kemper Games, Inc.

The latest sales figures were very encouraging. Laura Kemper, president and chief executive officer of Kemper Games, Inc., could not help but smile. After four essentially static years, Kemper Games looked like it was headed for an absolutely fantastic sales year that would carry the small company into the top ranks of the growing game industry. The reason behind this success was Kemper's latest entry in the market, "Bungle, The Game of Life Styles." Laura, an inveterate games player herself, had developed the game over the past two years with her husband, Peter, who headed the company's product development department. The company had introduced it about 6 months ago.

Bungle was an adult board game for two to six players. The game produced no winners—everybody lost. This outcome apparently struck a responsive chord in adults of every age and income group. The game's early success had set Peter Kemper to the task of developing a successor game, Deluxe Bungle, which would soon be ready for production. Deluxe Bungle was played by two-person partnerships (such as couples). In playing Deluxe Bungle, not only did everyone lose, but partners were encouraged to turn against one another. Peter had high hopes for the game, although no marketing study had been made as yet.

Bungle's success, however, was soon going to create capacity problems at Kemper Games. The factory could reliably produce about 900 sets per day in two shifts and inventory stood at 3000 sets. The magnitude of the problems anticipated depended in large measure on the particular forecasts for the game's sales (see Table 12-1). The range in the forecasts was due mainly to variations in the assumptions used. Low estimates derived from regarding Bungle as a fad that would soon pass. Higher estimates regarded Bungle as a long-running staple, like Monopoly.

The forecasts influenced not only the size of any additions to capacity but also the nature of those additions. Laura Kemper confronted several choices concerning capacity:

TABLE 12-1   Forecast of Bungle's sales by upcoming quarters (Kemper Games, Inc.)

QUARTER UPCOMING*	FORECAST 1[†] (sets/day)	FORECAST 2[‡] (sets/day)
1	600	600
2	750	750
3	1050	1050
4	1500	1500
5	1200	1500
6	900	1500
7	750	1500
8	500	1500

Source: Marketing Department

*The company plans for 62 days of production each quarter.

[†]Forecast 1 assumes that Bungle's sales are merely a fad that will peak in quarter 4 and gradually taper off.

[‡]Forecast 2 assumes that Bungle's sales will parallel those of Boggle and Pente. After rising like a fad, they will sustain themselves at a high level.

1. Kemper Games could expand on-site (enough space lead time were available to do so), or it could lease nearby space to satisfy forcasted requirements. Table 12-2 includes some cost estimates for the various space additions corresponding to the forcasted sales figures. The space could be used for any Kemper Games' products since nearly all of them required printing, plastic fabrication, gluing, and assembly. Costs for the additional equipment and working capital needed to reach various output levels are also included.

2. The company was going to operate two production shifts in the upcoming first quarter, doubling production from its current rate of 450 sets per day. A third shift was a definite alternative. Wage premiums would have to be paid, though, to attract a suitable workforce and supervisory personnel. In addition, productivity was expected to be less in the third shift, because the workforce quality would probably be lower than that prevailing in the first two shifts and because only a skeleton staff would supervise production. A 25 percent wage premium over first-shift

TABLE 12-2   Cost estimates for on-site expansion (Kemper Games, Inc.)

ADDITIONAL PRODUCTION RATE PER DAY OVER TWO SHIFTS	SQUARE FEET REQUIRED	ESTIMATED COST (PLANT)	ESTIMATED COST (EQUIPMENT) AND WORKING CAPITAL
300	15,000	$225,000	$ 75,000
450	20,000	300,000	100,000
600	24,000	360,000	120,000
750	27,000	405,000	175,000

Note: Present space allocated to Bungle production is 40,000 square feet. Current two-shift production capability is 900 sets per day.
A. Costs of on-site expansion
B. Costs of leasing the only nearby existing facility: 20,000 square feet or less—$7.00 per square foot per year. More than 20,000 square feet—$6.50 per square foot per year.

operations was expected as well as an effective output rate of 90 percent of the first shift. Labor constituted 40 percent of a set's variable cost of $2.40.

3. Kemper Games had so far manufactured all its games in its own facilities. However, the company could subcontract the manufacture of the entire game or of certain components (such as the printing of the board, the cards drawn as part of the game, the plastic pieces). The convenience of subcontracting was costly. According to one estimate, the 40 percent contribution margin that Kemper Games enjoyed on the $4.00 factory price of Bungle (the retail price was $7.95, just about a 100 percent markup) would be eroded to 20 percent through subcontracting the entire game.

It was also possible for Kemper Games to reduce or even to eliminate its need for additional capacity by pricing the game higher. However, Laura had decided to accept a lower margin than was the norm in the industry, and thus a lower retail price, in order to spur sales of Bungle and to spread the Kemper Games brand name. The initial marketing report for Bungle with estimated sales at different retail prices is reproduced as Table 12-3. Currently, sales for quarter 1 were running about 60 percent ahead of that market forecast.

Laura felt if Kemper Games was to have enough time to act on the on-site expansion option, she would have to act within the week. Any new on-site construction, begun now, was not likely to be available until the start of quarter 4. Inventory would have to be built up in the interim, which Kemper Games would be able to carry at a rate of 10 cents per unit per quarter.

TABLE 12-3   Marketing report on sales by price category (Kemper Games, Inc.)

RETAIL PRICE	CORRESPONDING FACTORY PRICE	SET SALES PER DAY*
$ 7.00	$3.50	420
8.00	4.00	360
9.00	4.50	315
10.00	5.00	285
11.00	5.50	260
12.00	6.00	240

*Set sales per day are estimated at 6 months after introduction.

impeding potential subcontractors. In gauging feasibility, then, it is largely a matter of testing the third shift and expansion alternatives.

Let us test the expansion alternative first. Any expansion on site begun now would not be ready for production until the fourth quarter. In the meantime, two shifts would have to suffice. Fortunately, two shifts begun now will be enough to carry Kemper Games until the fourth quarter, with some to spare. Inventory would be built up in the first two periods and depleted in the third. If the on-site expansion involves 24,000 square feet, the corresponding production rate is 600 — just enough to combine with the two-shift rate of 900 per day to meet a continued demand for 1500 units per day. Table 12-4 summarizes how demand would be satisfied under the on-site expansion alternative. As the table makes clear, immediate implementation of the second shift leaves the first quarter buildup as an inventory cushion of 18,000 units (calculated as 300 units extra per day times 62 days) — 12.5 percent of expected demand over the first three quarters.

Such a cushion is welcome and by no means too large, given the possibly explosive sales for the game.

Let us turn our attention now to the feasibility of a three-shift operation. The third shift is expected to be more costly and less efficient than either of the first two shifts, chiefly because of the quality of the labor and the skeleton staffing of supervisory positions. Specifically, only 405 units (calculated as 450 times 0.9) are expected during each third shift. With this in mind, we can construct an inventory buildup table like Table 12-4, using the larger forecast estimates. As is plain from Table 12-5

TABLE 12-4   Inventory buildup under the on-site expansion alternative (Kemper Games, Inc.)

QUARTER	DAILY DEMAND	DAILY PRODUCTION	DAILY INVENTORY BUILDUP OR DEPLETION	CUMULATIVE INVENTORY ON DAILY BASIS FOR QUARTER
1	600	900	+300	+300
2	750	900	+150	+450
3	1050	900	−150	+300
4	1500	1500	0	+300

TABLE 12-5   Inventory buildup under the third-shift alternative and forecast 2 (Kemper Games, Inc.)

QUARTER	DAILY DEMAND	DAILY PRODUCTION	DAILY INVENTORY BUILDUP OR DEPLETION	CUMULATIVE INVENTORY ON DAILY BASIS FOR QUARTER
1	600	1305	+705	+705
2	750	1305	+555	+1260
3	1050	1305	+255	+1515
4	1500	1305	−195	+1320
5	1500	1305	−195	+1125
6	1500	1305	−195	+930
7	1500	1305	−195	+735
8	1500	1305	−195	+540
9	1500	1305	−195	+345
10	1500	1305	−195	+150
11	1500	1305	−195	−45

the third shift can produce enough inventory early on, if implemented in quarter 1, to tide the corporation over until quarter 11. For the short term, then, a third shift is clearly feasible, although over the long term, if the second forecast prevails, a third shift would come up short of capacity. In that case, only on-site expansion would serve.

On the other hand, if forecast 1 prevailed, on-site expansion just for the Bungle game would be overkill because it would be necessary only for quarters 4 and 5. What seems to be suggesting itself is a wait-and-see approach to on-site expansion. A third shift appears to be the best way to meet short-term demand; a decision on on-site expansion can be deferred until the popularity and longevity of Bungle in the marketplace can be ascertained more precisely, perhaps even as late as quarter 7.

Although just this analysis of the feasibility of each of the options may be persuasive to many, a carefully documented decision is still far off. The economics of the decision have yet to be discussed.

Compared with the expansion alternative, both the subcontracting and third-shift alternatives leave the corporation with less contribution (i.e., dollars allocable to profit and to pay for overhead) per unit produced. By subcontracting, Kemper Games loses half of its 40 percent contribution margin on the $4 factory price of Bungle. Thus the contribution per unit from subcontracting declines to

$$\$4.00 \times \frac{0.40}{2} = \$0.80 \text{ per unit}$$

from the present $1.60 per unit.

By going to a third shift, the 25 percent labor premium Kemper Games has to pay will also cut into the contribution per unit produced. In this case, we need to calculate the third-shift increase in labor costs (composed of variable costs, labor content, and the shift premium) and subtract this amount, along with the variable costs for the first and second shifts, from the factory price:

$$\$4.00 - \$2.40 - (\$2.40 \times 0.40 \times 0.25)$$
$$= \$1.36 \text{ per unit}$$

Given that the third shift is feasible for the short term and that its contribution per unit produced is much higher than the subcontracting option, it is clear that the subcontracting option should be dropped from further consideration. Of course, the subcontracting of just a part of the Bungle game is a possibility, but that option has not been developed sufficiently within the corporation to address it here.

It remains to investigate the economies of expansion versus third-shift operation. After all, if adding capacity is cheap, Kemper Games may be better off holding excess space for a while, until Deluxe Bungle or some other game can fill up the space, no matter which Bungle forecast prevails. As noted, expansion can come either by building an addition to Kemper Games' original structure or by leasing some space very close by. Let us resolve the build-or-lease issue before examining the economics of the expansion/third-shift decision.

Because a build-or-lease decision would possibly prevail only for a demand pattern like forecast 2, let us assume forecast 2 for the purposes of the analysis. To meet that forecast, an additional 600 units per day will be needed, implying a space requirement of an additional 24,000 square feet. The estimated cost to build and equip this area of the plant comes to $480,000 — $360,000 for the plant and $120,000 for the equipment. In return for this investment comes the ability to earn a stream of contribution from the manufacture of Bungle. At a contribution of $1.60 per unit and production of 600 units per day, the investment pays itself back in 500 working days.

Although it is nice to see a payback so short, the most persuasive way to analyze the economics of capacity choice is to use net present value (NPV) analysis, sometimes referred to as discounted cash flow analysis. Almost all of the largest and most sophisticated companies employ the net present value technique in their capacity planning decisions.

Using NPV to Assess the Build/Lease Choice at Kemper Games, Inc.

For simplicity's sake, let us pursue the Kemper Games situation by following the convention of using real, constant dollar cash flows and real rates of discount.

To build an on-site expansion of 24,000 square feet requires an initial outlay of $480,000. For this investment, the company expects a contribution four quarters out of

$$\$1.60 \text{ contribution/unit} \times 600 \text{ units/day}$$
$$\times 62 \text{ workdays/quarter}$$
$$= \$59,520 \text{ contribution/quarter}$$

and a stream of annual contributions after that of $238,080 (4 × $59,520) per year in real terms. Although different assumptions can be made about the relevant discount rate on the investment and the proper time horizon, we may want to use expecially high rates of discount (in real terms) and a short time horizon because the games business is faddish and uncertain. Arbitrarily, then, let us select a real discount rate of 15 percent (very high) and a time horizon of only 5 years. The cash flows involved in this investment, excluding taxes from consideration, are shown in Table 12-6.

The net present value of this stream at 15 percent is $162,800. Kemper Games can thus expect to make money on this investment. But we still must compare it with the leasing option to see which would bring the higher expected increase in the company's worth.

Leasing 24,000 square feet of space at $6.50 per square foot per year comes to an annual expense of $156,000. This cost cuts into the $238,080 annual contribution, although leasing eliminates the plant expense of $360,000.

TABLE 12-6   A comparison of cash flows (Kemper Games, Inc.)

	CASH OUTLAY (−) OR RECEIPT (+)	
YEAR	ON-SITE EXPANSION	LEASING
0	$−480,000	$−120,000
1	+59,520	+82,080
2	+238,080	.
3	.	.
4	.	.
5	+238,080	+82,080

The cash flows involved with leasing (again excluding taxes from the calculation) are also shown in Table 12-6.

The net present value of this stream of cost and contribution at a discount rate of 15 percent is $155,150. Because this figure is less than the $162,800 present value of the build on-site option, the economics say build rather than lease, although there is not much difference between the two.

Third-Shift Operation Versus Building On Site

We can now turn to the economics of third-shift operation versus building on site. Suppose that forecast 1 prevails. A three-shift operation can meet the sales demand for the five quarters Kemper Games will need to determine whether forecast 1 or forecast 2 correctly gauges the market. Thus for the first year and for the first quarter of the second year, there will be a stream of contribution from the units sold. There will also be a stream of inventory carrying charges (given as 10 cents/unit/quarter) for the units actually produced. It is the sum of the net present values of these streams that we are interested in.

Note that because forecasted needs in excess of 900 units per day, which is the two-shift capacity, occur only in quarters 3, 4, and 5 of forecast 1, the contribution to be reaped from the third shift occurs only in those quarters. Nevertheless, because third-shift production should begin in quarter 1 and last through quarter 5 before being dismantled, inventory will accumulate all through that period. Accumulated inventory must be financed and this is a charge to be considered. This reasoning suggests that both the second and third shifts should be discontinued in quarter 6 so that the excess inventory can be drawn down. This may well be an optimistic action, expecially if demand continues at a somewhat higher rate than anticipated by forecast 1. More than likely, only the third shift would have to be discontinued, but for purposes of the calculation, let us stick strictly to forecast 1. We can summarize these flows of contribution and costs as shown in Table 12-7.

The net present value of this stream of cash flows at a 15 percent discount rate is $25,900. This figure confirms that even with the high inventories that are maintained under the option of instituting a third shift, the decision is a profitable one for Kemper Games. It is not hard to see that if forecast 1 prevails, building a plant just to get over the peak demands of quarters 3, 4, and 5 is not advisable. Contribution is raised from $1.36 to $1.60 per unit, but the volumes demanded are so low that total contribution, undiscounted, comes to barely over $100,000 and the total investment required is almost $500,000. Thus, if forecast 1 prevails, a third-shift operation is preferred.

Suppose forecast 2 prevails, however. How different are the results? We already know that three shifts can provide only enough inventory to last through quarter 10 and that, with a lead time of three quarters necessary for the build option, a decision on building a plant expansion can be delayed only until quarter 7. Since Kemper Games must build at some point if forecast 2 prevails, the question then can be rephrased as "If forecast 2 prevails, should Kemper Games build now or later?"

If the corporation builds now, it faces an immediate outlay of $480,000, contribution of $59,520 in quarter 4, and a stream of $238,080 per year in contribution thereafter, as we saw in Table 12-6. The net present value of this stream of cash flows is $162,800.

If the corporation builds later, say beginning in quarter 7, three shifts would be maintained through quarter 10 and inventory carried throughout that period. The extent of the required inventory buildup is already documented in Table 12-5, and the inventory carrying expense for any quarter is simply the

TABLE 12-7 Contribution and cash flows (Kemper Games, Inc.)			
	CASH OUTLAY (−) OR RECEIPT (+)		
YEAR	CONTRIBUTION	INVENTORY CARRYING CHARGES	TOTAL
0	$ 0		
1	+63,240	$−29,760	$+33,480
2	+25,296	−22,630	+2,666
3	0	−7,965	−7,965

quarter's cumulative inventory on a daily basis multiplied by 62 days per quarter and the 10 cents/unit/quarter inventory charge. Table 12-8 develops the contribution attributable to sales from third-shift operation in the first year, namely $12,648 in quarter 3 and $50,592 in quarter 4. Because demand remains constant at 1,500 units per day in forecast 2, the $50,592 figure is maintained in all subsequent quarters.

In summary form, these flows of contribution and cost, including the cash outlay for the plant expansion in year 2 and the resulting cash flow from it, are shown in Table 12-9.

The net present value of this stream of cash flows at a discount rate of 15 percent is $200,850. Because this figure is higher than the NPV for the build-now option, the economics argue for deferring the on-site expansion and

TABLE 12-8   Contribution attributable to sales from third-shift operation (Kemper Games, Inc.)

YEAR	QUARTER	CONTRIBUTION FROM UNITS SOLD FROM THIRD SHIFT	INVENTORY ON DAILY BASIS FOR QUARTER, FORECAST 1	INVENTORY CARRYING CHARGES ON THIRD SHIFT PRODUCTION
1	1	$ 0	+705	$ 4371
	2	0	+1260	7812*
	3	12,648[†]	+1515	9393
	4	50,592	+1320	8184
2	5	25,296	+1425	8835
	6	0	+975	6045
	7	0	+675	4185
	8	0	+575	3565
3	9	0	+475	2850
	10	0	+375	2325
	11	0	+275	1705
	12	0	+175	1085

*Calculated as (1260 cumulative inventory held per day of that quarter, from Table 12-5) (62 days/quarter) (10 inventory carrying charge per unit per quarter)

[†]Calculated as [(1050 − 900) = 150 units/day of third-shift production sold (62 days/quarter)($1.36 contribution/unit)

TABLE 12-9   Contribution and cost flows from build-later option (Kemper Games, Inc.)

YEAR	CONTRIBUTION	CASH OUTLAY (−) OR RECEIPT (+) INVENTORY CARRYING COSTS AND INVESTMENT COST	TOTAL
0	$ 0		
1	+ 63,240	$ −29,760	$+ 33,480
2	+202,368	−500,646	−298,278
3	+220,224	−3,069	+217,155
4	+238,080	0	+238,080
5	+238,080	0	+238,080

filling the gap in the meantime with a third-shift operation.

Under both forecasts, a strategy of using three shifts until at least quarter 10 is preferred. Not only is it feasible but the economics, under the assumptions made, are superior.

Are there any arguments for a build-now strategy? Although none of the noneconomic arguments may be strong enough to cause the abandonment of immediate three-shift operation, there are several reasons why on-site expansion may have to be undertaken sooner than quarter 7:

1. If Deluxe Bungle is introduced to the market, and if its sales parallel those of Bungle, Kemper Games may need considerably more capacity than three shifts could ever hope to provide.

2. The marketing department seriously underestimated Bungle's appeal, and there is no assurance that marketing's sales estimates are correct now. Of course, estimating is a difficult task, at best. If sales are understated again, however, Kemper Games could quickly fall short of capacity.

3. The economics advise establishing a third shift at the same time as a second shift is being established. The addition of so many new workers so quickly is bound to have a serious impact on productivity. The company's managers simply cannot be expected to train two new shifts of workers without considerable startup difficulties and monumental exhaustion. The capacity plan should allow for the especially low productivity both the second and third shifts can be expected to suffer for two or three quarters.

These three considerations ought to temper our enthusiasm for the third-shift alternative. Although they may not deter the establishment of a third shift in the near term, they should keep us alert to the need for additional production space.

Lessons from Kemper Games, Inc.

It should be clear from this discussion that even a relatively simple capacity situation, such as that faced by Laura Kemper, can be surprisingly complex in its analysis. No wonder then that most corporations take months developing and reviewing capacity plan proposals!

As we have seen, there are some critical features to the typical capacity plan and its analysis: the forecast, gauging capacity, the transition to increased space, economic analysis, dealing with risk, and noneconomic considerations.

The Forecast

The driving force behind the typical capacity planning exercise is a forecast or group of forecasts of sales in the upcoming quarters and years. In most companies, the forecast itself is not the particular responsibility of the production or operations managers, although they might have a hand in it. More often, the marketing department for a particular product or product line is charged with forecasting sales.

Forecasting the long term is frequently very uncertain, especially if the market is not well defined or the product is new or highly dependent on the fortunes of a hard-to-predict national economy. Over the years, forecasting techniques have become increasingly sophisticated, although one must admit that an entire arsenal of mathematical techniques may fall short of a single person's "gut feel" or refined intuition. (See the Chapter 6 Appendix discussion of forecasting.)

Gauging Capacity

As we discovered earlier in this book, production capacity can be very ambiguous, subject to a host of influences. Nevertheless, capacity management requires some appraisal of present capacity and what can be done to increase capacity by a given amount.

The Transition to Increased Space

As Kemper Games' situation makes clear, it is one thing to recognize the need for increased space in the long term and quite another to manage the transition from the present to that long-term goal. Capacity increases often come in lumps, and in such circumstances managers must devise short-term policies to ease the operation into that capacity increase. At Kemper Games that meant a third shift, but in other situations it may mean subcontracting, an inventory buildup, price adjustments to modify demand, overtime, line rebalancing, or something similar.

The nature of this transition, coupled with the needs of the forecast, often determines the timing of a capacity increase. As we saw in Kemper Games, the ability of a third shift to meet demands in the short run enabled the corporation to delay the construction of an on-site expansion until the sales pattern for the admittedly difficult-to-predict Bungle game was known with considerably more certainty.

Economic Analysis

Finding an economic justification for a particular capacity proposal is an important step in advancing that proposal through approval and implementation. Because capacity plans are almost always multiyear in outlook, the time value of money becomes an important consideration in any economic analysis of a project.

In any analysis of the economics of a project, assumptions have to be made. It is usually worth exploring how the economics vary under different assumptions. This kind of sensitivity analysis can be applied to (1) the time horizon of the project, (2) the cash flows expected in

any year, and (3) the rate of discount to be applied to the cash flows. The sensitivity of a project's net present value to such factors is important because the risks involved with a project may be great; whether it should be chosen may rest on how "robust" its economic value is, given a host of different assumptions. Sensitivity analyses help to identify the projects that entail high risks as well as high returns.

Dealing with Risk

A corporation's capacity decisions are among the more uncertain it faces. Although there is no sure-fire way to handle riskiness, it usually is helpful to know whether adding capacity is cheap or expensive and what penalties the corporation faces by having either too much or too little of it. If, for example, capacity is rather cheap to add and falling short of it would forgo substantial contribution, the corporation is probably better off to carry an excess. Many job shops and batch flow processes involve relatively little capital investment and considerable labor intensity. In such processes, it is often the case that some excess capacity can usefully be kept for peak periods of demand and left idle the rest of the time. Its use during the peak periods generates more than enough contribution to finance it for the remainder of the year.

Noneconomic Considerations

Although the economics of a capacity project are important, it is a narrow-minded manager who bases his or her decision solely on the economics. Three nonquantifiable (or at least difficult to quantify) factors deserve special mention:

1. *The impact of particular choices of capacity and facilities on existing operations.* The transition to enlarged capacity or to a new

facility can be time consuming and troublesome; it is seldom as smooth as either hoped for or planned. It often pays, however, to make some allowance for the impact of a new project on the existing operation. For example, in the Kemper Games situation, the introduction of the second and third shifts simultaneously is bound to involve an extra amount of chaos. How much is problematic, but its presence should be considered in the decision.

The nature of these impacts on existing operations is varied. Frequently, a capacity or facilities change will alter the character and extent of materials handling and movement within the company. Marketing and/or sales may have to adjust to such a change as well. Often, too, the startup of the new capacity saps engineering and management attention from the existing operation. Other impacts abound as well, many peculiar to the company or the investment itself.

2. *The competitive reaction of others in the industry to the capacity change.* Many capacity projects cannot be investigated in a vacuum, since their desirability rests in part on how competitors meet a company's particular decision on new capacity. This is particularly true of oligopolistic industries and those that are sensitive to transportation costs. How a competitor reacts may greatly affect the market in a geographic area or the market for a particular product line.

3. *Flexibility and new capacity.* In fast-paced industries, where new products, markets, and marketing approaches are routine, production must stay flexible. The choice of a capacity project in one form or another may hinder or help the company's ability to keep abreast of the latest developments. Flexibility is very difficult to quantify, but its importance to a firm's continued success can be immense.

Common Pitfalls in Dealing with Capacity

The capital appropriation request procedure as outlined at the beginning of this chapter has recently come under increasing attack.[3] The procedure is frequently seen as too rigid in practice and too narrow and confining in its outlook. How ironic it is that purchasing, which accounts for half of the cost of goods sold in many companies, and hiring, where lifetime decisions are often made, are management activities that are taken almost casually in many companies, whereas plant managers are often required to jump through hoops to justify capital expenditures for equipment that may be fairly inexpensive when compared with a worker's capitalized stream of earnings.

Some companies do tend to look at manufacturing choices involving large expenditures as isolated decisions, not as interrelated parts of an overarching manufacturing strategy (see Chapter 15 for a discussion of manufacturing strategy). This approach narrows the focus for many capital appropriations. This narrowed focus, coupled with the need for appropriation requests to win approval by the corporation's financial department, leads to several problems with these procedures:

1. Many corporate procedures for capital appropriations are typically used, or even intended, for the study of additional capacity, not replacement capacity. As a result, what needs to be done to shore up existing operations and to keep them fully competitive is sometimes overlooked or not examined very critically.

2. The procedures too frequently stimulate a weak and/or narrow list of alternatives to be investigated. Although the stated corporate policy may be to promote fresh, even radical, ideas, in practice the options that are considered tend to be conservative and may

even be specifically slanted to what the general manager thinks (or is believed to think by the group making the study).

3. Capital appropriation plans are typically excursions into uncertainty. Benefits from new plant, equipment, or process configurations — especially radical ones — tend to be more elusive to quantify than are costs. As a result, options that are safe (such as on-site expansions as opposed to new plants, known technologies rather than fledgling technologies) often have an inherent advantage, although they may be the worst choices for the company in the long run.

4. To get around the biases of narrowness and uncertainty, project teams that really believe in change often end up fudging their data. Games are then sometimes played between project teams and the corporate level managers who must decide on the appropriations. It is understandable then that capital appropriations that have to meet a certain hurdle rate (rate of discount) often end up with project proposals whose rates of return are just over those hurdle rates. Some companies deliberately set their hurdle rates high as a way to insist that new capital projects end up better than existing capital projects. Too often, however, setting these high rates simply stimulates more creative fudging of the data.

From my point of view, capital appropriation projects should be evaluated in the context of the company's formal manufacturing and competitive strategies. Thus individual projects would be seen as involving interrelated pieces of manufacturing, capacity, and technology. The key question for the decision would be how the individual project fits with the manufacturing plan. This is not to say that the economics of net present value should be ignored. Rather, this point of view argues for all the

right kinds of benefits to be included in the calculations. It also argues for the hurdle rates to be realistic and for postaudits of projects to be done so that project teams and top management can recognize the numbers problems that one can fall victim to in developing capital appropriation requests.

CAPACITY IN A SERVICE INDUSTRY

Most services are provided (1) on demand, (2) with interaction between the customer and someone who conveys, and often provides, the service, and (3) without any inventories to help smooth out the supply task. Under such circumstances it is not difficult to understand why capacity planning and choice are a critical element in the management of service industries. As background for a discussion of capacity management of services, consider Situation 12-2.

Discussion of Quaker City Squash

The situation Dick Murnane finds himself in illustrates many of the unique problems with which service firms have to wrestle, most of which are doggedly difficult to quantify.

The first-cut economics of this situation are actually quite straightforward. For each court, Quaker City Squash now receives $8 per half-hour or $16 per hour of use. There are 5 hours of each weekday when all the courts are occupied; for the remaining 10 hours per day and for all of Saturday only half of the courts are occupied. These figures imply a weekly revenue of $5520 ($920 per court) and a yearly revenue of $287,040, assuming the same level of use as now exists. We can use this revenue figure together with the investment sum required ($575,000) and the expected operating costs of $115,000 per year to analyze the economics of another downtown location. The cash flows in

SITUATION 12-2

Quaker City Squash, Inc.

So far it had seemed easy, but now Dick Murnane, president of Quaker City Squash, Inc., faced the first tough decision of his young business career: should he expand operations and, if so, how? For the past 2 years, Quaker City Squash, located in the business district of Philadelphia, had been selling half-hour blocks of time on its six squash courts to anyone willing to pay the $8 fee. Squash is a racquet sport generally played indoors by two people and is similar in style to handball. It had been enjoying growing popularity, especially at universities. Once out of college, however, players typically had to join downtown clubs to gain access to squash courts. The club fees were invariably high and covered more than just squash.

Sensing this latent demand, Dick (a former college varsity player himself) used some of an inheritance to purchase a building downtown and convert it into six air-conditioned squash courts, locker and shower rooms, several offices, and a bar and grill area where a bartender/short-order cook served drinks and a limited selection of sandwiches. Business boomed. Soon after opening, Dick extended the hours of operation so that now play began at 7 A.M. and continued to 10 P.M. every day but Sunday. Further, Dick hired a teaching pro who was now drawing a fine salary from teaching and from restringing squash racquets. The squash pro was regarded as a definite asset to the operation. Even the bar and grill was making money by doing a brisk luncheon and cocktail business.

From all Dick could tell, Quaker City Squash attracted mainly young (under 40) professionals who worked in Philadelphia's business district. Most had been introduced to squash while in college or in graduate school and were looking for a way to stay in shape. Over 90 percent of the clientele was male. The peak hours lasted from 11:30 A.M. to 2 P.M. and from 4:30 P.M. to 7 P.M.; it was rare when a court was open during those times. Average usage outside these hours ran about 50 percent. The same 50 percent usage prevailed all of Saturday as well. Standard practice was to reserve a court at least a day in advance, although no reservations except permanent ones (a year's worth of time at a particular hour and day of the week) were accepted for any more than a week in advance. There were no assigned or permanent lockers; players brought their clothes and racquets with them.

With business going so well, Dick naturally pondered the opening of another Quaker City Squash Center. Two different options presented themselves:

1. Open another center in the business district, much like the one currently in operation. A building suitable for the necessary remodeling was available in another portion of the downtown area. It could be purchased for $300,000 and the remodeling for six courts could be done for $275,000. Operating expenses (salaries, utilities, necessary maintenance, taxes) could be expected to run about $115,000 per year, the same rate as at the existing squash center. This new squash center would cater to the same market as the old one; but because it would be located in a different part of the business district, it would appeal to a different group of young professionals.

The available building was somewhat larger than the existing center, and that posed some special problems. Dick could add another court, but that would mean paring some space away from the locker rooms and the bar and grill. Or, Dick could expand the locker room space (some of the permanent reservation holders had been clamoring for their own lockers at the existing facility and would gladly pay for them) and/or expand the space of the bar and grill, making it fancier, perhaps adding a waiter, and permitting the kitchen to expand its limited menu in hopes of attracting more players after their games or of catering to a small band of squash "addicts" who might use the squash center as their own "club." The costs seemed roughly comparable; that is, $40,000 more would pay for either the additional court, or the expanded locker room ($10,000) and an expanded bar and grill (between $20,000 and $35,000, depending on decor and modifications to the kitchen).

2. Alternatively, Dick could open a squash center in one of Philadelphia's suburbs. Demand in the suburbs was more spread out and so only a smaller center, say three courts, could be supported at any one location. Building in the suburbs was likely to be considerably cheaper than remodeling a downtown building. For a suburb along the fashionable Main Line, construction estimates, including land, averaged about $200,000 for a three-court squash center. Operating expenses were also likely to be proportionately lower, the best estimate being $48,000 per year.

Much less predictable was suburban demand for the courts. Dick had no good feel for how strong any one of several trends would be and what each would mean for the operation:

- Would a suburban location shift demand to more women players than was true of downtown? Was that demand strong enough so that Quaker City Squash should offer nursery services?
- Would the peak demand period shift from lunch to, say, the early evening, as was true for many indoor tennis clubs?

- Would the time pattern of play support opening up later than 7 A.M. and extending play to 11 P.M. or even midnight?
- Three courts would not support a bar and grill, as in the downtown squash center, but Dick wondered whether some vending machines ought to be substituted.
- Three courts would not support a full-time teaching pro. Dick wondered whether a part-time pro would be necessary, say in the morning, teaching mainly women.

A further concern of Dick's, no matter which alternative was finally chosen, was how competition would affect all of his operations. Not only was there the threat of new squash centers opening up in either the business district or the suburbs, but there was also the threat of the downtown clubs introducing "squash only" privileges at special rates. Dick wondered whether adhering to the same strategy and being first in the market was sufficient, or whether he should try to "lock in" his clientele by offering memberships like the clubs.

real terms for that investment over, say, 10 years are shown in Table 12-10.

The net present value of the investment (10 years at a high real rate of interest of 15 percent) is a handsome $288,430. We can also calculate the average usage level that must be sustained if the investment is to have a net present value of at least zero. The annual revenues would have to be $229,570 per year to have an NPV of zero, and this in turn implies an average usage rate of 51 percent. Any rate higher than that and the NPV turns positive.

In similar fashion, the economics of the suburban location can be calculated. The investment comes to $200,000 with expected annual operating costs of $48,000. If the revenue per court remains the same as downtown with 5 hours of peak play and 10 hours of 50 percent

usage, weekly revenues will total $2760 and the yearly revenues will come to $143,520. The cash flow pattern is shown in Table 12-10.

The net present value of this suburban investment (10 years at a real rate of interest of 15 percent) is $279,390, which is almost as high as for the new downtown location. In fact, shortening the time horizon or raising the discount rate makes the suburban location relatively more attractive. The economics, then, do not argue much in favor of one location over the other.

What must guide the decision are a number of other considerations, most of which are exceedingly difficult to quantify. They are what make the capacity decisions of service industries, in many cases, relatively more troublesome than those of manufacturing enterprises.

TABLE 12-10   A comparison of cash flows (Quaker City Squash)

YEAR	NEW DOWNTOWN LOCATION			SUBURBAN LOCATION		
	CASH OUTLAY(−)	RECEIPT(+)	TOTAL	CASH OUTLAY(−)	RECEIPT(+)	TOTAL
0	$−575,000			$−200,000		
1	−115,000	$+287,040	$+172,040	−48,000	$+143,520	$+95,520
2	.	.	.	.	.	.
.						
10	−115,000	+287,040	+172,040	−48,000	+143,520	+95,520

Among these considerations are the mesh of marketing with operations, competitive reactions, demand modification, and nonspace capacity adjustments. These four considerations can all shape thinking about the need for and the design of capacity in service industries. Although their economic impact is frequently difficult to assess, their importance to the success of any service firm's operation can be crucial. At the same time, deciding what should be done about each is quite a burden. As we shall see, the course Dick Murnane should follow with Quaker City Squash is far from evident.

The Mesh of Marketing and Operations in Planning Capacity

Should Quaker City Squash's courts be centrally located in the business district or should it decentralize its capacity? The juxtaposition of marketing and operations is critical in this regard. Is growth to be found in the suburbs? Which is more convenient: squash near home or squash near work? How much a part of the game is camaraderie, and is camaraderie more likely to flourish at a downtown location? Will significant numbers of women be attracted to the game if courts are situated in the suburbs?

Because capacity is often the "product" in a service business, or at least intimately linked to the service provided, knowledge of why a customer purchases the service is a critical item in the plan for the design of new capacity. This is something Dick Murnane must ponder; the answers are by no means obvious.

Related to the issue of increasing the right kind of capacity is the concern about whether new capacity will undercut the service provided by already existing capacity. At Quaker City Squash, will a new downtown location divert players away from the existing downtown site, or will players refrain from downtown play during the day to play at the suburban location in the evening? To answer these questions, one needs knowledge of the customers, where they come from, and why.

Another aspect of the close marketing/operations bond in a service firm's capacity planning is the need to keep balance in the mix of elements which make up that capacity. For example, at Quaker City Squash there is the opportunity to build a seventh court, although at the expense of the locker rooms and grill. Alternatively, a private locker room could be installed and the bar and grill expanded.

If one assumes that the new court will attract use just like any of the other courts, the economics are much more likely to support a squash court than to support private lockers or added space in the grill room. What must be questioned is whether a new court, coupled with smaller locker and grill room space, would actually add enough revenue to warrant its construction. It is conceivable that usage would adjust itself not to the number of courts provided but rather to the locker room and/or grill room space. In that case, adding a seventh court would be a disastrous move since total patronage would actually fall.

Competitive Reactions

We have already discussed the importance of gauging competitive reaction for capacity planning in manufacturing companies. Because capacity decisions (how much, of what type, and where) are so critical to a service firm's competitive stance, the reactions of competitors are usually even more significant in service industries. At Quaker City Squash, for instance, competitive reactions may take the form of other squash centers being opened in downtown Philadelphia or they may be policy changes at some of the existing clubs (say, the inauguration of separate squash memberships for special fees). In any event, competition can quickly change the earnings potential of all of Quaker City Squash's operations.

Modifying Demand

When a bottleneck is reached in the provision of a service (e.g., during the noon hour at Quaker City Squash), often that bottleneck can be relieved, at least temporarily, by manipulating demand in various ways through pricing policies and service policies.

Pricing can be used as both carrot and stick to shift the extent and the timing of demand. High prices can shut off demand or shift it into other time periods. Low prices can stimulate demand. Whether a price cut will actually lead to more revenue as well as greater patronage depends on how sensitive the demand by patrons is to price changes. If changes of X percent lead to changes in the quantity demanded of more than X percent, the demand is said to be price elastic and revenues can be increased by cutting price. If an X percent change in price leads to less than X percent change in the quantity demanded, the demand is termed price inelastic and revenues can actually be increased by raising price. Ascertaining whether demand is in fact price elastic or price inelastic is often hit or miss, however, and it is usually under pressure—severe lack of capacity or embarrassingly excess capacity—that prices are altered significantly.

The extension of existing services and/or the provision of ancillary or complementary services can also have a substantial impact on demand. In terms of the Quaker City Squash situation, the teaching pro, the bar and grill, the private lockers, nursery care, the reservation system, and other extras may all have important implications for demand. One or more of them may be instrumental in attracting demand or in shifting demand from peak times to off-peak times. The reservation system is clearly a tool to apportion demand among peak and off-peak periods, and the timing of the teaching pro and nursery services could also have a bearing on broadening demand away from the peak periods.

Adjusting Capacity Without Altering Space Requirements

At Quaker City Squash, space means capacity, but there are a variety of other service industries in which capacity can be increased modestly by employing several techniques:

- Using part-time workers for peak times (e.g., think of temporary employment services).
- Increasing efficiency via methods improvements, labor specialization, labor cross-training, and the like.

- Inducing consumers to provide some of the service themselves (e.g., salad bars at restaurants).

- Sharing equipment or other resources with firms or organizations in the same or a related industry.

ON-SITE EXPANSION, BRANCH ESTABLISHMENT, RELOCATION

Once a company's managers agree that new capacity ought to be erected to satisfy expected future demand, they still face deciding how that new capacity ought to be implemented. Should on-site expansion, the opening of a new plant, or the relocation of an existing one prevail? This decision lies at the heart of Situation 12-3.

Discussion of Tighe Printing Company

Growth is forcing the Tighe Printing Company to seek more production space, and Ruth Tighe has to choose from among three means of adding that space: expanding locally, placing some operations in a new branch, and relocating to a larger facility. Of these options, expanding locally is the lowest cost one, at least in terms of initial dollar outlay. A building of 55,000 square feet can be purchased for $700,000 or leased for about $125,000 a year. That much space in new construction would cost about $1 million or more to build and over $200,000 a year to lease. Relocating would cost an additional $200,000.

Although a local expansion may be inexpensive in the short run, it may be frightfully expensive in the long run, especially if it locks Tighe Printing to outmoded technology, high materials handling costs, and other detriments to an effective operation. There are indications in the situation that some of this is occurring already. What then are the relative advantages of on-site expansion, new branch establishment, and plant relocation?

The Relative Advantages of On-Site Expansion

On-site expansion is by far the most popular way of adding production space. Its low cost (often involving no new land acquistion costs), the relatively short time lag associated with it, and its status as a "known quantity" make it appealing. By expanding on-site a company is assured of keeping its labor force intact and it does not have the difficult task of weeding out products or pieces of the production process for isolation in another plant.

Often, too, on-site expansion is viewed as a way to achieve the benefits of economies of scale. This is an elusive concept, however, and it merits some discussion of its own.

Economies of Scale[4]

It is frequently stated that large companies enjoy a competitive edge over small companies because of the inherent advantages of economies of scale. The phrase has a strong appeal, and large companies with sizable plants have a certain aura about them. General Electric or IBM is no stranger to the observation that its continued cost competitiveness is the result of its size. The phrase draws nods of recognition, yet few people can define it and fewer still have thought critically about it.

All too often you hear a company president nonchalantly say something like, "If we increase the size of our operations in this plant, we'll naturally enjoy some economies of scale." There is nothing natural about economies of scale and certainly nothing to be nonchalant about.

Confusion about the term is understandable because it serves as an umbrella for a number of real but quite distinct concepts. Because it is an umbrella term, economies of scale often loses its usefulness in making management decisions on plant size.

Tighe Printing Company

"Maybe sometimes you can be too successful," Ruth Tighe mused to herself. The Tighe Printing Company, of which Ruth was president, had enjoyed substantial growth. Ruth was increasingly concerned, however, that continued sales increases simply could not be accommodated without increased space.

The Tighe Printing Company specialized in high volume printing of magazines, catalogs, and business forms. The regional sales of magazines like *Newsweek* and *McCall's* were run on Tighe presses and constituted the bulk of company sales. Timeliness, as could be expected, was essential to the company's success. The company was located in downtown New Haven, Connecticut, in a six-story structure built in 1923. Founded by Ruth's grandfather 10 years before then, Tighe Printing was the building's first tenant. The company bought the structure in 1935.

As the company grew after World War II and thus required more space, it started displacing other tenants. By the late 1960s, the company had taken over the entire building, all 75,000 square feet (11,500 square feet on each of the six floors and 6000 square feet in the basement). In 1972, again pressured by space needs, the company leased 15,000 square feet in the building next door and knocked out a wall in between. Now even this space was fully utilized.

Ruth Tighe was convinced that she would have to decide soon how she was going to structure the company's growth. The possibility of starting afresh with either a new branch facility or a relocation was appealing, albeit more expensive. Ruth would have to weigh these relative benefits and costs.

Ruth Tighe confronted a choice from among three alternatives:

1. *Continued expansion in the neighborhood.* A building two doors down from Tighe Printing and on the other side of the street was up for either sale or long-term lease. The building was four stories high with a basement, occupying 55,000 square feet. A likely rent would be $2.25 per square foot per year. The asking price for the building was $700,000.

2. *Establishment of a branch plant.* Rather than keep all its production space close together, Tighe Printing could locate a satellite plant elsewhere in the city or in the suburbs. This new branch could house a new press or two and could concentrate on a certain type of printing.

3. *Relocation of the entire operation.* Rather than break up existing operations into two distinct locales or continue a more or less haphazard expansion around the company's main building, Tighe Printing could sell its old holdings and equipment and move the entire operation in one piece to larger quarters, purchasing new equipment in the process. Exclusive of new machine purchases, a move itself was likely to cost $200,000.

Newly constructed space for either a branch or a move would cost $18 per square foot or more, depending on location. Rents for suitable, existing space could vary from $2.20 per square foot per year to $3.50 per square foot per year. The higher values would secure space in newer, one-story structures.

As Tighe Printing was now, operations there left much to be desired. In particular, the movement of goods and information was worsening. The growth of the enterprise through all seven levels of the main building separated departments that logically should have been placed side by side. Work-in-process had to be handled almost twice as often as necessary, and, besides being time consuming, led to more routing errors than Ruth Tighe, at least, felt should be the case.

Intra-factory transportation was slow, limited to a single elevator. Because there were only two loading docks, the delivery of paper and other raw materials was often held up to give priority to shipments out of the building. Deliveries often had to be taken at odd hours. In addition, more and more materials and output were being stored outside the plant, sometimes expensively.

The constraints of the building also meant that many of the latest innovations in the printing industry such as phototypesetting and the faster and larger electronic web presses had to be forgone. Furthermore, space constraints made it impossible to reap the advantages of running machines close to one another in tandem and sorting entire magazines or forms on the spot—advantages in terms of production scheduling, production control, and workforce assignments.

Volume, Capacity, and Process Technology

How do economies of scale relate to plant size? The standard definition declares that economies of scale are reaped whenever higher output volumes lead to lower unit costs. However, the definition only seems to be clear.

Consider four different plants, all in the same industry. Plants A and B are physically identical in every respect. They differ only in that plant B produces just a fraction of the volume of plant A, perhaps because company A's marketing efforts are much more effective than company B's. Plant C is similar to plant A in layout and equipment but is twice the size. Plant D, on the other hand, incorporates an entirely different process technology from the other three and can produce twice what plant A can produce.

These examples are drawn to distinguish among the concepts of volume, capacity, and process technology. As Table 12-11 shows, plants A and B share the same capacity and the same process technology, although their actual volumes differ. Plants A and C share the same process technology, although not the same capacity; plants C and D share the same capacity but not the same process technology.

These concepts of volume, capacity, and process technology are important because they all relate to economies of scale. We will discuss these three specific types of economies.

Economies of Volume

The higher-volume plant A will enjoy lower unit costs than plant B because it can spread its fixed costs (overhead, capital costs, machine setup costs) over a greater number of units. If by scale we mean volume, then the difference in unit costs between plants A and B can be an economy of scale.

For some people, however, this "spreading of fixed costs" seems too trite an example to be labeled a "scale economy," so they dismiss it. These people then modify the definition of a scale economy to exclude economies that are really economies of volume.

Economies of Capacity

As the table shows, plant C's unit costs are lower than plant A's and thus much lower than plant B's. Plant C's increased capacity permits

PLANT	CAPACITY		VOLUME PRODUCED		PROCESS TECHNOLOGY		MANUFACTURING COSTS PER UNIT PRODUCED
A	100	Difference in capacity; same volume to capacity ratio	100	Difference only in volume	Type X		$1/unit
B	100		40		Type X		Higher than $1/unit
C	200		200		Type X	Difference only in process technology	Lower than $1/unit
D	200		200		Type Y		Much lower than $1/unit

TABLE 12-11   Defining the differences in unit costs

SOURCE: R. W. Schmenner, "Before You Build a Big Factory," *Harvard Business Review* 54, no. 4 (July-August, 1976): p. 102.

it to carry proportionately less raw materials inventory. This may be due to the familiar "economic order quantity" result that optimal inventories need increase only as the square root of volume and not proportionately with volume.

On the other hand, because plants A and C share the same technology, plant C is not likely to enjoy a proportionately lower work-in-process inventory, but its finished goods inventories may well be proportionately lower. For example, suppose that plants A and C both make two different products. If plant A has only one production line that it must change over from one product to the other, it will have to build up proportionately more finished goods inventory to cover its demand than plant C, which can afford to manufacture each product on a separate, dedicated line. Plant A needs to carry enough of an inventory of its first product to tide itself over while it makes its second product. Plant C need not worry about this problem.

In addition to these inventory-associated economies, plant C will have an advantage over the smaller plant A, as its increased capacity allows the luxury of more spare equipment and maintenance capabilities or of additional and useful overhead functions. If scale means "capacity," then any differences in unit costs between plants A and C can be appropriately called economies of scale.

Economies of Process Technology

As we can see, plant C enjoys some cost advantages over plant A. However, plant D enjoys even lower costs—for two major reasons: capital-for-labor substitution and labor specialization. Plant D has automated more (substituted more capital for labor) than plant C. Using more, better, and different equipment, plant D has been able to produce as much as plant C, but it employs fewer workers. As a result, its costs are lower. The increased auto-

mation at plant D may or may not be an advance in the state of technology, but the increase in capital certainly alters the process in plant D as compared with plant C. Moreover, often a company cannot make small additions to its plant's stock of equipment and space; large additions (frequently termed "indivisibilities") have to be made and are perhaps the classic explanation for economies of scale. Such indivisibilities are usually substantial alterations to a plant's process technology and so are more than just scale changes. They are changes in process technology, and should be recognized and managed as such.

One particular substitution of capital for labor merits special mention. The geometry of processes that deal with free-flowing materials (chemicals, molten metals) often permits output to vary according to the volume of its capital equipment, while the costs (construction and/or operating costs) vary according to surface area.

This is the frequently quoted "6/10 rule," which argues for large-process technologies. The rule is so named because it was found that for chemical plants a doubling of volume leads to only a 6/10 increase in surface area and thus costs. An oil refinery is one such example. Output depends greatly on the sizes of tanks and pipes, but construction and maintenance costs depend more on surface area of the material that surrounds the oil held inside. For most other (non-free-flowing materials) operations, however, such cost advantages are much less significant.

Plant D also has lower unit costs because it has altered its process technology to specialize its labor for particular tasks. In order to make its process more continuous, jobs are "deskilled." That is, instead of large numbers of highly skilled workers each doing a number of operations to form the product, the process is organized to link together less skilled workers doing a small number of specific operations.

The time and responsibility any one worker devotes to a particular product is reduced in an effort to increase productivity through repetition and specialized competence. The pace of the production process shifts from worker discretion to management option. The car assembly line is a classic example of labor specialization and management (that is, mechanical) pacing.

To reiterate, both capital-for-labor substitution and labor specialization alter process technologies. Thus, if by scale we mean "process technology," then we can appropriately call these two means of cost reduction "economies of scale."

Avoiding Ambiguity

As pointed out in the preceding paragraphs, the term *economies of scale* suffers from irredeemable ambiguity. It can take a variety of meanings depending on one's interpretation of *scale* as volume, capacity, or process technology. Even then, there are a variety of ways to effect the economies. Instead of inviting confusion by using the term *economies of scale*, people should think of volume, capacity, and process technology separately.

Plant Size Decisions over Time

The plant size decision depends on the careful balancing of a number of alternative technologies, costs, and risks. Over time, these alternatives become more and more blurred. Technological advance is unpredictable, and it is not clear whether new capital equipment with smaller or with larger break-even volumes will predominate. Lots of new technology is sophisticated, expensive, and large scale; and yet microcomputers and some machine tools are persuasive examples of some small-scale technological change.

In sum, "economies of scale" is vague enough to provide easy justification for any number of decisions on plant capacity. Because it is so vague, its usefulness to managers is minimal. Instead, two areas should be carefully scrutinized: (1) the cost reductions a company can achieve through specific changes in a plant's volume, capacity, or process technology and (2) what those changes mean for management control, logistics, inventories, and the ability to respond to product or process innovations.

Return to the Issues at Tighe Printing

These cautions about economies of scale notwithstanding, on-site expansion sports some very attractive features. Nevertheless, it is not an all-purpose remedy for production space shortages and there are a host of circumstances where either a new branch or a relocation would be preferred.

The Disadvantages of On-Site Expansion: Building the Case for Branching and Relocation

Although frequently attractive, on-site plant expansion can usher in a host of diseconomies as well, particularly if on-site expansion has been a repeated practice. For example, as more and more production space is added on-site, the layout of the plant typically becomes less and less optimal. Rarely is an entire plant reconfigured during an expansion; rather, only portions of the plant get shifted around. The result, over time, is that departments, once close together, become separated. Materials handling and storage become more difficult, with more chances for delay or error. Managers find themselves isolated from one another and/or from the work groups they are supposed to oversee. In short, intraplant transportation and communication become strained and this strain is likely to have detrimental consequences for product delivery and quality.

Staying at the same site often postpones the introduction of new process technology as well. Old equipment is kept in use, old methods are followed, and the advantages of new equipment and techniques are forgone, with consequences for both future costs and product innovation.

Continued on-site expansion means that more and more workers and, often, more and more products, must be managed. Such a layering of expanded responsibilities creates real complexities for managers at all levels. The existing cadre of managers may be asked to supervise more than they are readily capable of, thus lessening the attention certain problems should receive. With more products and output from the same plant, decisions on the levels, composition, and uses of inventories are likely to become more difficult and prone to error. Decisions on production control—what and when to produce, how to sequence it through the factory—are likely to become vastly more complex as well. As more and more products are added to the factory, the cost accounting system is likely to become more arbitrary and thus less helpful.

With more products in the plant, management runs the risk not only of complicating supervision, inventory, and production control systems and the like, but also of placing incompatible demands on managers, workers, and systems. For example, products produced in low volumes, but with special performance features, demand a different mode of management, worker effort, and control systems than products produced in high volumes with few special features. If both are manufactured in the same facility with common management, workforce, and operating policies and systems, the likely result is that both products will suffer in the individual dimensions (price, performance, delivery) that make a product competitive.

Besides this incompatibility problem, the addition of more workers to an existing site is apt to require the increasing formalization of the workforce–management relationship. The workforce is less apt to identify strongly with the company, and labor relations within the plant may become strained. Old management

concessions to the workforce may come back to haunt operations. With increasing size, the plant is more likely to become a target for unionization, if it is not already organized.

For these reasons, continued on-site expansion becomes less and less desirable. The alternatives—the establishment of a new branch plant and plant relocation—can avoid many of these long-term and frequently subtle pitfalls of on-site expansion, although their abilities to surmount certain of these pitfalls differ. Table 12-12 outlines some of the relative advantages of new branches and relocations.

New branch establishment, for example, is at a relative advantage if the plant's problems run more to product proliferation, workforce size, or meeting expected future growth. By branching, a company can avoid overloading one plant with either too many products or too many workers. At the same time, the new branch can exploit the latest production technology and the most sensible plant design. The operating policies and systems of the branch can also be meshed carefully with the product(s) chosen for manufacture and with the competitive priorities attached to them.

Plant relocation, on the other hand, is at a relative advantage if the plant's problems are more involved with plant layout, materials handling and storage, new process technology, production and inventory control, and lack of management depth. Relocation, by definition, means closing one facility and opening another at roughly the same time, which means that relocation can readily scrap old capital, technology, and policies for new. Thus relocation gains in standing when the plant's problems are less related to large size and more to process technology and control.

Return to Tighe Printing Company

From the description of Tighe Printing's problems, it would seem that relocation is best able to remedy them. The company's present site

TABLE 12-12   Advantages of branches and relocations vs. on-site expansion

PROBLEM AREA	NEW BRANCH PLANT	PLANT RELOCATION
Plant layout and materials handling	Radical improvements possible with operations placed in branch; some possiblities of improving base plant as operations are placed in branch	Radical improvements possible
New process technology	New technology for branch possible; likely that base plant will keep much of old technology	Scrapping of old plant, equipment, and methods possible; new technology can supplant it readily
Production and/or inventory control	Can mean radical change to production control procedures and policies in new plant, though not much change to be expected for old plant; inventories can build up by adding branch	Can mean radical changes to production and inventory control; inventory levels not likely to be affected
Managerial impact	Additional managers needed to open and run branch: staff demands increased to coordinate plant interactions	Old set of managers can generally run new plant without stretching themselves too thin
Product proliferation	Can easily manage new products, especially if branch plants are organized as product plants	New products less easily managed
Size of workforce	Keeps workforce levels at all plants under desired ceilings	Little or no effect on workforce size
Financial burdens	Extra overhead demanded to cover more than one location, new plant start-up expenses	Moving costs, new plant start-up expenses
Ease of meeting future growth	Relatively easy; geographic growth met best with new market area plants, product introductions with product plants, and vertical integration with process plants	Not easy; shares many future capacity problems with on-site expansion alternative

and building are contributing to poor materials and information flow, excessive outside warehousing, and delayed introduction of production process improvements. The time appears appropriate to abandon the company's history of incremental expansion on site in favor of the more radical approach of relocation. New branch establishment appears to be a less attractive solution since some economies are possible by running nearby machines in tandem. Production scheduling seems to be facil-

itated by bunching the printing presses at the same site.

Tighe Printing has not perfomed any quantitative study of the likely opportunity costs of maintaining the existing plant in the face of what the company could do at another location. Some such study would be warranted. However, the implications of the qualitative aspects of the situation, which have already been discussed, are unlikely to be reversed, as the study may show (see Situation 12-4).

SITUATION 12-4

Tighe Printing Company (Reprise)

After considerable thought, Ruth Tighe decided to relocate her growing printing company. Only by abandoning the old six-story structure the company had grown up in, she reasoned, could the company keep abreast of the latest technological changes that were making over the printing industry. New and sufficient single-story space appeared to be the answer. What remained to decide was where the space ought to be located. After some preliminary scouting by the company's vice-president for operations, four options presented themselves (excluding, of course, the ever-present option of continuing to search for sites). A summary of site-related costs is included in Table 12-13.

1. *Long Wharf Industrial Park in New Haven.* Tighe Printing could maintain its home in New Haven by moving less than a mile to some vacant land in the nearly complete Long Wharf

Industrial Park. Only one parcel of land of suitable size was left, a 5-acre site. Assuming single-story construction, having only 5 acres would restrict the plant space to 110,000 square feet. (Property covenants in the park restricted plant size to no more than half the area of the site.) Nevertheless, the site provided free parking for Tighe's 220 employees and the certainty that none of the plant's workers would leave as a result of the move. Long Wharf was served as well by bus as was the company's old plant. Construction was likely to take 15 to 18 months.

2. *Relocate to an existing plant site in East Haven.* In East Haven, about 4 miles east of the old factory, Tighe Printing could move into a factory building that had been vacated 6 months before by a machine tool manufacturer. Some renovation was going to be necessary, estimated at

TABLE 12-13   Site-related costs for plant relocation (Tighe Printing Company)

| COST ITEM | SITES | | | |
	LONG WHARF	EAST HAVEN	HAMDEN	NEW HAMPSHIRE
Land price per acre	$70,000/acre (for 5 acres)	$18,000/acre (for 10 acres)	$12,500/acre (for 12 acres)	$6,000/acre (for 12 acres)
Estimated cost of building plus land	$2.8 million (110,000 square feet to be built)	$2.1 million (Asking price for 140,000 square feet to be built)	$2.3 million (110,000 square feet to be built)	$1.9 million (110,000 square feet to be built)
Estimated remodeling costs		$.3 million		
Labor costs		Essentially invariant across sites		
Transportation costs		Essentially invariant across sites		
State income tax	9.1%	9.1%	9.1%	0%
Local property tax (% of real value)	4.0%	3.5%	3.2%	3.3%
Insurance per year—fire, etc. (now $10,000/year)	$15,000	$15,000	$17,000	$21,500

$300,000. The company could move in within 5 months. The plant housed 140,000 square feet on 10 acres of land. Thus, although the plant was more than large enough for the company's needs now, there was also the possibility of expanding on site in the future.

The plant was not served by bus, but parking was ample. Ruth Tighe would have to think about alternative means (company bus, car pools) by which all of the plant's present workers could get to work. Ruth expected that some workers might not want to commute across the Quinnipiac River Bridge to East Haven. The area, however, was as densely populated as most neighborhoods in New Haven itself, and reports indicated that labor availability in the area was excellent.

3. *Relocate to available land in Hamden.* Another site was available about 11 miles north of the old factory in Hamden. A 12-acre site was available there in the Hamden Industrial Park, with all utilities provided. Construction would take between 12 and 15 months, it was estimated. Bus service was available to the plant, but it was very sporadic, coming only every 60 minutes. Parking would be no problem, and the proximity to two expressways in the area would shorten commuting for at least some workers. There was surely a number, say 15 to 25 percent, who would not relocate with the company.

4. *Relocate to a 12-acre site near Manchester, New Hampshire.* Because most of Tighe Printing's business involved magazine or business forms and catalog contracts with national concerns, and because the transportation involved was the U.S. mail, the company was not neces-

sarily tied to a location within the New Haven metropolitan area. Of course, the company would lose some of its local clients, but with the continuing growth of southern New Hampshire, there was apt to be enough new business to pick up whatever slack existed.

New Hampshire was alluring for two reasons: (1) New Hampshire levied lower taxes—no income tax and comparable property and other taxes, and (2) New Hampshire was closer to Ruth Tighe's ski lodge and summer home in Vermont.

Picking up stakes and moving so far from New Haven would not be without cost, however. Few of the plant's workers could be expected to move to New Hampshire; some of the plant's managers might also jump ship. Labor was available in southern New Hampshire, but the recent word that Ruth Tighe had heard was that labor was becoming increasingly tight, especially for skilled positions. There certainly were not going to be any labor cost savings.

Moving would not be easy for Ruth Tighe herself or for her family. Her husband, an accountant, could find a new job in New Hampshire fairly easily, but a move could be more disrupting to her two teenaged daughters.

Moving to New Hampshire would probably cost an additional $75,000 over the local move estimate of $200,000. Construction time for the plant was estimated at between 12 and 15 months.

Ruth Tighe wondered which of these four sites, if any, she ought to choose. Net sales in the present year were about $12.7 million and net income before taxes stood at about $675,000. Tighe Printing's present site was valued at roughly $1 million.

Discussion of Tighe Printing Company (Reprise)

The economics of this location decision are very similar to the economics of the Kemper Games situation. The time horizon is long, and varying cash flows will be associated with the different years of each of the alternatives. Table 12-14 portrays some cost flows and their illustrative net present values. A time horizon of 20 years is assumed as well as a discount rate of 15 percent. Note that the cost flows represent marginal changes. That is, only costs that differ

TABLE 12-14   Cost flows and net present values (in millions of dollars) (Tighe Printing Company)

| | INITIAL INVESTMENT | | | YEARLY EXPENSES THAT DIFFER ACROSS SITES | | | |
| | | | | INCOME | PROPERTY | | PRESENT VALUE OF COST |
SITE	LAND	BUILDING	OTHER	TAX	TAX	INSURANCE	STREAMS*
Long Wharf	$ 0.350	$ 2.8	$ 0	$0.061	$0.126	$0.015	$4.414
East Haven	0.180	2.1	0.3†	0.061	0.090	0.015	3.619
Hamden	0.150	2.3	0	0.061	0.078	0.017	3.426
New Hampshire	0.072	1.9	0.075‡	0	0.065	0.0215	2.588

*Present value calculations assume real cash flows; a real discount rate of 15 percent, and a time horizon of 20 years.

†Represents the cost of renovating the available East Haven building.

‡Represents the incremental cost of moving to New Hampshire. All relocations will cost at least $200,000 but the one to New Hampshire will cost $275,000.

across locations are included in the calculation. The present values of the cost streams, then, are useful only to compare with one another. A more thorough inclusion of costs and expected revenues could have been done; in that case, the present values could be compared not only with one another but with zero as well, giving some indication of the worth of the investment as well as the relative desirability of each site.

Table 12-14 reveals that New Hampshire would be the lowest-cost site by quite a bit, $3.426 M − $2.588 M, over $800,000 of present value. Ruth Tighe must ask herself whether the abandonment of the company's present workforce and market area would be less costly over the long pull than this site-specific cost advantage. For sure, developing a new workforce and searching out at least some new sales contracts will cost the company something. For example, a training cost of $4000 per employee for each of the company's 200 plus job positions would itself wipe out the cost advantage of the New Hampshire site.

Because Tighe Printing has been thriving in New Haven, there are no strong pressures to cut costs, and there is every incentive to keep the company's workforce and customer base intact. There is not much to choose between the East Haven and Hamden sites. Both are measurably cheaper than the Long Wharf site and would retain most of the company's employees. The Hamden site offers potentially more space for expansion, but the East Haven site could be available 7 to 10 months earlier. It probably would be better to move earlier rather than later, but either Hamden or East Haven would be well worth the investment. The Long Wharf site should probably be avoided. Not only is it expensive, but its possibilities for expansion are modest at best, and building on the site would take even longer than building in Hamden.

Tighe Printing Company's situation is representative of many companies, as will be discussed in the next section.

THE NATURE OF LOCATION DECISIONS[5]

Plant relocations are overwhelmingly made by small, growing plants (often independent of particular suppliers, markets, or labor sources) that are pressed for more production space. They move to larger, modern quarters and in the process alter their production technology, sometimes in fundamental ways. The vast majority of relocations are over short distances (less than 20 miles), which help to ensure labor force continuity and the retention of customer and supplier contacts. To a lesser degree, relocations also occur to consolidate two or more plants into a single new facility, and to escape from high site costs (wages, land values, taxes). As one might expect, however, it is the plant whose profits are hurting the most that sees relocation chiefly as a means to lower costs. These plants are also the ones who are most likely to move further than 20 miles in search of these lower costs.

Plant relocations are trauma-filled experiences for many managers and so are apt to be avoided if at all possible. Only about a third of the relocations seriously contemplated are actually carried through; actual relocations occur at a rate of only 3 percent per year for manufacturing establishments of all sizes, and less than 1 percent per year for plants of at least 100 employees.

More common, especially within larger companies, is the establishment of new branch plants. Each year, on average between 3 and 6 percent of the existing stock of plants is added as new branch plants; of branches contemplated, more than two-thirds are actually established, double the rate of relocations.

New branches start out small—only 40 percent, on the average, of the size of their sister plants—and are simply organized. They are less apt to be unionized and more likely to enjoy simple logistics. While likely to be located in modern facilities with the latest technology, they are also frequently dependent on the corporate services provided by the plant or plants from which they were spun off. The products they make are commonly mature ones, technically well established, with few engineering

changes necessary, although for a quarter to a third of new branches the products and/or technology is brand new.

Multiplant Manufacturing Strategies

Analysis of new branch plants and of the base plants from which they typically spring reveals that the new branch plant fits into a prescribed place in a multiplant company's scheme of things. Four general types of multiplant manufacturing strategies seem to prevail in the operating division of large companies. Behind each one are some compelling cost or managerial reasons.

The Product Plant Strategy—Perhaps the most popular strategy is the product plant strategy where distinct products or product lines are manufactured in separate plants, each plant serving the company's entire domestic market area.

The product plant strategy permits each plant to concentrate on a limited set of products, generally within a well-defined market niche. This concentration has the advantage of permitting the plant management to select the process technology, equipment, labor force, manufacturing policies, and organization that are consistent with the particular competitive priorities (cost, performance, product flexibility, speed of delivery) associated with the plant's products. In this way, the company can avoid much of the complexity and congestion that plague many oversize, multipurpose factories. In addition, if there are any economies of scale to be reaped, a product plant strategy can take advantage of them. Product plants can also take advantage of any raw materials or worker expertise that are specific to particular geographic areas.

A product plant strategy is likely to correspond to a decentralized manufacturing organization with a relatively small staff at the corporate level. Plant locations may be far-flung, but more often they are clustered within one or two broad regions of the country. Within such a strategy, a significant challenge to management lies in recognizing when a plant has become too large. What constitutes "too large" varies from industry to industry, technology to technology, and company to company; but the most frequently quoted figures lie between 500 and 1000 employees, with few companies stating figures in excess of 2000.

Companies in many industries can divide operations according to a product plant strategy because the products manufactured are many and varied. About 60 percent of the Fortune 500 companies follow a product plant strategy. For smaller companies, the product plant strategy is even more prevalent.

The Market Area Plant Strategy — Under this strategy, plants are designated to serve particular subnational market areas. The plants themselves manufacture all or most of the corporation's product line. The market area plant is perhaps the classic notion of the branch plant. When freight costs are important because of high product weight or volume relative to value, it makes sense to spread plants apart geographically. This is all the more true if products are consumed over wide areas, and if the market requires a quick response by manufacturing.

A market area plant strategy is likely to require more corporate coordination than the product plant strategy. The corporate staff is likely to be larger and to carry considerable clout. By the same token, plant managers are less likely to be able to act autonomously. A different management challenge confronts the market area plant—namely, the sequencing and regional authority of new plants. For instance, should an east coast company's second plant be in the midwest, west, or south, and how should the market be split between

plants? If the second plant is placed in the West, where and when should a third plant be sited?

The national breweries are classic examples of market area plants, as are many glass, can, paper converting, food, and building products companies. All involve products consumed in quantity everywhere and subject to significant transport costs as a fraction of product value. About 30 percent of the Fortune 500 companies follow a market area plant strategy.

The Process Plant Strategy — Rather than separate their products into individual plants, some companies, notably those with complex products, separate their production process into various plants. These plants are often viewed as feeders to one or more final assembly plants. A process plant strategy is less prevalent than the others, occurring in less than 10 percent of Fortune 500 companies, for example.

Like the product plant, the process plant exists to simplify an inherently complex and confusing managerial situation. For complex products like automobiles, large machine tools, and computer systems, a number of plants become involved in making components of the completed product. The manufacturer typically faces a rash of make-or-buy choices for many of these components, but to be able to produce one or another of these components competitively may require different raw materials, labor skills, control systems, or management skills and organization. This difficulty, coupled with the already discussed diseconomies introduced by large size (Table 12-12), argues for a division of the complete manufacturing process into stages, with separate plants for each stage. For any one stage, however, there may be economies of scale. Diverse manufacturing requirements explain why some plants may be located in the south or the Far East (for lower labor costs), or in resource- or expertise-rich areas (such as the Silicon Valley in California), or merely in a separate location, to provide surrounding plants with a special service that would be uneconomic for them to provide for themselves. This stage-by-stage division may lead to many feeder plants shipping to one or more assembly plants, or to one or a few feeder plants (for a critical component, say) shipping to many other manufacturing plants. In any event, the concept of plant separation to simplify operations persists.

The process plant strategy is even more demanding of high-level corporate coordination than the market area plant strategy. The process plant strategy is generally accompanied by a manufacturing organization that is highly centralized, technically well versed, and responsible for the control and coordination of materials and product between plants. For this reason, process plants are often located within an easy commuting distance of one another.

The General-Purpose Plant Strategy — Some companies do not establish specific plant charters. Rather, plants are prized for their flexibility in adapting to constantly changing product needs. Defense contractors, among others, are typical companies following a general purpose plant strategy.

The general-purpose plant strategy demands a considerable degree of centralized control. Coordination of plants is a real management challenge, as is the smooth staging of transitions in plant use and in employee assignments.

Site Selection

Once a multiplant manufacturing strategy has been decided on and plant size picked, site selection follows. The multiplant strategy frequently can say a lot about the choice of region. For example, clustering of plants in a particular region of the country is most apt to occur under the process or general-purpose plant strategies

and is least likely under the market area strategy. The choice of where within a region to locate, however, is sometimes very straightforward and sometimes baffling.

Many people, including some location consultants, try to simplify the decision making by introducing elaborate rating schemes that quantify everything imaginable about a particular location. Much of this is false rigor. There are, of course, a number of costs that can usefully be estimated — among them expected labor costs, construction, rental, or remodeling costs; taxes and other government payments; transportation cost savings or penalties for both inputs and finished goods; expenditures for needed services such as energy, pollution control, roads, sewerage, water, and parking; insurance costs; moving costs; and expected plant start-up inefficiencies or time delays due to start-up or to governmental approvals. Often one or just a few of these costs are so important that they can control the entire site selection process. The six controlling considerations that occur most often (in about two-thirds of the cases involving major manufacturers) are labor costs, avoiding labor unionization, proximity to markets, proximity to suppliers or resources, proximity to another company plant, and quality of life (variously defined) in the area. Although costs such as these may be important to evaluate (and linear programming can help assess the impact of transportation costs), they do not tell the complete story and sometimes they do not differ enough to ground a location choice strictly on their merits. A company should not expect any quantitative analysis to isolate a single area or site that stands alone as clearly optimal. Rather, a company should expect that a number of sites will show more or less the same cost structure.

Often the next phase of site selection is an exploration of the intangible and qualitative features of a location that could be expected to contribute to the company's competitive success. Although these factors may be difficult or even impossible to quantify, they are no less real; companies should resist the temptation of letting hard numbers drive out reasoned but qualitative analysis. The intangibles can be of many varieties: risks associated with any of the quantitatively evaluated costs or the sales potential of the site; the area's prevailing "business climate" (which means different things to different people but mainly long-term competitiveness); educational and training strengths of the area; attitudes of the workforce toward productivity, change, and unionization; the aesthetic and cultural attributes of the area (important aspects for attracting and holding managers); the cooperation of the local and state government for resolving public service or other public matters faced by industry; the commuting distances of workers and managers; and the impact of other, perhaps competitive, industries in the area. Frequently, a careful point-by-point comparison of these difficult-to-quantify factors against the real demands a particular product, area, or process will make on the manufacturing function can argue decisively for a particular site. A site need not rate high on all factors, but it should rate high on those that truly make a difference for the plant's competitiveness.

The company should be prepared as well for location analyses that, at the end, do not favor one site over another. If careful analysis reveals no preferable option, a company should not feel guilty that a seemingly inconsequential item tips the scale toward one site. After all, in such a case the company stands to gain or lose little by the location choice itself.

The location decision making for service firms is akin to that of manufacturing ones.[6] Indeed, services are less "local" in character than many believe; a majority of service firms characterize themselves as serving an area beyond the metropolitan (or rural) area they happen to be located in.

There are typically several factors that "control" the site selection process for the service firm, both the choice of broad, general area in which to locate and the choice of a particular site within that broad area. The critical considerations for the broad area choice appear to be infrastructure (roads, communications, services), proximity to customers, and the ability to attract qualified labor. For the choice of a particular site, the chief influences are parking, an attractive building, attractive rent (or cost), and, where applicable, specialized space needs.

Plant Closings

An inescapable and always unpleasant fact of corporate life is the plant closing. It is one of the most difficult decisions that operation managers face. Closings can be forced on companies in a number of ways, but the most common are poor production technologies, layouts, and materials handling; lack of sales volume; and competitive pressure on price. Many plant closings cannot be avoided, but some may be put off by careful attention to the roles the plant can play at various stages in its corporate life (start-up, maturity, decline). Planning for the plant's various roles well ahead of time can save much anguish.

Tools of the Trade: A Recap of Lessons Learned

When the various means by which capacity can be augmented in the short run turn out not to be sufficient, then the company must think hard about how "bricks and mortar" ought to be added. The management of additional capacity (or its unpleasant sibling concept, the downsizing of existing capacity) is typically triggered by a forecast of demand into the future. Naturally, companies are better off if their capacity choice is a "robust" one, whose economic desirability is not compromised much by swings in the forecast itself. Some capacity is relatively cheap to hold compared to the alternatives (e.g., stocking out of a product and thus forgoing its contribution margin) and thus one wants to have excess capacity; other capacity is expensive to hold and thus excess capacity is less desirable. Given this, sensitivity analysis is often indicated.

Timing is also important to capacity decisions because many such decisions take the form of evaluating the option of "act now" versus the option of "wait, do this a while and then act". For this reason, the time value of money, and the technique of net present value, is a useful tool for evaluating any capital appropriations request. One needs to know how long a particular alternative dominates another, and thus when a course of action ought to be reevaluated. This quantitative analysis notwithstanding, a number of non-economic factors come into play as well because capacity affects both internal operations and the firm's competitive situation.

On-site expansion is the usual base case considered for additional capacity. There are frequently some economies to increasing the scale of activity at a site. On the other hand, there are also some diseconomies. One has to be very careful to understand what constitutes any economies of scale touted for a particular location. On-site expansion, especially repeated on-site expansion, can wreak havoc with items like layout, materials handling, and production and inventory control. On-site expansion can also compromise the use of new technology and cause problems in coping with product proliferation, managerial constraints, and plant size issues. When on-site expansion falls down, then the company should consider relocation or the opening of new, branch plants. These alternatives often can solve the problems posed by on-site expansion.

If a company is big enough to have multiple sites, then the sites typically fit into a strategy. Perhaps that strategy is different products at different sites, or different market areas served by different sites, or different portions of the production process mastered by different plants. No matter what the strategy, however, site selection is generally dependent on the multiplant strategy of the company. It is typically clear what the major controlling considerations are for a plant's site selection, and how they relate to the multiplant strategy. Similar things can be said of service firm location as well.

World-Class Practice. Progressive companies are very clear about the multiplant strategies they have in place. What is more, what contributes to the strategy can change. In Europe, for example, the move to a Single Market within the European Community has dramatically altered the multiplant strategies of numerous companies, reducing the prevalence of the market area plant (for individual countries) and increasing the prevalence of product plants operating for all of Europe.

Progressive companies are also very clear about the life cycles of their plants, knowing what each plant is to be responsible for and standing firm on the kinds of changes that would, or would not, be permitted to occur.

Net Present Value (NPV)

The technique of net present value grows out of that pervasive concept of opportunity cost. For an investment of such and such a size, a corporation expects to generate a stream of earnings that extends into the future. But, by investing so many dollars in that particular investment project, the company forgoes the opportunity of earning a return on those same dollars by putting them in some money market instrument to earn interest or by using them to finance another income-generating project. In order to decide, then, whether to undertake an investment, a corporation needs to ascertain whether that investment will likely generate a stream of income whose value over time would be greater than bank interest earnings or the return from other investment projects not undertaken.

Put in other words, because income-generating investment opportunities exist, there is a time value to money. That is, having a dollar today is worth more than having a dollar tomorrow because today's dollar enjoys the opportunity of being put to work to finance an investment that will generate a stream of dollars into the future. How much more is today's dollar worth than tomorrow's dollar (apart, of course, from inflation, which is another matter

altogether)? It depends, as we might expect, on the return that can be earned from investments.

This rate of return, usually called the discount rate, typically varies from company to company and industry to industry, because some companies and industries are generally more profitable than others. In any case, the discount rate is at least as high as the interest rate available from the financial markets, because all companies have the opportunity of placing their cash in some money market instrument.

How can the discount rate be used to value the worth of a dollar tomorrow? Suppose that r is the discount rate. A dollar (the principal) invested today at the day-to-day rate of return r would be worth

$$\$1 + (r \cdot \$1) = \$1(1 + r)$$

tomorrow. Put another way, tomorrow's dollar is worth only $1/(1 + r)$ of today's dollar; it has a "present value" of

$$\$1\left(\frac{1}{(1 + r)}\right)$$

How much is a dollar worth 2 days from now? Even more than today's dollar is worth tomorrow because it has the opportunity of

445

earning interest on tomorrow's interest as well. More formally, a dollar invested today at the day-to-day rate of return r would, in 2 days, be worth

$1	+ $(r \cdot \$1)$	+ $r[\$1 + (r \cdot 1)]$
(principal)	+ (tomorrow's re-turn on today's dollar)	+ (the return 2 days from now on tomorrow's total)

$$= \$1 + 2r + r^2 = \$1(1 + r)^2$$

Put another way, a dollar received 2 days from now is worth only $\$1/(1 + r)^2$ of today's dollar. Its "present value" is

$$\$1\left(\frac{1}{(1 + r)^2}\right)$$

As we should expect, a dollar received 3 days from now is worth $1/(1 + r)^3$ of today's dollar, and so on for successive days.

How can we use the discount rate to value a stream of income, such as would be generated by an investment? Suppose that the investment returned a dollar tomorrow and a dollar the next day. How much should we value the investment? Essentially we have already solved for this value. It is merely the sum of the present value of a dollar received tomorrow and the present value of a dollar received 2 days from now. Hence the value of the stream of dollars received tomorrow and 2 days from now equals

$$\$1\left(\frac{1}{(1 + r)}\right) + \$1\left(\frac{1}{(1 + r)^2}\right)$$

It is easy to see then that the value today (the present value) of a dollar received in each of n days is simply

$$\$1\left(\frac{1}{(1 + r)}\right) + \left(\frac{1}{(1 + r)^2}\right)$$
$$+ \left(\frac{1}{(1 + r)^3}\right) + \cdots + \left(\frac{1}{(1 + r)^n}\right)$$

What would be the value of a dollar received beginning on Day 3 and extending through Day n? Obviously, it is

$$\$1\left(\frac{1}{(1 + r)^3}\right) + \cdots + \left(\frac{1}{(1 + r)^n}\right)$$

but the ability to add and subtract present values implies that this figure can be computed as the present value of a dollar received from Day 1 through Day n less the present value of a dollar received tomorrow and the next day. This ability to add or subtract present values comes in handy when computing them.

It is important to understand how the present value calculation is affected by changes in the discount rate and the time horizon chosen:

1. The higher the discount rate, the lower the present value of any stream of income. Income received in distant periods is particularly hard-hit by high values of r, since $(1 + r)$ is raised to successively higher powers.

2. The longer the time horizon over which income is to be received, the higher the present value of any income stream. As the horizon stretches on, however, less and less is added to the present value, other things being equal.

We have discussed net present value and the discount rate in terms of days. Of course, usually the discount rate is quoted on an annual basis, and the cash flows associated with the investment are grouped into yearly accounts.

In calculating net present values one can choose between two conventions:

• Nominal (currently prevailing) rates of discount and nominal cash flows. Under such a convention, the cash flows in more distant years are swollen by inflation; even in a situation of no growth in unit volumes, dollar volumes show growth at the rate of expected price inflation for the product. The rates of discount, in turn, can be regarded as the sum of some inflationless "real" rate of discount and a premium reflecting the expected rate of general price inflation.

• Real (inflation-adjusted) rates of discount and real (constant dollar) cash flows. Under such a convention, inflation is netted out of both cash flows and discount rates.

In most simple situations, either convention will yield the same qualitative results. In some cases, however, when particular cash flows can be expected to grow over time according to different rates of inflation, only the first convention (nominal discount rates and cash flows) is flexible enough to allow divergent inflation expectations. What one wants to avoid, of course, is the use of real, constant dollar cash flows and a nominal discount rate, because then future cash flows will be needlessly penalized.

The actual calculation of net present values is accomplished by evaluating the expression

$$\text{cash flow}_0 + \frac{\text{cash flow}_1}{(1 + r)} + \frac{\text{cash flow}_2}{(1 + r)^2} + \cdots$$
$$+ \frac{\text{cash flow}_n}{(1 + r)^n}$$

where the subscripts refer to the year (or period) of interest. This calculation can be done term by term (the only option when the cash flows are all dissimilar) or many terms at a time if a cash flow remains the same over a number of periods and can be factored out. Many calculators are programmed to calculate present values, and tables exist with (1) the present value of a dollar received in a certain period given such and such a discount rate and (2) the present value of a dollar received each period for so many periods at such and such a discount rate.

In investment decisions, if the calculated net present value is positive, the corporation can expect to increase its worth by undertaking the investment. If the net present value is negative, the investment should not be undertaken, at least on economic grounds. (There may be other factors to consider in making such a decision.)

The use of present value calculations is clearly superior to the simple, and still widely used, payback calculation. Payback is simply the breakeven computation of when a project's income exceeds the investment. Unfortunately, the payback calculation ignores (1) the time value of money (i.e., the fact that income earned in later years is not equivalent, dollar for dollar, to income or expenditures incurred beforehand) and (2) the pattern of expenses and income over time. Thus wherever the pattern of income and expenses becomes at all irregular, the payback calculation is apt to yield wacky results.

For more subtle reasons, net present value is also preferable to a sophisticated technique called internal rate of return (IRR). The IRR method solves for a rate of discount that yields a net present value of precisely zero. As might be expected, for most purposes the NPV and IRR methods yield identical results, but there are some offbeat cases where the IRR method can lead to erroneous conclusions and where NPV is considered superior. Finance texts describe the particular cases where the two methods diverge; all that the operations manager has to know is that the net present value technique will always yield a proper figure by which investment projects can be compared. That figure is the gain in the corporation's worth that is likely to result by undertaking the investment.

An economic analysis of capacity should be an incremental one, comparing the proposed change to a base case. The base case, in particular, should be what is expected to transpire without the proposed change taking place. What makes this statement important is that too often what is taken as the base case is the status quo. The status quo should always be suspect, however. Market share and profitability can easily be eroded by what competing firms do. So often competitors' new investments in technology or capacity toll the death knell for companies that have blithely assumed that the present's status quo would go on forever. Managers have to be careful that the base case is well thought out and realistic. It is frequently very easy to decline bold, new investments in favor of doing nothing, or very little. But after a while, the company loses its spunk: its products ease into obsolescence, its costs edge up ceaselessly, and its managers stagnate. Easy assumptions about the status quo need to be avoided.

QUESTIONS

1. Why is Kemper Games described as being in a joyous predicament? What are some of the tactics Kemper explores for getting out of its predicament? How might Kemper's experience be generalized for other companies' benefit?

2. Define and discuss the following concepts:

(a) Net present value.

(b) Discount rate.

(c) The effect of changes in the discount rate and the time horizon on the present value calculation.

(d) The difference between real and nominal rates. How do these concepts fit into Kemper's procedure for deciding whether to build or to lease?

3. What role does long-term forecasting play in the capacity plan and its analysis?

4. Summarize and discuss the main lessons that can be learned from Kemper Games' dilemma.

5. For service industries, what are the main factors involved in decisions about capacity?

6. The term *economies of scale* covers a multitude of concepts. What are some of the more important concepts? How do they relate, if at all, to issues of capacity and expansion?

7. Review three of the relative advantages of branches and relocations vis-a-vis on-site expansion.

8. "Maybe sometimes you can be too successful," comments Ruth Tighe about the progress of her printing company. Are there times when this can be true for a company? In considering the four distinct options, does Tighe look at all aspects of the decision? Support your answer.

9. Summarize and discuss the importance of the various multiplant strategies mentioned in this chapter.

PROBLEMS

1. Hal operates an automotive repair shop. He has the opportunity to purchase another auto repair shop across town. This shop has been generating about $60,000 in annual profit after taxes and he feels that with the new demand and better management, he can increase this to $68,000 or $70,000. The present owner is asking $210,000 for the business. To help him decide whether this would be a profitable purchase for him, Hal has made the following estimates:

Life of the business	20 years
Internal cost of capital	10% real rate
Extra annual expenses Hal foresees	$20,000 for another manager's salary

Analyze Hal's problem and make a recommendation.

2. The Calway Cab Company currently operates a fleet of 27 cabs in a medium-sized midwestern town. The firm is considering expanding the fleet. They feel that each additional cab will generate less clear profit annually than the last additional cab. They have estimated that the twenty-eighth cab will generate $4000 per year and that each additional cab will generate 5 percent less than the previous one. Each cab has a purchase value of $11,000 and a trade-in value of $5000 at the end of 2 years. The company has an internal cost of capital of 15 percent. Ignoring taxes, how many new cabs would you recommend that they purchase?

3. Emil Gaines owns and operates Consolidated Salvage yards. Emil has a scrap compressor that does not have the capacity to maintain the level of production required. In deciding whether to buy a new compressor, Emil has compiled the following data:

Depreciated value of old compressor	$17,000
Market value of old compressor	$ 1,000
Price of new compressor	$65,000
Installation costs	$ 4,000
Economic life	10 years
Salvage value	$ 1,000
Required rate of return	12%
Estimated annual cost savings	$13,000

Ignoring taxes, answer the following questions:

(a) Should Emil purchase the new compressor?

(b) What is his actual rate of return?

4. The Cal Bodkin Electronics firm manufactures an assortment of consumer electronics, including CB radios. The firm has a standardization program and they have standardized 95 percent of their control knobs. Each of these knobs costs 39 cents when bought at an annual volume of 800,000. The firm's managers have completed a make-or-buy analysis and feel they can make the knobs at a cost savings of 13 cents each. They are currently analyzing two different potential processes:

	PROCESS A	PROCESS B
Installed cost	$35,000	$45,000
Annual operating cost	$ 9,000	$ 6,000
Economic life	4 years	5 years

Ignoring taxes, what is your recommendation, and why?

5. Discuss the appropriate management level that should gather information and make the decision regarding the three "brick and mortar" choices for additional capacity presented in the chapter.

6. Discuss the length of the review and planning procedures for various industries. Do those who use longer planning periods have better forecasting methods?

7. What kinds of alternatives should a planning group consider for meeting shortfalls?

8. Lake Country Diversified's Chicago plant, which made drive-train components for off-highway trucks, had performed at a level below expectations for several years. Many workers expected the plant to close. In investigating the profitability of the plant, which was evaluated as a profit center, management had divided up the plant's production into three groups. Some data for each group is at the top of the next column (in thousands of dollars). Of particular concern to management was the high pension expense that the plant carried. Being an old plant, there were many retirees on the books, so the plant carried what it thought was an unfair burden of $648,000 a year in pension expenses (included in the fixed-overhead allocation). Management had several things to contemplate:

(a) Should the group C products be discarded?

(b) How could the pension expense burden be lightened?

(c) Should the plant be closed?

	GROUP A	GROUP B	GROUP C
Sales	$6944	$4629	$ 5787
Direct labor	296	258	815
Materials	3195	2170	2910
Variable overhead	630	551	1735
Fixed overhead	2125	1415	1774
Profit (loss)	$ 694	$ 231	$(1447)

SITUATION FOR STUDY 12-1

CENTRAL LOUISIANA OIL—HANKS C WELL

The Hanks C well had been shut in for 45 days. This gas well in central Louisiana was leased to Central Louisiana Oil. This well was old and the gas pressure in it had declined significantly over the years. At present the pressure was too low to lift the water that was also in the well and so the gas could not be recovered by conventional means. Under the terms of the lease, Charlie Clotfelter, the production manager for the field where the Hanks C well was, had only 15 more days to get the well going again. If time ran out before production could be resumed, Central Louisiana Oil's rights to the lease would terminate and the owners of the land and its mineral rights could negotiate with another production company.

Before the well had been shut in, Clotfelter had tried all of the conventional means of increasing gas pressure to no avail. He was now faced with three major options involving larger capital expenditures:

1. Smaller-diameter tubing could be dropped down the well's shaft, causing the gas to travel faster up the tubing and thus probably to lift the slug of water. Such a scheme would cost $43,000 and could be expected to work, but perhaps not for long. Eventually, the pressure in the well would decrease even more and the smaller-diameter tubing would not be able to lift the water.

2. Natural gas could be forced into the well between the existing tubing and the outer casing of the well shaft. This gas would join with the natural gas from the well to create enough pressure to lift the slug of water. This operation promised to keep production going longer than that with the smaller tubing and it cost only $33,000. To ensure that operations would start up within the next 15 days, a right-of-way to a compressor unit on a neighboring lease would be needed immediately, but the two leases were not clear about whether this right-of-way could be obtained easily. Clotfelter was hopeful that it could be, but he was not certain.

3. A beam pumping unit could be installed on the well. The pump would lift the water out of the well and thus free the well to produce

TABLE 12-15   Expected cash flows (Central Louisiana Oil—Hanks C Well)

OPTION	INVESTMENT NOW		CASH FLOW DIFFERENCES FROM BASE CASE		
			YEAR 1	YEAR 2	YEAR 3
Small-diameter tubing	$43,000	(costs)	Base case	—	—
		(sales)	Base case	—	—
Gas lift	33,000	(costs)	$ 44,000	$44,000	$46,000
		(sales)	150,000	80,000	60,000
Beam pumping unit	74,000	(costs)	44,000	45,000	47,000
		(sales)	150,000	80,000	60,000*
		(salvage)	—	—	49,000

*This option will generate higher sales than projected here.

its natural gas. This operation would cost $74,000 in capital expenses, some of which could be salvaged after, say, three more years of well operation. This option also had the potential for generating more gas than either of the other options. Whether, in fact, it could produce more gas depended on what kinds of geologic structures were below ground.

The expected cash flows over the next three years are given in Table 12-15. Charlie Clotfelter wondered what actions he should take to resume the operation of the Hanks C well.

SITUATION FOR STUDY 12-2

MIDWEST BRANDS, FAST FOOD

Joel Huber had recently joined the corporate staff of the Fast Food Division of Midwest Brands. The company operated about 300 restaurants in the Midwest. Rather than standardize these restaurants, management had allowed about 30 different sizes and layouts to develop over the years. Joel's immediate concern was with a restaurant where the local management wanted to add 40 seats without expanding the kitchen. The dining area currently seated 112, and the restaurant was doing a booming business during lunch and dinner peak hours. Management at this division had been rather lax. Few systematic data were kept on the operation and use of the restaurants, and there were no guidelines on the steps to take in expanding a restaurant's seating capacity. It was Joel's job to study this request for expansion, using some systematic analysis. He wondered, first, how the decision ought to be structured, and second, what data he ought to collect to analyze the problem properly.

REFERENCE NOTES

1.   For more commentary on the management of productive capacity, see:
     Baden-Fuller, C. W. F., ed., *Managing Excess Capacity*, Cambridge, MA: Blackwell, 1990.
     Barnett, F. William, "Elastic Capacity and Skin Tight Costs: Low Budget Production Improvements," *Sloan Management Review* 31, no. 3 (Spring 1990): 65–72.
     Bower, Joseph L., *Managing the Resource Allocation Process*, rev. ed. Boston: Harvard Business School Press, 1986.
     Koster, M. B. M. de, *Capacity Oriented Analysis and Design of Production Systems*, New York: Springer-Verlag, 1988.
     Lieberman, M. B., "Strategies for Capacity Expansion," *Sloan Management Review* (Summer 1987): 19–27.
     Porter, Michael E., *Competitive Strategy: Techniques for Analyzing Industries and Competitors*, New York: Free Press, 1980 (Chapter 15).
     Swamidass, Paul M., "A Comparison of the Plant Location Strategies of Foreign and Domestic Manufacturers in the U.S.," *Journal of International Business* 21, no. 2 (Second Quarter 1990): 301–17.

2.   For an extended description of the decision-making mechanism for capacity expansion or contraction, see Joseph L. Bower, *Managing the Resource Allocation Process*, Homewood, IL: Richard D. Irwin, 1972.

3.   See, for example, Robert H. Hayes and William J. Abernathy, "Managing Our Way to Economic Decline," *Harvard Business Review* 58, no. 4 (July–August 1980): 67–77, and Robert L. Banks and Steven C. Wheelwright, "Operations vs. Strategy: Trading Tomorrow for Today," *Harvard Business Review* 57, no. 3 (May–June 1979): 112–20.

4.   This section on economics of scale is adapted from an article of mine entitled "Before You Build a Big Factory," *Harvard Business Review* 54, no. 4 (July–August 1976): 100–104.

5.   Much of this section is derived from Roger W. Schmenner, *Making Business Location Decisions*, Englewood-Cliffs, NJ: Prentice-Hall, 1982. Another useful reference on multiplant manufacturing is F. M. Scherer et al., *The Economics of Multiplant Operation*, Cambridge, Mass.: Harvard University Press, 1975. See also Roger W. Schmenner, "Every Factory Has a Life Cycle," *Harvard Business Review* 61, no. 2 (March–April 1983): 121–29.

6.   See Roger W. Schmenner, "Service Firm Location Decisions: Some Midwestern Evidence," Indiana University School of Business Working Paper, December 1991.

CHAPTER 13

DEALING WITH TECHNOLOGY AND DESIGN

The management of technological change may well be the least-developed aspect of operations management. Companies of wildly different philosophies and organizations have been successful innovators. No one model of technology management has so far been recognized as head and shoulders above the others. And yet everyone realizes that technological advance is critical to a country's continued economic growth. It is estimated that technological innovation has been responsible for almost half of U.S. economic growth. In addition, high technology companies grow quickly in both sales and employment and are often looked to as engines of a country's growth.

This chapter introduces some of the ideas that people have put forward as important to successful innovation and some of the issues that continue to be debated. What is technological change? There is debate even on this. Are new products evidence of technological change? How about minor modifications to existing ones? Do new machines constitute technological change? How about bigger (or smaller) revisions of existing ones? We can readily recognize new, far-reaching process technological changes (such as Henry Ford's moving assembly line and computer process controls in papermaking), but what about small alterations in an existing process that are more on the scale of line rebalancing? Merely trying to answer these questions can give us some feel for the debate.

The ambiguity associated with technological change is sometimes resolved by recognizing that both product and process changes can be either radical or incremental. The debate itself over which product innovation is significant or which process modification is radical is evidence of the belief many share that radical or significant technological change is somehow different and that it demands different modes of thinking and management to make it work. Hence, trying to identify what is or is not radical or significant is a worthwhile endeavor. Even though we may think of technological change as a continuum from "irrelevant" to "momentous," successful innovators, so some people argue, should treat one end of the continuum differently from the other. The significance of the distinction between radical and incremental change laps over into the chief managerial issues that companies confront about technology.

Several questions loom about technology and design:

- What are the sources of technological change?
- How can new products and processes be better introduced?
- How should new technology (product or process) be evaluated and implemented?
- What are some of the latest developments in manufacturing technology?

These questions form the themes of this chapter.

SOURCES OF TECHNOLOGICAL CHANGE

When we think of technological innovation, we often distinguish between product innovation (new product introductions) and process innovation. While conceptually distinct, the two are often related. Abernathy and Utterback have suggested one such relationship between product and process innovation.[1] Figure 13-1 captures the major points of their idea.

In the early portion of a product's life, product design is critical. A product's early users are almost always more interested in product performance than in price. Considerable product redesign is undertaken to make the product even more useful and desirable for its users. Abernathy has referred to this early phase of product technology as the search for the "dominant design."[2] Dominant designs are those products that "make a market," such as Ford's Model T car, the DC-3 airplane, the Xerox 914 copier, Kodak's "Brownie" camera, VHS-type video cassette recorders, and the IBM PC. In this early going, the production process is most apt to be a job shop or a near-job shop.

As acceptance of the dominant design proceeds, however, cost reduction becomes increasingly important. Process innovation—geared primarily to lowering costs, increasing yield, and gaining production speed—commands management attention. Changes become less and less radical as the product, the process, and the organization become more and more standardized. The production process edges closer to the continuous flow end of the process spectrum. At the same time, both product and process become increasingly vulnerable to a radically different offering of similar function by some producer outside the traditionally defined industry (e.g., Polaroid cameras, an electronics company like Texas Instruments introducing digital watches). In this view, then, product and process innovation are both aspects of broad technological advance and of the shift in market characteristics, over time, that usually accompanies such advance.

New products spring from someone's research and development (R&D), but the question is whose and how can such R&D be stimulated. Some companies are renowned for generating new products from within (du Pont, Intel, 3M, Sony, Corning Glass), and the management emphasis in such companies focuses on how the research labs can keep up their rich traditions. Other companies are forced (or perhaps choose) to generate their new product offerings from different sources—sometimes copying others' new products, sometimes acquiring the rights to the product, sometimes acquiring the innovating company itself.

Assessing the Technological Fortunes

Spotting the product designs that will be commercially successful is an uncertain business at best. It has been suggested that a new "technological winner" must score high on a mix of separate criteria.[3]

- *Inventive merit.* How important are the scientific advances in the product?
- *Engineering merit.* Does the engineering of the product fully exploit the scientific advances made? Must many changes be made?

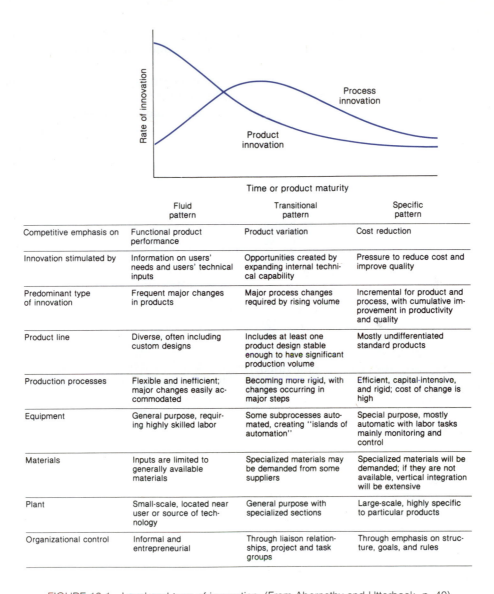

	Fluid pattern	Transitional pattern	Specific pattern
Competitive emphasis on	Functional product performance	Product variation	Cost reduction
Innovation stimulated by	Information on users' needs and users' technical inputs	Opportunities created by expanding internal technical capability	Pressure to reduce cost and improve quality
Predominant type of innovation	Frequent major changes in products	Major process changes required by rising volume	Incremental for product and process, with cumulative improvement in productivity and quality
Product line	Diverse, often including custom designs	Includes at least one product design stable enough to have significant production volume	Mostly undifferentiated standard products
Production processes	Flexible and inefficient; major changes easily accommodated	Becoming more rigid, with changes occurring in major steps	Efficient, capital-intensive, and rigid; cost of change is high
Equipment	General purpose, requiring highly skilled labor	Some subprocesses automated, creating "islands of automation"	Special purpose, mostly automatic with labor tasks mainly monitoring and control
Materials	Inputs are limited to generally available materials	Specialized materials may be demanded from some suppliers	Specialized materials will be demanded; if they are not available, vertical integration will be extensive
Plant	Small-scale, located near user or source of technology	General purpose with specialized sections	Large-scale, highly specific to particular products
Organizational control	Informal and entrepreneurial	Through liaison relationships, project and task groups	Through emphasis on structure, goals, and rules

FIGURE 13-1   Level and type of innovation. (From Abernathy and Utterback, p. 40)

- *Operational merit.* What will the product mean to company operations and organization? Will major revamping be required?
- *Market merit.* Is the product innovation what the market wants and will pay for?

To the extent that a product's inventive merit is strong, its marketing features attractive, and any required engineering and operational changes either slight or fully anticipated, a new product is likely to be a commercial success. A failure in any of these dimensions can spoil an otherwise promising product.

Pinpointing a promising technological advance is unavoidably subjective, but some research suggests that many of the uncertainties involved can be usefully quantified.[4] A number of these approaches to forecasting the success of a new product use the Delphi method, combining an expert panel with probability measures. Each panel expert is asked to indicate his or her own view of an R&D project's success. These opinions are collected, tabulated, and usually reviewed again by the expert panel, which can then revise its initial probability selections. Such assessments can be made at regular intervals over the years required to research and develop a new product.

Such assessments of the probability for success of various R&D projects can be important inputs to the management of these projects, identifying those that need attention and governing in part how resources can be controlled and apportioned among projects.

NEW PRODUCT INTRODUCTION

For most operations, there are new products to be made or new services to be delivered nearly all the time. Making the introduction of a new product or service go smoothly, however, takes some planning and effort. The merit of the new product or service should be assessed so that priorities and resources can be properly assigned, and engineering or design should make a smooth and timely transfer of the new product or service to the operations people. This section addresses such requirements and how companies have coped with them.[5]

The Product Development Cycle

No matter what the source of new product inspiration, new products typically follow a standard progression from design engineering to volume production in the factory:

1. Basic research, where the initial concept for the product is explored and decisions are rendered about its feasibility.

2. Development work, where different designs are investigated and a particular one chosen. Historically, these first two phases of the product development cycle have been the exclusive province of engineering, but this has been changing in many companies in recent years. (More on this later.)

3. Prototyping, where the initial versions of the product to be manufactured are created and tested for performance, and any necessary engineering changes made to the product. This has historically been manufacturing's first crack at the product and how it might best be made.

4. Pilot runs, where the manufacturing process for the product is examined and debugged, and where tooling, bills of material and purchasing needs, equipment needs, layouts, and methods are developed, as required. Engineering changes in the product often result from this stage as well. Often, the product design is "frozen" after the pilot runs.

5. Release to the factory and "ramp-up" for high-volume production.

There has been increasing concern in the United States that American firms are not quick enough to market with new products, particularly when compared to the Japanese. Such concern straddles numerous industries: autos, semiconductors, copiers, and pharmaceuticals, among others. The slowness of new product introduction is in part attributable to the attitudes held by engineering and manufacturing about each other. Too frequently, the two functions have been at odds with one another. Ensuring cooperation between engineering and manufacturing is frequently no mean trick; distrust and animosity between the two functions is too often the rule. Resentment of engineering by manufacturing may be caused by real, or perceived, incidents such as the following:

- Engineering is later than scheduled with the design, but manufacturing is still expected to get the first units to the market according to the initial schedule. Engineering is perceived by manufacturing as not being punished for taking too long.

- Engineering goes over budget at will, but manufacturing is held to its budget with religious fervor.

- Manufacturing is not consulted about the new product while it is in the design phase and yet has to accept a design that may please the engineer but is needlessly difficult to fabricate, assemble, and/or test.

- Engineering changes to designs, parts, and methods are needlessly frequent; it is as though engineering staff cannot make up their minds. Yet manufacturing is expected to react instantly.

- Engineers seem to have a penchant for specifying items that are hard to purchase and/or require a long lead time, although a little ingenuity could have made use of off-the-shelf items that are easily obtained.

For their part, engineers may gripe (and rightly) about the manufacturing group's inertia, singleminded attention to cost factors, needless devotion to existing equipment and methods, and so on. For new product introduction to be smooth, this chasm of mistrust must be bridged.

Fortunately, there have been a variety of techniques that have brought manufacturing and engineering closer to one another, techniques that have speeded up the flow of innovation to the marketplace. Many of these tech-niques smack of the JIT philosophy, and rightly so. The same kind of attention to time and quality that is so valuable to the factory floor is valuable to the new product development effort.[6]

1. *Overlapping development phases (concurrent engineering).* Historically, product development has been a sequential process. The development proceeded in linear fashion, much as a baton is passed from one runner to the next. R&D or engineering would design the product and then "throw it over the wall" to manufacturing for the subsequent steps of process and tool design, pilot runs, and ramp-up. Increasingly, however, development is proceeding in parallel, with phases of development overlapping one another. Not all product development is totally dependent on what went on before, so that overlapping phases can substantially cut down on the time spent in development.[a]

This is usually termed *concurrent engineering* and involves the creation of a team (typically between 6 and 25 in number) drawn from a variety of disciplines: design engineering, manufacturing or process engineering, manufacturing, and often a number of others such as marketing, cost accounting, quality control, purchasing, the model shop, and tool and die fabrication. More and more, suppliers are sitting in with these development teams as well, providing advice, design and manufacturing capability, and shortening the time needed to bring out the product. The team leader is often a design engineer but everyone on the team has knowledge of and input to the design process. This is more time-consuming early in the development process than the conventional mechanism but the result is often a

---

[a]The "fast track" construction described in Tour J can be considered another example of concurrent engineering.

better, more stable design for lower cost, and is accomplished in a shorter period of time. Process design and tool design can be completed more speedily, and in conventional development these steps can bog down a new product introduction dramatically. By these means, the "revisiting of mistakes" that too often seize the conventional development cycle is avoided.

2. *More value-added in development phases.* To the extent that needless duplication and iteration can be removed from a product's design stage, then the time devoted to that stage can be reduced. Several techniques are spurring on this trend, including the disaggregation of design and test to subassemblies of a product, the expanding use of simulation, and computer-assisted engineering (CAE). CAE enables the engineer to do many things to help manufacturing early in the design cycle. For example, CAE can "test" parts without the need for a prototype, thus eliminating a multitude of engineering change orders (ECOs), and it can specify the tooling for a product even as it is being designed. This means less need for prototyping, and both quicker and better designs.

3. *The quality of project leadership.* New product development projects have been organized in any number of different ways. Although there are no hard and fast rules, development projects are typically better managed (more quickly accomplished; containing more robust designs, easier to manufacture and costing less) when there is a strong project coordinator or leader who is part champion of the project in the upper ranks of management, part communicator of ideas and problems among the team members, and part critic of the design and advocate of the end user. The project coordinator must have the respect of all and needs considerable authority to push the project team's ideas and decisions through the company's bureaucracy.

4. *Incremental development.* The tendency that many people have when dealing with technology is to think first of the breakthrough and to concentrate on it. Leapfrogging the competition is a powerful vision, although only rarely achieved.

Less satisfying, perhaps, but more often successful, is the small-step approach to product development and innovation. This incrementalism, part of what is termed *kaizen* in Japanese, stresses the development of product variants from the core, or parent product. By constant, small-scale advances, a cadre of engineers is always improving and thinking about the product, and often can progress as far and as fast as development teams that always seek to leapfrog the competition. Incremental development does not preclude major advances in performance or cost reduction; it merely gets there in a different way. After all, good baseball teams can score lots of runs via singles, walks, and stolen bases. The home run is not always needed.

5. *More meaningful, periodic prototyping.* One of the ways that the time for this sequence of steps can be contracted is by making the prototyping step more meaningful for the manufacturing function. This has been accomplished in some companies by purposefully building the prototype in the same way that the company contemplates manufacturing it, and also by developing prototypes for use and test by manufacturing as well as by engineering. The prototype thus takes on the role of problem-solving tool for both manufacturing and engineering, and is not seen exclusively as a model for further production to copy slavishly.

In addition, generating prototypes on a regular, periodic basis has real advantages.

Such prototyping keeps interest up in the development, hones skills of both design and manufacturing, more visibly portrays the problems that have to be solved to improve the design or the manufacturability, and serves to rally the project team around a common "vision" for the product.

6. *Manufacturing prowess itself.* Clark and Fujimoto, in their study of the worldwide automotive industry, have shown that a significant portion of the Japanese time-to-market advantage over both the U.S. and Europe derives from the ability of Japanese companies, particularly tooling suppliers, to develop tooling, dies, or other needed production aids (fixtures to hold parts, jigs to guide tools) quickly, so that the project does not languish.[7] They talk of "manufacturing for design," where the benefits of JIT manufacturing help the speed of the design effort. This occurs because prototypes can be built more quickly and with more fidelity to the design, and tools and dies can be fabricated more quickly and for less cost (e.g., Japanese auto prototypes in 6 months vs. 11–12 months in the U.S. and Europe, Japanese die manufacturing in 6 months vs. 14–16 months elsewhere). Such firms deal with tool and die makers is much the same spirit as JIT firms deal with their single source suppliers, as part of the family, with frequent, clear communications of needs and decisions, and an appreciation for what the supplier can contribute to the design itself.

Manufacturing prowess also helps with pilot runs. In Japanese auto companies, for example, pilot runs are made on existing lines by spacing out the new models among the mixed models run on those production lines. The pilot runs are thus less disrupting to existing production and can be accomplished in several short spurts that can be very helpful to the development effort.

Manufacturing prowess also helps with ramp up. The prevailing Japanese philosophy is to ramp up with many hands helping the effort so that problems can be spotted easily and dealt with immediately. This effort involves both workforce and engineering staff and contrasts with the more Western approach of gradually expanding the numbers of workers involved with the ramp up, where the notion is that gradualism enables the old hands at it to teach the new hands what to do with the new product. By contrast, the Japanese notion is to get their broadly trained workforce on the problem right away so that learning by all can be done quickly and problems with the new product identified readily and ironed out on the spot.

With such thinking and techniques, the time to market with a new product can be slashed significantly, and even with fewer engineers on the job. The list of policies or attributes in Table 13-1, derived from research of new product introductions, focuses on some additional policies or attributes that are associated with quicker new product introductions.[8]

Improving Product Design

It has been advocated by a number of industry observers that between half and three-quarters of a product's cost is locked in during the design phase. Thus, to expect to reduce product costs much without recourse to product redesign is often asking too much of the factory. Naturally, this puts a lot of pressure on the design function to design a cost-effective product from the start. This is all the more important given the fact that changes to a product becomes more and more costly, and the "fixes" more and more constrained, as time progresses and as the product moves on

TABLE 13-1   Promoting Quicker Product Development Cycles

Motivation

* There is nothing like the urgency brought on by competitive challenges to existing products to speed product development.

Workings of Teams

* Full-time rather than part-time team participation appears to facilitate speedier product development. Specialized skills required in different phases of the projects may, however, make full-time participation difficult to implement.
* One benefits from a clear-cut coordinator for the team, preferably one that works full-time on the project.
* Concurrent engineering within a team does work better than sequential hand-offs among non-team members.
* It is very useful to have costing on the team, especially as it can feed back data to marketing.
* Marketing has to be involved early and fully because marketing volume and price numbers are key to certain design and manufacturing trade-offs and if they are not forthcoming, speed and team morale suffer.
* Regular formal meetings (say, weekly) are good, but the content of the meetings should neither be too general nor too particular.
* Locating close to one another is extremely beneficial to the speed of the project. A "war room" that is used for meetings and which houses documents, messages, and "bulletin board stuff" is also useful to team members.

Outside Influences

* Using single sources from the start, and having the important vendors of the group sitting in with the team on its meetings can quicken the project's pace and effectiveness.
* It is attractive to have physical, prototype models being developed in continual, incremental fashion to keep interest up in the project, the design and technical skills of the staff honed, and to keep the project moving ahead.
* It is helpful to have the model shop under the team's control, or, at least, for the team to have the ability to determine its priorities.
* Buying manufacturing's cooperation with the promise of more in-house work is often effective.
* Letting the work force know early what's coming and getting their input can be helpful to the design and its manufacturability.

Project Control

* Having a control document of some kind speeds up the project. Critical path charts are good, but other network charts, Gantt charts, etc. can also be employed. These work best when they are fairly detailed and publish dates due for each piece of the development work.
* The team is best left alone with few formal reviews but with the knowledge that top management is interested in them. Top management sends a useful signal to the team when it shows speed and decisiveness in its own spending and choice decisions related to the new product.
* The use of CAD for the design work, both in-house and with vendors, can speed up the project.
* Reducing time delays in the early stages of a project offers opportunities for further reduction in development time.

from the conceptual phase to the design engineering phase to the prototyping phase and then on to product start-up. The better the work done early on, the more cost-effective the product.

What makes a product "cost-effective" involves a variety of things: design for manufacturability, design for serviceability, and design for assembly, among others. Making this happen, of course, requires teamwork among a number of company functions and adherence to the notion of concurrent engineering, as introduced above.

There are a variety of "rules" that can help almost any new design's costs. Consider the following, some of the tenets of design for manufacturing and design for assembly:

1. *Part Reduction.* If one reduces the number of parts, and uses more commonly available ones, costs can decline markedly. Not only do parts require assembly, but they also require such tasks as purchasing; receipt; handling; inventory control; entry into bills of material, engineering databases, and other documents; and inspection. Reduce parts and you reduce such "extra" costs. Part reduction can be accomplished by designing one new part to do the work of more than one old part, such as parts that are structural as well as mechanical or electrical in function. Or, it can be achieved by eliminating moving parts or by using the same part in different locations in the product. One can go too far with this notion, however. Sometimes the creation of single parts to replace a series of other, simple parts, is more costly, if not to design, tool, and build, then to replace or service.

2. *Less Precise Specifications.* What often causes fits for the factory, and thus raises costs, is trying to hold to tight tolerances in the manufacture of a product. Sometimes, the product's function and performance does not require that particular tolerances be kept quite so tight. Engineers are frequently accused of specifying tolerances by whim, without understanding which ones are easy to hold and which are tough. One of the advantages of concurrent manufacturing is that it opens up dialogues between design and manufacturing people that increase understanding and reduces situations where tight tolerances add only cost, and not performance, to the product. A corollary is to use processes that are robust in themselves, well understood, and are not subject to lots of variation.

Engaging suppliers (typically sole source suppliers) in the design process can often be beneficial. Designs given to capable suppliers can be less precise in their specifications than is conventional. The specs could, for example, deal only on outside space dimensions or weight and with certain functional requirements, but not include the blueprint exactitude of much of the conventional approach to supplier efforts for new products. The rewards to the customer company can be substantially lower costs, speedier delivery of the new part(s), and better product performance from a clever supplier whose ideas had not been sought before.

3. *Use of Modular Design.* To the extent possible, the building blocks of a product should be bunched together. This keeps more of the full product line looking the same, and being manufactured in the same way, until late in the process. This "sameness" contributes to better quality and lower costs. It also helps when the product needs to be redesigned or updated or serviced.

4. *Design for Easy Assembly.* Assembly is easier if it can be built from the bottom up. A useful admonition is to make the design simple enough for a robot to position everything. It is also helpful to design the product's assembly so that there is only one way for the pieces to fit together, namely, the quality way, and with little of no adjustment required. If an assembler has a choice as to how two pieces fit together, the designer has invited problems into the assembly process. Thus, non-symmetric, no-adjustments-needed designs for pieces that fit together are advantageous.

5. *Reduce the Processes and Process Steps Used.* Employing the simplest of processes that can do the job, often brings advantages.

Processes that have many steps to them and that demand significant skills from the workforce often are the processes that generate the most quality problems, and the most cost. To the degree that the design requires simple, straightforward processing, the lower the costs can be. It is often useful as well to design the product with the machines in mind to make it. To the extent that general purpose machines are used, with capabilities that are much greater than the demands that the product will place on them, quality is likely to be enhanced.

Many companies have instituted programs (e.g., those of Boothroyd Dewhurst Inc.) that have helped their engineers learn these skills.

Value Engineering

The previous set of "rules" and attitudes towards new product designs are often associated with a task known as *value engineering* (sometimes also called value analysis). Value engineering often, although not exclusively, refers to product redesign efforts as a panel of managers and engineers looks at a product from the standpoint of the end user and asks what changes in design or materials are possible, what those changes would be worth to the user, and how such changes can be made economically. Value engineering is a useful complement to much of methods engineering, which focuses more on the process and the worker's contribution to it. Productivity gains can occur in many ways, and value engineering helps guard against "overengineering" a product.

Value engineering asks these sorts of questions: Does this feature of the product provide value for the consumer? What other features or designs could be substituted at lower cost that could do the same thing? Can components of the product be standardized? Or eliminated? Can materials be changed? Which tolerances are truly the ones that must be held to tightly?

How reliable is the product or its components? Is it too strong or in some other way overengineered? Does it have excess capacity or excess weight?

Another useful value engineering approach is to study all of the parts or products of the company that might be considered similar, with the object of identifying useful simplifications, design changes, or materials substitutions among a family of products. Still another means of value engineering is to query suppliers about the changes or improvements they see can be made.

Value engineering is very much a "habit of mind," but it is one that can pay tremendous dividends.

PROCESS INNOVATION

The sources of process innovation are similarly perplexing. Some innovations stem from a company's own work (occurring in conjunction with new product innovation), some stem from advances made by equipment or materials suppliers, and some result from copying or acquiring the technology developed by others.

Although the sources of process innovation are often baffling, the character of process innovation is less so. Moreover, an understanding of the character of process innovation carries with it some important lessons about how productivity can be enhanced.

A useful touchstone for a discussion of the nature of process innovation is the empirical phenomenon typically referred to as the "learning curve."

The Learning Curve[9]

The notion underlying the learning curve derives from the actual experience of manufacturers in some diverse industries. The concept of the learning curve first came to prominence before World War II when aircraft manufacturers observed that their costs of assembling an airplane decreased steadily and in a predictable fashion. Since that time, the universality of the learning curve has been demonstrated in a variety of industries, recently and most notably in the electronics industry.

The predictable decline of manufacturing costs is usually expressed in percentage terms that correspond to a doubling of the cumulative volume of the product in question. Thus, an 80 percent learning curve means that as the total number of units produced over time grows to twice any given number, manufacturing costs per unit can be expected to decline to only 80 percent of the cost per unit incurred before the doubling. If 100 airplanes of a new type have been manufactured to date, under an 80 percent learning curve the cost of the two hundredth airplane produced should be only 80 percent of the cost of the hundredth plane. The hundredth, in turn, should have been only 80 percent of the cost of the 50th plane, and so on. A 70 percent learning curve shows more cost cutting over the doubling of cumulative volume than an 80 percent curve; a 90 percent learning curve shows less.

Because inflation can be markedly erratic, learning curves are usually expressed in real (inflation-adjusted) dollars. Sometimes they are expressed in labor-hours, which gives the same kind of inflation-independent result.

Graphically, an 80 percent learning curve looks like Figure 13-2. This is a logarithmic relationship, which is more commonly portrayed as a straight line on a log-log scale grid such as Figure 13-3.

For the mathematically inclined, the learning curve relationship can be expressed as

$$(1) \qquad y = ax^{-b}$$

where x is the cumulative volume of units produced from the first and y is the cost to produce the xth unit. The coefficient a represents the cost of the first unit and the exponent b indicates the sensitivity of unit cost to cumulative volume.[b]

The straight line representation of the learning curve can be seen by taking the logarithm of both sides of equation 1, as follows:

$$(2) \qquad \ln y = \ln a - b \ln x$$

which is the equation of a straight line on a log-log scale with intercept $\ln a$ and slope $-b$.

It should be made clear that b is not 0.8 for an 80 percent learning curve; there is, however, a relatively simple correspondence between b and the percentage figure that labels the learning curve. This relationship can be derived as follows.

Let $y_1 = ax^{-b}$ and let y_2 be the unit cost for double the cumulative volume of x. The fraction y_2/y_1 would be 0.8 for an 80 percent learning curve, 0.7 for a 70 percent learning curve, and so on.

$$(3) \qquad \frac{y_2}{y_1} = \frac{a(2x)^{-b}}{ax^{-b}} = \frac{(a2^{-b}x^{-b})}{ax^{-b}} = 2^{-b}$$

[b]This exponent b is referred to by economists as the elasticity of unit cost with respect to cumulative volume. Elasticity has special meaning, representing the percent change in one variable for a given percent change in another. The price elasticity of demand is a commonly used example; it indicates the percentage increase in sales that a given percentage decrease in price will stimulate, all other things being equal. That b is in fact an elasticity measure is seen by taking the logarithm of equation 1 and differentiating with respect to $(\ln x)$ as follows:

$$y = ax^{-b}$$
$$\ln y = \ln a - b \ln x$$
$$\frac{(\partial \ln y)}{(\partial \ln x)} = -b = \frac{(\partial y/y)}{(\partial x/x)}$$

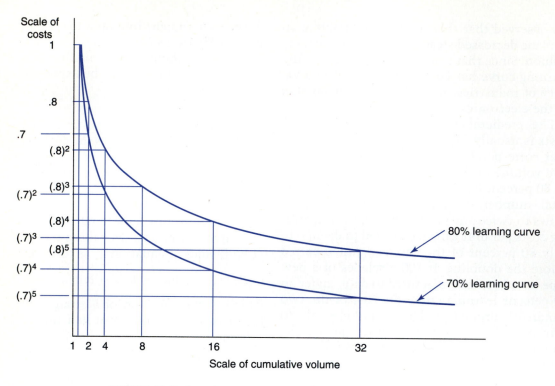

FIGURE 13-2   Learning curves portrayed on a conventional graph.

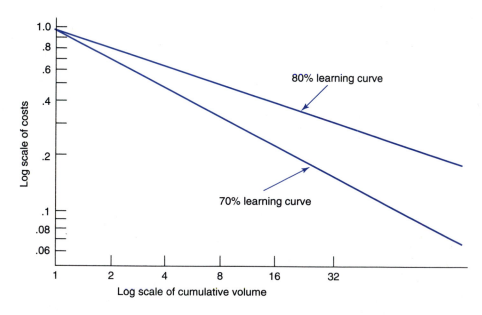

FIGURE 13-3   Learning curves portrayed on a log-log scale graph.

This equation gives the percentage learning curve for every choice of elasticity, b. We can easily turn this equation around to give the elasticity, b, that corresponds to every percentage level of learning curve.

$$\frac{y_2}{y_1} = 2^{-b}$$

$$\ln\left(\frac{y_2}{y_1}\right) = -b \ln 2$$

$$b = \frac{-\ln (y_2/y_1)}{\ln 2} = \frac{-\ln (y_2/y_1)}{0.6931}$$

For example, the b corresponding to an 80 percent learning curve is

$$b = \frac{-\ln(0.80)}{0.6931} = \frac{+0.2231}{0.6931} = 0.3219$$

This result can be checked in Figure 13-2 as the slope.

Composition of the Learning Curve

So much for the mathematics. What does the learning curve really represent and how can it be of use? Many, many factors stand behind the learning curve, and all of them can systematically reduce costs over the life cycle of the product. These factors can be grouped into major categories: workforce-related factors, process modifications, and product modifications. The workforce-related factors are more short run in character and distinguish, for some, the learning curve from the broader notion of the "experience curve." The experience curve is also a relationship between cost and accumulated volume, but it is generally associated with longer term factors dealing with process and product modifications as well as factors related to the workforce. Nevertheless, the terms "learning curve" and "experience curve" are often used interchangeably.

Workforce-Related Factors

• *Worker learning on the job.* This was the initial explanation for the results of the pre-World War II aircraft manufacturers, and for some people, this is what the learning curve represents. Cost decreases attributable to gains in worker efficiency through practice are generally quickly exhausted. That such cost reductions can be sustained over time is dependent on maintaining a stable workforce—high worker turnover can annihilate efficiency gains—and also a workforce that is interested in increasing productivity. As we have discussed before, worker attitudes toward productivity can depend on which compensation program is adopted. Some programs, like the Scanlon plan, by purposely tying compensation to overall productivity gain can do very well at sustaining cost reduction in the workforce.

• *Employee effort and management pressure.* Plain, old-fashioned hard work by workers and managers alike can keep costs down. Sometimes it is overlooked how important management pressure can be not only for worker learning but also for the full exploitation of any of the other factors in this list. Learning curve results frequently have an uncanny knack of matching learning curve expectations, and so we should not discount the impact of management pressure.

• *Workforce organization.* Akin to worker learning and effort is the organization of the workforce. A systematic training program, a thoughtful division of labor, and a considered plan for job mobility and advancement all have a role to play in lowering costs and in keeping them low over time.

Process Modifications

• *Methods improvements.* Included as methods improvements would be shortcuts in machine setup or operation, changes in the sequencing of a product through the factory, improvements in production and/or inventory control, and the introduction of production aids such as jigs, fixtures, or patterns.

These and other changes in methods can sometimes improve the yields of "good product" in the process and reduce the scrap involved.

- *Capital-for-labor substitution.* This is a classic means of lowering costs — substituting machine operations for tasks formerly performed manually.

- *Capital-for-capital substitution.* This means of reducing costs involves the replacement of existing plant and equipment with plant and equipment of more recent vintage. The new equipment is frequently more versatile, faster, more durable, and/or cheaper to operate with capabilities that exceed the existing models. The new equipment may be either larger or smaller than that which it replaces and for this reason more desirable.

- *Vertical integration.* By broadening the span of its process, a manufacturer can often lower its costs, although this is not always the case. This topic is discussed in more detail in the next chapter.

- *"Pure" technological change.* By "pure" technological change I mean the more radical redesigns of the production process that are brought about by breakthroughs in engineering or basic research. Examples abound in almost every industry — continuous casting in steel, computerized typesetting in printing, process control equipment in many industries, freeze drying methods in food processing, to name a few. For these changes new equipment must be designed, and so "pure" technological change usually includes capital-for-labor and capital-for-capital substitution as well. But pure technological change is usually much more than these kinds of substitution, for it involves a rethinking about how a particular product is made. Generally it triggers a host of changes throughout the production process, from changes in worker training and methods, and in production and inventory control, to changes in the company's handling of capacity requirements.

Product Modifications

- *Product design.* A frequent source of cost reduction is product redesign. Often a product can be simplified in its design so that its manufacture becomes significantly trouble free. As was discussed in Chapter 5, the term "value engineering" has been coined to describe this critical look at a product's construction and how it can be altered to lower manufacturing costs. Products as diverse as automobiles, television sets, and digital watches have all undergone significant value engineering in recent years.

  Not only does product redesign reduce manufacturing cost; it often enhances a product's performance and this, in turn, is valued in the marketplace. If product performance is what the market buys, this kind of product redesign can even be placed in the category of cost reduction. Semiconductor memory for computers is an example where product performance improvements have been a significant feature in the cost and price decline of the "product" purchased — the storage of a given unit of information in computers and related products.

- *Materials substitution.* Akin to product redesign is the use of different, possibly new, materials in place of other materials in the product. The substitution of aluminum for steel, plastic for glass or metal, metal for wood, and so on can lower the total costs of manufacture with relatively few design changes to many products.

Strategic Use of the Learning Curve Concept

This concern for the applicability of the learning curve phenomenon to short-term, tactical decisions of a corporation extends as well to longer term, strategic decisions. The strategic impact of the learning curve is nowhere more evident than in the electronics industry — historically with the growth and behavior of one company, Texas Instruments, in the 1970s. Texas Instruments then had a reputation as a tough competitor whose strategy was to generate huge sales volumes by repeatedly cutting prices. The company sought the sales leadership in all the markets it entered, using low prices to generate volume and using volume, via learning, to generate cost (and price) reductions. Texas Instruments was reputed to relish industry shakeouts, and such companies are generally feared as competitors.

The tactic that companies such as Texas Instruments have used with success to deter the entry of newcomers to a market and to stimulate a continual shaking out of an industry is preemptive pricing. Preemptive pricing is based on the expectations of the learning curve. Prices are set and announced at low levels, often with slim margins, and it is expected that volume production will stimulate enough cost reduction that total profits will become acceptably high. This kind of pricing is in distinct contrast to the practice of maintaining, for as long as possible, high prices in the face of declining costs, perhaps inviting a massive shakeout as more and more competitors enter the market under the existing price umbrellas.

These divergent strategies are represented pictorially in Figure 13-4. Note that although the initial profits are higher under the price maintenance umbrella strategy, because prices are higher the market is not likely to grow rapidly. Under the preemptive pricing strategy, although margins are often thin, volume is high, sales growth is rapid, and total profits can be quite acceptable.

In many industries, preemptive pricing has a distinguished history. For example, the so-

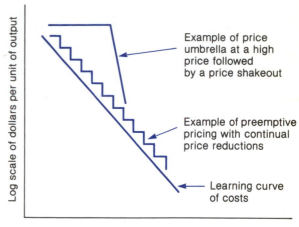

FIGURE 13-4   Pricing strategies and the learning curve.

called "trusts" of the late nineteenth century grew quickly and grew large by keeping prices low and using technology to cut costs. Rockefeller's Standard Oil refineries were models of the state of the art—the largest and fastest around. Carnegie's Edgar Thomson Steel Works in Pittsburgh was a triumph of rational plant layout and technology. The same stories could be told of sugar, tobacco, and the other trusts. In each of these cases, the first producer to get volume up and to force technological advance in process design was the one that traveled down the learning curve the first, forcing others out of the industry by pricing low and accumulating profits all the while.

It is no coincidence that all of the trusts of the nineteenth century were in commodities. Commodity manufacturers do not have to worry about new product introductions or product modifications undermining sales of their particular products. Commodity manufacturers can concentrate on cost reduction without fearing what competitors will market to counter their own product lines. The power of the learning curve phenomenon is greatly weakened by continual product model changes or new product introductions. These throw the company onto new learning curves or back up old ones. Consumer taste-sensitive industries are thus much less able to compete successfully with preemptive pricing than are more taste-invariant industries.

This fact suggests that the strategic implications of the learning curve are best exploited by particular classes of companies. Whether a company can adopt preemptive pricing and exploit the learning curve phenomenon successfully depends to a great degree on how it values a dollar received today versus a dollar tomorrow. Preemptive pricing defers its rewards more than traditional price maintenance because it must wait to foster growth in the market and because it stimulates an immediate and ongoing shakeout in the industry. Thus, preemptive pricing is relatively more attractive when some or all of the following features are true:

1. The expected product life is long.
2. The product is standardized, like a commodity, with little risk that consumer tastes will suddenly turn against it.
3. The expected growth of the market is rapid.
4. The apparent slope of the learning curve is steeper, since the learning curve in that instance can constitute a substantial barrier to the entry of other firms into the industry.
5. The product is not patent protected.
6. The company is not particularly strapped for cash.

To the extent these features are true, preemptive pricing (implied by the learning curve) can act to gain market share in a growing market and thereby command considerable profits.

Innovation in the Production Process

The introduction of process innovations is even less well delineated than new product introduction. As mentioned at the start of this chapter, incremental change is often thought to offer different management problems from radical change. It is frequently felt that continual, small improvements in the process technology can be stimulated by company reward systems or company grants for the pursuit of new ideas, such as are found at companies like 3M. Or, they can be induced by increased management pressure, or through work restructuring programs to call forth worker modifications, or by compensation schemes like the Scanlon plan, which ties pay increases to productivity gains. The factors cited earlier as influencing the learning curve can all account for incremental changes in process technology. Further, implementing incremental change can often be done without placing great demands on management's coordinating abilities.

Radical changes to the process technology are more problematic. More money and greater risk are involved, as well as the need to coordinate many people in various company departments. What leads to success in such an endeavor is not known with certainty, but analyses of some radical process innovations seem to up hold these general principles:

- The suggested process change must fit the company's business goals and thereby foster the commitment of the company's top management.
- The company must have a realistic idea of what the innovation can do well and should not get swamped by trying to innovate everywhere at once.
- Some attention should be given to how the innovation may be resisted by workers, consumers, or government.

- Any test should be made in a setting that is representative of the company's business and environment.
- A strong individual — sometimes termed the "change champion" — should act as both catalyst and critic for the innovation and its implementation. The change champion is usually a respected engineer or manager who pushes both the innovating team and the company itself. He pushes the team to develop something of true worth and technical distinction, and he pushes the company to spend the resources and time needed to make the innovation a success.

Naturally, it is important for companies to understand both incremental and radical process change.

We are all well aware that new discoveries are constantly making old knowledge obsolete. We take for granted that new technologies will someday, somehow transform whatever operation we care to name. The certainty of technological advance, however, does little to quell the uncertainties many managers feel about coping with it. Too often, managers abandon technology and the issues surrounding it to the engineers and scientists who are with it day to day. They fear that they need to understand the technology in all its details in order to manage it, and that fear can be paralyzing.

Managers need not be paralyzed by technology. Indeed, if technology is not to run the business, they must learn to manage technology. According to Wickham Skinner's insightful views, some inquisitiveness and the willingness to reason by analogy can be exceedingly helpful in coming to grips with process technology change.[10] The kind of inquisitiveness required to begin understanding technology is precisely the same kind of inquisitiveness that runs through the plant tours at the end of this book: What happens to the material during the process? How do machines or workers transform it? How does the new technology differ from the old? Where is the new technology vulnerable to breakdown or error? The manager need not fully understand the science behind the new technology, but he or she should be able to understand it by analogy or in some other way visualize what the technology does.

Armed with this knowledge, the manager can pose some basic, but tough, questions whose answers are critical to the effective management of new technology.

1. What will it cost? This, of course, is a favorite question of managers. Costs involved with new technologies can include procurement cost, installation, debugging, maintenance, materials handling, operating costs (labor, utilities, material), and overhead assignable to it.

2. What will it do? There are several key issues riding with any process change, among them: capacities involved, range of products affected, tolerances held, reliability offered, product characteristics performed, scrap/rework and energy saved. These issues, in essence, define the potential for the new technology.

3. What will it require? What will adoption of the new technology mean for direct labor skills, maintenance, setup, quality control, scheduling, inventory levels, materials needs, space, layout, and production control? These are the skills and the systems that support the existing process and must also support any new technology.

4. How certain are the costs, potentials, and requirements of the new technology? This, too, is a familiar managerial question that can be useful in defining the "bounds" of the technology. Here is where some sensitivity analysis can be very informative.

Skinner's questions help make the management of technology more systematic. They can often lead the way to an understanding of exactly what management must do well to have the new technology succeed, and that is valuable information indeed.

The remainder of the chapter addresses recent developments in manufacturing technology.

RECENT DEVELOPMENTS IN MANUFACTURING TECHNOLOGY[11]

During the past few decades, manufacturing technology has changed tremendously. There is now a dizzying array of sophisticated, electronics-based equipment to choose from. Consequently, factories of the future are likely to look very different from the factories of the present. The changes involve nearly every aspect of factory operation:

- *Materials.* New plastics, alloys, composites, powdered metals have been developed.

- *Inventory control.* Bar coding of parts and work orders, computerized parts storage and retrieval are commonplace today.

- *Materials handling.* Automatic guided vehicles, new kinds of conveyor systems, many with scanning equipment for bar codes are some recent changes.

- *Testing.* Automated testing equipment using X-rays, ultrasound, or radioactivity; automatic gauging, including robotic gauging are two types of equipment innovations.

- *New hardware.* The plethora of new, computerized systems is described next.

- *New means of controlling and linking hardware together.* Some techniques for improved control and communications are also described next.

- *Expert systems.* Expert systems are computer-based decision-making heuristics that simulate how trained, expert humans would examine evidence. Expert systems can be applied in either manufacturing or service sector businesses to do such tasks as life insurance underwriting or equipment maintenance planning and troubleshooting.[12]

Hardware Advances Affecting Continuous Flow and Line Flow Operations

As we have already discussed in this book, continuous flow and line flow operations are more capital-intensive than other kinds of processes. Historically, that capital intensity has been in conveyors for materials handling and special-purpose equipment designed expressly for the process and plant being considered. With the march of technology, this equipment has become faster, more versatile, and more capable of producing quality products. Conveyors now have the ability to scan products and packaging for bar code identification and route items to different places by code. Equipment has been filled with sophisticated sensors and process controls that permit increased speed of operation while maintaining quality control. Automatic inspection equipment also acts to maintain quality.

Robots have been introduced in many line flow operations, often to perform under particularly nasty conditions, such as heat, noise, or danger. Welding robots, painting robots, and "pick-and-place" materials handling robots are the most frequently observed types. Some of the more advanced robots, those capable of some assembly operations and those able to "see" and "feel" the work they handle are less prevalent, but still increasing in numbers.

Some technological proponents are so enthralled with the latest, flexible automation that they disparage the traditional, inflexible, "hard" automation that has characterized continuous flow and line flow processes for years. Yet if a company can live with inflexibility, as is the case with most companies employing continuous or line flow processes, then "hard automation" and "transfer lines" (a linked series of machines that work on material that is clamped into position) are often the cheapest and most reliable equipment to operate.

This debate between adherents of new, computerized, flexible equipment and those who value the specialized hard automation of many line and continuous flow processes is sometimes couched in terms of "economies of scale" versus "economies of scope." Scale economies are tied to the older, hard automation; given its relative inflexibility, its design best fits running large volumes of product over the equipment. With flexible equipment, however, the argument is that economies abound because of the greater scope (variety) of products that can be run over such machines. This is a debate for which both sides are right. Scope can provide economies but scale can still provide economies as well. Given high volumes for a product, managers should not shy away from the hard automation that could serve the process well.

Hardware Advances Affecting Job Shop and Batch Flow Operations

Some of the most intriguing recent changes in production technology have affected job shop and batch flow processes, which have not traditionally been the objects of technological innovation. These changes have, in many instances, altered these processes so that they function more like line flow or continuous flow ones.

Let's tour some of the major innovations and the alphabet soup that their introductions have cooked up. Of comparatively long standing is the numerically controlled (NC) machine, which is capable of altering workpiece feed, workpiece position, and tool speed and orientation through the use of prepunched paper or plastic tape. NC machines are particularly prevalent in metal working functions like turn-

ing, milling, drilling, grinding, punching, and the like. NC machines can often perform jobs faster and more accurately than manually operated machines, but the tapes require programming, sometimes at considerable expense.

Much more recent has been the development of the CNC (computer numerically controlled) machine, which allows the operator to punch in the various specifications and controls for the job and eliminates the need for prepunching a tape. A CNC machine is thus more flexible than an NC machine. In addition, many CNC machines can change tools automatically, selecting the next tool from a rotating inventory of tools. The next generation of machines are DNC (direct numerical control) machines. DNC ties together a CNC machine with a computer (known as a controller) that directs its movements, and usually the movements of other DNC machines.

The capabilities of CNC and DNC machines constitute one pillar of CIM (computer-integrated manufacturing). The ideal for CIM is the creation and design of new parts on a computer, the translation of those designs into instructions usable by CNC and DNC machines, and the activation of actual production using those instructions.

Other pillars for CIM are CAD (computer-aided design) and CAM (computer-aided manufacturing). CAD capabilities have progressed rapidly in recent years. In essence, CAD hardware and software enables the design engineer to do mechanical drawing without a drafting table, and to do it better. With CAD, parts can be created, their dimensions changed easily, various geometric views checked out, and the parts tested for fit, stress, and other features without the necessity for any expensive, physical prototype. This can be an enormous saving of time and expense.

CAM (computer-aided manufacturing) has been less widespread. CAM is also a more ambiguous subject. For some, CAM has a fairly limited scope, applying only to the use of the computer to generate instructions for CNC or DNC machines or for designing tools, fixtures, and layouts or for preparing or analyzing quality control data. For others, CAM is broader, even to the point of taking the CAD-generated design and following it through to the finished piece.

To date, linking CAD with CAM and the potential of CIM are only beginning to be realized in many manufacturing situations. Also in its infancy is flexible manufacturing system (FMS) technology. FMS combines the group technology layout and parts family concepts discussed in Chapter 3 with DNC technology. With FMS, a series of DNC machines are grouped together and linked with automatic materials handling equipment, all under the direction of a controller. In essence, then, FMS is a more fully automated example of group technology. A part is started in one machine, perhaps necessitating an automated tool change. Once finished, the part is automatically removed and transferred to the next operation. The part proceeds from one machine to the next, all without human intervention — even for setup. Such a system is capable of running independently, say at night, without workers to monitor it. All actions are taken as result of the controlling computer program. A key advantage of an FMS is that it can switch to making completely different parts. Small lots, or even single parts, can often be easily run through the system. There are still comparatively few true FMS's in operation throughout the world, although their numbers are growing rapidly. To date, they have not been as flexible as their proponents have trumpeted. Often the effort to get the system up and running on the initial small family of parts is so exhausting for the engineers and managers involved that they are reluctant to broaden its role. Nevertheless, many FMS adopters are pleased with the technology and flexibility is increasing.

We are already seeing glimpses of the factories that these technological advances may lead to in the future. A number of companies, often led by those in the machine tool and defense industries, have succeeded in linking CNC and DNC equipment together, often with automated materials handling, and all under the direction of a computer controller. Such operations automate the movement of workpieces between and within machining centers, the changing of tools, the operation of the machines themselves, and sometimes the inspection of pieces and the self-correction of a machine's operation (say, because of tool wear). Several dozen machines have been linked in this way. Even more sophisticated systems are being devised that make use of computer-based "expert systems" that can help guide products through production, do diagnoses of problems encountered, and initiate steps to correct those problems.

The most dramatic of these factories permit unmanned operation. Operators can load the factory in the evening and return the next morning to collect the night's production. This technology has enabled companies to increase significantly the efficiency of the machining they do, cut lead times for delivery, reduce space needs, and reduce their dependence on skilled operators.

Companies that are pioneering the "factory of the future" effort are, in general, advocating several principles. Among them:

- Design the product for automation. Sometimes this requires significant rethinking about the product form, materials, and assembly. Modular designs are often desirable in this regard.
- Make sure that the interface between worker and equipment is absolutely transparent. Otherwise, the full potential of the equipment will not be realized, inadvertently hindered by both labor and management.

- Implementation should occur in stages.
- Each individual workstation should have as much autonomy as possible, and there should be as little interaction or dependency across workstations as possible. This principle acknowledges that the most difficult technical problems are often those encountered in getting one machine to "talk" to others.

Technologically advanced factories are now found all over the globe, most prominently in Japan. In the United States, aerospace companies like Northrop, Lockheed, and others are leaders as are electronics companies like Motorola. However, these advances are also made in the more prosaic industries. For example, Allen-Bradley, Inc., of Milwaukee has developed a $15 million, completely automated assembly line for making contactors and relays (starters and controllers for industrial electric motors) in various models. The assembly line consists of 26 stations where nearly 200 different parts are installed in just 45 minutes. Each hour 600 units are produced, but the lot run can easily be as small as a single unit. Bar codes help the system differentiate one product model from another. The line can make products in two different sizes and with up to 999 possible combinations of parts, without slowing for changes. Quality is aided in part with 3500 automatic inspection steps. Not surprisingly, Allen-Bradley became the lowest-cost producer in the world.

Not all factories will become so automated, but increased automation does seem to be a clear trend. It is likely that with time, many factories will replace more and more of their direct labor with automated equipment.

Managing the New Technology

Day-to-day manufacturing problems are becoming more the problems of white-collar workers (engineers and managers) than solely the problems of blue-collar workers. Machines will increasingly do the dirty work. The workforce will then act mainly to changeover equipment, if need be, and to maintain it.

This is not to say that manufacturing is likely to become easy to manage well. Rather, we can expect that a different set of problems will demand attention. More processes are likely to need sophisticated engineering skills on a regular basis. This may mean shortages of engineers or of technical assistants such as programmers. The heightened need for engineering may also trigger more conflicts within the company—clashes between engineering and manufacturing, purchasing, marketing, and finance. There may be more conflicts between different kinds of engineering (such as design versus process engineering).

With technological advance, management's responsibilities are apt to be modified in various ways:

- The means by which new technologies (such as CAD/CAM) are evaluated financially may need to be changed, because present procedures (including net present value techniques) tend to discriminate against the new technological advances. Although the costs are all captured, the benefits of such advances are often so diffuse and hard to quantify that they escape traditional procedures.[13]

- Because companies are more likely to have to live with technological change on a continuing basis, a company must become more attuned to feedback on its technological choice. No longer may companies simply study and choose technology as if it were a one-shot deal.

- Because of the recent advances for job shop or batch flow processes, many companies are going to have to reconceptualize these processes and the planning and control systems for them. Particular attention will have to be paid to the management of transition within the company. Items like training and dealing with workers who feel threatened by technological advance may take on added emphasis. These issues have already been faced by many continuous flow process companies; their experiences have and will serve as models for others.

- At a strategic level, a key decision is the choice of which technologies to pursue and when. Many companies have failed to detect the change of technological winds — vacuum tube makers that discounted semiconductor advances, chemical companies that discounted materials changes, pinball machine makers that discounted video games. The antennae must always be up and searching for new developments, if merely to defend the company's interests.

In regard to the last-mentioned change, a company must be well aware of the limits to its own technological prowess so as not to be blinded when the promise of something better rolls around. Ten signals that can tip off a company to the imminent approach of a technological limit have been identified.[14]

- Top managers have an intuitive sense that the company's R&D productivity is declining.
- R&D is missing more and more of its deadlines. This trend is often misinterpreted as a signal that the department is losing effectiveness, when in reality improvement is becoming more difficult to achieve.
- Typically, as limits are approached, process improvements are easier to attain than product improvements (as we noted at the start of this chapter).
- The creativity of R&D is perceived as having declined.
- There is disharmony among the R&D staff.
- Replacement of R&D leaders or staff brings little or no improvement.
- Profits come from increasingly narrow market segments.

- Market share in a specialized market niche is lost. This signal is particularly important when market share is lost to a smaller competitor.
- There is little difference in returns despite spending substantially more — or less — than competitors over a period of years. When significant variations in expenditures produce no significant differences in performance, the technology is near its limits.
- Smaller competitors are taking radical approaches that "probably won't work." If they do work, larger competitors may suddenly become smaller ones.

The company that is alert to such signals can start to assess more radical revisions to its product offerings and process capabilities.

Tools of the Trade: A Recap of Lessons Learned

Clearly, in the future a manager will have to spend more time dealing with technological change. No single model for technology management has emerged as superior, however, although many of the principles of just-in-time manufacturing lap over into the management of new technology: focus on time, work in teams (and put suppliers on those teams), feedback information to those involved, work to get problems visible and then quickly analyzed and solved, engage work in parallel fashion (concurrent engineering), work incrementally, practice (as with periodic prototypes), and drive non-value-added steps out of the process.

Process innovation shares much with product innovation, and has a variety of sources, some involving pure technology, some involving the capital stock, and some involving labor and management. The learning curve is a helpful construct for thinking about such change. Importantly, however, learning involves design and not simply the operations floor. Design and manufacturing are more complementary than many think. Good design for manufacturing or assembly can reduce costs markedly, while good "manufacturing for design" can reduce the time it takes to develop new products (quicker tooling, dies, prototypes, pilot runs, ramp up).

Factories of the future will certainly look vastly different from the factories of the past, but blind faith in new technology is to be avoided. There is simply too much waste in existing processes that can be removed easily for us to jump to the expense of new process technology too quickly. There is still a lot of "blocking and tackling" yet to do before the machines take over. World-class companies recognize this and do not place undo stress on the importance of new technology to their competitiveness.

MORE ON THE LEARNING CURVE

The learning curve phenomenon is founded on a wealth of strong influences. What is not so clear is how the regularity of the learning curve can be used as a planning and decision tool. Consider Situation 13-1.

Discussion of Nashville Fire Equipment Company

It is easy enough to calculate the learning curve implied by the brief list of data in Table 13-2. The learning curve calculated is depicted in Figure 13-5. Mal Getz could use this finding of an 85 percent learning curve to grant an "improvement factor" to the company's workforce on, say, an annual basis payable at the end of the next year.

Calculating an Improvement Factor

One procedure for calculating an improvement factor is as follows:

1. Calculate how many labor-hours will be spent on the product over the course of the next year.

2. Estimate how many more units of the product are expected to be manufactured during that number of labor-hours.

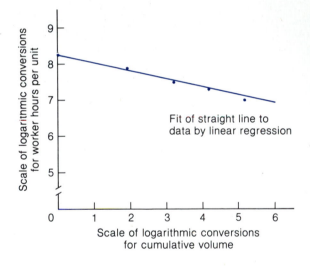

FIGURE 13-5   Learning curve for Nashville Fire Equipment Company's "Squirt."

3. Using the formula, compute the expected labor-hours for the unit that will be manufactured in exactly one year from now.

4. Compare the labor-hours per unit then with the present labor-hours per unit. The labor rate can then be adjusted to reflect the lower labor-hours that will be spent on each unit in a year's time.

476

SITUATION 13-1

Nashville Fire Equipment Company

Mal Getz, president of the Nashville Fire Equipment Company, was in the midst of contract negotiations with the business agent of the union that had recently won the right to represent his company's employees. The union, as one of the points it wanted to negotiate, had broached the concept of regular wage increases as so-called "improvement factors." The union argument was that because workers over time produce more with the same amount of human effort (due to such things as technological progress, better tools, methods, processes, equipment, and worker cooperation), they ought to be rewarded with regular wage increases.

Mal recognized this union point of negotiation as one based on the learning curve principle, and different in character from the typical wage increase to keep up with the cost of living. Mal had to admit to himself that the improvement factor was founded on sound reasoning. Moreover, Nashville Fire Equipment had been benefiting from the learning curve over the last several years, and Mal knew that the union would almost certainly be able to document that fact. It seemed likely then that management would have to agree to negotiate "improvement factors" for the next 3 years, the term of the contract.

The difficulty was that Mal was at a loss as to how to estimate what labor savings the company could legitimately expect over the next 3 years, and thus what improvement factors the company ought to consent to.

The company had recently been rejuvenated by the introduction of an articulating hydraulic boom (known within the industry as a "squirt"), a new product and technique for fire fighting that permitted the quick, pinpoint spraying of water into otherwise inaccessible places. The squirt had been winning increasing acceptance in the fire departments that served as the company's customer base and now accounted for most of the company's sales. Nashville Fire Equipment's squirt was a patentable modification of a similar product that the Seiver Company had licensed Nashville Fire Equipment to make. Table 13-2 documents various times, production volumes, and worker hours associated with the squirt. Over the next year Mal Getz expected to continue to employ 50 workers on the squirt. He wondered whether the table data, coupled with the learning curve concept, could form the basis for a recommendation on suitable improvement factors.

TABLE 13-2   Data on Hours to Produce a Unit

WEEK	CUMULATIVE UNITS PRODUCED AS OF THAT DATE	HOURS NEEDED TO MANUFACTURE FIRST UNIT OF WEEK
2	1	4000
10	7	2550
38	25	1850
70	65	1500
150	180	1170

For the Nashville Fire Equipment Company, following this procedure entails these calculations:

1. 50 weeks x 50 workers x 40 hours/week = 100,000 labor-hours for the next year.

2. To get an estimate of how many more units can be made in 100,000 labor-hours, we have to calculate what the learning curve is by fitting a straight line to the data.

If we did this, we would find that the 100,000th labor-hour was spent on the 270th unit. That is, exactly 1 year from now, if the Nashville Fire Equipment Company continues on its 85 percent learning curve, its workers will be working on the 270th squirt.

How can we fit a straight line to these data? Consider the following Aside.

ASIDE

Fitting a Straight Line to the Company's Data

Fitting a straight line to the data of Table 13-2 involves converting those data to logarithmic form (Table 13-3) and then applying the technique of linear regression to the converted data. Linear regression, part of most computer software statistical packages and even found on many calculators, is merely a procedure for choosing the straight line that minimizes the sum of the squared vertical distances between the data points and the fitted line.

The equation of the fitted straight line to the converted (natural logarithm) data, as found by linear regression, is

$$(1) \quad \begin{pmatrix} \text{natural} \\ \text{logarithm} \\ \text{of hours} \\ \text{needed} \end{pmatrix} = 8.293 - 0.237 \begin{pmatrix} \text{natural} \\ \text{logarithm of} \\ \text{cumulative units} \\ \text{produced} \end{pmatrix}$$

or put in the nonlogarithmic terms of $y = ax^{-b}$:

$$(2) \quad (\text{hours needed}) = 4000 \left(\frac{\text{cumulative units}}{\text{produced}} \right)^{-0.237}$$

Converting this estimate of b (0.237) into the appropriate label for the learning curve is accomplished by equation 3:

$$(3) \quad \frac{y_2}{y_1} = 2^{-b} \qquad 2^{-0.237} = 0.85$$

Thus the data of Table 13-3 imply an 85 percent learning curve.

Each successive unit will require slightly less labor. We could generate a table (Table 13-4) like Table 13-3. Using Table 13-4 we could sum up the individual times, starting with unit 181, until their sum reaches 100,000. By referring to the table, we could then discover the corresponding number unit on which the 100,000th labor-hour was spent.[c]

TABLE 13-3 Conversion to logarithms of data in Table 13-2			
CUMULATIVE UNITS PRODUCED	NATURAL LOGARITHM CUMULATIVE UNIT PRODUCED	HOURS NEEDED	NATURAL LOGARITHM of HOURS NEEDED
1	0.00	4000	8.29
7	1.95	2550	7.84
25	3.22	1850	7.52
65	4.17	1500	7.31
180	5.19	1170	7.06

---

[c]Fortunately, we do not have to perform this calculation, although it takes a little integral calculus to avoid it. The labor time spent between the mth unit and the nth unit is approximated by the integral of the learning curve function evaluated between m and n. Since the learning curve function is a monomial, the integral is easily evaluated. In particular, labor time between mth and nth units is equal to

$$(1) \quad \int_{m}^{n} y \, dx$$

TABLE 13-4   Estimate of labor-hours needed to produce a particular Squirt (Nashville Fire Equipment Company)

NUMBER OF UNITS PRODUCED	ESTIMATED LABOR-HOURS NEEDED TO PRODUCE THAT UNIT*
1	4000.0
180	1170.9
181	1169.4
182	1167.9
183	1166.4

*From learning curve estimated in Figure 13-5.

where y is the learning curve estimate of labor time needed to produce any unit x. We can then use the formula for the learning curve to substitute for y:

(2)
$$\int_{m}^{n} ax^{-b}\, dx$$

Then we factor out the constant a:

(3)
$$a \int_{m}^{n} x^{-b} dx$$

Taking the integral, then, yields

(4)
$$\frac{ax^{1-b}}{1-b}$$

which can be evaluated at n and m and subtracted to reveal the labor time spent between units.

We can apply this formula to figure out how many squirts the Nashville Fire Equipment Company can be expected to make in a year. We know that 100,000 labor-hours will be spent during the next year. The company has already manufactured 180 squirts according to the learning curve that was derived as

$$y = ax^{-b} = 4000x^{-0.237}$$

Stepping through equation (4), we have

$$100,000 \text{ hours} = \int_{180}^{n} 4000x^{-0.237}\, dx$$

where n, the as yet unknown number of the unit to be worked on in exactly 1 year's time, equals

$$\frac{4000\, n^{1-0.237}}{1 - 0.237}$$

We evaluate at n and 180 and subtract as follows:

$$\frac{4000 n^{0.763}}{0.763} - \frac{4000(180)^{0.763}}{0.763} = 5240 n^{0.763} - 275{,}410 = 100{,}000$$

3. Over the course of the year it will have produced 90 (270 − 180) squirts. The 270th unit will require only

$$y = 4000 \, (270)^{-.0237}$$
$$= 1060 \text{ hours to complete}$$

This situation shows the phenomenal power of learning. It took the Nashville Fire Equipment Company 3 years to produce its first 180 squirts; in a single year's time, if its 85 percent learning curve is maintained, it can expect to produce almost half that amount.

4. At present, 1170 labor-hours are needed to produce the latest unit, the 180th. From the calculation above, in a year's time only 1060 hours will be required, which is only 90.6 percent of the present labor input. With this productivity advance, if the price of the squirt remains the same (a rather big if), the company can afford to pay its workers 1.10 (1/0.906) times as much in a year's time and still enjoy the same profit margin it now does. This figure of 1.10 thus provides a ceiling on the improvement factor that Mal Getz has to decide upon.

Applying the Learning Curve

The forgoing series of steps demonstrates how the learning curve can be used as a planning device. Here the learning curve was used to calculate an "improvement factor" for labor, but it could also have been used to figure out by what date a given number of units will have been produced or to guide labor hiring practices to fulfill a certain commitment by a given date. The learning curve can be particularly useful in tracking and in planning new product introductions.

However, a larger issue than the mere mechanics of the learning curve is whether the concept can be applied at all to situations like that of the Nashville Fire Equipment Company. What are the weak points in using the learning curve for such situations?

As a glimpse back at Figures 13-2 and 13-3 reveals, the learning curve is keenly sensitive to position and slope, which are determined respectively by the coefficient a in equation 1 and by the exponent b. We ought to ask the question then whether the first squirt made by the company in week 2 of Table 13-2 should be viewed as the absolute first unit or not. As was mentioned in the situation, Nashville's squirt is a patentable modification of a similar product that the Seiver Company had licensed Nashville Fire Equipment to make. It could easily be argued that the first unit should not be the squirt itself but the Seiver Company's licensed product. The impact of such a change

---

Rearranging terms and dividing through yields

$$\frac{(100,000 + 275,410)}{5240} = n^{0.763}$$
$$71.64 = n^{0.763}$$

Taking the natural logarithm of both sides enables us to solve for n:

$$\ln (71.64) = 0.763 \ln n$$
$$\ln n = 5.6$$

which implies that

$$n \approx 270$$

can be dramatic. Suppose, for example, that Nashville Fire Equipment's first squirt is viewed as the 31st unit for purposes of computing the learning curve. The latest unit would thus be the 210th. The refigured learning curve would be

$$y = 21,850x^{-0.570}$$

which translates to a 67 percent learning curve and drastically different results.

Perhaps the company's early experience with a similar licensed product brought down its labor-hours per unit much more quickly than could otherwise be expected; what is learned from producing one product can frequently be transmitted to those working on another product within the company, or even from within one company to another. In that case, the learning curve the company could anticipate over the next year would be much less steep than the currently calculated 85 percent curve.

Moreover, what learning curve rate would be appropriate may be very unclear, especially at the outset. Much of the issue of "technological forecasting" revolves around the appropriate choice of the learning curve elasticity, b, for various new products and which influences on that rate are likely to be important.

These doubts about the position and slope of the learning curve are not easily resolved and argue for great caution. Mal Getz may well be able to grant an "improvement factor" to his workers, but one somewhat less than 110 percent of the current rate seems prudent.

Situation 13-1 underscored the conflicts that can arise in the use of standard times. Technological change and learning will often improve worker times and erode the standards established. Making adjustments, especially if wages are tied to performance relative to the standard, can be tricky.

QUESTIONS

1. What do you understand by the term *technological change*?

2. What are the mathematical assumptions behind the learning curve? What does learning represent for manufacturing?

3. What is linear regression and of what relevance is it to the Nashville Fire Equipment Company? Give three examples of other situations in which the concept would be useful and applicable.

4. What are some uses of the learning curve in long-term strategic decisions?

5. Analyze and discuss the relationship between product innovation and process innovation. Provide a concrete example from your own reading or experience.

PROBLEMS

1. Summarize the procedure for calculating a learning curve. Of what use is this procedure?

2. It has been observed that preemptive pricing strategies are more useful in certain situations than in others. What are the characteristics of these situations?

3. How can changing the product design result in a technological change?

4. For each of the following elasticities, what is the implied learning curve percentage?

(a) 1.00.

(b) 0.75.

(c) 0.50.

(d) 0.25.

5. If the Nashville Fire Equipment Company (Situation 13-1), having just produced its 180th squirt unit, continues on its 85 percent learning curve, in half a year's time, how many units will have been produced?

6. Time has passed, and more data can now be added to Nashville Fire Equipment Company's set of statistics (Situation 13-1). The new data are as follows:

WEEK	CUMULATIVE UNITS	HOURS NEEDED TO MANUFACTURE
200	270	1050
250	365	1000

What is the learning curve implied by all the data now available?

SITUATION FOR STUDY 13-1

KID KLOTHES, INC.

Bob Stifler is president of Kid Klothes, a small family-owned apparel manufacturer for children. He was contemplating the acquisition of some new technology in an effort to keep his company competitive in the cut-throat world of apparel. Children's clothes were highly cost sensitive; mothers did not wish to spend a lot of money outfitting their children, yet children's clothes contained all of the pieces and seams to sew as large clothes, effectively constraining many of the cost savings possible. Labor was a significant variable cost in the manufacture of children's clothes, accounting for as much as 35 percent of the cost of goods sold. Any savings on the direct labor making up a child's garment would help make that clothing more price-competitive in the marketplace.

Bob was especially interested in applying some of the latest technology to the fabrication of children's clothes, particularly the use of automated sewing machines and of automated fabric cutters. The investments in this new technology were sizable. An automated cutting system of the type Bob felt was necessary cost $300,000, and programmable automated sewing machines with edge guides cost $15,000 each. The best estimates predicted the automated cutter could replace 14 full time cutters. The automated sewing machine could, on average, double the output from a conventional sewing machine. The average shop wage was $6 per hour.

Bob wondered what he should know before making this decision. What were the things he should be leary of about this new technology, and how should he assess it? He had to act quickly or else forfeit the chance to use this new technology for the next seasonal peak in production.

SITUATION FOR STUDY 13-2

MEDICAL HOSPITAL

Medical Hospital is a large regional hospital located in Weatherford, Oklahoma. Included in the hospital is a large pharmacy. Recently, the pharmacy has been experiencing long lead times for filling prescriptions, and the pharmacy manager, Patricia Pritchard, is anxious to find a solution to the problem. The main difficulty occurs during night hours and weekends when the regular pharmacists are off duty.

During the evening shifts (3 P.M. to 11 P.M.) and weekends, the pharmacy is staffed by technicians. The technicians are trained in pharmacy procedures but are not pharmacists. They are allowed, however, to fill prescriptions. Due to the time of the shift and the requirement of working every other weekend, it has been very difficult to find highly motivated and trainable people to work full-time. Most people preferred to work days and wanted weekends off. Many who were hired were slow to learn the job.

An alternative to hiring full-time technicians was to hire students from a nearby pharmacy school on a part-time basis. Each student worked about 20 hours a week, so twice as many employees were needed to fill all the slots. Most pharmacy students could be ex-pected to work a maximum of 2 to 3 years because once they finished their pharmacy program, they were not happy to be technicians. Also, some of the students would tie their work to the school calendar, and turnover was high in May, August, and December. The result was that from two to five technicians (an average of three) were new every 4 months. Usually, technicians were brought in at a rate of two per week and were trained on the job as well as by a course taught by the pharmacists.

Patricia had noticed not only that productivity was low, but also that the cost of training technicians was high. She had heard that the learning curve concept had been applied in manufacturing and wondered whether it could be applied to the pharmacy's product, a filled order. Pat defined an order as having three medications, on average. If an order had six medications, it was counted as two orders. On an average night 55 orders were processed. The normal maximum speed for an employee was around 20 to 24 orders per hour.

Patricia decided to study, for a period of several weeks, the speed with which 10 technicians filled orders. The shift supervisor collected

TABLE 13-5   Orders filled per hour (Medical Hospital)

EMPLOYEE	WEEK 1	WEEK 2	WEEK 3	WEEK 4	WEEK 5	WEEK 6
A	8	12	12	15	18	20
B	6	11	12	14	16	18
C	7	12	14	16	16	16
D	8	10	10	14	14	18
E	12	18	20	21	21	22
F	5	8	10	11	12	14
G	6	8	10	13	15	16
H	8	10	13	15	17	19
I	9	10	12	12	14	16
J	10	12	15	17	19	21

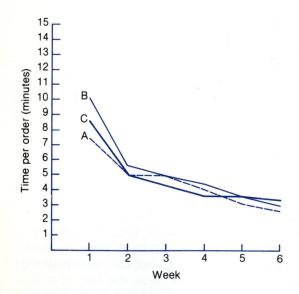

FIGURE 13-6   Learning curves for technicians
A, B, and C.

information on orders filled correctly by these technicians. It was easy to identify each technician's work, as they were required to initial each order filled. Table 13-5 shows the results of this study in orders filled per hour, and Figure 13-6 shows the learning curves for 3 of the 10 employees.

Now that Patricia had this information, she was uncertain how it could be used to help reduce the cost of training new employees, and how she could help to reduce her turnover problem.

REFERENCE NOTES

1.   Abernathy, William J., and James M. Utterback, "Patterns of Industrial Innovation," *Technology Review* 80, no. 7 (June–July 1978): 40–47.

2.   Abernathy, William J. *The Productivity Dilemma.* Baltimore: Johns Hopkins University Press, 1978.

3.   White, George R., and Margaret B. W. Graham, "How to Spot a Technological Winner," *Harvard Business Review* 56, no. 2 (March–April 1978).

4.   Hundy, B. B., and D. J. Hamblin. "Risk and Assessment of Investment in New Technology," *International Journal of Production Research* 26, no. 11 (1988) : 1799–1810.

5.   For some additional information on new product development, see:
Adler, Paul S., Henry E. Riggs, and Steven C. Wheelwright, "Product Development Know-How: Trading Tactics for Strategy," *Sloan Management Review* (Fall 1989): 7–17.
Clawson, Richard T., "Controlling the Manufacturing Start-Up," *Harvard Business Review* 63, no. 3 (May–June 1985): 6–20.
Frand, Erwin A., *The Art of Product Development*, Homewood, IL: Dow Jones-Irwin, 1989.
Rosenau, Jr., Milton D., *Faster New Product Development*, New York: AMACOM, 1990.
Souder, William E., *Managing New-Product Innovations*, Lexington, MA: Lexington Books, 1987.
Wheelwright, Steven C., and W. Earl Sasser, Jr., "The New Product Development Map," *Harvard Business Review* 67, no. 3 (March–April 1989): 112–25.

6.   Many of these points are explained very well in the fine book by Kim Clark and Takahiro Fujimoto, *Product Development Performance: Strategy, Organization, and Management in the World Auto Industry*, Harvard Business School Press, (Boston: 1991). See also Chapters 10 and 11 of R. H. Hayes, S. C. Wheelwright, and Kim B. Clark, *Dynamic Manufacturing*, Free Press: New York, 1988, for a discussion of some of these points.

7.   See Kim B. Clark and Takahiro Fujimoto, op. cit.

8. Adapted from V. A. Mabert, J. F. Muth, and R. W. Schmenner, "Collapsing New Product Development Times: Six Case Studies," *Journal of Product Innovation Management*, June 1992.

9. For some additional information on the learning curve, see
    William J. Abernathy and Kenneth Wayne, "Limits of the Learning Curve," *Harvard Business Review* 42, no. 5 (September–October 1974): 109–19.
    Belkaoui, Ahmed, *The Learning Curve: A Management Accounting Tool*, Westport, CT.: Quorum, 1986.
    David L. Bodde, "Riding the Experience Curve," *Technology* Review 78, no. 5 (March–April 1976): 53–59;
    Winfred B. Hirschmann, "Profit from the Learning Curve," *Harvard Business Review* 42, no. 1 (January–February 1964): 125–39;
    Majd, Saman, and Robert S. Pindyck, *The Learning Curve and Optimal Production Under Uncertainty*, Cambridge, MA: National Bureau of Economic Research, 1987.
    Smunt, Timothy L., "A Comparison of Learning Curve Analysis and Moving Average Ratio Analysis for Detailed Operational Planning," *Decision Sciences* 17, no. 4 (Fall 1986): 475–95.
    Teplitz, Charles J., *The Learning Curve Deskbook: A Reference Guide to Theory, Calculations, and Applications*, New York: Quorum Books, 1991.

10. This discussion draws heavily from Chapter 6, "Technology and the Manager," in Wickham Skinner, *Manufacturing: The Formidable Competitive Weapon*, New York: Wiley, 1985. Another excellent source is Robert A. Burgelman and Modesto A. Maidique, *Strategic Management of Technology and Innovation*, Homewood, IL.: Irwin, 1988.

11. Information on recent developments in manufacturing technology is sometimes available in the popular press. Other articles on the management of technology include:
    Barton, Leonard, and William A. Kraus, "Implementing New Technology," *Harvard Business Review* 63, no. 6 (November–December 1985): 102–10.
    Beatty, Carol A., and John R. M. Gordon, "Preaching the Gospel: The Evangelists of New Technology," *California Management Review* 33, no. 3 (Spring 1991): 73–94.
    Betz, Frederick, *Managing Technology, Competing through New Ventures, Innovations and Corporate Research*, Englewood Cliffs, NJ: Prentice-Hall, 1987.
    Brody, H., "Overcoming Barriers to Automation," *High Technology* (May 1985): 41–46.
    Darrow, William C., "An International Comparison of Flexible Manufacturing Systems Technology," *Interfaces* 17, no. 6 (November–December 1987): 86–91.
    Das, Sidhartha R., and Basheer M. Khumawala, "Flexible Manufacturing Systems: A Production Management Perspective," *Production & Inventory Management* 30, no. 2 (Second Quarter 1989): 63–67.
    Dean, James W. Jr., *Deciding to Innovate: How Firms Justify Advanced Technology*, Cambridge, MA: Ballinger, 1987.
    Dunn, Alan C., "Considering Robots?" *Quality Progress* Vol. 17 (June 1984): 10–13.
    Ettlie, John E., and Stacy A. Reifeis, "Integrating Design and Manufacturing to Deploy Advanced Manufacturing Technology," *Interfaces* 17, no. 6 (November–December 1987): 63–74.
    Foulkes, Fred K., and Jeffrey L. Hirsch, "People Make Robots Work," *Harvard Business Review* 62, no. 1 (January–February 1984): 94–102.
    Gilbert, James P., and Peter J. Winter, "Flexible Manufacturing Systems: Technology and Advantages," *Production & Inventory Management* 27, no. 4 (fourth Quarter 1986): 53–60.
    Groover, M. P., and E. W. Zimmers, Jr., *CAD/CAM: Computer-Aided Design and Manufacturing*, Englewood Cliffs, NJ: Prentice-Hall, 1984.
    Hayes, Robert H., and Ramchandran Jaikumar, "Manufacturing's Crisis: New Technologies, Obsolete Organizations," *Harvard Business Review* 66, no. 5 (September–October 1988): 77–85.
    Hayes, Robert H., and Ramchandran Jaikumar, "Requirements for Successful Implementation of New Manufacturing Technologies," *Journal of Engineering & Technology Management (Netherlands)* 7, no. 3,4 (March 1991): 169–175.
    Jaikumar, Ramchandran, "Postindustrial Manufacturing," *Harvard Business Review* 64, no. 6 (November–December 1986): 69–76.
    Kusiak, Andrew, ed., *Flexible Manufacturing Systems: Methods and Studies*, North Holland, Amsterdam: Elsevier, 1986.
    Liberatore, Matthew J., ed., *Selection and Evaluation of Advanced Manufacturing Technologies*, New York: Springer-Verlag, 1990.
    Martins, J. G., and M. Svensson, *Profitability and Industrial Robots*, New York: Springer-Verlag, 1988.
    Meredith, Jack R. "Implementing New Manufacturing Technologies: Managerial Lessons and the FMS Life Cycle," *Interfaces* 17, no. 6 (November–December 1987): 51–62.

———, ed., *Justifying New Manufacturing Technology*, Atlanta, GA: Industrial Engineering and Management Press, 1986.

———, "The Strategic Advantages of New Manufacturing Technologies for Small Firms," *Strategic Management Journal (UK)* 8, no. 3 (May–June 1987): 249–258.

Meredith, Jack R., and Marianne M. Hill, "Justifying New Manufacturing Systems: A Managerial Approach," *Sloan Management Review* 28, no. 4 (Summer 1987): 49–61.

Noori, Hamid, *Managing the Dynamics of New Technology: Issues in Manufacturing Management*, Englewood Cliffs, NJ: Prentice-Hall, 1990.

———, and Russell W. Radford, *Readings and Cases in the Management of New Technology*, Englewood Cliffs, NJ: Prentice-Hall, 1990.

Noro, Kageyu, ed., *Occupational Health and Safety in Automation and Robotics*, Philadelphia, PA: Taylor and Francis, 1987.

Schreiber, Rita R., "Assembly Robots Build Quality," *Manufacturing Engineering* 106, no. 6 (June 1991): 47–51.

Schroeder, Dean M., C. Gopinath, and Steven W. Congden, "New Technology and the Small Manufacturer: Panacea or Plague?" *Journal of Small Business Management* 27, no. 3 (July 1989): 1–10.

Schroeder, Roger G., Gary D. Scudder, and Dawn R. Elm, "Innovation in Manufacturing," *Journal of Operations Management* 8, no. 1 (January 1989): 1–15.

Suresh, Nallan C., and Jack R. Meredith, "Achieving Factory Automation Through Group Technology Principles," *Journal of Operations Management* 8, no. 1 (January 1989): 1–15.

Tidd, Joseph, *Flexible Manufacturing Technologies and International Competitiveness*, London: Printer Publishers, 1991.

Tushman, Michael L., and William L. Moore, eds. *Readings in the Management of Innovation*, 2nd ed. Cambridge, MA: Ballinger, 1985.

Vonderembse, Mark A., and Gregory S. Wobser, "Steps for Implementing a Flexible Manufacturing System," *Industrial Engineering* 19, no. 4 (April 1987): 38–48.

12.   See the article by Marc H. Meyer and Kathleen Foley Curley entitled "Putting Expert Systems Technology to Work", *Sloan Management Review*, Winter 1992 (Vol. 32, No. 2), pp. 21–32.

13.   Gold, Bela, "CAM Sets New Rules for Production," *Harvard Business Review* 60, no. 6 (November/December 1982): 88–94. See also Robert S. Kaplan, "Must CIM Be Justified by Faith Alone?" *Harvard Business Review* (March–April 1986): 67–95.

14.   These signals were identified by the consulting firm of McKinsey and Co. as reported in *Business Week* (May 24, 1982): 26, 28.

CHAPTER 14

DEALING WITH OPPORTUNITIES FOR VERTICAL INTEGRATION

The sequence of activities that bring a good or service into being and then to market is frequently a long one. Very often, the companies involved in the sequence confront some intriguing choices about how much or how little of that sequence to involve themselves with. Take gasoline, for example. There are at least six broad stages involved in manufacturing and marketing gasoline: (1) exploring for and purchasing the mineral rights to the land under which the oil lies, (2) drilling the well and pumping oil from it, (3) transporting the crude oil from well to refinery, often by pipeline or tanker, (4) refining the crude oil into a multitude of petroleum products, (5) transporting refined gasoline from refinery to storage tanks (tank farms) and thence to service stations, and (6) selling gasoline at the service station pump. A company can decide to engage in just one of these six activities or to straddle two or more of them. Indeed, there are companies that engage in just one activity and there are companies that engage in all of them

At issue in this chapter is how to decide the extent of a company's commitment to portions of the full sequence of production and distribution activities for any good or service. In the common terminology, it is a decision about the degree of vertical integration that a company

should seek. The decision is complex, influenced by a number of factors.[1]

FACTORS IN THE VERTICAL INTEGRATION DECISION

As an introduction to some of the factors that can guide vertical integration decisions, consider Situation 14-1.

Discussion of Rich Clover Dairy

Ralph Maynard faces several challenging decisions. Traditionally, the Rich Clover Dairy has been vertically integrated (at least with respect to guernsey milk). Of concern now is whether the dairy ought to remain so. The potential for either selling the land to developers or relocating the bottling plant forces this decision upon Maynard.

Given its current degree of vertical integration, a company usually needs to make three decisions about which way to go:

1. Should the company stay put or should it move toward more or less vertical integration? If the latter, should it change its degree of *backward integration* (integration toward the source of raw materials) or *forward*

SITUATION 14-1

Rich Clover Dairy

Perhaps it was time to close this chapter in the dairy's history. All Ralph Maynard knew was that it would be an emotional ending, if it came to that. Ralph, president of the Rich Clover Dairy and son of its founder, faced the decision whether to close the dairy's own farm. The farm, located in the lush, rolling countryside north of one of the city's reservoirs, had been the dairy's showpiece for over 40 years. Rich Clover's prize herd of guernsey cows was pastured there and the dairy had used this fact to advantage in marketing itself as "the dairy with cows." (Most other dairies had no cows of their own; they bought milk from dairy farmers, processed and bottled it, and then sold it under their brand name. Rich Clover bought most of its milk from area dairy farmers but produced its own guernsey milk.) Many of the area's families had made a point of bringing their children to the dairy to see the cows and bulls, the twice-daily milkings, and the egg-laying chickens. The farm no doubt contributed not only to the dairy's image among local people of being devoted to quality but also its position as the area's second largest dairy.

Times were changing, however, and Rich Clover was not immune. Although the dairy's sales had continued to grow with the development of the area, the nature of those sales had changed drastically. Where 20 years ago, two-thirds of the dairy's sales were from home deliveries, now they barely accounted for 20 percent. The slack had been taken up by increased sales to grocery stores (Rich Clover was a leading supplier to local chains and Mom-and-Pop independents) and by the founding of the dairy's own convenience stores, which were open from 7 A.M. to 11 P.M. and sold a variety of grocery items in addition to dairy products.

Other changes had some significant implications for the farm itself. The area's continued growth had spread population and housing farther and farther from the downtown. The farm, once a considerable trek out into the country, was now much more within the fringe of the area's growth. Land prices were escalating and the dairy could reap a handsome profit by selling its 1500 acres.

In addition, the farm raised only guernsey cows, which were known for their very rich milk. The dairy's "top-of-the-line" milk (8 percent of sales), "Golden Guernsey," was supplied almost entirely by the farm's herd. With changing life-styles, however, milk rich in butterfat was steadily losing market share to nonfat and low fat milk. Ralph Maynard could foresee the day when the farm's herd would be forced to supply milk for other than the Golden Guernsey line—the dairy's highest margin product. As it was, 20 years ago the herd had been able to produce only half the dairy's need for guernsey milk; the milk from other guernsey herds had been purchased then to make up the other half.

Ralph could foresee other changes at Rich Clover. The dairy's receiving and bottling plant, located in the city, was over 45 years old. Although periodically modernized, the bottling plant was neither as efficient nor as fast as the latest plants of other dairies. Moreover, traffic around the plant had become increasingly snarled, in part because the plant's loading operations were more suited to traditional, small milk trucks than to the larger, refrigerated trailers that could better supply grocery and convenience stores. The relocation of the dairy's bottling operations was a choice Ralph could sense he would have to face in the years to come. First, he thought, the farm's future should be addressed.

integration (integration toward the final consumer)?

Rich Clover Dairy is more vertically integrated than most dairies: backward integrated into its own herds and forward integrated into its own convenience stores as well as home delivery. The decision the dairy faces regarding its farm is whether to stay put or decrease its backward integration.

2. How far backward or forward should the company go in its vertical integration?

Rich Clover Dairy decided years ago to move backward all the way to ownership of cows and land. It could have chosen to stop short of so deep a backward integration by not owning cows or land or both.

3. What balance among activities should the company seek? How complete should its backward or forward integration be?

Although Rich Clover Dairy became backward integrated into the production of guernsey milk, it did not produce all the milk it needed; it still bought heavily from area dairies. This incomplete vertical integration (sometimes referred to as *partial* or *tapered integration*) can provide many of the advantages of vertical integration but without some of its disadvantages. Although partial integration is sometimes deliberately chosen by companies, it often occurs simply because capacity increments can be "lumpy," inevitably causing an imbalance in capacities among the segments of a process.

An important feature of a reasoned analysis of these decisions, especially their long-term aspects, involves an assessment of the economics of vertical integration. That is, where in the entire span of activities that a company can choose to perform are the rates of return to invested capital high and where are they low?

This important question is often very difficult to answer. We can get a handle on it by taking a brief excursion into the realm of economics and the concept of an "economic rent."

Scarcity and Vertical Integration

A sure way to riches is to offer a product or service that is both in demand and scarce in supply. In such situations the price people are willing to pay is often very much higher than the costs of bringing the product or service to market. The difference between the two is termed an *economic rent*.

Land is the classic example of an economic rent. Certain categories of land, say land with oil underneath it, are in at least relatively fixed supply. That is, even over several years, the known quantity of oil land does not vary greatly (although vary it does). As long as oil companies bid higher for the use of the land than housing developers or commercial interests, owners would be willing to offer their land for drilling. A higher bid brings no further change in behavior from the landowners; all they can say, no matter how high the price, is "Yes, go ahead and drill." What they receive from the oil company is an economic rent.

To aid our understanding, we can graph this situation (Figure 14-1). The quantity *OS* is the land available. *SS* is the supply curve, which is vertical, indicating that a higher price offered for the use of the land will not induce any more production. Production is fixed at the *OS* acreage level. *DD* is the demand curve, indicating the price the oil company would pay for various quantities of land offered. *OP* is the per acre price that the company is willing to pay for the services of *OS* acres of oil land. The *OP* price is the economic rent in this situation of a production input that is in rigid supply. The landowner reaps a total economic rent of *OSP'P*, which can be a considerable sum.

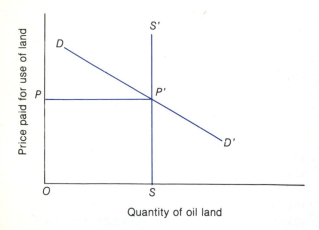

FIGURE 14-1   Economic rent.

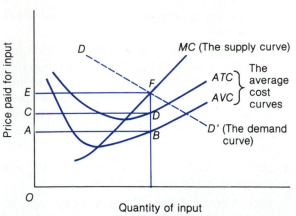

FIGURE 14-2   Economic quasi rent.

Few inputs are in rigid supply, of course, but many come into temporarily rigid supply. Owners of these inputs can in these short-lived cases earn what is called a *quasi rent*. Figure 14-2 illustrates this case. Here, increases in the price offered will induce more of the input to be supplied, as we can tell from the rising supply curve, marked *MC* because it is really the marginal cost curve for the input. The demand curve intersects the supply curve at *F*, thus establishing the level OE as the price for the input. The input could be purchased, however, for any price higher than *OA*, because any price higher than *OA* would cover the average variable costs of the owner of the input. In the long term, however, we could not count on having the input supplied if the price did not exceed *OC*, which is the level at which average total costs are covered. The difference, of course, is that some fixed costs must be covered in the longer run, if not in the short run.

Quasi rents are pure rents in the short run, although in the longer run they must go to pay off fixed costs; otherwise the input would have to be withdrawn. In Figure 14-2, the total quasi rent is *ABFE*. The quantity *ABDC* represents the fixed costs that must be paid off in the

longer run, whereas *CDFE* represents pure economic profit.

What do rents and quasi rents have to do with the economics of vertical integration? Rents and quasi rents are important because they often indicate where rates of return are high. Because scarcity brings about rents and quasi rents, it often pays to ask where scarcity lies in an industry. Find scarcity, and you often find a road to riches.

Returning Home with the Cows

Back to Ralph Maynard's responsibilities at the Rich Clover Dairy. Where does scarcity exist in dairying? Where do the bucks and the high rates of return lie? According to Maynard, dairy cows are not scarce, and so we should not expect large rates of return from owning a herd of guernsey cows, at least not for their milk production. (Their marketing clout because of the slogan "the dairy with cows" is a different question.) This is an argument then for selling the farm.

If scarcity and high rates of return do not lie with owning cows, where do they lie? The answer to this is less than clear-cut, but the

Rich Clover brand name and its quality image appear to be what is "scarce." The industry is one where the retailing end of the business is likely to be a much more important factor than the origin of the milk. Whatever else, a dairy wants to maintain its strength as a bottler and marketer.

This seat-of-the-pants case against a dairy's owning its own herd is insufficient by itself to decide the matter, because the immediate issue is more concerned with whether it is better to hang onto the farm for a few more years in anticipation of even higher land prices in the area or to sell the farm now and invest the cash in some other income-generating alternative. Once again, the decision involves a marginal analysis. Will the land price gains outweigh the expected net benefits of relocating the dairy and adopting a more efficient bottling technology? The strict economics of the decision rest on these calculations, which are less a matter of vertical integration and more a matter of investment analysis.

This divergence of project economics (hold land or relocate the bottling plant now) from industry economics (where the economic rents are) is an intriguing one, common to many vertical integration situations. The industry economics are likely to govern the long-run earnings of a company, but individual decisions in the short run are likely to be made on narrower grounds. In this instance, Rich Clover Dairy is likely to be better off without its own guernsey herd in the long run, but land speculation in the short run may well advise keeping the farm at least a few more years. What managers need to be vigilant about is that the sequence of short-run decisions made is consistent with the long-run economics of the industry. We shall return to this theme in further discussions about vertical integration.

Another purely operational concern with implications for vertical integration centers on the balance of capacities between stages of the production process. Over the years at the dairy, the herd of guernsey cows had been insufficient to supply all the Golden Guernsey milk demanded. If recent trends persist, however, the herd will soon provide more than enough milk to meet the declining demand for Golden Guernsey milk. The farm's excess milk would then have to be mixed with other, lower profit margin milk, or else production would have to be cut back. Thus, just because the product mix demanded of the dairy has changed, the attractiveness of vertical integration, already suspect, is eroded further. The capacities of the farm and the bottling plant simply do not match as well as they used to, and the dairy risks forgoing contribution because of this mismatch. It was mentioned earlier that the Rich Clover farm had, at least historically, aided the company's marketing efforts and its quality image. This remains an argument for maintaining the farm. There may be other ways, of course, of promoting a quality image. For instance, assuming the dairy is relocated, the site and advertising of a new bottling plant with all the latest, fastest, and cleanest equipment could well make up for whatever tarnishing occurs to the Rich Clover image as a result of selling the farm.

THE ADVANTAGES OF VERTICAL INTEGRATION

The arguments brought to bear against vertical integration at Rich Clover Dairy notwithstanding, vertical integration can be very attractive. For one thing, integration either backward or forward may lead to the location of the profit-enhancing economic rents in the industry.

Let us pursue our oil industry example further and delve into the history of the industry in the United States.[2] The oil industry did not begin fully integrated but rather integrated backward and forward in piecemeal fashion. In fact, the Standard Oil Company, and later the Standard Oil trust, did not seek vertical integration until it was forced to.

After the discovery of oil in western Pennsylvania in 1859 came a decade of remarkable advance in refining technology that led to low-cost, high-speed continuus flow refineries. John D. Rockefeller's Standard Oil Company in Cleveland came to the fore in the early 1870s because it was the largest and fastest refinery. At that time, the industry was being racked by excess capacity and price cutting. Industry associations sprang up but were essentially toothless in enforcing prices on their members.

In trying to add discipline to the loose cartel of refiners, Rockefeller struck a bargain with the railroads that gave Standard Oil and its associates significant price breaks in exchange for the steady use of large numbers of rail tank cars. This arrangement was beneficial to both Standard Oil (lower transport costs) and the railroads (better scheduling of railcars). The lower rail rates were enough to discourage future investments in refineries and to keep existing refineries in the cartel. Rockefeller had succeeded in discovering a way to protect the economic rents he and others in the Standard Oil group enjoyed—namely, by ensuring that refining capacity remained scarce.

Some oil producers soon found a way around Rockefeller's railroad arrangement, however. Using the innovation of the long-distance oil pipeline, some crude oil producers (Standard Oil did not own any crude oil production) in western Pennsylvania were able in the late 1870s to bypass the railroads to bring crude oil to the east coast for refining, and at a lower cost. These producers then integrated even further forward by building their own refineries because pipelines were even better than railroads at maintaining an even flow of crude into continuously operating refineries. It did not

take long for Rockefeller and Standard Oil to realize that the scarcity (and the economic rents) in the industry had shifted. The scarcity of refining capacity was no longer protected. What was scarce was low-cost transportation of crude oil from oil field to refinery. Standard Oil moved quickly to build its own pipelines and then to relocate its refinery capacity in the markets that could now easily be served by pipeline. By the early 1880s, low-cost crude oil transportation was no longer scarce. Standard Oil's economic rents from refining were once again secure.

In a few years, these economic rents were again threatened, this time by the oil wholesalers to whom Standard Oil sold its product. Once small and widely dispersed, the wholesalers began to build large storage facilities, selling large quantities over wider areas for lower unit costs. The larger scale of individual wholesalers gave them increased market power and began shifting the scarcity away from refining capacity to storage and marketing. Again, Standard Oil reacted swiftly to this threat and protected its economic rents by creating its own national sales organization and buying out the wholesalers.

This action brought protection for only a few years until it was realized, with the decline in production in the western Pennsylvania fields, that crude oil production was itself the source of possible scarcities and economic rents. Standard Oil then moved to purchase oil-producing properties to ensure crude oil supplies and any economic rents to be gathered due to periodic constrictions in supply. By the early 1890s, then, Standard Oil was a completely integrated company, able to protect itself against the threat of any scarcities.

The point of this brief recapitulation of the fascinating early history of Standard Oil is to spotlight the importance of scarcity and economic rent to vertical integration decisions. So long as no scarcities existed, Standard Oil was content to remain less than fully integrated. But when scarcities arose, Standard Oil was forced to dispel them or face an erosion of its own profitability. Standard Oil's history raises another interesting point: Scarcity and economic rents can be tremendously mobile. That fact became painfully obvious in the oil industry once again with the Arab oil embargo of 1973. The Arab oil states reaped stupendous economic rents. Since then, however, there have sometimes been gluts of crude oil, with the likely result of throwing scarcity and economic rents back on refining capacity.

Indeed, Ashland Oil thought that this was the case. As the chairman of Ashland was quoted:[3] "If our refineries don't have crude oil, no industry runs in those areas and no commercial enterprises operate.... The government will not allow such a situation to exist." Ashland Oil cut back its oil exploration and production but increased its refining capacity, precisely because it felt the economic rents lay in refining. Other oil companies disagreed, of course, with this assessment and stepped up their exploration programs because they felt the economic rents lay in the production of crude oil.

In sum, a major attraction of vertical integration is its ability to control the reaping of economic rents and quasi rents. It does this by assuming all the relevant profit margins in the chain of production and distribution activities that bring a good or service into being and off to market. By controlling the key features of an entire production/distribution system, the vertically integrated company can cut its costs and at the same time control scarcities so as to help maintain prices and reduce the threat of new capacity to the industry by raising any barriers to entry.

There are some other strong motives for vertically integrating that relate to exercising control over the industry or process:

1. Providing better control over the quality of suppliers and/or end products.
2. Providing better control over the delivery performance of supplies and/or end products.
3. Possibly protecting the company against the exploitation of any proprietary technology it may have that may not be patentable.

These three motives are defensive ones for a corporation, much as Standard Oil's actions were defensive. Companies that pride themselves on quality, delivery, or technology are sometimes compelled to provide their own supplies or take their products closer to the consumer. Vertical integration is often seen as one way to maintain reliability in quality, delivery, or technology, although it may not be the only way. Suppliers, however, can frequently be managed successfully to provide consistently dependable deliveries of quality products. Much recent management attention has been devoted to this task, as was indicated in Chapter 9 of this text and will be underscored in the next chapter. Maintaining reliability probably contributed to Rich Clover Dairy's establishment of its farm, a motive borne out by the dairy's successful cultivation of a quality image.

Other motives address cost advantages that can sometimes be enjoyed because of vertical integration:

1. The transactions costs between process stages can be saved, as can middleman profits. This cost-related point is allied with the notion of economic rents, because those activities that post high middleman profits or for which transactions costs are high are also likely to yield impressive economic rents.
2. Process planning, especially when it encompasses technological change, can encompass two or more process stages, leading to advantageous product designs and process designs.

3. By linking process stages directly, materials handling can be improved.

4. By linking process stages directly, production scheduling and control can often be made smoother and better coordinated. Inventories are less likely to build up unnecessarily between process stages.

5. The company can assure more complete use of company resoures or by-products that might otherwise be difficult to sell (such as using agricultural waste as fuel in an electricity cogeneration project, where a boiler produces heat to produce process steam and drive an electric generator).

6. The extra costs incurred by a company for inducing a supplier to take on a risky endeavor (termed a "risk premium") may be reduced to the extent that the demand on an in-house supplier can be at least as steady, if not more so, than the demand on an outside supplier.

7. Information about the market may be easier to gather. Marketing may be more effective, product differentiation more successful, and the prices charged higher.

THE PERILS OF VERTICAL INTEGRATION

As attractive as vertical integration can be, there are also a host of perils associated with it. The situation at Rich Clover Dairy uncovered a couple of them:

1. Integration into some links of the production/distribution chain may be uneconomic because economic rents are not likely to reside there. For instance, the dairy's backward integration into the ownership of guernsey cows was not likely to provide the dairy with control over scarcities, and so

high rates of return on such an investment were unlikely.

2. Vertical integration may induce problems in achieving a balance of capacities among portions of the production process, particularly if the product mix is subject to considerable change. At Rich Clover Dairy, the farm was coming into greater and greater imbalance with the bottling plant since Golden Guernsey milk was passing out of favor.

There are other possible perils with vertical integration, as may surface in Situation 14-2.

Discussion of Matheson Semiconductor Corporation

The allure of high and steady sales volumes for the bubble memory is certainly a strong one. The same sort of allure attracted many semiconductor manufacturers like Texas Instruments, Rockwell International, and National Semiconductor into the hand-held calculator market in the early 1970s. Not all of those companies would admit that the experience was a good one, however. There are many perils to vertical integration, and the type of situation Matheson Semiconductor finds itself in is indicative of many of them.

The prominence of the learning curve concept in industries like electronics has fostered a chicken–egg phenomenon: innovations like the bubble memory cannot break into new markets unless their costs are lower, and their costs will not be lower unless their markets are large. The pressure, then, is to seek out as many applications as possible to get production volumes up. Vertical integration has been among the means employed, despite the fact that the scarcity almost certainly lies in the expertise and production of the innovation (the bubble memory) and not in the application.

The alternative is to manage around the vertical integration. That is, instead of integrating

SITUATION 14-2

Matheson Semiconductor Corporation

Matheson Semiconductor Corporation, located in the Silicon Valley of California, was a young but growing firm with a remarkable reputation for innovation. Rod Matheson, the company's founder, was a near-legend in the electronics industry for his ability to crack seemingly impossible engineering problems.

Recently, Matheson had done it again, solving much of the yield problem associated with the "bubble memory," a revolutionary way to store electric charges by using layers of garnet crystal. Bubble memory was an alternative way to store charges; RAM (random access memory) silicon chips or CCDs (charge-coupled devices) were other ways. Matheson Semiconductor was not the only company working on bubble memories; Texas Instruments, Intel, National Semiconductor, and IBM were among the other, larger companies that had research and development projects devoted to them. With this breakthrough, however, Matheson Semiconductor was clearly in the lead in what promised to be a very large market.

Heretofore, Matheson Semiconductor had been a supplier of semiconductor devices to the original equipment market. It did not manufacture products for the end-user.

The top tier of management at the company was concerned that Matheson Semiconductor keep its lead in bubble memories. Given the well-documented effects of the learning curve, management stressed the importance of quickly establishing sales for the product. At issue was how best to stimulate sales. A substantial number of end-users, to be sure, would order the bubble memory immediately. However, it was not at all evident that the large mainframe computer manufacturers (IBM, Unisys) or even the minicomputer makers (Digital, Data General, Hewlett-Packard) or personal computer makers (Apple, Compaq) would switch immediately to bubble memories. Some of these companies were tied up with their own R&D projects for bubble memories, CCDs, or RAM, and they might be reluctant to jettison them for Matheson Semiconductor's version. Others might want to wait out the technology a while, refraining from redesigns of their own equipment and eagerly awaiting the arrival of still newer technology that might dominate bubble memories.

It had been suggested that Matheson Semiconductor might best help its own cause by going straight to the end-user market with add-on memory units compatible with either IBM or Digital Equipment Corporation computers, or both. This step into forward integration would take some engineering work and some entirely new manufacturing expertise, but the market potential was very enticing.

Management was split on the issue. Marketing, and to a lesser extent, manufacturing, were behind the move to integrate forward. Marketing was eager to conquer new territory; manufacturing, though somewhat cautious about the new demands expected of it, was eager to see costs drop as volume rose. Engineering, on the other hand, was dead set against the proposal, largely because it relished its state-of-the-art position in semiconductor electronics and did not want to be bothered with the more "pedestrian" engineering involved with computer applications.

It looked more and more like the issue was Rod Matheson's alone to resolve.

backward, the company can try to manage its suppliers better; and instead of integrating forward, the company can try to manage its customers better. What does this mean? In the case of Matheson Semiconductor, it might mean a stepped-up campaign to locate and to convince a computer hardware manufacturer that the bubble memory would be an ideal means of adding on memory to an existing computer, perhaps enticing the manufacturer more with some of the cost savings from learning curve advances. In the case of backward integration, managing suppliers better might mean deliberately seeking out multiple suppliers to ensure both delivery and reasonable cost or it might mean selecting a single supplier and working to coordinate production planning, process design, and quality needs between the company and the supplier. Of course, a company will not be able to manage around vertical integration all the time—companies often have to settle for limited product applications—but it is an attractive alternative.

What about managing vertical integration should a company be sensitive to?

The Flexibility Issue

Vertical integration will almost surely make a company less flexible to react to changes in products, production processes, and production volumes. By layering another full step of production or distribution on the company's existing operations, it is nearly certain that the coordination of change will be more difficult and the evaluation of managerial performance less objective. It is to be expected that as vertical integration broadens, the probability increases that at least some product, process, or volume changes cannot realistically be made at all. The company locks itself in more and more.

At Matheson Semiconductor, for example, forward integration into add-on memory units may constrain the company to produce bubble memories of a certain type for a lot longer than a company so eager to stay on the technological frontier may want to. The classic example of how vertical integration, coupled with a constant drive to lower costs (the learning curve), combined to halt product flexibility is the Model T car of Henry Ford.[4]

Ford was fully integrated in 1927, with processes specifically designed to manufacture almost all of the Model T's parts. From 11 cents of capital investment per sales dollar in 1913, Ford integrated backward and forward to the point where in 1926 each dollar of sales was backed by 33 cents of capital investment, representing such vertical integration as a steel mill, a paper mill, a railroad, a glass plant, and others. Starting in the early 1920s, however, car buyers' tastes began changing perceptibly. The Model T, an open car design, was being beaten in the marketplace by heavier, more comfortable, closed body cars. Ford tried to modify the Model T to include these features being pioneered by General Motors but could not do so successfully. At last, Ford bit the bullet as its sales shrunk. In 1927, it had to close its Highland Park plant for remodeling, but the inflexibility of Ford's process at the time was so great that the remodeling lasted almost a year! From that time on, Ford's leadership in the automotive industry was relinquished to General Motors.

Vertical integration can influence process change as well. With so much more invested in the process, integrated producers are less likely to scrap a production technology and somewhat more likely to "patch up" the existing one—not so much because they are ignorant of sunk costs but because individual pieces of the process tend to come up for review and action rather than the entire process itself.

The question for a company contemplating vertical integration is whether inflexibility really matters or not. For many companies, products and demands are standard enough and stable enough that a loss of flexibility is of little concern. For other companies, however, flexibility is of central importance, and for most of these companies significant vertical integration carries with it many perils.

Effect on the "Core" Business

How does a move toward vertical integration affect the nature of the company's "core" business? Increasing the span of production a company takes upon itself raises a couple of questions about the management tasks in operations.

1. Will the move to more vertical integration tend to draw a company out of the particular niche in which it finds itself most competitive? This concern is related to the balance of capacities issue that surfaced in the Rich Clover Dairy situation. Often there exist imbalances in the capacities between stages of a production process. Since management abhors excess capacity almost as much as nature abhors a vacuum, pressure for "doing something" with the excess capacity arises. This "something" may well turn out to be incompatible with the company's core business, the reason for the vertical integration in the first place. For example, a pharmaceutical company's backward integration into chemicals may leave the company with excess chemical production capacity. The company may then seek to sell other chemicals on the side. Over time, the side business may grow, leaving the pharmaceuticals division at odds with the chemicals division for the capacity it was originally supposed to have by right. At this point, the company is no longer a pharmaceuticals company—it has been pulled out of its niche.

2. Will managing a move toward greater vertical integration place new costs or risks on the company? If so, are these costs or risks acceptable? Often, an increase in vertical integration is quite costly and it raises the level of output at which the company breaks even. With a higher breakeven volume, the company faces increased pressure for sales and greater vulnerability to a downturn in the market or to cyclical sales. The company must assess what the higher breakeven volume may be and whether it poses any real problems.

Other Perils

Besides financial cost and risk, vertical integration can place strain on a manager's capabilities and on the management organization. The coordination of an integrated operation is frequently different from and more difficult than the coordination of operations within a narrow portion of the process. Concern for production control, logistics, and inventories tends to be much more intensive. Authority must cut through all layers of the operation. This typically places managers in roles different from what they are used to and strains the prevailing organization of those managers. Frequently, a different set of managerial skills must be mastered and the organization modified before vertical integration can be made to work well. These changes can be costly.

To return to Situation 14-2, Matheson Semiconductor has good reason to fear forward integration in this instance. It is probably a company ill-prepared to handle a great deal of coordination across manufacturing, although the engineering-manufacturing coordination is probably well oiled. What is more, vertical integration is likely to weaken its position on the frontier of semiconductor electronics by drawing talent away from the most technically demanding tasks and lessening the company's flexibility.

Vertical integration raises both barriers to entry in an industry (protecting economic rents as mentioned earlier) and barriers to exit (imply-ing that a company cannot easily get out of the business if it wanted to). The expense of verti-cal integration, which can be substantial, is thus somewhat risky if the ability to exit the industry quickly is valued highly.

Tools of the Trade: A Recap of Lessons Learned

Before making some closing observations about vertical integration, it is useful to sum-marize its attractions and perils.

Attractions of Vertical Integration

Control-Related Attractions

- Vertical integration can insulate that which is scarce in the span of possible company ac-tivities, and thus what is apt to yield desir-able economic rents. This is a defense against technological or other shifts of bargaining strength that can work to the disadvantage of the nonvertically integrated company.

- Barriers to entry into the industry may go up as a result of an integration move, thus dis-couraging potential competition. (This point is closely allied to the previous one, being an "offensive" strategic move rather than a "de-fensive" one.)

- A vertically integrated company may im-prove its control over the delivery and qual-ity of suppliers.

- Vertical integration can protect against ex-ploitation of proprietary (but not patented) technology.

Cost-Related Attractions

- Closer linking of process stages can aid ma-terials handling.

- Process planning, especially with respect to technological change, can encompass both process stages, leading to advantageous prod-uct designs and process designs.

- Production scheduling and control can be made smoother, more coordinated.

- Transactions costs and middleman profits can be saved; "contracts" between process stages may be simplified.

- Supplier risk premiums can be reduced be-cause demand on in-house suppliers can be expected to be at least as steady, if not more so, than demand on outside suppliers.

- Information about the market may be easier to gather; marketing may be more effective, product differentiation more successful, and the prices charged higher.

- More complete use can be made of company resources or by-products that would other-wise be difficult to sell (e.g., cogeneration of electricity instead of selling agricultural waste).

Perils of Vertical Integration

Control-Related Perils

- An integration move can be away from scar-city and desirable economic rents into oper-ations with a lower rate of return to invested capital.

- Vertical integration can lead to an imbalance of capacity among process stages. Trying to cope with this (such as by raising the return on the capacity investment) can draw a company out of its competitive niche in a market.

- The sometimes equally effective choice of managing suppliers or customers better (de-manding required aspects of delivery, quality, or price; or communicating needs better about schedules, technology, product change, etc.) can be overlooked.

- Flexibility for change in materials, products, technology, process design, or production volumes can decrease—rendering a company less nimble, or even less willing to compete. As a result, a company can fall behind technologically.

Cost-Related Perils

- Trying to overcome any barriers to entry that existed in the industry being integrated into can be expensive.

- The breakeven volume of production may go up, leaving the company more susceptible to economic fluctuations.

- Exit barriers for the company may go up, implying that the company cannot easily get out of the business if it wanted to.

- Management talents and organization can be strained by the demands of vertical integration for coordination. The costs of overhead, planning, and logistics can rise markedly. Efforts and talents may become fragmented if coordination is not attended to.

- Performance evaluation of process stages may become less objective as process stages become more "captive" to one another.

World-Class Practice. World-class companies appear to be following two major themes for their decisions about vertical integration. On the one hand, they jealously guard the technologies that they consider to be their competitive advantages; they do not permit others to gain experience in these technologies. On the other hand, world-class companies are doing a better and better job of managing their carefully selected suppliers as if they were part of the company itself. These suppliers, typically single sources, are invited into new product development and are linked closely with the company's production planning.

Vertical Integration in the Life Cycles of Products and Processes

When we first examined vertical integration in Chapter 1 of this book, we noted that it is most prevalent in products that are standardized and produced in large volumes by processes that have rigid patterns of flow. In terms of the product-process matrix (Figure 1-2), vertical integration is characteristic of the lower right-hand corner.

In observing the behavior of industries, we should expect to find then that changes triggering both product standardization and process simplification are apt also to trigger vertical integration in the industry. The firms that are fully integrated are likely to hold a tremendous edge over their nonintegrated competitors.

Watches

A fascinating example of this tendency has recently been played out in the watch industry. Historically, the watch industry was extremely fragmented. Separate groups of companies fabricated watch movements, cases, and other components, assembled them into watches, and distributed them. The Swiss watch industry, for example, consisted of many small components fabricators and assemblers.

Enter the age of the quartz movement. This innovation at once improved time-keeping accuracy and simplified watch construction. Quartz movement watches, powered electronically, have fewer parts than mechanical watches. This fact in turn forced watchmaking away from labor-intensive assembly and permitted the introduction of automated equipment. Getting the most from the automated equipment, however, generally meant scheduling longer production runs than the watch industry was used to; this, in turn, meant standardizing designs. There was a risk, of course, that even the modest standardization that occurred would have been resisted by the market, but the advent of the quartz movement electronic watch apparently has been novelty enough. Bulova, for example, reduced the number of basic model series it manufactured from 18 to three.

The acknowledged leader in quartz watches, Seiko of Japan, was a vertically integrated company from the start. It was therefore able to exploit the quartz movement invention (originally Swiss) and develop it further under its own R&D group. Seiko's success has forced a rash of vertical integration moves within the industry that have simplified watchmaking's logistics, cycle times, and labor input. Bulova is integrated and the Swiss industry, historically so fragmented, has clustered into large watch companies.

The Swiss companies like Omega and the U.S. company Bulova still are impressive trademarks in the industry and are relatively the strongest names in jewelry and department store sales. Whether the trademark will remain so influential (that is, whether it will be the source of economic rents) is a matter of some debate. Some believe that the rents are now to be had in production.

Fiber Optics

Vertical integration's impact on the fledgling fiber optics industry is also likely to be fascinating. It is now technologically possible to manufacture telephone cables that use strands of pure glass to transmit light pulses (instead of copper wires that conduct electrical impulses). The industry's manufacturing requirements can be divided into three major segments: (1) manufacture of the optical fiber itself, (2) manufacture of telephone cable using this optical fiber, and (3) manufacture of the electronic equipment that sends and receives the light pulses sent along the cable (e.g., transmitters, multiplexers, repeaters).

This division of the industry is analogous to the division of the traditional telephone transmission business: copper producers, cable producers, and transmission equipment makers. In this traditional business, there are companies that concentrate on one portion (such as just copper producers or cable producers) and some that overlap two portions (e.g., Anaconda produces both copper and cable, AT&T Technologies produces both cable and transmission equipment). Furthermore, the traditional copper cable telephone transmission industry is so well established that even small independent telephone companies can engineer their own systems, choosing equipment and cables from a number of different suppliers.

As time goes by, this diversity of suppliers may also apply to fiber optics telephone transmission, as telephone companies of all sizes begin to master the fiber optics technology. However, some observers feel that the only strong competitors in this new industry will be those that are vertically integrated. In the United States, only AT&T Technologies is totally vertically integrated. Due to some fortuitous cross-licensing of technology between AT&T and Corning Glass Works, AT&T Technologies is able to manufacture its own optical fiber as well as cable and electronic transmission equipment. Several Japanese companies have combined, under government auspices, to accomplish a similar degree of vertical integration. There still remain, however, some companies that are not vertically integrated: Corning Glass Works makes only optical fiber, although under substantial patent protection; Siecor, a joint venture of Seimens of West Germany and Corning Glass Works, makes only cable; and an assortment of companies manufacture the electronic equipment needed.

At issue as this industry develops is whether the successful companies will be those that are vertically integrated or those that are not. Where will the economic rents lie, and what will it take to protect a company's position in this industry? There is much speculation now. Time, as always, will tell.

QUESTIONS

1. Define and discuss the relationship between vertical integration, economic rents, quasi rents, and scarcity.

2. Summarize and give examples of the main advantages of vertical integration.

3. How does the dilemma of the Matheson Semiconductor Corporation illustrate some of the dangers that can attend vertical integration? Can you think of some of your own examples?

4. Vertical integration almost always affects the nature of a company's core business. Why is this?

5. Think of a life cycle of products and processes similar to that of the watch industry. What part does vertical integration play in that life cycle?

PROBLEMS

1. What market conditions should exist for a firm to consider vertical integration?

2. What are the methods of vertical integration? Discuss the relative merits of each.

3. How does the management coordination task change with greater vertical integration?

4. Give an example of a situation where the reverse of more vertical integration would probably be the better strategy.

5. "Vertical integration will almost surely make a company less flexible to react to changes." Discuss this statement, supporting your points with examples from your reading and/or your own experience.

SITUATION FOR STUDY 14-1

DEVINE NUTS REVISITED—THE SHELLING PLANT DECISION

Two years ago, the profit margin Devine Nuts received for grading and drying peanuts was cut in half by inflation and other factors. The past year was even worse, causing Frank Coyne to wonder why he had chosen this line of work. Frank did notice, however, that the shellers to whom he sold his loads of graded and dried peanuts seemed to be doing well enough, even during the past two years. This observation sparked him to assess the possibility of building his own shelling plant. If Frank were to do this, Devine Nuts would be the sole middleman between the farmers who brought harvested peanuts to the company for grading and drying and the peanut processing companies (e.g., peanut butter manufacturers, candy companies) that purchased bagged shelled peanuts of particular sizes (jumbos, mediums, splits) from shellers.

Devine Nuts' grading and drying operations were located in three south Texas towns within about 30 miles of each other: Devine, Pearsall, and Charlotte. Pearsall was the most centrally located of the three and the logical choice for a shelling plant. Devine Nuts could easily purchase some land contiguous to its holdings in

Pearsall on which to build the new plant. Having a shelling plant in Pearsall would be particularly attractive to area farmers. Currently, peanut growers had to haul their harvests 250 miles or more north to shellers at a cost of about $26 per ton in transportation expense. (The prevailing average price of a ton of farmer's stock was $550.) Since a ton of farmer's stock consisted of only 75 percent peanut kernels, some of the transportation expense of hauling peanut shells, stones, and debris 250 miles could be removed by having a shelling plant for the local area. Given this transportation advantage, Frank Coyne was confident that the shelling plant could easily attract 10,000 to 12,000 or more tons of peanuts each year. The peanut growing season in south Texas was the longest in the nation; even so, Frank expected that 80 percent of the year's peanuts would be delivered within a 6-week span.

Spurred by the potential for a shelling plant in the area, Frank had obtained some estimates of the costs of erecting and running a new shelling plant. The total costs (land, plant, equipment) for a sheller with a rated capacity of 10 tons per hour (and an effective capacity probably higher than that) were estimated to be $1.25 million. The expected variable costs per ton of farmer's stock were as follows:

Labor	$16
Bags	$20
Utilities	$ 3
Insurance, taxes, inspection fees, lab fees, and other overhead	$16
Total	$55 per ton

Frank felt that, for the time being, a fee of about $90 per ton could be charged for the shelling. Devine Nuts, of course, could and did buy farmers' harvests and then resell them to peanut brokers or peanut processing companies. To Frank Coyne's way of thinking, however, it was imperative that Devine Nuts not speculate in peanuts. Even if it meant having to settle for a few pennies less per pound, Frank wanted to be sure that every ton Devine Nuts purchased from farmers was quickly sold to someone else.

Before he could commit to building the shelling plant, Frank had some other concerns that needed to be addressed. One was simply the operation of the plant itself; neither he nor any of his sons had managed an operation like it. Another concern related to customer deliveries. Frank knew that some peanut processors bought their tonnage in large batches but wanted delivery stretched out over several months. If such purchases were the rule, Devine Nuts would have to store shelled peanuts, a somewhat perishable commodity, for longer than they had ever stored peanuts before, with all the added costs of carrying them, insuring them, and storing them. Finally, and not least, came the matter of financing this venture. The local Pearsall bank seemed willing to lend, but only for a maximum 5-year period and at rates that were expected to be two percentage points above prime (currently running at 11 percent).

Frank wondered whether he should go on with the sheller venture or simply hope for better profit margins in the existing business. How many tons would Devine Nuts have to shell to break even every year? He needed to decide soon or risk missing the next shelling season altogether.

REFERENCE NOTES

1. For some additional commentary on vertical integration, consult:
    Caves, Richard E., and Ralph M. Bradburd, "The Empirical Determinants of Vertical Integration," *Journal of Economic Behavior and Organization* 9 (April 1988): 265.
    Harrigan, Kathryn Rudie, *Strategies for Vertical Integration*, Lexington, MA.: Lexington Books, 1983.
    Kumpe, Ted, and Piet T. Bolwijn, "Manufacturing: The New Case for Vertical Integration," *Harvard Business Review* 66, no. 2 (March–April 1988): 75–81.
    Langlois, Richard N., and Paul L. Robertson, "Explaining Vertical Integration: Lessons from the American Automobile Industry," *The Journal of Economic History*, 49 (June 1989): 361ff.
    Lieberman, Marvin B., *Determinants of Vertical Integration: An Empirical Test*, Stanford, CA: Hoover Institution, Stanford University, 1990.
    Porter, Michael E., *Competitive Stategy: Techniques for Analyzing Industries and Competitors*, New York: Free Press, 1980 (Chapter 14).
    Rodgers, Robert C. "Saturn: A Vertically Integrated Automotive Complex," *Automation* 37, no. 7 (July 1990): 50–52.
    Walker, Gordon, "Strategic Sourcing, Vertical Integration, and Transaction Costs," *Interfaces* 18, no. 12 (May–June 1988): 62–73.

2. For the history of the U.S. oil industry, see Alfred Chandler, Jr., *The Visible Hand: The Managerial Revolution in American Business*, Cambridge, MA.: Harvard University Press, 1977.

3. *Business Week* (September 4, 1978): 99.

4. See William J. Abernathy and Kenneth Wayne, "Limits of the Learning Curve," *Harvard Business Review* 52, no. 5 (September–October 1974): 109–19.

CHAPTER 15

OPERATIONS STRATEGY

It is time to draw together the topics we have discussed so far into a unified view of operations and how operations can be used as an effective competitive weapon for the corporation, one that can help to give it "world class" status. This chapter contains two parts. The first deals with manufacturing strategy and the second with service operations strategy. If a company is to achieve world-class performance and status, it will, of necessity, have to think deeply and clearly about its operations. This process of thinking is what operations strategy is about. Without such thinking, the company risks inconsistent results and a future that is full of struggle.

PART ONE

MANUFACTURING STRATEGY

OVERVIEW

Manufacturing strategy is best viewed as an ongoing process rather than as a document such as a 5-year plan. It is a process that feeds off the strategic planning of the company for each of its business units. It cannot exist in a vacuum. Moreover, given the importance of the customer and what the customer is willing to buy, it is marketing, and not manufacturing, that typically drives the strategy process. Manufac-

turing's influence, while critical, plays a largely supporting role.

Figure 15-1 portrays the strategy process. As shown there, the process begins with the development of a business unit strategy that is then "translated" into a statement of manufacturing tasks. The statement of manufacturing tasks reflects the company's beliefs about what manufacturing has to do well for the business unit to succeed, and, sometimes implicitly, what it need not have to do as well. Once the

504

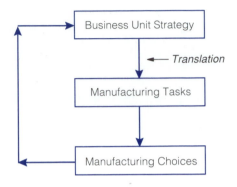

FIGURE 15-1   The manufacturing strategy process in perspective.

manufacturing tasks are set, a range of manufacturing choices must be made that will implement the decisions made about the tasks. If those choices are good ones, strengths will be fed back to the business unit strategy. If they are poor ones, weaknesses will be fed back, and the strategy will likely have to be redrawn.

Good manufacturing strategy is undergirded by two pillars or principles: consistency and continual improvement. Effective manufacturing choices are those that are consistent with one another and consistent with the set of manufacturing tasks set. The tasks, themselves, should be consistent with the adopted business unit strategy. Moreover, the sets of manufacturing tasks and choices are always evolving and changing. The effective manufacturer must be continually improving its capabilities so that it can do more and more. The manufacturing strategy must recognize this need for learning.

MANUFACTURING TASKS AND CHOICES: CONSISTENCY, THE FIRST PILLAR

As should be clear by now, manufacturing managers have a wealth of choices in the means by which they can influence the manu-

facture of goods. Chapter 1 presented an entire list of them, which, by this point in this text, should be reasonably well understood. Table 15-1 reviews the matter by providing the major headings for the list in Chapter 1.

These choices—and all operations must make them either explicitly or implicitly—define to a great degree what the corporation's production process looks like and how it operates. Furthermore, these choices are a significant explanation of how well an operation is running or could be expected to run. This is an important point; let us consider it in greater depth.

In the course of this text we have discovered that for each broadly defined type of process (job shop, batch flow, line flow, continuous flow), certain characteristics of the process hang together and define what that process is. For example, one would expect a job shop to be characterized by such operations choices as general-purpose equipment, broad work content of the jobs workers perform, production to order, and lots of information flowing within the shop. If these features were not all present, one would have good reason for suspecting that the particular process being investigated was not a pure job shop. Similarly, one would expect a continuous flow operation to be characterized by operations choices such as special-purpose equipment and capital intensity, production to stock rather than to order, machine pacing of production, and at least some dependence on the forecasting of demand. Deviations from this list would also raise questions about the purity of a continuous flow operation.

We can proceed a step further. Not only do certain broad characteristics of any process hang together, but they *should* hang together. The operations choices outlined above should be carefully matched up to be consistent with one another. In most instances, deviations

TABLE 15-1   Operations Choices

Operations choices can be segregated into three broad categories: (1) technology and facilities, (2) operating policies, and (3) operations organization. Let us review these categories and choices in turn.

Technology and Facilities

These choices frequently involve large capital expenditures and long periods of time. These are the big decisions that do much to define the type of process employed.

1. Nature of the process flow
2. Vertical integration
3. Types of equipment
4. Degree of capital or labor intensity
5. Attitude toward the process technology
6. Attitude toward capacity utilization
7. Plant size
8. Plant charters and locations

Operating Policies

Once the process technology and facilities have been selected, management must still decide on a host of features concerning how the process technology is used. Three broad segments of such operating policies present themselves: loading the factory, controlling the movement of goods through it, and distribution.

Loading the Factory

1. Forecasting
2. Purchasing
3. Supply logistics
4 Raw materials inventory system
5. Production planning

Controlling Movement Through the Factory

1. Production scheduling and inventory control
2. Pacing of production
3. Production control
4. Quality control
5. Workforce policies

Distribution

1. Distribution
2. Logistics

Operations Organization

1. Operations control, measurement, and evaluation
2. Talent

from the model and consistently defined process should be changed because they run a serious risk of acting at cross purposes to the other elements of the process. For example, a "job shop" whose workers are limited by their training or by management edict to perform only a few, very specific tasks risks losing a great deal of flexibility and scheduling ease compared to a job shop with broad work content to the job. Such a process choice risks establishing workers as well as machines as regular bottlenecks to shop capacity. Clearly, in this case, worker tasks ought to be expanded.

To cite another example, a line flow operation that permits many or significant engineering changes on a regular basis risks losing the benefits of production speed, lower labor value in the product, and low work-in-process inventories. The products of a line flow process should be fairly standard; if they cannot be, the other elements chosen for the process ought to be seriously examined and the process altered to make it less line flow and perhaps more batch flow.

Consistency in the choice of process characteristics is the first principle of what has come to be known as manufacturing strategy. Let us explore the concept in more depth.

To say that certain broad characteristics of any process should hang together as a consistent whole is not to say that every operations choice mentioned above can be unambiguously assigned to a specific process. Picking the process does not at the same time mean that all of the operations choices we discussed above are automatically determined. A company typically can exercise a great deal of latitude in selecting the specific elements of its production process. The job shop and the continuous flow process, which lie at either end of the process spectrum, are probably more constrained to specific choices of process elements than are batch or line flow operations. In other words, the job shop and continuous flow processes must, in general, be consistent over a greater range of process element choices than either a batch or line flow process. Nevertheless, considerable choice is possible in defining almost any process. Recall also from Chapter 1 that different challenges to management are inherent in different types of production processes and in the process elements that make them up. Here, too, the job shop and continuous flow processes are more clearcut in what they demand of management. For the job shop, scheduling workers and machines, bidding for new jobs, handling materials, and maintaining the flexibility to manufacture a great variety of goods are paramount to management. For a continuous flow process, management's challenges are completely different.

What is demanded is care with the planning of capacity and new process technology and with the management of materials from suppliers to the plant and from the plant to the customers. The challenges to management for batch or line flow operations are less well delineated, since both must be concerned with the balance of capacities within the process, product flexibility, worker motivation and training, and product design. There are shades of stress on these challenges, but the distinctions between them are more blurred and less clear-cut.

Selecting Consistent Elements of a Production Process

Selecting the component elements for a manufacturing or service operations company is not a trivial task. Situation 15-1 offers some practice in describing how an operation ought to be organized.

Discussion of Porterfield Glass Company

The preceding section on operations choices provides a framework for selecting a production process suitable for Porterfield Glass Company's Glass Container Division. Let us then follow the choices outlined in that section.

Technology and Facilities

Although the glass containers made by the division are most likely specialized (shape, quality glass, color, printing) for the individual customer, each bottle goes through identical steps in the process. Different glass recipes or bottle molds may be used, but the process itself does not vary by customer order in any significant way. The pattern of the process flow can be rather rigid then. As suggested in the situation itself, the equipment is special purpose. Combined with the rather rigid process flow, this feature suggests a considerable degree of capital intensity to the process.

SITUATION 15-1

Porterfield Glass Company, Glass Container Division

The Porterfield Glass Company's Glass Container Division was almost exclusively a manufacturer of bottles for beer, soft drinks, and other beverages. The technology of glass bottle making was well known and not subject to radical advances, although bottle-making machines were continually becoming faster and able to produce larger and more intricate bottles. Basically, all that was required to fabricate a bottle was to (1) melt down sand, lime, and soda ash, usually with natural gas, (2) purify the resulting molten glass mixture, (3) suck up a portion of the molten glass into the molds of a rotating bottle forming machine, (4) treat the molded bottle in a tunnel-like annealing oven, and (5) do any required printing on the bottle. All the equipment was specific to bottle making.

There were substantial cost advantages with increased size because of energy saving in melting and because larger, more efficient bottle-molding machines could be dedicated to specific bottles without wasting any time with mold resetting. Different colors or glass types had to be manufactured in separate batches.

The company's major customers (breweries and soft-drink bottlers) were scattered across the country but were concentrated in the major metropolitan areas. These bottlers did not like to inventory bottles, primarily because they were bulky, and so demanded at least daily shipments—even shipments three or four times a day to feed their typically multishift bottling operations. For such service these bottlers were willing to enter into long-term contracts and to provide lead time of a month on all orders.

The division could well be vertically integrated to a great degree, with the company owning and controlling its own sand pits and limestone quarries, particularly if demand is large. On the other hand, the raw materials that go into making glass are so common, so readily available, and used in so many other ways that it is hard to believe that backward integration into sand, lime, or soda ash supply would be very attractive. The rate of return to the company's capital is almost certainly as high or higher in bottle-making as in the supply of raw materials for glassmaking. The delivery of such commonly available materials is surely reliable enough to remove delivery as an argument for the division's maintaining control over its own raw materials.

The fact that supplies are so readily available and easier to ship than bottles themselves is an argument as well for locating close to customers and not suppliers. Because customers demand a steady stream of bottles into their plants, it is clearly advantageous for the company to locate its plants close to its largest customers, if that can be arranged. Plant size should be determined in large measure by customer needs. The plant should be large enough to incur some of the cost advantages there are in energy savings and mold resetting, but after some threshold level, additional cost advantages from increased size are not likely to be large. As it is, the plant should be constructed purposely with capacity in excess of long-term customer contracts. Otherwise, the division risks poor delivery performance during a customer's demand peak, which may endanger the entire contract. The division should track customer markets fairly closely so as to anticipate needed capacity expansions. Technological changes are likely to be more incremental than radical in nature, as suggested in Situation 15-1 — apart, of course, from a possible redefinition of the business to encompass plastic bottles, too. This implies that the division could successfully follow, rather than lead, process innovation in glass bottle manufacture.

Operating Policies

The loading of the factory is likely to be dominated by the continual output demands of the long-term customer contracts. Raw materials supplies, which are bulky, are probably best shipped by rail or barge on at least a weekly basis, and perhaps more often. The inventory of raw materials is apt then to be replenished on a regular basis, the quantity of each replenishment being enough to bring the desired buffer stock up to some desired level (in a periodic reorder system). The buffer inventory itself is likely to be large because (1) stockouts of materials could jeopardize any contracts and (2) the raw materials are of low value anyway. Heavy investment in forecasting is probably not required; the long-term customer contracts are great aids in determining plant material needs with enough lead time so that deliveries are not jeopardized.

Production should be to customer order and a "chase" type strategy should be followed in planning production over the short term. With long-term customer orders and customer willingness to provide sufficient lead time, production planning of workforce size, hours, and schedule should be fairly straightforward. The schedule, in practice, should not vary much from month to month. There are likely to be advantages in batching orders that use the same glass, though in different shapes or sizes. Changing products — glass or molds — should be a routine matter requiring a very standard directive from management. The quicker and more routine the product changeovers, of course, the more profitable the entire process. Except for changeover periods, the pace of production is determined solely by machine speeds. Information is largely one-way, from management to the workforce, except for sounding the alarm when a machine breaks down.

The workforce is likely to divide into two skill groups. Highly skilled, highly paid labor is likely to be required for troubleshooting equipment repair, for maintaining the equipment, and perhaps for controlling the glass making itself so that it conforms to the "recipe." Other tasks are likely to be unavoidably narrower in content and therefore suitable for lower skilled, lower wage labor. With the equipment largely pacing production, incentive pay schemes are inappropriate. Rather, all workers should be paid by the hour. Opportunities for either cross-training or worker advancement outside the rather narrow and well-defined departments making up the process are relatively scant. Job advancement is likely to be mostly shift changes or relief work. Unionization is likely to make little impact on operations.

Quality of the product is largely a matter of the proper functioning of the equipment. Quality control is best accomplished by sampling output and using control charts to identify problems or potential problems in the process.

Manufacturing Organization

Within a division like the Glass Container Division, authority is best left centralized. Plants are not autonomous, but are much like one another. The central staff should control the larger decisions on technology and facilities, leaving the plant responsible for dealing with day-to-day operating problems, personnel, and equipment maintenance. The plant is best evaluated as a cost center and its measures should be overwhelmingly physical, as opposed to financial.

Within the plant, management talent in the plant engineering department is critical. At the division level, management talent should be directed to the choice and development of equipment and process technology, because the division's future rests largely on these.

Summary

Glass container manufacture, as should be clear by now, is largely a continuous flow process and it shares a host of traits with other continuous flow or near-continuous flow hybrid processes. In fact, the bottle-making process is very similar to the bottling process itself, as the description of the Stroh brewery operation in Tour E at the back of this text makes evident. There simply is not a lot of room in such processes for much deviation in the process characteristics and challenges.

The Impact of Competitive Priorities on Consistent Choice in Operations: Manufacturing Tasks

Too many managers falsely believe that manufacturing's goal should be low-cost production in every instance. While being the low-cost producer is often a very advantageous position, there are many other ways to compete successfully in most industries. Products can be differentiated from one another too readily and markets have too many niches in which companies can position their products. In one sense, this marketing diversity takes the heat off operations people to cut costs continually; but in another sense, it broadens the role of operations within the company, because operations must now react to different sets of competitive demands.

What is more, the competitive demands that can be placed on manufacturing are diverse and numerous. These eight competitive demands, sometimes termed manufacturing tasks, can be organized into three distinctly different groups. These groups are (1) product-related concerns, (2) delivery-related concerns, and (3) cost concerns.

Product-Related Concerns

1. *Product performance.* Whether the product's design or engineering permits it to do more or better than comparable products.
2. *Product reliability and workmanship.* Apart from differences in product design, whether the quality of materials and workmanship enhances the product's value and increases its durability and reliability.
3. *Product customization.* How adaptable the operation is to meeting special customer specifications.
4. *New product introduction.* How readily the operation can bring out product variations or completely new products.

Delivery-Related Concerns

1. *Speed of delivery.* The time between order taking and customer delivery.
2. *Delivery reliability.* Apart from the speed of delivery, how close actual delivery is to any quoted or anticipated delivery dates.
3. *Volume flexibility.* How readily the operation can switch production rates on some or all of its products.

Cost Concern

• *Cost to produce.* The traditional burden on manufacturing to become the low-cost producer, generally related to higher volumes and less customization.

Which of these competitive demands takes priority at a company depends on several forces: the economics of the industry, particular competitive pressures, government mandates and incentives, the company's own resources, and the company's culture and attitudes. What is important to recognize is that operations can be subject to different, and changing, competitive demands. Determining priorities for competitive demands frequently is challenging. Marketing managers typically want the company to compete along all aspects. They can recognize competitors that cater to each "manufacturing task," although no one competitor competes across all dimensions. Determining the competitive demands the firm will concentrate on (essentially, translating the business unit strategy into clear manufacturing tasks) often is very difficult.

Manufacturing has more to offer the company than what some business strategists thought, not so long ago, were the strategy elements of manufacturing: economies of scale and the learning curve. For these strategists, manufacturing benefits a company when it can increase its scale and lower its cost, and when through experience, manufacturing can move down the learning curve ahead of other companies. What is significant here is that the notions of economies of scale and learning curve are both cost related. What the strategists neglect are the many other ways by which manufacturing can be strategically important to the company. Good manufacturing strategy can highlight the importance of these non-cost-related factors.

Translation: The Toughest Thing

From my experience, the toughest aspect of the manufacturing strategy process is making the translation from business unit strategy to a set of manufacturing tasks. The translation has to be in plain language so that manufacturing can discern what its true priorities are, and also

understand how it is to be measured against those priorities. The sad fact of the matter is that the numerous manufacturing choices listed early in this chapter will be made by the company by one means or another. If both manufacturing and general management do not become involved in setting tasks and in making these choices, they will be made by people lower in the organization, including workers and foremen. If companies do not engage in a dialogue about these priorities, they risk inconsistencies all across the board. For general management, the question is a very real one: Whom do you want to run the company? Do you want general management to run the company, or do you want foremen's and workmen's decisions on strategic matters to hold sway?

Not falling prey to this problem takes concerted action. What is needed is a determination from top management of what corporate strategy means for manufacturing—an evaluation of the importance of the eight "manufacturing tasks" cited earlier. Once manufacturing is clear about what is demanded of it, it can go about choosing the technology, facilities, operating policies, and organization that are at once internally consistent and consistent with the declared corporate strategy and competitive priorities. The importance of these consistencies has been argued forcefully and persuasively by Wickham Skinner.[1] For Skinner, good manufacturing strategy must start with an explicit statement of the corporation's objectives and business unit strategy. This statement must then be translated into "what it means for manufacturing," and the existing operation must be examined, element by element, in a kind of manufacturing audit of the existing facilities, technology, and operating policies. Only then can the corporation think of altering any of the elements that do not mesh with the explicit statements of corporate strategy and its meaning for operations.

The goal is to have all of the operation—right down to the first-line supervisors and all of the workforce—pulling in the same direction, the direction implied by the proclaimed corporate strategy. Manufacturing should be well coordinated with marketing, finance, personnel, engineering, R&D, and other departments. Naturally, this takes a substantial amount of communication and discussion among functions and between levels of the organization.

CONTINUAL IMPROVEMENT: THE SECOND PILLAR

The Limits and Promise of Operations: The Dilemma

In almost any industry you could name, there are some companies that compete primarily on one set of manufacturing tasks while other companies compete on entirely different ones. Scattered among the low cost producers are others that emphasize product performance features (Porsche cars, Hewlett-Packard pocket calculators) or product workmanship (Rolls-Royce cars, Ethan Allen furniture) or any of the other characteristics mentioned. What is both interesting to recognize and crucial for the development of strategy, however, is that no one product and no single production operation truly competes across all competitive dimensions with equal vigor. A company typically chooses which competitive dimensions it wants to rank high and which must necessarily rank lower. It competes in a niche. Even the low-cost producer competes in a niche.

These competitive niches aside, it is becoming increasingly evident that to become "world class," manufacturing must do many things well: produce output of consistently high quality, on time, at low cost, and often with a measure of flexibility—diverse models, fluctuating volumes, and new product introductions. The important word here is "and," because with world-class competitors we no longer have the option of trading these attributes off against one another. We have learned, painfully, that quality and cost are not substitutes for one another but are, instead, complements (the discussion of JIT in Chapter 11 reinforces this point). This has driven us to consider quality, cost, delivery, and flexibility all as complements. This may not be so in every instance, but it is clearly the trend.

This thrusts us into a dilemma. On the one hand, we acknowledge that companies typically do not stress all of the competitive priorities (manufacturing tasks) at the same time, but on the other, we note that world-class companies are seeing cost, quality, delivery, and flexibility all as complements. Are these two points of view at all compatible?

Figure 15-2 provides a possible reconciliation. The figure graphs a "production possibilities frontier" that is similar to, but not exactly the same as the production possibilities frontiers that one finds in some economics textbooks (i.e., where one trades off the production of "guns" versus "butter"). The curves in Figure 15-2 (such as Curve A) trade off one manufacturing task (say, product mix flexibility) versus another task (say, cost). In the short run, lowering costs may involve decreasing the process's ability to offer greater product mix flexibility. However, in the longer run, through the implementation of technology, better management, improved worker training, and so on, one could expect the curve to shift outward (Curve B) so that costs can be lowered and flexibility enhanced at the same time.

This is what world-class companies do. They push the production possibilities frontiers outward so that over time they accomplish lower costs, more timely delivery, improved flexibility, and better quality, all at the same time. The name of the game is continual improvement or

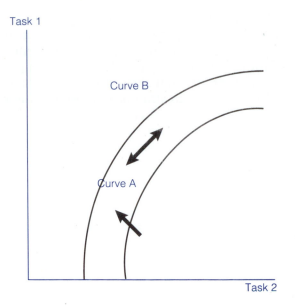

FIGURE 15-2   A production possibilities frontier for manufacturing strategy.

learning. World-class companies do a better job at it than others, and therein lies their distinctive and strategic advantages over their competitors. It has been advanced by Hall and Nakane, knowledgeable observers both, that there is not a trade-off of competitive priorities (i.e., quality, cost, delivery, and flexibility), but rather a progression in these "capabilities," namely first quality, then delivery, cost, and finally flexibility.[2] Thus, quality is the base upon which other manufacturing capabilities are built. As people and the organization learn, these capabilities are enlarged and broadened until truly world-class performance is achieved.

The Contrast: Seeing Operations as a Trade-off vs. Seeing It as a Learning Experience

The operating manager who views the operation as a trade-off, as something that constantly demands that he choose one priority over another is likely not to perform as well over time as the manager who views the operation as one big, continuing experiment to learn from and tinker with.

Consider the following scenario. The manufacturing operation is humming along nicely until the marketing department starts reciting so-called "customer complaints" about product delivery or quality. The manufacturing manager feels compelled to react quickly to this pressure, and he or she makes certain adjustments to the process without really examining the longer-term implications of the short-run actions. After a time, the finance department confronts manufacturing with the "corporation's" desire to reduce inventory carrying costs or capital expenditures. In an effort to appease the finance people, manufacturing shaves inventory or tables some spending plans. Step by step, as manufacturing gets tugged and pulled by such forces, whatever consistency existed among the numerous operations choices discussed above starts to unravel. Not only does the operation suffer from lack of consistency in its choices, but the time of the manager is spent fighting fires and not spent improving the process. For this manager, the process exists to be manipulated and not necessarily to be studied and improved. The manager spends time trying to earn a "hero" button and not trying to make long-term life easier.

The more effective operating manager works hard to stabilize the process, to study its capabilities, to improve them and the quality those capabilities can produce, to study and to experiment with the process. This often means resisting some of the pressures to fight the constant fires that seem to seek out line managers on the factory floor. This is naturally harder to do in practice than to state ahead of time, but there are many managers who have altered the way they approach "life with the process."

What is thus strategically important for companies is to define ways by which managers can enter a culture of continual improvement. A number of these ways have proved successful in helping a manufacturing organization learn more and act more effectively.

Tools for Continual Improvement

If continual improvement is to be the rule for the company, certain foundations must be in place. Many of these have been discussed already, particularly in the presentation of the just-in-time manufacturing philosophy in Chapter 11, and in the presentation of quality and quality improvement processes in Chapter 4. It is helpful to review them briefly here.

Quality Improvement Programs.   Improvement often involves quality and/or the steps one should take to improve quality. Steps include the analysis of problem situations, the recommendation of countermeasures, the testing and experimentation with them, and then the standardization of any changes made. The elements of good quality management, as outlined in Chapter 4, are an essential feature of any effort to improve continually.

Cycle Time Reduction and the Identification of Waste.   While the rigor of a quality improvement program is essential for eliminating variation from the process, a quality improvement program, by itself, does not guarantee that all of the waste in a production system can be identified. Even more important, adherence solely to a quality improvement program jeopardizes the introduction of radical ideas for totally altering the design of the process. A quality improvement program can help with incremental improvements (and that is a significant achievement), but it may ignore more substantial rethinking of the process. For these reasons, other tools are required.

As was discussed in Chapter 11, the use of throughput time (cycle time) can be a tremendously useful tool. It is, perhaps, the most useful tool a company can deploy.[3] When process and information flows are clearly identified (by the application of process and information flow charting) and the times for each elements of them are tracked, waste can be identified. Waste is found where delays occur. By asking why those delays occur and thinking about what can be done to keep the product (or the information) in motion, the process can be altered, and in dramatic fashion. Moreover, this is a tool not only for the production process itself, but also for the white-collar, information-intensive, overhead areas of the company. Companies need to look at their total cycle times, including order entry, complaint resolution, new product introduction, customer delivery, customer service, as well as the manufacturing process itself.

The use of time can lead to more radical change. When companies realize how much time it takes to react to customer wishes relative to the value-added time (the actual time it takes to make the product or to handle the information, without any delays) they are often stunned. The ratio of total cycle time to value-added time may well be 50 or 100 to 1. It can be only 2 or 3 to 1. Such a dramatic drop in that ratio, however, frequently takes a radical restructuring of the process that not only eliminates the wasted time, but also eliminates the activities of the process that do not add value at all and which can be replaced by other means, or totally dispensed with. Such activities might include management reviews or inspections, or having suppliers add the value rather than wasting time to transport the material elsewhere. If these non-value added activities are eliminated, the entire process flow diagram gets redrawn along with dramatically altered layouts, the mixing of more activities, and the inclusion of more, different kinds of people (management, engineering, support staff, direct labor) right there with the product. Time, more than any other metric, has this ability to foster radical thinking in how the product is made or how information flows.

People Development. To accomplish any of this takes people, people who understand what is at stake and who "buy in" to the tools. People development takes commitment to training at all ranks; constant dialogue among managers and workers; the development of teams, often involving multiple functions within the company; sharing of information and data that, in many companies, has too long been thought to be management prerogative; and truly participative management on the factory floor.

Benchmarking. Increasingly, companies have been seeing the benefit of benchmarking, systematic investigation of the capabilities of other firms, not necessarily in the same industry for the purpose of devising targets for improvement. Xerox has been justly cited for its work in the early 1980s in benchmarking its position and then taking steps to bring its performance in line with the benchmarks established.

Thus, a company that is "good" at manufacturing strategy has several characteristics:

1. Its manufacturing tasks are well stated and understood, and its manufacturing choices are consistent among themselves and consistent with the manufacturing tasks, and the business strategy that spawned them.

2. More than that, the company engages in continual improvement, so that it can push out the "production possibilities frontier" that it faces. It typically uses a variety of tools to foster continual improvement:

a. Process and information flow diagrams to nail down the particulars of materials flows and the information surrounding the process. Bottlenecks can often readily be identified in this way, as well.

b. Tracking of throughput times and other times in the process (e.g., order entry to delivery, leadtimes to change the product mix or the specifications of the product) so that the places where delays occur can be noted and studied, for there lie the wastes in the process, and the seeds for radical restructuring of the process and information flows themselves.

c. Use of a formal and systematic quality improvement process to study the root causes of and to develop countermeasures for the identified problem areas of the process. Those problem areas are those where quality problems have cropped up or where waste has been found. The techniques of the quality improvement process include, among others, data gathering, Pareto analysis, brainstorming of causes, experimentation, data analysis to test hypotheses about root causes, and standardization of potential remedies.

d. Benchmarking, not only of themselves against their past, but also of other companies and plants to keep company sights set high.

Simplifying the Learning I: Factory Focus

There are other tools that companies have applied to help them improve in continual fashion. One tool that has been advocated since its debut in 1974 in an article by Wickham Skinner is that of factory focus.[4] In the light of our twin pillars for manufacturing strategy, the concept of factory focus is best seen as a means of simplifying the life of the factory so that competence and learning are enhanced.

Factory focus is typically defined as the isolation of diverse products or processes, products or processes that have distinctly different manufacturing tasks (in the sense defined above), in their own plants or "plants-within-the-plant." Focus is designed to avoid situations where manufacturers have geared up to do well what the corporation does not need done and then do poorly on the task on which the corporation's success really depends. Or, commonly, too many conflicting tasks are asked of the same operation, caused, for example, by product proliferation or by a blind acceptance of "professionalism" in functions like engineering, inventory control, finance, or quality control. The result too often is that no tasks are accomplished very well. Factory focus tries to remove the conflicts by simplifying the tasks and thus spurring the learning. Consider Situation 15-2.

Discussion of Matthew Yachts

This situation with Matthew Yachts is a common example of product proliferation that has begun to sabotage an otherwise smoothly operating production process. How does the apparently innocuous addition of one more type of yacht to a boatyard that makes all kinds of yachts anyway sabotage operations?

To understand this situation, let us trace through some plausible consequences of adding the fixed design yacht to the yard. First of all, we should recognize that the custom work and the fixed design work were very different from one another. The competitive priorities for the two diverge greatly. For the custom work, competitive demands such as product customization, product performance, and product reliability and workmanship are critical; cost and delivery are much less important. For the fixed design craft, on the other hand, cost and delivery carry relatively more weight, and features like product performance are not nearly so important. These differing competitive priorities place conflicting demands on the production process.

The traditional custom work requires a skilled workforce, broad job content, a tremendous amount of information transmittal through the process, and great attention to the scheduling of work, among other things. The boatyard is a job shop. Into this job shop has been thrown a product that is more standard, sold in higher volumes, and susceptible to price competition. Ideally, it should not be manufactured with a job shop type of process but with one that is more like line flow. Such a process would use more and more special-purpose equipment, narrower worker tasks balanced against one another to stabilize the pace of work through the yard, a materials inventory and handling system that removes most responsibility for worker ordering or handling of materials, a lower skilled workforce, and more explicit quality checks in the process, among other things.

When the two different kinds of yachts are manufactured in the same boatyard, costs on the fixed design yacht naturally climb over expectations. The workers are obviously treating its construction with the same care and methods that they use for the custom yachts. Deliveries too tend to be long and to grow longer, because the present process is not organized to reduce work-in-process inventories or manufacturing cycle times and the lower margins on the fixed design yachts relegate them to second priority whenever the going gets tough.

The pressing space problem at the boatyard indicates that now is a convenient time for seeking "focus" at the yard by splitting the two product lines and manufacturing them in separate facilities with separate equipment, workforces, and controls. The new, fixed design

SITUATION 15-2

Matthew Yachts, Inc.

Matthew Yachts, located in Montauk, Long Island, manufactured sailing yachts of all descriptions. The company had begun by building custom-designed yachts for a largely New York-based clientele. Custom-designed yachts still accounted for three-fifths of Matthew's unit sales and four-fifths of its dollar sales and earnings. Over the years, as Matthew Yachts' reputation for quality design and workmanship spread, sales broadened to cover all of the Eastern Seaboard.

In an effort to capitalize on this increased recognition and to secure a piece of the fastest growing market in sailing, Matthew Yachts began manufacturing a standard, fixed design craft. Matthew attacked only the high end of this market, as the boat measured 37 feet long. Nevertheless, even this end of the market was more price sensitive and less conscious of performance than Matthew Yachts' custom design customers were.

All of the company's yachts were manufactured at the Montauk plant, and shared the same equipment and skilled labor force. Custom designs were given priority in scheduling, and the new boat was rotated into the schedule only when demand slackened. As sales of the fixed design boat increased, however, scheduling the new boat on a regular basis became necessary.

Matthew Yachts were built basically from the bottom up. Fabricating hulls was the first step. Increasingly, fiberglass hulls were demanded for their speed and easy maintenance. Afterward came the below-decks woodworking, followed by the fiberglass and woodworking on the deck itself. The masts were turned and drilled separately. Masts and hull were then joined and the finish work completed.

Over the past year, as the fixed design craft continued its steady increase in sales, costs and deliveries began to slide precipitously, especially on the fixed design yachts. During this period, when push came to shove, construction of the fixed design craft always yielded time and resources to the higher profit margin custom designs. As a result, many fixed design yachts were strewn around the yard in various stages of construction. Moreover, space in the existing shipyard was becoming scarce, and a plant expansion of one sort or another appeared inevitable.

The company wondered whether it should stay in the business of building fixed design yachts and, if so, how it should continue.

---

yacht is probably best produced in a new boat-yard, possibly located a considerable distance from Montauk, where a new workforce and more specialized equipment and controls can be devoted to the task. Space problems could be relieved at the same time focus is achieved.

The separation of production lines and manufacturing facilities (perhaps in plants-within-the-plant) so that only one set of competitive demands is addressed at a plant is a tremendously compelling idea. For Matthew Yachts it means maintaining the match of a job shop process to the custom yachts it was originally designed to manufacture. It also means the match of a new

process, more like a line flow, to the standard design, higher volume yachts. In principle, the need for separation to achieve "focus" is clear, and yet there may be good reasons to resist separation, even perhaps in the case of Matthew Yachts. It is to these resistances to focused manufacturing that we now turn.

Resistance to Focused Manufacturing

Although focused manufacturing is an enormously appealing and "clean" concept, it can meet with some resistance. Focus is rooted in the many inevitable trade-offs managers must make about their facilities, technology, and operating policies. It states that there is no one "right" set of policy decisions; choice depends on how one competes. Focus manufacturing's strength lies in the recognition that simplification and concentration of one's efforts can lead to better and faster learning, and thus a more powerful production system. What is often overlooked, however, is that focused manufacturing is itself just one choice among the many that companies confront; while appealing and powerful, it is no cure-all.

Consider these cases where focused manufacturing may be resisted by managers, perhaps justly so:

- A single factory produces two distinct product lines, each serving a different type of market. One product line's sales are buoyant and growing, while the other's are shrinking, increasingly vulnerable to competitive attack. Within the factory, minimal separation seems warranted, but splitting the lines cannot be done easily by installing a plant-with-the-plant because (1) production space on site is already cramped and (2) the existing and rampant job bumping between product lines would be difficult, if not impossible, to eliminate without totally disrupting the workforce. Focused manufacturing thus recommends a completely new plant for a product line that is in trouble. Moreover, the new plant will incur some one-time startup costs and an ongoing overhead structure.

  Here lies the resistance. Do the benefits of focused manufacturing actually outweigh the known costs of new plant construction, start-up, and staffing? And for a product with a possibly shaky future?

- Two different products are manufactured in the same factory. Sales for each are seasonal and offsetting; one product's demand peak nearly matches the other's trough. Nevertheless, the competitive demands on each product are vastly different. One must be produced to rigid specifications, and so worker attention to detail must be paramount. For the other product, quality is not crucial; speed of delivery is. The workforce, inculcated with the importance of caring for product quality, cannot change its ways sufficiently to make the factory a very successful manufacturer of the second product.

  Here lies the resistance. Do the benefits of focused manufacturing outweigh the costs not only of establishing a new plant but of carrying seasonally slack capacity in both plants?

- A new plant is being started up to specialize in the high-volume products of a high-technology company. The plant's capital-intensive component supply department could be constructed specially for the products to be placed in the factory—in a sense, focused for just those products. If this were done, the department would forgo the opportunity to supply other, more sophisticated products of the company. But a more general-purpose design of the department would add 25 percent to capital expenses. Product lives in the industry are short, and nobody knows what the next set of high-volume products will be.

  Here lies the resistance. Do the benefits of designing the plant to be focused, including the capital expense savings, outweigh the technological rigidity that focus may imply for the factory's future?

Similar manufacturing situations are not uncommon, and they offer some perplexing options for managers. Evaluating the forces pushing either for or against focused manufacturing is not a clear-cut endeavor, since a variety of difficult-to-measure factors must be balanced against one another. In fact, assessing what factory focus really means in a given situation can be vexing. Nevertheless, merely understanding the pushes and pulls on focused manufacturing may help to ease manager's burdens in confronting this type of situation.

Analyzing Factory Focus

How does focus help? What should a manager look for as predictable consequences of focusing a factory? Focus can help in a variety of ways (although not in all instances):

1. *Improving the flow of materials and product within the company.* Often the first sign of confusion in a factory shows up in plant logistics—late or misdirected materials or product, incorrect order taking or order filling of either materials or product, and increasingly complex production scheduling. In such a situation, focusing by product or product group often makes sense, because by organizing in that way, plant logistics can be simplified and thus improved.

2. *Reducing throughput times in both production and support activities.* Often, by separating products and at the same time pulling together portions of the production and support processes, a focused factory can reduce its throughput times. This fact in turn implies reduced work-in-process inventory, quicker recognition of quality problems and quicker resolution of them, fewer rush orders and disruptions to ongoing operations, less expediting asked of suppliers, greater potential capacity, and perhaps, lower finished goods inventories.

3. *Increasing job specialization and product identification.* By concentrating on a limited number of products or processes, workers and managers may more easily find ways to redefine procedures and methods to smooth the flow of manufacturing even more. Especially if the workforce can identify with and take pride in the specific product manufactured, the company may benefit from more suggestions for product and cost improvements. Breakthrough thinking and more radical changes to the process may become easier to accomplish.

4. *A clearer cost accounting system.* By streamlining operations, the cost accounting for a product or process is simplified, and the simplification often means better pricing decisions and improved capacity utilization.

5. *Better reaction to production gone awry.* Factory focus means ingrained operating routines and corrective measures that can simplify and smooth the factory's reactions to nonstandard, potentially troublesome developments.

All of these are evidence for improved learning and more consistent improvement of the process and its capabilities.

In any individual case, of course, these advantages to focus may be present in greater or lesser degree. Furthermore, the ability to measure their impact varies widely. Nevertheless, managers may do well making some estimates of their strength so as not to fall victim to the syndrome of hard numbers driving out reasoned, but qualitative analysis.

As the examples introduced above make clear, there are a number of arguments against focused manufacturing. These, too, vary in the ease by which they can be quantified. A company's ability to focus its manufacturing may be constrained by:

1. *Idle machines.* In separating products that use the same machinery, a company risks running certain machines at much less than their rated capacities. Thus a lower volume of output must absorb the fixed costs of the machinery. Fortunately, the sensitivity of

total costs and earnings to this increase in fixed cost is relatively easy to calculate.

2. *The savings of placing differently cyclical or seasonal products of diverse characters within the same factory.* As with idle machinery, the costs of focusing operations through product separation, measured in terms of increased inventory building and possibly increased hiring and firing of labor, can frequently be handled quantitatively.

3. *New plant construction with increased overhead.* A more difficult constraint to analyze involves a focusing of operations that, for one reason or another, demand a totally new factory with an accompanying overhead structure rather than the designation of plants-within-plants. Such a radical change is sometimes mandated because of growth of the product lines affected and insufficient room to expand on site. Sometimes it is mandated by the need to develop an entirely different kind of workforce or to get out from under outmoded work rules and practices that could not be modified easily at the old plant. In a sense, then, new plants are mandated generally only when the penalties of maintaining production in a single facility are great.

   Even with the plant-within-plant option, overhead will probably increase because of separate supervision, production control, and materials handling. These costs too can be deterrents to focused manufacturing.

4. *"Deserting" an existing workforce.* In separating products or processes for their own plants, some layoffs of an existing labor force may be necessary, at least temporarily. This is never a happy prospect, even if improvement in workforce practices is one of the reasons behind such a move to focused manufacturing.

5. *Production rigidity.* A more subtle, but nevertheless important, possible constraint on the focusing of manufacturing is the threat of rigidity it poses for production. By segmenting the operations for each of a string of products, for example, one runs the risk of losing the ability to react quickly to some new product or new process innovations. The new independence of each product may lock the company into a much less fluid structure than what it was used to. It makes sense then for companies that experience a large share of product or process innovation to be leery of a strict segmentation of products or portions of the production process. In fact, focused manufacturing would itself argue for a concentration on the creative, technological function in these kinds of companies anyway, because that is how many such companies compete. This is an important point. Focused manufacturing, and good manufacturing strategy, do not argue necessarily for a segregation of products. It may make more sense to segregate portions of the process or particular skills within the company. The guiding principle is always consistency with the acknowledged manufacturing tasks.

A common threat that emerges from these possible constraints on focused manufacturing is that change—reflected in sales growth, decline, or uncertainty—is an enemy of factory focus. When products with different competitive demands are continually introduced, when the mix of products sold frequently shifts, or when seasonality or cyclicality abounds, then the costs of complying with factory focus become high. Thus, the firm that is growing in sales and product offerings and is in constant need of cash is often the one that confronts the balancing of factory focus against other aspects. Since the costs of focused manufacturing may outweigh its benefits, managers must be careful to resist the enticement of hard number costs overshadowing a more nebulous evaluation of benefits. Further, they must be vigilant in continually testing whether focused manufacturing can fit into their situations; for, as easily as focused factories can lose their focus with change, nonfocused factories can acquire the need for it. Indeed, while change makes focus more difficult to accomplish, it is precisely through focus that a manufacturing organization can help itself learn more. In our judicial system, we have found that it works best if one is presumed innocent until proven otherwise. In manufacturing, it can be argued, it is better to seek focus until proven otherwise.

Technological Advance and the Focused Factory

Some observers of the recent change in manufacturing technology discussed in Chapter 13 have remarked that technological advance may soon make the focused factory concept obsolete. These observers take note of (1) how successfully the Japanese have maintained product flexibility while moving down the product-process matrix diagonal and (2) the promise the latest technologies (e.g., CAD/ CAM, flexible manufacturing systems, DNC) hold for improving the product cycle times through job shop and batch flow processes. With such change, they reason, the factories of the future will be able to make customized products as easily as mass-produced ones. No longer then, according to this view, will factories have to be focused for specific products or process types.

This is an interesting proposition, and one that will bear keeping in mind as the years go by. The new technologies, coupled with better production scheduling techniques, are likely to close some of the gap between job shop and continuous flow processes. New technologies will not, however, be able to close all of the gaps between the various process types distinguished in this book. As long as there are relative differences between the various process types, the notion of factory focus will remain useful. If focused factories are able to outperform more general purpose ones, even by a small margin, it will pay to adopt them.

Simplifying the Learning II: Manufacturing Performance Measurement

Perhaps the easiest way to thwart both consistency in manufacturing choices and continuous improvement is by imposing some counterproductive performance measures on the process. "Scorecards" too often measure the wrong things. For example, many companies manage by establishing variances between actual performance and established standards. Managers then focus on keeping the variances small, and, it is hoped, positive (actuals better than standard). Progressive companies are increasingly abandoning such management by variances, essentially, because variances do not tell you how to improve the process or even what the root causes of problems may be. Visual measures of the process, such as JIT provides, and measures of quality, time, productivity, cost, and performance to schedule are all more apt measures for a company and ones that can better aid learning.

More than variances are under attack. Labor efficiencies and machine utilizations are other measures of long standing and wide use. They, too, are being abandoned as ineffective measures that do little to improve the process. Labor efficiencies, it will be remembered, look at the ratio of standard hours "earned" by the workforce to the actual hours spent. A ratio greater than 1.0 (greater than 100 percent) is regarded as desirable, evidence of hard work put in by the workforce. Unfortunately, such a measure tells little about how the factory as a whole is doing. Several deficiencies crop up in the measure:

1. *What the established standard is.*   An unrealistic standard can lead to an unrealistic labor efficiency value.

2. *How the earned hours and the actual hours are measured.*   Supervisors whose performance evaluation depends on labor efficiency can easily play games with the measurement of hours.

3. *More fundamentally, labor efficiency is conceptually flawed as a measure of the entire factory's performance.*   Whether labor is working efficiently or not does not measure whether the right materials are moving through the factory well. Efficient labor can create inventory and scrap as easily as it can produce good, needed output. Indeed, "efficient labor" can really be inefficient in the sense that it follows wasteful (time-consuming, non-value-adding) practices.

The same kinds of gripes can be leveled at machine utilization. Just because machine utilization is high, for either a single machine or an entire group of machines, means little for the productivity of the factory as a whole. There are issues of quality, changeover time, and waste. As noted in the early chapters of this text, and underscored in the tours at the end, one should expect that machine utilization differs for different processes; a job shop's machine utilization should be lower than that of a continuous flow process, given the increased variances of demand and supply and thus, if a job shop were to exhibit high machine utilization, the likelihood is that wait time will be excessive and delivery thus impaired. It

is only with the continuous flow process, which acts in a sense like a big on-off switch, that machine utilization really means something. There, the integration of machines is often so tight, that utilization is a good proxy for the bottlenecking of the entire process. This does not routinely occur with other processes. Bottleneck operations need high machine utilizations for the factory to be productive, but not non-bottleneck operations. Therefore, for most factories, machine utilization measures can be exceedingly misleading.

The Problem with Cost Accounting

Not only do variances, labor efficiencies, and machine utilizations have problems; cost accounting has come under increased fire recently. Even some within the profession itself have asserted that cost accounting is not providing manufacturing with the information that it requires in a day and age filled with technological advance, declining direct labor in the factory, and exploding overheads. Cost accounting is seen as preoccupied with the wrong things. The major complaints from manufacturing people are centered on the allocation of overhead to products. These allocations, which in some companies become part of the product costs against which managers are evaluated. More and more, when the direct labor content of a product is what is used to allocate overhead, managers are crying foul. Direct labor content is not correlated well with how products actually consume overhead resources such as plant space and equipment value, engineering time, and the time of other overhead functions such as quality control, production scheduling and control, materials handling, the stockroom, and others.

For this reason, other mechanisms have been adopted in a number of companies, and the list grows longer all the time. Here are the chief alternatives to the traditional cost accounting systems, alternatives that are more in tune with continual process improvement:

1. *Direct costing.* Direct costing does not assign overhead to specific products but instead, focuses on contribution margins. Decisions about product profitability can be made, but they require the development of specific plans and pro formas. This remedy has been espoused for decades by a small, but devoted band of followers. It has remained a minority position, however, largely because many people want fully allocated product costs for "easier" product decision making and fear the ambiguity that lies in dealing only with contribution calculations.

2. *Activity-based costing.*[5] Activity-based costing (ABC costing) tries to get a much firmer grip on the real cost "drivers" for the various overhead accounts. With activity-based cost accounting, overhead allocations that depend only on direct labor, or some other simple measure, are replaced by numerous, more precise ways of allocating overhead. In this way a more precise cost for a product is developed.

   For example, the cost of maintaining a stockroom (an overhead expense) can be tied to particular products by noting that the number of shipments received by the stockroom is a plausible metric (cost driver) for how the stockroom allocates its "activities" of receiving and storing material. The more shipments received, the more space and people the stockroom will demand. Knowing how many shipments of materials were devoted to a particular product could then be a fine way upon which to allocate the expenses of the stockroom. The more shipments devoted to a product, the greater that product's share of the activities of the stockroom and thus the greater the allocation should be that

represents the stockroom's piece of the total overhead to be allocated to it.

While activity-based costing is conceptually very appealing, many firms resist it because of its complexity. For some, the cost of collecting and using additional information required to allocate overhead more precisely overwhelms the advantages such an approach enjoys. Still, numerous companies are either adopting or piloting ABC costing, although the use tends to be for strategic decisions. The ABC system typically does not replace the existing systems to capture costs but rather stands off to the side, providing input for strategic decisions about products.

3. *Throughput time accounting.*[6] For those seeking a simple system, yet one without the biases inherent in allocating overhead exclusively according to direct labor, some companies (and my own sentiments) advocate substituting a different rationale as the basis of the allocation, namely, the value of production throughput time. In this instance, the faster a product goes through the factory, the less overhead it picks up per dollar of product value. This introduces an incentive to managers to improve the flow of products in their area because reducing a product's throughput time has the added benefit of reducing its cost, as computed by the cost accounting system.

   Furthermore, reducing throughput time may well stimulate a reduction in overhead. How might this occur? Many of the tasks of overhead personnel in a factory are addressed to a minority of products. The old 80/20 rule applies here, just as it does elsewhere in business; 80 percent of overhead may actually be spent working on just 20 percent of the products. Which 20 percent? My candidates are the 20 percent of products that languish in the factory. These are the products that need more materials handling, more production and inventory control resources, more scheduling, more accounting, more quality control, and more engineering changes. If materials flows are streamlined and throughput times are contracted, some of the time devoted to these overhead activities can be shed.

No doubt that over time more changes will be made to the cost accounting systems. Cost accounting remains an area of intense interest and much recent innovation.

Insights into Performance Measurement

More and more, observers of manufacturing are advocating the divorce of cost accounting and performance measurement.[7] Indeed, financial measures of the factory are only important for the highest levels of management. Good managers shield the lower levels of managers and workers from those kinds of measures, in favor of measures that are physical and easily comprehended: measures of delivery performance, quality, throughput time, actual (not standard) costs over time, and so on.

Performance measures should be linked to the company's strategy. If "what gets measured, gets managed," then performance measures *have* to be integral to the strategy: consistent with it and with each other, simple to understand, and attuned to what customers require. It stands to reason, then, that performance measures should change as the strategy changes. Keeping the same performance measures in place for years and years (perhaps with the idea of comparing operations over time) should, in fact, be avoided. If strategies are expected to change, then performance measures should change right along with them. The same set of measures for years on end is simply an indication of an inconsistent manufacturing strategy. Performance measurement should be as organic as any other manufacturing choice that managers confront.

Simplifying the Learning III: The Manufacturing Organization[8]

In the discussion of manufacturing strategy earlier in this chapter, a recurrent theme has been the search for operations choices that were at once consistent with one another and consistent with the perceived competitive demands that manufacturing must meet if the corporate strategy is to be advanced. Simplicity has also been valued—no conflicting goals for a factory, single-mindedness of purpose and of worker and management tasks— in part because it can foster learning and continual improvement.

These rubrics for the selection of a plant's operating characteristics follow through as well in the definition of the management organization that oversees and coordinates operations. Consistency and simplicity should anchor the manufacturing organization. How can an organization be consistent and simple? To begin to answer this question, let us analyze two distinct ways to organize an operation's management: the product-centered organization and the process-centered organization.

The Product-Centered Organization

At its simplest, the product-centered organization resembles a traditional plant-with-staff organization, which is repeated at higher and higher levels to control groups of plants and then groups of products and/or product lines. Customarily, individual products or product lines are placed in their own plants and managed exclusively by a product-oriented group of managers. Figure 15-3 offers a schematic representation of a product-centered organization.

Authority in this kind of organization is spread among the products and forced down within each product organization. Each product group is a small but independent company.

So far, this type of organization has been associated exclusively with products and product lines. Equally valid would be an association

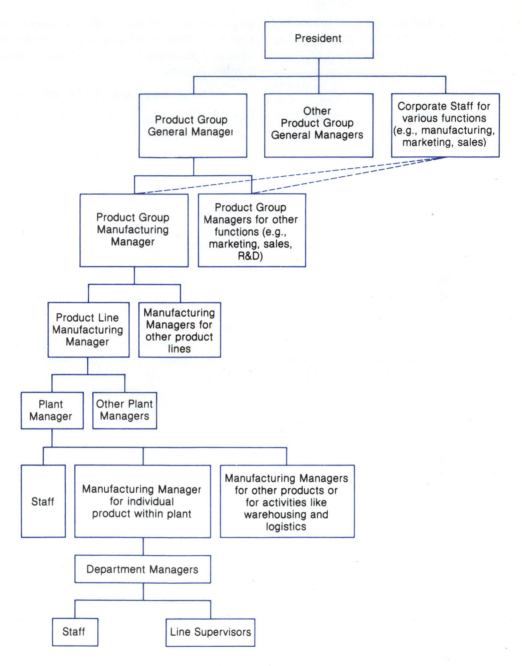

FIGURE 15-3   Organization chart typical of a product-centered manufacturing organization.

with geographic market areas. Here, instead of independent product companies, the company is independent within the boundaries of a certain geographic area. Most commonly, these separate geographic divisions of manufacturing involve foreign operations; but for some companies, domestic operations are organized with considerable authority placed at the regional level. Although it is more complete to describe such organization as product/market-centered, for ease of exposition I will continue to call it a product-centered organization, leaving the market-centered variation implicit. A product-centered organization is most often found in less complex, perhaps more labor-intensive production processes that do not rely on vast economies of scale or on the latest or most sophisticated equipment. Rather than be dominated by cost reduction, tight controls, and formal planning cycles, the product-centered organization is much more attuned to product mix flexibility and innovation. Consumer goods companies are often product-centered in organization.

The product-centered organization is appealing because it is "clean;" responsibilities are well delineated. Performance within the organization is easy to evaluate, because usually product groups are established as profit centers. Plant managers are important in the company and are often charged with responsibility for decisions about process technology, plant equipment, product development, and capacity as well as more routine concerns such as personnel, maintenance, and production control and scheduling.

The product-centered organization demands —and breeds—a special kind of manager: independent, entrepreneurial, well rounded. Junior level managers frequently move within the organization and must be tracked continually as they take on new or different responsibilities.

The corporate staff is relieved of much of its duties, particularly day-to-day operations. The corporate staff in the product-centered organization is often small and left to coordinate policies and personnel across product lines or geographic regions.

The Process-Centered Organization

Instead of breaking apart product lines for their own plants and manufacturing staffs, the process-centered organization breaks the process apart into distinct pieces. In this type of organization, most often several plants combine to manufacture the final product, although a multiplant operation is not a necessary feature of the organization. It is necessary to delineate responsibilities in manufacturing not by product line but by segment of the complete manufacturing process—hence the appellation "process-centered."

Because an entire series of plants or process segments must be coordinated to ensure smooth changes in the mix and quantities of products manufactured, the process-centered organization must be strongly centralized. This centralization of authority contrasts sharply with the decentralization that prevails in the product centered organization (see Figure 15-4).

In addition to providing coordination, the centralized authority is often absorbed with decisions on needed capacity and its character, on the balance of capacities in the entire process, on logistics, and on technological change and its impact on the process. These decisions are often the key challenges to management when products or processes are complex, as when the company is oriented to particular material (such as, oil, chemicals, agricultural products) or to advanced technology (such as, computers or other electronic products). Because such important decisions are retained centrally, plants within a process-centered organization are nearly always evaluated as cost centers.

Such a centralized, process-centered organization is not conducive to rapid new product introduction nor to vast swings in volume be-

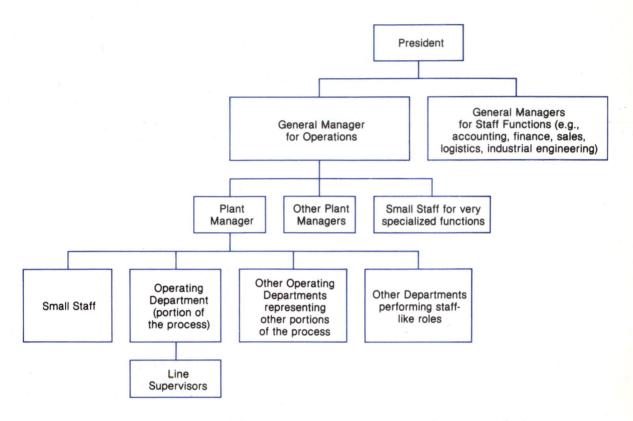

Figure 15-4   Organizational chart typical of a process-centered manufacturing organization.

cause the pipeline momentum and geographic separation between plants often lead to sluggishness. Nevertheless, if aspects of a process can benefit from large-scale or particularly specialized and rapidly advancing technologies, then a process-centered manufacturing organization offers a substantial contribution.

Because authority in the organization is kept centrally, the character of the managers who "grow up" in such an organization tends to be very different from the character of product-centered managers. Whereas the product-centered manager is likely to be an entrepreneur and a generalist, the process-centered manager tends to be a technician and a specialist. Because technologies within a process-centered organization are frequently so com-

plex and because decisions involving them are so expensive, the young manager in a process-centered organization is typically removed from the substantial production decisions that are made. Instead, the young manager usually undertakes a rather technical and frequently long apprenticeship. At the end of the apprenticeship, however, lies the opportunity to make decisions of a substantially larger magnitude than most product-centered managers make.

Table 15-2 lists the major operating differences between product centered and process-centered manufacturing organizations.

There are no strong arguments for consistently favoring one type of organization over the other. In terms of total manufacturing costs, neither organization could always be

TABLE 15-2   Differences between product-centered and process-centered manufacturing organizations

COMPANY	PRODUCT-CENTERED	PROCESS-CENTERED
Profit or cost responsibility; where located	Product groups	Central organization
Size of corporate staff	Relatively small	Relatively large
Major functions of corporate staff	Review capital appropriations requests	Coordinate decision making with marketing
	Communicate corporate changes and requests	Make facilities decisions
	Act as clearinghouse for personnel information, management recruiting, purchasing, used equipment, management development programs	Set personnel policies Set purchasing policies Review logistics-inventory management
	Evaluate, reward plant managers	Coordinate production schedules
	Select plant managers & manage career paths—possibly across product group lines	Decide whether to make or buy and whether to vertically integrate Recruit future plant managers Review plant performance, cost center basis
Major responsibilities of plant organizations	Coordinate decision making with marketing	Use materials and facilities efficiently
	Make facilities decisions (subject to marketing)	Recruit production, clerical, and lower management workers
	Make purchasing and logistics decisions	Train and develop future department and plant managers
	Schedule production and control inventory	Respond to special requests from marketing within limited ranges
	Decide whether to make or buy	
	Recruit management	

SOURCE: R. H. Hayes and R. W. Schmenner, "How Should You Organize Manufacturing?" *Harvard Business Review* 56, no. 1 (January–February 1978): 112

expected to dominate. The process-centered organization, for example, can best manage economies of scale or situations where tight controls are desirable; but at the same time, the process-centered organization is likely to increase overhead and logistics costs. It may also experience longer production cycle times and larger inventories, because it is less flexible in general than the product-centered organization.

Because of the staff specialization within a process-centered organization, technological change is often fostered best under such an organization. The central staff is apt to be highly expert, aware of technological alternatives, trends, research, and the experiences of other companies in the industry. This is a clear advantage of the process-centered organization, because the exploration of technological advance in a product-centered organization may be hit or miss. Some corporate staffs may take on the responsibility, or a separate research division may sponsor the needed research and development. In any case, major technological issues are often removed from the line operations of the typical product-centered organization.

A process-centered organization may also enjoy a relative advantage when it comes to managing the purchasing function, because such an organization can often win volume discounts. These economies may or may not accrue to a product-centered organization. In some product-centered companies, purchasing is split between the product lines and a central purchasing staff for the very reason of securing volume discounts.

Choosing the Organization: Product, Process, or Some Combination?

As they have been introduced so far, the product- and process-centered organizations are polar opposites in some important respects, and they can be expected to place fundamentally different demands on the company—different policies and practices, measurement and control systems, managerial attitudes, kinds of people and career paths. The question arises then in any given situation as to whether a product-centered organization should be put in place, whether a process-centered organization should be preferred, or whether some combination of the two ought to prevail. To explore this question consider Situation 15-3.

Discussion of STE Electronics Corporation

We can picture the manufacturing organization at STE Electronics as in Figure 15-5. As Rich Charles is only too aware, this kind of organization places distinctly different demands on him and his staff. The profit-center product divisions view the central staff as umpires and as little else. They do not approach the control staff for much day-to-day support, but rather remain in their nearly autonomous worlds. The San Jose plant makes much more use of the central staff, mainly because the central staff cannot afford to risk seeing San Jose "on

its own." The plant is too important to the smooth operation of the entire company.

For this kind of situation, Hayes and Schmenner argue for either a product-centered organization or a process-centered one, but not the combination that is evident at STE Electronics. The corporate staff cannot function properly when so many conflicting demands are placed on it. Just as the role or charter for each plant must be clear and simple, so must the role or charter for the central manufacturing staff. Hayes and Schmenner propose a test for such "organizational focus," as they put it: How easy would it be to fragment manufacturing, say, if forced to by the antitrust division of the Justice Department? If the operations could be fragmented easily, organizational focus has been achieved. If too many parts of the organization become entangled so that fragmentation is difficult, they argue, the organization is not focused. The organization will not be able to learn as well from its mistakes. Continual improvement is put at risk.

What then should Rich Charles do at STE Electronics? Clearly, the suggestion to break up the San Jose plant and to toss a chunk of it to each of the product divisions is sound, at least organizationally. Apart from that, Charles may be well advised to formalize the relationship between San Jose and the product divisions. This organization can be accomplished by insisting that transactions between each product division and the San Jose plant be carried out at "arms length;" San Jose could enter into contracts with each division and could maintain the right to refuse compliance with a division request, just as the divisions would be free to seek outside suppliers. This approach argues for an organization that is more subsidiary than divisional. To maintain this organizational formality, the corporate staff could be divided into two parts, one aiding and overseeing the San Jose plant and the other coordinating the two

STE Electronics Corporation

Rich Charles, vice-president for operations at STE Electronics, swore that the San Jose plant was built on quicksand and that any week now, on a visit there, he would be sucked under with no more than a muted gulp for his last breath. It was a constant wonder to him why he wasn't having nightmares about the plant.

STE Electronics Corporation, headquartered in San Francisco, was a small but growing company manufacturing specialty electronics equipment for the process machine industry and for the Department of Defense. The company was organized around two divisions: the industrial products division with its manufacturing facility in Livermore, and the defense products division with its manufacturing facility in Bakersfield. Supplying both of these divisions was the San Jose plant, which made power supplies and printed circuit boards.

As Rich Charles saw it, the San Jose plant was a battleground over which the industrial products and defense products divisions waged an intermittent war. Charles himself was some sort of umpire who would periodically award the spoils of victory to one division's "army" or the other's. There were continual skirmishes between the two divisions concerning the allocation of the San Jose plant's capacity. Each division wanted priority to be given the delivery of its power supplies or printed circuit boards. Both products were critical components of the company's range of electronic equipment, and the failure to deliver on time could seriously interrupt each division's production schedule. Nevertheless, both divisions were playing games with orders on the San Jose plant. In an effort to ensure delivery, both divisions would over-order from San Jose and at the same time pretend they needed every unit they ordered, and quickly at that.

To make matters worse, the defense products division continually harassed San Jose about product quality. This was understandable, too, for the defense products division had to produce to exceedingly precise specifications. The industrial products division was not under as stringent pressure.

Delivery and quality concerns flared up periodically as skirmishes, but there was a pitched battle every year at transfer price setting time. Each division was evaluated as a profit center but the San Jose plant was evaluated as a cost center. Because materials costs were a substantial fraction of product cost, each division had the incentive to keep the transfer prices they paid for their power supplies and printed circuit boards low. But, because San Jose's total costs had to be covered, there were fevered sessions where the allocation of San Jose's overhead expenses were argued over, each division trying to throw them onto the other. Rich Charles did not enjoy these sessions.

Charles had a measure of compassion for the San Jose plant and its management. They were always caught in the crossfire, and it was only the concerted efforts of the cadre of old engineers who ran San Jose that kept things moving as smoothly as they did. However, in not too many years most of the San Jose management were slated for retirement, and Rich Charles was concerned that no young heirs apparent seemed to be rising to take charge of the plant. Charles knew that the young talents in the company were shying away from the San Jose operation, both because of the divisional crossfire and because the product divisions were viewed as the "fast track." Charles himself had come up mainly through industrial products and his predecessor had headed defense products. It was no wonder that STE's young managers viewed San Jose as a dead end for their careers.

Rich Charles had pondered closing the San Jose operation and spinning off the power supply and printed circuit board production to each division, but he had shied away from this option. San Jose did realize some economies from the scale of its operation and its breakup would be costly and would spread the existing expertise too thin. What Charles really wanted to think about was some alternative organization of manufacturing that might lighten his burden and that of his staff in dealing with San Jose and its interactions with the product divisions. As it was now, San Jose housed only about a quarter of the company's manufacturing employees, but it accounted for well over half of his time and the time of his staff. The plant needed persistent guidance on matters such as forecasts, production control, and inventories.

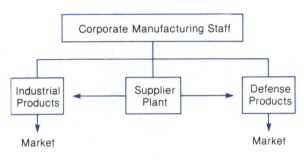

FIGURE 15-5   The manufacturing organization
at STE Electronics.

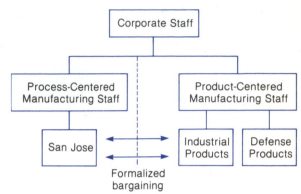

FIGURE 15-6   A revised organization chart
for STE Electronics.

product divisions. By seeking this separation, the two organizations assume a more equal status and eliminate the conflicting roles that the corporate manufacturing staff had to play.

A revised organization chart might look like Figure 15-6. As for Rich Charles himself, he might have to choose sides, occupying a role either at the head of the product-centered staff or at the head of the process-centered one. He, too, should not have to occupy the middle ground.

REVISITING THE PRODUCT–PROCESS MATRIX

These notions of manufacturing strategy can be reinforced by drawing upon the product-process matrix that was introduced in Chapter 1. As will be recalled, the product-process matrix related the character of a plant's products—one of a kind, commodities—to the character of the process flow—jumbled versus rigid, loosely linked versus tightly linked. Figure 15-7 reproduces the product-process matrix as it was introduced in Chapter 1 (Figure 1-2).

As it was first discussed, the product–process matrix was devoid of any dynamic features. It was merely a handy box in which to categorize the range of production processes we have con-

sidered in this book. As should be evident by now, the typical manufacturing operation must deal constantly with change and with the threat of change. It is natural, then, to expect a number of companies to undergo considerable movement within the product-process matrix. The learning curve, for example, represents a constant pressure to move along the matrix diagonal down and to the right.

Movement within the product–process matrix is not likely, however, to be precisely along the diagonal. Such a smoothly modulated movement is difficult because many process and product changes tend to be abrupt; at least, that has been true of much change in the past, particularly in the continuous flow process industries. A smoothly modulated diagonal movement is also made difficult by the fact that companies tend to concentrate their efforts at any one time on either a large product change or a large process change. Movement along the diagonal requires simultaneous product and process change, which is a complex management enterprise. Thus, rather than smooth movement along the diagonal, companies undergoing change are more apt to demonstrate steplike

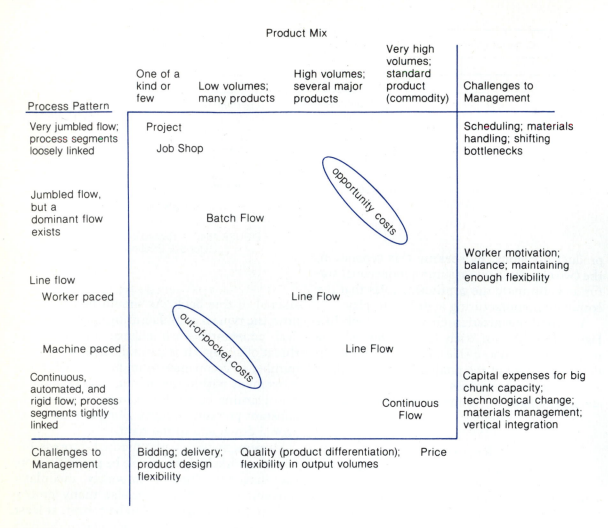

FIGURE 15-7   A product–process matrix.

movements within the matrix, such as depicted in Figure 15-8.

In charting their movement over time within the product-process matrix, companies can generally choose to remain above or below the diagonal, since it is unlikely that they can remain on it. Is a path over time that lies predominantly above the diagonal any better than a path that lies predominantly below it? The question is a meaningful one and aims at the heart of corporate as well as manufacturing strategy.

As noted in Chapter 1, the area of the product–process matrix that lies below the diagonal is characterized by out-of-pocket expenses. This area is where the rigidity and capital intensity of the process are relatively further advanced than the acceptance (the in-

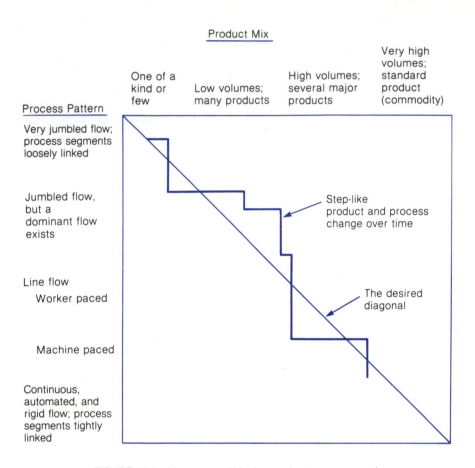

FIGURE 15-8   Movement within the product–process matrix.

creasing commodity aspects) of the process's products. A company whose decisions position it below the diagonal places significant pressure on its marketing people to drum up the sales volume needed to sustain its relatively further advanced process characteristics. Otherwise, the high costs incurred in financing a process change that is significantly capital-intensifying are likely to lower the company's profits. What such a company would like to see is continuing sales growth that brings the company's position within the matrix back to the diagonal, for only the diagonal represents a perfect match of process and product characteris-

tics. In a sense, the diagonal represents those "patches" of the matrix where manufacturing consistency is best achieved.

The area above the diagonal, as also noted in Chapter 1, is characterized by opportunity costs. Here, the process's characteristics are relatively less advanced than is product acceptance. The company's profits suffer, not because high investment in plant and equipment must be paid off, but because process advances could manufacture the product for less. By having its process lag behind its product acceptance, the company forgoes the opportunity to earn more. By positioning itself above the diagonal, the company places pressure on its manufacturing people to lower costs.

A strategy that consistently positions a company above the diagonal of the product–process matrix is essentially a conservative one; the company risks the loss of dollars that could have been made rather than the loss of dollars already earned. As we observed before, opportunity costs are every bit as real as out-of-pocket ones, but the conservative firms would rather lose potential profits by lagging behind in production technology than incur certain, out-of-pocket expenses for advances in the pro-

cess. While the road to high profits has no shortcuts and lies in keeping both out-of-pocket and opportunity costs low, the shortcut to bankruptcy lies in incurring out-of pocket expenditure. Incurring opportunity costs is simply a longer and more tortuous road to bankruptcy, as the company gets left in the dust by faster moving companies.

An analogy to one of my favorite sports may serve as a useful characterization of the "strategy paths" companies can choose to follow. Think of the company as a golfer teeing up in the upper left of the matrix (see Figure 15-9).

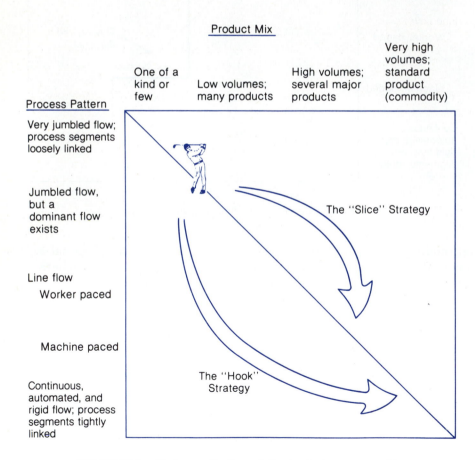

FIGURE 15-9   Strategy paths through the product–process matrix.

The company/golfer is pondering a shot down the diagonal of the matrix, the center of the fairway for our purposes. Like all golfers, the company/golfer cannot expect to hit the ball straight; he's probably going to put some hook or slice on the ball. Knowing what swinging a certain way will do to the ball, our company/golfer compensates. If he expects to hit a slice, he aims left of the center of the fairway; if he expects to hit a hook, he aims right. It just so happens that the physics of golf dictate that a hook shot is apt to travel further than a slice, since the hook puts an overspin on the ball that makes it roll more when the shot strikes the fairway. In contrast, the slice puts an underspin on the ball that dampens the roll of the ball. The hooked shot, then, is apt to travel farther, but is also more apt to run into the rough if not properly struck. The sliced shot is safer because its roll, once on the fairway, is less, but the shot sacrifices distance.

The "slice" is the conservative company's path through the matrix, risking only opportunity costs. The "hook" is the more adventurous path through the matrix, which risks out-of-pocket expenses. Both kinds of "shots" can be successful, ending up with the company/golfer lying on the diagonal (center of the fairway). The hook, while riskier, is also a more likely way to leave other companies/golfers in the dust, far back up the fairway.

The masters of the hook strategy in this era of manufacturing have been the Japanese. The leading Japanese companies have made significant inroads in all international markets, primarily at the high-volume-product end of the product mix spectrum. These companies have led with their manufacturing capabilities that stress high-quality conformance at low cost; the pressure in these companies is on their marketing arms.

The manufacturing philosophy lying behind the hook strategies of the major Japanese companies is the topic of the following section.

Two Issues Affecting Manufacturing Strategy

1. Productivity

Since 1970, U.S. productivity, measured as real GNP (gross national product) per worker, has not been advancing at its historical (post-World War II) pace. This falloff in the rate of growth of productivity has puzzled many observers, and no definitive explanation has yet emerged. A number of possible explanations have been advanced, however. Among them are some traditional, macroeconomic explanations:

- Low levels of business investment, caused perhaps by increased uncertainty in the economy.

- The phenomenal increase in energy costs, causing many companies to turn to less energy-intensive production technologies but with no compensating increase in output.

- More pervasive government regulations, which have required huge "nonproductive" capital outlays for pollution control equipment.

- Shifts in employment away from agriculture and manufacturing and toward the service and government sectors, where productivity increases may be harder to come by.

- Increased reluctance to lay off workers when business turns down.

- Decline in the willingness to work hard, caused perhaps by high marginal income tax rates and/or too "socialist" a welfare system.

- Demographic shifts toward younger, less-skilled workers.

Some people say that these macroeconomic explanations do not hold much water because, among other reasons, they do not explain cross-country variations. These critics put forth other explanations:

- American management has been too enamored of short-run return and has failed to plan effectively for the long run. It has, according to this view, failed to invest in new capital and technology and has given up on new product innovations.
- There is a tension between productivity (as exemplified by the learning curve) and the customization, product variety, and new product introduction that are common to many marketing strategies. Marketing demands for flexibility, so this argument goes, cut into the national rate of growth of productivity.
- The data on productivity are unreliable, especially for service industries. Output in many cases may not be valued properly, and there may be some problems in accounting for the quality and learning of the workforce.

How this debate will be resolved is unclear. We have much yet to understand about how the national figures on productivity are affected by changes at the plant level. At the plant level, concern for declining productivity is much less evident than it is in Washington. For many plants, productivity gains are being reaped regularly, although there are questions that could be asked about the reliability of plant data, just as there are questions about the reliability of the national data.

Productivity Measures

In Chapter 5, where we discussed workforce management, the widely used labor efficiency measure (standard hours/actual hours), which is often used also as a productivity measure, was introduced. This efficiency measure and its cousin, the effectiveness measure, are not classic productivity indices. That is, they do not represent a ratio of output to input, which is the standard definition of productivity. In addition, labor efficiency is a partial measure of productivity; it relates only to labor and not to capital

or materials. Better measures are the more direct ones that measure units produced per labor-hour, or tons produced per day, or something similar.

There are many other measures of productivity in use. Most of these are partial measures. The inventory turn ratio, for example, is a measure of the productivity, broadly construed, of dollars invested in inventory. Value added per employee, gross margin, and ratios of indirect labor are other widely used measures.

While one can gripe that partial measures of productivity do not balance labor and capital productivity, partial measures do have their place. They are usually easy to calculate and readily understandable by workers and managers alike. When the measure indicates that performance is off, people can generally explain what happened easily. Because people typically pay attention to things that get measured and rewarded, the choice of productivity measure to track is not a casual one. The measure should capture something that managers feel is important to company competitiveness. Nonetheless, the measure chosen is less important than the company's prevailing manufacturing strategy. If the strategy adopted is a good one, productivity cannot help but improve.

Productivity Policies

Productivity is a complex phenomenon. There are a host of anecdotes about what contributed to a particular factory's or company's productivity advance, and many stories differ completely. For some people, productivity means hardware advance. For others, it means motivating people. For some it means simplifying the organization, while for others it's synonymous with quality management. Some say it's value engineering. Yet others say it may be methods improvements and industrial engineering. Thus, the plant manager who is faced with a mandate to improve productivity is confronted with a great diversity of options. They can be dizzying and confusing, and the plant manager who attempts to undertake all of them at once risks doing none of them well.

At issue, then, is which of the many potentially conflicting policies the factory should adopt. Are there some policies or programs that are more fundamental than others for a factory of a certain character? Simplicity has long been a cardinal management tenet, so it would be advantageous for plant managers to be clear on what should be stressed in order to improve productivity.

Some recent research on this matter shows some promise. Its key suggestion is that reducing throughput time—the time between when materials are ready to be worked on and when the product is finished, inspected, and ready to be shipped—is a key explanatory influence on productivity. We have seen where time has a strong claim as an aid to continuous improvement. This research highlights just how important such as influence can be on a factory's productivity performance. Throughput time is thus a strong candidate as a simple, unifying "banner" for plant managers to proclaim.

We should not be prepared to discard completely, however, more historical explanations of factory productivity gains. The research also points to the limited but discernible influence of investment in new technology as a spur to productivity gain and also to human resource management themes, as in Chapter 5, that spread participation in decision making more widely than before. It seems clear, however, that just-in-time manufacturing, and the means that broad heading offers for continuous improvement, do seem to be supported in the results as a clear way to achieve productivity gains.

2. The Paradox of Flexibility

Mentioned earlier in this chapter was the term *world-class manufacturing*, which has come to mean those who can seemingly do all that may be requested of manufacturing, be it low cost, quick delivery, high conformance quality, or flexibility. In Chapters 4 and 11 we addressed the importance of high conformance quality as a key means to lower the total costs of manufacturing. Much of the basis for just-in-time production rests on this relationship. Quick delivery and throughput time reduction have just been discussed and shown to be intimately related to improved productivity. What remains to fit into this mosaic of manufacturing strategy is flexibility.

To my mind, flexibility in an operation is linked to permissiveness. Importantly, however, permissiveness should be shunned in an operation; it is an evil to avoid because it breeds rigidity in the process. This seems paradoxical. How does it occur? When the process is permissive—when rescheduling is done all the time, when expediting is permitted, when engineering change orders are rampant—paralysis can easily set in. The process may become unable to put anything out on time and it can lose its grip on quality. As has often been observed, when everything in the factory is tagged red for immediate action, that's as good as saying that nothing is tagged red.

On the other hand, it is discipline that breeds flexibility. This is the paradox of flexibility. Consider the job shop. A job shop is tremendously flexible in the mix of products that it can produce—large quantities, small quantities, all with different characteristics. However, a job shop only operates well if it has an absolutely rigid flow to the information and recordkeeping within it. Tooling, setups, and the load on particular workstations can all change dramatically in a job shop, but the flow of information—the paperwork, the recordkeeping—has to run like clockwork or the job shop becomes hopelessly snarled.

Consider another paradox—that relating to quick-change setups. Quick-change setups are the quintessence of flexibility. But think what it takes to be able to change setups quickly: prescribed methods for making them, similar tooling designs, well-conceived workstation layouts, and strict machine maintenance schedules and procedures. Discipline breeds flexibility.

Different Kinds of Flexibility

Now, of course, there are different kinds of flexibility that one may want to foster. There is the product mix flexibility, best characterized by the job shop. There is also the flexibility of new product introduction: being able to add products easily and keep up with the state of the art. There is also volume flexibility: being able to ramp up production quickly or, alternatively, cut it off quickly, so as to capitalize on fads or to react to severe changes in marketplace conditions. There is also customization, being able to produce precisely what the customer requires. These different kinds of flexibility demand different kinds of discipline.

Product mix flexibility requires quick changeovers, and as mentioned above, quick changeovers demand their own rigidities. For product mix flexibility, it is helpful for the workforce to be cross-trained, to be able to operate multiple machines at the same time, and to be able to move from one bottleneck operation to another. It it also helpful if production schedules are frozen for periods of time, and expediting is not permitted. Product mix flexibility also demands discipline in the sharing of information, both within the factory, with the sales staff, and with vendors.

New product introduction, on the other hand, requires a different set of disciplines. Some of the companies that do new product introductions best make sure that there is a rigid set of reviews and sign-off procedures that must be adhered to at various stages. In addition, there are frequently set procedures through which all of the engineering drawings must pass, as well as set procedures and people through which engineering changes must go, so that nothing falls between the cracks. Good new product introduction typically often also requires engineering follow-up. Engineers know in advance that they are to be responsible not only for the design, but for initial production as well.

With volume flexibility there are still other disciplines that have to be fostered. Typically, the most important is making sure that there is a cushion of capacity which resists any management wishes to see it fill up with unnecessary product. Without a cushion of capacity, being able to increase production quickly becomes much more difficult. Volume flexibility is enhanced with equipment that is general purpose and convertible, with layouts that are easily changed, with subcontracting of jobs, with temporary hires, and with cross-trained workers who are willing to go from one task to another. Effectiveness in volume flexibility may also require low breakeven levels. This may mean a more labor-intensive operation than exists for operations that do not value volume flexibility.

The paradox of flexibility is this: The extent and the nature of the discipline that you are willing to impose on the factory will, in fact, decide for you what kinds of flexibilities you will be able to enjoy. Being permissive will only cause the factory to gum up and to lose the flexibility that permissiveness seeks in the first place.

PART TWO

SERVICE OPERATIONS STRATEGY

Part One of this chapter opened by reviewing a whole series of operations choices faced by manufacturing managers. Service operations managers, of course, face a similar list of choices. Table 15-3 outlines some major ones.

Furthermore, just as manufacturing choices work best when they are consistent with each other and with the competitive priorities implied by the business strategy, so do service operations choices. Service business competition can be as cutthroat as any, perhaps more so, because barriers to entry are frequently lower for service businesses. The service business with inconsistent operations choices is often quickly weeded out.

Service businesses can no more be all things to all customers than can manufacturers. The concept of factory focus has easy parallels in the service arena. A number of the criticisms leveled at various service companies reflect concern over too many different kinds of services offered or a failure to achieve the operational synergies touted when the merger or acquisition was first undertaken. One cannot assume that the "formula" that has been so successful in one service business will necessarily carry over to another service business, even if that business is merely a segmentation of the old one.

SERVICE TASKS, SERVICE STANDARDS, THE SERVICE DELIVERY SYSTEM, AND THE SERVICE ENCOUNTER

The first part of this chapter, in dealing with manufacturing strategy, discussed how good manufacturing strategy rests in consistency between the manufacturing task, as derived from business unit strategy, and the numerous choices that a manufacturer must make. Service is no diffferent. Consistency between the service task and the service delivery system is the hallmark of what is a good service operations strategy.

The service task[a] is a statement of what the consumer really values about the service provided. It states what the operation must do well for the customer to be satisfied. It can serve as a "creed" for those who provide the service, a creed that constantly reinforces what it is that leads the customer to this service rather than some other one.

It is then up to the service delivery system to provide the service that fulfills the service task. There are many choices to make, as the foregoing section recounts. Consistency among them is not assured. Service managers must be diligent in their pursuit of consistency, especially given the very fast-paced nature of much of the service sector.

Service standards are the guardians of the service task. They act to define and constrain the service delivery system. They specify what the service is, much as product specifications and bills of material specify what a manufactured item is.

The relationship among the service task, service standards, and the service delivery system is represented in Figure 15-10. The three are

---

[a]The service task is sometimes also referred to as the service concept.

TABLE 15-3   Key choices for service operations

I. Technology and Facilities

Type of facility and equipment (Used by customers themselves? How? How attractive do they have to be? How are they used during a peak versus a nonpeak situation? General-purpose versus special-purpose equipment? Anything proprietary about equipment or its use?)

Layout (Job shop-like? Fixed position? Other?)

Location (How critical to attracting customers?)

Size of the facility

Geographic spread of any multiple facilities

Degree of labor intensity

Attitude toward capacity utilization (Facility for the peak demand or not?)

II. Operating Policies

A. Planning the operation

Forecasting (Extent required? Type used?)

Logistics and inventory system used for materials employed

Manpower planning

Schedule setting (Can service provision be "leveled" in any way?)

Demand management for peak and off-peak times

B. Controlling the operation

Labor issues—hiring, training, skill levels required, job content, pay, advancement, unionism

Accounting controls used

Checklists developed

Foolproofing designed into the layout and the equipment

Quality control audits and policies

What triggers provision of the service and the pace of the operation (Customer? Forecast?)

Production control (How does the information flow within the operation? What is on track? What is not? How can anything gone amiss be fixed? How can any changes be implemented?)

C. Operations organization

What is kept at the individual unit level, and what is centralized?

Where is talent held in the organization?

---

shown to surround, and support, the "service encounter," what the customer encounters when he or she interacts with the service delivery system. The service encounter embodies what Jan Carlson of SAS terms "moments of truth" about the service operation, when one actually confronts how well the operation can satisfy customer needs and wants.

It has been said that profitability in services is directly related to customer satisfaction, and that customer satisfaction follows a formula:

Customer Satisfaction = Delivery − Expectation

Naturally, the development of the service delivery system is an effort to build the capability of the service process to deliver a service that more than meets customer expectations. This formula also indicates that managing customer expectations is another important endeavor. Pleasantly surprised customers and profits both follow those services that over-deliver and under-promise.

That other pillar of manufacturing strategy, namely continuous improvement, is also as at home with service operations as is the concept

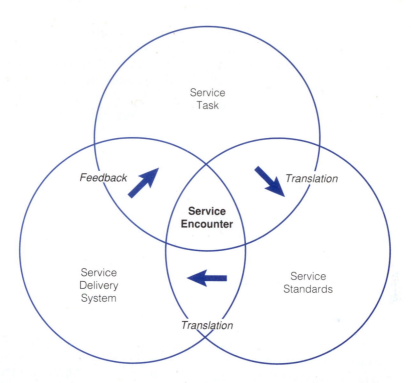

FIGURE 15-10   The relationship among service task, standards, and service delivery system.

of consistency. Thus, the tools of process and information flow diagrams, tracking of time, quality improvement programs, removing waste, and benchmarking are equally valid for services and manufacturing. Progressive companies, either in their service operations per se, or with the overhead operations that support the company's manufacturing, have pursued these points successfully, providing both good service and profitable service at the same time.

REVISITING THE SERVICE PROCESS MATRIX

The classification, in Chapter 1, of service businesses into four categories — service factory, service shop, mass service, and professional service — can be used to investigate what has happened strategically to service opera-

tions over time. The most salient development in many service businesses has been vast segmentation and diversification. Services that were once clearly service shops or mass service firms are no longer so clearly labeled. Service firms are spreading themselves out across the service matrix. Some examples are instructive:

1. One of the classic changes has involved the development and evolution of fast food. The traditional restaurant could be positioned as a service shop with relatively high customization and interaction for the consumer and a middling labor intensity. The elegant gourmet restaurant may even be classed as professional service. On the other hand, with the advent of fast food, interaction and customization for the consumer has been dramatically lowered, as has the labor intensity. For the classic fast

food companies, the change has been from a service shop to a service factory. This has left the restaurant business with a wide diversity of operations types.

2. Another interesting innovation within the service shop quadrant has involved hospitals. The new kinds of hospitals developed by Humana, Hospital Corporation of America, and others are different in type from the traditional community hospital or university medical center. The traditional hospital (and especially the university medical center) is geared to treat any disease, and thus invests in all of the latest equipment and technology for diagnosis and treatment. This new breed of hospital customarily deals with the more routine kinds of medical treatment; often such care facilities do not include intensive care units and other high expense units for very sick or dying patients. Those patients are referred to larger and more well-equipped hospitals. For its part, the new type of hospital offers much lower cost service that is convenient for the consumer. The new hospital thus offers less customization but also a higher degree of labor intensity.

3. Another series of changes has occurred in some of the services characterized as "mass service." Retailing offers some interesting examples. The expansion of catalog stores (such as Best's or Sharper Image), warehouse stores (such as Toys 'R' Us), mail order sales (such as L. L. Bean), and brandname discounters (such as Loehmann's) has broadened the traditional retailing operation toward a lower degree of labor intensity, largely by providing less than department store-type "full service." On the other hand, the proliferation of boutiques and specialty operations within stores like Bloomingdale's is evidence for a different kind of change, one where interaction and customization are stressed, often with higher labor intensity (more than "full service"). By being more "professional" (frequently with salespeople on commission), such stores hope to convert more browsers into buyers.

4. The deregulation of commercial banking and financial services has also created some intriguing strategic changes for operations within the mass service quadrant. Automation in commercial banking (automatic teller machines, electronic transfers, and other new technological advances) has pushed aspects of commercial banking to lower labor intensity. Indeed, credit card operations and check clearing are now placed in their own facilities, often at quite a distance from the commercial banks themselves, and they do essentially the kind of work that one would expect in a service factory. This kind of change is evident in some other financial service companies. One of the justifications given for the acquisition of Lehman Brothers Kuhn Loeb by Shearson/American Express was the fact that the trading operations of Lehman Brothers could be easily absorbed by slack capacity in the backroom operations at Shearson.

However, even as technological advances have helped commercial banking become less labor intensive, there is a move to customize some services even more. The "unbundling" of financial services brought on by such deregulation milestones as the removal of interest rate ceilings on certificates of deposit, the cessation of fixed brokerage commissions, the demise of Regulation Q (which placed ceilings on interest payable by commercial banks), and the first steps toward interstate banking has promoted this development. Many of the services that have been acquired by the "financial supermarket" companies are essentially services that will give those companies greater interaction with and customization for consumers. The increasing menu of services from the traditional brokerage houses may cause the oldtime broker, who tried to be all things to his or her client, to become an anachronism.

Moving Toward the Diagonal

Given the importance of focused operations and the quickening pace of segmentation and diversification of service businesses, understanding the dynamics of service process change can be very helpful. There are several observations about these dynamics that bear attention. The first observation is that many of the segmentation moves that service businesses have made, as were just discussed, have been toward the diagonal that runs from the service factory to the professional service firm. Figure 15-11 illustrates this point. But why is this the case? What is it about the diagonal that is so attractive for existing services? The unifying thread in these moves toward the diagonal seems to be better control, although the kinds of controls needed differ from the service shop side of the diagonal to the mass service side.

On the mass service side, control often relates to labor costs and efficiency; these services are constantly trying to get a grip on their labor scheduling and productivity. With mass service, plant and equipment are rarely constraints. In retailing, for example, labor is a critical variable cost, so scheduling labor is an important activity. The recent increased use of point-of-sale terminals has permitted sales tracking of different items throughout the day. Such information, in combination with selling cost (wages paid divided by sales for any period—a productivity measure) is tremendously useful to workforce scheduling. Such information also helps inventory control, of course. Moves toward more customization,

such as is illustrated by the department store boutique and its commission salespeople, can also be understood as moves toward increased control of the selling situation itself with concomitantly higher revenues, profits, and productivity.

On the other side of the diagonal, the service shop frets about control of the service itself. In the service shop, plant and equipment are frequently constraining. Thus there are concerns for how frequently unpredictable jobs (such as autos for repair, health care patients) can be sequenced through areas with expensive capital equipment. Control is affected as well by great uncertainty about when and how people know that the service is satisfactorily rendered. Hospitals, for example, have a high proportion of fixed costs and thus worry a great deal about capacity utilization. Current debates about whether administrative staffs or medical staffs should dictate the utilization of hospital resources (particularly given changes in reimbursement practices) are, at core, debates about resource control. One can thus understand pressures to provide less customized and/or interactive service (fewer tests, more ambulatory care).

The service factory and the professional service firm suffer less from loss of control. This is not to say that control is not an issue for both of these service types, but with the professional service firm, the high degree of interaction and customization is at least matched with a high degree of labor intensity and a tremendous amount of skill in the workforce. Control is a much more individual affair, relatively free of constraints of plant or equipment. The service factory, on the other hand, can plan its "process" to foster more control; in this regard, the service factory shares many of the benefits that manufacturing operations enjoy. The process defines the service and the flow of the process is relatively smooth. The labor needed is well

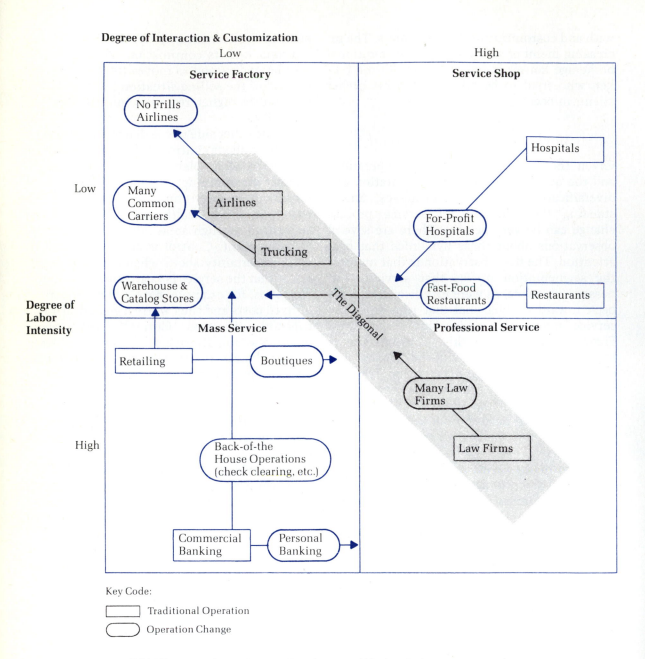

Degree of Interaction & Customization

FIGURE 15-11   Strategic operation changes within the service process matrix.

known for given levels of demand, and scheduling of labor, plant, and equipment is fairly straightforward. Even though there are forces on existing operations to move toward the diagonal, we should not be alarmed that all service shops or mass service operations will become extinct. Many operations will be able to adhere to their traditional operations choices, and marketing pressures for increased customization and the generation of completely new services are likely to replenish the supply of service shops or mass service operations that are pressed to transform themselves. Witness, for example, the upsurge of luxury airplane travel or luxury hotel accommodation. New services are being rendered to particular market niches. The change recently in how computers are sold, with more mass outlets as opposed to individual salespeople, is another example.

Moving Up the Diagonal

A second observation on the dynamics of service processes centers on some of the service business already located on the diagonal. The professional service firm and the service factory are not immune themselves from strategic changes. Of those services that have changed their positions within the service process matrix, most have moved up the diagonal. Consider the changes that have occurred within many law firms over the last decades. The growing staffs of paralegals and other lower-cost labor and the increasing specialization of what many law firms do has driven many firms within that professional service industry toward lower labor intensity and less customization (that is, less full service). Similarly, other professional service firms have invested in equipment, much of it for word processing or data processing.

In a similar vein, the deregulation of trucking and air travel has brought significant change to those service factories. Deregulation of trucking has forced many of the old line common carriers to invest dramatically in breakbulks and additional freight terminals so as to do better with less-than-truckload shipments. While these capital investments have meant increased barriers to entry by competitors to these trucking firms, these firms can no longer afford to offer so customized a service for their clients. The firms are more specialized and offer fewer services. Their pricing structures have been changed to encourage the clients to ship in particular ways. This has meant movement up the diagonal for most of the common carriers.

A similar story can be told of the airlines. Deregulation encouraged their operations to shift to hub and spoke systems. All of the major airlines have changed their operations in this way so that a number of "pushes" can be made from their major airport hubs during any day. Where once competition was on time of day and the elegance of service, competition these days is more on price and less on the number of flights and their timing during the day. Thus for most airlines, the result has been both less customization in what they offer the consumer and lower labor intensity, because significant investments have been placed in the hubs.

Just as movement toward the diagonal does not necessarily mean that service shops and mass service firms will cease to exist, the move up the diagonal does not necessarily mean that professional service firms will all become small service factories. There will always be some firms that will be able to remain successfully with high labor intensity and high interaction and customization, and there will be new professional services rendered that, at least initially, will demand a combination of high labor intensity and high interaction and customization.

A Concluding Comment

This text, with both its chapters and its tours, has strived to describe and interpret, *from the inside out*, the fascinating and ever important world of the operations manager for you, the reader. If you had started the journey of learning this subject with some trepidation, I hope that it strikes you now as friendlier terrain upon which to travel. It is certainly everchanging terrain, and that fact makes us all students.

QUESTIONS

1. Three categories of decision making or policymaking define to a great degree what the corporation's process looks like and how it operates. Summarize the main aspects of each category.

2. Why is consistency valued so highly in the production process choice? To defend your position, provide examples that you know about from your own experience or reading.

3. Compare the main operational features of the Porterfield Glass Company with those of Matthew Yachts.

4. What are four of the main competitive demands that can be placed on manufacturing?

5. For a corporation, what are the principal advantages and disadvantages of factory focus? What are the main areas of resistance to focus?

6. "Movement within the product–process matrix is not likely to be precisely along the diagonal." Discuss the theory behind this statement, paying particular attention to what it means to be above or below the diagonal.

7. Name the two clear ways to organize an operation's management. What advantages does each offer? For what kind of operation? How does one choose? Provide an example for your discussions.

8. Discuss the factors or concerns that should be considered in selecting the elements of a production process.

9. Discuss the competitive demands on manufacturing. Can a manufacturing manager expect to perform equally well in all of the demand aspects? Why?

10. What is meant by the term focused manufacturing, and why is it an important concept?

PROBLEMS

1. "Inflexible, special-purpose machinery can be as much a part of a flexible manufacturing system as flexible, general purpose machinery." Comment on this statement.

2. In its Puerto Rico plant, which assembled display terminals, Hewlett-Packard adopted a concept that the company termed "focused flow." Instead of building many different terminal models at the same time, it sequenced production serially so that at any one time the factory of 275 employees assembled only one model of terminal. What do you see as the advantages to such a strategy? What are its disadvantages?

3. In Situation for Study F-1, Tom Stoddard is attempting to use production and operations management in the house construction industry. Where is his operation located in the product–process matrix? Relate the theoretical concepts of the product–process matrix to Tom's actual process.

4. The home construction industry has for many years tried to get form the top left-hand corner to the bottom right-hand corner of the product–process matrix. What progress in this direction has been made in your local area? Why is this so?

5. Bob Leone's company manufactures a material that significantly lengthens the life of fluorescent bulbs. This material is patented, but the patent will run out in 5 years. Demand for the material is growing at a rate of 25 percent a year as the major fluorescent bulb manufacturers use it in more and more of their product offerings. What manufacturing choices would you recommend to Bob to help him cope with this situation?

6. U.S. auto manufacturers are faced with some monumental production challenges—the need to produce more small cars, the need to increase gas mileage on all their cars, and the need to meet emissions control standards. What kind of manufacturing changes would you suggest to meet these challenges and still keep autos affordable? Be sure to consider such items as product design, product line breadth, purchasing, vertical integration, process technology, capacity balance, and also workforce policies.

7. You are a maker of Nintendo-compatible games, one of the latest crazes in the United States. What manufacturing choices would you make to serve this market, which is expected to grow by 450 percent over the next year? Support your choices. What manufacturing strategy underlies your analysis?

8. What parallels do you see in the dynamics of the product–process matrix and its service counterpart? What clear distinctions exist?

SITUATION FOR STUDY 15-1

STAR TRUCK CORPORATION

Tom Horn had just been appointed vice-president of operations for the Star Truck Corporation and had moved to the corporate headquarters in Atlanta. Tom had most recently been plant manager of the Star Truck plant in Little Rock, Arkansas. The firm had decided to build a new truck plant in the Midwest and had taken an option on 20 acres in an industrial park just outside Dayton, Ohio. Tom's first job as vice-president was to coordinate all the design aspects and managerial concerns to make sure the new plant operated in an efficient and effective manner and made a high contribution to the firm. Tom thought he could rely heavily on his experience at the Little Rock Star Truck plant, plus his education, to bring this assignment to a successful conclusion.

The trucks that were built at the Little Rock plant, as well as those to be built at the Dayton plant, were the diesel tractor portion of highway tractor-trailers (the "18-wheelers" of CB and TV renown). The tractors were the "cab over"—flat in front and no hood. Basic production followed this format:

1. The running gear (including engine, transmission, axle, brakes, and suspension systems) was purchased from name brand suppliers and

adapted to suit the particular requirements of the buyer.

2. The frame rails were purchased with pre-punched bolt holes in the front. The remaining holes were drilled on semiautomatic machines with patterns, again to suit the configuration ordered by the buyer.

3. The cab was the single component that most distinguished the Star Truck from every other manufacturer's. The cab was entirely assembled in the plant from components both made in the plant and purchased.

4. There were a number of variations on the truck design from which a buyer could choose to individualize or customize a truck.

Star manufactured to order and felt comfortable with a 90-day backlog—much shorter, and the firm felt uneasy about sales; much longer, and they felt that buyers would go to the competition. Although the firm did occasionally sell trucks to dealers for floor models, the usual sales were to the individual owner-driver through a local dealer. The firm would negotiate lower prices for stripped-down models to fleet operators when necessary to absorb slack capacity. Tom would allow no more than three fleet buyer's trucks to be scheduled in the plant at one time. He thought it best to intersperse these multiple orders in with the typical single-unit orders. Star Truck's reputation as the "limousine" of trucks was enhanced by the fact that the buyer had such a wide selection of designs and accessories from which to choose, compared with the limited selection offered by larger volume manufacturers. The truck sales price ran from $70,000 to $120,000—typically higher than that of the competition. However, this higher price apparently did not deter buyers.

To get a better grasp of the whole, Tom made the following attempt to identify, define, and categorize the various departments and work centers.

- *Machine shop.* Did precision machining of component parts, usually in batches to inventory. Occasionally built and repaired simple tools, jigs, and fixtures for both itself and other areas. Would also, at times, do precision machining for the maintenance department. Employees were highly skilled machinists.

- *Maintenance department.* It kept the plant and equipment operating; also included the janitorial function; assembled and installed new machines and processes. With the exception of janitors, the maintenance crew was made up of highly skilled craftsmen such as electricians, maintenance welders, and millwrights. On many of the jobs, more than one trade or skill was involved. Maintenance jobs were scheduled in order of priority: (1) breakdowns, (2) building or installing a new process (once project funds were approved, the investment was charged to the plant; Tom thought it was not too smart to have a $100,000 machine sitting idle in the receiving area), and (3) preventive maintenance (coordinated with machine/process availability and maintenance department slack).

- *Press department.* This contained a variety of brake and punch presses, metal shears, metal rolls, spot welding, metal routers, and metal saws. The machines were used to fabricate a large assortment of component parts from aluminum cab "skin" to support brackets. The parts were made in batches to inventory. The employees were semiskilled machine operators.

- *Production welding area.* Workers assembled and welded various component parts, which were run in batches to inventory requirements. The welders were skilled production welders.

- *Paint areas.* These were scattered throughout the plant.

The smaller component parts were hung on a conveyor line, which transported them through an automatic washer and "static" paint process for undercoat paint. The parts were run in batches through the identical process. The one employee required for this process was very low skilled.

Each finished cab was primed and baked, then given its unique set of colors and trims by painting and baking each color or trim stripe; each color or trim required a separate trip through the paint booth and oven. Each cab was scheduled to match up with the correct truck chassis.

The truck wheels were painted in a separate paint booth near the truck line. Again, each set of wheels, usually numbering 10, was painted to order, although white predominated. The wheels were also scheduled to match the production build schedule.

The chassis or remainder of the truck was painted and baked on the assembly line. The line passed through a paint booth and oven. The painters in these operations were skilled production painters.

- *Frame rail drilling.* All the common holes on the front three-foot portion of the truck rails were prepunched by a vendor. The remainder of the holes (about 75 percent) were drilled on two semiautomatic drilling machines. Both rails of a set of two had the same pattern with the exception of a few holes. The rails were thus clamped together and drilled two at a time. The rails were sched-

uled and drilled to suit the plant build schedule. The workers were all skilled machine operators.

- *Assembly processes.* These included frame assembly, suspension assembly, transmission to engine assembly, drive shaft assembly, cab assembly, cab trim, door assembly, and wheel tire assembly. These minor assembly operations fed into the major truck assembly operations. Each of these operations performed the same basic set of tasks on each unit, as on an auto assembly line. The major difference at Star Truck was the fixtures and tooling that were necessary for each set of possible combinations. For example, the axle/suspension/brake assembly required an inventory of welding fixtures. Another oddball operation was the drive shaft assembly. The product engineers emphatically declared that they could not consider all the variables that would enable them to predetermine drive shaft lengths. Therefore, each drive shaft was custom fit as the truck chassis was pulled along the assembly line, thus precluding many of the economies that would have been possible otherwise.

Figure 15-12 is a schematic representation of most of the productive processes with which Tom had to be concerned in trying to design the new plant. Ancillary functions such as shipping, receiving, and materials management Tom neglected for the moment. He felt that if he could get a firm grasp of his problem with the processes described above, then he could fit in the others.

1. What concerns or considerations would you have if you were Tom?
2. What do you suggest that he do?

SITUATION FOR STUDY 15-2

HERRIESTON CORPORATION

The plastics division of Herrieston Corporation had sales of $53 million and produced a variety of plastic parts out of 12 plants scattered throughout the south and the southeast. In the late 1960s researchers at DuPont Corporation developed a new bottle made of plastic that was shatterproof, 11 times lighter than the comparable glass bottle it could replace, and capable of storing liquids for three months without loss of flavor or carbonation.

The plastics division, seeing this development as an extremely attractive growth oppor-

tunity, seized upon this idea by proposing to purchase a number of machines that could utilize this technology. The costs of this technology were significant. If the division were to place a 2-liter bottle-making machine in each of the 12 plants, the cost would be $13 million in capital outlay. Having machines in each of the geographically scattered plants was attractive from the standpoint of servicing local markets but unattractive from the standpoint of introducing new technologies to plants that performed many different kinds of operations.

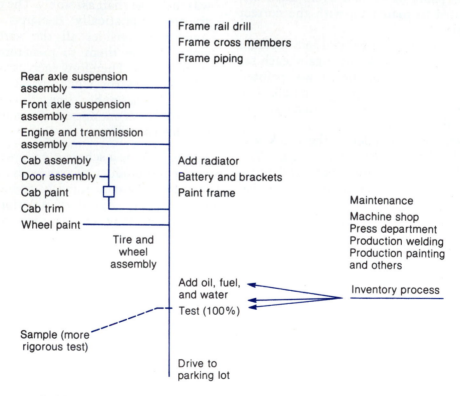

FIGURE 15-12   Major processes to be incorporated into the new Star Truck plant.

Relatively few of the plants shared any commonalities with the new technology and those commonalities were simply the manufacturing of milk bottles. One decision facing the division was how big an investment to make and whether the new plastic bottle-making machines ought to be introduced to the existing plants of the division or whether new plants in selected areas ought to be constructed.

Most of the costs of bottle making were in materials. Plastic resins, being oil derivatives, were much more expensive per pound than the raw materials of glass fabrication. Glass making, on the other hand, was much more energy-intensive than the making of plastic. The 2-liter plastic bottle was a good choice to start off producing because its ratio of product volume to plastic was much higher than the similar ratio for smaller bottles. Still, as the plastic bottles were currently engineered, materials costs were of concern relative to glass bottles.

What should Herrieston's strategy be? How big an investment should it make in two-liter plastic bottles? Where should this investment be made? What strategies for manufacturing, especially in the future, should be adopted?

SITUATION FOR STUDY 15-3

Cowles Computer Corporation

Cowles Computer was a small but rapidly growing maker of medium-size digital computers (minis and superminis, in the trade's jargon, rather than personal computers or mainframes). Sales were doubling every year and prospects for sales growth in excess of 50 percent a year remained strong. Cowles manufactured computers for so-called end use, that is, for use by customers without any further modification. The end-use market could be contrasted with the original equipment manufacturer OEM market, which dealt with computers that would be included in other products (such as large machine tools). Customers typically wanted different special features added to the basic computer that they purchased. Thus, part of any computer sold looked like all others, and part was unique. In its product development, Cowles stressed the development of software (for example, programs, operating language) rather than hardware. Thus, although technically its computers were not particularly advanced or fast, they were extraordinarily versatile and could communicate directly with all other models of Cowles computers. Cowles had been introducing more and more advanced computer systems at a rate of about one new model a year, a fast pace even for the industry.

The assembly of a computer from its component pieces and its subsequent testing, although a complex task, did not require vast scale or particularly sophisticated equipment. The secret to performance lay in the engineering.

What should be Cowles Computer's manufacturing strategy?

REFERENCE NOTES

1. See Wickham Skinner, *Manufacturing: The Formidable Competitive Weapon*, New York: Wiley, 1985, which brings together a number of Skinner's articles that appeared first in the *Harvard Business Review.*

2. See Robert W. Hall, *Attaining Manufacturing Excellence*, Dow Jones-Irwin, 1988.

3. See Roger W. Schmenner, "The Merit of Making Things Fast." *Sloan Management Review* 30, no. 1 (Fall 1988), pp. 11–17. Also consult Schmenner, "Behind Labor Productivity Gains in the Factory." *Journal of Manufacturing and Operations Management* (Elsevier), Winter, 1988, vol. 1, no. 4, pp. 323–338; Schmenner, "An International Comparison of Factory Productivity," with Rho, Boo Ho, *International Journal of Operations and Production Management* (MCB University Press), vol. 10, no. 4 (1990), pp. 16–31; and Schmenner, "International Factory Productivity Gains," *Journal of Operations Management*, vol. 10, no. 2 (April 1991).

4. Wickham Skinner, "The Focused Factory," *Harvard Business Review*, 52 no. 3 (May–June 1974), 113–121.

5. See Robin Cooper and Robert Kaplan, "Measure Costs Right: Make the Right Decisions," Harvard Business Review 66, no. 5 (September–October 1988): 96–103.

6. See Roger Schmenner, "Escaping the Black Holes of Cost Accounting, *Business Horizons* (January–February 1988): 66–72. A worked example is included in this article.

7. For an excellent book on the subject, see J. Robb Dixon, Alfred J. Nanni, and Thomas E. Vollmann, *The New Performance Challenge: Measuring Operations for World-Class Competition*, Homewood, IL: Business One–Irwin, 1990.

8. This section is based on Robert H. Hayes and Roger W. Schmenner, "How Should You Organize Manufacturing?" *Harvard Business Review* 56, no. 1 (January–February 1978): 105–118.

Some additional sources of information on manufacturing strategy include the following:

Adam, Everett E., Jr., and Paul M. Swamidass, "Assessing Operations Management from a Strategic Perspective," *Journal of Management* 15, no. 2 (June 1989): 181–203.

Anderson, John C., Gary Cleveland, and Roger G. Schroeder, "Operations Strategy: A Literature Review," *Journal of Operations Management* 8, no. 2 (April 1989): 133–59.

Anderson, John C., Roger G. Schroeder, and Gary Cleveland, "The Process of Manufacturing Strategy: Some Empirical Observations and Conclusions," *International Journal of Operations & Production Management* 11, no. 3 (1991): 86–110.

Bower, Joseph L., and Thomas M. Hout, "Fast-Cycle Capability for Competitive Power," *Harvard Business Review* 66, no. 6 (November–December 1988): 110–118.

Buffa, E. S., *Meeting the Competitive Challenge: Manufacturing Strategies for U.S. Companies*, Homewood, IL: Irwin, 1984.

Chase, Richard B., and W. J. Erickson, "The Service Factory," *Academy of Management Executive* 2, no. 3 (August 1988): 191–96.

Chase, Richard B., and David A. Garvin, "The Service Factory," *Harvard Business Review* 67, no. 4 (July–August 1989): 61–70.

Clark, Kim B., and Robert H. Hayes, "Recapturing America's Manufacturing Heritage," *California Management Review* 30, no. 4 (Summer 1988): 9–33.

Cleveland, Gary, Roger G. Schroeder, and John C. Anderson, "A Theory of Production Competence," *Decision Sciences* 20, no. 4 (Fall 1989): 655–668.

Cook, William J., "Ringing in Saturn," *U.S. News & World Report* 109, no. 16 (October 22, 1990): 51–4.

Cross, Kelvin F., "Making Manufacturing More Effective by Reducing Throughput Time," *National Productivity Review* 6, no. 1 (Winter 1986–1987): 35–47.

De Meyer, Arnoud, Jinichiro Nakane, Jeffrey G. Miller, and Kasra Ferdows, "Flexibility: The Next Competitive Battle—The Manufacturing Futures Survey," *Strategic Management Journal (UK)* 10, no. 2 (March–April 1989): 135–44.

Ettlie, John E., Michael C. Burstein, and Avi Fiegenbaum, *Manufacturing Strategy*, Boston MA: Kluwer Academic Publishers, 1990.

Ferdows, Kasra, and Arnoud De Meyer, "Lasting Improvements in Manufacturing Performance: In Search of a New Theory," *Journal of Operations Management* 9, no. 2 (April 1990): 168–84.

Ferdows, Kasra, Jeffrey G. Miller, Jinichiro Nakane, and Thomas E. Vollmann, "Evolving Global Manufacturing Strategies: Projections into the 1990s," *International Journal of Operations Management (UK)* 6, no. 4 (1986): 6–16.

Ferdows, Kasra, and Wickham Skinner, "The Sweeping Revolution in Manufacturing," *Journal of Business Strategy* 8, no. 2 (Fall 1987): 64–9.

Gunn, T. G., *Manufacturing for Competitive Advantage: Becoming a World Class Manufacturer*, Cambridge, MA: Ballinger Publishing Co., 1987.

Hax, Arnoldo C., and Nicholas S. Majluf, "The Concept of Strategy and the Strategy Formation Process," *Interfaces* 18, no. 3 (May/June 1988): 99–109.

Hax, Arnoldo C., and Nicolas S. Majluf, *The Strategy Concept and Process: A Pragmatic Approach*, Englewood Cliffs, NJ: Prentice-Hall, 1991.

Hayes, Robert H., "Strategic Planning—Forward in Reverse?" *Harvard Business Review* 63, no. 6 (November–December 1985): 111–9.

Hayes, Robert H., and Steven C. Wheelwright, *Restoring Our Competitive Edge: Competing Through Manufacturing.* New York: Wiley, 1984.

_____, Steven C. Wheelwright, and Kim B. Clark, *Dynamic Manufacturing: Creating the Learning Organization*, New York: Free Press, 1988.

Hayes, Robert H., Steven C. Wheelwright, and Kim B. Clark, "The Power of Positive Manufacturing," *Across the Board* 25, no. 10 (24-30).

Hill, Terry, *Manufacturing Strategy*, Homewood, IL: Irwin, 1989.

_____, and Stuart Chambers, "Flexibility—A Manufacturing Conundrum," *International Journal of Operations & Production Management (UK)* 11, no. 2 (1991): 5–13.

Jackson, Dave, and Roland Toone, eds., *The Management of Manufacturing: The Competitive Edge*, New York: Springer-Verlag, 1987.

Kaplan, Robert S., "Yesterday's Accounting Undermines Production," *Harvard Business Review* 62, no. 4 (July–August 1984): 95–101.

Klein, Janice A., *Revitalizing Manufacturing: Text and Cases*, Homewood, IL: Irwin, 1990.

Kotha, Suresh, and Daniel Orne, "Generic Manufacturing Strategies: A Conceptual Synthesis," *Strategic Management Journal (UK)* 10, no. 3 (May–June 1989): 211–231.

Krafcik, John F., "Triumph of the Lean Production System," *Sloan Management Review* 30, no. 1 (Fall 1988): 41–52.

Marucheck, Ann, Ronald Pannesi, and Carl Anderson, "An Exploratory Study of the Manufacturing Strategy Process in Practice," *Journal of Operations Management* 9, no. 1 (January 1990): 101–23.

Meredith, Jack R., "The Strategic Advantages of the Factory of the Future," *California Management Review* 29, no. 3 (Spring 1987): 27–41.

Miller, J. G., and A. V. Roth, "Manufacturing Strategies," Executive Summary of the 1988 Manufacturing Futures Survey, School of Management, Boston University.

Moody, Patricia E., ed., *Strategic Manufacturing: Dynamic New Directions for the 1990s*, Homewood, IL: Dow-Jones Irwin, 1990.

National Center for Manufacturing Sciences, Craig Giffi, Aleda V. Roth, and Gregory M. Seal, *Competing in World-Class Manufacturing: America's 21st Century Challenge*, Homewood, IL: Business One Irwin, 1991.

Peters, Tom, "Part One: Get Innovative or Get Dead," *California Management Review* 33, no. 1 (Fall 1990): 9–26.

_____, "Part Two: Get Innovative or Get Dead," *California Management Review* 33, no. 2 (Winter 1991): 9–23.

_____, *Thriving on Chaos: Handbook for a Management Revolution*, New York: Alfred A. Knopf, 1987.

Ruwe, Dean M., and Wickham Skinner. "Reviving a Rust Belt Factory," *Harvard Business Review* 65, no. 3 (May–June 1987): 70–6.

Schonberger, Richard J., "Frugal Manufacturing," *Harvard Business Review* 65, no. 5 (September–October 1987): 95–100.

_____, *World Class Manufacturing: The Lessons of Simplicity Applied*, New York: Free Press, 1986.

Schroeder, Roger G., John C. Anderson, and Gary Cleveland, "The Content of Manufacturing Strategy: An Empirical Study" *Journal of Operations Management* 6, no. 3/4 (May–August 1986): 405–415.

Shapiro, Harris Jack, and Teresa Cosenza, *Reviving Industry in America: Japanese Influences on Manufacturing and the Service Sector*, Cambridge, MA: Ballinger, 1987.

Shetty, Y.K., and Vernon M. Buehler, eds., *The Quest for Competitiveness*, New York: Quorum, 1991.

Skinner, Wickham, "A Strategy for Competitive Manufacturing," *Management Review* 76, no. 8 (August 1987): 54–56.

Stalk, George, Jr., "Time—The Next Source of Competitive Advantage," *Harvard Business Review* (July-August 1988): 41–51.

Starr, Martin K., *Global Competitiveness: Getting the U.S. Back on Track*, New York: W.W. Norton & Co., 1988.

Swamidass, Paul M., "Manufacturing Strategy: A Selected Bibliography," *Journal of Operations Management* 8, no. 3 (August 1989): 263–77.

———, "Manufacturing Strategy: Environmental Uncertainty and Performance: A Path Analytic Model," *Management Science* 33, no. 4 (April 1987): 509–524.

———, "Manufacturing Strategy: Its Assessment and Practice," *Journal of Operations Management* 6, no. 3/4 (May–August 1986): 471–484.

Thompkins, James A. *Winning Manufacturing*, Norcross: Institute of Industrial Engineers, 1989.

Wheelwright, Steven C., and Robert H. Hayes, "Competing through Manufacturing," *Harvard Business Review* 63, no. 1 (January–February 1985): 99–109.

Whybark, D. Clay, "Evolving the Manufacturing Strategy," *Engineering Costs & Production Economics (Netherlands)* 12, no. 1–4 (July 1987): 243–250.

"Explaining Productivity Differences in North Carolina Factories," with Randall Cook, *Journal of Operations Management* (APICS) (May 1985): 273–279.

"The Merit of Making Things Fast," *Sloan Management Review* 30, no. 1 (Fall 1988): 11–17.

"Behind Labor Productivity Gains in the Factory," *Journal of Manufacturing and Operations Management* 1, no. 4 (1988): 323–338.

"International Factory Productivity Gains", *Journal of Operations Management*, Spring 1991.

Other articles on productivity and plant improvement include:

Adler, Paul S., "A Plant Productivity Measure for 'High-Tech' Manufacturing," *Interfaces* 17, no. 6 (November–December 1987): 75–85.

Cross, Kevin, *Manufacturing Planning: Key to Improving Industrial Productivity*, New York: Marcel Dekker, 1986.

Edosomwan, Johnson Aimie, *Integrating Productivity and Quality Management*, New York: Marcel Dekker, 1987.

Hayes, Robert H., and Kim B. Clark, "Why Some Factories are More Productive than Others," *Harvard Business Review* 64, no. 5 (September–October 1986): 66–73.

Harmon, Roy, and Leroy Peterson, *Reinventing the Factory: Productivity Breakthroughs in Manufacturing Today*, New York: Free Press, 1990.

Kobayashi, Iwao, *20 Keys to Workplace Improvement*, Cambridge, MA: Productivity Press, 1990.

Schmenner, Roger W., and Boo Ho Rho, "An International Comparison of Factory Productivity," *International Journal of Operations & Production Management* 10, no. 4 (1990): 16–31.

Schroeder, Roger G., John C. Anderson, and Gary D. Scudder, "White Collar Productivity Measurement," *Management Decision (UK)* 24, no. 5 (1986): 3–7.

Shetty, Y. K., and Vernon M. Buehler, eds., *Productivity and Quality Through People: Practices of Well-Managed Companies*, Westport, CT: Quorum, 1985.

———, *Productivity and Quality Through Science and Technology*, New York: Quorum, 1988.

Skinner, Wickham, "The Productivity Paradox," *Harvard Business Review* 64, no. 4 (July–August 1986): 55–59.

Timpe, A. Dale, ed., *Productivity*, New York: Facts on File, 1989.

Shingo, Shigeo, *The Sayings of Shigeo Shingo: Key Strategies for Plant Improvement* (Japanese Management Series), Cambridge, MA: Productivity Press, 1987.

Tours A–J
Tour Summary

A CONTINUOUS FLOW PROCESS
International Paper Company
Androscoggin Mill
Jay, Maine

The Androscoggin Mill, situated along the Androscoggin River in central Maine, was one of 28 domestic pulp and paper mills of the International Paper Company. It was one of the largest, built originally in 1965 with significant additions in 1968 and 1977. The plant occupied 478 acres of land, had 20 acres under roof, and represented a book value investment of nearly $400 million.

The mill, part of International Paper's Pulp and Coated Papers Group, produced three distinctly different kinds of paper: (1) forms bond, envelope, tablet, and offset paper for office or computer use; (2) publication gloss for magazine printing; and (3) specialty papers such as microwave popcorn and fast-food french fry grease-proof papers.

Androscoggin was a fully integrated mill; that is, it produced all the wood pulp it needed to make its paper. The mill was laid out to receive logs or chips of wood at one end and, about a mile farther down, to ship packaged "logs" of paper from the other end. (See Figure A1 for a layout of the mill; the production flow is from left to right.)

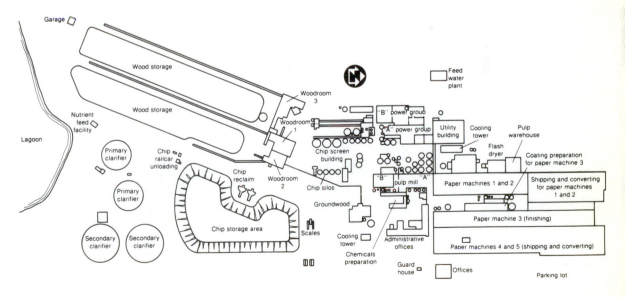

FIGURE A1   Layout of International Paper Company's Androscoggin Mill.

PART ONE

PROCESS DESCRIPTION

A Brief and Simplified Description of Papermaking

The paper we use today is created from individual wood fibers that are first suspended in water and then pressed and dried into sheets. The process of converting the wood to a sus- pension of wood fibers in water is known as pulpmaking, while the manufacture of the dried and pressed sheets of paper is formally termed papermaking. The process of making paper has undergone a steady evolution, and larger and more sophisticated equipment and better technology continue to improve it.

In the woodyard, wood being unloaded. (*Courtesy of International Paper*)

The Woodyard and Woodrooms

The process at Androscoggin began with receiving wood in the form of chips or of logs 4 or 8 feet in length. From 6 A.M. to 10 P.M. a steady stream of trucks and railroad cars were weighed and unloaded. About 40 percent of the deliveries were of wood cut from International Paper's own land (over 1 million acres in Maine), while the other 60 percent were supplied by independents who were paid by weight for their logs. The mill also received wood chips from lumber mills in the area. The chips and logs were stored in mammoth piles with separate piles for wood of different species (such as pine, spruce, hemlock).

When needed, logs were floated in flumes from the woodyard into one of the mill's three woodrooms. There, bark was rubbed off in long, ribbed debarking drums by tumbling the logs against one another. The logs then fell into a chipper; within seconds a large log was reduced to a pile of chips approximately 1 inch by 1 inch by 1/4 inch.

The chips were stored in silos. There were separate silos for softwoods (spruce, fir, hemlock, and pine) and hardwoods (maple, oak, beech, and birch). This separate and temporary storage of chips permitted the controlled mixing of chips into the precise recipe for the grade of paper being produced.

The wood chips were then sorted through large, flat vibrating screens. Oversized chips were rechipped, and ones that were too small were collected for burning in the power house. (The mill provided approximately 20 percent of all its own steam and electricity needs from burning waste. An additional 50 percent of total electricity needs was produced by harnessing the river for hydroelectric power.)

Once drawn from the silo into the digesters, there was no stopping the flow of chips into paper.

Pulpmaking

The pulp made at Androscoggin was of two types: Kraft pulp (produced chemically) and groundwood pulp (produced mechanically). Kraft pulp was far more important to the high-quality white papers produced at Androscoggin, accounting for 80 percent of all the pulp used. Kraft pulp makes strong paper. (Kraft is German for strength. A German invented the Kraft pulp process in 1884.) A paper's strength generally comes from the overlap and binding of long fibers of softwood; only chemically was it initially possible to separate long wood fibers for suspension in water. Hardwood fibers are generally smaller and thinner and help smooth the paper and make it less porous.

The groundwood pulping process was simpler and less expensive than the Kraft process. It took high quality spruce and fir logs and pressed them continuously against a revolving stone that broke apart the wood's fibers. The fibers, however, were smaller than those produced by the Kraft process and, although used to make newsprint, were useful at Androscoggin in providing "fill" for the coated publication gloss papers of machines 2 and 3, as will be described later.

The chemical Kraft process worked by dissolving the lignin that bonds wood fibers together. It did this in a tall pressure cooker, called a digester, by "cooking" the chips in a solution of caustic soda (NaOH) and sodium sulfide (Na_2S), which was termed the "white liquor." The temperature in this cooking process reached as high as 340°F, and the pressure was as great as 11 atmospheres. The two digesters at Androscoggin were continuous digesters; chips and liquor went into the top, were cooked together as they slowly settled down to the bottom, and were drawn off the bottom after about three hours. By this time, the white liquor had changed chemically to

A chip unloading station. (*Courtesy of International Paper*)

A groundwood log conveyor. (*Courtesy of International Paper*)

A debarking drum. (*Courtesy of International Paper*)

"black liquor"; the digested chips were then separated from this black liquor.

In what was known as the "cold blow" process, the hot, pressurized chips were gradually cooled and depressurized. A "cold liquor" (170°F) was introduced to the bottom of the digester and served both to cool and to transport the digested chips to a diffusion washer that washed and depressurized the chips. Because so much of the lignin bonding the fibers together had been removed, the wood fiber in the chips literally fell apart at this stage.

The black liquor from the digester entered a separate four-step recovery process. Over 95 percent of the black liquor could be reconstituted as white liquor, thereby saving on chemical costs and significantly lowering pollution. The four-step process involved (1) washing the black liquor from the cooked fiber to produce weak black liquor, (2) evaporating the weak black liquor to a thicker consistency, (3) combustion of this heavy black liquor with sodium sulfate (Na_2SO_4), and redissolving the smelt, yielding a "green liquor" (sodium carbonate + sodium sulfide), and (4) adding lime, which reacted with the green liquor to produce white liquor. The last step was known as causticization.

Meanwhile, the wood-fiber pulp was purged of impurities like bark and dirt by mechanical screening and by spinning the mixture in centrifugal cleaners. The pulp was then concentrated by removing water from it so that it could be stored and bleached more economically.

By this time, depending on the type of pulp being made, it had been between 3 1/2 and 5 hours since the chips had entered the pulp mill.

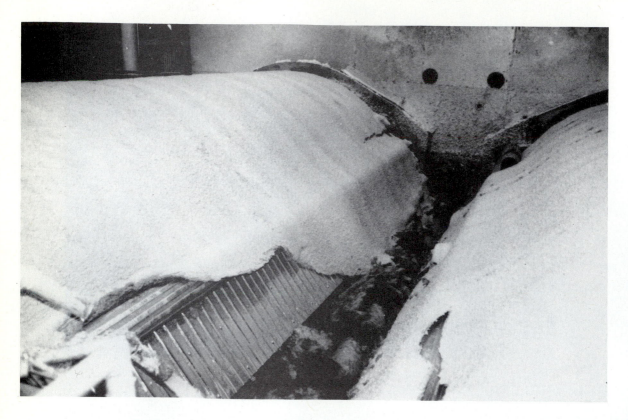

A pulp washer. (*Courtesy of International Paper*)

All the Kraft pulp was then bleached. Bleaching took between 5 and 6 hours. It consisted of a three-step process in which (1) a mix of chlorine (Cl_2) and chlorine dioxide (ClO_2) was introduced to the pulp and the pulp was washed; (2) a patented mix of sodium hydroxide (NaOH), liquid oxygen, and hydrogen peroxide (H_2O_2) was then added to the pulp and the pulp was again washed; and (3) chlorine dioxide (ClO_2) was introduced and the pulp washed a final time. The result was like fluffy cream of wheat. By this time the pulp was nearly ready to be made into paper.

From the bleachery, the stock of pulp was held for a short time in storage (a maximum of 16 hours) and then proceeded through a series of blending operations that permitted a string of additives (for example, filler clay, resins, brighteners, alum, dyes) to be mixed into the pulp according to the recipe for the paper grade being produced. Here, too, ''broke'' (paper wastes from the mill itself) was recycled into the pulp.[a] The pulp was then once again cleaned and blended into an even consistency before moving to the papermaking machine itself.

[a]It made a difference whether the broke was of coated or uncoated paper, and whether it was white or colored. White, uncoated paper could be recycled immediately. Colored, uncoated paper had to be rebleached. Coated papers, becasue of the clays in them, could not be reclaimed.

Continuous digesters at the pulp mill. These work like pressure cookers, dissolving the lignin bonds between wood fibers. (*Courtesy of International Paper Company*)

Papermaking

The paper machine was a simply awesome engineering creation, stretching hundreds of feet. At Androscoggin, there were five of varying size. The paper machine had a wet end and a dry end. The pulp entered the machine at the headbox, which released the heavily diluted pulp through a slit onto a moving belt of synthetic fabric. This belt was called the "wire" because originally it was made of bronze mesh.

As the wire with its deposit of pulp stock moved away from the headbox, water drained through it. Suction was also applied. Within 60 feet after the headbox slit, so much water had been removed that it could be said the wire carried paper, rather than a suspension of pulp in water. From the wire, huge woolen felts picked up the paper "web" and started it through a series of rollers and steam-heated drying drums, which pressed and evaporated even more water out of the web. If the paper was to be coated, the coating would be applied first to one side of the paper and then to the other and more drying done. If the paper was to be shiny, it would be pressed on smooth, shiny rollers called calenders. Drying and pressing having been done, the paper was essentially finished and was picked up from the machine on winders.

The speeds at which all this happened were incredible. The fastest machines (machines 2 and 3, devoted to publication gloss) produced over a mile of paper every 2 minutes. This great speed was made possible by continual improvements made by the machine's builders and by computerized process controls, which constantly adjusted the machine's settings based on the values registered by numerous electronic sensors scattered along the length of the machine.

Once on the winders, the "logs" of paper could be slit directly into widths the customers had ordered, or they could be slit into two logs of equal width and stored until needed. These stored logs, called "parent logs," were then placed on rewinders for successive slitting to customer order. Once cut to order, rolls were packaged, labeled, and then shipped.

MAINTAINING THE ENVIRONMENT

Being a relatively new mill, Androscoggin had environmental control equipment designed into it. Pulp and paper mills were large users of water. Androscoggin took water from the river, filtered it, and then used it in the process. Before being released back to the river, the water was treated in three stages to remove the effluent it had picked up.

1. A primary clarifier (two large cone-shaped basins) allowed particles to settle to the bottom. A slow-moving rake forced the settled sludge to the middle, where it was removed to be burned.

2. In a 37-acre lagoon the clarified effluent was aerated by a collection of churning fountains, and a special collection of microorganisms broke down the effluent over the course of 1 1/2 to 2 days.

3. A secondary clarifier (two more large basins) settled more solids (including dead bacteria from the lagoon) from the water.

The entire treatment process took 2–3 days.

Air pollution control was another area where great strides had been made since the early 1960s. Gas collection systems to collect and burn noncondensible gases from the digester and evaporators had largely eliminated the rotten egg smell that had been characteristic of paper mills. Scrubbers and precipitators on the stacks at Androscoggin had gone even further in removing sulphur and other particulates from the smoke and air at the mill.

THE DESIGN OF THE ANDROSCOGGIN MILL

Vertical Integration

The Androscoggin Mill was vertically integrated; that is, it combined pulpmaking and papermaking at the same site. Not all mills, especially not the older ones, were designed to combine both activities. Several advantages argued strongly in favor of vertical integration:

1. Integration removed the transportation costs of transporting pulp between a pulp mill and a paper mill. Vertical integration eliminated the double handling of pulp, a cost saving in any event. In addition, because the mill shipped all over the country and because it specialized in lightweight grades of paper, transportation costs for finished paper were not affected significantly by the mill's location; and so a location near the supply of pulp was highly desirable. Pulp drying and reslushing costs were also eliminated.

2. Integration provided for better quality control. Since the quality of the paper produced depended heavily on the quality of the pulp, vertical integration permitted hour-by-hour control over the quality of pulp entering each paper machine. Such control greatly reduced the chances for paper breaks on the machine and it also eliminated any quality problems caused by old pulp or other deficiencies.

3. The wood preparation and pulpmaking operations generated sawdust, bark, chemicals, and wood chips that could be burned as fuel for the whole plant, thus cutting down on plant-wide energy costs.

Although the Androscoggin Mill integrated pulpmaking and papermaking, the extent of the mill's vertical integration could have been even broader. For example, only about 40 percent of the wood used at the mill was cut from land owned by International Paper. The company could have integrated backward even more by raising that 40 percent figure. It chose not to do so, however, because the 40 percent figure, and modest variations around it, (1) permitted good forest management of International Paper's own land and (2) meant that the company could add directly to the area's economy by purchasing from independents. There were no substantial cost differences between independent and company tree harvesting. Only if independent loggers could not supply the mill's needs would the company have harvested more from its own land.

By the same token, the operations at Androscoggin could have been extended by forward integration, so that the plant converted more of its paper into end uses (for example, sheets of bond paper rather than rolls of it). The company chose not to do this as well. For one thing, adding converting facilities would have greatly increased Androscoggin's already large size and would have diluted management's attention from the mission of producing quality paper efficiently. Converting operations (and International Paper operated many of them) were also best located closer to markets, since the nature of conversion often placed the paper in bulkier and more awkward packages that were harder to ship and were often shipped in smaller quantities than Androscoggin's full truck and rail carloads. Splitting conversion from papermaking, at least for white papers, meant lower overall transport costs, whereas there were no real economies in the linking of the two operations.

Capacity

Over time, the size and capacity of individual paper machines had grown enormously. There were distinct advantages to this large scale of operation, since the largest and latest machines could produce more tons of paper per worker, thus reducing costs. At Androscoggin, for example, before 1981, machines 2 and 5 both manufactured carbonizing tissue. Machine 2, built in 1965, could produce paper 208 inches wide at a speed of 1750 feet per minute. Machine 5, in contrast, built in 1977, could produce paper 230 inches wide at a rate of 2000 feet per minute. In only 12 years, the capacity of the machine per unit of time had increased by over 26 percent. The capacity difference between

machines 1 and 4 was even more striking — 40 percent. The increases in machine size and speed enabled the newer mills to house fewer paper machines and employ fewer workers than older mills and yet produce more paper.

At its current size, the Androscoggin Mill was about as large as it was likely to grow. In 1980, the company rebuilt paper machine 2 and converted its production from carbonizing tissue to lightweight publication gloss. The project, which cost $71 million, included the addition of some groundwood pulping, supercalenders, and coating preparation processes. The output rates of its five paper machines (over 500,000 tons per year, up from approximately 410,000 tons per year before the conversion of machine 2), its two continuous digesters, its groundwood mill, and its three woodrooms were all reasonably well balanced. As it was, the pulp mill had been deliberately built larger so that another mill down the river and other International Paper mills could be supplied with pulp.

Location

The Androscoggin Mill's general location in central Maine was attributable largely to three factors:

1. The production of strong, good-quality white office papers demanded long wood fibers. This demand favored a Maine location because cold winter weather favors the production of longer fibers in both the softwood and hardwood trees that Maine and neighboring Quebec grow in abundance.

2. A paper mill uses so much water that it has to be located on either a lake or a river, and preferably a river because of environmental factors.

3. This part of central Maine enjoyed the availability of a good supply of labor.

Paper machine 5, built in 1977. (*Courtesy of International Paper Company*)

Technology

In pulpmaking and papermaking, keeping abreast of the latest changes in process equipment and technology is critical to a company's continued success. At International Paper the main responsibility for keeping up with technology rested with the corporate staff. The staff was also instrumental in the selection of major new equipment or process changes. The plant, however, did play an important role by providing managers (such as the plant engineer or the paper mill superintendent) who joined the corporate staff in studying any new capacity addi-

tion or process technology change. This team was responsible for developing the project's capital appropriation request and its supporting documents.

LOADING THE FACTORY

Production Planning

Orders for any of the white papers produced at Androscoggin would be taken by the company's central order department in Memphis, Tennessee. Orders, as they were received, were

Paper machine 4. (*Courtesy of International Paper Company*)

placed in the next available open time in the planned weekly production cycles. One week before the start of the next weekly cycle, the manager of operations control planned the production schedule for that cycle. This production schedule allocated the types and weights of paper to be produced at each mill by each machine for every day of the weekly cycle. These machine assignments and the orders they represented were then transmitted to each of the company's plants.

Each quarter a linear program was run to develop the optimum schedule. Marketing and customer constraints were factored into the quarterly plan by operations control. This plan was reviewed with product line managers and was used to establish weekly cycles.

The assignment of types and weights of paper to each paper machine was a sophisticated enterprise. Because the company's machines were all slightly different from one another, due primarily to their different vintages, their capabilities differed. That is, they were often relatively more successful and/or more cost efficient at making certain types and weights of paper than others. In assigning products to machines, therefore, the corporation took care to make the assignments in such a way that the company's total contribution to profits and overhead was maximized. This assigning

A paper machine coater. (*Courtesy of International Paper*)

A paper machine heated dryer "cans." (*Courtesy of International Paper*)

required not only information on the mix of papers demanded for the period and on machine capabilities, but on distribution and transportation costs as well. Thus for some grades, colors, or weights of paper, depending on the situation, Androscoggin might have filled an order that might otherwise have been filled by a mill much closer to the customer.

At present, Androscoggin's five paper machines were likely to be assigned the following "paper machine products":

- Machine 1 — various types of bond or offset paper in white or colors
- Machine 2 — coated publication gloss
- Machine 3 — coated publication gloss
- Machine 4 — various types of bond, ledger, envelope, and tablet paper (all white)
- Machine 5 — specialty grades

Since machine 4 was both larger and faster than machine 1, it was devoted to fewer different kinds of paper and thus to longer production runs than machine 1. In that way, the mill could produce more total paper than if machine 4 continually had to be interrupted and reset for different colors, grades, or weights. Better to interrupt machine 1 for such changes.

Demand for paper products was cyclical in nature. Volume requirements tracked the general economy very closely, although lagging it slightly both upward and downward. In good economic times, Androscoggin and other mills could be running flat out, with sales so strong that customers had to be put on allocation and orders monitored very closely. This situation was far different from some recessionary times when mills took any order and still had to cut back production.

While the company's central order department (operations control) decided what would be produced on what machine on which days, the mill still had its own considerable production planning problems. Customer orders not only specified the type of paper but also the size (width) to be cut, the quantity (the diameter of the roll), and some other features. In planning for any specific run on a paper machine, the mill wanted to group together all those orders calling for the same diameter and in such a way that the entire width of the paper machine's output was accounted for. The mill did not like to leave any waste, since even an inch of paper unaccounted for could cost the company foregone revenue of several thousand dollars. The mill used a computer program to optimize paper machine trim.

The mill also wanted to schedule slitting the rolls to minimize additional handling. This goal meant slitting as much off the machine winder as possible; small widths, however, should be cut from the center portion to reduce the risk of distortion in slitting.

Two production planners scheduled orders on the mill's five machines. They had to remain flexible enough to allow at least some changes in customer orders as late as the Wednesday before the start of the next week's production. The mill also had to schedule trucks and railcars for order shipments.

Inventories and Purchasing

Most of the wood that Androscoggin Mill used was provided by contract for 10,000 to 100,000 cords per year. International Paper did not cut its own wood, but contracted that task out as well. The company's woodlands division was responsible for purchasing the mill's wood and managing the woodyard at economical inventory levels. In addition, the mill stood ready to purchase, by weight, the wood of small, independent loggers without prior commitments being made. The mill did, however, deliberately seek the building up of wood in inventory in the fall and winter so that the spring thaw and the mud it brought did not disrupt the mill as it disrupted logging. It was not uncommon

A paper machine take-up reel. (*Courtesy of International Paper*)

to have 30 days of production in pulp wood inventory. All the work-in-process inventories were small and temporary, used mainly to permit the mixing of different types of wood chips or pulp and to permit production to continue in case some portion of the process went down temporarily.

The extent of finished goods inventory varied. Comparatively few finished goods were held for the products of machines 2, 3, and 5. Some were kept to permit stray order fulfillment or order amendments, but most paper production was soon headed out the door, waiting only long enough for all items in the order to be gathered together. The parent log and finished goods inventories for machine 1 (bond and offset papers) were very much larger, typically about 15 days of production. Such a large inventory was required to meet customer needs on any of the 48 paper machine products that machine 1 was qualified to manufacture, while keeping to a minimum the number of machine setups over the course of a year.

At Androscoggin, all responsibilities for production planning, loading machines, product distribution to customers, and finished goods inventories rested with the planning, scheduling, and distribution department because it was felt that all of these items affected one another at the plant and should be controlled by the same authority.

THE WORKFORCE

The Androscoggin Mill employed about 1400 people, about 250 of whom were salaried. The 1100 or so hourly employees (paid according to the hours worked), were represented by three international unions. The United Paperworkers International Union (UPIU) represented the woodyard, woodroom, pulp mill, paper mill, maintenance, finishing, and shipping workers; the International Brotherhood of Firemen and Oilers represented the power house employees and plant electricians; and the Office and Professional Employees International Union represented many of the office and clerical workers at the mill. The agreements were specific to Androscoggin and had been negotiated by the plant's management.

Labor relations at the plant had been soured by an extended strike that lasted from June 1987 to October 1989. The UPIU, in an effort to reverse a series of concessions that it had granted other manufacturers, stood firm against the company's contract offer. The company, faced with increasing competition, felt obliged to reduce what it felt were costly concessions it had made in earlier contracts, specifically the pay of double time for Sunday and holiday work and the mandatory shutdown of the mill at Christmas. (Other companies had successfully altered these items in their new contracts with the UPIU.) The union finally accepted the company's offer, but not until all of the hourly employees then striking were replaced with other workers. Because of subsequent attrition, 250 of the 1100 who went out

on strike were back working at the mill. A petition from the replacement workers to decertify the UPIU was, at the time of this tour's writing, under review by the National Labor Relations Board.

Work at the mill was defined as either day work or tour (shift) work. Most work was in fact tour work. At the suggestion of the workforce, the mill operated a modified 12-hour shift (see Figure A2). Workers worked 16 out of every 28 days. Every day, except Thurdays and Fridays, workers worked either 7 A.M. to 7 P.M., or 7 P.M. to 7 A.M. On Thursdays and Fridays, they worked more traditional 8-hour shifts. Workers worked no more than 5 days in a row and no more than 3 12-hour shifts in a row. This implied that workers worked 2 out of every 4 weekends, but got 3 days off for one weekend and 6 days off for the other weekend in every four.

Even though one crew or another was at the mill 24 hours a day, there were still some occasions that warranted overtime pay. Overtime

	Monday Tuesday Wednesday Thursday Friday Saturday Sunday		Monday Tuesday Wednesday Thursday Friday Saturday Sunday		Monday Tuesday Wednesday Thursday Friday Saturday Sunday		Monday Tuesday Wednesday Thursday Friday Saturday Sunday
7:00 A.M.— 3:00 P.M.	A A A A A D D		D D D D D C C		C C C C C B B		B B B B B A A
3:00 P.M.—11:00 P.M.	(1) D D D D C C C	(2)	C C C C B B B	(3)	B B B B A A A	(4)	A A A A D D D
11:00 P.M.— 7:00 A.M.	C C B B B B B		B B A A A A A		A A D D D D D		D D C C C C C
OFF DAY	B B C C D A A		A A B B C D D		D D A A B C C		C C D D A B B

Notes:
1. Schedule repeats every 4 weeks.
2. Three out of every 4 weeks each shift works 5 days and is off 2.
3. One out of every 4 weeks each shift works 6 days and is off 1.
4. Average hours worked, 42.

FIGURE A2   Shift schedule for International Paper Company's Androscoggin Mill.

was paid on a daily basis past either 8 or 12 hours of the regular shift and after 40 hours a week of regular time.

All of the jobs in the mill were classified and assigned different rates of pay according to the hierarchy of knowledge and responsibilities required. When job openings came up, usually precipitated by a worker's transferring or terminating for one reason or another, workers in the next lower position were advanced according to seniority, assuming they demonstrated the abilities required of the new job. When a high-level job became vacant, a cascading of open positions was triggered, as workers moved up in the hierarchy. Transfers within the mill were permitted, in addition to promotions within the same department. Seniority governed transfers as well. For most hourly workers, the typical career path would entail transfers and promotions up to the highest paying jobs.

With the adoption of the latest contract, however, and its concept of more teamwork and more cross-training of the workforce, the number of specific classifications for the workforce had been significantly reduced (e.g., on a paper machine from 7 to 4 classifications, and in the technical department, from 9 to 2) and this shortened the lines of progression and at the same time broadened the content of jobs in the mill. This helped to keep work balanced across workers and fostered increased productivity.

Worker complaints were relatively few and, if not adjusted on the job, could be adjusted by a four-step grievance procedure involving successively higher levels of union and management officials. The fourth and final step was arbitration, which was seldom invoked — perhaps once a year.

Over the years, as papermaking equipment grew in size, speed, and sophistication (particularly with the advent of computerized process controls), papermaking became less and less an art and more and more a science. This change left its imprint on the workforce. Not only could a new paper mill be operated with fewer workers than an old one, but the requisite worker skills were less manual and more cerebral. The latest breed of papermakers were more highly educated, in general, and more analytical. As one would expect, some of the older workers were frightened by the automation and its demands on them.

Management had to be understanding of this, and considerable resources were expended on training. The adoption of a quality improvement process (QIP) that used teams of workers and managers to solve particular problems, many necessitating considerable data-gathering and mathematical and statistical analysis, had also required substantial training. Happily, the QIP had been a great and continuing success that had helped the plant to enjoy significant productivity gains that had stood it in good stead even as the economy turned downward. The mill, unlike many others, was not forced into shutdowns because of a lack of orders; its competitiveness had assured it of constant operation, and thus of a constant string of paychecks for its workers.

CONTROLLING THE OPERATION

The entire papermaking process at Androscoggin was designed to manufacture paper with as little downtime as possible. Equipment maintenance could generally be scheduled during the periodic changing of the paper machine's "clothing" (i.e., changing the worn wire and felt, accomplished every 2-3 months). Process control equipment and workforce skills were geared to react instantly to production disruptions, such as paper breaks, or to product changes in weight or color. The time it took to correct a disruption such as a paper break varied enormously (from, say, 3 minutes to an hour). But the average break time was in the 10- to 15-minute range. A color change normally took about 10 minutes, and a careful weight change could be done without any downtime.

Paper was continually being tested for quality, and the feedback to the workforce was swift so that any needed adjustments could be made. The workforce also kept their own process control charts that statistically tracked the process' ability to make quality paper and that signaled when changes to the process were called for.

In general, operations at the mill could be described as quiet and watchful. Because the elements of the process were so interdependent, skilled managers and workers were required in every phase of the mill's operations. Indeed, many of the mill's supervisors had been promoted out of the ranks and were thus intimately familiar with the mill's operations.

Changes in the process, routine capital acquisition and maintenance were the province of plant engineering. Plant engineering was constantly engaged in projects. Here is a sampling:

- A new, larger storage tank for broke (waste paper) for machine 4.
- A proposal to dry bark with exhaust gases.
- Installation of equipment to incinerate the sludge created as part of the effluent treatment process.
- Replacing a worn section of a debarking drum.

- Installing steam meters.
- Overseeing the repair of a roof.
- Laying out the specifications for a new gas scrubber.
- Replacing winders and supercalenders with state-of-the-art equipment.

EVALUATING MILL PERFORMANCE

The mill was evaluated as a cost center; it had no control or authority over prices, markets, or revenues. The mill operated to a budget. Given a sales forecast from the marketing function at headquarters, prices, and a product mix, the mill developed standard costs for producing the quantity of each paper product forecasted. The budget reflected these standard costs and the mill was held to the budget. However, if the product mix changed, the resulting cost changes were charged to marketing, not the mill. The mill was accountable only for those costs over which it had control.

While the efficient production of paper was an important goal, it was by no means the only aspect of mill operations that was evaluated. Others played an important role: employee safety; management and worker training; commitment to environmental controls (not only meeting present standards, but keeping the environmental control equipment balanced in capacity with the rest of mill operations and advancing in technology); expenditures for new capital appropriations and for maintenance; and industrial relations.

PART TWO

DISCUSSION

The International Paper Company is in no way responsible for the following views and presentation. They remain solely the responsibility of the author.

THE PROCESS FLOW

The process of making paper, while a frightfully complicated endeavor, follows a clear cut and rigid pattern. All of the paper that the Androscoggin Mill produces—however different in appearance, weight, and feel—proceeds through essentially the same production steps, from logs of wood in the woodyard to "logs" of

paper at the shipping dock. This kind of production process can be readily portrayed in a diagram like Figure A3, which is commonly called a process flow diagram.

The process flow diagram of Figure A3 is a fairly general one and could be made considerably more detailed. Whether more detail is desirable depends, naturally, on the use to which a process portrait like Figure A3 is put. Several points about the diagram ought to be noted:

1. Actual processing operations are usually distinguished from storage points in the process. In the diagram, processing operations are indicated by rectangles and inventories

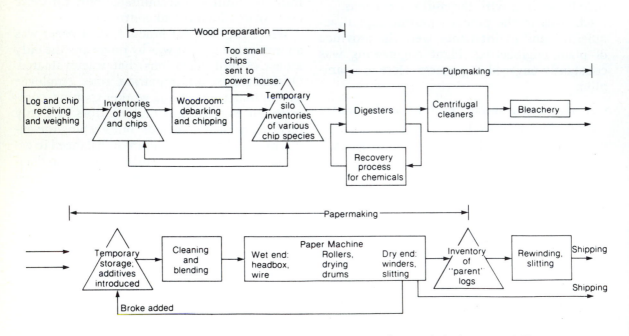

FIGURE A3   Process flow diagrams for International Paper Company's Androscoggin Mill.

by triangles. (Rectangles are used here because it is easier to fit the descriptions in them. However, typically in industrial engineering, circles are used to indicate processing operations and rectangles are used to indicate tests and inspections. See Figure 5-1 for the standard conventions.)

2. Several operations could be bypassed and are indicated by two arrows emanating from one operation and pointing to others. For example, all of the woodroom's output need not have gone directly to the silos but could have been placed in the woodyard for storage. The rewinders were superfluous for those orders that could be slit directly from the main winders.

3. The continuous nature of the process is evident by the very low level of inventories and by the designation of the silo chip and pulp additive inventories as temporary.

Information in the Process

A production process is more than a series of operations performed on a collection of materials. What a process flow diagram can depict—the sequencing of process steps, the choice of equipment and technology, the capacity of process steps, the tasks required of the workforce—while critical, is only part of the story. Another part of the story involves the procedures that have been put in place to direct the process flow. We can usefully think of a companion to the process flow diagram—namely, an information flow diagram. Figure A4 provides an example of what might be placed in such a diagram. Note how the actions of different layers of managers and workers are distinguished in the diagram and how information is fed back up the channels of communication.

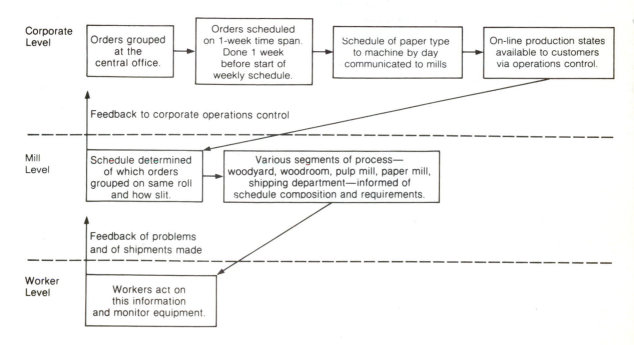

FIGURE A4   Information flow diagram for International Paper Company's Androscoggin Mill.

Most of the information flow in this continuous flow process is directed from the top down. Feedback is needed only to acknowledge receipt of information and to signal significant problems in the actual workings of the process. The process is designed with such care and the workforce so carefully trained that workers do not have to be in repeated touch with management to do their jobs well. The information needs of the process simply are not great, although the thought and effort standing behind that information (such as scheduling tasks of both the corporate office and the mill itself) are considerable and sophisticated.

CAPACITY

Capacity in this continuous flow process is fairly well defined and can be spoken of in physical terms—namely, tons of pulp and paper manufactured and cords of wood consumed. Likewise, there is a straightforward meaning to the term capacity utilization. In fact, all one needs to do is check to see whether the paper machines are running. If they are, capacity (all of it) is being used. Simply compiling the time the paper machines have run over some period of time and dividing that sum by the total available machine-hours gives a splendid indication of capacity utilization. Indeed, only when sales are insufficient or when the machine is broken or down temporarily for a setup or a change of "clothing" would we expect capacity utilization that was less than 100 percent.

Despite the relative ease with which we can talk of capacity and capacity utilization in this continuous flow process, on closer inspection of the concept a number of disquieting ambiguities surface. The ambiguities surface largely because the capacity of the mill is dependent on the number of factors. Among them:

1. *Product mix.* Some grades and weights of paper are more difficult to produce than others. Newer machines, of course, are apt to run at faster speeds than older machines, given the same kind of paper. But, given similar vintage paper machines, thin papers and high-quality finish papers have to run at slower speeds than others. Thus comparing the capacity of the Androscoggin Mill with that of other paper mills or with its own in other years can only be done with some understanding of the product mixes involved.

2. *Run lengths.* Every time a paper machine must be set up for a different grade, weight, or color of paper, capacity is lost. Although some changes are inherently more timeconsuming than others, every change implies at least a small reduction in theoretical capacity. The capacity of machine 1 was somewhat lower than that of machine 4 at least in part because machine 1 was scheduled to produce 48 different paper machine products in white and in colors, whereas machine 4 was devoted primarily to white.

3. *Maintenance.* A paper machine might also lose capacity because it was being repaired or because its "clothing" was being changed. The machines that could be kept running until only regularly scheduled maintenance interrupted production were those that lost the least capacity.

4. *Slitting schedule.* Slitting paper "logs" from the winders and rewinders was purposely scheduled so that as little waste as possible resulted. The more successful the schedule, the less waste there was.

For these reasons, the terms capacity and capacity utilization were not as unambiguous as might have been anticipated. Nevertheless, the process was extraordinarily well balanced and smooth running, and this made discussion of capacity and capacity utilization easy.

Demands of the Process on the Workers

Over the years, papermaking, like other continuous flow processes, has become a process where increasingly machinery "does it all." Pulpmakers and papermakers have been systematically removed from playing their traditional direct "hands-on" roles in the process. Instead, papermakers are more and more "indirect" labor-setters, monitors, and repairers of equipment that makes paper much faster and more reliably than any crew of traditional papermakers could using the older technology and their own skills. Where papermaking was "art," it is now "science." The former lengthy apprenticeships from sixth hand on a paper machine to machine tender are no longer justified by the need to learn the art. Despite the change to science, the level of skill required in the process remains high. The workforce may have been removed to gazing at control panels, adjusting knobs, and throwing switches, but the technical demands of the process are even higher in a modern, as compared to traditional, paper mill. Training is therefore an important consideration and an apprenticeship, while different from before, is still required. The lines of worker progression in the mill are clear (in fact, they were diagrammed for each department in the labor agreement), and workers are paid accordingly.

What comprises any job in the mill, while somewhat restricted, has been increasing steadily, however. The team concept, job rotations, and the collapsing of the traditional classifications into many fewer, had helped to broadened the definition of jobs at the mill. The new labor contract permitted those changes that encourage productivity and kept everyone well occupied.

Skilled workers are needed everywhere in the process, largely because the process is so interdependent. If one aspect of the process falls down (e.g., quality of the pulp), another aspect is likely to suffer (increased risk of a paper break).

Demands of the Process on Management

The interdependence and capital intensity (a high ratio of plant and equipment value to labor payroll) of a continuous flow process like papermaking place tremendous demands on management, especially in the realms of coordination and of the choice and care of equipment. Furthermore, these demands are made not only of the mill's own management but of the corporation as a whole. Profitability in such processes largely rests on (1) assuring that the proper technology is selected, (2) balancing the capacities of all segments of the process so that as little capacity goes to waste as possible, (3) scheduling the use of that capacity as completely as possible, (4) keeping the equipment running up to speed and up to quality standards; and (5) an ever-continuing effort to improve in all areas of the operation. Let us review these points in turn.

Choice of Technology

The march of technological change is inexorable. While the output of continuous flow processes seldom changes by much, there can be upheavals in how that output is manufactured. The introduction of process control equipment to papermaking is a case in point. Management must be constantly aware of equipment advances across the industry, and savvy manufacturers engage in regular dialogues with equipment makers so that their own ideas and needs can be tried out in new equipment designs. The impact of technology is so fundamental that the corporate staffs of continuous flow process industries are usually charged with monitoring and selecting major new technology for all the plants of the company. Individual plants may have representatives on any plant and equipment choice studies, but corporate-level managers are apt to take the lead in the study and decision-making process. Many of the latest technological advances had involved changes spurred by pollution control issues or energy savings.

What is true for technology decisions is also true for decisions on how vertically integrated the process should be. This is generally also a corporate decision and one that is the province of corporate staff.

Balancing Capacities

The papermaking process, as we have seen, can be broken down into distinct segments (as in the process flow diagram); associated with each of these segments are machines and other equipment, often very large. Frequently, these machines and equipment are manufactured by different companies and do not come in just any size. It is management's responsibility to select the equipment for each process segment that represents both the suitable technology and size for the contemplated plant or plant expansion. One segment's capacity should be balanced, as well as possible, against that of other segments, so that as little extra capacity as possible has to be financed. After all, in an integrated, continuous flow process the capacity of the entire process is determined by the lowest capacity segment.

In reality, balancing process segment capacities is a difficult chore, and choices have to be made as to which process segments are to be assigned whatever excess capacity may exist. Often such a choice entails an investigation of equipment costs, with the relatively cheaper equipment being assigned any excess capacity.

In papermaking, for example, the big bucks get chewed up in financing paper machines and bleacheries. Woodrooms, digester, and rewinders are relatively less expensive. Thus, frequently, the spare capacity is to be found at the ends of the process (woodrooms, rewinders) rather than in the middle (bleacheries, paper machines). Having spare capacity at the rewinders increases the flexibility of the process as well to modifications of the product mix or order specifications. With excess capacity in place, unusual orders can be serviced without undue strain or delay.

The need for, and the expense of, maintaining a balance of capacities in a continuous flow process like the Androscoggin Mill are dramatically highlighted by the 1980 rebuilding of paper machine 2 and its conversion from carbonizing tissue to publication gloss. That conversion speeded up the flow of paper from about 20 miles per hour to nearly 35 miles per hour and raised the mill's net yearly tonnage by about 100,000 tons. To accommodate a change of that magnitude meant altering the pulpmaking capabilities of the mill as well as its papermaking capabilities. When process segments are tightly coupled, capacity changes echo throughout the entire process.

Scheduling

Large, high-speed continuous flow processes are geared to high-volume production of standard items. Typically, they are low-cost processes, but they often sacrifice the ability to respond quickly to changing customer specifications without introducing a lot of waste and thus destroying their low-cost character. In order, then, to satisfy customer orders at low cost, continuous flow processes must schedule their capacity well in advance, offering their customers longer lead times than may be common in other types of processes or else filling orders out of in-process or finished goods inventories.

At International Paper, as we saw, the broad "paper machine products" scheduling proceeded one week in advance for production and the schedule devised was of a 1-week period. The customer orders that were produced in each run, however, were set at the plant level only 2 days in advance of the production cycle. As many customer orders as could be slit right off the machine winder were scheduled, to eliminate rehandling and converting of paper logs. More exotic orders for special colors, grades, or weights from machine 1 were often filled out of finished goods inventory to avoid having to schedule special production runs. Scheduling at International Paper, as these observations indicate, was a sophisticated enterprise.

The idea, of course, behind such sophisticated scheduling is to keep the process flowing as continuously as possible. Generally, the process could be more responsive to customer orders, but to do so would necessarily mean interrupting production to set up equipment. Capacity would be reduced and costs increased. By stretching out deliveries or by keeping finished goods inventories, the continuous flow process can keep its costs down.

Equipment Maintenance

Most continuous flow process plants are evaluated as cost centers. As with the Androscoggin Mill, such plants are not given authority over revenues (no sales forces are tied to them), and so the plant is judged by how well it can adhere to a budget. The plant has an incentive then to keep its equipment well maintained; if it does not, it risks assuming costs that can be very high (equipment breakdowns force high repair costs and expensive makeup work.) Given such an evaluation scheme, most managers would opt to spend all of their budget for maintenance to avoid the chance, however slim, of suffering a huge cost increase and an instantly bad reputation for plant management.

QUESTIONS

1. The changeover of machine 2 from carbonizing tissue to publication gloss involved changes in other parts of the process. Why was this so? Speculate in as much detail as possible about the kinds of changes that would have had to occur as a consequence of such a paper machine changeover.

2. Much of the "higher-level" production planning for Androscoggin Mill occurs in International Paper's Memphis and New York headquarters, whereas "lower-level" production planning occurs at the mill itself. Why is there this division of responsibilities? What would happen if Memphis or New York handled all the production planning or if Androscoggin did it all?

3. If you were a production planner at the Androscoggin Mill, what might you be most likely to consider important? Why? How would these priorities affect your production plans?

4. Why did the Androscoggin Mill choose vertical integration for its design? What would be some of the advantages and disadvantages of increased forward integration for the mill?

5. How is the workforce organized at the mill? What are some of the operational implications of this workforce composition?

6. As a manager, which aspects of a continuous flow process appeal to you? Which aspects might not?

BEACON GLASS WORKS

Beacon Glass Works, located in Marysville, West Virginia, manufactures hollow glass tubing for use in catalytic converters. Beacon is a large division of one of the leading glass manufacturers, and was founded when the Environmental Protection Agency strongly urged the development of antipollution devices for trucks and automobiles. The hollow glass tubing manufactured at Beacon comes in around 30 different size combinations. The lengths are from 6 to 12 inches, and the inside diameter ranges from 1/32 to 1/16 inch.

These tubes are produced by extrusion. The molten glass is first forced through a die that determines the inside diameter of the tube. The tube then is cooled and cut to the proper length. After the cut edges have been finished so that they are smooth, the product is sent to a holding area where it waits to be packaged. Once the tubing is packaged, it is again stored, this time awaiting shipment to Beacon's customers, the four major domestic motor vehicle manufacturers.

The extrusion process is basically a make-to-stock operation driven by production scheduling. Based upon anticipated stock shortages, Jose Torrez, the production manager, schedules the next week's production, and the necessary dies are readied. Each die is used until about one month's demand has been produced. Then, the dies are changed and another product with a different inside diameter is run. Actually, since several lengths have the same inside diameter, several products can be extruded with one extrusion run.

Once the product cools to the proper cutting temperature, the cutting area cuts the proper lengths, again based upon the stock needs determined by the production manager. The product is then stored in a holding area waiting to be packaged. Packaging is usually done to customer order, as different customers want different quantities and want them packaged differently.

Sometimes, however, Jose will schedule the packaging department based on what he believes will be the customer orders, so that he can use the packaging machinery efficiently. This scheduling sometimes results in excess inventory for one customer while another customer's order is backordered because of a stockout.

Jose is concerned that his process scheduling is not as good as it could be and thinks that possibly the first step he should take is to diagram the production flow and the information flow for the process.

1. Diagram the production flow.
2. Diagram the information flow.
3. Discuss possible problem areas for Beacon's production process.

SITUATION FOR STUDY A-2

SUNMEADOW DAIRY

Sunmeadow Dairy produced and distributed a variety of products including whole milk, skim milk, 2 percent milk, ice milk, fruit drinks, buttermilk, eggnog, chocolate milk, and many different mixes used to produce ice cream. The production of milk, however, was 90 percent of the operation.

From 8 A.M. until 5 P.M., raw milk (unpasteurized milk direct from the farm) arrived by refrigerated tanker truck at or below 42°F. The raw milk was pumped from the tanker (capacity 5500 gallons) to one of three refrigerated storage silos (capacity 70,000 gallons total) adjacent to the receiving area. During this transfer, the milk flowed through a plate heat exchanger to lower its temperature to 37°F. From these silos, the milk was pumped to the second floor as needed for production and was stored in four of six storage tanks. The empty tanks were used for mixing and diluting the raw milk, which was 3.5 to 4.0 percent butterfat.

Because whole milk contains 3 to 5 percent butterfat (depending on its source) and some skim milk products contain as little as 0.5 percent butterfat, a certain percentage of butterfat had to be removed from the raw milk. A separator was employed to remove the excess butterfat. (A separator is a centrifuge in which raw milk is spun at high speed, forcing the separation of the butterfat from the heavier milk and allowing the isolation of the two fractions.) One by-product of this process contained 40 percent butterfat and was pumped off (at a rate of 1000 to 1500 gallons per day) into nearby holding tanks and shipped to a butter factory once a week. The other by-product was 0.5 percent milk; it was pumped into an empty tank and mixed with 3.5 percent raw milk to obtain the desired butterfat level for the product being packaged next in the plant. This mixing was called "standardizing" the product. From an information chart on obtaining the percentage of butterfat needed for each product, the separator operator determined the exact quantity of each input (3.5 percent milk and 0.5 percent milk) to be mixed in the empty tank.

From the holding tanks the standardized product was pumped into one of two pasteurizers (capacity 1500 and 4000 gallons per hour). (Pasteurization is the process of heating raw milk to 161°F for 16 seconds to kill harmful bacteria.) The pasteurized milk was pumped into one of two homogenizers, which blew the milk through a very small opening to break up the fat globules and create a uniform, or homogenous, mixture. The pasteurized, homogenized milk was directed into any one of seven pasteurizing holding tanks to await packaging on the first floor. Because packaging might start as early as 5:30 A.M. pasteurization usually began at midnight-to build a buffer inventory in the holding tanks for the first and second products to be run that day.

Milk could be pumped into one of five packaging processes:

1. The filler filled plastic gallons and half-gallons at a rate of 70 per minute. The plastic containers were made in the blow-molding department and arrived at this machine by overhead conveyor. This conveyor continued through the completely automated processes of filling, capping, and storing the filled containers in metal cases for shipment. These metal cases were stacked six high and sent to the cooler for order filling.

2. The H-75 machine produced paper half-gallons (capacity 75 units per minute) and automatically filled, sealed, and stacked them. Manual attention was needed only to take the wax cardboard containers from the adjacent stockroom and load them into the feed mechanism. Each carton was automatically unfolded, heat-sealed at the bottom, and filled with the product being produced in this run; the excess foam and overfill were drawn off the top, and the carton was heat-sealed. The cartons were then automatically stacked into metal cases and put on the conveyor for the short trip to the cooler.

3. The Q80-110 machine operated the same way as the H-75 but was for half-pints, pints, and quarts (capacity 80 quarts, 110 pints, and 110 half-pints per minute).

4. The NEP 210 machine (capacity 210 half-pints per minute) also operated the same way as the H-75. The unique feature of this process was that it produced only for schools.

5. Using the bag and box filler was a totally manual process for filling 5-gallon plastic bags inside boxes. These were shipped to local schools and hospitals for use in self-service milk dispensers.

While each process was running, there was at least one operator with each machine at all times. The high volume of production demanded constant attention and minor adjustments, especially to keep the product from becoming caught in the moving parts. Once packaged, the product was carried by conveyor from the packaging areas to the 38°F cooler, where it was stored according to how soon it would be used to fill orders on the daily trucks. The cooler operated on a first-in, first-out (FIFO) basis; rotating the product this way kept the buffer stock as fresh as possible. The product was moved to the loading areas on the same conveyor that circulated through the packaging and cooler.

Orders to the dairy were called in by the branches and other customers, usually 2 or 3 days in advance. The salesclerks wrote the orders on computer-generated, preaddressed order sheets for each account. Special phone orders and adjustments could be made until shipment, because of the buffer inventory in the cooler.

Local orders were written by the drivers of the route trucks while visiting each customer on their daily schedule. The day before shipment, the cooler supervisor scheduled and consolidated all orders for filling based on the usual arrival times of the incoming trucks. Two order copies stayed in the sales office, two copies were sent with the order, and one stayed with the cooler supervisor. After delivery, the signed order copy was compared with the original order and all necessary information was entered into the computer. This update triggered the daily report on actual production, cooler inventory, actual units loaded on all trucks by product line, and total amount needed for the next day's production. The plant manager received this printout early the following morning. Two 10-hour shifts were needed to accommodate each day's scheduled production. The same order of production was run each day (such as 2 percent milk first and chocolate last). To accommodate the trucks that started arriving at 4 A.M., the buffer inventory in the cooler at the start of the day for the last two products run might be as high as 60 percent of total production.

The quality control supervisor was responsible for product quality throughout the plant. Quality control was a major factor in the production of milk. As the tankers arrived from the farms, three quality checks were done to make certain the raw milk was within the acceptable range of temperature, bacteria count, and acidity. Except during pasteurization, milk was maintained at 38°F throughout the plant. In addition to the standardization checks for butterfat content, two tests were done after pasteurization to check the bacteria count and type of bacteria still in the milk. Raw milk arrived with a bacteria count of 10,000 to 20,000 per milliliter and was shipped out at about 500 per milliliter. Each worker was acutely aware of the importance of quality control, and all machinery and equipment were thoroughly washed at least once a day to ensure cleanliness. Most of the equipment—including tankers, silos, and storage tanks—was stainless steel to facilitate cleaning and reduce the chances of contamination.

1. Diagram the process and information flows.
2. What do you see as the challenges that the management of the dairy faced?
3. Recently the dairy lost a large contract for supplying milk to a chain of grocery stores. How do you think the loss of sales will affect the operations of the dairy? Be as specific as you can.

A JOB SHOP
Norcen Industries
Jersey Shore, Pennsylvania

It had been years since Joe Gehret finally gave in to his desire to control his own company and resigned his position as general foreman in the machine shop of the Litton Industries plant in Williamsport, Pennsylvania. In July 1967, Joe and a partner (who had since left the firm) began Norcen Industries by selling stock and taking over an old garage in Jersey Shore, about 15 miles west of Williamsport. They initially intended for Norcen to be a plastics distributor and fabricator, but it soon became apparent that both their experience and the demands of industry in north central Pennsylvania dictated a change of course. Norcen quickly became mainly a metal-working machine shop, and the machine shop accounted for almost 90 percent of gross revenues.

The company had experienced reasonably steady growth. In 1981, Norcen moved into a new, 21,000-square-foot building on the outskirts of town. Sales in 1991 had grown to over $2 million and employment to 40.

PART ONE

PROCESS DESCRIPTION

PRODUCTS, SALES, AND ORDER HANDLING

As a general machine shop, Norcen was capable of producing a seemingly endless succession of metal and plastic parts that a host of companies typically assembled into machines and other products. Norcen specialized in close tolerance work for the electronics and aerospace industries. Almost all of Norcen's customers were manufacturers, but 95 percent of this business Norcen had to win by submitting low bids. The purchasing departments of Norcen's customer firms generally requested Norcen and at least two other machine shops to "quote" the work they wanted done. The request always specified (1) the number of pieces desired, which varied enormously, from 5

Joe Gehret beside a recently purchased lathe. (*Courtesy of Norcen Industries*)

pieces to over 1000, although most lots were less than 250; (2) the nature of the material required and whether it was supplied by the customer, (3) the design of the piece (a blueprint would be sent), and (4) the date by which the order had to be received. During peak years, Norcen bid on as many as 250 such requests each week, knowing that on a long-run average its quote would be accepted about 15 percent of the time. (In the past this figure had sometimes been as high as 40 percent.)

Joe Gehret was responsible for deciding all metalworking quotes. Naturally, it was easier for him to quote jobs that Norcen had done before. Not only did he know the hourly charge-out rate (currently, $31.20 per hour for general machining and $40–50 per hour for electronic discharge machining (EDM)), which would cover both direct labor and overhead expenses, but he had a past record of what the piece had actually cost Norcen to make on all previous occasions. (See Figure B1 and the discussion on record keeping below.) In addition, Norcen would already have a blueprint of the piece and a "process sheet" (Figure B2), which would outline the steps the shop had previously taken to manufacture it and the time standards for each of those steps. Deciding quotes for pieces Norcen had not previously made was more difficult. In such cases, Joe had to ponder the blueprint and develop, at least in his mind, a rough-cut process sheet.

Sometimes the request for the quote included a process sheet developed by the company itself, and this was a great aid. Other factors such as delivery dates (rush orders, being more trouble, commanded higher margins), and the prevailing load in the shop (the more

INFORMATION:

Material Furnished

Sheet #3

COMPANY & ITEM 53147 Transformer

VARIANCE	PRICE UNIT	TOTAL	HOURS	COST MATERIAL	LABOR	SERVICE	TOTAL	VOL.	COMP DATE 84	JOB INVOICE	DATE 84	P.O. NUMBER
	9 20	607 20					13 41	66	2/14	12107 17503	1/4	53368
	9 20	708 40					53 87	77	1/16	11813		
	9 20	579 60					747 43	63	1/12	11788		
	9 20	322 00					263 14	35	3/19	12419		
	9 20	956 80					37 00	104	3/28	2585		
	9 20	883 20	1.369				60 13	96	4/16	12107 17583	84	
	8 70	453 70					483 36	51	6/12	3355	2/8	55590
	8 70	685 30					— 77		6/20	3410		
	9 10	191 10					— 21		8/8	3896		
	9 10	172 90					719 76	19	8/6	3883		
	9 10	464 10					— 51		8/7	3884		
	9 10	336 70					214 57	37	8/5	3917		
	9 10	873 60					222 20	96	8/23	4004		
	9 10	455 00					13 75	50	9/27	4368		
set up	182 40	2.181				59 23	1.26	10/27	4451			
	10 24	1.382 40					612 72	135	1/25	5521	10/25	60938
	10 00	1180 00	88.5	made on 18980			121 79	118	2/8	5668	2/14	61673
	12 24	722 16	26				259 89	59	10/2	2004	9/6	65630

Notes:

1. This record card is for a so-called "transformer" (not what we would ordinarily term a transformer) for a company whose name has been withheld. This part is distinguished at Norcen by its part number, 53147, which also corresponds to a blueprint and a process sheet (see Figure B2).

2. The material to work on is furnished by the customer company.

3. This is a part that Norcen has machined for years. This record card is sheet #3. On this sheet, we can see at the top of the columns that the first order was placed on January 4, 1984, under purchase order (P.O.) number 53368. That job was assigned job number 17503 and consisted of 6 separate shipments of varying shipment dates (such as 2/14, 1/16) and corresponding and separate invoices to the customer (such as 121074, 11813). The shipment quantities varied as well (for example, 66 units, 77 units. 63 units). The total for the entire order is 441 pieces.

   The latest order (#65630), dated 9/10 and shipped 10/2, involved just 59 pieces. It carried job number 20414 and was billed on invoice 18004.

4. The quoted price for the first order on the card (job number 17503) was $9.20 per piece. The total billing for the first invoice (#12107) was $607.20. The total billing for the entire order was the sum of the six relevant invoices, $4,057.20.

5. A total of 136.9 hours of labor were spent fulfilling the first order. The total labor cost of the job was $1,175.44 (the sum of the cost total column).

   For the latest job, a total of 26 hours of labor was expended at a cost of $259.89. The total billing was $722.16 for the 59 pieces shipped, quoted at a price of $12.24 each.

6. Norcen Industries made its charge-out rate in the latest completed job. The labor hours of 26 times the prevailing charge-out rate of $24 per hour equals $624, which is below the actual revenue for the job of $722.16.

7. Two special notations are worth explaining: (1) The latest entry for P.O. number 55590 is a charge of $182.40 for an extra setup. The customer company had not provided the material on time, requiring Norcen Industries to incur an additional, unscheduled setup. Norcen charged the customer for that setup.

   (2) P.O. numbers 60938 and 61673, involving job numbers 18980 and 19206, were able to be run with a single setup. This was advantageous to the shop (denoted by the notation "made on 18980") and resulted in a lower quote for P.O. number 61673.

FIGURE B1   A sample record card for a particular item manufactured at Norcen Industries.

MATERIAL DESCRIPTION	OFHC Cu				COMPANY Some company	PART NAME Transformer	PART No 512864			
LINEAL INCHES	AREA (SQ. IN.)	DIMENSIONS			COMPUTATION-MTL REQD	PREPARED BY	PG OF	REV		
EQUIPMENT					OPERATION		ESTIMATED			
FIXTURES & TOOLS	MACH.	SPEED	FEED	No.	DESCRIPTION		No OF PIECES PER HR	TIME		
								PER PC	SET UP	
Arbor, collet				10	Rough face to turn O.D. to size (.972-.974) on Arbor					
				20	Face waterfall drag down off of open end					
				30	Face off other end to .400 long					
				40	Deburr					
Collet, template				50	Dress radius & grind radius in part. holding length to size (.381)					
				60	Deburr					
				70	Inspect					

Notes:

1. A collet holds the workpiece. An arbor is another device to hold the workpiece.
2. A template was used to help guide the grinding.
3. The waterfall was excess metal, a humpline bulge, that had to be machined off the part.
4. Various dimensions were given in steps 10, 30, and 50. They were also available on the blueprint of the part.
5. In the material description, OFHC Cu means oxygen-free copper, which is a very pure copper.

FIGURE B2   A sample process sheet used at Norcen Industries.

slack in the shop, the lower the margin) also influenced the quote.

The acceptance of a Norcen quote typically came by phone and was confirmed later by letter. Upon notification of acceptance, Joe's wife, Lillian, who was responsible for all bookkeeping, wrote up two copies of a job sheet (Figures B3 (and B4). One copy was kept in a record book, and the other was attached to a blueprint and placed in a special tray. Joe was then responsible for looking over each job sheet to see whether any materials not provided by the customer needed to be ordered and whether any tooling needed to be done. Job sheets that awaited materials or tooling were kept in a sep-

arate tray. When enough materials were on hand to complete the order, the job sheet and blueprint were placed in a special drawer. This action released the order to the shop.

PLANT AND PERSONNEL

The machine shop operated on two shifts: 6:50 A.M. to 3:20 P.M. and 3:30 P.M. to midnight, with a half-hour break for lunch or dinner and two 10-minute break periods. The day shift consisted of 24 hourly workers, and the evening shift currently ran with only 9 workers. [This shift operated primarily the company's direct

ORDER No. _60938_ JOB _18980_

PART No. _53147_ LOT SIZE _____

MATERIAL _Furnished_ 245

DELIVERY _2/10_

Checked By	Date	Name	Operation	Qty.	Hrs.	Acc. Hrs.
	1/14	RPH	255			
DM	1-16	J.C.	Lathe		8.0	8.0
	1-17	JL	LATHE		8.0	16.0
	1-18	JL	"		8.0	24.0
	1-19	JL	"		5.0	29.0
	1-21	JL	"		2.5	31.5
	1-21	Doyle	Deburr		3.9	35.4
	1-22	Doyle	Deburr-Buff		4.2	39.6
DM	1-22	D M.	Set up & Gr.		2.0	41.6
	1-22	O.R.	Gr.		0.7	42.3
JJ	1-23	O R.	Gr.		8.0	50.3
DM	1-24	O.R.	Gr.		8.0	58.3
JJ	1-25	JJ.	Deburr		2.0	60.3
	1-25	Doyle	wipe - Buff		1.5	61.8
	1/25	JJ.	DeBuff, Insp		1.5	63.3
	135	PCS	To Ship	1/25	JJ	
	1-25	O R	Gr.		8.0	71.3
JJ	1-26	O R	Gr.		2.5	73.8
DM	1-28	o R	Gr.		4.0	77.8
	1-29	O R	Gr.		4.5	82.3
	2-7	DOYLE	CLEAN Deburr-Buff		5.2	87.5
	2-8	JJ	INSP		1.0	88.5
	118	Pcs	To Ship	2/8	JJ	

Notes:

1. This job sheet accompanied job number 18980 around the shop from the initial receipt of materials from the customer (255 pieces on 1/14) until shipments on 1/25 and 2/8 of 135 and 118 pieces, respectively.

2. The lot size on this job changed. It was increased to 245 (253 were actually shipped) when P.O. number 61673 came along, and the shop was able to do both jobs with a single setup.

3. Dates worked, worker initials, operation done, time spent, and total time to date on the order are filled in by each worker after he or she completes working on the job. Periodically, others have checked the order's work (RTM and JJ).

4. The accumulated labor hours are 88.5. This figure is also displayed on the record card.

FIGURE B3 A sample of the job sheet used at Norcen Industries.

ORDER No. _____609-38_____   JOB _18980_

PART No. ___53147___   LOT SIZE _245_

MATERIAL _____

_____

DELIVERY ___2/10___   _10.24_

Checked By	Date	Name	Operation		Qty.	Hrs.	Acc. Hrs.
	1/13	8 L.	246	50			29
	1/20	GG.	33	25		3.5	32.5
		8L.	21	25		2.5	35
		OR	231	20		27.2	62.2
		JM	19	00		2	64.2
		DD	71	52		9.6	73.8
			622	72			
		chgd	155.29				
	1/27	OR.	72.25			8.5	82.3
	2/3	DD	40.04			5.2	87.5
		GG	9.50			1	88.5
			121.79				
		chgd.	156.68				

Notes:

1. This job sheet was compiled by Lillian Gehret after job 18980 was completed. The information for the job sheet was derived from (a) the job sheet (Figure B3) that follows the job around the shop, (b) worker time cards (Figure B5), and (c) information on worker hourly rates.

2. The top portion of this job sheet contains the same information as Figure B3 and includes as well the quoted price on the job ($10.24 per piece).

3. Worker times are gathered by the week, with the weeks being dated from their start on a Sunday (1/13, 1/20). Workers' hours spent on this job in each week are listed in the Hrs. column. The cumulative hours for the job as a whole are totaled in the Acc. Hrs. column. In the center columns are listed the worker pay for the week due to the job. By dividing this pay by the hours, one can see that workers have different hourly wages. GJ has an hourly wage of $9.50 ($33.25/3.5).

4. The total wage bills for the job are $622.72, charged to invoice 15529, and $121.79, charged to invoice 15668. Recall that job 18980 combined two purchase orders and ran them on the same setup.

FIGURE B4   A sample showing Lillian Gehret's office copy of a job sheet.

numerical control (DNC) lathes and mills and was also used to expedite rush jobs.] Norcen's hourly wages (the highest paid by any of the machine shops in the area) started at $8.50 per hour for unskilled workers and went to $12.75 per hour for the most demanding work. This pay was augmented in two ways:

1. Overtime pay at time and a half

2. A bonus plan, payable at Christmas and in the summer

The size of each bonus depended on the fortunes of the company over the previous 6 months. Currently, bonuses averaged about $1500 per employee, greater by far than the dividends paid to the company's shareholders. In leaner years, of course, this bonus was much lower. The size of the bonus for each worker, like the number of vacation days, depended on seniority. The workers recognized this system as an incentive for continued good work. In the words of a lathe and drill press operator: "The better we are to the company, the better the company is to us."

A rough layout of the plant, showing the general groupings of machines, is included as Figure B5. An inventory of the most important machines in the shop, along with their weekly average hours in use, is provided in Table B1.

THE WORKINGS OF THE SHOP

Although Joe Gehret spent a good deal of time on the shop floor, responsibility for shop operations rested with Rob Thomas, the production manager, and Dan Steinbacher, the shop foreman. Their duties were varied, but the first concern for the smooth running of the shop was to see that all workers were assigned tasks and that they understood these tasks well, both their nature and their time standards. Moreover, Rob and Dan knew that it was important to the success of the company that tasks be assigned in a way that saved Norcen money. Cost reductions from such

The shop as seen from a balcony. (*Courtesy of Norcen Industries*)

The shop as seen from a balcony. (*Courtesy of Norcen Industries*)

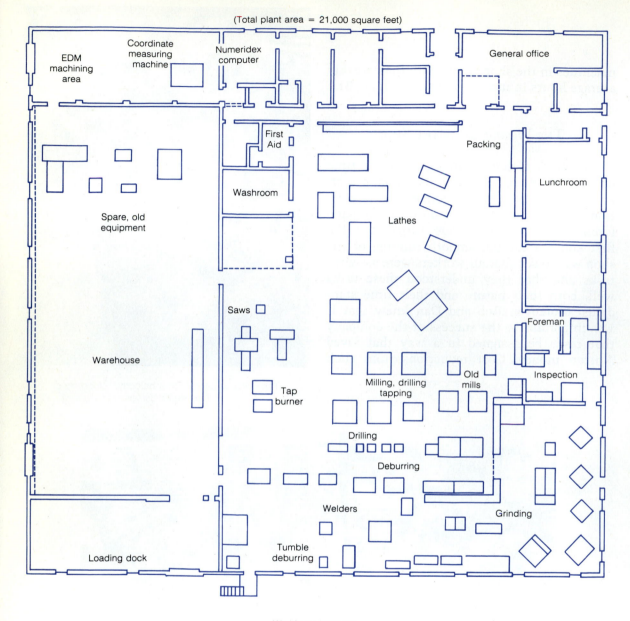

(Total plant area = 21,000 square feet)

Workforce (40 total)

1 general manager
1 production manager
1 QC manager/project engineer
2 foremen
1 bookkeeper
1 computer programmer

26 machine operators
1 full-time inspector
1 inspector/operator
1 maintenance man
1 shipper/receiver
2 utility people (clean-up, etc.)
1 manager of the plastics business

FIGURE B5   Plant layout and workforce at Norcen Industries.

TABLE B1   Inventory of Major Machines at Norcen Industries

Machine and Number Owned		Percent of Scheduled Time Machine Is Run
1	Hardinge HXL	100
1	Hardinge Conquest superprecision lathe	100
5	Hardinge CHNC II lathes	100
2	Large slant bed lathes	100
3	Hurco machining centers	100
2	Matsuura machining centers	100
1	Fadal machining center	100
1	Kitamura Mycenter machining center	100
2	Mitsubishi traveling wire EDM machines	75
1	Sodick die-sinker EDM machine	30
1	Blanchard-type grinder	40
1	Do-All automatic cut-off saw	40
1	Sheffield coordinate measuring machine	50
2	J&L 20" optical comparators with power feed & digital read-out	75
2	O.D., I.D. universal grinders	50
1	Super-precision universal tool & cutter grinder	30
1	Heat treat facility (size limitations, muffled main furnace, nitrogen induction)	75
1	Special roll grinding unit	50
1	110 ton Cincinnati hydraulic punch press	30
1	20 ton Perkins crank press	70
	+ surface grinders, manual mills, old lathes used for very small runs or for tool and die work or in support of the highly used machines (i.e., for creating fixtures, tooling)	25

scheduling and advance planning could come about in several ways:

1. Before an operator could machine any of the myriad jobs Norcen was capable of performing, the necessary machine had to be set up. The setup varied from about a half-hour for simple jobs to 4 hours for the most complex. An average setup was about 1.5 hours. Since, once set up, the machine could produce any number of pieces with only minor adjustments, it made sense to run full lots whenever possible. In this way, setup costs were spread over all the units in the order.

2. Frequently, especially with automatic equipment, the running time of the machine per piece produced was long enough so that, by staggering operations, a single operator could attend two different machines at the same time. That is, while one piece was being machined, the operator could be working at another machine, typically inserting new material or removing a finished piece. To exploit these possibilities,

A Hurco machining center down for a changeover
after machining some white plastic parts.
(*Courtesy of Norcen Industries*)

foremen to keep these special capabilities in mind in his assignment of jobs.

4. Cost reductions could also occur through improvements in the process sheets that were kept in special binders. Often, changes in the sequence of drilling, cutting, milling, grinding, tapping, threading, deburring, and so on, could have profound effects on the total time Norcen had to allocate to a particular job. Rob and Dan had to be alert for possible improvements they could make in the process.

5. The probability of successfully reducing costs was naturally greater for more complex operations and for repeat business. Furthermore, cost reductions could lie in more than process sequence changes. To cite an example, Norcen machined a part composed of a thin molybdenum strip bonded to steel. The customer's initial process specifications called for cutting two semicircles into the part using a lathe, but Joe Gehret recognized this technique as a high-risk, low-yield way to machine this part. By changing the process to use a specially adapted grinder, Norcen was able to machine the part successfully with much higher yields than would otherwise have been the case. Such dramatic success was not to be found often, but its significance was great enough to demand a good deal of the foremens' thoughts.

Rob and Dan had grouped similar kinds of machines together on the shop floor. It was to Norcen's advantage for Dan to identify and to match together those jobs that permitted this kind of labor savings.

3. As might be expected, some operators could perform certain machining tasks relatively better than others and thus held a comparative advantage over other workers in the shop for some jobs. It was worth it for the

Even more important to the foremen than assigning tasks in money-saving ways was seeing to it that the delivery schedule was met. Norcen could not miss deliveries and still expect to receive repeat orders. The foremen were constantly aware of the promised delivery dates, and they attached high priorities to imminent deliveries due. The scheduling of jobs was further influenced by Norcen's cash flow

A Hurco machining center. (*Courtesy of Norcen Industries*)

The Hardinge Conquest superprecision lathe. (*Courtesy of Norcen Industries*)

needs. Other things being equal, and particularly at the end of the month, small jobs were given preference so that billing could proceed at once.

The foremen dealt also with worker morale, training, and development, with materials handling, and with quality control. But job scheduling was the most demanding on the foremen's energies. They tried to stay one day ahead of the workers, so that they knew precisely which job would go next to every machine, who would work that job, and when they would start on it. As Joe Gehret put it: "Scheduling is the most difficult function we have around here. It's the easiest thing in the world to say no to a customer, that we can't fit

The Sheffield coordinate measuring machine.
(*Courtesy of Norcen Industries*)

his job into the shop. But after a few noes, you may not have a customer."

During the recession period of 1990–91, scheduling was even more difficult than usual. The drop in orders then meant that the foremen had fewer chances to group compatible orders together to avoid bottlenecks or to create labor-saving opportunities. In busier times, the foremen could choose among the many orders in the shop, spreading out demand for particular pieces of equipment and taking advantage of the economies possible.

Frequently, planned schedules had to be interrupted because of customer desires for expedited delivery, because materials were received late from the customer, or because of the breakdown of the machine or its tooling. In these cases quick, remedial action had to be taken. Often, this remedial action took the form of adding a new job or rerouting an existing job through the shop. Norcen had enough equipment on hand, typically, to permit a new or rerouted job to be set up on an unused machine. Joe Gehret purchased new machines for the shop with this kind of compatibility in mind. Not all jobs, of course, could be rerouted through the shop if things went awry, but it was possible with a sizable number, especially those that were not too complex. Some jobs could even be machined in an alternative sequence (e.g., milling could be done first rather than third), and this helped when remedial action was necessary. In general, the foremen preferred to have lathe work performed first, then milling, and only then grinding; but they were willing to abandon this sequence if necessary.

Some aspects of just-in-time manufacturing had repercussions for machine shops such as Norcen. Customers were now placing orders that required multiple, time-phased deliveries of smaller-than-usual lots. Even though many of Norcen's new machines were capable of running small lots economically, this trend did serve to complicate scheduling.

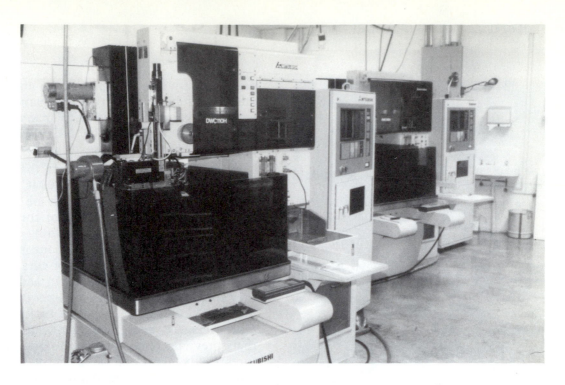

The Mitsubishi traveling wire EDM machines. (*Courtesy of Norcen Industries*)

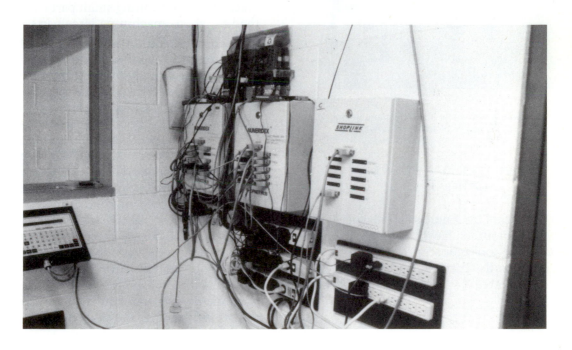

The control box for the fiber optic cables of Norcen's CAD/CAM network. (*Courtesy of Norcen Industries*)

Joe Gehret with some inspection equipment.
(*Courtesy of Norcen Industries*)

Technological Advances and Other Trends

Over the years, technological advances, especially in computers, have been revolutionizing the machine tool industry. The old fully manual lathes, mills, grinders, and other tools had been supplanted by a succession of newer tools:

1. The first set of newer tools were NC (numerically controlled) machines that could repeat operations exactly by reading punched paper tapes. The paper tapes had to be specially pre-

pared and thus were typically suitable for only high-volume, repetitive jobs. Naturally, in job shops like Norcen with many low-volume, one-of-a-kind jobs, NC machines could find only limited use.

2. The next generation of equipment, the computer numerically controlled (CNC) machines, found more of a home in job shops like Norcen Industries. Much of this machinery was programmable, enabling operators to enter, at the computer console, the depths, speeds, and feed rates of cuts to be made along X-, Y-, or Z-axes. In addition, they could operate from tapes (paper or magnetic). These newer machine tools were more versatile, more accurate, and faster than previous ones. Their automatic features easily permitted an operator to run more than one machine at a time, although this doubling up was often less feasible for machine shops running small parts (Norcen) than for shops machining large parts.

3. The latest generation of equipment, the direct numerical control (DNC) machines, could be run directly by a computer, eliminating the need for the preparation of tapes or the case-by-case programming of the CNC machine. All of Norcen Industries' were now DNC controlled. These machines made use of Norcen's CAD/CAM equipment and a fiber-optic cable network that linked the computer to the machines it controlled. Parts could now be designed using CAD computer software (Norcen used CADKey). The design could then be translated by CAM software (Norcen used IGES translation and the Anvil 1000 MD CAM program) into the precise settings for each machine's operation, and then downloaded to that machine via the cable network. Norcen kept a library of its past jobs, so that any repeat job could be pulled from the library and the machine set in exactly the way it was done before, without any extra work.

All of Norcen's lathe operators programmed their own jobs as did some of the machining center operators. The computer programmer and the foremen helped to troubleshoot particularly difficult jobs.

Norcen Industries was continually buying new, trading in old, and upgrading existing equipment. Joe Gehret remarked that the introduction of new technology meant that Norcen could produce some parts for costs that were less now, in current dollars, than those same parts cost in 1967, when the company was started.

Doing more and more machining were some of the latest machines to be added to the shop: three electrical discharge (EDM) machines. These machines used wire electrodes of various, and often small, diameters to machine parts of any conductive material, such as metal. The conductive material would be charged positively and the wire electrodes negatively and the material would be "burned off" by the travelling wire. EDM machines produced no burrs (thus, no need for grinding) as they machined parts, and they were capable of very precise machining, even in previously hard-to-work corners of the material. This equipment was kept in its own air-conditioned "clean room."

Norcen did not have any flexible machining centers (FMS) that one could sometimes find in larger fabrication areas nor did it operate any machining cells. FMSs linked several machines together and used sophisticated conveyors to move pieces from one machine to another. FMSs did not make sense for a job shop like Norcen because the tooling needed was too great and expensive, and jobs could be moved around more efficiently by the existing workforce. Similarly, machining cells were not employed because the continually changing nature of the product mix in the shop precluded the creation of product families with similar routings for those machining cells.

Nevertheless, the advent of the new technology permitted the shop's management to devise a number of cost-saving procedures. For example,

1. With carbide cutting tools and with EDM machining, Norcen could now harden metals by heat treating and could then machine them afterward. Previously, parts were machined, heat treated, and then re-machined (e.g., ground).
2. The new machines had multiple tool holders so that more facets of the job could be done while the part was held in position.
3. Parts could now be held in programmable indexers which could automatically rotate the part so that different sides could be machined without operator intervention.
4. Some machines had multiple vices on them so that 2 or 3 pieces could be set up at once.
5. Some of the lathes had automatic bar-feeds.

The movement for just-in-time (JIT) manufacturing and concern for quality and inventory levels were affecting operations at Norcen Industries. Customers were less lenient about scrap, no longer providing an extra 10 percent of material as a matter of course to cover potential scrap. Customers were also requiring multiple deliveries of their orders, forcing Norcen either to hold the inventory or to set up the job more than once. Concern for rapid delivery had also recently led the company to purchase a 110-ton punch press so that it could make its own blanks quickly for many of its parts. The press had also permitted the development of some press work for outside customers.

QUALITY CONTROL

Quality at Norcen was pursued by following what the quality control manager termed "the

Norcen way." Typically, these steps were followed in working on a job:

1. The foreman wrote the process sheet.
2. The foreman would then discuss the process sheet with the assigned operator and make any changes.
3. Each operator was responsible for his own setup.
4. The first piece would be checked, first by the operator, usually with the shop's coordinate measuring machine, and then by an inspector.
5. The job would be run in lots of 5 (usually) and each 5th piece would be inspected. If it did not pass inspection, all the previous 4 pieces would be checked 100 percent, and the next 5 pieces would be inspected 100 percent as well.
6. Quality was the joint responsibility of the operator and of quality control.
7. All finished jobs would be sampled by QC for final inspection.

The QC manager noted that the keys to quality were following the proper steps for the job and performing some steps with such tight tolerances that subsequent work, which relied on the previous machining, all fell within acceptable limits.

RECORDKEEPING

When an operator completed his assigned task on any job, he completed the job sheet (Figure B3) attached to the blueprint. On this sheet he filled in his name, the date, the operation he performed, the quantity he completed, and the length of time it took him to complete the task, including setup time. In

CHARGE NO	SUN	MON	TUE	WED	THUR	FRI	SAT	Total Hours
18980	✓	4 -	4 5					8 5
18593	✓	4 -	2 -			- 3		6 3
18976	✓		3 5	1 5				5 -
19363	✓			1 2		1 -		2 2
19191	✓			3 6	1 5			5 1
19192	✓			3 7	1 5			5 2
19363	✓				2 5			2 5
19116	✓				4 5	2 5		7 -
19215	✓					6 2 5 -		11 2
TOTAL	8 -	10 -	10 -	10 -	10 -	5 -	53 -	

Notes:

1. This is Orie Reading's time card for the week beginning January 24, 1985. This was a busy time at Norcen, and Orie worked 53 total hours during the week, including two hours of overtime from Tuesday through Friday and five hours on Saturday.

2. During the week, Reading worked on nine different jobs. He worked on the job we are most interested in, #18980, on Monday and Tuesday, for 4 and 4.5 hours, respectively. On Monday he also worked 4 hours on job 18593, and on Tuesday, he worked on job 18976 as well.

3. By totaling the hours by each worker on each job as recorded on these time cards, Lillian Gehret has another, independent estimate of the hours worked on any job. These estimates can be compared with the hours totaled on the job sheets.

FIGURE B6   A sample of an operator's time card used at Norcen Industries.

addition, at the end of the day, each operator completed a time card (Figure B6), listing the time spent on each job during the day.

In this way, Lillian Gehret had two records of the time each operator spent on each of the jobs done in the shop. Knowing the times and the wages of each operator involved in any job permitted Lillian to calculate the labor cost of that job. She made this calculation weekly and took care to see that any discrepancies in the two time records were resolved. Usually, the daily time cards were more accurate than the job sheets.

Knowing this information, at the close of every job Lillian would complete the record card (Figure B1), which Joe relied on for making quotes on repeat business. As a rule of thumb for calculating the profitability of a job, Lillian multiplied total labor-hours by the relevant charge-out rate ($24 to 30 per hour) and com-pared that total with the total price Norcen had quoted. She alerted Joe to any major deviations in that comparison, either positive or negative. Subsequent bids could be based accordingly.

In the summer of 1991, Norcen Industries committed to the purchase of an IBM A/S 400 minicomputer so that it could run some job shop-specific job control software. This software would automate much of the bookkeeping work now done by hand. Job histories would be kept on the computer, by customer and by part number so that quoting would become easier. It would handle order entry of jobs; billing and shipping; financial tasks such as payables, receivables, and the general ledger; inventories; and job tracking in the shop itself. Information for tracking jobs would come from the daily information now kept by hand. There were plans eventually to use bar-coding to collect the shop information.

PART TWO

DISCUSSION

Norcen Industries is in no way responsible for the following views and presentation. They remain solely the responsibility of the author.

THE PROCESS FLOW

Compared with the continuous flow process at International Paper's Androscoggin Mill, the job shop process at Norcen Industries is strikingly loose and ill-defined. A process flow diagram becomes difficult to draw in any precise or meaningful way. For example, we might sketch a process flow diagram such as in Figure B7. Although many of the parts made at Norcen Industries would have passed through exactly this sequence of operations, many more would have required a different set of operations and a different order. Some of the differences might be minor, but there are a host of parts for which the differences are major. At best, Figure B7's flow diagram can be termed a dominant flow or, perhaps, a preferred, simplified flow. For the most part, the job shop process exhibits great product flux and flexibility. Work-in-process can be routed anywhere within the shop so that even

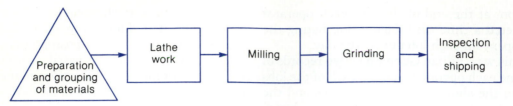

Notes:
1. Other tasks could be sequenced before, in between, or after the three middle tasks.
2. Work-in-process inventories exist between each operation.

FIGURE B7   A dominant, or preferred, process flow diagram for Norcen Industries.

extraordinary machining requirements can be met.

THE INFORMATION FLOW

What is not flexible but is almost totally rigid in the job shop is the flow of information through it. In stark contrast to the diversity of paths a machined part can take through the shop, information flows in a prescribed way (see Figure B8). Record keeping is done in the same manner for every order; the responsibilities of the workers, bookkeeper, and managers toward information in the process never vary. Everyone has fixed information tasks to perform: quoting, job sheet and blueprint preparation, release to the shop, filling out of job sheets and time cards, labor cost calculations, updating of quote record cards, and signaling of any variances.

The reason for all this rigidity, of course, is that the job shop lives and dies by its ability to process information. Significantly, too, the information flows in the job shop are as much from worker to management (job sheets, time cards, process suggestions, machine breakdowns) as from management to worker (job and machine assignments, schedules, quality control checks, troubleshooting, training). With-

out suitable records, there would be no clear or readily available means of routing an order through the shop or of specifying exactly what should be done to satisfy it. Without suitable records, the job shop's managers would have little idea how to bid for various jobs. Without suitable records, advances in productivity would be more sporadic and less well retained for future use. Without suitable records, managers would not be able to load the shop effectively. Information and the responsibility of everyone in a job shop to maintain its accuracy and smooth flow constitute the glue by which this type of process is held together.

CAPACITY IN THE JOB SHOP

In the job shop at Norcen Industries, capacity is as ambiguous as the process flow's pattern. No single measure of capacity makes complete sense. While a paper mill can measure its capacity rather straightforwardly in tons per day or some similar measure, a machine shop like Norcen Industries cannot readily do the same thing. The large and constantly changing mix of products in the typical job shop ensures that a simple count of units produced is a meaningless way to gauge what the shop's effective capacity really is. We must avoid that simple

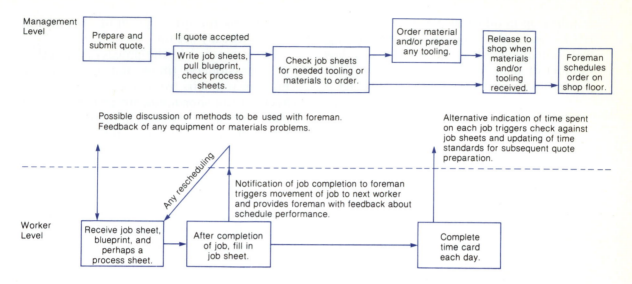

FIGURE B8   An information flow diagram for Norcen Industries.

measure in favor of one that transcends the product mix problem. The easiest remedy is to measure capacity in dollar terms, but that still leaves us with a variety of options. Dollars of typical output per unit of time? Dollars of output per worker? Dollars of output per machine? Dollars of output per dollar of machine value (at cost? at replacement cost?)? All of these measures grab a piece of what we mean intuitively when we say capacity, and yet no one of them fully describes the concept. Only over the long term, when we can feel comfortable that the mix of products has been representative, can we point at differences in shop output over time or output per worker or output per machine as valid measures of high or low capacity. In the short run, no capacity measure is clearly appropriate. Any measure of the shop's capacity is dependent on a host of factors, such as:

1. *Lot sizes.* The larger the lots ordered, the fewer setups in any one day and the greater the number of pieces produced, other things being equal.

2. *Complexity of the pieces worked on.* The more complex the piece, the more likely it will require a large succession of operations and thus the more likely its manufacture will demand a succession of time-consuming setups and difficult scheduling decisions. Of course, this factor is relatively more important in accounting for differences in the number of units manufactured than it is in accounting for differences in dollar value measures of capacity.

3. *Mix of jobs already on the shop floor.* The number and nature of orders already released to the shop floor affect the capacity of the shop in at least two ways: shifting bottlenecks and worker–machine interference.

   Many of the orders in the shop may require the services of particular machines. We can expect, then, that some orders may back up while awaiting a particular machine or operation. Further, we can expect to see such bottlenecks occurring from time to time all over the shop. That is, we can expect to see work-in-process inventories building up in different places and at different times in the shop. A smooth-running, well-scheduled job shop will have a low number of such "shifting bottlenecks," but given the diversity of output and run lengths within a job shop process, they are absolutely unavoidable.

In a typical job shop, there will be many machines that require the constant attention of a worker when they are in operation. Increasingly, however, automatic equipment is entering the shop with its ability to perform without the constant "hands-on" attention of the workforce. This advance is not without a challenge of its own. While automatic equipment frees up worker time, only the (sometimes fortuitous) scheduling of two or more jobs to the same worker actually leads to greater worker productivity apart from any speed advances built into the automatic equipment itself. If the scheduling cannot mesh two or more jobs together, the machine can be said to "interfere" with the worker, and capacity in the shop drops relative to the situation where the worker can easily operate two or more pieces of equipment at once. For this reason, Joe Gehret and his managers positioned machinery within the shop to maximize the possibilities for reducing worker-machine interference.

4. *Ability to schedule work well.* As Joe Gehret himself put it, "Scheduling is the most difficult function we have around here." The matching of workers to machines and of workers and machines to jobs often separates a profitable job shop from an unprofitable one. Good scheduling lessens shifting bottlenecks and worker–machine interference. Poor scheduling introduces more work-in-process inventory to the shop than is necessary. In particular, if too many rush orders are permitted in the shop, their scheduling will become difficult, often necessitating the interruption of runs on machines already set up.

As was noted in the process description, in recessionary times management has many fewer jobs from which to choose to keep all the shop's machining centers busy and to avoid bottlenecks. In better economic times, management can search through the released orders for jobs that can help smooth the flow of work through the shop.

5. *Process improvements.* Any advances in the methods employed in producing a part at Norcen Industries permitted the shop to increase its capacity and thus its profitability. With so many orders passing through the shop, there are many opportunities for improving the process: resequencing of operations, different use of machines, quicker setups, special jigs or fixtures for increasing speed and/or accuracy, possible redesign of the piece ordered. Because any of these improvements takes time to work through, it is likely that only the higher volume, repeat business will benefit from such attention.

6. *Number of machines and their condition.* It is obvious that, even without expanding the workforce, the addition of equipment to the shop is likely to increase capacity detectably. For one thing, rush orders will be less likely to necessitate the dismantling of existing machine setups before completion of the run. Having more machines also increases the probability of finding favorable combinations of orders to lessen worker–machine interference. Fewer bottlenecks, too, are likely to occur with the addition of more equipment. It is evident as well that machines in good condition are less likely to break down and thus demand attention both for themselves and for the rescheduling of operations through the shop.

Technological advances, notably the evolution from manual to NC machines to CNC machines and then to DNC machines, have greatly influenced the shop's capacity. Setups are often easier with the more advanced machine tools; more time can be spent actually cutting metal. Moreover, the newer tools permit fancier machining (such as curved shapes) to be done with greater accuracy, higher speed, and less scrap. And given that the machining was accomplished via tape or computer instruction, the new equipment does not tie up workers as long as the older equipment did.

7. *Quantity and quality of labor input.* Another obvious set of factors affecting the capacity of a job shop is drawn from the labor force itself. Overtime and second shift work is a standard way to augment capacity with the same stock of plant and equipment. Employing inexperienced workers and having to train them is a drain on capacity, however.

Obvious in identification but subtle in design and application is an incentive system for the workers. At Norcen Industries, the bonus plan that was dependent on company profits was viewed by the workers as a fair and reasonable spur for continued good work. But it is not the only payment system that could be installed; different systems might have different effects on worker effort and thus on capacity.

Increasing the capacity of a job shop in the way outlined here, at little or no cost, is important to the shop because that is how it makes its money. Of course, the shop must be flexible enough to bid on a tremendous variety of jobs, but the shop's ability to earn any profit, once granted the business, is linked fundamentally to its capacity and its knowledge of that capacity. As we have discussed, these two items are influenced by a number of factors, many of which require the accurate processing of information around the shop.

THE ROLE OF STANDARDS AND INCENTIVES

Time standards for all of the operations to be performed at Norcen Industries are an integral feature of its cost estimation and bidding responsibilities. Some of the standards are developed internally, mostly through past experience doing the same thing; but others are supplied by the customer when bids are solicited, having been worked out for or by the customer.

The standards are a useful guide for both workers and managers. For the workers, the standards (written on the job sheet) provide continual feedback on how well they are doing the job. For managers, the standards provide information on how long certain jobs should take and thus how they might be scheduled. The standards also provide management with a yardstick for worker performance that is useful not only for making advancement/layoff decisions but also for determining which tasks each worker does relatively better.

Other than by furnishing feedback for each worker, the time standards are not tied formally to any incentive system. Incentives for good work and/or for speedy work are provided either through knowledge that such work often contributes to better company profits (although such work is not the only determinant of profits, by far) or by the prodding and cajoling of the foreman. In job shops like Norcen Industries, the foreman carries much of the responsibility for pacing work through the shop and for ensuring that the quality is satisfactory.

QUESTIONS

1. Write up a hypothetical job sheet for an order and briefly explain each column entry. How do the functions of the job sheet and the record card differ?

2. Can you make any generalizations about cost reductions from the example of Norcen Industries? Do these generalizations apply to any other process with which you are already familiar?

3. Why might scheduling be "the most difficult function" in a job shop?

4. "The job shop lives and dies by its ability to process information." Discuss this comment and compare the information flow in the job shop with that in another type of process.

5. Discuss three of the factors that influence a job shop's capacity. What might be "the perfect set of circumstances" in a shop foreman's eyes?

6. As a worker, which would you prefer: the calm predictability of the continuous flow process or the frequently frantic unpredictability of the job shop? As a manager, which would you prefer? How might your attitude affect the setting of standards and incentives in each process?

7. Norcen's foreman places particular emphasis on meeting customer due dates. Due dates thus significantly affect the scheduling that is done. Suppose, however, that due dates were not as important to you as the foreman. What other factors could take precedence? How could they serve as the basis for scheduling jobs through the shop? Under what circumstances might different factors take precedence in establishing scheduling rules?

8. Suppose that you were forced to justify to upper management the purchase of new technology, such as a CNC machine, for an existing job shop. Where would you look for the benefits of such an addition to the shop? How would you measure the extent of those benefits?

9. What advantages do you see to a layout such as that used at Norcen Industries? What disadvantages do you see?

SITUATION FOR STUDY B-1

OWENS, INC.

Fred and Ralph own and operate a small machine and metal fabrication shop. Most of their business involves small orders from local industry. These orders are usually ones that the larger customer firms farm out because they do not have the excess capacity. In other words, Fred and Ralph provide "slack" capacity to the larger firms. Figure B9 shows the plant layout.

Fred and Ralph feel that if they can once get an order in their shop, they can learn how to make the part and then underbid on future orders. Moreover, they think they can make most parts more cheaply than the customer can in house.

They are currently considering bidding on part 273. Figure B10 is an engineering sketch of this part. For the first order the customer will supply a coil of 1 5/8 cold-rolled steel (CRS), but Fred and Ralph are investigating to see whether other processes might use other sizes of raw material more efficiently. The proposed order is for 500 pieces, with possible future orders of 500 per month. The proposed bid must be submitted in 2 weeks, the bid

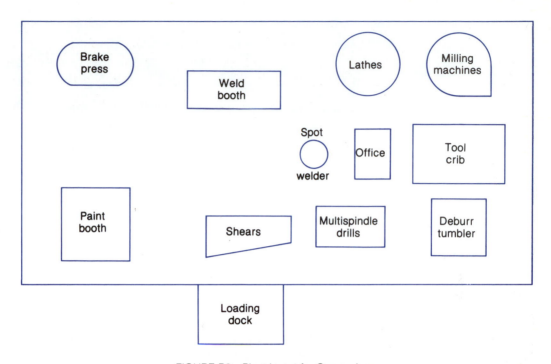

FIGURE B9   Plant layout for Owens, Inc.

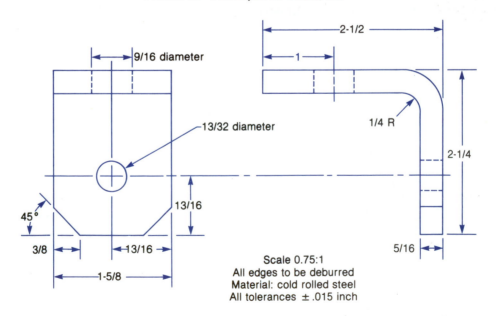

Scale 0.75:1
All edges to be deburred
Material: cold rolled steel
All tolerances ± .015 inch

FIGURE B10   Engineering sketch for part 273, a bracket.

Table B2   Process sheet for part 273 (Owens, Inc.)

OPERATION SEQUENCE	MACHINE	OPERATION	SETUP TIME (mintues)	OPERATION TIME PER UNIT (minutes)
1	Shear 3	Shear to length	5	0.030
2	Shear 3	Shear 45° corners	8	0.050
3	Multispindle drill presss 1	Drill both holes and deburr	15	3.000
4	Brake press	Bend 90°	10	0.025
5	Tumbler	Deburr	5	a
6		Pack in boxes		30.000

a The deburr operation can be left unattended, so the only labor required is to load and unload.

award determined within 1 week, and the 500 units delivered 1 month after that.

Fred has generated the estimates shown in Table B2. The cost estimates used for bids are $10.55 per hour for labor and $18.50 per hour for machines — or $29.05 per hour for one worker/ one machine operations. Any time "sold" at this rate makes Owens a margin of 25 percent with which to cover indirect costs such as administrative costs.

1. If you were Fred and Ralph, what other aspects of the shop, the market, and the future would you consider regarding the bid for part 273?

2. Analyze the current proposal and then generate a bid to the customer and a rough shop schedule.

SITUATION FOR STUDY B-2

STREETER DIE & STAMPING COMPANY

Streeter Tool & Die Company was founded in 1956 by entrepreneur Jack Streeter. For 20 years the small firm flourished, furnishing tools, dies, jigs, and fixtures to metal stamping firms. In 1976 the firm became Streeter Die & Stamping Company (SDS) and began a move toward a new business. Today SDS has two plants in Toledo, Ohio, furnishing metal stampings and assemblies primarily to the major U.S. automakers. In the most recent year, General Motors (GM) accounted for 85 percent of SDS's sales.

The main plant (plant 1) had 14 presses, ranging in power from 20 to 1000 tons, that

stamped out metal parts from raw steel. Production was usually based on firm customer orders, but occasionally some forecasting of future orders was done in an attempt to get longer, more efficient production runs.

The secondary plant (plant 2), representing a type of forward vertical integration for SDS, was more labor intensive than plant 1. Plant 2 specialized in the assembly and welding of parts stamped at plant 1. Extra space at plant 2 was used as a finished goods warehouse for any parts made at plant 1 or 2 that would not be shipped within 1 week.

The orders that SDS normally produced to were contracts that SDS had won by submitting quotes to the customer. In the case of GM contracts, it was known at SDS that price was not always the deciding factor; SDS's record for quality and timeliness played an important role in winning many of these contracts.

Raw materials arriving at SDS, almost all of them coils weighing 1 to 5 tons, were stored in a warehouse adjacent to plant 1. When the steel was needed in the shop, a forklift picked it up and brought it to the press where it was to be used. From this point, the steel could go through many different combinations of presses and operations, depending on the specific part being made.

In general, the first operation involved running the steel through an automatic press. Steel coils were hoisted onto a reel and fed into the press automatically. The stamped parts fell off the press into bins, which the press operators occasionally checked to see whether the parts were being stamped properly. When the order had been completed, the remaining part of the steel coil was removed from the reel and stored against the shop wall. For several parts this operation was the only one necessary; a random sample of these parts was inspected by quality control and, if the parts were good, they were taken to the loading dock and prepared for shipping.

After passing a quality inspection, bins of more complex parts were moved to nonautomatic presses for the rest of their stamping operations, such as piercing, drawing, and forming. These operations were much slower, because the parts had to be placed manually into the dies by the press operators. These operations completed some parts. Other parts were sent to a vendor for sandblasting, painting, heat treating, and/or deburring before being shipped to the customer; still others were sent to plant 2, where several component parts were welded and/or assembled into the finished product.

The finished goods inventory at SDS was relatively small. Parts that were to be shipped within one week were stacked along the walls of the loading docks at both plants so that forklift drivers could easily spot which parts must be loaded onto a certain truck.

Smooth operations at SDS were extremely dependent on a coordinated, accurate information flow. The information flow followed a cycle that began and ended with engineering submitting a quote for a contract. If information was not accurate at one point in the cycle, the results could have a domino effect on the various departments at SDS. To help improve information flow, all departments at SDS used an IBM System 34 computer extensively.

After analyzing customers' blueprints and past SDS results on similar parts, engineering quoted prices per piece for lot sizes ranging from 100 to 600,000 pieces per year. Reflecting the fixed costs of initial tooling, the price per piece decreased as the lot size increased. After winning a contract, SDS had 6 to 8 months before its first shipment was due on a new part and 3 to 4 months on a carryover part.

During the 6 to 8 months before the first due date, engineering was responsible for getting the shop ready for production. This included ordering the tooling from a vendor, drawing up process sheets for each subassembly, determining what presses could or should be used to stamp each part, assigning part numbers to new parts, and entering all pertinent data about each part into the computer. Engineering also sent a copy of the contract and process sheets to production planning so that they had a record or parts that had to be scheduled in the near future.

Production planning needed no further information until 13 weeks before the first shipment due date. The data systems manager had designed a computer program that received information directly from GM's computer each Monday morning and printed out a list of SDS's week-by-week requirements over the next 13 weeks; the first 2 weeks were required by contract to be firm. When new parts showed up on the list, production planning had to (1) check with engineering to see how the tooling was progressing, (2) order materials, and (3) schedule a sample run. No full production runs could be shipped to GM or other top customers until samples had been approved.

The Monday morning printout listed other important information for each part, such as the number of pieces on hand, the week in which the number on hand would be insufficient to cover cumulative requirements, and the amount of components and raw materials on hand for each part. The production scheduler used this report to determine which parts would have to be made 2 weeks into the future. The scheduler filled out a small card for each operation that must be performed and posted the card on a schedule board that enabled him to schedule work on each press for the present week and the next 2 weeks.

While the scheduler was busy loading the presses, a production planning assistant used another copy of this report to determine raw material requirements over the entire 13 weeks. Any materials that SDS would not have enough of to meet these requirements were written on a material requisition, which was delivered to the purchasing agent for further action.

Each morning, after surveying the progress on the shop floor, analyzing the previous day's production summary, and updating the schedule board, the scheduler brought to the shop foreman (1) a schedule of parts to be stamped in each press, (2) routing cards for each part, and (3) process sheets with pictures of each operation. Press operators used counters on the presses to count the number of pieces stamped each shift, each day, and each production run. When the total count for the run reached the number on the routing card, the operator signed the routing card, wrote on his activity card his shift count and his hours spent on the part, and signaled the foreman for a new job. At the end of each shift, the foreman gathered all the activity cards and summarized the data into the daily production summary. In addition to the scheduler's use of this summary, the inventory control specialist combined this information with shipping and receiving data to update inventories on the computer files daily. Keying in this inventory data and some additional cost data also updated financial and job cost data, thus completing the information flow cycle and ensuring that accounting and engineering had access to the information necessary for preparing monthly financial statements and accurate quotes on parts contracts.

A somewhat separate line of information flowed between the shipping department and the billing clerks. Each day shipping filled out a list of the parts, with quantities, that it had shipped. This list was delivered to the billing clerks who, at the end of each day, wired these shipments into the customers' computers. With this method, the customers' procurement agents knew exactly from day to day which parts were in transit, and both SDS and its customers had a common reference date if questions arose about a particular shipment.

Often, the information flow was less formal when customers' procurement agents called and asked for early delivery or when SDS was behind schedule on a part. Information flow became a matter of someone from production planning personally expediting these orders through the shop and arranging special rush transportation.

1. To the extent that you can, diagram the process and information flows.
2. Sales at Streeter Die & Stamping recently increased and bottlenecks emerged in the

operation, although the exact cause of the bottlenecks was not immediately obvious. How would you determine where any bottlenecks were? Be specific.

3. How would you control the levels of inventory? How would you monitor and, perhaps, change the status of jobs in the company? What things could the computer do?

A BATCH FLOW PROCESS
Jos. A. Bank Clothiers
Hampstead Manufacturing Operations
Hampstead, Maryland

Jos. A. Bank Clothiers was a maker of high-quality, traditionally styled men's and women's clothes—primarily suits, sport coats, slacks, and skirts. It sold these clothes along with shirts, ties, and other accessories through 40 retail stores[a] and an extensive mail-order business. Compared with retailers of similar clothes, the company used lower price mark-ups and had a lower overhead. These features, combined with manufacturing efficiencies, permitted Jos. A. Bank to market its high-quality clothes at prices significantly lower than those of its competitors. In early July 1981, Quaker Oats purchased Jos. A. Bank Clothiers from the Bank family, who had owned and operated the company since its founding in 1902. Quaker Oats, in turn, sold the company in late 1986 to a group of investors in a leveraged buyout.

The company employed 625 people in manufacturing, 34 of whom were salaried managers. There were three manufacturing plants in Maryland—one in Baltimore and the other two in rural Hampstead. The Hampstead plants did all of the company's cutting, and produced men's coats and all the women's clothes. The Baltimore plant, situated on North Avenue, produced men's coats and all the pants. All warehousing and distribution were accomplished in Hampstead.

The Hampstead plants were located within a few miles of each other. The larger of the two, a former Black and Decker manufacturing site that was acquired in November 1986, housed both the distribution center and the cutting room. As a result of its acquisition, an older, multistory operation in downtown Baltimore was closed and the work and workforce relocated to either the North Avenue or Hampstead facilities.

---

[a]Two stores were in the Baltimore area (on Light Street in downtown Baltimore and in Towson, north of the city). Other stores were located in Atlanta, Birmingham, Boston (2), Buffalo, Charlotte, Chicago (4), Cincinnati, Cleveland, Columbus, Dallas, Denver, Detroit (2), Houston (2), Indianapolis, Los Angeles, Louisville, Memphis, Minneapolis, Nashville, Philadelphia (2), Pittsburgh, Raleigh, Richmond, Rochester, St. Louis, Stamford, Summit (NJ), Washington, D.C. (3), and Winston-Salem.

PART ONE

PROCESS DESCRIPTION

The making of a high-quality sportcoat or suit coat was a complex endeavor, encompassing 140 to 150 distinct operations (see Figure C1). The coat was an assembly of various parts, including sleeves, backs, fronts, facings, collars, and pockets. Each of these parts, in turn, combined the basic fabric for the coat itself (known as piece goods) with 40 to 50 trim items. These trim items included linings, fusings (the firm material fused by heat to various parts of the piece goods to help them retain their shape and resist wear), pocketing, stays, tape, thread, and buttons. The assembly of pants and skirts was less complex.

The Cutting Room

The assembly of a suit or coat began when a bolt of fabric (the piece goods) was withdrawn from raw materials inventory and examined. Any flaws were marked so that they could be trimmed out during the cutting process. The examined fabric was then carefully laid on long cutting room tables. This task was known as "spreading" the cloth; depending on the particular order, the number of layers spread ranged from 1 to 110. Once spread, a paper pattern known as a marker was placed over the layers of fabric.[b] The coat, vest, pants, and/or skirt pattern pieces were carefully arranged on it so that as little fabric as possible was wasted. This

[b]Only occasionally would a pattern have to be chalked on fabric by hand.

arrangement was generated by one of two special computer-based marker-making instruments (AM-5s, made by Gerber Garment Technologies) and was printed out, ready to be used. The AM-5 could easily adjust sizes and arrange the pieces in the most efficient pattern. For example, certain suit sizes (such as 40 regulars and 41 regulars or 39 regulars and 43 regulars) might be combined efficiently on the marker pattern. The markers were stapled to the layers of fabric and the fabric was cut using either electric cutters or manual shears, depending on the number of layers and the outline of the pattern.

The pieces that composed the same coat and/or pants were cut from the same layer in the spread fabric. After being cut, they were matched together and assembled (this assembly was termed "fitting") into various groups of materials—one group, for example, for the pants, one group for the coat. These matched and assembled cutouts of fabric were then stapled with special identifying numbers by the Soabar marking and ticketing machine (see Figure C2). These numbers were critical in ensuring that material cut for the same suit would stay together throughout the process. After being tagged, the pieces destined for a coat, pants, skirt, or vest were subdivided and assembled, or "fitted", into special bundles with an average of 12 units in each bundle. Bundle sizes ranged from 1 to 22. Bundles were fitted for various segments of the coat, such as sleeves, fronts, backs, collars, facing, and linings. (Flaps and welts (pockets), being movable items, were fitted in the coat shop itself).

continued on p. 617

```
                    MODEL MASTER LISTING                          RPT-PAY032

MODEL-S361BCSSHC PLN

    OPER              DESCRIPTION        WRKCTR     PRICE      S-A-M      SINGLE     SUB-FLAG

SUB ASSEMBLY 00023-SLEEVE

      27          JN 1ST SLEEVE          00401      .1029      .8542      .1132
      76          FUSE CUFF              00401      .1073      .0000      .1181
      77          MAKE CUFF              00401      .1292      .0000      .1422
      75          JN ELBW SEAM           00401      .2704      .0000      .2974
     128          UNPRE FIR SLE          00401      .0865      .7835      .0909
     129          UNPRE SEC SLE          00401      .0898      .9624      .0940

SUB ASSEMBLY TOTAL                                  .7861     2.6001      .8558


SUB ASSEMBLY 01017-SLEEVE-LINING

     226          MK SLEEVE LIN          00402      .1533     1.0259      .1685
     273          MK BTN SHANK           00402      .0360      .3190      .0376          N
     277          FEL CUF1TURN           00402      .1585      .9938      .1744
     281          SEW 3BUT1TURN          00402      .1525     1.5159      .1679          Y
     376          TK LN1TURN             00402      .1253     1.0853      .1315
     427          MATCH SLEEVE           00402      .0465      .4607      .0488

SUB ASSEMBLY TOTAL                                  .6721     5.4006      .7287


SUB ASSEMBLY 04058-BACK

     744          MK BK FUL LIN          00405      .5091     3.9938      .5601
     777          FUSE BK TAPE           00405      .0463      .3321      .0510
     822          PR BK FUL LIN          00405      .0986      .2128      .1036
     834          TK VT SM FL            00405      .0880      .5365      .0924
     836          FIT BK FUL LN          00405      .1351      .8169      .1419
     902          MATCH BACK             00406      .0475      .4285      .0498

SUB ASSEMBLY TOTAL                                  .9246     6.3206      .9988


SUB ASSEMBLY 05001-CANVAS

     954          CUT GORES              00406      .0522      .3187      .0547

SUB ASSEMBLY TOTAL                                  .0522      .3187      .0547


SUB ASSEMBLY 06043-FACING-LINING

    1031          JOIN LINING            00407      .1986     1.4979      .2185
    1215          PREPARE PKT 2          00407      .0622      .5304      .0683
    1055          REECE IBP              00407      .1389     1.1577      .1527
    1080          FIN 2IBP WFS           00407      .2378     1.9717      .2615
    1085          BANK LBL IBP           00407      .0626      .4668      .0657          Y
    1088          SW XTRA LABEL          00407      .0675      .4622      .0708          N
    1111          SEW CR LB IBP          00000      .0478      .3576      .0501
    1155          PRESS FACING           00407      .1260      .9002      .1385
    1202          PAIR IN FACING         00407      .0475      .4285      .0498

SUB ASSEMBLY TOTAL                                  .9889     7.7730     1.0759


SUB ASSEMBLY 08036-FLAP
```

Notes:

1. This portion of a Model Master Listing details the specific operations to perform, in sequence, for various subassemblies for the coat of suit model 361, a plain fabric Kent Model. These subassemblies include the sleeve, sleeve lining, back, canvas (fusing), facing-lining, and so on. The entire listing contains 13 different subassemblies.

2. Each line consists of an operation number, a description of the operation, the workcenter number where it is performed, the piece rate price for a standard bundle, the standard allowed minutes (not applicable to the Hampstead coat shop), and the piece rate price for a single unit. The Sub-flag column denotes exceptions to the standard that are made for selected coats.

FIGURE C1   Model master listing extract for a man's coat.

The Gerber AM-5 marker making machines. (*Courtesy of Jos. A. Banks Mfg. Co.*)

Spreading cloth in the cutting room. (*Courtesy of Jos. A. Banks Mfg. Co.*)

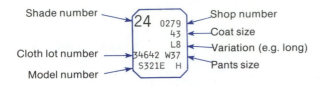

FIGURE C2   A sample Soabar ticket used by
Jos. A. Bank to identify cut fabric.

Each bundle (batch) of materials to be processed in the coat shop would have the same operations performed on each of the pieces in the bundle. For the most part, workers in the shop did the same things to successive bundles of materials. A record of what was to be done to each bundle and how many items were in each traveled with the bundle.

The Coat Shop

Most operations in the coat shop consisted of sewing pieces of the fabric together and/or attaching various trim items to the fabric. The bundles representing parts of the coat were worked on separately in various areas of the coat shop. Periodically, some were brought together to make larger subassemblies before these were joined in the assembly that was the coat itself. At two key places in the shop and at several minor locations, these parts were matched before being sewn into a major subassembly. These matching stations served to control the process, ensuring that all pieces in each bundle were accounted for and that only those pieces of fabric that had been cut together would be assembled into a coat. Once the coat was fully assembled—linings in, welts and flaps sewn, facings fabricated and assembled to fronts, and sleeves assembled to the rest of the coat—it was ready for finishing. In finishing, basting (temporary stitches) was removed, various parts of the coat were pressed, and the coat was thoroughly inspected. Finish pressing for all coats was accomplished at the central pressing facility at the North Avenue factory. There, each part of the coat was pressed with specific equipment designed to finish it in an impeccable manner. Once inspected, the coat was transported to the matching warehouse in Hampstead. At the matching warehouse, pants and coats for the same suit were placed together and taken to the adjacent distribution center. Orders were filled from the distribution center.

The Skirt Shop

Skirts underwent similar, although simpler processing (see Figure C3). Much as with coats, the cut piece goods for a skirt had been fitted into bundles of from 1 to 20. The cloth was then serged to prevent fraying. The processing that followed could differ from one model to another, but in general, these steps were followed. Cloth for any pleating was sent outside to a vendor with special equipment for putting pleats into skirts. Once returned, the pleats were stitched down. Pockets and zippers were sewn in and any vents were constructed. Then sides were seamed, linings attached, and the waistband was sewn in. The seams were underpressed and the waistband was top stitched. The skirt was then ready to be hemmed. Button holes were cut, buttons sewn on, and finish pressing was accomplished. Inspection concluded the process and the completed skirts were then sent on to the matching warehouse. The skirt shop had a capacity of 225 skirts per day.

THE WORKFORCE AND THE PIECE-RATE SYSTEM

Except for some nonunion clerical workers, all the workers in the Hampstead manufacturing facilities were members of the Amalgamated

MODEL MASTER LISTING   RPT-PAYO32

MODEL-K12XBKS5HS PLN

OPER	DESCRIPTION	WRKCTR	PRICE	S-A-M	SINGLE	SUB-FLAG
SUB ASSEMBLY 21014-WAISTBAND						
9000	FIT PNL W/BND	00601	.1974	1.4885	.1974	
SUB ASSEMBLY TOTAL			.1974	1.4885	.1974	
SUB ASSEMBLY 22133-ASSEMBLY						
9025	FUSE W/B	00602	.0608	.3435	.0608	
9027	PIPING WB	00602	.0540	.0000	.0540	
9052	SRG PNLS SHRD	00602	.2054	1.5368	.2054	
9100	CUT ZIPPER	00602	.0286	.1618	.0286	
9285	MK FRNT PLTS	00602	.0803	.0000	.0803	
9425	MK BACK GORES	00602	.0671	.8748	.0671	
9474	JN 1CNTR BKSM	00602	.0897	.8465	.0897	
9510	SET ZIPP&FAC	00602	.1899	.8207	.1899	
9526	OV/LK 3SEAMS	00602	.1196	1.5476	.1196	
9531	JOIN SIDESM 2	00602	.1676	1.5808	.1676	
9549	OVERLOCK PKT.	00602	.0827	.8925	.0827	
9554	ATT TP PKT WB	00602	.3143	.0000	.3143	
9577	SW BND BTM VT	00602	.1405	1.1191	.1405	
9604	HEM BOTTOM LN	00602	.1114	1.0178	.1114	
9614	MK VENT ZIPPER	00602	.0814	.8082	.0814	
9615	MAKE VENT LNG	00602	.0814	.8082	.0814	
9620	MK FRTVT CRNS	00602	.1233	1.0854	.1233	
9625	PRS VENT	00602	.0594	.6093	.0594	
9647	ATT LNG WST	00602	.2720	2.1815	.2720	
9700	SEW WAISTBAND	00602	.1758	2.0291	.1758	
9741	SEW 2 LABELS	00602	.0463	.4044	.0463	
9745	MK EXT & CRNR	00602	.1483	.8302	.1483	
9777	T/ST W/B	00602	.1143	1.4745	.1143	
9786	LOCK ZIPPER	00602	.0848	.6536	.0848	
9803	PRS OPN S/S 6	00602	.1131	.0000	.1131	
9810	HEM BOTTOM	00602	.1269	1.2973	.1269	
9811	TACK VENT BTM	00602	.0381	.0000	.0381	
9911	BARTACK ZIPP	00602	.0339	.0000	.0339	
9951	MARK&MK 1 BH	00602	.0473	.3651	.0473	
9961	SEW 1 BUTTON	00602	.0486	.5051	.0486	
9975	OFFPRS.TOPPLT	00602	.1938	1.4340	.1938	
9982	OFPRS BDYW/VT	00602	.2098	1.9258	.2098	
9990	CLEN.EXAM HNG	00602	.2581	2.5879	.2581	
SUB ASSEMBLY TOTAL			3.9685	29.7415	3.9685	
MODEL TOTAL			4.1659	31.2300	4.1659	

Notes:

1. This portion of a Model Listing details the specific operations to perform, in sequence, for the two subassemblies, waist band, and assembly making up skirt K12XBKS5HS, a plain fabric model.

2. Like the Model Master Listing for the coat, each line indicates operation number, operation description, workcenter number, standard bundle piecerate price, standard allowed minutes, single unit piecerate price, and the sub-flag (exceptions) indication.

FIGURE C3   Model master listing for a woman's skirt.

Some hand cutting of cloth. (*Courtesy of Jos. A. Banks Mfg. Co.*)

A roll of cloth on the inspection equipment. (*Courtesy of Jos. A. Banks Mfg. Co.*)

The Soabar ticketing machine. (*Courtesy of Jos. A. Banks Mfg. Co.*)

Clothing and Textile Workers Union. These workers were paid an incentive wage, although the type of incentive differed across departments. Workers in the coat and skirt shops were paid piece rates. These piece rates were the prices in dollars per piece that the company paid for each piece of sewing, pressing, matching, or inspection.

When workers finished their assigned tasks, they detached the appropriate cardboard coupon from the perforated sheet of coupons (Figure C4) that traveled with the job. On the coupon was an operation number that was uniquely identified with the piece rate for performing the indicated job. By accumulating these coupons and turning them in at the end of the day, workers were paid for exactly what they had done. There was no minimum payment and no maximum payment. In actuality, the wage payment averages in the coat and skirt shops ran about 110 percent of the established standard, with a range for experienced workers from 80 percent of the standard to 160 percent. The coupons that workers turned in were read by a laser wand so that payroll and work-in-process were tracked automatically.

This piece rate system was modified somewhat for the cutting room workers. Before about 1978, workers in the cutting room were paid by the hour. At that time, however, it was felt that the output of the cutting room could be increased if it, too, were put on some incen-

Notes:

1. This is a portion of a page of coupons that had been placed onto a sheet of paper as a record of what a piecerate worker (B. Fuhrman) had done during the day. Whenever she completed a bundle, she detached the relevant coupon. The payroll department then wands the top row with a laser scanning wand and creates this record page.

2. The top row is used for payroll purposes only and indicates the shop number, bundle number, and operator number.

3. The middle row provides a description of the operation and, where appropriate, indicates what trim items are to be used.

4. The bottom indicates the bundle number, the cut number (same as shop number), the operation number, and the quantity in the bundle.

5. The Coupons Entered—50 and Count Entered—130 are merely check codes for the payroll office.

FIGURE C4   Coupons.

tive plan. A consulting firm was brought in, and standards were set for all operations in the cutting room—spreading the fabric, cutting the patterns, and other tasks. The standards set were called "standard allowed minutes" (S-A-M) for doing particular jobs. All the jobs in the cutting room were studied and rates were set. Thus, workers in the cutting room were paid for the standard allowed minutes that they acquired rather than for the pieces of work that

Seamstresses in the shop. Note the bundle in the foreground.
(*Courtesy of Jos. A. Banks Mfg. Co.*)

A seamstress in the dress shop.
(*Courtesy of Jos. A. Banks Mfg. Co.*)

The major matching station.
(*Courtesy of Jos. A. Banks Mfg. Co.*)

they were able to finish. However, a pay floor was established; workers who did not accumulate 8 hours of standard allowed minutes during the day earned a set base rate. Workers who accumulated more than 8 hours of standard allowed minutes were paid for those minutes. In fact, the average in the cutting room ran between 120 and 125 percent of the established standards, with the range being between 100 and 160 percent for experienced workers. Workers were paid by filing incentive declarations that documented start and stop times for various operations. There were no coupons to detach as in the coat or skirt shop. This incentive system did, in fact, increase the productivity of the cutting room dramatically.

There were times when workers were not paid piece rates but were paid the average of their work. This occurred whenever a worker was taken off his or her standard job to do something else. For example, if a worker was asked to fill in for an absent colleague, that worker was taken off the piece-rate system. Similarly, if a worker was asked to work on something new or something special, piece rates were removed.

Once a piece rate had been established, it could not be changed without some change in model, materials, methods, or machinery. There were literally thousands of piece rates that could apply in the shop, and typically less than 10 percent of them were changed each year. Some had remained unchanged for years. Many changes were, in fact, triggered by a worker's retirement. (Turnover was only 2 to 3 percent a year.)

Management recognized that the piece-rate system caused some wage inequities in the factory, but there seemed to be no clear solutions. It was known, for example, that some piece rates were tight whereas others were loose; but if no changes were contemplated to those jobs, management could not remedy those inequities easily without union consent. Spotting such inequities was not always simple. Variations in performance across operators doing the same job could range from 90 percent of the standard to 150 percent. Given such variations, it was often hard to say that a particular piece rate was loose when for at least some workers it might be considered tight.

Setting New Piece Rates

Establishing new piece rates was the responsibility of the factory's industrial engineer, who spent about 60 percent of his time determining piece rates and the rest of his time calculating the factory's unit costs and following up on special projects (such as studying new equipment purchases). The first step the industrial engineer took in setting a new piece rate in the coat or skirt shop was to meet with the foreman in charge to discuss the procedure that the foreman thought was required to do a good-quality job. An operator was then brought in and informed of the new task. He or she was given some time to become familiar with the work. After that time the industrial engineer observed the operator for about an hour and a half, informing the operator that this study would result in the establishment of the piece rate.

During this observation period, the industrial engineer used an electronic time study board—a clipboard to which was attached a special digital readout stopwatch that kept track of various times. By pressing a lever the engineer could indicate the beginning of one element of the job (jobs were composed of 2 to 30 elements) or the end of another. The time study board would then freeze the duration measured so that the engineer could write it down while at the same time accumulating the time of the next element of the job. In this way the industrial engineer isolated particular movements and tasks. The engineer could also evaluate whether the operator was working as

Pressing a coat. (*Courtesy of Jos. A. Banks Mfg. Co.*)

efficiently as possible and just where improvements could be made.

Setting piece rates was complicated by an agreement the company had with the union. That agreement specified that workers would be given the opportunity to maintain the same average wage no matter where they worked in the shop. Abiding by this agreement meant that new piece rates for established workers in the company had to be constructed specifically for those workers. As could be expected, new jobs with new piece rates were generally assigned to new workers, since that was one means by which costs could be kept under control.

This company–union agreement made setting piece rates in the coat and skirt shops somewhat more difficult than setting rates in the cutting room. To comply with the agreement in the cutting room, all that needed to be changed was the base pay of the operator. The standard allowed minutes for a task remained the same. However, for average wages in the coat and skirt shops to be maintained, some juggling of the piece rates normally assigned to a worker had to be done. The old piece-rate system was not as "clean" as the newer standard allowed minutes system. (In fact, the North Avenue factory and the Hampstead distribution center had adopted the standard allowed minute system from the start.)

Once the piece rate was established, the industrial engineer informed the worker of it.

The union was also notified. Often this was all that was needed, although many times some negotiation with the union ensued. Sometimes a restudy of the job was done to dispel any disagreements. Some restudies lasted as long as 8 hours. When a piece rate was first changed, it was deemed "unapproved," but usually after a month the union and management would agree that the piece rate was fair. Then the rate was formally approved until it had to be changed because of changes in model, materials, methods, or machinery. The company wanted to set rates that could be beat if the workers worked hard—but not so tight that the workers felt they were not earning a fair incentive wage. In 9 out of 10 cases there was little problem in establishing a piece rate, although company and union both had to compromise their initial positions to some degree.

The Nature of the Workforce

Most of the workforce had had years of experience in the apparel industry. The union was a great help in recommending skilled workers to the company. Thus, there was little problem in educating new workers to tasks. They already knew fairly well what was required of them; with a little training and help from supervision, these workers were able to meet Jos. A. Bank's high quality standards.

For the most part, once a new operator was assigned a job, it was likely that he or she stayed at that job until retirement. Almost all of the workers spent many years at the same level in the company; only about 10 percent attempted to elevate their grade. Advancement was possible, and workers could request transfers to higher-rate jobs through a formal job posting system for vacancies.

The company's relations with the union were good. It helped that the company was growing in an industry that, for the most part,

had suffered severe employment declines. In the last 20 years there had been only one strike, and it was resolved very quickly. Worker complaints were few—only two or three informal complaints each week, and these were usually resolved quickly. In general, complaints related to earnings, such as:

- A piece rate was set too low.
- Bundles were not large enough. With bundles of less than 12, the operator suffered some inefficiencies, so a high number of low-count bundles sometimes sparked complaints.
- Too many plaids were put through the factory in a week. Plaids required a great deal of pattern matching and slowed up work in the factory considerably.
- Changes to a new model should require a new piece rate.

As many as 90 percent of these complaints were resolved by the factory's first-line supervision. There was, however, a formal grievance procedure, which involved supervision at the first level, the plant manager at the second level, the vice-president of manufacturing at the third level, and then arbitration at the fourth level. There had been only two arbitration proceedings in the last 15 years. Most of the complaints taken to higher levels of the grievance procedure related to firings, which typically occurred for absenteeism or poor quality of work. For the most part, union involvement could save a worker from the first threat of firing; but if the worker did not improve, management could fire the worker without union intercession. The only instant and uncontested dismissals were for theft, gross insubordination, or fighting in the factory.

Dismissal, of course, was the most radical form of discipline that the company used. Other levels of discipline were usually used before a dismissal. The most minor form of discipline was a verbal warning, and in most cases two or three verbal warnings were given before a written warning was issued. This written warning would give the reason for the warning, suggest a way for the worker to correct matters, state what the next step would be if the worker did not improve, and indicate at what future time the worker would be reviewed again. If the worker did not improve, there might be another written warning, and after that would come suspension from work. Some suspensions carried definite starting and ending dates, while others were of indefinite duration. Only if these measures did not resolve matters would the worker be dismissed.

Worker pay depended on the task and the shop. Average hourly wages in the Hampstead coat shop averaged $8.34; in the cutting room, $8.25; in the skirt shop $8.05; and in the distribution center, $7.21.

In addition to these wages, Bank paid about 41 percent of base salary in benefits. These included medical and disability insurance, life insurance, a pension plan, an eye care plan, a prescription plan, and holiday and vacation pay. All of these plans were administered through the union.

It was the responsibility of the factory's industrial engineer to monitor costs and piece rates once they were set. Foremen also helped monitor tasks, since they had to sign off on all worker summary sheets and were required to calculate the percentage efficiencies for each worker. Thus, the workers who were consistently making well above average on their piece rates were known. If their high pay was a surprise, their jobs and/or quality were sometimes studied to make sure that they were not taking shortcuts.

PRODUCTION AND QUALITY CONTROL

Jos. A. Bank's quality standards were rigid. Not only did the company not tolerate sloppy work in the factory, but it also wanted to make sure that a 40 regular one year exactly matched a 40 regular in every other year.

Quality control started at the beginning of the process. Numerous areas of the factory had their quality checked by the appropriate foremen on every coat or skirt processed. The chief final inspection area, however, was in pressing, after the coat or pair of pants was fully assembled.

The foremen, who supervised between 30 and 35 people each, were responsible chiefly for checking quality and for instilling in the workers an appreciation of what quality is and a desire to flag quality problems themselves. It was the foreman's task as well to approach workers with any rework. Workers were responsible for the quality of their own work, and any rework on their portion of a coat or pants was done without pay. It was to the worker's advantage to do things right the first time, every time. The chief role of first-level supervision at Bank was to check and maintain quality. Nothing else took so much time and care.

The control of production, as opposed to quality, was chiefly the responsibility of the production managers to whom the first-level supervisors reported. These production managers were as well known to the workforce as the foremen. It was as if factory supervision had been split in two, with the foreman responsible largely for quality and the production manager responsible largely for materials and production control.

The factory could produce 550 coats per day, of which about 25 were usually special orders, single coats that were processed separately. The rest were standard orders batched together into the standard bundles. It generally took 5 to 6 weeks for a coat to travel through the process, from the laying out of fabric in the cutting room to the transport of the finished coat to the distribution center. The primary concern of production control was to make sure that all the parts for a coat moved smoothly through the factory and arrived at the various matching stations at about the same time so that they could be joined with specific other parts for further processing. Naturally, some parts took more labor than other parts to complete. For example, backs might be sewn in 5 minutes, while fronts might take an hour. To ensure a steady flow of goods through the factory, the workforce on fronts was 12 times larger than the one on backs. If parts were not ready, production had to shift to another lot, and operations were not as smooth as they could be.

Matching

Two critical matching stations helped control the flow of parts through the process and uncover any shortages. The first matching station, located about a quarter of the way through the process, was where backs were matched with fronts. The other key matching station was where sleeves were matched with the front and back subassembly; this occurred when the coat was about 75 percent complete. Matching was accomplished by looking at the tickets that had been attached to the materials near the start of the process. As we saw in Figure C2, the ticket numbers supplied various information: the model processed, the cloth lot to which it belonged, the size of the coat being fabricated, and the shade (a number for the bolt of cloth, which was the same for all pieces belonging to the same coat). The matching stations collated pieces by using these numbers.

If there were quality problems or if particular parts could not be located, fabric sometimes had to be recut. Naturally, Jos. A. Bank wanted any recuts to be from the same bolt of cloth that the original had come from; thus, lot numbers and shade numbers, as well as shop numbers, were recorded for any item that had to be recut. One day was allotted to do a recut and to rework the materials so that they could rejoin their bundle in the standard flow of work through the shop. Production control did not want to break apart any bundles of material except for super-rush jobs, which occurred about once a week. For all other jobs the bundles were kept the same.

Formal records were kept showing which lots were started in the shop through the fitting operation, and where the bundles were in the shop. Using the laser wand on the worker coupons provided the information needed to keep this tracking up to date.

Because of rework needed or production problems that might be encountered, the capacities of the upstream operations tended to be slightly higher than the capacities of the downstream operations. To avoid the accumulation of work-in-process inventory, the factory wanted to make sure that it could easily recoup any extra time spent in the early stages of the process. Some models were more difficult to make than others—for example, game coats with gusseted bi-swing backs and bellows pockets. When and how many of these coats to make at one time had to be carefully evaluated. This concern for extra capacity in the early stages of the process was particularly evident when plaids were done. As mentioned previously, plaids were the most difficult fabrics to work with because of pattern alignment for pockets, collars, back seams, and so on. This pattern alignment led to a 10 to 15 percent drop in production and a good bit more rework. (Rework in plain fabrics amounted to only 2 or 3 percent of any production lot, but in plaids the rework figure generally climbed to between 5 and 8 percent.) With this in mind, the target for the easy control of work through the factory was to have 60 percent of the work in plain fabrics, 20 percent in stripes, and the remaining 20 percent in plaids. This grouping was the ideal; sometimes, of course, the percentage of stripes or plaids would have to be increased to meet demand.

Supervision

As noted earlier, the first-level supervisor's prime responsibility was quality. Another of his or her tasks was to deal with absences. The first 15 minutes of the morning were spent determining whether everyone was there (absences averaged only 5 percent) and finding replacements for those absent. A number of the workers were cross-trained to fill in on other tasks; when such a switch occurred, they were paid their average earnings rather than the piece rates for these jobs. (There were also some specified utility workers who were paid their choice of a flat rate or a piece rate for the job.) The workers who were put on absentee work generally came from those who were ahead of the pace of the factory as a whole and could thus afford to take time off from their jobs.

The flow of information to the worker from the foreman generally concerned rework or methods. The flow of information from worker to foreman could be for a variety of problems: machine malfunctions, damaged goods, poor quality upstream. The production manager handled all grievances about piece work and about change in the operation. Sharing the complaint load was another advantage to the split of supervision between the foreman and the production manager. The foreman would, however, handle gripes about the size of the bundles for particular workers.

LOADING THE FACTORY

The specific planning for, say, the next year's spring fashions began in March. From March through June, the merchandising division reviewed planned sales for both retail stores and the catalog. As a result, manufacturing received a production "layout" for spring deliveries. From the layout, purchase orders for goods were sent on to manufacturing, authorizing production.

Orders for particular bolts of cloth (piece goods) were then placed with various mills. Jos. A. Bank had long had dealings with many mills and knew exactly which mills it wanted to produce particular fabrics. For many piece goods, about 10 percent extra was purchased so as to give the company some flexibility if demand exceeded the forecast. Since Jos. A. Bank fashions changed in only subtle ways from year to year, any surplus material could probably be used the following year; being caught with more fabric than required was not nearly as much a problem as being caught with too little. On those rare occasions when there was not enough fabric for a particular model, the merchandising division sometimes was able to reorder fabric from the mills. The fabric ordered for the next year's spring fashions arrived during the summer, any necessary examining or sponging (preshrinking) was done, and cutting began in August. Delivery to the distribution center started in October and delivery to the stores in December. From December to June or July the spring line was sold.

Manufacturing had no responsibility for the purchase of fabric, the factory's single most important raw material. All responsibility and authority for selection and purchasing rested with the merchandising division. However, manufacturing was charged with the purchase of the numerous trim items that were combined with the piece goods to make coats or skirts. Manufacturing also kept track of the status of the piece goods inventory: quantities remaining and their locations. For trim items such as shoulder pads, fusings, and sleeve heads, there were always two sources of supply; the split between the vendors depended upon price. Commitments to vendors typically extended 6 months in advance, but the shipment of items was specified for much narrower time periods; linings, for example, were delivered every week or two, buttons every 6 weeks.

Many of the vendors had been dealing with Bank for 25 years or more. Only if a vendor's price got out of line would the company threaten to reduce the amount of orders placed with it. New vendors were occasionally chosen. Vendor qualification was handled by design/quality control, which tested the item's physical characteristics (such as stretching and pulling) as well as the way it behaved when sewn or dry cleaned. If a vendor was deemed qualified, the trim purchasing manager negotiated the price. Typically, new vendors started with small orders as a further check on quality and delivery capabilities and on the stability of the pricing structure. Only if these tests were passed would the order be enlarged for subsequent years.

Production Planning

The production plans began with broad, large-scale decisions, which were subsequently refined and made more detailed. The broadest plans involved ranges—aggregate counts of fabric needs by type. Thus, there were ranges within plains, within stripes, within herringbones, and within plaids. These ranges were then broken down into particular quantities of specified fabrics required; they helped define what was cut and assembled during the entire season. A particular week's production plan was part of a 6- to 8-week batch of work from the merchandising division. The merchandising division balanced finished goods deliveries to meet catalog and retail store sales. It was the merchandising division's responsibility to monitor sales and to adjust subsequent batch "purchase orders" accordingly.

The production schedule set was typically for an 8-week period. The first week was allocated to planning, the next 2 weeks to cutting, and the ensuing 5 weeks to the shop. What was cut in a particular week depended on three things: the expected finished goods inventory, the attempt to develop an appropriate mix of cloth (the 60/20/20 plain/stripe/plaid breakdown mentioned earlier), and piece goods availability.

Production Planning Paperwork

Once the merchandising division provided the initial batches (batches contained specific purchase orders for defined finished goods deliveries), the batches were matched to the available piece goods inventories. The paperwork was then initiated to direct and control the transformation of fabric into suits, coats, and pants. The production plan for any week triggered several actions in the cutting room office. One of them was the computerized issue of a cloth pull sheet (Figure C5), which authorized the withdrawal of fabric from inventory. Authorization to cut the cloth was also provided by the computer. It created a cutting slip and the accompanying cutting chart (Figure C6), which indicated, for each lot number, the quantities and the size combinations for the markers or patterns that were actually used to cut the fabric. Cutting was done in groups of fabrics compatible for sewing and for machine setup and operation (see Figure C7). The cutting slip and the cutting chart traveled with the job. After the materials were cut and fitted, the Soabar report (Figure C8) was issued by the cutting room office. This report indicated not only the lot number but also the make, model, and size for all the bundles of fabric to be processed. The data entry for the Soabar report generated, via a special computer program, the fitter's report, which indicated the lot, bundle, and quantities, by size, that were to be run through each shop. The fitter's report in turn generated the coupons that traveled with the job and were critical to the operation of the piece-rate system.

continued on p. 633

25⁴

1, 001 UNITS
CLOTH PULL SHEET

PAGE NO. _142-67_
WK 18
75

MODEL: _S365T/S33XT_   _Coat & Pant_ TYPE OF CLOTH: _Plain - Cotton/Poly_

DELIVERED BY: _____   RECEIVED BY: _____   DATE: _1-17-89_

Lot No.	Ydg. Rec'd.	Shade	Ydg.	Loc.	Shade	Ydg.	Loc.	Shade	Ydg.	Loc.	Shade	Ydg.	Loc.
35530	3.40	R											
1123 R	286/972	114.	60.	609.	115.	59.	609.	117.	60.	152.	119.	58.	549.
1124 L	214/728	124.	60.	152.	125.	60.	152.	128.	58.	151.	130.	52.	609.
		140.	59.	055.	141.	60.	055.	142.	60.	145.	145.	60.	055.
	(1,700)	146.	59.	145.	151.	31.	610.	153.	42.	609.	154.	26.	611.
		158.	20.	610.	161.	57.	145.	163.	60.	146.			
		L											
		89.	55.	055.	99.	55.	145.	101.	32.	044.	102.	23.	044.
		105.	29.	145.	106.	30.	145.	107.	4.	055.	109.	43.	044.
		110.	54.	055.	116.	60.	152.	118.	59.	152.	121.	55.	152.
		126.	59.	055.	127.	58.	152.	131.	60.	152.	132.	59.	152.
35531	3.40	R											
1123 R	292/993	97.	59.	694	99.	60.	054.	100.	56.	053.	101.	59.	053.
1124 L	208/707	102.	58.	054.	109.	27.	705.	118.	60.	705.	119.	59.	705.
		120.	59.	053.	123.	58.	705.	125.	60.	705.	126.	59.	705.
	(1,700)	127.	57.	069.	130.	58.	697.	134.	57.	054	136.	59.	054.
		137.	58.	697.	140.	59.	053.						
		/												

Notes:

1. This is the cloth pull sheet for group 259, coats and pants for model S365T/S33XT.

2. The sheet indicates the yards of fabric cut for each lot, the shade (bolt of fabric), and where that bolt of fabric was located (the bin number).

FIGURE C5   Cloth pull sheet.

CUTTING CHART A   *#4· Cutting Shee*

CUT = _____   MODEL *C 355 E   SPORT COAT*   *589* _____   MAKE = ___*001*___

GROUP = *260* _____   UNIT ___*PLAIN – wool*_____   GOODS *R. 60.*
I. 60,
Sto. ¹/Yd. 59

LOT NO.	SHADE NO.	40 R 44R	41 R 43R	42 R 42R	39 R 46R	38 R 48R	37R 46R		45R					
35910		30	28	21	15	8	3		4					
		42L 44L	43L 46L	41L 48L	39L 46L				40L					
35910		29	19	12	4		8							
		38S 40S	43XL 44XL	39S 42S	42XL 48XL				42S	46XL				

Notes:

1. This is a portion of the cutting chart for an order for group 260, York sport coat model C355E. The cutting chart indicates which size combinations will be used for the markers on this order and how many of each lot and shade of fabric are to be cut. The group number and shop number are given in the cutting office.

2. These coats will be cut from a plain wool fabric.

FIGURE C6   Cutting sheet used by a spreader.

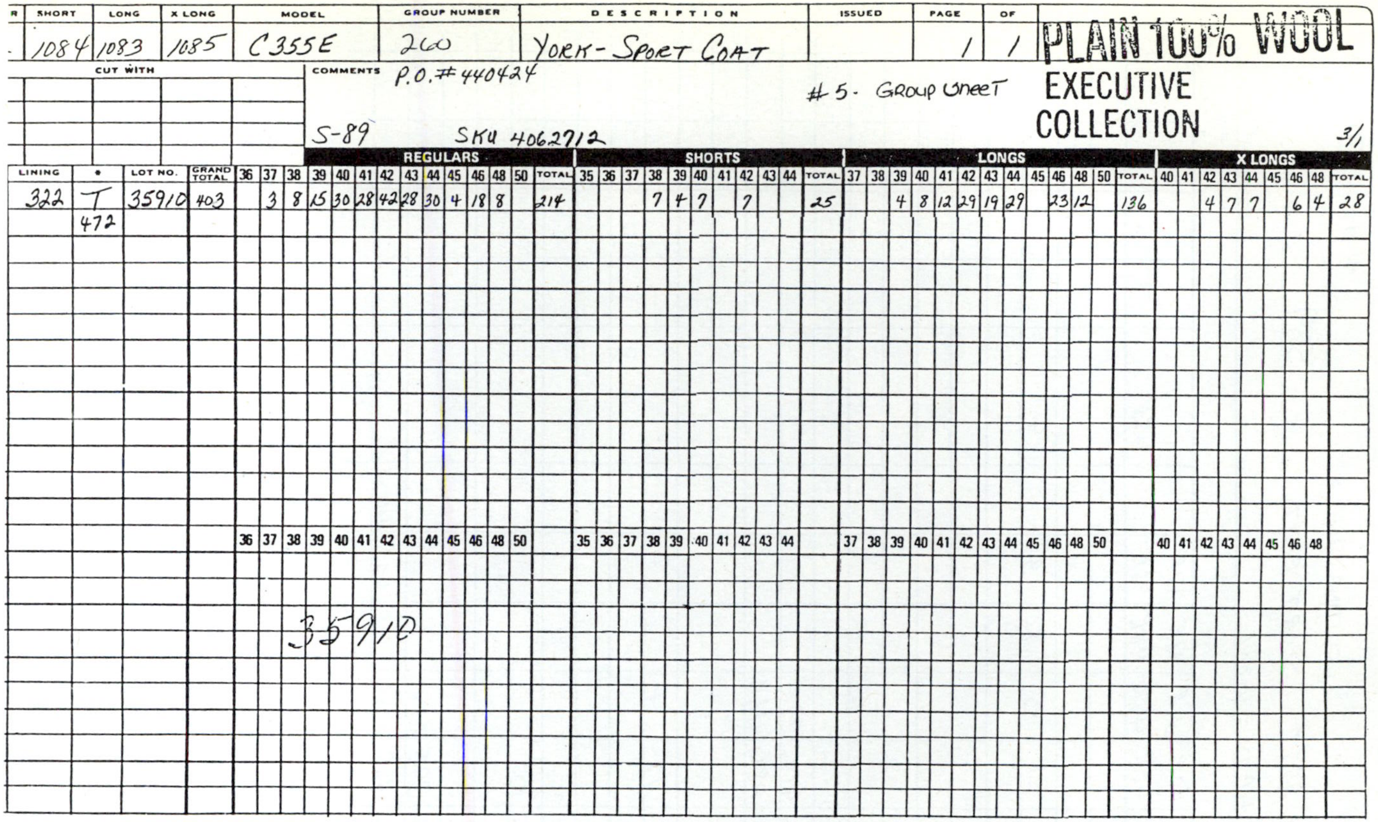

PLAIN 100% WOOL

R SHORT	LONG	X LONG	MODEL	GROUP NUMBER	DESCRIPTION	ISSUED	PAGE	OF
1084	1083	1085	C355E	260	York- Sport Coat		1	1

CUT WITH

COMMENTS: P.O. # 440424 #5 - Group Sheet

EXECUTIVE COLLECTION

S-89   SKU #062712

REGULARS

	GRAND TOTAL	LOT NO.	36	37	38	39	40	41	42	43	44	45	46	48	50	TOTAL
LINING 322 * T 472	403	3590	3	8	15	30	28	42	28	30	4	18	8			214
		3590 D														

SHORTS

35	36	37	38	39	40	41	42	43	44	TOTAL
				7	7	7	7			25

LONGS

37	38	39	40	41	42	43	44	45	46	48	50	TOTAL
	4	8	12	29	19	29		23/21				136

X LONGS

40	41	42	43	44	45	46	48	TOTAL
	4	7	7		6	4		28

3/1

Notes:

1. This is the group sheet for a York sport coat, model C365E. The group number is 260 and the shop numbers are 1083 through 1085. After the shop number has been assigned, five copies of this sheet are given to the cutting room floor and one copy to customer service.

2. There is one lot in this group (35910) calling for various quantities of regulars, shorts, longs, and extra longs.

3. The colors to be used for UC (undercollars), linings, and * (fusings) are indicated by the numbers or letters in those columns at the left.

FIGURE C7   Group sheet.

MEN'S SOABAR REPORT

SEASON: S	YEAR: 89	LINING: 0	WIDTH: 60	DESCRIPTION: STRIPE-WOOL	DATE: 9/26/88

SHOP# 0279	COAT MODEL S321E	PANT MODEL S32XE	VEST MODEL	MAKE 001	GROUP 87	VAR 3	DUE DATE 11/4/88	DATE SOABARED

TYPE:1 STOCK
K:1 000
2 2PC SUIT
3 3PC SUIT

UNIT COST MFG *HC* RETAILER *B*
UNIT COST MFG *NP* RETAILER *B*
UNIT COST MFG RETAILER

VAR: 1 REGULAR   6 PORTLEY SHORT
2 SHORT   7 PORTLEY LONG
3 LONG
4 X LONG
5 PORTLEY

LOT #	M A R K E	T Y P E	COAT	VEST	PANT	29 / 35	30 / 36	31 / 37	32 / 38	33 / 39	34 / 40	35 / 41	36 / 42	37 / 43	38 / 44	39 / 45	40 / 46	41 / 48	42 / 50	43 / 52	44 / 54	46	48	PO #
		1																						
34653	2	1	57		57											5	5	4	5					44030
		1																						
34642	2	1	140		140											5	11	11	5					44030
		1																						
34631	2	1	59		59											5	5	4	5					44030
		1														5	21	19	15					
		1														15								
		1																						
		1																						
		1																						
		1	*Debbie - Pants 10-3-88*																					
		1																						
TOTALS		1	256		256				15	19	21	21	43	29	38	15	21	19	15					

HAMPSTEAD

Notes:

1. This is the Soabar report for shop number 0279, for group 87, coat model S321E and pants model S32XE, soabared on September 26, 1988 and due out of the shop on November 4, 1988. The HC and NP entered for Mfg. indicate the Hampstead coat shop and the North Avenue pants shop, respectively.

2. There are three separate lots. The report indicates the exact sizes and how many of each should be made.

FIGURE C8 Fitters or Soabar report.

The paperwork just described served to follow up on the production planning by initiating work in the cutting room and the several shops of the factory. Other paperwork served to help control the movement of materials through the factory and to cope with irregularities (such as rework, recutting, and failed matchings).

TECHNOLOGICAL INNOVATION

Apparel manufacture, especially the manufacture of high-quality merchandise such as Bank's line, still required operator involvement and manual handling of suit parts. Innovations such as automatic thread trimmers and needle positioners, and work aids to improve operator efficiency and to reduce handling time had been implemented. There were some automatic machines for pocket cutting and other tasks, but most of the factory's sewing machines were strictly manual.

There had been other investments over the years in improved technologies. The AM-5 machines for automatically generating markers for the cutting room was among these. They eliminated waste in laying out the piece goods, and they automatically increased sizes, saving considerable time in the design of the markers themselves. New equipment for piece goods inspection, for fusings, and for steaming cloth had also been purchased. Productivity had improved each year in all product types.

Bank reviewed new technology on a regular basis, always with an eye to maintaining or improving quality. Chronically under review for application to Bank's high standards were high-speed automated spreading and computerized cutting equipment that were compatible with the AM-5 machine.

PART TWO

DISCUSSION

Jos. A. Bank Clothiers is in no way responsible for the following views and presentation. They remain solely the responsibility of the author.

THE PROCESS AND INFORMATION FLOWS

In comparison with the job shop of Norcen Industries, the flow of production at Jos. A. Bank is better defined. While there are some understandable differences between the coat shop and the pants shop, we can compose a process flow diagram for each of those shops that adequately represents the process, at least in general terms (see Figure C9). A more detailed process flow diagram, one that encompasses the actual operations described on the coupons, would reveal that there are still substantial differences between various products (such as suit coat versus sportcoat), models, and materials. Variations in any one of these lead to the inclusion of certain process steps and the exclusion of others. In general, however, the flow of materials through the process is fairly smooth and easily directed. Concern for materials routing is much less critical in this batch flow process than in a job shop.

Because of the wide variety of materials, models, and sizes, the factory must deal with a vast array of different batches. These variations underscore the still very strong need of the batch flow process for information and control. In their own way, the information flows of the batch flow process are inflexible, although compared with the job shop, the flow of information between workers and management is somewhat reduced. While workers are responsible for notifying management of machine malfunctions and quality problems, the major information flows to and from workers revolve mainly around the coupons that move with each bundle, as shown in Figure C10. In essence, the coupons serve as both process sheet and job sheet, to borrow language from the Norcen Industries example. The coupons inform

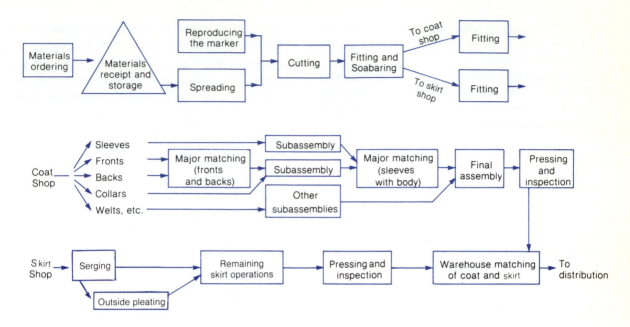

FIGURE C9   A simplified process flow diagram for a Jos. A. Bank suit.

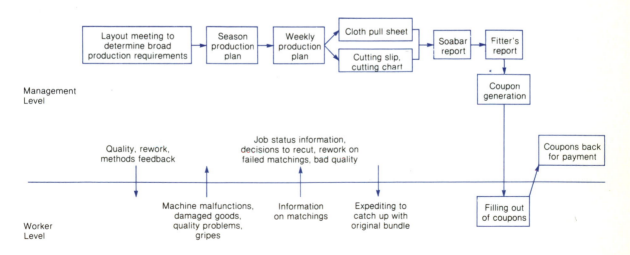

FIGURE C10   A simplified information flow diagram for Jos. A. Bank.

the worker of what must be done to the batch, and they also keep track of who did what to each batch of work.

The increased stability of the batch flow process reduces the amount of information that must be communicated from management to workers. The foreman typically has less troubleshooting and expediting to do and fewer scheduling changes to make. Therefore, it is often possible to supervise more people in a batch flow operation than in a job shop. The foreman at Jos. A. Bank has time to track quality and improve methods rather than having to make schedules and troubleshoot.

The need for continuous flow of information is reduced in the batch flow process, primarily because its product selection, while large, is still limited and well known in advance. Furthermore, it seldom has to bid for contracts, and so precise, order-specific cost information is not required.

CAPACITY IN THE BATCH FLOW PROCESS

The notion of capacity is only slightly less ambiguous in the batch flow operation than in the job shop. Nevertheless, the wide arrays of products that can be produced in any week make it difficult to specify a batch process's capacity in specific units of output without pegging it first to some standard mix of products.

The capacity of the batch flow process is influenced by many of the same factors that determine the capacity of the job shop.

1. *Batch (or lot) sizes.* Although the setup time for any bundle of materials is low at Jos. A. Bank relative to Norcen Industries, certain procedures are necessary (change of thread and needles) before a new batch can be processed. As mentioned in the process description, the larger the batch, the more

satisfied is the worker, too many low-count batches spur complaints from workers.

2. *Complexity of the products run through the operation.* As noted, plaids and stripes are markedly more difficult to handle than plain fabrics; when the production schedule includes a higher-than-normal percentage of these materials, the capacity of the entire coat shop declines.

3. *Nature of the jobs already on the factory floor.* Batch flow operations, like job shops, are subject to shifting bottlenecks and worker-machine interference. However, the definite limits to product diversity in batch flows generally mean that the severity and frequency of these two capacity thieves are reduced.

4. *Scheduling.* Scheduling is still an important consideration with the managers of the batch operation but the task is generally simpler than in the job shop. Again, the reduction of products and capabilities offered accounts for the bulk of this difference. With the batch operation, scheduling begins to shed some of its critical importance to the success of the operation as a whole. Given a reasonable composition of orders for the week (e.g., the 60/20/20 split of plain/stripe/plaid fabric), scheduling becomes much more routine.

5. *Process improvement.*

6. *Number of machines and their condition.*

7. *Quantity and quality of labor input.*

The last three major influences on capacity, like many of the other influences cited, take on less importance for the typical batch operation than for the typical job shop. This is because the batch flow process itself is better defined, with workers more apt to be assigned to particular machines and less free to float among machines. The transition from job shop to batch operation accounts already for considerable process improvement (such as specific methods, worker aids at the workstation, quicker setups). Considering its routing tickets (coupons), assigned machines, and specialized training, the batch operation is more likely to treat machines and workers as one producing unit. Thus, capacity can be enhanced significantly by increasing both in similar proportions. This is not to say that equipment improvement cannot be made in the typical batch flow operation. Jos. A. Bank, for example, was making substantial investments in new cutting tables and was monitoring advances in sewing machines.

THE ROLE OF STANDARDS AND INCENTIVES

At Jos. A. Bank, the role of standards takes on a different function than at Norcen Industries. At Norcen, you will recall, standards were used to estimate times and costs for bidding on jobs and also to provide information for scheduling jobs through the shop. Although the second function for standards is retained at Bank, standards are not used for job cost estimation because there is no bidding for new orders. Rather, standards in the form of piece rates and standard-hour (i.e., the standard-allowed minutes system) incentives are used explicitly to determine worker compensation and thus to spur worker effort and output. In a sense, the piece-rate and standard-hour incentives act to pace work through the factory.

As the process description made plain, the setting of incentives at Jos. A. Bank is serious business. Great care is taken to see that they are both accurate and fair. Considerable time and expense are spent setting new piece rates, observing workers trying out new piece rates, and testing modifications in methods. The procedures for setting and using incentives and for resolving disputes over them are well established and routine.

DEMANDS OF THE PROCESS ON WORKFORCE AND MANAGEMENT

The demands made on the workforce center around the incentive system. The system, rather than foremen, acts to pace production; thus, the demands for worker output are mainly triggered by the incentive system itself. Demands for quality output, on the other hand, come from the foremen. The foremen's most important job is to check the work that has been done and make sure that the methods being followed are proper.

At Jos. A. Bank, the ambition of workers is not great, and so not a lot of effort is expended to provide career paths for the workforce. Some other batch flow operations are more concerned with such issues. Of course, there is cross-training in a batch flow operation like Bank. Not only does cross-training help cope with absenteeism, but it also helps break bottlenecks, particularly chronic ones.

Clearly, one of management's chief concerns in a process like Bank's is the incentive system. The incentive rates should be fair to the workers but not so loose that costs get out of control. Striking a suitable balance is an important endeavor.

Apart from the incentive system, management must also be concerned with keeping the flow of work smooth. This means grouping work well and staying within guidelines for the materials mix. Not only must management be wary of the policies by which it loads the factory, but it must also track jobs through the process and act immediately to remedy any difficulties. While the flow of information is not as rigid and sophisticated as in a job shop, management must still pay considerable attention to the information pertinent to the flow of materials and jobs.

QUESTIONS

1. Compare and contrast the characteristics of the information flow in the batch flow process used by Jos. A. Bank with the information flow characteristics of the two other processes discussed in Tours A and B. Which process requires the least in the way of information flow from its managers? From its workers? Does anything replace the flow of information to workers or managers?

2. The typical duties of first-line supervision at Jos. A. Bank are divided between two people. What are the strengths and/or weaknesses of such an arrangement, as you see them?

3. Discuss the difference between the piece-rate and standard-hour schemes at Jos. A. Bank. Could the cutting room have been put on a piece-rate system? Had that been done, what complications would you have anticipated?

4. Under what conditions do you have the most confidence in the factory's ability to produce 550 coats a day? Under what conditions do you have the least confidence? Why?

5. How would you describe the market for Jos. A. Bank? What relationship, if any, does this market have to the production process chosen by the company?

6. In what ways is the union a help to Jos. A. Bank? A hindrance?

7. Could the low rate of movement of workers among jobs in the factory be related to the activities of the industrial engineer? Elaborate.

SITUATION FOR STUDY C-1

BROWN, SMITH & JONES (BS&J)

Paul Alvarez has just been hired by BS&J as superintendent of the fabricated tank division. Paul's previous experience had been with a leading manufacturer of home heating systems where the processes were highly automated. At the previous firm, the goods were also processed in a continuous manner, with a product line being set up for a year or more of production.

The products manufactured by the fabricated tank division were typically oil storage tanks made from sheet steel, which were sent to the customer much like a kit and were erected in remote oil fields. Occasionally, the firm sold the tanks in various sizes to other customers for other uses, such as to the military for the storage of diverse liquids like water and jet airplane fuel. Figure C11 is an example of a tank and some of the parts that must be fabricated.

Table C1 is a partial process sheet for these parts. Figure C12 is a layout of the relevant processes.

Many of the smaller parts are common to most tanks, and they are made on a continuous basis to inventory requirements. There are three parts that are unique to each tank size: the tank side parts called staves, the triangular roof, and the bottom sections. The smallest tank, 1500-gallon capacity, requires five staves, five roof sections, and five floor sections. A 55,000-gallon tank, five rows high, requires 13 roof sections, 13 floor sections, and 65 staves. The typical order is for three tanks of the same size, but an order can be as small as one tank and has been as large as 15 tanks. The tanks are made to order, but BS&J has a backlog estimated at 1 1/2 months' production.

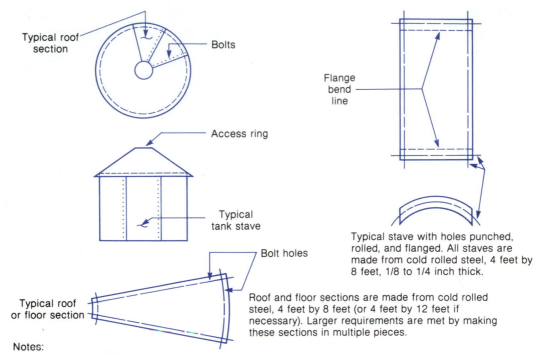

Typical roof section

Bolts

Flange bend line

Access ring

Typical tank stave

Bolt holes

Typical roof or floor section

Typical stave with holes punched, rolled, and flanged. All staves are made from cold rolled steel, 4 feet by 8 feet, 1/8 to 1/4 inch thick.

Roof and floor sections are made from cold rolled steel, 4 feet by 8 feet (or 4 feet by 12 feet if necessary). Larger requirements are met by making these sections in multiple pieces.

Notes:

1. A typical tank stave (1) is trimmed (if needed), (2) has holes punched on all sides, (3) is rolled to the approximate shape (depending on the tank diameter) on the pyramid roll, and (4) is flanged on both ends in a large brake press. The press is set up with a flanging that also causes the stave to be formed to the final shape according to the tank diameter. There is a flanging tool for each diameter tank.

2. The roof and floor sections are made from material similar to the tank staves, ranging from 1/8 inch to 1/4 inch, per the engineering design. If necessary, these sections can be made from material 4 feet by 12 feet. BS&J uses standard mill sizes to avoid the extra costs involved in having special sizes available; also, pieces that are too big would create construction problems. Two sections are usually cut from a single sheet:

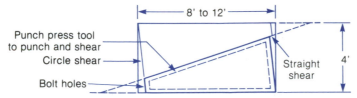

8' to 12'

Punch press tool to punch and shear

Circle shear

Bolt holes

Straight shear

4'

(The view is somewhat exaggerated to demonstrate the method.) If a top or bottom piece were too long, it would be made from two pieces:

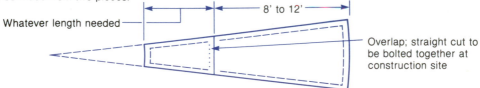

Whatever length needed

8' to 12'

Overlap; straight cut to be bolted together at construction site

FIGURE C11 Typical tank and parts.

TABLE C1   Partial outline of the process (Brown, Smith & Jones)

STAVE		ROOF AND FLOOR SECTIONS	
OPERATION	PROCESS	OPERATION	PROCESS
Shear edges true	Sheet shear	Shear edges and split each sheet on diagonal	Sheet shear
Punch bolt holes in both ends	Brake press setup with punches	Punch/shear small end*	Punch press
Punch bolt holes in both sides	Brake press setup with punches	Shear large end	Circle shear
Roll to tank diameter	Pyramid rolls	Clean and paint	Paint line
Flange both ends	Brake press setup with correct flange	Pack and ship	
Clean and paint	Paint line		
Pack and ship			

*For typical tank sizes there were punch press tools for this. Others had to be laid out, circle sheared, and single punched.

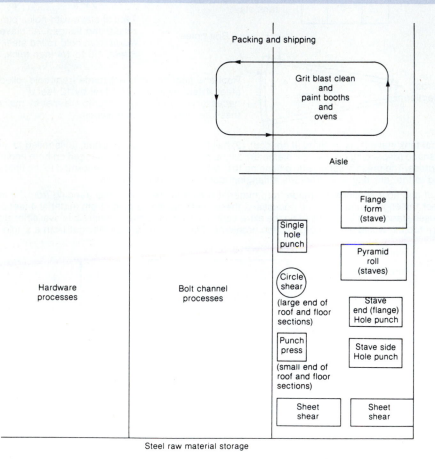

FIGURE C12   Fabricating process layout for a side stave, roof section, or floor section.

Paul was hired because the fabricated tank division has been unable to meet scheduled shipping dates, quality problems are cropping up, and the workers are unhappy about never being able to earn any incentive pay. Paul thought he would focus on the manufacture of the side staves, roof, and floor sections. What information does he need next? What information should be included on parts routing sheets?

1. Draw a process flow diagram for the two major tank parts.

2. What information would you need in order to set up a production schedule for this shop?

3. What approaches would you suggest to Paul to help him with both his quality problems and his incentive pay problems?

SITUATION FOR STUDY C-2

WELCH'S ICE CREAM

Ice cream production took place every weekday at Welch's, a shop that sold ice cream to the public. The entire process of making the ice cream was performed by Brad Gentry, an hourly employee. In addition, Brad had sole responsibility for inventory maintenance, quality control, and production scheduling. He worked about 4 hours a day during slow periods and about twice that during the peak season. Because he knew more than anyone else at the shop about production needs at any given time, and because he was decoupled from the sales portion of the business by an inventory of finished ice cream, Brad set his own hours. He usually worked from about 10 A.M. to 2 P.M.

The shop owned one ice cream machine with a 40-quart batch capacity. (The capacity refers to the finished volume rather than the volume of ingredients.) The manufacturing process involved rapid agitation and aeration of the ingredients, which nearly doubled the batch's volume. Hence, a 40-quart batch required approximately 20 quarts of mix.

The machine itself resembled a stainless steel 20-gallon water heater, lying on its side on a waist-high stand. The cylinder had a refrigeration layer on the outside, through which re-

frigerants circulated when the machine was in use. At one end of the cylinder was the motor unit, which drove the refrigerant pump and the agitation blades inside the tank. At the other end was a lower spout, through which the finished ice cream was pushed by the rotation of the blades. A table was positioned here so that buckets on it could catch the ice cream. Above the spout were two built-in funnels, through which extracts and flavorings could be added. On top of the cylinder was a stainless steel tub, which held a little over 20 quarts of liquid.

The ice cream machine was located in a small room called the process room, which also contained also a few shelves of supplies, reusable plastic buckets, scales, and a large sink. Three doors in the process room provided access to a walk-in refrigerator where the raw mix was kept, a walk-in freezer ($-30°F$) called the hardening box, and the rest of the premises.

The ingredients used in production were of three general types. The major type was the "mix," which was milk containing sweetener and a high cream content. An order of mix was delivered daily by a local dairy. This order consisted of five 50-pound cartons. Welch's placed an order with the dairy every day for delivery the next day, even though a different quantity was rarely ordered.

The second type of ingredient was the flavor bases: the various syrups and extracts that were added to the mix during agitation. The third type consisted of nuts, cookies, and some syrups that were "marbleized" into the ice cream. (Marbleization means that the ingredient is not beaten into the mix, but is poured into the stream of finished ice cream as it comes out the spout.)

The major vendor for the second and third types of ingredients was a company that usually delivered within 7 to 10 days after an order had been placed. This supplier paid freight costs but required a 15-case minimum. Brad took inventory on flavor bases once a month, and on marbleizing ingredients every 2 weeks. After taking inventory, Brad gave his boss a list of supplies needed and the boss did the ordering.

A production run began with sterilization of the ice cream machine and enough plastic buckets (a bucket equals 10 quarts) to hold the day's batches of ice cream. Each batch required 20 quarts of mix, and Brad usually made six batches a day, so Welch's received about 125 quarts of mix every day. Because the mix doubled during processing and the other ingredients also took up space, Brad used about 25 buckets per day.

While the machine and buckets dried, Brad checked the stock of ice cream in the tempering box, a walk-in freezer (adjacent to the hardening box) kept at 6°F. Ice cream in the hardening box was too cold to scoop, and so a sufficient inventory in the tempering box was necessary to ensure that Welch's did not run out of servable ice cream. After seeing which flavors were running low, Brad rotated those flavors from hardening to tempering box via a doorway connecting the two freezers. A map of ice cream locations was posted in the tempering box so that people who worked out front could easily find whatever flavor they needed. Maintaining this map was important because there were usually about 150 buckets ranging over 40 flavors stocked in this freezer. Welch's complete inventory comprised the contents of (1) the tempering box, (2) the hardening box (also an average of 150 buckets), and (3) the glass-topped refrigerators located in the service area (usually 32 buckets).

After rotating the stock, Brad took from the stock room the flavor bases and any other ingredients he needed that day. On Monday and Friday mornings, he took inventory of the hardening and tempering boxes and on the basis of this decided which flavors to make on a given day. Because he did not work weekends and Welch's sold 65 to 75 percent of their ice cream from Friday to Sunday, Brad had a target inventory for each flavor that he wanted to have on hand by Friday. The Monday inventory informed him how much of each flavor was sold over the weekend and therefore how much he needed to make above daily usage to reach Friday's target. The Friday inventory was a check that enough of each flavor was on hand for the weekend in the tempering box. By comparing usage figures from week to week, Brad could schedule flavor production to reflect demand. The target inventory was not a specific figure but rather a rough number that Brad carried in his head and altered according to his judgment. He did keep track of what he produced during the week, and he used this list to avoid duplicating batches or forgetting needed flavors. It was usually necessary to make two batches of vanilla a day, because about a third of all sales were vanilla.

Brad scheduled the batches from lightest taste and color to heaviest taste and darkest color. This way he avoided having to rinse the machine after every batch and saved 5 to 10 minutes, plus the ice cream that stuck to the blades. Vanilla was always first, because what little stayed in the machine blended well with the following batch. Brad could usually run three different flavors before having to rinse the machine.

Brad dumped 20 quarts of mix into the tub and allowed it to drain into the tank. He then closed the valve between the two and poured another 20 quarts into the tub, ready for the next batch. He turned on the agitator and added a measured amount of vanilla flavoring and cream tone. Welch's provided a complete index of recipes and instructions, but Brad had those for the most popular flavors memorized. He timed the batch—the first batch of the day for 10 minutes and the subsequent batches for 8 minutes. The first took longer because the machine was not yet cold.

When the time was almost up, he took a sample and weighed it. The batch was done if one cup weighed 90 to 95 grams, though chocolate flavors ran a bit denser. When ready, the ice cream was run out into buckets, covered with waxed paper, labeled with flavor and date, and placed in the hardening box, where it stayed for at least 12 hours before being rotated to the tempering box. Ice cream usually stayed in the hardening box for more than 12 hours; it would keep indefinitely at that temperature. When the end of the batch would not fill most of a bucket, Brad filled a few pint and quart cartons which, after freezing, were stored in a confection box at the front of the restaurant and sold. Although not a large business segment, this procedure avoided waste.

Brad repeated this batch process until he had either used up the five cartons of mix or made all he wanted to for the day. Normally he used all the mix, although it could be kept in the refrigerator for several days. When finished, he again sterilized the machine and washed the walls and floor of the process room before leaving. The cleanup took about 45 minutes.

1. Diagram the process and information flows for this process.

2. How would you plan production for such a process (i.e., how would you schedule which flavors would be produced and when, both during the day, the week, and the month)?

3. What controls on ice cream production would you use in this ice cream parlor? Why?

A MACHINE-PACED LINE FLOW PROCESS
General Motors Corporation
Chevrolet–Pontiac–Canada Group
Oklahoma City, Oklahoma

The Oklahoma City plant of the Chevrolet–Pontiac–Canada Group (C–P–C), one of the 28 domestic assembly plants operated by General Motors, was situated in the southeast part of Oklahoma City. The plant, built in 1979 on a 436 acre site, was huge, having 3 million square feet of space under roof. The plant employed a total of 5300 people, of which 430 were salaried. The nearly 4900 production workers were employed on two production shifts. (See Figure D1 for a layout of the plant.)

The plant currently assembled two nameplates of the body type A car, the Oldsmobile Cutlass Ciera and the Buick Century, in both sedan and station wagon models. In the past several years, however, as many as five different models (Buick Century, Oldsmobile Cutlass Ciera, Chevrolet Celebrity, and two versions of the Pontiac 6000 [front-wheel drive and all-wheel drive]) had been assembled there.[a] The plant delivered truckloads and rail carloads of these cars to every domestic auto dealer carrying those nameplates. The Oklahoma City plant was one of only two plants to assemble the A car, the other being in Ramos Arizpe, Mexico.

The Oklahoma City plant assembled cars of world-class quality. In the previous two years the plant had been rated as either the third or fourth best plant by J. D. Power, an independent company widely recognized as the arbiter of car quality for North America. In the latest year, the Pontiac 6000 produced in Oklahoma City tallied only 78 defects per 100 cars. This rating was exceeded only by the Lexus and surpassed that of the Toyota Camry and the Infiniti. The Buick Century's rating, 91, and that of the Oldsmobile Cutlass Ciera, 97, were better than that of the Honda Accord and of all other nameplates sold in North America.

[a] Although still selling well, the A car "platform," as it was termed, was aging and for the latest model year, the Chevrolet Celebrity and the Pontiac 6000 had been discontinued.

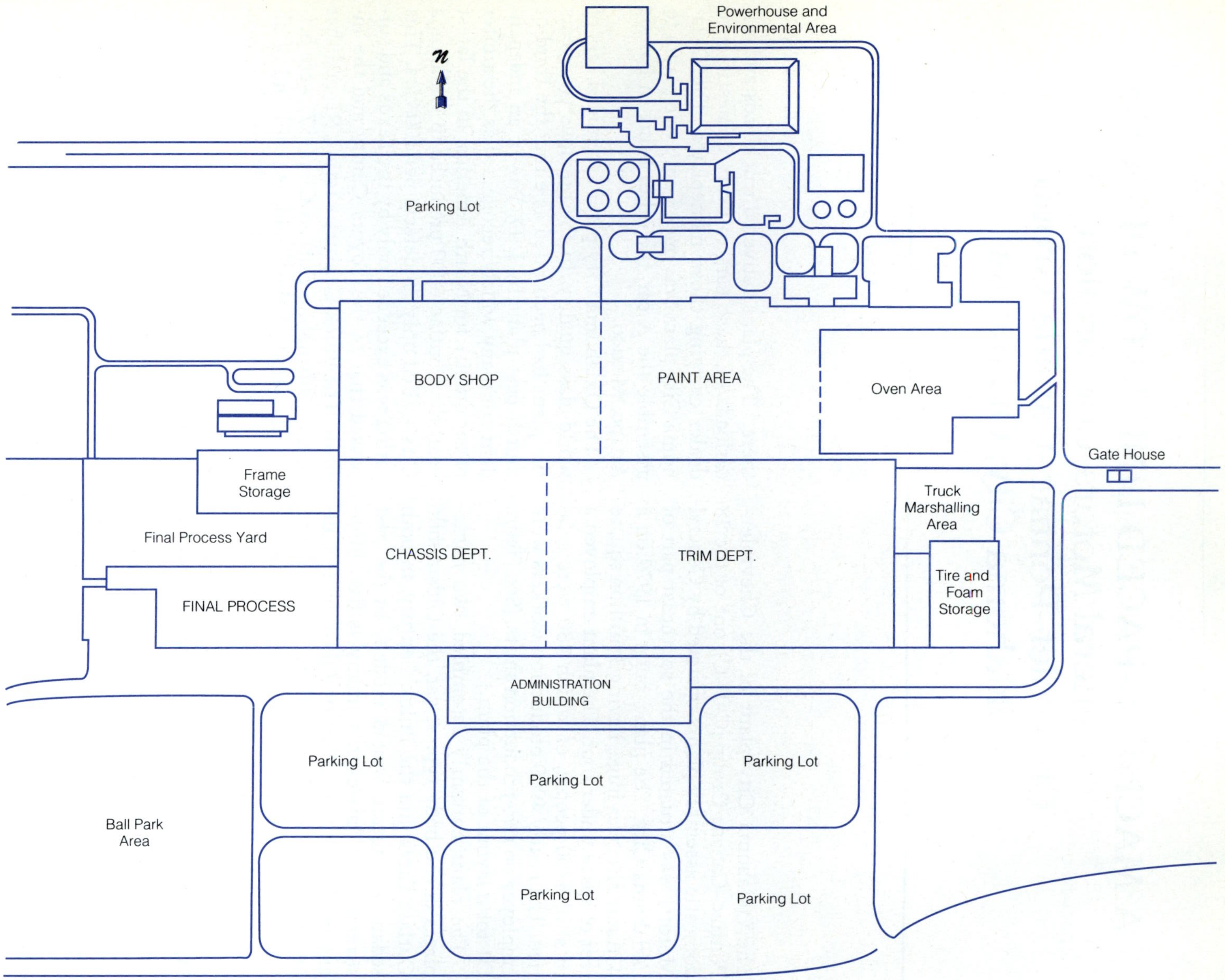

FIGURE D1   Map of the plant site. (*Courtesy of General Motors, Oklahoma City Plant*)

PART ONE

Process Description

HOW A CAR WAS ASSEMBLED: A SIMPLIFIED DESCRIPTION

The production process at C–P–C Oklahoma City was a classic, but modern, example of the moving assembly line so closely associated with Henry Ford and the Model T. The essence of the process was to build the car bit-by-bit by having workers perform the same tasks on each car as it moved through their work stations on a conveyor system. The A body car, with its front-wheel drive, could be viewed as the marriage of two large subassemblies: one for the body of the car and one for the engine cradle (the engine, its support structure, the front axles, and the trailing exhaust system and rear-wheel brake lines). The assembly plant was organized to build up the body and the engine cradle separately, to "marry" them, and then to finish the car's assembly and construction. The body and engine cradle lines were both fed by smaller subassembly lines. The plant was laid out with a flow that went clockwise, starting with the body shop, then the paint shop, the trim line (known as General Assembly I), and the chassis line (known as General Assembly II). See Figure D2. A more detailed version of the steps to build a car would require more space and explanation than is merited.

The entire line was composed of about 1800 cars with a single car taking about 28.5 hours to progress through the complete assembly process, a rate of about 1 car per every 50 seconds per work station. A typical work station consisted of 2 workers, 1 on each side of the line, some space in which to do the work, equipment specific to the job, and in many stations, some totes or bins stocked with the parts to be assembled onto the car. A great deal of variety existed from one work station to another. For example, a typical work station in the General Assembly 1 area might have the car moving at ground level, while at other positions in the General Assembly 2 area, the car was raised for work underneath it. An inspection station might have intense lights and raised platforms so that inspectors could easily identify any flaws and mark them for correction. Some work stations in the body shop were completely mechanized, with either robots or fixed automation. About the only consistent feature across work stations throughout the plant was that work averaged about 50 seconds per car.

For some options, such as occurred with station wagons, a worker might have to take somewhat over 50 seconds to do the job. This deviation was termed being overcycled. Naturally, a worker could not be continually overcycled without falling behind the pace of the line. To keep up, overcycled jobs had to be balanced off quickly by undercycled jobs for the workers affected. Given the existence of overcycled jobs, then, the sequencing of cars along the line, to provide a balance of overcycled and undercycled work, was an important endeavor. For example, the larger labor content station wagons were never scheduled back-to-back on the line.

Much of the overcycled work occurred along the General Assembly 1 and General Assembly 2 lines, where many of the multitude of options were added to the cars. Workers were advised what options to include on any car by reading the "broadcast," or "manifest," an instruction

continued on p. 650

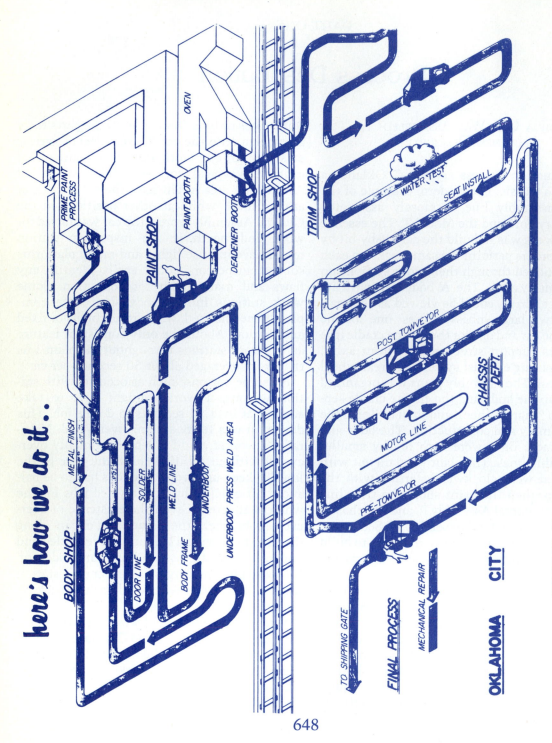

here's how we do it...

BODY SHOP

METAL FINISH

SOLDER

DOOR LINE

WELD LINE

BODY FRAME

UNDERBODY

UNDERBODY PRESS WELD AREA

PAINT SHOP

PRIME PAINT PROCESS

OVEN

PAINT BOOTH

DEADENER BOOTH

TRIM SHOP

WATER TEST

SEAT INSTALL

POST TOWVEYOR

CHASSIS DEPT.

MOTOR LINE

PRE-TOWVEYOR

FINAL PROCESS

MECHANICAL REPAIR

TO SHIPPING GATE

OKLAHOMA CITY

FIGURE D2   How a car is assembled. (Courtesy of General Motors, Oklahoma City Plant)

The body shop's wheelhouse subassembly area. (*Courtesy of General Motors, Oklahoma City Plant*)

Calling for replenishment in the body shop's wheelhouse subassembly area.
(*Courtesy of General Motors, Oklahoma City Plant*)

sheet attached to the car, usually on a window. The broadcast was the primary information by which the worker determined what part or component was to be placed on a particular vehicle. In an effort to reduce error, the plant generated some "mini-manifests," which highlighted in large bold print the options that were called for in a particular department. The mini-manifests required more paperwork, but they enhanced quality by simplifying the information flow to workers along the line.

LOADING THE PLANT

Planning Production

Because cars and the parts to make them were both expensive and bulky, considerable thought was given to limiting inventories in the plant. No finished goods inventories were kept. All the cars assembled were trucked or shipped by rail to individual dealers within a short period of time after coming off the production line. Therefore, all of the cars destined for a particular dealer had to be scheduled for completion at roughly the same time to avoid significant delays.

Furthermore, all of the cars were produced to order as to make, model, color, and options. C–P–C Oklahoma City, on its own, was not permitted to ship any dealer a car that the dealer had not ordered. Roughly 5 percent of a dealer's order represented cars that were already sold to particular customers. About 25 percent were orders for fleet sales, and the remainder represented the dealer's speculation about what the dealership would be able to sell. Such production to order, given the wide range of options permitted in a car, placed tremendous demands on the plant's materials function to schedule the proper mix of cars through the line and to secure enough of the proper parts to fill the order.

The scheduling and material procurement functions were characterized by successive tiers of forecast orders and due dates that represented ever more precise refinement. At the broadest level, the corporate office in Detroit determined a rough production schedule for the year. The schedule served as a target for the company's outside suppliers and for its own internal supply groups such as engine plants and transmission plants. As dealer orders came in, the schedule became increasingly firm. The C–P–C central office in Warren, Michigan, scheduled production in the plant in order to meet a customer promise date known as *target build*. When a dealership received confirmation of an order, it also received word of exactly which week each car ordered was scheduled to be built. Meeting this commitment drove the production plan and the materials scheduled from a myriad of suppliers.

This target build program represented a change from the production planning used until the early 1980s and was part of the company's just-in-time production principles, termed "synchronous manufacturing." The plant operated with a firm car-by-car production schedule for 10–15 days in advance; it was known as the "stable schedule process." Any changes to production had to be accomplished beyond the three-week limit set by this stable schedule process. With this schedule, 98 percent of the target build sequence was accomplished when expected. Moreover, that production schedule persisted throughout the entire plant. It was only rarely that a car had to be taken out of the production sequence because problems could not be fixed on line. Thus, a car begun in the body shop maintained its order in the line throughout the paint shop and the General Assembly areas. Some cars, like station wagons, were periodically removed for extra work, but then they were repositioned to the same place in line. This schedule was announced on Mondays, so that every Monday there were 15 days of firm car-by-car schedules.

Although the line was flexible in handling many make/model and option variations, there were certain limitations. A special computer program, known as "auto sequence," helped to take account of them. The restrictions varied. For example, the plant wanted to batch colors as much as possible, but every day, every color was made, although in a specific sequence that went from white, then to light blue, dark blue, medium blue, red, brown, and finally silver, before returning to white. Never, for example, was a red car scheduled next to a white car because of the trouble in purging the paint lines. There were also restrictions caused by the number of side gates in the body shop used to hold side panels. There were 54 gates and only so many of particular models. This meant, for example, that the plant could run no more than 6 Oldsmobiles in a row. In addition, they could run no more than 7 station wagons in every cycle of 54, and these wagons had to be spaced out because of the overcycle condition they provoked in General Assembly 1. Similarly, the auto sequence program spaced out the options of power doors and windows as much as possible.

This system was very much appreciated by the factory's supplier base. The visibility this gave suppliers and the level aspects of the demands placed on them permitted some suppliers to reduce their costs, and this in turn was passed on to the plant in lower prices. The auto sequence program that created the stable schedule was communicated electronically to all of the plants' suppliers on Mondays. This communication was called the production point-of-use. Every Monday, as well, suppliers received a 20-week planning schedule from the Warren Central Office which included the firm 3 weeks that matched the production point-of-use sequence, but also included forecasts for the subsequent 17 weeks. In the future, it was planned for these two documents to be merged into one document that the supplier would receive from the plant.

This change in production planning both removed a good deal of expediting and improved quality at the supplier and at the plant. Although the sequence was known in advance, no material was shipped from the suppliers in sequence. Oklahoma City did its own sequencing simply because it was more economic to do so. All of the plant's part numbers were scheduled in this way.

Purchasing and Raw Materials Inventory and Control

There were 4046 active parts and 493 suppliers that were used by the Oklahoma City plant. These figures represented reductions from the years when all four nameplates were manufactured at the plant and the active parts had totaled 7059. The reduction in the number of suppliers from as many as 783 suppliers to the current 493 was deliberate. The reduction in suppliers often improved both quality and price. For example, the 7 film suppliers previously used by the plant had been collapsed into a single vendor. The C–P–C Central Office in Warren negotiated all of the long-term contracts for the plant. The plant's purchasing people purchased non-production items, paints, sealers, and vehicle fluids, such as fuel, brake fluid, etc.

To avoid excess materials inventory, the delivery of materials was very tightly controlled. The plant had moved to provide reservations for both the truck and rail deliveries. It was essential that suppliers were reliable in these deliveries. For the most part, only between one or three days' worth of production items were kept in inventory at the plant, with one and one-half days being typical. For the relatively expensive items (engines, radiators, alternators), less than one day's production items were inventoried. Such tight control helped greatly, because the most expensive 350 parts accounted for 80 percent of the dollar value of inventory. Even small items like screws were stocked for only two to three days. How many days of inventory were held depended on:

1. *Volume.*   The more regular a part's use, the tighter the schedule.
2. *Monetary value.*   The higher the value of the part, the tighter the schedule.
3. *Physical size.*   The larger the part, the tighter the schedule. And,
4. *Transportation.*   The farther away the supplier, the greater the days of inventory held.

Many of the deliveries at the plant involved "milk routes," where a truck or train was filled with deliveries from several vendors and moved on a regular daily basis to the plant from as far away as the Midwest. Oklahoma City unloaded 50 rail cars and 90 truck trailers a day.

With materials delivery scheduled tightly, it was necessary to monitor the entire supplier network constantly. Failure to have on hand a key part could shut down the plant. Constant vigil was kept on the number and location of all supplies coming to the plant. Every day a list of critical parts that might turn up short or out of stock for the next two days was developed. The plant notified the suppliers to expedite the parts sought. Usually expediting involved securing a part from either the supplier or a sister plant and switching delivery to a faster and generally more expensive form of transportation.

Monitoring materials delivery involved not only the materials department at Oklahoma City (staff to track shipments, unload and handle materials) but also line workers and their supervisors. Even transportation carriers had electronic access to the materials system and knew exactly what the plant's needs were. However, the plant was careful not to incur extra transportation costs just to carry lower inventory levels. Calculations were made that traded off the extra transportation costs against the inventory carrying costs that might be incurred. And, often, more inventory was kept in order to keep the total cost of procurement lower.

As a result of these improvements in production planning and in "synchronous manufacturing" (discussed in more detail below) inventory turns at the plant had risen consistently over the years. From levels in the mid-1980s of 30 to 40 turns, the plant's inventory turns had risen to a level in excess of 56 turns for 1991, with a peak month level of over 79 annualized turns. What is more, these high turn figures included the fact that the plant owned inventory in transit as well as the inventory that was physically at the plant.

SYNCHRONOUS MANUFACTURING

The Oklahoma City plant was well along with a C–P–C division goal to develop synchronous manufacturing in the plant (what many other companies call just-in-time production). Synchronous manufacturing involved a coordinated set of strategies (17 were identified; see Figure D3) that were designed to help identify and eliminate the non-value-added activities in the factory. Thus, there was an effort to rid the plant of unnecessary activities and controls, to eliminate excess inventories, to contract the working space, or work envelope, so that workers could take fewer steps, to simplify the presentation of parts to the worker so that mounting parts on the car was easier, and to help the worker ergonomically so that less stress and strain was placed on hands, wrists, arms, backs,

A car entering the ELPO bath. (*Courtesy of General Motors, Oklahoma City Plant*)

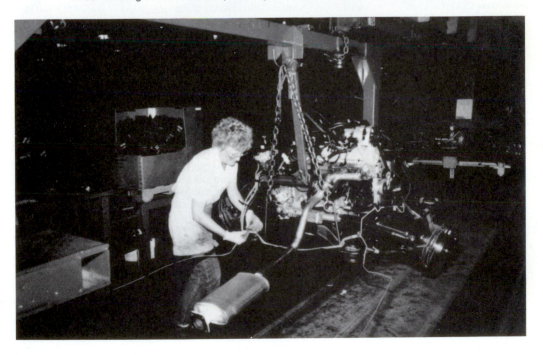

Along the chassis line. (*Courtesy of General Motors, Oklahoma City Plant*)

The marriage of the body and the chassis. (*Courtesy of General Motors, Oklahoma City Plant*)

FIGURE D3 Synchronous Manufacturing	Attachment F
KEY SUCCESS FACTORS	**DESCRIPTION**
1. Lead time reduction.	Method to identify, measure and eliminate waste.
2. Supplier involvement.	Extension of the process.
3. Reduction of variation.	Method of decreasing deviation from a target.
4. Pull system.	Replenishment based on consumption.
5. Leveling.	Providing stable & smooth flow.
6. Quick set-up.	Rapid preparation and changeover.
7. Total preventive maintenance.	Proactive planned upkeep.
8. Process/operator control.	Prevent problems from being passed on.
9. Problem solving.	Logical thought process to identify & eliminate problems.
10. Standardized systems.	Consistent and repeatable operations.
11. Small lot production.	Mimimum material quantity and flow.
12. Capable systems.	Equipment/process within desired range.
13. Facility/equipment layout.	Effective layout of worksites (I.P. "U" shaped cells, etc.)
14. Workplace organization.	Orderly place for everything.
15. Audio visual controls.	Make problems noticeable/status at a glance.
16. Error proofing.	Proactive problem prevention.
17. Flexible systems.	Equipment and process adaptability.

SOURCE: Courtesy of General Motors, Oklahoma City Plant.

and legs. In line with this approach to manufacturing, the Oklahoma City plant was relatively "low tech" compared to a number of auto assembly plants. It had only between 30 and 40 robots vs. perhaps 150 in a more automated plant. The plant's management pursued a strategy of more incremental automation.

Supporting this movement for synchronous manufacturing was the development of a support network to generate ideas for improvement, to analyze them, and then to implement them. Also supporting synchronous manufacturing was a change in the way materials were ordered and packaged for the line. On the line there were switches that turned on lights on a special panel to indicate to materials handlers that they were needed to fetch small bins and boxes of parts for particular positions on the line. In a standard assembly plant, in contrast, large baskets of many parts would be placed in the vicinity of the line, taking up space and forcing the workers to walk considerable distances to pick the parts needed for a particular car. No special signals would be devised either. Under synchronous manufacturing, parts were placed within easy reach of workers and would be replenished much more frequently by the use of worker-initiated signals. This was called the "pull card" system that was used to "pull" materials through the plant (as opposed to "pushing" materials onto the factory floor in anticipation of their use). Cards were used to identify parts, where they came from, where they were used on the line, the standard quantities in the totes and bins are used, and when the worker should signal for more (See Figure D4). This system forced the coordination of materials handling to the needs of the line worker, and placed demands on both the process and the supplier network to produce high quality parts so that the line could function well. In addition, all kinds of worker aids and assists had been created for synchronous manufacturing. Special hoists and other ergonomically friendly tools and devices eliminated much of the effort that traditional assembly lines placed on workers.

The areas of the plant that had been converted to synchronous manufacturing were provided with only four hours of parts on the line. These parts were picked from so-called "supermarkets," where larger bins of the parts were kept and where materials handlers replenished the totes and small bins of parts that were found near the line. The supermarkets were laid out in mirror image to the line so that the materials handling fork lifts and cart trains could easily move in to collect the parts that the pull card or "smart light" systems had indicated were needed.[b] Similarly, synchronous manufacturing had also led to the realignment of the line so that subassembly areas were greatly reduced and placed closer to the point of their use along the main assembly line.

The move to synchronous manufacturing affected both supervision and design. The changes in the process gave supervisors "cradle-to-grave" responsibility for distinct segments of the process, say, for example, all of the assembly for a rear door. This increased the identity that supervisors, and their workers, felt for the product and the process. Contributing to this was a plant program (termed "Design for Assembly") to foster redesigns of the car itself so that assembly could be easier and of higher quality. From time to time groups of design engineers came to the plant and spent time learning from the workers how to do the job, actually doing the job on the line, discovering for themselves the problems that production workers had in doing the job, and leaving with numerous ideas for altering the way that the car was designed.

---

[b]The hand deliverable parts were on the pull card system; the bulk or fork truck deliverable parts were on the "smart light" (Synchronous Manufacturing Andon Retrieval Transport) system.

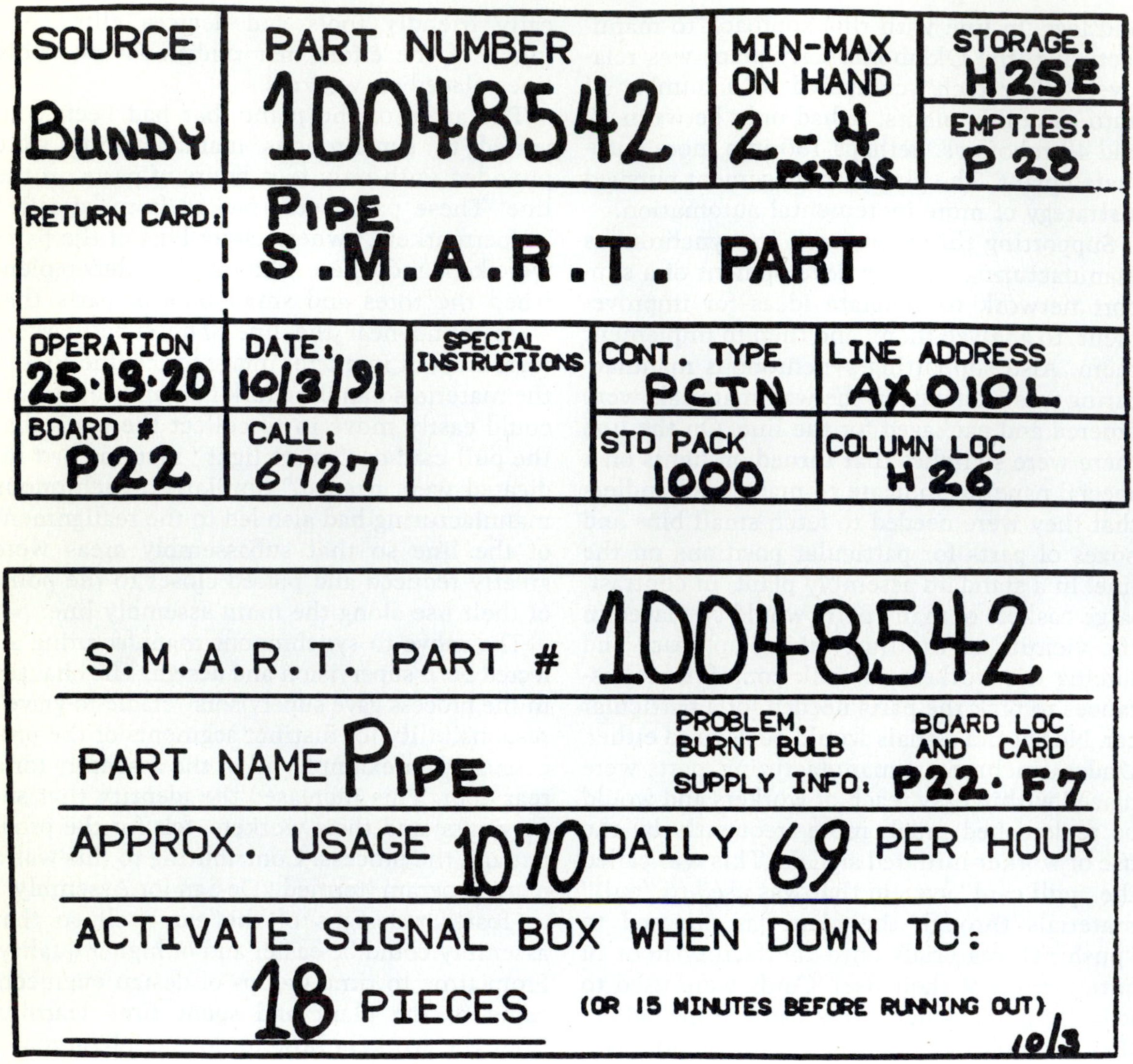

SOURCE	PART NUMBER	MIN-MAX ON HAND	STORAGE:
BUNDY	**10048542**	2 - 4 PCTNS	**H25E**
			EMPTIES: P 28

RETURN CARD:	PIPE
	S.M.A.R.T. PART

OPERATION 25·19·20	DATE: 10/3/91	SPECIAL INSTRUCTIONS	CONT. TYPE PCTN	LINE ADDRESS AX010L
BOARD # P22	CALL: 6727		STD PACK 1000	COLUMN LOC H26

S.M.A.R.T. PART # 10048542

PART NAME **PIPE**   PROBLEM, BURNT BULB   BOARD LOC AND CARD
                          SUPPLY INFO: **P22·F7**

APPROX. USAGE **1070** DAILY **69** PER HOUR

ACTIVATE SIGNAL BOX WHEN DOWN TO:
18 PIECES   (OR 15 MINUTES BEFORE RUNNING OUT)
                                                        10/3

FIGURE D4   A card for the "pull" system. (*Courtesy of General Motors, Oklahoma City Plant*)

Synchronous manufacturing did not lead to any workers losing their jobs. Workers removed from direct labor tasks on the line were reassigned to such tasks as repacking parts in the supermarkets, recycling cardboard and other materials, and doing prototype work on future models.

Synchronous manufacturing had resulted in tremendous gains in productivity, space saved, inventory dollars saved, and quality enhanced.

For example, in the body shop a synchronous manufacturing project reduced the square feet of space from 32,400 to 7593. Work-in-process inventory was reduced from 2076 pieces to 54, and the variable cost declined. Along the instrument panel feeder line, square footage diminished from 30,730 square feet to 11,340, work-in-process declined from 98 pieces to 24, and the number of jobs on the line at any one time dropped from 81 to 44. Similar results were being achieved all over the factory, and new opportunities to apply synchronous manufacturing concepts were being revealed all the time.

Underlying the plant's accomplishments with synchronous manufacturing concepts was a commitment by the plant's top leadership to be the instructors for 13 four-day workshops held at the plant in the previous 16 months. The well-designed workshop format was created by experienced manufacturing personnel utilizing concepts observed at the GM/Toyota (NUMMI) factory in Fremont, California.

REVISING AND CONTROLLING THE OPERATION

Although the plant did not have much leeway in the purchase of materials or the mix of makes, models, and options it was to assemble, it did have considerable leeway in the design and management of the assembly line itself. Key roles in this effort were played by quality control, the industrial engineering staff, and the plant engineering staff.

Quality Control

The move to synchronous manufacturing put pressure on the plant and its suppliers for quality. The system would only work when everyone had high confidence in what was being sent to the plant. There was no receiving inspection of items; parts went right to the line.

The quality control department coordinated a number of specific functions: inspection, reliability, audit, quality statistics, and customer acceptance criteria, commonly called the "voice of the customer."

It was the plant's aim to try to match key characteristics of the product, as made evident by the customer, to the capabilities of the process. This was known as "matching the voice of the customer to the voice of the process." Outside data were used to provide information on customer desires. Warranty information, data from outside independent surveys and consultants such as Camip and J. D. Power, and surveys done by Buick, and those by the plant itself (called "Sunvisor Surveys") brought the plant in contact with customers. In one year, for example, the plant, on its own, called 10,000 of its car buyers to solicit their opinions and to evaluate their complaints to find out exactly what customers thought was wrong or right with their cars. The plant was beginning to be "obsessed with the customer" and would stop at nothing to remedy a customer problem, even if it meant sending people and replacement parts to dealers anywhere in the country. With this focus on the customer, discrepancies had dropped to less than one per car, which was world-class performance.

While the plant still did some inspection, the number of inspectors had dropped considerably from a level of about 370 in 1986 to about 200 in 1991. Under a concept called "Build-In-Station," people on the line were paid a higher rate to assume responsibility for their own inspection with the intent of integrating inspection into the building of the car. There were inspection stations along the line, as are typically found in auto assembly plants, but these inspections stations were fewer than ever before. Inspectors marked on both the car and the inspection tickets that travelled with the car exactly what corrective work needed to be done farther down the line in certain repair areas.

The quality control department's reliability group was charged with a number of tasks. They were the overseers of driver and passenger safety in the car. They also dealt with incoming quality by working with suppliers. They were also responsible for engineering change orders and for leading any product changes that had been mandated.

The audit group was responsible for grading the production quality every shift. Eighteen cars were investigated every shift, selected at random. Four of those cars were checked for everything. Fourteen others were checked systematically for the most prevalent problems. Once the more prevalent problem areas were studied, and preventive countermeasures arranged for them, then the inspection sampling for those problems was reduced and other problem areas were sampled more intensively. Three cars a month were subjected to a comprehensive evaluation and were measured in every way possible so that "variables," as opposed to "attributes," data[c] were collected on things like noise, fit, finish, and other important criteria for the customer. In addition to these three cars, a competitor's car was also evaluated in the same way.

The daily audit included a special 18-mile road test and a 12-hour cold soak. These processes allowed the auditors to check for cold starts, squeaks, and rattles. The number of defects per car was assessed. In addition, about ten times a year, unannounced and at random, a quality audit team from Detroit would perform their own audit of 20 cars. This procedure served as a check on the plant's daily audit.

By matching the voice of the customer to the voice of the process, the plant developed some statistical measures that provided the information needed to improve the process. Currently, the plant had between 70 and 80 quality control charts, about the same number of trend charts, and very many more charts of raw data that supplied insight into process improvement. Problems that had shown up repeatedly were assigned to special cross-functional teams that were charged with mapping the process with a process flow diagram and then indicating where production variability might be caused. This systematic investigation, in conjunction with some experimentation, led to effective countermeasures aimed at prevention of these problems.

In addition, any employee could shut down the line if a quality problem cropped up. Such actions were seldom taken, however. The factory had programs whereby both hourly and salaried employees acted on specific problems that had been raised by customers. Employees were also sent to supplier companies to educate suppliers on exactly what the assembly plant's requirements were. Numerous quality aids had been introduced in the process. In some cases these aids were protective devices for the car itself, such as to avoid scratches or other damage, and in others they were worker production aids that decreased effort or fatigue (ergonomics) or increased precision or reliability.

Industrial Engineering

The Industrial Engineering Department was charged with: (1) translating the engineering design of an automobile into a step-by-step procedure for assembling it; (2) laying out the line, work stations along the line, and any production aids such as equipment and fixtures for the work stations; (3) assigning work to each worker on the line and measuring that work so that it was appropriate; (4) establishing the authorized level of work and monitoring that

---

[c]For this distinction, consult Chapter 4 on quality management.

The point on the line where the engine is lifted up and married with the body.
(*Courtesy of General Motors, Oklahoma City Plant*)

Ergonomic fixtures in the door pad subassembly area just off the main line.
(*Courtesy of General Motors, Oklahoma City Plant*)

Small-part delivery trucks in a "supermarket." (*Courtesy of General Motors, Oklahoma City Plant*)

level; and (5) devising methods improvements or other ways to lower costs and/or increase productivity. The Department spearheaded many of the synchronous manufacturing initiatives at the plant. Ideas for changes typically sprang up from the workforce, supervision, or industrial engineers themselves. Conceptual drawings for any change were done by the Industrial Engineering Department.

As mentioned earlier, the line was designed in such a way that all workers averaged about 50 seconds' work on each car. Given any major model changes, the line would have to be thoroughly redesigned so that workers assigned possibly very different work tasks from what they were used to still had about the same amount of time to work on each car. This is

what line rebalance was about. With the adoption of synchronous manufacturing, line rebalance occurred all the time as tasks were changed and jobs (or job links) were altered, non-valued-added tasks removed, and value-added tasks reshuffled so as to put fewer workers directly on the line.

In greater detail, the rebalance of the line followed a number of stages:

1. Given the car's engineering and design, the industrial engineer described what had to be done to assemble the car. This description involved painstaking detail; for example, one could not simply specify, "mount headlight" since the task may take either shorter or longer than the desired cycle time. Mounting

Call board for large parts. Part of synchronous manufacturing.
(*Courtesy of General Motors, Oklahoma City Plant*)

a headlight would involve a number of separate actions such as walking 5 feet to the supply bins, reaching for a headlight and screws, walking back to the line, positioning headlight in the socket, and using a power screwdriver to screw in two mounting screws.

2. Each of these separate actions (job elements) was assigned a time. These times were primarily determined by reference to standard time data established by General Motors, although sometimes they were determined by a special stopwatch where an engineer would time a worker performing a task. The standard time data typically would derive from numerous stopwatch studies, films of workers, and other sampling studies of specific worker tasks.

3. Once times were assigned to specific elements, the industrial engineer would review them with the relevant supervisor on the production line to determine whether they were reasonable. The changes would then be made.

4. Once an agreement was reached, the task, its elements, element time estimates, and whether the task was performed on all cars or only a few, were placed in computer-readable form. A computer program then calculated the work schedule, trying to bal-

ance work loads among the workers, such as by taking job elements from one worker and placing them with another, or otherwise shifting assignments to distribute the work evenly.

5. Generally, the initial computer schedule left some workers with either too much or too little to do. What was to be avoided was putting workers in an overcycled condition where they would consistently have too much work to perform for the time cycle decided on. If an imbalance in the line were serious and could not be remedied easily, the task was sometimes taken off the main line and performed as a separate activity. This course of action was considered a last resort, however.

On the basis of the computer-run, tasks and times were adjusted and a subsequent computer run made. This run was studied for reasonableness, and again adjustments were made. Numerous iterations of this adjustment process were generally performed before the line was satisfactorily balanced, at least on paper.

6. The paper balance of the line was then ready for trial on the factory floor. Industrial engineers, supervisors, and workers all became involved in this activity. The danger was that the predicted and actual times would not mesh. If this in fact occurred, the first thought was to help the worker improve the time by changing the layout of the work station, adding fixtures or other equipment, or changing methods. Such action was usually sufficient, but, if need be, another operator could be added or a modest rebalance of the line effected.

Plant Engineering

The Plant Engineering Department directed all the construction, maintenance, repair, recy-

cling, and environmental work of the Oklahoma City site's land, buildings, and equipment (except tooling, which was handled by the assembly engineering group). All the plant's requests for major expenditures and capital appropriations were coordinated by Plant Engineering.

Plant Engineering was constantly on the alert for technological improvement and was involved in all of the plant's initiatives for synchronous manufacturing. The Plant Engineering Department did the detailed engineering on equipment involved in the synchronous manufacturing effort. It did the costing and the approvals for any capital dollars required. And, it was responsible for the installation. Oklahoma City tried to do as much of its own installation as it could, contracting out, as a matter of course, only concrete work and roof work. Installations were meant to be flexible and easily repositioned. There was no bolting of machines to the floor unless safety concerns so dictated.

The Plant Engineering Department was involved in technological advances such as the use of turbo bells for painting and the use of robots for applying adhesives to the windshield or to the door paper. A new dynamic vehicle testing area was the responsibility of Plant Engineering.

The Oklahoma City plant prided itself on being able to make most line and equipment adjustments on the fly, without disrupting production.

THE WORKFORCE AND THE PERSONNEL DEPARTMENT

The plant's non-supervisory factory floor personnel were paid by the hour. All supervisory and clerical people were paid a salary. The plant currently operated two shifts of production with an hour of scheduled overtime each shift, a total of 18 hours of production a day. In addition, the plant operated two Saturdays a month. The first shift began at 7 A.M. and ended at 3:30 P.M., with a half hour break for lunch. The second shift began at 5:00 P.M. and ended at 1:30 A.M. During each half of each shift, the entire line shut down for 23 minutes so that workers could rest. This was termed "mass relief" and replaced the individual-by-individual tag relief that had proceeded it. Mass relief was seen as improving quality.

The hourly employees were represented by the United Auto Workers (UAW) union. Most features of the agreement between GM management and the plants workers were spelled out in a 597-page national agreement covering a wide range of topics, including wage payments, fringe benefits, seniority, grievance procedures, overtime, layoffs, safety production standard procedures, and numerous other topics. There was also a smaller 194-page additional agreement between the local UAW organization and the plant's management. This separate agreement provided detailed information on wages, layoff procedures, health and safety measures, and the like.

Plant relations with the union were generally cordial and constructive, fostered by daily dialogue between the parties. There were no secrets at the plant. Any union member could attend any meeting at the plant except for those relating to human resource issues. A union representative was always present at staff meetings.

Most worker complaints were resolved without recourse to the grievance procedure. Grievances, such as overwork conditions, were generally resolved at the first of four levels of the grievance procedure. Few grievances reached the final arbitration level.

The plant's personnel department was an essential factor in plant management. Not only did the personnel department oversee labor relations with the UAW, but it was also charged with training responsibilities and payroll.

The personnel department had spearheaded the drive for increased competitiveness and worker involvement with a program called the "Voluntary Input Process" (the VIP process). Teams of workers, typically between six and ten, met for half an hour every Wednesday morning to discuss improvements that could be made to the process. About 75 percent to 80 percent of the plant was involved in this effort. For their involvement, workers were paid an additional fifty cents an hour. A VIP operator had a number of responsibilities, including attendance at the weekly meetings, knowing all of the jobs that were done within the unit, agreeing to rotate to retain proficiency in those jobs, helping to maintain the cleanliness and good housekeeping of each unit area, providing training for others, and working to improve the materials scrap processing and efficiency of that area. VIP operators worked to build quality into the car in their areas through the use of statistical process control (SPC) and self-inspection. The VIP process had resulted in an increase in quality and a number of helpful suggestions for the increased efficiency of the plant. It fit well with the drive for synchronous manufacturing. In addition to the VIP operators, the program also had duties for the support people assigned to every unit along the line (see below) and for the group coordinators or supervisors that oversaw the area.

C–P–C Oklahoma City offered the hourly employee a number of possibilities for advancement and work preferences from the simple movement from one shift to another, to transfers between departments and promotions to higher-pay classifications. Hourly employees could also move to management by becoming supervisors.

Supervision

Work on the line was directly overseen by first-line supervisors assisted by three to four group leaders, called support people. Typically, the supervisor was directly responsible for between 30 and 35 workers on a shift. There was one support person for about every seven-to-ten member unit. Supervisors were thoroughly familiar with all aspects of work within their own sections of the line and of what workers could or could not be expected to do, largely because of once having been one of them. At the start of the shift, the supervisor's immediate concern was manning the production line and making sure material and tools were available.

Once workers were at all of the line positions, the supervisor spent much of the remainder of the shift in two activities: promoting and checking quality and troubleshooting any problems on the line, such as equipment malfunc-

tions, pending parts shortages, or defective materials. Very little of the supervisor's time was spent at a desk. For most of the shift, the supervisor walked the line, talking with the line's workers, trying to solve any problems they might have, and promoting quality.

Four or five supervisors in turn reported to a general supervisor, and the general supervisors reported to one of four superintendents, one each at the body shop, the paint shop, General Assembly 1, and General Assembly 2. (See the organizational chart in Figure D5.)

The superintendent's job included many aspects of the first line supervisor's, plus other responsibilities. The superintendent was concerned primarily with quality and safety, but also spent considerable time and effort on cost

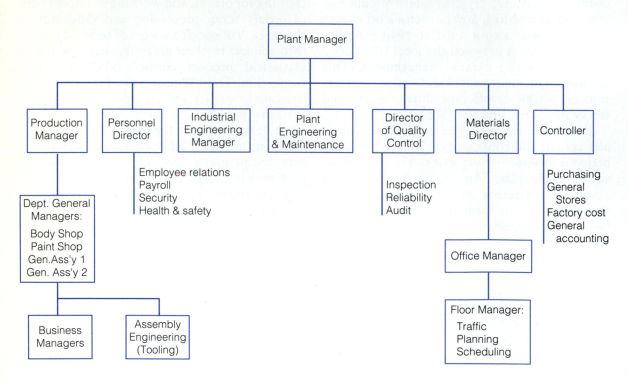

FIGURE D5   Organization chart for the C–P–C Oklahoma City plant.
(*Courtesy of General Motors, Oklahoma City Plant*)

control, human relations, and housekeeping chores. Some of these tasks were routine, such as meeting with union committeemen, checking absenteeism, and conferring with other su-perintendents. But there were always special plans or meetings and unexpected problems on the line to overcome.

PART TWO

DISCUSSION

The GM C–P–C Group is in no way responsible for the following views and presentation. They remain solely the responsibility of the author.

plies a great deal of communication between workers and management in the plant and between the plant, corporate level managers, and outside suppliers.

THE FLOW OF THE PROCESS AND OF INFORMATION

The process flow in a factory such as C–P–C Oklahoma City is among the most complex one can encounter. A blizzard of different tasks, equipment, and skills are required to assemble an automobile. Yet the process flow can be diagramed in a very straightforward way. Figure D2 represents a rough-cut process flow diagram. A more detailed version would consist of recounting all of the workstations along the production line; it is these workstations through which every car passes. Thus, compared with a job shop or a batch flow process, the classic assembly line has a very well-defined process flow.

With a process as complex and as closely scheduled as an automobile assembly plant, it is not surprising that the information flow is complex as well. Figure D6 portrays an information flow diagram for C–P–C Oklahoma City. While the information flow is essentially oneway, from the top down, the need to track materials both inside and outside the plant im-

CAPACITY

The notion of what capacity is in a machine-paced line flow process like C–P–C Oklahoma City is as clear as it ever gets. At C–P–C Oklahoma City, one car rolls off the end of the production line every minute. Everything in the process is geared to making that happen. It is as though the process were a huge on-off switch: when it is on, one car a minute is produced; otherwise, nothing is produced.

Similarly, the notion of capacity utilization is relatively unambiguous. When the plant is up and running, capacity utilization is 100 percent. Otherwise, it is zero. Simply by tallying the hours the plant actually runs and dividing this total by a count of hours the plant could be open gives an idea of what capacity utilization is. How many hours the plant could be open is subject to some debate, however. For example, it may be very contrary to corporate policy to schedule a third shift of production, and so a capacity utilization measure should be based on two shifts only. Another problem may lie in the determination of how much overtime is the

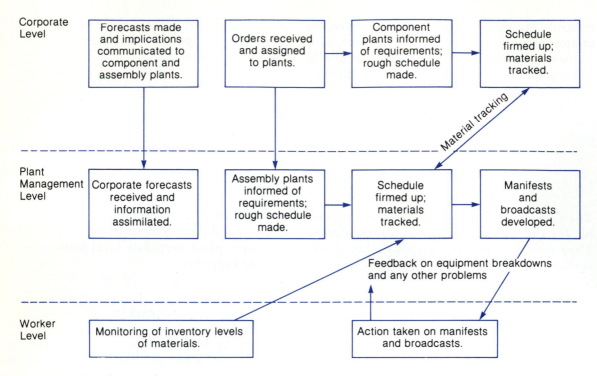

FIGURE D6   An information flow diagram of the Chevrolet–Pontiac–Canada Group, Oklahoma City.

maximum the plant could schedule without antagonizing workers. Is it an hour a day? Two hours a day? Every Saturday? Questions like these make a capacity utilization calculation fuzzy, but it is clear how much the process is capable of producing when it is up and running.

In a machine-paced line flow process, capacity in the short run is severely constrained. It can be modified only by scheduling more or fewer hours. In the medium run, however, capacity can be substantially modified (without huge, new additions of plant and equipment) by adjusting the balance of the production line. Thus while the machine-paced line flow process cannot modulate its capacity smoothly, as can occur in job shops and batch flow processes, capacity can be adjusted in abrupt steps and by varying amounts.

DEMANDS OF THE PROCESS ON THE WORKFORCE

A machine-paced assembly line places special demands on a worker. In a sense, the worker is a "slave to the iron monster," which is the moving assembly line. He or she cannot simply decide to take a break from work and wander about chatting with co-workers. (But then neither can the second-grade teacher.) Moreover, the assembly line worker must perform the same, fairly routine task repeatedly throughout the workday. Many people view that as monotonous or degrading or worse.

But not all people hold to that view, and therein lies a fundamental point about assembly lines and assembly line workers. Happily, people are very different from one another in tolerances for certain activities and in expectations about their jobs. One class of personality savors proximity to power and influence and will forgo higher salaries to do so (consider many staff workers on Capitol Hill), while another will sacrifice salary for considerable freedom of schedule and the chance to be "their own boss" (such as many college professors). If the assembly line worker can be typed in this way at all, he or she has forgone diversity in the job for relatively high wages and freedom from the anxiety many people suffer when they are confronted with decision making. For many people, that is not a bad trade to make.

Nevertheless, despite the matching of personalities to job types that goes on continuously in our economy, progressive companies like General Motors constantly search for ways to improve the quality of work life for their workers.

We have dealt more with quality of work life in Chapter 5. Suffice it here to observe further from the process description (1) the importance that the personnel department at C–P–C Oklahoma City enjoyed and the mandate it bore for worker training, safety, and promoting worker involvement in the process; (2) the challenge supervisors felt for recognizing worker achievement and integrity and for assisting their workers in solving problems both on and off the job; and (3) the care with which the industrial and plant engineers studied the tasks that make up automobile assembly and devised ways to eliminate or simplify more difficult or disagreeable ones. Within most machine-paced line flow processes, the career path of the worker is much less well defined than it is in many continuous flow processes or in the batch flow process. Lines of progression are not spelled out as directly, and most workers earn about the same pay. Most job changes involve transfers from evening or night work to day work, or from one department to another with what may be perceived as easier or more exciting work. Some promotions in job categories are frequently possible as well.

Demands of the Process on Management

The machine-paced line flow process places a number of taxing demands on management. The products it produces are typically consumed in high volumes; therefore they are likely to compete on price as well as product performance, workmanship, and perhaps even the reliability of delivery. This is a broad front on which to compete, and it means that line flow managers must devote their attention to cost reduction measures (items 1 through 4 in the following list) and to product performance and workmanship measures (items 5 and 6):

1. *Balance.* The definition and balance of the line, as discussed previously, is a critical aspect in ensuring that as little labor as necessary is placed on the line.

2. *Materials management.* Like the continuous flow process, the line flow process places very regular and steady demands on its suppliers. Because of this regularity, the line flow process can often avoid drawing on large raw materials inventories and paying the necessary finance charges, but only by carefully managing the purchase and logistics of supplies. Such materials management involves the establishment of superb information and control systems so that (1) parts in imminent danger of falling short are identified and noted and (2) parts from suppliers are tracked thoroughly and delivered on time. The process description noted how recent changes have contributed to coordinating materials more tightly.

3. *Technological change.* In processes as complex as most line flow ones there are always ways to improve their workings. Most such changes are incremental in nature, but periodically more significant steps can be taken either to speed up the process or to eliminate some stations through the introduction of new equipment. Computers,

more versatile machine tools, and increased automation have often made significant contributions to process improvement.

4. *Capacity planning.*   The managers of a process as rigid in capacity as the machine-paced line flow process must be very careful to plan diligently for its future capacity needs. Line rebalance may be called for or more drastic modifications to the design of the line. Design changes must be thought through carefully from the beginning of any planning cycle.

5. *Product design.*   The high-volume products of the typical line flow process must be designed to be manufactured; the more exotic the demands the product places on the design of the line and the tasks workers perform, the more likely the competition will bury the product on price. Yet some of the best assembly line products give the impression that they were custom-made, there being so many options for the customer to choose from. Automobiles are often this way. The trick, of course, is to design flexibility into the product so that the same general worker task results in a different appearance or performance in the product. This is where the Oklahoma City plant's program for "design for assembly," where design engineers are brought to the plant to be educated by the workforce on what is easy to assemble and what is hard, is so important.

6. *Workforce management.*   In so many line flow processes, quality is fundamentally dependent on the workforce. Fostering pride in workmanship and creativity in the identification and solution of problems, when the tasks performed are as repetitious as they are, is a challenge, but an essential one.

QUESTIONS

1. Why is the production process at C–P–C described as "classic but modern"?

2. It has been said that industrial engineering is the most important element of control in a machine-paced line flow process. Do you agree? Why or why not?

3. What are the connections between the workforce and its supervision and product quality at C–P–C Oklahoma City?

4. Discuss in detail the relationship between the notion of capacity in a machine-paced line flow and the technique of line balance.

5. Based on different operations you have so far studied, which operation would you rather work in? In which would you rather be a manager? In which would you like least to be a worker? a manager? Support your answers with specific references to the processes.

6. Various changes in materials management were mentioned in the process description for C–P–C Oklahoma City. Trace the changes in both process and information flows that you would expect to result from these changes.

7. Suppose that you had just been designated the plant manager for a new plant about to be completed. You are in charge of assembling your own staff (such as in quality control, plant engineering, supervision, production planning). Based on the process descriptions to date, how would the composition and relative strengths of your chosen staff differ from one type of process to another?

8. Suppose C–P–C Oklahoma City has been designated to produce a new General Motors car. What aspects of the plant's operations do you think will be most affected by this change?

A HYBRID (BATCH/CONTINUOUS FLOW) PROCESS
Stroh Brewery Company
Winston-Salem, North Carolina

In 1982, the Stroh Brewery Company of Detroit acquired the Jos. Schlitz Brewing Company of Milwaukee. In the process, Stroh acquired Schlitz's Winston-Salem brewery, which had begun production in June 1969. At the time, the Winston-Salem brewery was the largest brewery ever built at one time. It occupied 1.1 million square feet on 150 acres and cost $130 million. Originally constructed to produce about 5 million barrels of beer per year (a beer barrel is equal to 31 gallons), the brewery was now capable of producing about 6 million barrels of beer per year.

PART ONE

PROCESS DESCRIPTION

THE BREWING AND PACKAGING OF BEER

Beer is a malt-and-hops-flavored drink that is fermented and carbonated. It has been popular for thousands of years and, in one form or another, exists in almost every society. Although known and enjoyed for so long, the chemistry of brewing is very complex and is still not understood in all its particulars. Brewing remains very much an art as well as a science, with rapid innovation over the past 20 to 25 years; developments in brewing automation and control have reduced the need for manual labor by 75 percent and placed much of the control of the process on instrument panels.

The brewing process itself can be divided into two main stages: (1) the production of natural cereal sugars, called wort production, and (2) fermentation and finishing. (The process is depicted in Figure E1.) At Stroh, a batch of beer typically took about a month to brew and age.

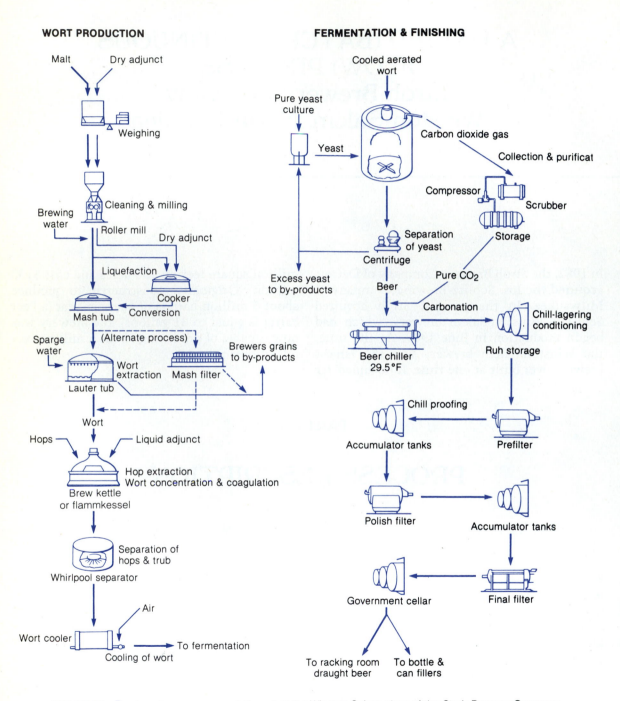

FIGURE E1   The brewing process as followed at the Winston-Salem plant of the Stroh Brewery Company.

A control panel in Stroh's Winston-Salem brewhouse. (*Courtesy of Stroh Brewery Company*)

Wort Production

Wort production starts with the weighing, cleaning, and milling of the malt, which gives beer its flavor. The malted barley and an adjunct (corn, rice, malt, and/or liquefied corn) are prepared separately. The adjunct is used because malted barley can convert more than its own weight of starch to fermentable sugars. To the milled or crushed barley, malt, and adjunct is added the brewing water, which is specially treated by filtering and pH adjustment to remove impurities.

This addition of water to ground malt is known as *mashing*. This occurs in large, stainless steel vessels in which the ground malt and brewing water can be stirred together and brought to a boil.

The Stroh's brand and its Stroh Light extension brand are famed for being "fire brewed." Instead of the wort being cooked by heat from steam coils, as was the common practice, the wort for Stroh's and Stroh Light is cooked using direct flame in the flammekessel. This step gives Stroh and Stroh Light its distinctive taste.

After mashing, the mixture is transferred to a lauter tub, which separates the liquid (the wort) from the undissolved solids of the mash. The lauter tub is a large, circular tub with a false bottom and a rotating set of "rakes" that can be raised or lowered. These rotating rakes smooth out the mash as it enters the lauter tub and makes an even filter bed. Undissolved particles in the mash settle to the false floor of the lauter tub. Once settled, the liquid wort can be

A brew kettle in Stroh's Winston-Salem brewhouse. (*Courtesy of Stroh Brewery Company*)

drawn off with the aid of hot "sparging" water, which rinses the wort off the grain. The undissolved mash is then collected and sent to the nearby Miracle Feed Company, where it is processed into a highly nutritious livestock feed.

The drawn-off wort then enters the brew kettle, where it boils for 2 hours. This boiling stabilizes the wort by killing off bacteria, deactivating enzymes, and coagulating any still undissolved particles. During this boiling, hops are introduced to the wort to enhance its flavor by extraction of desirable hop components. Boiling also increases the concentration of the wort through evaporation. Once boiled, the hops and trub (coagulated sediment) are separated out by whirlpool action in a special vessel. The hot wort is then cooled and aerated.

Fermentation and Finishing

Only after yeast has been introduced to the cooled wort and fermentation is complete can the product be called beer. The yeast that is introduced to the 50-foot-long fermentation tanks transforms maltose, glucose, and other sugars in the wort to alcohol and carbon dioxide. Stroh's fermentation is accomplished by a patented process called "accurate balanced fermentation." During fermentation, carbon dioxide gas is collected and purified. After fermenting has finished and the yeast has been separated from the beer by centrifuge, the carbon dioxide is reintroduced to the beer (carbonation), and the beer is chilled.

Once fermentation is complete and the yeast separated, the remaining finishing consists of a string of storage and filtering steps. The chief aging occurs in the ruh cellar, where the beer is held for between 15 and 21 days. Once properly aged, the beer is filtered several times through diatomaceous earth. Final storage occurs in what is called the "government cellar," where the next day's production is inventoried and where the quantity of beer subject to federal tax is monitored. Beer typically spends less than 7 days in the government cellar, the time largely dependent on packaging needs.

Out of the government cellar, the beer flows to the keg, canning, and bottling lines.

Canning, Bottling, and Kegging

The Winston-Salem brewery operated six canning lines, three bottling lines, and a keg line. Each line performed essentially the same functions: (1) filling and capping and (2) pasteurizing (for the non-draft beers). Cans and bottles were also boxed into six-packs and cases, and cases were placed on pallets for shipping by truck or rail. The process from filling to palletizing ran continuously, with no significant buildup of in-process inventories unless a piece of equipment in the line broke down. The key piece of equipment on the line was the filler (an investment of $400,000) — the limiting element for capacity. All of the machines that followed had rated capacities in excess of the filler; so any equipment breakdown after the filler did not necessitate shutting down the entire line. Small reservoirs of work-in-process were held at several places along the line in order to keep things moving in case one part of the line had to be shut down temporarily (such as a shutdown to remove a broken bottle).

The can lines could run at 1600 cans per minute, the bottling lines at 900 bottles per minute, and the keg line at 340 half-barrels per hour. With six canning lines, several brands brewed at the plant could be canned at the same time. Alternatively, the same beer could be packaged in two different size cans at the same time. Three bottling lines were needed not only to match the demand for bottles but also to permit bottles of different shapes and sizes to be run at the same time. In addition, returnable bottles were separated from new bottles for filling and packaging. A canning line shift required 6 workers, while a bottling line shift required 15.

Warehousing

The Winston-Salem plant preferred to load the beer directly from the line into railcars or trucks rather than warehouse it. (See the plant layout in Figure E2.) In fact, about half of the plant's output was shipped without any storage in the warehouse. The warehouse's fully enclosed railroad docks could store fifty 50-foot freight cars. In addition, 15 truck loading docks were available. The warehouse was designed so that output could be shipped by either rail or truck. The warehouse stored some brands that were not produced at the plant. Also, 10-ounce cans were not filled here. These items had to be supplied by other Stroh plants.

Loading the Plant

Order Taking

Stroh's Winston-Salem brewery served approximately 200 wholesalers stretching over the entire eastern seaboard from New England to some parts of Florida and inland to Michigan, Indiana, Kentucky, and some parts of Tennessee. A wholesaler generally handled one of the major national brands and perhaps a smaller, regional brand. Each wholesaler serviced a range of retail accounts (such as liquor stores, bars, grocery stores) within a specified geographic area.

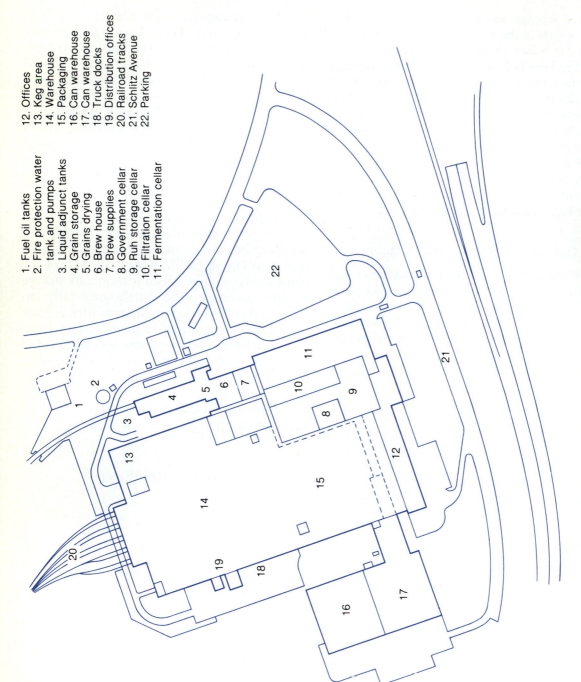

1. Fuel oil tanks
2. Fire protection water tank and pumps
3. Liquid adjunct tanks
4. Grain storage
5. Grains drying
6. Brew house
7. Brew supplies
8. Government cellar
9. Ruh storage cellar
10. Filtration cellar
11. Fermentation cellar
12. Offices
13. Keg area
14. Warehouse
15. Packaging
16. Can warehouse
17. Can warehouse
18. Truck docks
19. Distribution offices
20. Railroad tracks
21. Schlitz Avenue
22. Parking

FIGURE E2   Layout of the Winston-Salem plant of the Stroh Brewery Company.

At the middle of each month, every wholesaler placed an order for delivery next month. This procedure meant that the brewery had lead times of 2 to 6 weeks for each order. The order, entered either directly by computer or mailed in, specified item, quantity, and date (sometimes hour) for delivery of the beer ordered. The brewery offered a remarkable diversity of items.

Seventeen different brands of beer were brewed at the plant (Stroh's, Stroh Light, Old Milwaukee, Old Milwaukee Light, Old Milwaukee Non-Alcoholic, Old Milwaukee Draft, Schlitz, Schlitz Light, Schlitz Malt Liquor, Piels, Piels Light, Goebel, Goebel Light, Schaefer, Schaefer Light, Red Bull, and Silver Thunder). Old Milwaukee accounted for 37 percent of production and Stroh for 22 percent, with the other brands dividing up the remaining 41 percent of the brewery's capacity. This beer could be packaged in cans, bottles, or kegs of various sizes, with different lids and in different kinds of cardboard cartons. In all, there were 759 beer and packaging variations that a wholesaler could order. Wholesalers could pick up their order by truck, paying the expenses themselves, or they could have the plant load a special Stroh-assigned rail freight car and direct it to the wholesaler's own rail siding. The truck option offered greater speed and reliability of delivery than rail, thereby cutting down on the inventory the wholesaler had to carry, but it cost considerably more. About 20 percent of the plant's volume was shipped by rail and 80 percent by truck.

The kegging line at Stroh's Winston-Salem plant. (*Courtesy of Stroh Brewery Company*)

The Winston-Salem plant monitored the inventory positions of each of its wholesalers, and it followed trends in the marketplace as well (such as promotional campaigns and their effect). The plant sometimes made suggestions to its wholesalers on what and how much to order, especially if it felt strongly that a wholesaler might not have ordered correctly.

Production Planning

Production planning was intimately related to order taking and was performed in the same department. (See the organization chart in Figure E3.) The role of the production planners was to figure out how the brewery could best fill the orders placed by its wholesalers. A good production plan was one that satisfied all of the wholesalers' demands with (1) little repositioning of the requested delivery dates, (2) few changes in the workforce, and (3) full usage of the existing equipment.

Planning and scheduling production for each month meant determining (1) how much beer of which type was required and the timing of its delivery and packaging and (2) the precise sequencing of lines and beer/packaging combinations on each line.

Because it took a month to brew beer, the brewmaster had to produce to a forecast of sales rather than to customer order. This production to forecast was a key reason why the plant kept close tabs on retail activity and made buying suggestions to its wholesalers. The packaging of beer, on the other hand, could be, and was in fact, done to customer order.

In planning packaging for any given month, it was advantageous to:

1. Group runs of the same beers together, such as packaging Old Milwaukee on several lines at the same time.
2. Group similar packaging sizes and types together, such as running all quart bottles at

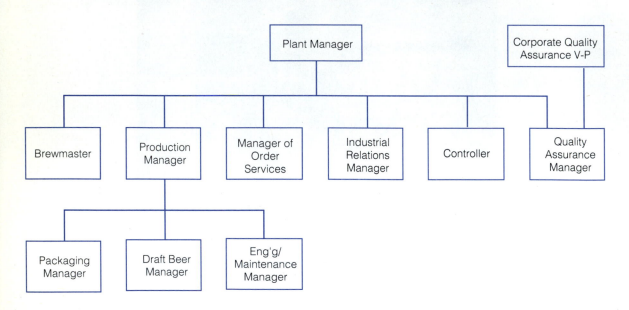

FIGURE E3   Organization chart for the Winston-Salem plant of the Stroh Brewery Company.

the same time rather than interspersing them among other production runs.

3. Run similar lines at the same time, such as running two can lines in tandem. To run a second line, given the first was running, required only four more workers rather than the eight it took to run the line by itself.

4. Run a canning line together with the keg line, since the joint workforce for such a combination could then be shifted en masse to a bottling line if need be. In that way, the groups working on any line would not have to be split up and reassembled as often.

Once the next month's wholesaler orders were received, the production planners went to work to schedule the production to fulfill that demand. This scheduling was accomplished manually, using trial and error but with reference to decision rules like the four mentioned here. After the production plan became "final," the wholesalers' delivery dates and order quantities were acknowledged on a week-by-week basis, giving the wholesalers 2 weeks' notice. About 70 percent of the wholesalers' original orders were acknowledged without any changes in dates or quantities. For the other 30 percent, all quantities were filled with timing dislocations that were always less than 2 weeks. Fully 99 percent of all the acknowledged orders were shipped on time.

Wholesalers could request changes in the schedule after their initial orders had been placed. This meant calling a production coordinator who could change the production plan or arrange swaps between wholesalers (since the brewery was aware of all wholesaler inventory positions). Major modifications of the production plan rarely had to be made. Minor adjustments were made about twice a week, but these occurred normally for production problems such as a machine breakdown. The production plan was declared fixed as to labor content one week in advance of production. It was declared fixed as to packaging variations only one day in advance of production. About one shift's production was kept as a safety stock so that snafus in demand, production, or logistics would not necessarily affect delivery.

The production plan had another role somewhat different from the weekly scheduling of machines to match orders exactly. This role had to do with anticipating the seasonality of beer drinking. There was a definite seasonal trend to beer sales. Demand peaked from May through September, with the height of the peak accounting for sales roughly double those of the winter trough. Because Stroh's pasteurized beer had no living organisms in it (milk has 500 per milliliter), the beer did not spoil. (The draft beer was specially filtered so that it, too, had no living organisms in it.) The brewery could therefore cushion itself somewhat for the peak season demand by building up inventories, both in its own warehouse and in the warehouses of its wholesalers. In this way, the brewery did not have to hire or lay off as many workers during the period when demand was changing abruptly. The brewery and its wholesalers built up some inventories during the late spring and ran them down as summer progressed.

Because beer tasted best when it was fresh, it was company policy to hold no packaged beer at the brewery that was older than 21 days. The policy for wholesalers was not to hold any beer longer than 60 days. These policies helped to ensure that consumers enjoyed a fresh product.

The precise production schedule for the various lines depended in large measure on the time it took to accomplish various changeovers with the equipment. There were three key types of changeovers:

1. *Beer changeovers.* When a different beer was to be packaged, the pipes leading to the

line from the government cellars had to be cleaned out by a "blow back" of beer to the cellars, using carbon dioxide. Only then could the new beer travel the pipes to the filling machine. For most beer switches, this changeover could be accomplished in 15 minutes, quicker than that for lines close to the government cellars and longer than that for lines far from the cellars. Because of the significantly different composition of light beers, a changeover to or from light beer involved, in addition to a "blow back" with carbon dioxide, flushing the pipes and the filling machine bowl with water. Thus, when light beer was run, the changeover entailed 20 or 25 minutes to prepare the line for the beer and a similar amount of time to clean up the line once it was run.

2. *Packaging changeovers.* On the canning lines, the chief packaging changeovers involved either a change of can lids or a change of case carton. The "ecology lid," where the tab was pushed down into the can and remained attached, was used exclusively, but these lids could carry special tax-paid statements for various states, and thus can-lid changes had to be made. There were three major carton possibilities: (1) a standard case of 24 made up of four six-packs, (2) a fully-enclosed box of 24 loose cans, and (3) a fully enclosed box of 12 loose cans. Other carton possibilities were possible, however. A change of lids took 5 to 10 minutes on average, while a change of case cartons took about 5 minutes.

The only packaging change for the bottling lines involved a switch of cartons for the 7-ounce bottles, a changeover that averaged 10 minutes.

3. *Container changeovers.* The most complex and time-consuming changeover involved a switch of container sizes—for example, from 14-ounce to 16-ounce cans or from 7-ounce bottles to quarts. (Four can sizes were possible—12, 14, 16 and 24 ounces—and four bottle sizes—7, 12, 32, and 40 ounces.) Such changes involved resetting the filling equipment, the lid placement or bottle crowning equipment, the labeling equipment, and the cartoning equipment. Typically, an entire shift was devoted to these changeovers.

On a typical day, for example, a 12-ounce can line would make 8 to 10 lid or beer changeovers, a 16-ounce can line would make six such changeovers, and a quart bottling line would make three such changeovers.

The extent of container changeovers depended on the season. There were two basic types of can changeovers and two basic types of bottle changeovers. In the summer, the cycles shown in Table E1 prevailed for these can and bottle lines.

Given changeovers of various sorts and mechanical and other problems, can and bottle lines could not operate all the time. For the most part, the line efficiencies ran about 80 and 85 percent. Changeovers accounted for between 5 and 8 percent of the difference between these efficiencies and 100 percent.

Purchasing

The purchasing function for the plant was housed in the same department that handled order taking and production planning. This arrangement made sense since the production plan implied, in a straightforward way, the material needs for the entire process.

All of the brewery's major materials inputs (malt, adjuncts, cans, bottles, cardboard) were purchased on long-term contracts from major suppliers. These contracts were all negotiated by Stroh's headquarters staff in Detroit. The Winston-Salem plant negotiated on its own only those contracts for materials and services

TABLE E1   Container changeovers for can and bottle lines during summer (Stroh Brewing Company)

LINE	SHIFTS PER CYCLE
Can Line	
Type A (12-ounce cans)	Line devoted to 12-ounce cans: no changes
Type B	
16-ounce cans	15 shifts
14-ounce cans	6 shifts*
Bottle Lines	
Type A	
12-ounce returnables	12 shifts
12-ounce nonreturnables	3 shifts*
Type B	
Quarts	20 shifts
7-ounce bottles	5 shifts
Stroh taper bottles (12 ounces)	12 shifts
32 ounce or 40 oz. bottles	6 shifts*

*Cycle then repeats.

that went into the maintenance of the brewery itself and were not an integral part of the final product. Still, there were plant-specific negotiations for 2500 contracts each year. Also, even though Detroit headquarters may have negotiated for a specific material, there was often a direct supplier-brewery relationship. This was particularly true for cans and bottles. Cans were supplied by Stroh's own can plant located near the brewery, and bottles were provided by three outside sources: Anchor, Foster-Forbes, and Owens-Illinois.

The brewery placed orders for these major materials in much the same way that Stroh's wholesalers placed orders with the brewery. For example, the July order for cans or bottles would be placed on June 10. At that time, the brewery would place a firm 4-week order and an estimate of the succeeding 4 weeks' demand.

The inventory levels for any one of these major materials depended crucially on the dependability of the supplier. If the supplier had proven reliable and no special circumstances intervened, the brewery would want to hold only enough of the material to last until the next delivery. As it was, special circumstances were continually intervening, and so the brewery often adjusted the amount it ordered to take advantage of volume discounts in price or seasonal differences in sales or expectations about a strike in the supplier's plant. Some typical inventory levels for major materials were:

Bottles	1–2 days of production
Malt	5–7 days
Cans	1 week
Cardboard cases	1 week
Bottle labels	1 month

THE WORKFORCE

The brewery currently employed about 600 people, producing on three shifts a day, 5 days a week. Of these, about 100 were salaried (managers and staff) and 500 were paid hourly. The hourly workers were represented by a union, the International Brotherhood of Teamsters.

Most of the workforce had been hired from within a 50-mile radius. Half of the salaried workers were drawn from other Stroh breweries, but none of the hourly workers came from other Stroh breweries, mainly because they had to be volunteers and had to pay for their own move. During plant startup a strong effort was made to hire persons so that there was a representative mix of ages, races, and sexes.

The hourly workers were divided into two main groups: the production workers who brewed and packaged the beer, and the engineering and maintenance workers who cared for and fixed the plant's equipment. The maintenance workers were paid about 5 percent more than the production workers, since their jobs required greater expertise and wider skills.

Once in a department (brewing, packaging, engineering/maintenance), workers tended to stay, often doing the same job. This status quo was frequently by choice, since workers who complained of boredom were generally shifted to more complex jobs or made relief workers (spelling other workers who were taking breaks). Advancement through the ranks and into management was possible, especially in brewing and packaging, although the odds were short. Most movement among the hourly employees was between shifts; such moves were based on seniority.

CONTROL AND EVALUATION OF THE OPERATION

Quality Assurance

Quality assurance (QA) at the brewery was an important activity, employing 33 people in three separate labs: microbiology, brewing, and packaging. The staff at those labs performed 1100 separate tests, many of them repetitions of the same test, on each batch of beer brewed at the plant. QA staff worked round the clock, seven days a week. Quality assurance had the authority to stop operations at any point in the process. The QA manager was responsible for dealing with and rectifying any customer complaints.

Here are two examples of the specifications that the QA staff tested for, one set by the corporation's brewing staff and the other set by the corporation's packaging staff:

1. The brewing staff had set a standard of 16 million yeast cells (plus or minus 2 million) per milliliter of beer during fermentation. Quality assurance tested for this standard twice for every fermentation tank's batch. The test had to be completed within the first four hours of fermentation so that any necessary corrective measures could be taken. The test was accomplished by taking a 4-ounce sample from the tank, diluting it in three steps, and then counting yeast cells by microscope.

2. The packaging staff had established a standard that the air content in a 12-ounce can of beer be no more than 0.66 cubic centimeter, because oxygen caused flavor instability in the beer. The QA staff tested for this standard by checking each canning line five times per shift, taking three cans from each line each time. If the cans were off specification, the QA staff would quarantine all the line's beer up to the last good check and then systematically inspect the quarantined lot until the beginning of the off-specification beer was encountered. The beer that passed inspection was released, and the beer that failed inspection was discarded. As it turned out, about 98 percent of any quarantined beer was eventually released as good product.

Rarely did any of the batches of beer fail to satisfy standards, but the quality assurance department prided itself on its vigilance. It viewed itself as the early warning station for detecting any encroaching degradations of the process's integrity.

Information Flows Within the Process

Most of the information flows within the process were directed from the top down. If the process was working smoothly, workers needed to be informed of only routine things, such as the changeover in a line from one beer to another or the shifting from one line to another. Most jobs did not vary that much anyway, and so the information that was transmitted could be sparse without any detrimental effect.

Top management was eager, however, for the thoughts of the workers on all matters of plant operations. The plant had begun a total quality management (TQM) program that was involving the workforce in process improvements. Of particular concern were breakdowns in the process. After worker signals of process breakdown, the engineering and maintenance force would be called in. These troubleshooting jobs were generally regarded as the most complex at the brewery and the most pressure-filled. A good engineering/maintenance department was a real asset and a chief way to keep costs low by maintaining high speed and high yields. The TQM program was also instituting supplier certifications to improve quality even more.

Evaluating Plant Performance

Each week management at the brewery developed an operations summary that listed goals and actual results for an entire list of performance measures such as productivity (barrels per worker), cost per barrel, packaging line efficiencies, material losses, beer losses, beer rejected by QA, deliveries made on time, shipment errors, wholesaler complaints, and worker absenteeism. Each week, too, the plant's beer was evaluated by taste tests in Detroit and elsewhere. More than any other test, this was the one the plant always wanted to pass.

In many respects, the plant's performance could be fine-tuned only from week to week. The major elements of plant performance were either already decided (plant design) or beyond the scope of plant management (the plant's sales volume). As the plant manager admitted: "The heaviest decision in our industry is capital equipment investment." Major cost and quality advances were very much a function of the equipment the company's design people in Detroit decided to incorporate into the plant.

Over the past 20 to 25 years, breweries had also been built larger than before. There were real cost savings implied by larger brewery size: vats and tanks could be larger and still make quality beer at lower cost per barrel, and packaging line speed and capacity could be used fully. The size for the Winston-Salem brewery was determined as a sort of "lowest common denominator" for different "lumpy" capital investments. Specifically, the plant's lauter tub could process 3 million barrels of beer a year, and the mandatory assortment of packaging lines (lines for kegs, cans, one-way bottles, and returnable bottles) could process 5.2 to 6 million barrels a year. With the addition of other canning lines and another bottling line, packaging capacity could then precisely match the output of two lauter tubs—hence a brewery capacity of 6 million barrels a year.

What the plant manager worried about depended a lot on volume. As he noted, "Volume solves everything." When volume was high, all the manager had to worry about was whether the workers would accept the necessary overtime

A worker inspecting cans of Schlitz along the Stroh's Winston-Salem canning line.
(*Courtesy of Stroh Brewery Company*)

and whether maintenance could hold everything together. At capacity, quality and meeting shipments were chief concerns, with product cost lower in priority. When sales volume dropped, however, meeting cost targets became relatively more important, and that meant tightening up on staff and laying off workers. The plant was operated as a cost center, since it had no responsibilities for revenue raising (e.g., no marketing at the plant) and its geographic market area was fixed. Plant management was concerned chiefly for the maintenance of the plant and the motivation of the workforce; with them came good-quality beer, high yields, and on-time shipments.

PART TWO

DISCUSSION

The Stroh Brewery Company is in no way responsible for the following views and presentation. They remain solely the responsibility of the author.

THE PROCESS FLOW

Much of the Stroh brewery's operation is reminiscent of the continuous flow process at International Paper's Androscoggin Mill. Even though there are about 420 beer and packaging variations, most of these modifications require setup times that are tiny in comparison with the run length. The process flow is well defined, and every product goes through the same steps in the process. (See Figure E1 for a process flow chart.) Over time, equipment advances have speeded up the process and have driven labor out of the product's value.

The job contents of the direct labor in the process are of two kinds, polar extremes. Most workers have well-delineated tasks, most of which are repeated time and again as the product is manufactured. These jobs, typically, are not very demanding. At the other extreme lies the brewmaster. Despite great strides in automation and control, the brewmaster's duties still lie within the realm of art. Brewing is still imperfectly known enough to require great experience to produce a quality product.

Requiring nearly comparable levels of skill and experience are the plant engineers. Plant engineering and maintenance are crucial to an operation such as a brewery, where the capital investment is enormous and the success of the process depends on meeting quality standards and delivery schedules and on maintaining high yields of output. All of these goals are influenced importantly by the design and upkeep of the equipment in the plant. Moreover, all of the equipment must be functioning smoothly for any of the product to meet its specifications. Breakdowns or below-specification performance of any of the equipment is likely to lead to poor quality, poor shipment performance, or excessively high costs (such as low packaging speeds).

Despite these important similarities to the continuous flow process at the Androscoggin Mill, the Stroh brewery is not strictly a continuous flow operation. This is true primarily because the brewing phase of the process is done on a batch basis. True, these batches are large, but it is only because of their sequencing and the accumulation of a work-in-process inventory just before packaging that the batch operation in brewing can supply a steady flow of beer to the continuous flow that is packaging. The work-in-process inventories in the ruh and government cellars separate the two different types of processes. Without such an inventory, even though the daily or weekly capacities of the two processes (brewing and packaging) may be matched perfectly, the lumpiness of the batch flow operation's output cycle would cause severe problems for the smooth-running continuous flow operation.

THE INFORMATION FLOW

The inventory between brewing and packaging does more than even out production. It also separates the information needs of the two processes. The brewmaster need only know how much beer of which kind to brew, and the packaging department need only to know what, when, and how much it is to package.

Moreover, the timing of the information flows is much different. The precise packaging department schedule is set as final only one day prior to packaging. Up to that time, adjustments to the schedule can be accepted. Needless to say, the brewmaster cannot react so quickly. Since it takes about a month to brew a batch of beer, the brewmaster can only work to an estimate of demand, not from firm wholesale orders. The work-in-process inventory between brewing and packaging, then, acts as a safety stock of beer as well, so that last-minute changes in demand can be accommodated. It also serves as the dividing line between production triggered by estimate (brewing) and production triggered by firm orders (packaging). This difference in information requirements, as much as anything else, sets aside the Stroh brewery as a hybrid (batch/continuous flow) process.

The separation of information needs and flows at the brewery is depicted in an information flow diagram (Figure E4). Note that the information basically flows in only one direction (top down) except for the usual signaling of equipment breakdowns and acknowledgments of orders shipped. In this way, the information flows are reminiscent of the Androscoggin Mill.

CAPACITY MEASURES

The capacity of the brewery is a firm and easily understood number, 6 million barrels of beer a year. Barring a major breakdown of equipment, the capacity figure is primarily what the existing plant and equipment permit. Neither product mix nor scheduling has much impact on capacity. Other than to maintain the equipment and keep quality up, the plant can pull only so many levers to increase the quantity of good product brewed and packaged.

By the same token, capacity utilization is a well-defined concept. The capacity of any piece of equipment is known because it was engineered that way, and the current capacity is also easily measured. Moreover, because the plant's designers want to leave as little waste or spare capacity unused as possible, the capacity utilization for any single piece of equipment is often very close to the capacity utilization figure for the process itself. Since the process's capacity utilization is determined by the bottleneck operation, one can say that all equipment capacities are likely to lie close to the bottleneck capacity.

STANDARDS AND INCENTIVES

The standards that matter at the Stroh brewery are not the type of standards that managers of

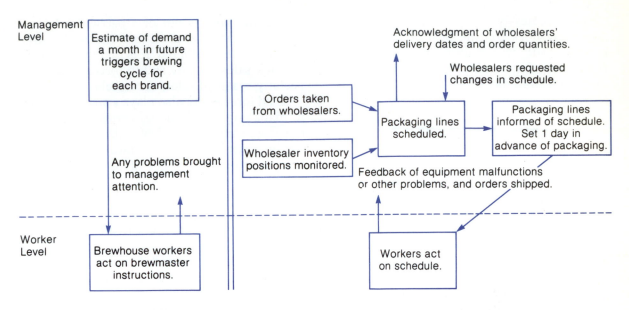

FIGURE E4   An information flow diagram of the Winston-Salem plant of the Stroh Brewery Company.

batch or job shop operations get concerned about. Labor standards (output per worker per unit of time) do not exist at the brewery. Rather, it is machine standards (units processed per hour) and quality standards that capture managers' attention and influence their behavior. These are the standards management strives to meet.

Similarly, labor incentives are absent from the brewery. The plant's output is so overwhelmingly related to machine performance rather than worker performance that worker incentive schemes make little sense.

DEMANDS OF THE PROCESS ON THE WORKERS

The process at this Stroh brewery places a variety of demands on a variety of different work groups within the brewery. Art, science, and routine are all found. For the brewhouse work-

ers, even with the panels of controls that now abound, the brewing process still requires art, a "feel" and a "taste" for what makes quality beer. In many breweries, it still is true that the brewmaster's skills have been handed down from generation to generation.

While science may not have overtaken all aspects of the brewing art, it rules the quality assurance operations of the brewery itself (although the Detroit taste test is still the single most important check on the brewery's quality performance). Science is evident everywhere in quality assurance, from the chemistry involved to the statistical sampling. The training for the quality assurance department can thus be long and rigorous.

Science and art are mixed in the plant's maintenance activities, which are an important and prestigious activity. The capital intensity of the process dictates heavy reliance on maintenance and the ability to troubleshoot problems quickly so as not to waste time and output needlessly. Only with the line jobs in packaging does routine best characterize the process. Still, switching from canning to bottling to kegging provides a measure of diversity in the job from day to day.

DEMANDS OF THE PROCESS ON MANAGEMENT

Stroh's Winston-Salem brewery and International Paper's Androscoggin Mill place many of the same kinds of demands on management. The high degree of capital intensity they both share dictate concern for (1) selection of the proper technology and (2) the balancing of capacities in all segments of the process. Over time, technological advance has dictated numerous advances in the process itself (centrifuges, stainless steel tanks, panels of process controls) and in the scale of the process (larger tanks than ever before). The design, choice, and matching of equipment are critical decisions for a brewer like Stroh.

Scheduling production, while a nontrivial matter, is not as critical to a brewery's success as it is to a job shop. As important, or even more so, as the positioning of orders within any work shift or within any week's production is the accurate forecasting of demand. A Stroh

wholesaler who is out of a particular beer or package risks that the consumer will merely choose a different brand. It is up to the brewery to track sales reliably and to forecast sales, especially seasonal sales, accurately.

Good forecasting also plays a significant role in managing suppliers' deliveries and inventories (e.g., cans and bottles) well. Since individual breweries do not negotiate price, delivery becomes the key aspect of brewery–supplier contact.

HYBRID PROCESSES

Hybrid processes, mixes of two or more of the purer process types (job shop, batch flow, line flow, continuous flow), are common occurrences. Numerous plants are organized with line or continuous flow processes on their front ends, say, to assemble a product or to package it. On their back ends, these plants often have components fabricated or materials processed by job shop or batch flow processes. Much of food processing, drug manufacture, many complex consumer items (such as, appliances and furniture), and other products combine production in lots with assembly/packaging in lines — the kind of hybrid process represented by the Stroh brewery.

QUESTIONS

1. What similarities are there between the Stroh brewery operation and the Androscoggin Mill operation? What dissimilarities?

2. What did Stroh consider to be the elements of the satisfactory production plan and the requirements for good package planning? What particular elements of Stroh's hybrid process planning might be applicable to the hybrid process in general? Why?

3. Describe how quality control may differ between the two main processes involved in a brewery.

4. Compare and contrast the demands of the process on Stroh workers and management with the demands on the same groups in two other processes you have studied.

5. It was stated that the brewery had a capacity of 6 million barrels of beer per year. If the plant

were to can, bottle, and keg 6 million barrels in a year, what would be the capacity utilization of the packaging equipment, given the speeds mentioned in the case? Feel free to make assumptions about the relative demands for cans, bottles, and kegs.

6. How could the capacity of the brewing process itself be increased without a physical expansion of the brewery? How attractive is such a change?

7. Given the chapter's information on the relative demands of different brands and the summer cycle of container changeovers, concoct a month's production schedule for each major Stroh company brand. In an average week, what fraction of time would be devoted to changeovers of various types?

SITUATION FOR STUDY E-1

COUNTRY GELATIN COMPANY

Thomas Brewer is the plant manager for Country Gelatin Company, maker of powdered gelatins for human consumption. The main component of gelatin is collagen, a fibrous substance that is mostly protein. Steerhide, a major source of collagen, is the main ingredient in the gelatin powder produced at Country's Kansas City, Kansas, plant.

Figure E5 shows the process that Country uses to produce its gelatin. First, the steerhides are treated (acid is added to remove the hair, then neutralized), weighed, and put into storage. Enough steerhide is treated at one time to provide 2 days' worth of production. The hides are then stored for about 2 weeks so that they begin to "break down" and can be blended with other ingredients. This storage is called "aging."

Once several batches of steerhides (1500 kilograms each) have aged properly, enough batches are ground (to about the consistency of hamburger) to run 1 week's production. These mixed batches (5500 kilograms each) are then aged again for a few days before they are mixed into a gel blend (8000 kilograms, of which about half is water). One gel blend is enough for 1 day's production of gelatin.

After being mixed and aged, the mixed batch undergoes microcutting, a process that cuts the material into very small pieces. It then is mixed with water and an acid to form the gel blend. Next the gel blend is forced through a dryer to make it a powder. Under proper conditions, the powder can be stored almost indefinitely, until it is needed for final production.

Final production consists of two stages. First, the powder is mixed with sugar and a powdered flavoring (unless the final product is to be an unflavored gelatin); second, the product is measured and packaged. Final production, although continuous, is done to customer order. All stages in the process are highly mechanized, and workers are needed only to monitor the machines.

Tom feels that quality control at Country can be adequately handled by three types of highly trained process control workers: solutions technicians, powder technicians, and line technicians. The solutions technician is responsible for the proper acid balance, salts, density, percent of solids, and the like of the gel blend. The powder technician's main responsibility is proper mixing of the final product—

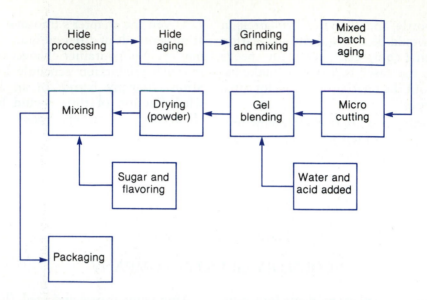

FIGURE E5   The production process at Country Gelatin Company.

that is, the proper proportion of powder, sugars, and flavoring. The line technician is responsible for packaging.

Tom is wondering how he should schedule production at the various stages in the process, including the scheduling of labor and machines. In addition, he is uncertain how much inventory should be carried, and in what form.

He feels he could begin to answer his concerns about scheduling and inventory control by (1) analyzing the characteristics of the process, and (2) asking himself what aspects of the process would help him in scheduling production and controlling inventories and which ones would hinder him.

SITUATION FOR STUDY E-2

BUCHANAN TOBACCO COMPANY

Buchanan Tobacco employed 2000 people in a plant that stretched 15 acres. Ten different cigarette brands were made at the site, each with its own blend of bright, burley, and Turkish tobaccos and various other flavorings.

After the tobacco (the whole dried leaf) had been purchased by the company's buyers at various tobacco markets, it was brought to the stemmeries. Here it was cleaned with moist air and dried, and the stems were removed. The tobacco was then cut into small pieces and stored for 2 to 3 years in hogsheads before being brought to the plant for blending. The hogsheads were loaded onto flatbed tractor-trailers, and the tobacco arrived at the west end of the building in the containers in which it was aged.

In the blending area, the tobaccos were blended and flavorants were added in order to

produce the correct blend for each brand of cigarettes. At each stage in the blending process,
the tobacco was exposed to certain relative hu-
midities and temperatures. At the completion
of each stage, the blended tobacco was aged
through drying. This aging allowed the tobacco
to absorb moisture and flavorants.

After being aged, the tobacco went through
the "casing and cutting" process, where certain
liquids (casings) were added and the tobacco
was shredded and blended for the particular
brands being made. A pneumatic vacuum system transported the blended tobaccos to one of
the 100 making machines (makers) designated
for its brand. Each maker required one operator.

The sealed vacuum tubes ran above the machines, and so the tobacco fell to the bottom of
the makers by gravity. The tobacco entered at
the rate of 10 pounds per minute. The action of
turning drums pulled the tobacco up through
the machine. This action also vacuumed out
dust and particles such as stems. As the tobacco came out of this section of the maker, it
entered a small tube at such a constant rate
that one continuous rod of tobacco was
formed; this rod stretched almost the length of
the machine.

At this point, the paper was wrapped around
the tobacco. The paper, which was produced by
a subsidiary, came in rolls over 6 miles long.
Each roll contained enough paper for 95,000
cigarettes. The paper was threaded through the
machine, and the cigarette brand name was
printed on it. As the paper was folded around
the continuous rod of tobacco, a wheel applied
glue to one edge. The paper was then sealed
around the rod of tobacco.

After sealing, the continuous cigarette was
cut by a revolving knife, which cut 4100 cigarettes per minute and was timed to cut the exact length required.

The cut cigarettes next went to a drum. Vacuum pressure held the cigarettes on the drum

as they slid into pockets. Simultaneously, the
filters for the cigarettes, which were made in
another area of the building and were precut to
the appropriate size, were blown over on a
tube. The cigarettes in the pockets were directly opposite each other, and a filter simply
fell between them so that a cigarette, double-
length filter, and another cigarette were all
sealed together. At this point, the double cigarette passed over a sealer, and the filter was
sealed to the paper on both ends. At the same
time, corking paper, which was fed from a roll
into the maker, was cut to the proper length
and wrapped around the filter by the action of
the drums. The "tipping" picked up enough
glue to cause it to adhere to the filter. The double cigarette was then moved to a knife and cut
into two 80-millimeter filtered cigarettes.

There were various automatic quality checks
during the making process. Any cigarette with
a defect would be automatically rejected. Common defects found at this point included air
leaks and pin holes. All defective cigarettes
were broken down and completely recycled
through the process.

Immediately after cutting, the cigarettes that
passed inspection were moved on a belt, with
cigarettes on the back part of the drum being
turned so that all were now facing the same
direction. This belt carried the cigarettes to a
special reservoir that created a buffer between
the making and packing processes.

This reservoir loaded cigarettes onto a conveyor for up to 17 minutes. If the maker went
down, the packer could continue running for up
to 17 minutes by using the reserve cigarettes
from overhead. If the packer went down, the
process simply reversed and cigarettes fed onto
the conveyor directly from the maker. If either
machine was inoperable for more than 17 minutes, both machines had to be shut down.

From this point, the cigarettes were fed by
gravity into the packing machine or packer. As

the cigarettes were fed into the machine, they automatically fell into three rows of seven, six, and seven cigarettes. A plunger pushed the 20 cigarettes into a foil pocket, which was formed as a series of turning drums took the foil that was fed off a roll and wrapped it around a form identical in shape and size to a pack of cigarettes. This foil was the interior of a pack of cigarettes.

Next, the label was taken from the front of the machine and formed into a pack and sealer was applied. Another plunger pushed the foiled pack into the label. A stamper picked up glue and applied a closure (stamp) to the pack. The pack slid on chains to the film machine.

The film fed off of a roll and was cut to the desired length to wrap a pack. Just before the pack was pushed into the film, the tear tape was laid across the film and sealed to it. Mechanical folders folded the film around the pack and electric heaters caused it to seal.

The packer packaged 210 packs of cigarettes a minute (4200 cigarettes); it was operating at a slightly faster rate than the maker. There were approximately 110 packers in the plant, each one operated by one worker.

The packaged cigarettes were then fed on conveyor belts past an inspector. The inspector looked for missing closures, missing tape, or any other defective packaging. Packs were fed to the inspector from two packers; therefore, an inspector saw about 25,000 packs an hour.

After the visual inspection, the packs proceeded on belts to the carton boxer machine. There were approximately 50 carton boxers, each one receiving from two packers; thus, two belts came together at the boxer. The cartons were fed down by a series of rollers and automatically opened. The packs from the two belts (five packs from each belt) lined up. The pressure from the last pack activated a switch, which in turn activated the plunger, pushing ten packs into a carton. The flaps of the carton were turned down by guides. When the carton

emerged at the back of the boxer, it was sealed. At this point, an electronic inspector (operating on the basis of the metal in the foil) was used to detect missing tape or packs. The carton boxer's speed was determined by the speed of the packers, but generally it boxed 200 packs, or 20 cartons, per minute.

Next, the cartons proceeded up the elevator to the carton conveyor and down to the case packer. Usually at least five packers sent cartons to a single case packer; however, a single case packer could receive from as many as 20 packers. There were five case packers at the plant.

Flattened cases were stacked and fed down through one end of the machine, where guides opened them automatically. The cases had been preprinted with the brand name, but print was now applied to the cases to specify the date and the name of the plant. Cartons came down overhead sides from the carton collector and were conveyed into the back of the machine. The cartons were stacked in layers. As soon as 10 rows of six cartons each had accumulated, a switch was activated to plunge the 60 cartons into one case. Thus each case contained 12,000 cigarettes. The filled case was then moved forward. Glue was applied from rollers to the end flaps. When the machine cycled again, these flaps were sealed down with pressure. The case stayed in this compartment until the glue was dried and sealed, which was less than a minute. Each caser had the capacity of packing approximately five cases per minute.

The sealed cases crossed an automatic scale before going up an elevator. This scale inspected for light cases; being underweight indicated that a pack or carton was missing. A heavy case was not rejected, because extra weight merely indicated the presence of collected moisture.

After reaching the top of the case elevator, the case traveled on a conveyor to the transfer room. Here all brands were assembled on pallets by brand. Each pallet had an identification number, which the computer married to the brand's code number to permit tracking these cases to the warehouse. After the cases had been palletized, elevators lifted the pallets to the conveyor to the shipping center. All of the shipping was done from the new, completely computerized central distribution center. The cigarettes were produced strictly to order. Everything produced during the day was shipped out at night so that there was virtually no work-in-process or finished goods inventory.

1. Diagram the process flow.
2. Speculate, as best you can, on the flows of information in the process.
3. Why might this operation be classed as a hybrid? What type of hybrid is it and how does it compare with the Stroh brewery?
4. What do you see as key challenges for management in this process?
5. What kinds of changes to the process would occur as new makers are purchased with speeds twice as great or more than those already in the plant?

A WORKER-PACED LINE FLOW PROCESS AND A SERVICE FACTORY
Burger King Restaurant
Route 37
Noblesville, Indiana

The Burger King Restaurant on Route 37 in Noblesville, Indiana, was one of over 6000 fast-food restaurants operated worldwide by Burger King Corporation, a wholly owned subsidiary of Grand Metropolitan, and by the corporation's franchisees. Service was available from 6 A.M. to 2 A.M. every day. The dining room was open from 6 A.M. until 11 P.M., Sunday through Thursday, and from 6 A.M. until midnight on Fridays and Saturdays. Breakfast was available from the opening until 10:30 A.M. and consisted of a choice of bacon, ham, and sausage bagels, biscuits, and croissan'wiches, with or without eggs or cheese, French toast sticks, mini-muffins, breakfast buddies, biscuits with gravy, hash browns, and a scrambled egg platter. The regular menu consisted of a wide selection of hamburger sandwiches (regular hamburger and cheeseburger; a special Whopper sandwich with or without cheese; hamburger deluxe; double burger, double cheeseburger, and double Whopper; and a bacon double cheeseburger) as well as some specialty sandwiches (chicken, ham and cheese, fish), BK broilers (broiled chicken sandwiches), chicken tenders, and prepacked salads. To accompany these selections, the restaurant offered various choices: french fries, onion rings, soft drinks, shakes, Breyer's frozen yogurts, Snickers ice cream bars, Kool Aid cool pops, apple pie, cherry pie, and lemon pie. Occasionally, there were promotional items such as Burger Buddies.

The restaurant, the highest sales volume Burger King in Indiana, was located on Route 37, a busy thoroughfare between Marion and Indianapolis. The restaurant was opened in May 1987 on the site of a former farm. The restaurant was owned and operated by a franchisee, Douglas Brucker of Marion.

The restaurant was rectangular, freestanding on 2 acres, and was constructed mainly of wood planks, bricks, and glass. The building was situated toward the front of the lot, freeing the sides and rear for 64 parking spaces. A "drive-thru" was operated on one side of the restaurant, with the order board at the rear and the pick-up window toward the front. Within the restaurant 96 seats (either benches or free-standing chairs) were arranged in groups. There were 16 tables of four, 1 table of three, 8 tables of two, and 13 single places within the restaurant. The eating area was tastefully appointed in oak wood, with a handsome color scheme, an atrium off to the side of the main eating area, ceiling fans, and many hanging plants. In the eating areas there were service counters for napkins, straws, and so on, and several large trash cans for customer use. The rest rooms were located next to the main door. The eating area occupied about half of the restaurant's square footage; the other half was occupied by the counter, the kitchen, and the storage area.

PART ONE

PROCESS DESCRIPTION

RESTAURANT OPERATIONS

Order Taking (The Front Counter)

After entering the restaurant, customers walked between the railings of the queuing line (or, as Burger King calls it, the "customer guidance system"). While awaiting their turns to state their order at the counter, they could consult the brightly lit menu above the counter area. At the right-hand side of the counter were two cash registers with screens (point-of-sale devices). An order-taker greeted each customer there and took the order. As the customer spoke, the order-taker keyed the contents of the order on a register (e.g., pressed the key for the Whopper—a large hamburger with lettuce, tomato, onions, pickles, ketchup, and mayonnaise), read the printed total for the order, took the customer's payment, and gave the customer change and a receipt that listed the order. Each register could accommodate one or two cashiers simultaneously and could handle up to three transactions at the same time.

At the other end of the counter, a printer printed a duplicate copy of the order. Using this copy, an expediter proceeded to assemble the order. Assembling the order meant going to the counter and the chutes that lay between the front counter and the kitchen to gather the sandwiches, fries, or other items that flowed from the kitchen. When shakes were ordered, the expediter was responsible for drawing them from the nearby dispensing machine. After assembling the kitchen items of the order, the expediter presented them to the customer either on a tray (for inside use) or in a bag (to go). The expediter also gave cups to the customer so that he or she could fill them with the desired soft drinks, tea, or coffee.

Several features of the front counter's operation were comparatively new and merit highlighting. The most important innovation was the split made in the duties of the order-takers and expediters. The system in use at the Noblesville Burger King was known as the "multiconventional" line-up. It was a throwback to the early years of Burger King when the "conventional" line-up was used. That system consisted of a single line of customers served by a single cash register where orders were taken. Expediters assembled the orders and presented them to the customers farther along the counter. The present system operated in much the same way, although up to six orders could be taken at the same time, and several expedit-

The front counter. (*Courtesy of Burger King Corporation*)

The dining area. (*Courtesy of Burger King Corporation*)

ers could handle those orders. The new, multi-conventional system was thus far faster than the old system.

The multiconventional line-up was a more radical change from the "hospitality line-up" that had replaced the old conventional system. With the hospitality line-up, cash registers were evenly spaced across the entire front counter and customers had to choose which of several lines they wished to wait in. Cashiers both took orders and assembled them. The hospitality line-up, although somewhat more labor intensive than the old conventional line-up, could handle peak hour demand more efficiently than the older system. With the advent of the multiconventional line-up, however, the labor intensity of the hospitality line-up was reduced without sacrificing its peak-period capacity advantages. Customers preferred the new line-up as well. With a single line, they did not risk becoming annoyed because some other line was moving faster than theirs. In addition, they were frequently better prepared to give their order to the order-taker, and this speeded up the process for everyone.

Another innovation involved the self-service drink dispensers. With their installation, customers could fix their drinks precisely the way they wanted (whether with lots of ice or no ice). By moving the drink dispensers out of the kitchen, time and some labor costs were also saved. Because free refills were possible, materials costs were somewhat higher, but not appreciably so.

Order Filling (The Kitchen)

Burger King differed from McDonald's and some other fast-food restaurants in that comparatively little finished goods inventory was kept; sandwiches were assembled continuously. While certain orders might not have been delivered as quickly as when larger inventories were kept, this approach offered the distinct advantage of producing to order when appropriate. As an old Burger King slogan put it, you can "have it your way," say, by ordering a Whopper with double cheese and mustard but no pickle, or a hamburger with extra onions.

Providing this kind of customer order variation with minimum customer waiting demanded a production system that was extraordinarily flexible. In fact, two kinds of flexibility were required: flexibility to meet special customer orders and flexibility to meet large surges in customer demand during lunch or dinner hours. Many aspects of the production system contributed to this flexibility.

THE "LINE": LAYOUT AND JOB DESCRIPTIONS

The process of making sandwiches and filling orders at Burger King was explicitly recognized as an assembly line. Production of the hamburger sandwiches (burgers, for short) followed a straight path from the back of the kitchen to the front counter. Along this path were a series of workstations (see Figure F1).

Any of the various burgers was begun either by taking a broiled meat patty and toasted bun out of an environmentally controlled holding compartment called a "steamer" or by placing a frozen meat patty (Whopper size or regular) and bun onto the chain drag at the feed end of a specially constructed, gas-fired broiler at the back of the kitchen. The meat patties were drawn from a freezer below the broiler. The broiler cooked the meat at approximately 800° F, allowing the grease to drip into a special compartment, and it also toasted the bun.

Next in the assembly line came the "board," where buns and meat were transformed into Whopper sandwiches, burgers, hamburger de-

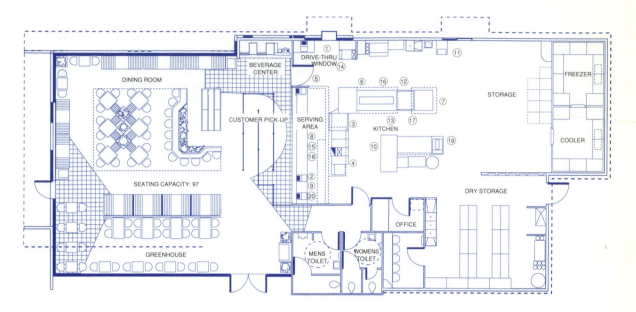

FIGURE F1   Layout of the Noblesville Burger King. The circled numbers indicate the sequence of additions of workers to the kitchen as demand increases.

luxe sandwiches, double cheeseburgers, and the like. This was the key portion of the line, where the burger could be assembled "your way." The board itself was a long, stainless-steel table in the center of which were bins of condiments (refrigerated cheese slices, bacon, pickles, onion, sliced tomatoes, shredded lettuce, mayonnaise) kept at room temperature. Below the table were racks for holding spare quantities of condiments and supplies and also places for waste disposal. There were two work areas, one on each side of the center inventory of condiments. Above each side were two microwave ovens that could be used for keeping assembled sandwiches hot, stacks of various wrappers into which the sandwiches were placed, and a special series of touch controls that were part of the information flow system of the kitchen (more on this later). Beyond the board, on the pick-up counter, were chutes that

held completed sandwiches ready for assembly into customer orders.

On one side of this main burger assembly line were the frying vats and the specialty sandwich board. The four frying vats were computer controlled, two just for french fries and two for other products (such as onion rings, chicken sandwich portions, chicken tenders, or fish portions). Near the frying vats were racks of thawed or thawing french fries (1- to 2-hour thaw times maximum). Behind the frying vats was the specialty sandwich board, which had its own assortment of condiments, buns, and wrappers. To one side of the specialty sandwich board were two warmers (one for items such as cooked chicken patties and the other for chicken tenders) and to the other side was a bun toaster.

On the other side of the main burger assembly line were the automatic drink machines for the drive-thru operation. A worker simply placed an ice-filled cup under the appropriate drink spout and pressed a button for small, regular, or large drinks; the machine then dispensed the proper quantity of soft drink, freeing the worker to do something else while the cup was being filled. The shake machine (for chocolate, vanilla, and strawberry) was located nearby, close to both the drive-thru window and the front counter.

Around the periphery of the kitchen were sinks and storage areas for food and supplies. There were also cooler and freezer rooms, an office, a training room, and a crew room where workers could congregate.

THE DRIVE-THRU OPERATION

Because the drive-thru at the Noblesville Burger King accounted for approximately 50 percent of the restaurant's business, its smooth operation was critical. Work assignments at the drive-thru depended on how busy the restaurant was. In very slow times, a single worker (order-taker/expediter) could handle the drive-thru alone. When business picked up, two people were assigned to the drive-thru — one as an order-taker and one as an expediter. At peak times, five workers were assigned to the drive-thru: order-taker, window pusher, expediter, drink assembler, and cashier. (The Noblesville Burger King had two windows in the drive-thru lane, the first to handle the money and the second to "push out" the order.)

PEAK VERSUS NONPEAK OPERATION

Employees were allocated to the workstations at the restaurant according to the pace of demand. When demand was slow, a smaller crew operated the restaurant: an order-taker/expe-diter at the front counter, a drive-thru order-taker/expediter, and two kitchen workers. One worker took the broiler and the burger board, while one worked at the fry station and the specialty board. At peak times, there were as many as 24 employees working in the restaurant plus two or three managers. Management knew from historical data exactly when the peak hours would be at the Noblesville restaurant and so could plan worker arrivals.

Before the daily peaks in demand at lunch and dinner, considerable preparations were made, such as stocking the freezer under the broiler with hamburger patties, taking frozen fries out for thawing, and slicing tomatoes. If things worked as planned, minimal restocking would be needed during the peaks.

As demand picked up, more workers were added, and job assignments became more specialized. In slow times, for example, one worker split duties between the fry station and the specialty sandwich board; during peak periods there were separate workers for these two work areas. During slow periods, one worker handled the broiler and all burger sandwich preparation; during peak times, one worker "fed" the broiler, one or two "caught" patties and buns and monitored the steamer, and up to four workers assembled sandwiches at the burger board. At the Noblesville Burger King, the burger board was divided during peak times into cheese and noncheese sides. (At other Burger Kings, this division might be Whopper-sized sandwiches on one side and small sandwiches on the other. Such a split frequently depended on the composition of demand at the particular restaurant.) During peak periods as well, two workers helped keep the dining area clean.

During very slow periods, much of the sandwich and drink production was triggered by the customer order itself. The burger given to the customer might actually be the one placed on the broiler chain in response to punching in the

From behind the front counter. (*Courtesy of Burger King Corporation*)

Slicing tomatoes in preparation for the noon peak period. (*Courtesy of Burger King Corporation*)

At the burger board. (*Courtesy of Burger King Corporation*)

The fry station. (*Courtesy of Burger King Corporation*)

The drive thru window. (*Courtesy of Burger King Corporation*)

order. Some items could be held in inventory. An inventory of the most popular sandwiches (such as regular burgers and Whoppers) were usually kept in the chutes, but only for 10 minutes; if no customer requested them, they were discarded. Each sandwich wrapper was marked with a "time to discard." The system used the numbers pointed to by the minute hand on the clock. Thus a Whopper wrapped at 20 minutes after the hour (the 4 on the clockface) would have a 6 marked on the wrapper, because at 30 minutes after the hour (the 6 on the clockface) its 10-minute hold time would have expired. French fries were handled in a special way. Fries took approximately 2 minutes and 30 seconds to cook—too long to have a customer wait. As a result, fries had to be waiting in finished goods inventory. Here, too, the restaurant kept track of the time fries were waiting. If no one claimed fries within 7 minutes, they were discarded. (The fry station computers helped by keeping track of the time between batches so that fries were not sitting out too long.)

As demand picked up, it became more and more likely that a finished goods inventory of the major burgers and specialty sandwiches would be kept and that the broiler feeder would be loading the broiler in response to the inventory situation rather than in response to the particular orders displayed on the video screens. To guide production and to maintain inventory levels as demand increased, the Noblesville Burger King operated "stock level charts." There were four stock level charts—one each for the broiler-steamer, sandwich board, specialty sandwich board, and specialty sandwich warmer. The chart featured a stock-level light indicator on which there were seven lights. Corresponding to each light was a volume of sales range to which were pegged inventory standards for the various items. Each change in the level (e.g., from level 4 to level 5 or from level 6 to peak) was announced by a

special bell and the light switched so that all workers could infer, for example, how many sandwiches of particular types should be in the finished goods chutes and how many patties and buns should be in the steamer. The workers at the board were constantly tracking how many sandwiches were in each chute as well as keeping an eye on the video screen for any special order variations entered by the cashiers. When a worker finished such an order, he or she pressed one of the special touch controls above the station (mentioned earlier), which removed the call for that sandwich from the video screen. The worker then marked the wrapper to identify it to the expediters.

For the manager, the choice of a "level" represented a trade-off between meeting service standards with quick deliveries, on the one hand, and keeping too much in inventory and risking waste on the other. The manager always wanted to raise the level just before a surge in demand and to lower it just before the surge evaporated.

COPING WITH BOTTLENECKS

Surges in demand or runs on particular items could strain the production system and cause it to miss its established service standards. Three service standards were routinely tracked: (1) door-to-door time—the time elapsed from the moment the customer entered a line to place an order until the customer was served, (2) drive-thru time—the time elapsed from the arrival of a car in line until the customer was served, and (3) drive-thru transaction time—the time elapsed between a car's arrival at the pick-up window and its receipt of the order. The Noblesville Burger King restaurant tried to keep these service standards at company-mandated levels: 3 minutes average for the door-to-door and drive-thru times and 30 seconds average for the drive-thru transaction time. This last time was tracked automatically by the store's computer.

To reach these goals consistently the restaurant's managers had to avoid bottlenecks in production. By providing both guidance to the crew and a set of spare hands, a manager could help ensure a smooth operation. (The Noblesville Burger King's sales volumes were high enough to warrant the creation of a "production leader" position for work in the kitchen, and a head cashier position. These workers were active in guiding the crew and anticipating bottlenecks.) During peak hours especially, it was not uncommon to see the production leader or a manager stepping up to the board to assemble a sandwich or drawing a shake from the machine or bagging fries or replenishing materials. The managers also encouraged crew team work. For example, workers with some slack time helped those who were overloaded — a cheese sandwich worker might assemble some non-cheese burgers. In fact, the manager's mandate for each of the workers was (1) to be aware of which "level" was being operated and what was on the video screen, (2) to be aware of any materials that would soon be out of stock, (3) to keep the workstation clean, and (4) to help anyone needing help. This mandate to keep production smooth and efficient placed special demands on a manager. The manager's job was not only to break any current bottlenecks but also to anticipate any potential ones. The manager constantly sought information from the workers on where bottlenecks were, how many cars were in line, and which materials were running low, and encouraged the workers to handle as much of any surge in demand themselves without intervention.

Adequate prepeak period preparation also meant a great deal in coping with bottlenecks. If the shake machine was not completely filled with ingredients or if the cash register tape was close to running out, the operation was in danger of some bottlenecks that might affect ser-

vice adversely. The restaurant manager had to check the status of items like these to be sure that incomplete preparation did not detract from the success of the operation.

PURCHASING AND MATERIALS MANAGEMENT

For any item, the quantity ordered was the amount needed to bring the restaurant's existing inventory up to a certain established level. This level was the quantity of the item the manager expected to be used between receipt of the order and receipt of the next order (3 to 5 days at a maximum) plus a safety stock amount equal to 20 percent of the expected usage. This safety stock helped ensure that the entire menu was always available. The expected usage was calculated from the previous week's demand but was adjusted to reflect special trends or such conditions as holidays, school vacations, and newspaper promotions.

Orders were typically placed 1 day in advance of delivery. Materials were generally received in the afternoon. All materials were dated on receipt, and the oldest materials were always used first. However, usage levels were so high that the restaurant never needed to worry about spoilage, even for produce or dairy products. Orders for meat, paper products, and certain other supplies were placed three times a week with the Chicago regional office of Burger King Distribution Services, Burger King's supply subsidiary. Fresh produce was ordered from either Burger King Distribution Services or an independent supplier 6 days a week. Baked goods were ordered 6 days a week and dairy products 5 days a week.

THE WORKFORCE

The restaurant employed 45 workers. A typical worker put in 35 to 40 hours each week, spread over 5 or 6 days. Most lived in the area of the restaurant. At night and on weekends, high school students and second-job workers were often employed. The area was considered one that was hard to find staff for.

The crew for the Noblesville Burger King was hand-picked by the assistant managers. Almost none had worked at a Burger King before. Workers were paid for the hours worked, with overtime given after 40 hours each week. The average hourly wage was $5.20, with some workers receiving a minimum wage of $4.25 and the best-performing worker receiving $5.80. The fringe benefit package was modest. Labor turnover at the restaurant ran at 40 percent.

A worker's hours were likely to differ from day to day, although there was a set crew for the breakfast shift. The schedule for any week, however, was worked out about a week in advance. The schedule did try to reflect worker preferences about the amount and timing of their work. Most work assignments were shuffled daily. The prime reason for shuffling assignments was to promote worker cross-training, which increased the operation's flexibility. A manager might also be forced to shuffle assignments because of worker scheduling or absentee problems. For these reasons, the order-taker knew how the kitchen operated, and the kitchen knew how to take orders. The cross-training also had the welcome benefit of heightening the tolerance any worker had for the momentary troubles of other workers.

The daily rise and fall of peak and off-peak demand required continual changes in crew size, so that neither too few nor too many workers worked at any one time. The flexibility needed in the schedule was achieved by using the part-time labor force and by scheduling their arrivals and departures at different times. Some, for example, reported to work just 15 minutes or so before the noon rush hour. Workers worked at least 3 to 4 hours at a time. A worker's departure from the restaurant was at the discretion of the manager. If sales were light, the manager could let some of the workers out early; if demand was heavy, it was understood that workers would be asked to stay past the scheduled departure time. The best crew, however, was generally scheduled for the peak times on Friday and Saturday.

The production leaders and the head cashier were responsible for training new workers. There were seven main stations on which to be trained [sandwich board, specialty sandwich board, fry station, broiler-steamer, order-taker/cashier, drive-thru, and hostess (clean-up)]. Training for any one of these stations took about 3 hours and consisted of reading some prepared material, watching a videotape on the station's work, doing a worksheet based on that video, reading a manual on procedures on the workstation, watching the trainer demonstrate procedures at the workstation, performing on a trial basis from 1 to 2 hours, taking a written test, and then having performance evaluated by the trainer. A new worker trained first on the broiler and steamer and then progressed to every station in the kitchen, finally winding up at the front counter and drive-thru operations. The fry station and the burger sandwich board were generally acknowledged to be the toughest tasks at the restaurant.

QUALITY

Periodically, about once a month, a team from the corporation would visit the restaurant unannounced to perform a quality inspection. They used a form for rating the food, the service, the appearance of the restaurant, and some other factors. In addition, the district manager for the franchisee made weekly visits to the restaurant to audit quality. Twice a year, management performed a 3-day restaurant operations consultation during which every aspect of the restaurant's operation on every shift was analyzed. The results of such audits were taken very seriously. In addition, much of the managers' time was spent in the dining room talking with customers and assessing how well the service was being delivered.

MANAGEMENT

The Noblesville Burger King was a franchise store. It was part of the company's Chicago area. Supervising the Noblesville Burger King were a manager and five assistant managers. One of these six was always at the restaurant. During peak times managers would overlap, so that two or three sets of extra hands and eyes were available for breaking bottlenecks. The five assistant managers worked 5 days a week, but on a rotating shift.

Any manager's primary responsibility was to ensure that a good-quality product was promptly served in a clean environment within company guidelines. Although a manager's ability to control costs was valued, meeting the corporation's service, quality, and cleanliness goals came first. Meeting these goals meant developing the capabilities of the production crew and maintaining its morale. Thus the manager was first and foremost a crew foreman, teaching new hires, guiding work assignments, checking quality, breaking bottlenecks, and providing an example for the crew. Layered on these responsibilities were others—ordering materials, receiving deliveries, checking and posting standards of performance (such as the door-to-door or transaction times), checking on the preparations for the day and for the peak periods, and scheduling the part-time workforce. Of the restaurant's five assistant managers, three specialized in the functions of ordering, scheduling, and breakfast operations.

Burger King headquarters provided restaurant managers with a number of aids. Chief among these aids for the week-by-week operation of the restaurant were the charts, formulas, and decision rules for scheduling the workforce, given the restaurant's particular configuration. The corporation had developed aids that showed, for different sales levels and hours of operation, how much labor each restaurant should have and where that labor ought to be assigned. These charts, formulas, and decision rules greatly aided the managers in controlling labor costs, which was the second highest cost that could be controlled.

The restaurant's POS system (computer/cash register network) was also a useful management tool. From the computer, the manager could obtain information, by half-hour increments, on sales, product mix, and discounts. The POS system also kept track of the restaurant's service standard performance.

FACILITIES AND TECHNOLOGY

This Burger King was a design called BK87. According to Larry Levensky, the franchisee's district manager, the BK87 was an excellent design for a high-volume restaurant. The layout was very efficient, workers did not waste many steps in performing the tasks required of them, materials were readily available, and there was ample storage space. Workers at nonpeak times could easily cover more than one workstation.

In addition, the Noblesville Burger King included several technological improvements that had been initiated by the corporation. For example, computerized frying vats had been installed with temperature probes that automatically adjusted the frying times so that french fries or other fried foods were cooked perfectly. Buzzers indicated when the fries should be shaken or removed from the vats. The automatic drink machines were another advance; these permitted workers to engage in other tasks while soft drink cups were being filled automatically. A new shake machine mixed shakes automatically as well. The addition of TV screens to indicate variations on the standard sandwiches was another advance; these reduced the cacophony that could strike the kitchen during peak periods. The kitchen included a new breakfast grill and hood, a specially designed unit that could be set up for the breakfast business and then transformed into a work area the rest of the day. The heat chutes for finished sandwiches were coated with Teflon and were enclosed to keep the sandwiches warmer. The extra window for the drive-thru was still another innovation, helping to cut 12 seconds off the line time for drive-thru customers. Burger King was always inventing new ways for delivering better service.

PART TWO

DISCUSSION

The Burger King Corporation is in no way responsible for the following views and presentation. They remain solely the responsibility of the author.

The Flow of the Process and of Information

In most service industries, the time delay between service provision and service consumption is necessarily very short. Put another way, one can rarely inventory a service, at least not for very long. Hotel room-nights cannot be inventoried, nor can timely tips on the stock market, nor can tasty hamburgers. Hence a whole degree of freedom is removed from the service manager, which heightens the importance of capacity choice in most service industries.

It is also frequently the case that services must be particularly flexible so that they can be customized to individual consumer needs. Think of the travel agent, the salesperson, the cab driver, and, yes, the fast-food restaurant.

These two basic features of many service businesses place substantial demands on the process design and information systems of an enterprise like Burger King. Flexibility in both product and volume is paramount, and Burger King has adopted some classic policies for yielding such flexibility.

Figure F2 is a process flow diagram for assembling an order. It is simple, and that is one of its advantages. Responsibilities are clearly demarcated, and yet all of the key tasks for customizing the hamburger or specialty sandwich rest with the worker who actually assembles the sandwich at the board. This fact, in turn, simplifies the information flow so that only one worker need pay strict attention to the punch-in of a special order. (An information flow diagram is found in Figure F3.) In this way, with a clear delineation of tasks and direct information flows, rapid product changes can be facilitated.

Flexibility in changing production volumes is achieved by a continual rebalancing of the sandwich production line. The time pattern of demand is well known from past experience, and this knowledge has permitted the staggering of work hours for the part-time workforce so that varying numbers can be on hand at any one time. The workforce ebbs and flows from

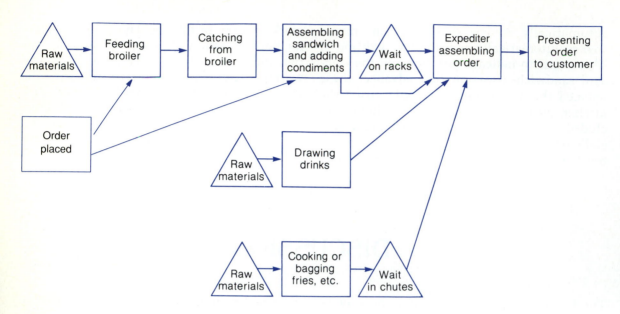

FIGURE F2   A process flow diagram (nonpeak period) for the Noblesville Burger King restaurant's kitchen operations.

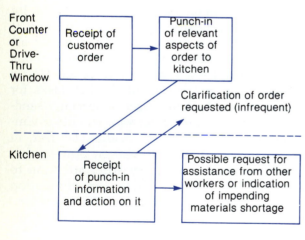

FIGURE F3   A process flow diagram (nonpeak period) for the Noblesville Burger King restaurant's kitchen operations.

More and more, production is keyed by the levels of the finished goods inventory and of the work-in-process inventory of broiled burgers and toasted buns. Instead of reacting to the order punched in, the broiler feeder reacts to the manager's posted list of burgers to be held as work-in-process, feeding the broiler so as to maintain the desired inventory level. Similarly, the board workers respond to the level of finished goods inventory, although these board workers must also respond to any special orders that are punched in. Special orders, of course, must take precedence over the maintenance of the finished goods inventory. The peak period changes in the process and information flow diagrams are pictured in Figure F4.

DEMANDS OF THE PROCESS ON THE WORKFORCE

The flexibility of the Burger King fast-food operation demands flexibility from its workers.

four during slack periods to 24 during peak periods. As more workers are added, the job contents narrow and, importantly, the flow of information changes. No longer is production keyed completely to the punch-in of the order.

Process Flow Diagram for Burgers

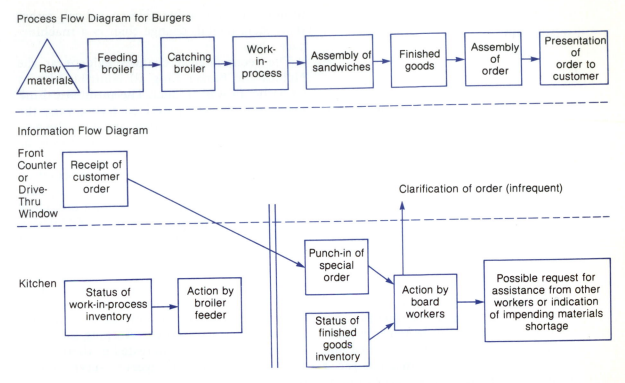

FIGURE F4   Process and information flow diagrams (peak period) for the Noblesville Burger King restaurant.

The job contents and the production pace vary markedly throughout any worker's shift during the day, requiring a special tolerance. Workers on a machine-paced assembly line get used to the rhythm of the conveyor. But, on a worker-paced line, and particularly one in a service industry, any rhythm may soon dissolve.

What is especially true of worker-paced lines is that the crew on the line views itself as a team, largely because they are so dependent on one another. This fact often permits worker-paced lines where the demand is steady (unlike at a fast-food restaurant) to be paid according to a group incentive scheme. Some standard of production—X units per time period—is established and the crew on the line, if it can better the standard, is paid extra for doing so. This

kind of group incentive pay scheme can tie a crew together as a team even more thoroughly than usual.

DEMANDS OF THE PROCESS ON MANAGEMENT

An operation like a Burger King restaurant places specific demands on management both because it is a service operation and because it is a worker-paced line flow process. As a line flow process, many of the issues that were discussed in Tour D with respect to General Motors surface:

• Balance of the line.

• Materials management.

- Technological advance.
- Capacity planning.
- Product design.
- Workforce management.

Managers of worker-paced lines should be sensitive to all these issues, since to fall down on any one could seriously jeopardize the entire operation.

The issue of technological advance is particularly acute, since such an advance may enable a smoother, more regular flow of the product. After all, a worker-paced line flow usually remains worker, rather than machine-paced, because there is as yet no easy way to guard against a succession of overcycle conditions striking a worker all at once (such as a run on "Whoppers, hold the pickles"). If the mix of product options can somehow be smoothed out or if particular advances can be made in product or workstation design, a worker-paced line can easily be transformed into a machine-paced line. In other words, the worker-paced line is somewhat more vulnerable to radical, rather than incremental, change than is a machine-paced one.

The technological changes evident at the Noblesville Burger King all contributed to a smoother flow of product. The automatic drink machines, the computerized frying vats, the TV monitors of information, and even the new BK87 layout all act to speed up and smooth the delivery process.

A Burger King's status as a service operation only heightens the importance of capacity planning, since service firms cannot ordinarily inventory their product. The new BK87 design is testament to the importance the corporation assigns to planning capacity. The importance of workforce management is also heightened, because the worker-customer interaction is part of the service. Ordinarily, one cannot hide the service process as readily as one can hide the manufacturing process. Keeping the workforce productive and interested in their jobs is a key challenge for managers in service operations.

QUESTIONS

1. Visit a fast-food restaurant with a friend to perform an experiment. While one of you times the other's entry and exit and the time from the placing of the order to its receipt, the other should observe how many workers are actually involved with the order. Visit the fast-food restaurant at a slack time and at a peak time, and compare the differences in time and number of workers.

2. Watch for bottlenecks the next time you are in a fast-food restaurant (or any worker-paced service operation). Where do they arise and why? How might they be remedied?

3. Suppose that Burger King were to introduce a new type of sandwich. What aspects of the restaurant's operations would be affected, and how?

4. Why is flexibility one of the key features for success in a worker-paced assembly line operation such as Burger King?

5. How would you define capacity at a Burger King? What factors influence the establishment of a limit to capacity?

6. Would you prefer to work on a machine-paced assembly line or a worker-paced one? Why? Which would you prefer to supervise? Why?

7. Visit a Burger King close to you. It is unlikely to be a brand new one with all the ad-

vances incorporated into the Noblesville Burger King. Consider the ways by which the Burger King you visit could be modernized. Consider as well what could be done at the restaurant to increase capacity as sales increase. Prepare plans of action that specify which things ought to be done, and in what order, to modernize the restaurant and/or increase its capacity. Defend your choices.

8. Consider an alternative kitchen layout to the one at the Noblesville Burger King. In what ways is your alternative better or worse than the one discussed in the tour?

SITUATION FOR STUDY F-1

LEGACY HOMES

Tom Stoddard had been employed part-time as a "rough carpenter" by Legacy Homes while he completed his degree at Interstate Tech University. After his graduation, Legacy hired Tom as a construction superintendent; this job combined the duties of foreman, scheduler, and expediter.

By building only to order, Legacy took a conservative approach to the ups and downs typical of the home construction industry. Legacy offered a limited number of home designs; Tables F1 and Table F2 show the work involved on most of them—usually about 90 days' worth. When a buyer arranged a home loan from a lending institution, the local practice was to allow the construction company to receive the money in partial payments (cash draws) depending on the degree of completion of the home. Thus, for Legacy to receive the money, Tom needed to finish a home as fast as possible.

Legacy had a nucleus crew of highly skilled workers that it wanted to keep working most of the time. Other workers could be hired to supplement this crew when necessary. Table F3 shows this crew and the various tasks they were capable of performing. Each crew member knew well a primary function such as plumbing but could be counted on to help with other functions such as rough carpentry. The crew who installed the plumbing, for example, also installed the heating and electricity. This overlap of skills was possible for two reasons: (1) most of the work on a Legacy house was rather straightforward and did not entail the total skill requirements of journeymen in any of the trades, and (2) Legacy hired nonunion workers.

Table F1 shows the minimum and maximum crew sizes for completing each task. For example, task 8 requires a minimum of two workers because one person working alone cannot position the furnace or handle the ductwork; two people working on separate houses would take much longer to complete their work on both houses than if they worked together on one house at a time. At the maximum crew size, of course, it can happen that workers interfere with each other's work. Tom assumed that additional workers could work at the same rate as workers already on the job; he therefore scaled the number of workers up or down proportionately. For example, in task 1 if two workers can do the job in 1 day (i.e., 2 labor-days are required), three workers can do the job in 2/3 day (2 labor-days divided by three workers available), and four workers can do it in 1/2 a day. All operations have been designed to require a low level of skill; thus, any worker can shift to another operation with only a negligible loss of efficiency. Usually one worker is used as parts chaser and relief. The supervisor can help with

TABLE F1　Crew size and times for various tasks (Legacy Homes)

TASK	DESCRIPTION	WORK TYPE*	MINIMUM/MAXIMUM CREW SIZE	DAYS TO COMPLETE (with minimum crew)
1	Concrete footer	A	2-4	1.00
2	Foundation	F	Subcontract	1.00
3	Grading	A	2-4	0.50
4	Framing	B	5-8	2.60
5	Roofing	F	Subcontract	1.00
6	Concrete	A/B	3-5	1.50
7	Wiring	C	2-4	1.00
8	Furnace and ducts	C	2-4	2.25
9	Plumbing	C	2-3	2.50
10	Insulation	B	1-3	2.25
11	Dry wall	F	Subcontract	1.00
12	Siding	F	Subcontract	5.50
13	Sewer line	A	2-3	2.00
14	Painting	F	Subcontract	1.50
15	Finish carpentry	D	2-4	2.50
16	Tile	D	1-2	5.00
17	Electrical trim	C	1-2	1.50
18	Finish plumbing	C	1-2	3.50
19	Heating trim	C	1-3	2.50
20	Carpeting	F	Subcontract	2.00
21	Cleanup	E	1-3	2.00

*A, excavate; B, rough carpentry; C, plumbing, heating, electrical; D, finish carpentry; E, part-timeclean-up; F, subcontract.

TABLE F2　Precedence relationship (Legacy Homes)

JOB	PRECEDING OPERATION(S)
1	
2	1
3	2
4	3
5	4
6	2
7	4
8	5
9	5
10	7, 8, 9
11	10
12	10
13	9
14	11
15	14
16	14
17	14
18	15
19	14
20	6, 12, 13, 16, 17, 18, 19
21	20

TABLE F3 Labor force (Legacy Homes)		
TYPE OF WORK	CURRENT LABOR AVAILABLE	TASKS PERFORMED*
A Excavate	3	1, 3, 6, 13
B Rough carpentry	5	4, 6, 10
C Plumbing, heating, electrical	4	7, 8, 9, 17, 18, 19
D Finish carpentry	2	15, 16
E Part-time cleanup	1	21
F Subcontract		2, 5, 11, 12, 14, 20

*See Table F1 for an explanation of these numbers.

the parts chasing but is not expected to take part in operations except as a troubleshooter.

Although the products (houses) do not move, Tom feels that the principles of worker-paced lines can be applied with beneficial results.

1. Design a process flow for Tom.
2. What is the minimum time requirement to build a house? How many houses should be built to keep the permanent crew busy? What is the maximum number of houses that should be under construction at one time?
3. Is the workforce balanced? If not, which crew sizes should be changed?
4. Design an information flow for Tom.
5. Set up a sequence for raw material flow for Tom.

SITUATION FOR STUDY F-2

SMALL CITY NEWSPAPER

Small City Newspaper (SCN) was a family-owned paper with a circulation of 43,000 in the morning and 22,000 in the evening. There were three editions of the morning paper and two editions of the afternoon paper; the earlier editions of each paper were for delivery further away.

Newspapers have two content inputs: news and advertising. For SCN, as with most daily newspapers, advertising determined the size of each issue, which in turn affected the length of the production run. To maintain an average of 43 percent news content and 57 percent adver-tising, the number of columns of news was allocated according to the number of columns of advertising sold for a particular issue. A minimum number of news columns (front page, sports, editorial, comics, and so on) was required for the morning paper (96 total columns) and for the evening paper (80 total columns). Information about the number of columns of advertising was communicated to the editors and the production manager by the advertising department.

Advertising was divided into three categories: display, classified, and national. Display was used primarily by local retailers and ad agencies, who often had camera-ready copy prepared for the paste-up department. On occasion, the advertising and art departments worked with the customer in preparing the advertisement. Three days' lead time was generally required of the client when placing a display ad; however, flexibility in this procedure allowed for changes to be made until press time. Classified ads had to be submitted by 5:00 P.M. the previous day and could be prepared for paste-up on the computer. National advertising copy was usually camera-ready and followed much the same procedure as the display ads. To expedite the use of both display and national ad copy, an inventory of these ads was maintained.

The morning news department had a staff of 40: city desk, copy desk/AP wire service, sports, style, editorial, and cartoon. The afternoon paper had a staff of 22 (no cartoonist). The library kept microfilmed copies of newspapers, publicity brochures, and a master file of news clippings (filed alphabetically) on important individuals. A personal computer that was tied into the Associated Press news retrieval system was also located in the library.

The managing editor assigned reporters and photographers to cover newsworthy events. Editorials as well as style and feature articles had the earliest deadlines (5:00 P.M. for the morning paper and 9:00 A.M. for the afternoon paper) and so were often written the day before they were to be published. Front page stories and sports coverage demanded up-to-date reporting; the deadlines for these articles were 11:00 P.M. for the morning paper and 11:00 A.M. for the afternoon paper.

Flexibility in the entire operation was a must, to allow for reporting late-breaking news. Because these stories could significantly disrupt the process, reporters were encouraged

to prepare as much of their other copy as far in advance of press time as possible. To facilitate this, the newspaper had adopted a flexible time schedule for its reporters, who were generally on the job from 3:00 P.M. to midnight (for the morning paper) and from 7:00 A.M. to 4:00 P.M. (for the afternoon paper). In spite of this effort, half of the newspaper was made up in the two hours before press time.

Generally, a reporter composed a story at a video display terminal (VDT). The story appeared on a screen (CRT) as it was typed. Errors could easily be corrected, and additions and deletions did not affect the rest of the story. Typewritten copy could enter the computer through the use of an optical character reader (OCR), which sensed typewritten copy if it was prepared in a compatible typeface. Once the reporter was satisfied with the story, a single key command put the story in line in the computer. If this was done close to the deadline and the reporter considered the story to be especially important, the reporter could supersede this queue by notifying the editor.

The editor had a list of all stories in the queue. He accessed the stories and made revisions using special keyboard commands. Once he was satisfied, the editor added the headline via the keyboard, specifying the type size of the heading (in points), and designated where the story should go in the paper. The story was then transferred by computer to page paste-up, where a hard copy was printed.

Syndicated features and wire service copy were transmitted to the paper in a form compatible with the VDT system, replacing the old teletypewriter and eliminating the need for rekeyboarding. At the copy desk these stories were reviewed and a determination was made concerning which stories to use; then the copy was forwarded by computer to page paste-up. The laser photo service was another external input. Photographs were beamed to the newspaper along with a subtitle and/or a short explanation. The photo came to the paper ready for paste-up.

Advertising copy, news stories, and photographs were pieced together on a page the size of a newspaper page. Reusable by-lines and photos of syndicated columnists were kept at the paste-up tables. Efforts were made to smooth the flow through paste-up by having advertising, feature articles, and as much of the news as possible to paste-up as early as possible. The paste-up operation for the morning paper normally ran from 5:30 P.M. to 11:00 P.M. Paste-up for the afternoon paper generally ran from 8:00 A.M. to 11:30 A.M. The time difference for the papers was due to the latter's smaller size and the advance preparation the morning paper paste-up had to do for its Sunday edition. Changes for the final edition of each paper could be made after the presses began the earlier edition(s).

Once the pages had been laid out, they were photographed. An aluminum-backed, polymer-coated plate was placed in an exposure unit beneath the page negative. After a 45- to 70-second exposure, it was put through a wash unit, which used water and a small amount of biodegradable defoaming agent to remove the nonexposed polymer areas. After drying, the plate had a raised mirror image surface, which was curved to fit the rotary press cylinder. One page could pass through the engraving process in 22 minutes standard time, though several plates could pass through simultaneously. Once the plates had been prepared, a press room employee picked them up on the fourth floor and carried them to the press on the first floor.

The current printing process replaced several steps of the old letterpress system. It was compatible with both the rotary press and the offset press systems. An offset press, although it wasted as much as 4 percent newsprint per run, offered computerized printing capabilities, better quality of print, and flexibility with the use of color.

Press runs for the morning paper were approximately as follows: first edition (outlying counties) 12:10 to 12:40 A.M., second edition (adjacent county) 1:00 to 1:25 A.M., and final edition (city) 1:45 to 3:20 A.M. Press runs for the afternoon paper were approximately as follows: first edition (all counties, except the immediate one) 12:50 to 1:30 P.M., and final edition 2:15 to 3:15 P.M. The capacity of the press was 80 pages (16 per deck) per run. When full color was used, only 64 pages could be printed per run, because two half-decks had to be devoted to overlap the four colors necessary for full color. To circumvent this bottleneck, the Sunday feature section was printed on Saturday and many of the inserts were preprinted during slack time. The Sunday comics, for example, were printed externally and were delivered early in the week.

The press cylinders were on the first floor. Rolls of newsprint were loaded on the press in the basement, and the cutting and folding mechanism was on the second floor. The newly printed newspaper was carried by conveyor to the mailroom.

Maintenance could be performed by disconnecting the individual decks; however, all five were operated simultaneously by a central drive system. A problem with the electrically powered drive unit, which had happened recently, could prevent the press from operating. Fortunately, this situation was corrected in time to get the paper out, but it pointed up a weakness in the functioning of the entire operation.

In the mailroom, the sections and inserts were assembled. The papers were counted, bundled, labeled, and dropped through a chute to an alley where they were picked up by carriers or were trucked to drop points.

Timely delivery was very important. The evening newspaper had to reach the customer before the evening TV newscast, or it would not be read. The morning paper had to be to the carriers (who usually held day jobs) in time for them to make deliveries before they went to work. In meeting these objectives, the earlier editions of the morning paper were to leave the headquarters by 2:00 A.M. and the final edition was to be to the drop points by 4:00 A.M. The

earlier afternoon edition was to be ready to go by 1:30 P.M., with the final edition due at the drop points by 3:30 P.M.

1. Diagram the process and information flows for this process.

2. In what ways is this a worker-paced line flow operation and in what ways, if any, is it not?

3. How do the concepts of line balance and rebalance apply to this situation?

A SERVICE SHOP
Ogle–Tucker Buick
Auto Service and Repair
Indianapolis, Indiana

Ogle–Tucker Buick was located on Keystone Avenue in northeast Indianapolis. It was located on 7.25 acres of land and occupied 43,000 square feet of space. The service department consisted of 27 stalls available in the service area, 24 stalls in the body shop, 6000 square feet dedicated to parts storage, and the associated office space for these operations.

The dealership was owned by Robert Ogle and Thomas Tucker, who had grown up on farms and had moved to Indianapolis after World War II. The dealership had been at its current location since 1969. It sold Buick cars exclusively.

PART ONE

PROCESS DESCRIPTION

WRITING THE SERVICE ORDER

The service operation began at 7:00 A.M., Monday through Friday. At that time, the two service advisors who were responsible for developing the repair orders began greeting the customers who had lined up for service. The service advisors' responsibilities were to detail (1) the symptoms of the problem experienced by the customer, and/or (2) work the customer wanted to have performed. This called for skill-ful listening by the service advisor and the knack for asking the right questions to glean the proper symptoms to explain to the mechanic assigned. The service advisor was also responsible for suggesting additional services the customer might want but had not thought of, such as oil changes, new brake linings, or front-end alignments. The service advisor was the operation's chief contact with the customer through the service process. If the customer needed to be informed about the progress on his or her car, or that additional work was needed, or that the car might not be ready in time, the service advisor communicated that information.

Upon greeting the customer the service advisor logged in necessary identification information to the service department's computer, and a customized repair order form was generated (Figure G1). When the customer finished detailing the work to be done on the car, he or she signed the repair order, giving the service operation authorization to work on the car. The service advisor then assigned a control number to the job and placed an oaktag control number on the rearview mirror of the car and also attached a tag with the same number on the keys to the car.

The repair order consisted of four paper "soft" copies plus an oaktag back "hard copy." The four soft copies were of different colors. The top sheet, white copy, was the accounting copy. The yellow sheet was the customer's copy. The pink sheet was kept by the service advisor. The green sheet was a control copy that was used by the person who followed up with customers on their repairs and was then routed back to the service manager whose secretary filed them. After the service advisor filled out the repair order, he detached the green sheet and sent the other soft copies, together with the hard copy, to the dispatcher in the "tower."

DISPATCHING

By keeping the soft and hard copies together,[a] Ogle–Tucker was able to reduce some double entry of information by the tower and yet still keep control over what was done, so that me-

chanics could not enter work that was not ordered by the customer. The service advisor sent the repair order into the dispatching office (the tower), the nerve center for the service operation. Here the dispatcher (the tower operator) logged in the control number, the name of the customer, the type of car, the time the service advisor promised the car for delivery, the repair order number, and the work that was to be done, together with the number of the mechanic who was supposed to do that work. Figure G2 shows a dispatcher's route sheet.

Jobs were classified by category. The categories included lubrication and filter changes, tune-ups, front-end work, rear axle work, transmission work, brake work, and the like. Ogle–Tucker's mechanics (termed "technicians" at the dealership) were specialized by task. For example, one did transmission work exclusively, two did front-end work, three handled tune-ups, while others devoted themselves to other tasks. When the tower operator logged in a repair, he noted which technician in which specialty was to do the job. When that job was completed, another technician might take over the car to work on a different aspect of its service or repair. An x-ed out entry on the route sheet meant that the technician had completed his work.

Routine items such as tune-ups and oil changes could be assigned to technicians without requiring a diagnosis about what was wrong. For a number of cars, however, some diagnostic work had to be done before the car could be fixed. Whether diagnostic work was needed was generally determined from the service advisor's wording on the repair order. Key words such as "hesitating sometimes," "occasionally stalling," and so on, were tip-offs. Technicians were responsible for diagnosing problems as well as fixing them. On particularly tough assignments, the service manager was consulted. The dealership referred to the

---

[a]Many dealerships separated the soft copies from the hard copy that went to the mechanic.

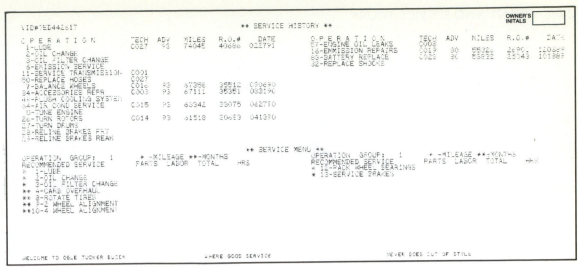

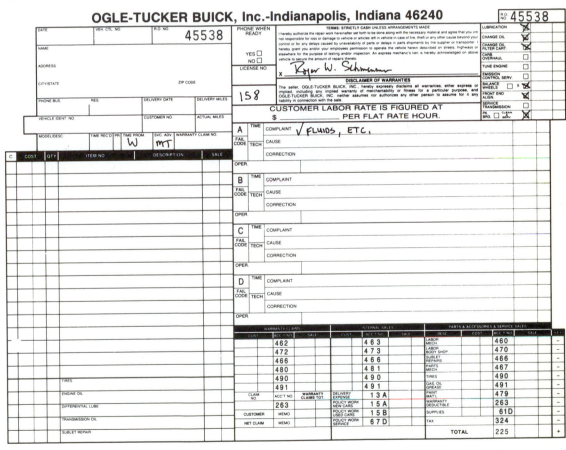

FIGURE G1   A repair order.

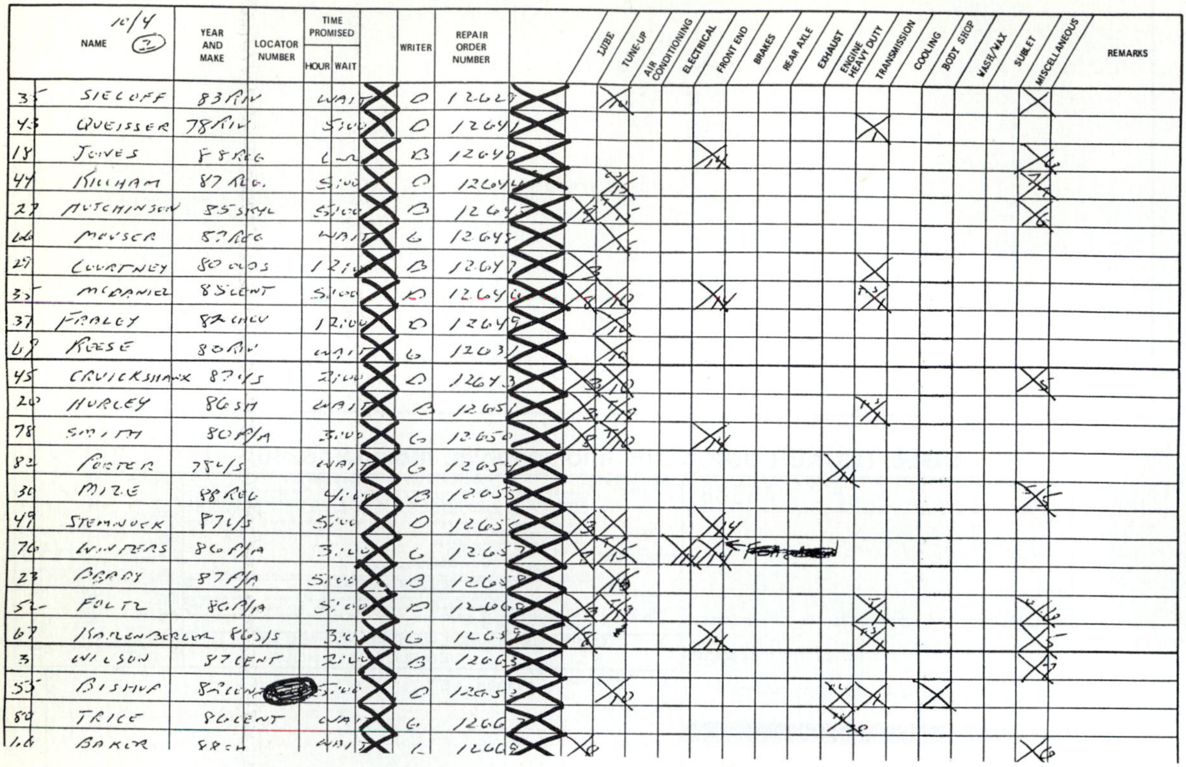

FIGURE G2 Dispatcher's route sheet.

"3 C's" of "complaint, cause, and correction" that were listed on the repair order. The service advisor was responsible for detailing the complaint and the technician was responsible for determining cause and performing the correction.

SHOP OPERATIONS AND THE WORKFORCE

The service floor had 27 stalls that were divided among the 13 technicians currently employed. Most technicians were assigned two stalls. Most of the shop's technicians (mechanics) had some post-secondary school education, generally technical school training, and they frequently had a rich family heritage of working on mechanical things. The better-established mechanics could make over $30,000 a year and carried at least $20,000 worth of their own tools. Established technicians earned a higher total income than younger technicians who did not have as many tools or as much knowledge. Younger technicians built up their collections of tools on a continual basis. Tool salespeople came to the shop every week to peddle their tools and collect payments from any technicians who were buying tools on credit.

The technicians were paid on a flat-rate basis. A manual indicated what the time standard should be for every particular kind of job. The technician was paid according to the standard for the job and not according to the time spent on it. Thus, if he spent less time than

The service entrance. (*Courtesy of Ogle–Tucker Buick, Indianapolis, Indiana*)

The dispatcher's tower to the left and the main desk for the service advisors.
(*Courtesy of Ogle–Tucker Buick, Indianapolis, Indiana*)

A line of service bays. (*Courtesy of Ogle–Tucker Buick, Indianapolis, Indiana*)

that indicated in the manual the technician could "beat the book" and earn increased wages. If he spent more time at the job than indicated, the technician would not earn as much. Any rework was directed back to the technician on an unpaid basis, and the shop's policy was that any rework had priority over other work. There were also some skill-level differences incorporated into the wage structure. The most inexperienced worker received $10.50 for every "flag hour" (standard hour), and the most experienced worker earned $14.50. Thus a technician could earn more annually in two ways: (1) by improving his skill level or (2) by being more efficient. When technicians had to wait for a part, it was to their advantage to go to the tower operator and get assigned some quick work that they could perform while waiting. In this way technicians were able to convert more of the time they were in the shop into flag hours for which they earned income.

After a mechanic finished working on a car, he road-tested it, and if satisfied, wrote the mileage on the back of the repair order's hard copy and filled in both the cause of the complaint and the correction he made. He then returned the repair order to the tower. There, the dispatcher reviewed what the technician had written, and in consultation with a standard reference, determined what the flag hours should be for the repairs made. The dispatcher then filled in and attached a gummed segment of the time sheet (Figure G3) to the back of the hard copy.

The repair order was routed to the cashier for payment and accounting, the pink copies were retained by the service advisors, and the hard copies filed. White and green copies were filed

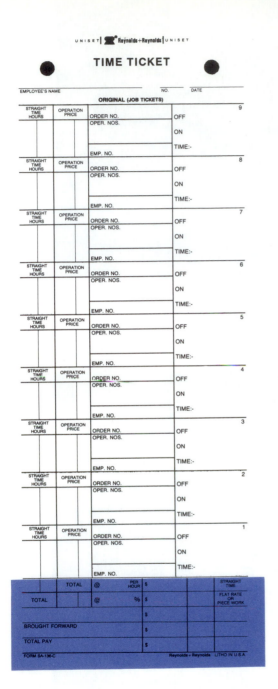

FIGURE G3   A worker's time chart. (*Courtesy of Ogle–Tucker Buick, Indianapolis, Indiana*)

by repair order number; hard copies were filed by car serial number. Customers returning for their cars went to the cashier to pay, and there received the yellow copies of the annotated repair orders and the keys to their cars. Within a few days of repair, customers were called by the dealership and asked to comment on the quality of their repair. If customers were not at home, a letter was sent. Postcard surveys were also left in each car repaired, with customers encouraged to mail their responses back in. The service manager viewed these customer surveys as an important aspect of the service department's operation.

Cars typically arrived during the morning. On an average day, 70 cars might be worked on, with a typical range of from 40 to 85. The day's chief period of stress came after 4:00 P.M., when most people came to pick up their cars. This was when customer complaints surfaced about cars not being ready or about the bill being too high. The service advisors were pulled in many different directions during this time.

QUALITY CONTROL

Quality was enhanced in several ways. There were, of course, the road tests done on each car. And because technicians had to do rework for no pay, there was an incentive to do things right the first time.

Shoddy workmanship was grounds for termination. If, in the opinion of the service manager, there were three instances in one month where poor workmanship was involved with no mitigating circumstances, the technician was terminated. The technicians, of course, were alert to this policy, and each instance of shoddy workmanship was discussed between the service manager and the technician.

From time to time, technicians were sent to training schools, usually either in Indianapolis

A service bay with an oil change in progress. (*Courtesy of Ogle–Tucker Buick, Indianapolis, Indiana*)

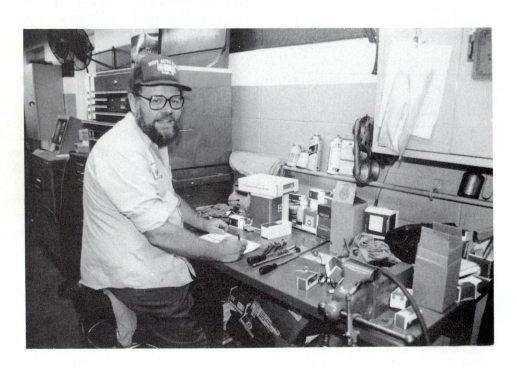

A technician updating a service order. (*Courtesy of Ogle–Tucker Buick, Indianapolis, Indiana*)

The parts department counter. (*Courtesy of Ogle–Tucker Buick, Indianapolis, Indiana*)

or in Cincinnati, where the zone's main training center was located. Most schools lasted between 2 and 4 days. The dealership paid for the school's fees plus food, lodging, and transportation. Technicians were also encouraged to take the twice yearly Automotive Service Excellence tests for accreditation in major aspects of automotive service and repair.

THE PARTS DEPARTMENT

The parts department employed six people: a manager, a computer operator, a truck driver, and three countermen. The parts department was charged with purchasing all the parts used by the technicians, maintaining that inventory, handling damaged parts, and filing claims and other returns. The department maintained an inventory of $320,000 worth of parts, an inventory that turned over three times a year. In a typical day perhaps 300 parts would be used to serve the needs of the technicians in the shop, but only two or three of those parts would be out of stock and have to be special ordered.

The parts department could place orders with General Motors Service Parts Operations (GM-SPO), outside jobbers, or other dealers. The parts department liked to turn a part around (sell it) every 30 days, but with an actual inventory turn ratio of around three they were only able to turn parts on the average every 120 days. Exactly which parts should be purchased at any time was determined jointly by the parts department and a computer program developed by Reynolds and Reynolds. The computer system and its software were reasonably standard for the industry. This computer software used past sales histories to generate suggestions for which items should be purchased to replenish the inventory. Each day, the parts department manager made sure that parts that had been used or sold were removed from the perpetual inventory record kept by the computer. In this way the computer knew what was needed for replenishment, although the parts department manager could override the suggestions made by the computer to take into account special circumstances.

There were several ways by which parts were ordered:

1. *Stock orders.* The stock order was the regular way to order parts. Parts were ordered every week at a set time (Tuesdays). Parts ordered through GMSPO by stock order received a 7 percent purchase discount and an 8 percent credit on returned parts. The credit was useful because it permitted Ogle–Tucker to return slow-moving parts at no cost, thereby improving the way the dealership's inventory dollars were spent. Stock orders were issued for regularly used parts and for parts whose demands were known with considerable lead time. There were also "target orders" (placed on Mondays) for the fast-moving parts. Extra discounts were available for these parts. Target orders could mean a 22 percent extra discount, and on special promotions the discount could be 27 percent plus 16 percent return privileges.

There were several other ways to order parts. These were more costly than stock orders, but provided faster service.

2. *Car Inoperative Orders (CIOs).* CIO orders were handled faster than stock orders and, consequently, no discounts or returns were given. For CIO orders, GM checked the six closest depots in that region for parts. (The distribution channel for parts went through three types of depots. The most local were termed Parts Distribution Centers (PDCs), the closest being in Chicago. The next level up, Field Processing Centers, carried more inventory. Lansing, Michigan, was the nearest Field Processing Center. The last resort were the parts plants themselves, controlled from Flint, Michigan. If the part could be located in any of the depots checked, it was sent. If the part could not be located, the dealership was so informed and would have to resort either to waiting for the part or giving it higher priority.

3. *VIP (Very Important Parts) Orders.* VIP orders called for all depots in the country to be checked for parts. The cost charged was $2.00 per item plus the dealer cost plus 5 percent.

4. *SPAC (Special Parts Assistance Center) special orders.* If all else failed, and if GM had a formal customer complaint, SPAC could be initiated. A SPAC special order called for action within 48 hours and dealers across the country could get involved.

5. *Partech.* For problem parts, Partech could be consulted. Partech helped with part number problems such as no part number or part number misprints in catalogs.

6. *Other dealers.* Ogle–Tucker also bought parts from other dealers through informal relationships. The parts manager sometimes phoned other dealerships and asked whether they had specific parts that could be shipped by bus or some other means. Typically, the cost was dealer cost plus 22 to 25 percent.

Within the first floor storage area of the parts department, there were both small bins and large bins. In the upstairs area, bulk materials such as sheet metal parts were stored. Parts were stored in numerical order by group number. For example, 2000 series numbers were for electrical parts, 3000 series numbers for fuel and exhaust, 4000 series numbers for transmission and brake parts, and so forth. Ordering parts, however, depended not on the group number, but on a specific part number of six to eight digits in length. The workers at each counter were responsible for identifying the proper part for the job, tracking it down in the storage room, and delivering it to the technician or the purchaser, with the proper paperwork done. At each counter, there were manuals for each model or body type of Buick. Each manual had both text and illustrations to aid the search. Prices were kept on the computer. When a part was billed out, the computer automatically deducted it from inventory and kept its accounting straight.

Body Shop

The body shop's major task was to repair cars that had been in accidents. These cars had smashed some sheet metal parts or even the car's frame itself. The shop employed nine: a manager, four bodymen, a painter, a painter's helper, a car washer, and a porter (who did waxing, detailing, and kept the shop clean). There were 15 stalls designated for the four bodymen and 9 stalls for the painter.

The body shop was a full-service body shop that could handle both ordinary sheet metal tasks and framework—the shop was equipped with the hydraulics, chains, and other equipment necessary to straighten automobile frames. The dealership provided benches, welding equipment, painting equipment, some special tools and the consumables such as sandpaper that were used in the course of the shop's work. Each bodyman, however, owned his own tools, which were valued between $15,000 and $20,000.

The bodymen typically worked on two cars at any one time, one that could be classified as a "heavy hit" and one that demanded much lighter work. Often the major job proceeded in fits and starts as parts became available. The smaller job thus served to keep him fully occupied. The painter often worked on six or seven cars in a day.

Within the body shop, the bodymen continually helped one another with problems, such as opening smashed doors or attaching frame straightening equipment. Even though they were paid for their individual work, they recognized the need for assistance when one of them ran into trouble. Within the paint shop, the painter was aided by a helper who did much of the required sanding and the taping of those parts of the car that were not to receive paint. The painter was primarily responsible for "shooting" the paint and ordering supplies. The other two were mainly cleanup people.

The typical car spent 1 or 2 days in body repair and 2 to 3 days in painting. (Painting required the application of a base coat and a clear coat, with time in between for drying, sanding, and rubbing out.) Some cars with major damage spent much longer in the shop. They required much more work and were sometimes delayed until parts were received. Typically, the body shop did not begin work on an automobile until all the parts were received. Thus, if the owner could still drive the automobile, he or she continued to drive it until all the parts were received and an appointment was scheduled. However, if the car had sustained major damage and could not be driven, it remained outside the shop until all the parts were available. With major damage, however, there was always a substantial chance that hidden damage would be revealed as the car was worked on. Hidden damage took even more time to fix and often required the purchase of additional parts. This took time, especially if

In the body shop with a "major hit." (*Courtesy of Ogle–Tucker Buick, Indianapolis, Indiana*)

the car was a foreign make. The body shop had two constituencies: car owners and their insurance companies. Almost all of the shop's work was paid for out of insurance claims, which meant that the shop had to agree with the insurance companies about how much a particular job was worth. The estimates by both the insurance companies and the shop itself were aided by reference to standard books that periodically updated prices and gave standards on the time it should take to remedy certain problems. The shop, in particular, followed the "Mitchell Crash Book" and had very few disputes with insurance company adjusters. Most of the disputes centered around which parts ought to be straightened rather than replaced. (It was standard practice to replace a smashed part with a brand new one on new cars, whereas on older cars, the existing parts were more often straightened or replaced with used parts of the same kind and quality as used on the car.) The shop contacted insurance adjust-

ers either by telephone or when adjusters came by to check jobs in progress or to obtain estimates for potential jobs.

Adjusters could not insist that repairs be made by a particular body shop, but they could recommend shops. Thus, keeping the insurance adjusters happy was an important aspect of the body shop's efforts. The body shop manager kept the adjusters up to date on what was happening and indicated when hidden damage was found. They could check this type of damage themselves and then reach agreement with the shop on a supplement to the original estimate. The shop tried to do its work quickly. Often the insurance company had to pay for rental cars for inconvenienced owners. Thus, the quicker work was done, the more likely the insurance company would not have to pay extended rental charges. To make life easier for the adjusters when they came to visit, the body shop maintained a small office and stocked it with supplies and a telephone.

The body shop manager also spent considerable time with car owners. Typically, there were many occasions for such customer contact. The first occurred if an estimate of the damage was requested. An estimate was made by systematically examining the car, in the same fashion as an insurance company adjuster, and by using the pertinent information in the crash book. Once the owner received the estimate, the body shop manager typically followed up by phone to answer any questions about the estimate or about the insurance process itself. Often in these follow-up calls the shop was informed that it or some other shop had been selected to do the work. However, sometimes it took two or three follow-up calls before a determination was reached about what to do with the car.

After the car was brought in, the body shop manager went over with the owner what was going to happen and what the probable time for repair would be. The body shop manager instructed owners to call the shop before the car was due, in case there were any schedule problems, hidden damage, or the like. When the car was ready, he went over the car in detail with the owner, telling the owner what had been done and letting the owner check the quality. The shop delivered cars to owners in as close to showroom condition as possible. Every car was road-tested and thoroughly cleaned.

How Workers Were Paid

Workers were paid in different ways. As discussed previously, technicians were paid on the basis of a flat-rate hour. The best technicians generally were the fast ones. They were the best at diagnosing problems, they did not squander time, and they knew exactly how to do each job. Technicians could also receive "spiffs," bonuses for working more than a set number of flat-rate hours. A bonus of 25¢ per hour was paid each week to technicians who had more than 35 hours time logged. A bonus of 50¢ per hour was paid to those with over 40 hours, and a bonus of 75¢ per hour went to those with over 45 flat-rate hours for the week.

The bodymen and painter were paid in a similar way. If they were able to beat the standards set for their work, they were able to make more money. If they could not meet the standard, they made less. Some standards naturally were looser than others. In the body shop, for example, sheet metal standards were generally looser than other types of work: thus, sheet metal jobs were the most prized. The parts countermen were paid a salary, although they received a bonus if sales reached some target level. Service advisors were paid commission. That is, they got a set percentage of the service business brought in. This was an incentive for them not only to do a good job so that people were satisfied, but it was also an incentive to suggest additional repairs that the owner might not have thought of. They also received spiffs for special sales (e.g., tires, antennas) that were documented on their pink job order sheets.

The Duties of the Service Manager

The service manager was constantly on the move. In the early mornings and in the late afternoons his main interactions were with customers. He helped the service advisors at the start of the day so that particularly concerned and/or aggressive customers could be handled individually by him. This saved time for the service advisors and contributed to a routine flow of cars into the shop. In the late afternoon, the service manager spent time with customers who complained about work or registered dissatisfaction with the size of the bill.

He listened to customers and then, typically, educated them as to the real problems with their cars and/or the reasons for the costs charged.

During the day the service manager made the rounds of the repair floor, the body shop, and the parts department. He checked the dispatching and the efficiency of technicians. He did diagnostic work and road tested completed cars. On heavy days the service manager was particularly stressed. The tasks were the same, but heavier days were more demanding on everyone, especially if the shop was short-handed.

PROMOTIONS AND SPECIALS

The flow of work into both the service area and the body shop was not even from day to day or month to month. People drove their cars more during the summer months and thus incurred more repair problems during that time. Periodically, in an effort to smooth the flow of work it received, the dealership publicized some service specials. These were typically cut-rate prices on services such as oil changes and lubrications, cooling system winterizing, or total paint jobs. These specials were timed so that they did not interfere with normally busy periods.

PART TWO

DISCUSSION

Ogle–Tucker Buick is in no way responsible for the following views and presentation. They remain solely the responsibility of the author.

THE FLOW OF THE PROCESS AND OF INFORMATION

Figures G4 and G5 display process flow and information flow diagrams for a car in the course of its repair at Ogle–Tucker Buick. Similar diagrams could be drawn for the operation of the body shop or indeed for the operation of the parts department. (The development of those diagrams are left as exercises at the end of this tour.) Note that the customer's car may wait at almost any point in the process. The only wait shown in the diagram is between logging in the job and its diagnosis; it is but one of many such waits that could exist. What is most striking about the flow of the process and

of information at Ogle–Tucker versus a service factory like Burger King is the extent to which there is communication between customer and management and between management and technicians. Because of the customization that is required in auto repair, the flow of information at Ogle–Tucker is much more frequent and less structured than that at a Burger King. Management is very much an intermediary between the customer and the technician. At the same time, management is also very much the source of control for the process.

One can see from this information flow diagram why peak days are so tough on management. There are no shortcuts in the information flows that must occur on busy days. The only source of help is to draw on others, such as calling on the service advisors to do road tests or to give diagnostic work to technicians themselves.

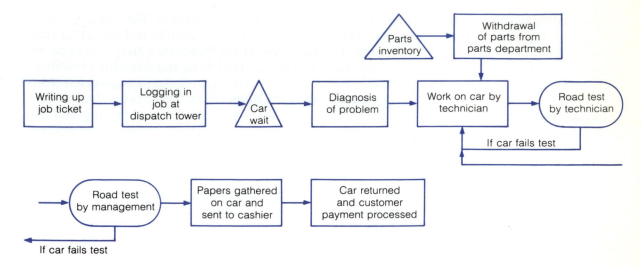

Note: The car may wait between any of the process steps shown and not simply between the log-in of the job and the diagnosis of any problem.

FIGURE G4   Process flow diagram that follows a car in the course of its repair at Ogle–Tucker Buick, Indianapolis, Indiana.

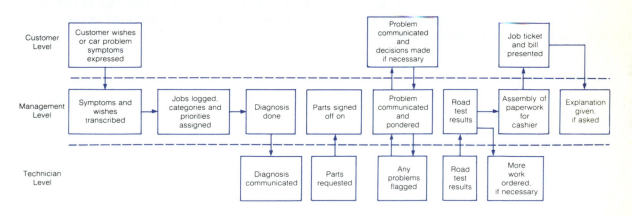

FIGURE G5   Information flow for a car in the course of its repair at Ogle–Tucker Buick.

Of interest is the fact that Ogle–Tucker represents an example of a fixed position layout, where materials are brought to a particular location. Chapter 2 had occasion to discuss this type of layout more and contrast it with other types.

DEMANDS OF THE PROCESS ON THE WORKFORCE

Most of the repair workers at Ogle–Tucker Buick are paid on an incentive basis. The technicians, bodymen, and painters all are paid according to the ratio of their performance relative to a time standard. The base wage against which that ratio is applied depends on their level of skill, which in turn depends on their willingness to learn and the experience they have accumulated. If a technician wants to get ahead and if he wants to be able to be promoted to more prestigious jobs in the shop, he must be willing to go to school. The increasing sophistication of cars demands this. The increasing sophistication of auto repair work is reflected in the fact that Ogle–Tucker's mechanics are not called mechanics but technicians. The use of that term is a reflection on how the workforce and management alike consider the demands of the job.

Working hard as a technician means more than simply beating the standard as set in the book. Working hard also means being able to juggle several different jobs at once so that little time is spent waiting for parts or otherwise removed from the actual repair job. In this regard, the technician is responsible for the pace of his own work. That pace is not dictated by the dispatcher nor by the flow of work along a line. The pace is dictated solely by the technician and his level of skill and effort.

The high level of individual responsibility evident for the technician job carries over to the considerable inventory of tools that everyone is expected to carry. Indeed, the capital-to-labor ratio implied by this tool inventory nearly equals 1 (about $20,000 in tools and somewhat over $20,000 in annual wages, for an average technician) without counting any of the plant, equipment, or parts that are made available to the technicians by the dealership itself. As owners of their tools, technicians are almost like independent operators rather than employees of the company. They have considerable freedom to come and go on their own, for they are paid only for what they do and not for how many hours they spend in the shop per day. Nevertheless, being part of a dealership's service op-

eration has its advantages. The work is steady, training is provided, and the technician does not have any of the headaches that would be involved in establishing and directing a business.

Ogle–Tucker represents a significant contrast to a Burger King. The jobs involved in auto repair are more individual and varying, and, despite some specialization within the workforce, there are fewer job niches. Although some team work is required, the element of team work at the dealership is much less than that which has to apply in a service factory.

DEMANDS OF THE PROCESS ON MANAGEMENT

As the information flow diagram leads one to suspect, auto repair places some significant demands on management. Some of those demands deal with the control of the process, which essentially means control over the rather independent operators. The elements of control include dispatching, diagnosis, the periodic checking of efficiencies, and the quality checks made for both part ordering and road testing. Control over the workforce is a continual endeavor, and one that is fairly loosely defined.

In addition, management must deal with complaints from technicians about their pay or what has been assigned to them. Management must also keep track of the training possibilities for the technicians and who should be assigned to which school.

Monitoring technician training is just one aspect of keeping up to date. Other aspects deal with equipment and information. The use of computers is now standard in most auto dealerships, and computer facilities are steadily being improved. This permits more of the paperwork and information needs of the dealership to be put on computer. Keeping up-to-date with the increasing sophistication of cars requires added investments in plant and equipment and in manuals and documentation. It is important for management to have the right kind of equipment and enough of it so that workers are not cramped for space or equipment and thus do not impede the capacity of the shop. An important aspect of the capital requirements of the dealership is the parts department. Dealers, of course, want to have enough of the right parts available so that they do not stock out too often, but they do not want to drown in inventory. The use of the computer helped to strike this balance, as did the discipline that was applied in keeping track of the parts on hand.

Another whole realm of management effort is devoted to dealing with customers. Managers were in repeated contact with customers—taking orders, handling complaints, and explaining the shop's actions. Senior managers, in particular, were safety valves for the service advisors so that when all the service advisors were occupied with customers, the flow of the process could continue unabated.

Management is also responsible for generating a steady flow of work into the shop. That means trying to perk up business in the off peak times of the year through promotions and specials. And, once a customer does come in, the service advisor informs the customer about other services that his or her car could really use.

QUESTIONS

1. How does the payment scheme for the shop's technicians compare with other incentive pay schemes discussed in previous tours?

2. How does the Ogle–Tucker Buick dispatcher compare with the foreman at Norcen Industries?

3. Why are the hard and soft copies of the repair order split the way they are?

4. What kinds of parts orders suggested by the parts department's computer program would likely be overridden by the parts manager?

5. Why was the parts manager so dedicated to updating each day the perpetual inventory records kept of each part?

6. In what ways was managing the body shop similar to managing the service shop? In what ways were they different?

7. In what ways is the Ogle–Tucker service manager's job like that of the Noblesville Burger King manager? In what ways is it different?

8. Develop process and information flow diagrams for the body shop.

9. Develop process and information flow diagrams for the parts department.

SITUATION FOR STUDY G-1

BIG CITY HOSPITAL EMERGENCY ROOM

The emergency room of the Big City Hospital served about 100 patients a day. Some patients' arrivals were unannounced. Others, however, were announced through communication with ambulance drivers or physicians. The pattern of admission was routine except for patients who were obviously in medical distress.

Each patient entering the emergency room was met by one of two clerks at the front desk. The clerk was the first to assess whether the patient needed immediate medical attention. The clerk would announce "patient here" over the loudspeaker and a triage nurse would take over, reviewing the paperwork that had been done, taking the patient's vital signs, and gathering additional information. The triage nurse then determined whether the patient had to be seen by a physician immediately or whether he or she could wait. If a patient arrived by ambulance, the communication from the ambulance would indicate an estimated time of arrival and some word of the patient's condition. If the patient could give some preliminary information on arrival to the clerk, that would be done; if not, the patient was wheeled directly into the emergency room treatment area.

There were several treatment areas to the emergency room. These included the acute trauma room; a large open area with seven bed slots with curtains to close off each of the slots; four other, separate holding slots; an ear, nose and throat room; an OB GYN room; a cast room; and a family room. The acute trauma room was kept always set up in anticipation of a patient with a severe problem. Up to 20 people could be mobilized to help stabilize a patient in acute trauma.

The open area with the seven bed slots was used for general admission patients. If the patient's condition warranted continual observation, he or she was placed near the nursing station. Patients not needing constant surveillance were placed at more of a distance. The ear, nose, and throat; the OB GYN; and the cast rooms were used for isolating patients with those conditions. The family room was used either for psychiatric patients or for the families of patients who were critically ill.

Once patients were seen by the triage nurse and then by the resident nurse, they were evaluated by a surgical or medical intern, and, if necessary, by an attending resident or physician. Upon examination, the patients were either treated and discharged or admitted to the main part of the hospital. The time goal for admitting, treating, and discharging a patient was 4 hours. Nurses handled any transfers to other areas of the hospital.

Each patient generated substantial amounts of paperwork: personal medical history, notes from physicians and nurses, diagnoses, orders and instructions, laboratory tests performed, and any supplies used. This paperwork was necessary for recording instructions to be used by other hospital personnel, or by departments for billing and for the creation of a permanent medical record. Given these various purposes, the record forms included space for personal data, the nature of the complaint, the physician's name, the type of insurance, notes from the physician, orders from the physician, diagnosis from the physician, and instructions for the discharge. Separate records were kept for nursing, the physicians, and insurance. Lab work was tracked using individual tickets that were kept grouped with the emergency room reports. After a specific time, these reports were entered in the hospital computer for storage.

1. Diagram the process flow.
2. Diagram the information flow.
3. How would you determine how the emergency room ought to be staffed with clerks, nurses, and physicians?

MASS SERVICE
Thalhimers—Cloverleaf Mall Store[a]
Richmond, Virginia

The Cloverleaf Mall store of Thalhimers was one of 25 that the company operated in Virginia, North Carolina, South Carolina, and Tennessee. Thalhimers was a general department store chain, begun in Richmond in 1842, and the fifth generation of the Thalhimer family was presently in the senior management of the corporation. Through the years, Thalhimers grew both by acquiring stores and by building new ones. In 1978, the firm merged with the Carter-Hawley-Hale group of stores, headquartered in California. By dollar volume, the firm's largest store was at Cloverleaf Mall, located in one of the fastest-growing areas in the Richmond market, and situated in one of the 15 fastest-growing counties in America. The store was built in 1972 and had been remodeled twice—extensively in 1978 when a second floor was added to the initial one-floor layout. It now occupied 130,000 square feet. As a general department store, Thalhimers carried a broad selection of merchandise for women, men, children, and the home.

[a]This tour describes operations at Thalhimers in the summer of 1988. Since that time Thalhimers was purchased by the May Department Stores.

PART ONE

SERVICE PROCESS DESCRIPTION

LAYOUT

The ambiance and design of the store was recognized as an important feature in attracting and appealing to customers and generating sales. A good layout accomplishes several things. It provides entertainment and excitement for the shopping experience. The customer is led in a logical pattern from one area to another. A good layout allows a store to

The cosmetics counter. (*Courtesy of Thalhimers—Cloverleaf Mall Store*)

present fashion merchandise in an accessible manner with proper adjacencies, to change the fashion statements each season, and to present the merchandise assortment in a meaningful manner and thereby maximize each area's sales potential.

Traditionally, department stores were designed with long central aisles that effectively divided the store into large quadrants of merchandise of a particular category. Through the years, the old-fashioned layout was supplanted by more architecturally pleasing designs. The Cloverleaf Store, for example, did not have the straight aisles. It was an open store with aisles that led the customer from one fashion world to another using soft wall treatments to designate particular areas or shops. The adjacencies of the areas and the merchandise presentations created the feeling that the customer was shopping in an area with a complete assortment of women's, men's, children's, or home merchandise. It incorporated shops—small, distinctly identified, sometimes partially enclosed areas that dramatized particularly important fashion statements. The Cloverleaf layout did not follow the "world concept." (In that concept of department store design, the consumer enters a well-defined arena or "world" of a particular category of merchandise—such as a "children's world," a "junior's world," or a "men's world." A world concept surrounds the customer with one category of merchandise and catches the customer's eye with dramatic wall treatments, color schemes, and fixtures.)

When Thalhimers designed a new store, space requests would be submitted by the Merchants as well as the Stores Division. For example, a request would be submitted for ladies' accessories for a certain square footage based on a history of results from other stores. Sim-

Signature Sportwear with the Liz Claiborne shop in the background.
(*Courtesy of Thalhimers—Cloverleaf Mall Store*)

ilar requests would be submitted for men's furnishings, the home division, and other areas, each broken down to the level of detail of individual departments within each area. Management then examined the current sales trends for the various segments of the business and the sales per square foot that were generated in each area in the current stores. Decisions were then made concerning space allocation for the new store. Together with the Thalhimers' architects, management then decided on the character of the store's environment; a store in Charleston, South Carolina, would probably be different aesthetically from one in Fayetteville, North Carolina. At this point, decisions were also made about which departments should be adjacent to one another, and what the proximity should be to the various store entrances. The different entrances generated different levels of traffic and it was important to keep that in mind in developing

the plan for the store. Plans and space allocations were changed from one store to the next, depending on the performance expectations of the various areas of the business.

There were substantial differences in the sales per square foot generated by the different departments. For example, fragrances enjoyed an annual sales volume per square foot that was 12 times the store's average. Men's ties produced sales per square foot of 7 times the store's average. Electronic sales were well above the average also. On the other hand, some areas, such as infants' clothing or men's robes, generated sales that were substantially less than the storewide average. Nevertheless, all elements of the store were important to the image of a department store with broad assortments of merchandise.

A tour of the store can help dramatize the importance of the layout. On the first floor, the Cloverleaf store had four entrances. The entrance

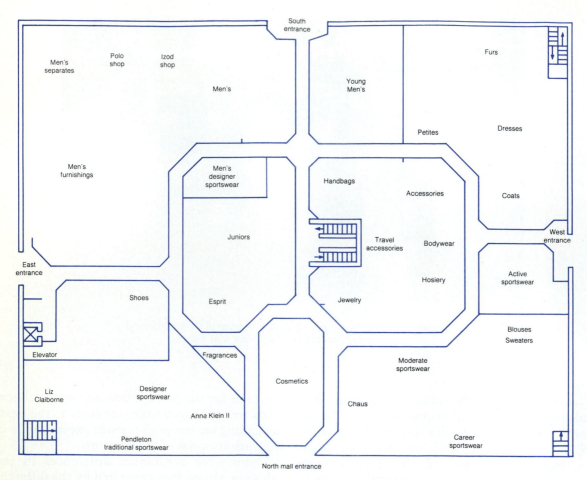

A. First floor layout.

FIGURE H1   Thalhimers—Cloverleaf Mall Store.

with the most traffic was the north or mall entrance. The cosmetics department was located there. The cosmetics counters produced high dollar volumes per square foot. It was an area that created excitement and generated impulse purchasing. To capture the attention of the customer entering the store, immediately adjacent and to either side of the cosmetics counters were a series of fashion shops featuring well-known designer sportswear (Liz Claiborne, Anne Klein

II, Chaus, and Pendleton). These designer fashion shops were appropriate complements to the excitement generated at the cosmetics counters.

As demonstrated by both the cosmetics counters and the designer sportswear shop, the interior design of the store was dramatic, creative, and yet very pleasing to the eye. One could look at the presentations of the merchandise and sense a statement of fashion. The distinctiveness of these statements of taste and

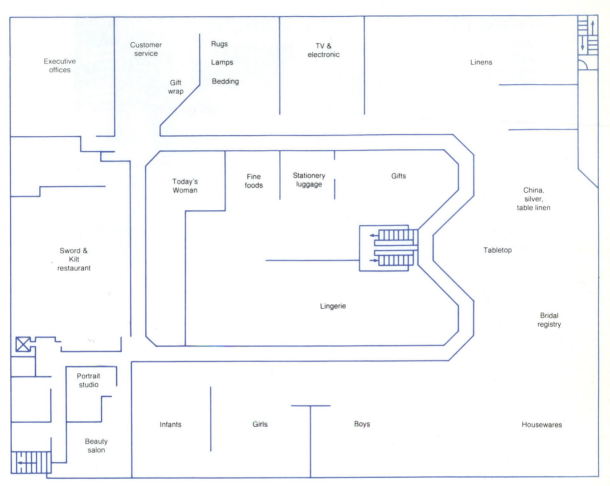

B. Second floor layout.

FIGURE H1   (*continued*)

fashion were important aspects of the sales performance of the store.

Moving west from the cosmetics counters and the designer sportswear shop, a customer would find both fashion and fine jewelry, and on the right, a large selection of moderately priced and career sportswear, blouses, sweaters, and active sportswear. In general, designer fashions and the more expensive labels were located closer to the front of the store. As one continued moving west toward the west entrance (the third most important entrance) (see Figure H1), one could see the bodywear and accessories departments on the inside of the aisle and women's coats, dresses, and petites' clothes on the other side of the aisle. Each of these areas maintained its own fashion statements.

The men's area was located between the south and the east entrances. The south entrance was

The Juniors area. (*Courtesy of Thalhimers—Cloverleaf Mall Store*)

The south entrance. (*Courtesy of Thalhimers—Cloverleaf Mall Store*)

The Polo shop for men. (*Courtesy of Thalhimers—Cloverleaf Mall Store*)

the second most important entrance, and the east entrance was the least trafficked entrance. As with the other areas on the first floor, the fashion statements were presented both on the aisle and with important wall treatments in the rear of the department, acting to draw the consumer's eye into that area of the store. The menswear area had its own brand name shops, including Izod and Polo.

Several areas in the store served as "swing" areas. Some, for example, offered sweaters for the fall and winter seasons and bathing suits for the spring and summer seasons. Other areas of the store were assigned even more temporary space, such as seasonal merchandise for Christmas. These trends and seasonal changes meant that the layout changed constantly.

The store's second floor was as fashion conscious as the first. On entering the second floor

the customer viewed a broad assortment of fine china, silver, and glass stemware from around the world in the tabletop area. To the right of this area was the gourmet housewares area, which dramatically presented a broad selection of both imported and domestic merchandise. Other areas on the second floor were the children's departments, gourmet foods, ladies' lingerie, large-size women's clothes, the restaurant, television and electronics, and the linen and bedding shops.

MANAGEMENT AND THE WORKFORCE

The organization chart for the Cloverleaf store is displayed in Figure H2. Reporting to the store manager were the assistant manager, the

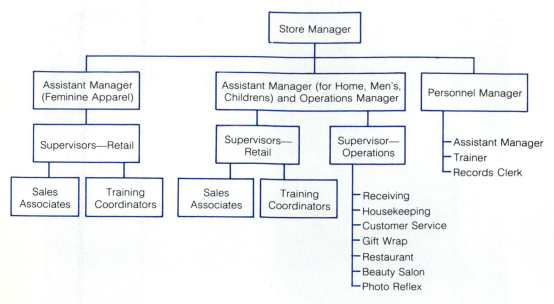

FIGURE H2   Organization chart for the Cloverleaf Mall Store of Thalhimers.

operations manager, and the personnel manager. The store manager also retained direct supervision of the home and children's areas. Reporting to the assistant manager were the sales managers for the other sales areas within the store. Reporting to the operations manager were the lease departments, the restaurant manager, the customer service supervisor, and the supervisors for housekeeping and merchandise handling. Each selling center (made up of multiple complementary departments) of the store was managed by a sales manager, and selling center sales associates reported directly to their sales manager. Also reporting to the sales manager could be an assistant sales manager (three intern-interim training positions for sales associates working their way up through the ranks).

Ensuring customer satisfaction was the primary focus of all associates. Each level of management periodically attended training seminars devoted to teaching, reinforcing, and perpetuating the culture of customer satisfaction of the Cloverleaf store. Monitoring and evaluating all elements of customer satisfaction for all associates was an ongoing function of management. Customer service was an element on which each employee was annually reviewed.

There were 250 employees in the Cloverleaf store, roughly half of whom worked full time and half part time. The store was open from 10:00 in the morning until 9:30 in the evening, Monday through Saturday, and from 12:30 to 5:30 on Sundays. With the store open so many hours during the week, employing part-time personnel was essential to providing maximum staffing. Full-time people were required to work two nights a week and to work every other Sunday. The part-time people worked a fixed schedule: mornings, evenings, and weekends.

During the Christmas season, additional temporary employees were hired (180 for the 1988 Christmas season). This temporary workforce was hired in late September and early October, trained at the end of October and began

The china and crystal area. (*Courtesy of Thalhimers—Cloverleaf Mall Store*)

The Market Place shop. (*Courtesy of Thalhimers—Cloverleaf Mall Store*)

work on the floor in early November. This hiring and training process ensured that every temporary employee has sufficient experience on the floor prior to Thanksgiving, the beginning of the strongest part of the Christmas buying season.

In addition to the store's part-time employees, there was an additional group of sales associates. The group was known as "flyers," and they formed the "flying squad." These substitute employees on the flying squad were called on short notice to work a particular day or period of time. Depending on their individual schedules, they would or would not be available to work. They were all fully trained to fill in in various departments throughout the store.

The store's employees were paid in different ways, although the same employee benefit package applied to all. Support personnel were paid between $3.35 and $7.00 an hour, based on experience and background. Sales associates, on the other hand, were paid a commission based on their individual sales. The commission rate varied from one department to another. The typical commission was 6.5 percent for most areas of the store, but it could reach 8.5 percent for accessories and 10 percent for shoes and the bridal gift areas. Each employee was paid a base hourly rate depending on experience and his or her ability to generate sales, which was applied against the commission earned. In other words, each associate was paid a commission on the merchandise he or she sold, which constituted his or her total earnings. The commission arrangement was new for the store except for shoes and electronics. It showed great promise for generating higher sales for the stores, encouraging individual effort, and for using product knowledge and individual selling expertise.

Both hiring and training were extremely important activities. The right sales associate for a particular department was critical to that department's sales performance. Different personalities were required for different departments. For example, cosmetics required sales associates who were aggressive and enthusiastic as well as highly knowledgeable about the product. China and silver, on the other hand, required sales associates of good taste and great patience, capable of serving the bride and the mother of the bride. Younger sales associates could interact more effectively with teenagers in the juniors' department. If certain clothes were selling well, it was advantageous to have sales associates who could relate to the merchandise and enjoy wearing that style of fashion.

For these reasons, the hiring decision was important and took on elements of theatrical casting. There were three required visits before a new employee could be hired, after which the employee was given 24 hours of training off the floor, 2 weeks of training on the floor, and then a 90-day period of evaluation. In addition, training sessions for sales associates (meetings and videotape viewings) were held weekly. Training was important not only for acquiring knowledge about the merchandise and the operation of the department, but also for learning selling skills. Thalhimers engaged in several innovative programs to train its sales associates. Everyone played the "Selling Game," a monopoly-like game specifically developed to teach selling skills. Through the play of the game, and by viewing accompanying videotapes, sales associates learned the best ways to approach the customer, to interpret verbal and nonverbal buying signals, to present the merchandise, and to close the sale, among a host of other things. Sales associates also attended product seminars that Thalhimers hosted from time to time. At these seminars, the store's vendors displayed the merchandise that Thalhimers had ordered for the next season and provided the sales associates with the latest product knowledge.

The sales managers were essential to successful operation of the store. Each year for a week they attended a sales leadership development seminar, focusing on the latest in store operations and in motivation and communication techniques. The 15 sales managers were responsible for a number of important activities, among which were:

1. Scheduling the sales associates in the department, both full and part-time, and determining when to use the flying squad.

2. Overseeing the training of sales associates.

3. Monitoring all sales transactions.

4. Receiving inventory, checking it against the paperwork, and reconciling any errors in its shipping or paperwork (explained later).

5. Adjusting prices and preparing the department for sales and promotions.

6. The store's management was very supportive of sales manager initiatives to make the Thalhimers shopping experience as enjoyable as possible. Fashion shows, cosmetics make-overs, special senior citizen promotions, focus group meetings with customers, and clientele books with customer preferences noted, among other things, were routinely employed to enhance Thalhimers' already enviable reputation in the community as the fun place to shop. Figure H3 displays two courtesy notes that Thalhimers' sales associates sent customers.

Some of the sales managers began their careers as sales associates. Others were management trainees, employed out of college, who served in the store for 2 years before entering the merchandising arm of Thalhimers as assistant buyers.

THE TRANSACTION — AND SUPPORTING IT

At the heart of the store's operations was the sales transaction itself, and the information it generated. Scattered throughout the store in the various departments were point-of-sale devices or registers. These registers were all tied into the store's computer, and via that computer, they were tied to the computer at Thalhimers in Richmond and the computers of Carter-Hawley-Hale in California. All transactions, once they were rung on the register, were immediately sent on to the computers in California and in the Richmond headquarters. Each transaction consisted of the following items of information: the type of payment to be made (cash or charge), the personnel identification number for the sales associate (the "pin number"), the department's number, the class of merchandise being sold, the stockkeeping unit number for the item (the "SKU number"), and the price for the item (see Figure H4). Once these bits of information were entered into the register, the sale could be completed, the drawer opened, and if change was required, change would be returned to the customer. With these registers, immediate credit information was available. Credit card numbers could be traced immediately and information relayed to the register on whether or not to complete the transaction. With the registers tied in to the company's computers, up-to-the-minute sales reports, or return reports, by department, could be generated at any time.

Supporting the store's transactions was a staff in the customer service area. The customer service department was responsible for

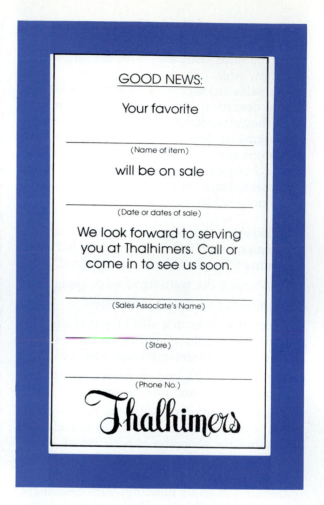

FIGURE H3   Courtesy notes sent to customers by sales associates.

opening up each register each day (including supplying it with $75 reserve change). Sales associates were required to log into the register upon arrival and to log out of the register when they had finished working for the day. At the end of the day the customer service staff was responsible for accounting for all cash and charges, and for checking any discrepancies. The money and the charge account media were taken to the registers by a sales associate and returned to the credit office in the evening by a sales associate. They were verified by the sales managers before being turned in to the customer service office.

The customer service department also ran the customer service window, through which it received payments, adjusted bills, sold tickets to various theater events, provided extra coins or cash to registers throughout the store, handled some customer satisfaction issues, and

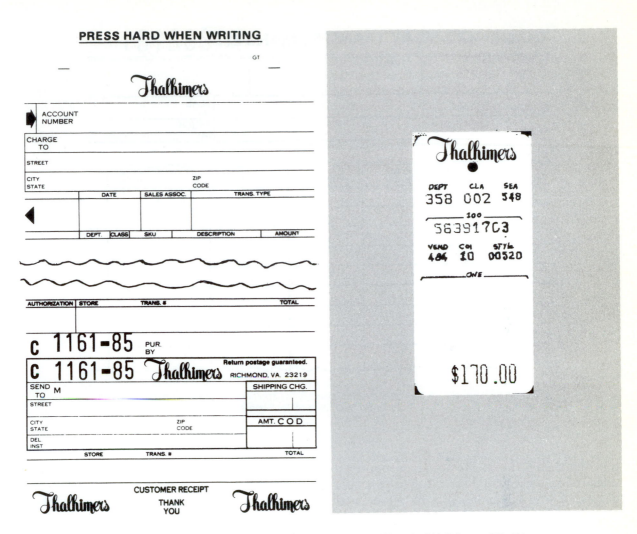

FIGURE H4   A price tag and sales ticket form used at the Cloverleaf Mall Store of Thalhimers.

cashed checks. Customer service was a vital center for the store's transactions.

MANAGING AND CONTROLLING THE INVENTORY

Providing the merchandise for the Cloverleaf store was the responsibility of Thalhimers' merchandising staff located at the Richmond headquarters. The buyers purchased the merchandise that would be sold at Cloverleaf and determined how much merchandise was to be purchased, when the merchandise was to be delivered, and for what price the merchandise would sell. Many of the items in Thalhimers were fashion sensitive, with lead times between ordering and receiving of 6 to 12 months. Thus

THALHIMERS RELAY TRANSFER MANIFEST 292101

DATE		TIME OUT		TIME IN		

DRIVER					SEAL No.	
FROM		TO			TRUCK No.	
DISPATCHER			RES. CLERK			

NO.	DEPT.	RECEIVING NO. OR ROSTN. NO.	HAMPER NO.	SEAL NO.	NO. OF CARTONS	NO. OF UNITS	Chk.
1							
2							
3							
4							
5							
6							
7							
8							
9							

FIGURE H5   A shipment manifest used in transporting merchandise at the Cloverleaf Mall Store of Thalhimers.

a buyer often had to make decisions months in advance on what would sell during a particular season. If the buyer bought the wrong merchandise or ordered too much, the store's sales and profits suffered. If the buyer erred on the other side and did not order enough merchandise, the store suffered a loss of potential revenue. The buying decision was thus a very important one, but one that was essentially outside the control of a branch store such as Cloverleaf. Of course, the store could provide feedback to the buyers on what merchandise was selling well, the type of customers who seemed to be buying it, and how it could be effectively presented, but the merchandising decision rested with those at the corporate headquarters.

Thalhimers' new computer system, dubbed IMIS (Integrated Merchandise Information System), was a key resource that helped relieve the store of its past labor-intensive burden of recordkeeping. IMIS kept up-to-the-minute track of every item in stock at all of the company's stores, including such information as the quantity on hand or on order, sales, returns, price changes, and transfers to other stores or returns to the vendor.

Even automatic reordering of merchandise was possible with the system to maintain inventories within a minimum-to-maximum

MOVEMENT LABEL

SERV. BLDG.

RICHMOND, VA.

DEPT.	Receiving Apron No.	RTW Pieces	SPECIAL INSTRUCTIONS
			This is of Cartons -Pieces
			Wheeler Container NO.
			Sent by | Location | Date

FIGURE H6   A movement tag that accompanies merchandise between the service center and the store.

range that was preagreed to between the buyer and the store manager.

Receiving

All items received at the store were sent from Thalhimers' Distribution Center (DC). At the DC, items were checked, sorted, priced, and marked. Deliveries to the Cloverleaf store were made between 4 and 6 A.M. The receiving crew started work at 6 A.M. and delivered all of the merchandise that was received that morning to the floor by 9 A.M. Several checks were made of

the merchandise received to ensure that the merchandise received and accounted for was what was expected and had, in fact, arrived. Each truck was physically sealed and the number on the metal seal had to match the manifest that accompanied the shipment (see Figure H5). After the seal was broken, the crew grouped those boxes that were shipped together and checked their numbers against the manifest. Hanging garments in canvas rollers, termed "hampers," were similarly checked. Each group of boxes or "hampers" had been labeled with a movement tag (Figure H6). The movement tag indicated the store department to receive the merchandise, the purchase order number, the hamper number, and the seal number (which indicated the truck shipment). The movement tags came in triplicate; the white tags stayed with the hamper or box, yellow tags were kept by the receiving store, and pink tags were kept by the service building.

After the truck was unloaded, the merchandise was moved onto the selling floor. The merchandise handling group knew exactly where each department received its merchandise, usually near its register. The merchandise moved out to the floor was accompanied by the white movement tag and the paperwork on the purchase, typically the purchase order (see Figures H7 and H8 for examples of this paperwork).

Every morning the sales manager and the sales associates in each department checked the merchandise delivered against the accompanying paperwork to ensure that all of the items they expected to receive were, in fact, received. Each department checked the seals on their hampers and boxes, circled the counts to indicate agreement, stamped the paperwork "received," and then filed that paperwork appropriately. The stock was then displayed in the appropriate place in the department.

The records of the inventory received were kept in a "journal room." The journal room was also next to the office of the inventory controller and the mailboxes for all the supervisory personnel. It was the nerve center for processing the merchandise paperwork for inventory.

VARIATIONS ON THE STANDARD RECEIPT OF MATERIAL

While most of the merchandise received at the store followed the standard procedure, there were a number of errors and changes that had to be recorded. These errors and changes were generally kept on the computer, although some hard-copy "books" were kept as well.

1. *Interstore transfers.* Often items of merchandise were transferred from one store to another. When this occurred, those stores involved entered the information on the store's computer and the appropriate records were updated (see Figure H9).
2. *Errors in shipment.* If there were errors in a shipment received by the store (what was received did not match the paperwork), these errors were noted.
3. *Report of changes in retail price.* This computer record tracked changes that had occurred to the retail price, for example, missed markdowns or markups, and month-end special sale liquidation of inventory.
4. *Merchandise transfer.* If merchandise had been incorrectly charged to the store or had been incorrectly charged to another department, these errors were corrected on the computer records.

Through the use of computer records for changes and errors, the store kept excellent track of exactly what inventory it had and that with which it should be credited. Price changes were also kept track of by the sales managers and the inventory controller. Figure H10 offers an example of a price change document.

SHIPPING ITEMS OUT OF THE STORE

It was Thalhimers policy not to ship items between stores or back to the service building unless the dollar value of the transfer justified the expense. In general, unless a shipment was worth $1000 or more, no transfers were made.

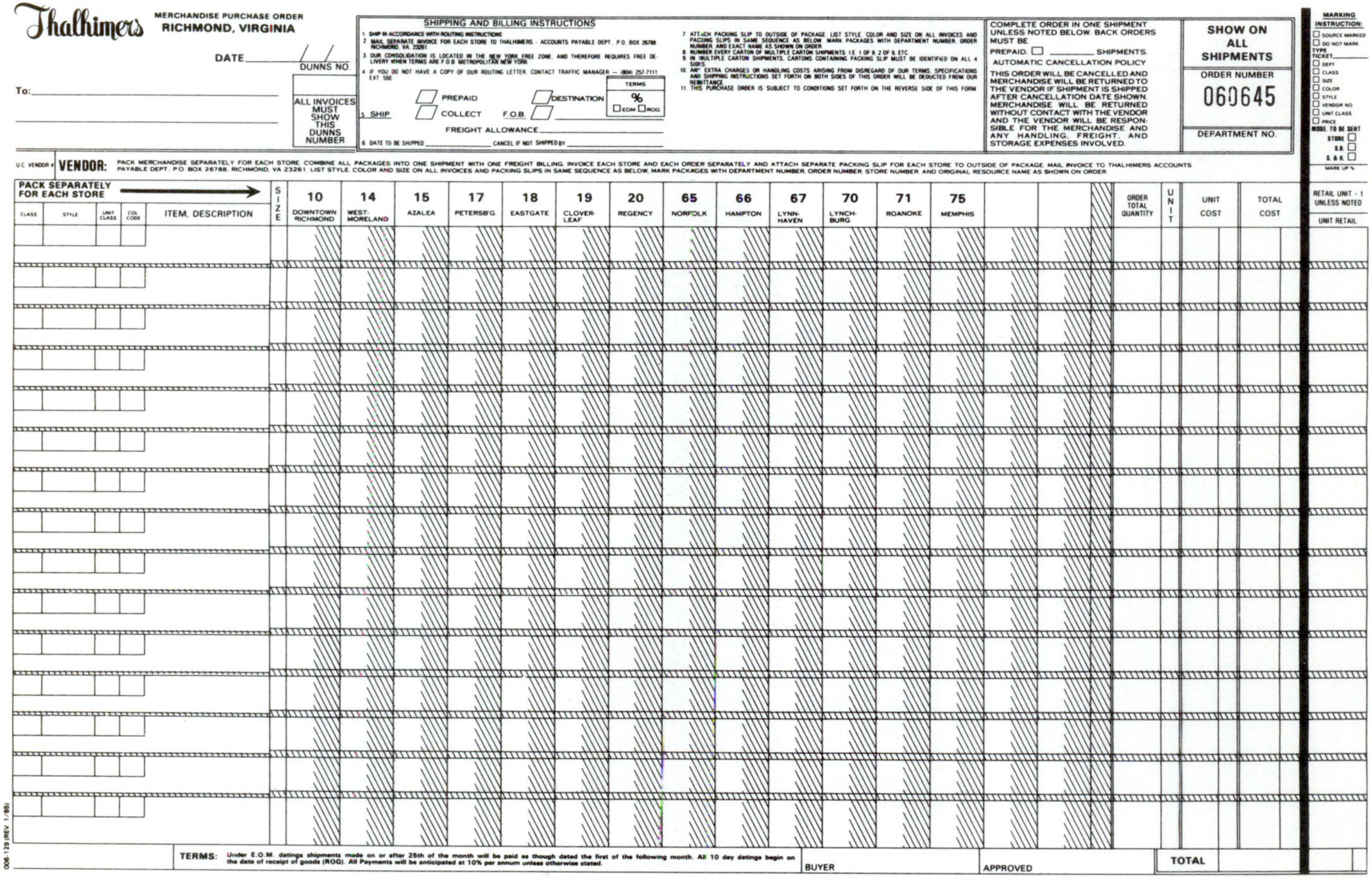
Thalhimers MERCHANDISE PURCHASE ORDER RICHMOND, VIRGINIA

DATE / /

DUNNS NO.

To:

ALL INVOICES MUST SHOW THIS DUNNS NUMBER

SHIPPING AND BILLING INSTRUCTIONS

1. SHIP IN ACCORDANCE WITH ROUTING INSTRUCTIONS
2. MAIL SEPARATE INVOICE FOR EACH STORE TO THALHIMERS - ACCOUNTS PAYABLE DEPT., P.O. BOX 26788 RICHMOND, VA 23261
3. OUR CONSOLIDATION IS LOCATED IN THE NEW YORK FREE ZONE, AND THEREFORE REQUIRES FREE DELIVERY WHEN TERMS ARE F.O.B. METROPOLITAN NEW YORK
4. IF YOU DO NOT HAVE A COPY OF OUR ROUTING LETTER, CONTACT TRAFFIC MANAGER — (804) 257-7111 EXT 550
5. SHIP: PREPAID / COLLECT; F.O.B.: DESTINATION; TERMS: % ☐ EOM ☐ ROG; FREIGHT ALLOWANCE
6. DATE TO BE SHIPPED; CANCEL IF NOT SHIPPED BY
7. ATTACH PACKING SLIP TO OUTSIDE OF PACKAGE. LIST STYLE, COLOR AND SIZE ON ALL INVOICES AND PACKING SLIPS IN SAME SEQUENCE AS BELOW. MARK PACKAGES WITH DEPARTMENT NUMBER, ORDER NUMBER, AND EXACT NAME AS SHOWN ON ORDER.
8. NUMBER EVERY CARTON OF MULTIPLE CARTON SHIPMENTS. I.E. 1 OF 9, 2 OF 9, ETC.
9. IN MULTIPLE CARTON SHIPMENTS, CARTONS CONTAINING PACKING SLIP MUST BE IDENTIFIED ON ALL 4 SIDES.
10. ANY EXTRA CHARGES OR HANDLING COSTS ARISING FROM DISREGARD OF OUR TERMS, SPECIFICATIONS AND SHIPPING INSTRUCTIONS SET FORTH ON BOTH SIDES OF THIS ORDER WILL BE DEDUCTED FROM OUR REMITTANCE.
11. THIS PURCHASE ORDER IS SUBJECT TO CONDITIONS SET FORTH ON THE REVERSE SIDE OF THIS FORM

COMPLETE ORDER IN ONE SHIPMENT UNLESS NOTED BELOW. BACK ORDERS MUST BE PREPAID. ☐ ______ SHIPMENTS.
AUTOMATIC CANCELLATION POLICY
THIS ORDER WILL BE CANCELLED AND MERCHANDISE WILL BE RETURNED TO THE VENDOR IF SHIPMENT IS SHIPPED AFTER CANCELLATION DATE SHOWN. MERCHANDISE WILL BE RETURNED WITHOUT CONTACT WITH THE VENDOR AND THE VENDOR WILL BE RESPONSIBLE FOR THE MERCHANDISE AND ANY HANDLING, FREIGHT, AND STORAGE EXPENSES INVOLVED.

SHOW ON ALL SHIPMENTS

ORDER NUMBER 060645

DEPARTMENT NO.

MARKING INSTRUCTION: ☐ SOURCE MARKED ☐ DO NOT MARK; TYPE TICKET; ☐ DEPT ☐ CLASS ☐ SIZE ☐ COLOR ☐ STYLE ☐ VENDOR NO ☐ UNIT CLASS ☐ PRICE; MODE TO BE SENT: STORE ☐ S.B. ☐ S.&H. ☐; MARK UP %

U C VENDOR # VENDOR: PACK MERCHANDISE SEPARATELY FOR EACH STORE. COMBINE ALL PACKAGES INTO ONE SHIPMENT WITH ONE FREIGHT BILLING. INVOICE EACH STORE AND EACH ORDER SEPARATELY AND ATTACH SEPARATE PACKING SLIP FOR EACH STORE TO OUTSIDE OF PACKAGE. MAIL INVOICE TO THALHIMERS ACCOUNTS PAYABLE DEPT. P.O. BOX 26788, RICHMOND, VA 23261. LIST STYLE, COLOR AND SIZE ON ALL INVOICES AND PACKING SLIPS IN SAME SEQUENCE AS BELOW. MARK PACKAGES WITH DEPARTMENT NUMBER, ORDER NUMBER, STORE NUMBER, AND ORIGINAL RESOURCE NAME AS SHOWN ON ORDER

PACK SEPARATELY FOR EACH STORE → CLASS	STYLE	UNIT CLASS	COL CODE	ITEM. DESCRIPTION	SIZE	10 DOWNTOWN RICHMOND	14 WEST-MORELAND	15 AZALEA	17 PETERSB'G.	18 EASTGATE	19 CLOVER-LEAF	20 REGENCY	65 NORFOLK	66 HAMPTON	67 LYNN-HAVEN	70 LYNCH-BURG	71 ROANOKE	75 MEMPHIS			ORDER TOTAL QUANTITY	UNIT	UNIT COST	TOTAL COST	RETAIL UNIT - 1 UNLESS NOTED UNIT RETAIL

TERMS: Under E.O.M. datings shipments made on or after 25th of the month will be paid as though dated the first of the following month. All 10 day datings begin on the date of receipt of goods (ROG). All Payments will be anticipated at 10% per annum unless otherwise stated.

BUYER APPROVED TOTAL

006-129 (REV 1/85)

FIGURE H7 A purchase order sheet that is filled out by a buyer and that accompanies merchandise to the Cloverleaf Mall Store. Note that Cloverleaf is Store 19.

```
   11      99999      11      99999    ••••• MARKING AND TRANSFER •••••     PRINTED 12/17/85  TIME 06:22    PAGE    1
  111     99  99     111     99  99
   11     99  99      11     99  99    MRKNG SEC/LOC _____ MRKNG LINE _____ MRKER ID ___ TKT PRT T3C@        OP ID TP00021
   11     99999       11     99999     RCV 12/16/85                  DOC : N          DOC TYPE P
   11        99       11        99
 1111        99     1111        99     VENDOR    878 . JOSIAH WEDGWOOD AND CO

 DEPT   281 # OF STYLES    2  SKUS     2
 RCVNG  281530                UNITS    2

 PO  20155396

    MFG-STYLE STYLE CD COLOR    SIZE   RCVD   REA LINE   SSKU     DESCRIPTION         RTV  CLASS   RETAIL UM   EXT-RTL

         6 00006 00                1    ___ 0601 56147314 OCEANSIDESANDWICHTRA        0     5     25.00 EA     25.00
         8 00008 00                1    ___ 0801 55948658 OCEANSIDETANKARD           0     5     17.50 EA     17.50
```

FIGURE H8   A computer-generated marking and transfer sheet that performs the same function as the purchase order sheet. It, or the purchase order sheet, accompanies goods to the store. It is generated by a buyer and is store-specific. Note the 19 in the upper left corner.

One day a month was allocated to transferring merchandise to other stores to balance inventories. There were some exceptions to this policy; vendor returns, damaged goods, alterations, and monograms could be returned to the service center or moved between stores on a daily basis. Shipping of this merchandise invoked the same care and controls that were shown for the receipt of the merchandise.

EVALUATING THE STORE'S PERFORMANCE

The store was evaluated as a profit center. The profit plan for the store involved the determination of sales for each department and a calculation of the proposed selling cost. (Selling cost was calculated as the compensation paid divided by the sales in the department for the time period under consideration.) The sales for each department were forecast by the merchandising arm of the company. The merchandisers were responsible for analyzing the sales trend for the store as a whole and any trends for individual areas within the store. A sales plan was developed for each department of the store and stock levels were planned to vary seasonally. Different "thrust" areas were also singled out for unusual attention; separate plans were developed for increased inventory, increased promotional activity, and additional personnel to achieve a substantially larger-than-normal increase in volume. There were daily, weekly, and monthly reports that monitored the progress of these departments and any variances from the projected sales and profit plan.

The store carefully monitored the "penetration" of particular departments. Penetration of a specific area was calculated by dividing the sales of the area by the total sales of the store. If this fraction for a particular area at Cloverleaf differed significantly from those of other Thalhimers stores, the store manager could then investigate reasons for that difference. In this manner, the store manager could move more aggressively to direct business opportunities.

The Cloverleaf store manager spent a great deal of time on the selling floor, observing the operation and communicating personally with the employees. His priorities for ensuring the store's profitability were clear: assuring complete customer satisfaction in all areas of the store, monitoring sales, selecting the proper personnel and the development of that personnel, presenting the merchandise in an appealing way to attract customers, and finally, monitoring expenses.

DIV: THALHIMERS
ON-LINE REPORT   ID: HNASO600   I N T E R - S T O R E   T R A N S F E R   S Y S T E M
                                            TRANSFER DOCUMENT
                                                                          PAGE: 1 OF: 2
STORE: 22   REPORT DATE: 09/23/88
                                                                          RUN DATE: 09/23/88
DEPT: 7521   TRANSFER NO: 10362192   X-REF:
REASON: 26 CENTRL STK VERSION: SEND-BY DATE: 09/23/88
MSG:

                                                    TO   11   99999
                                                    STR   111   99 99
                                                          11   99 99
                                                          11   99999
                                                          1111   99

CUST: PHONE: EMP:

DESCRIPTION:	CLASS:	VNDR:	STYLE:	*** COLOR *** NUM: DSCRPTN:	** SIZE * NUM: DESC	SSKU/UPC NUM:	MFG/STYLE:	RETAIL:	RQST QTY:	SENT QTY:	RCVD QTY:
LIBERTY 5 PC.PL	30	82	1207	0	0	59901524		105.00	1	1	0
STERL.COVE CUP	50	142	5702	0	0	72479750	5	24.50	1	1	1
STERL.COVE SAUC	50	142	5703	0	0	72479769	5	12.00	1	1	1
STERL.COVE SALA	50	142	5704	0	0	72479777	5	17.00	1	1	1
STERL.COVE BREA	50	142	5705	0	0	72479785	5	12.50	1	1	1
STANFORD CT.20	50	709	1708	0 0	0	59320432		340.00	2	2	2
STANFORD CT.OVA	50	709	1711	0	0	59278142		65.00	1	1	1
STANFORD COURT	50	709	1712	0	0	59287591	1	105.00	1	1	1
ROTHSCHILD 20 P	50	709	3308	0	0	59320351		278.00	1	1	1
CARLYLE 5 PC.PL	60	134	1007	0	0	70968029		198.00	12	12	12
CARLYLE 20 PC.S	60	134	1008	0	0	59910868		792.00	2	2	2
CARLYLE COMPLET	60	134	1009	0	0	70968037		520.00	1	1	1
CARNATION GRAVY	60	134	1119	0	0	71781364		48.00	1	1	1
SHERBROOKE DINN	60	134	1301	0	0	59623370	1	34.00	4	4	4
SHERBROOKE CUP	60	134	1302	0	0	59282026	1	24.00	4	4	4
SHERBROOKE SAUC	60	134	1303	0	0	59282093	1	18.00	4	4	4
SHERBROOKE OPEN	60	134	1311	0	0	59505874		78.40	1	1	1
SHERBROOKE COVE	60	134	1317	0	0	72471342		160.00	1	1	1

FIGURE H9 A transfer document.

```
BATCH REPORT HMMS5411-01              P R I C E   C H A N G E
   DIVISION: BOYLE 27-TABLE TOP
DEPT      PC NUMBER     XREF NUMBER    XREF TYPE     REASON
7521      5556503                                      12
                     J.Koles
COUNTED BY --------------------------DATE- 08/30/88
MESSAGE: NORITAKE PRICE INCREASE - PATTERN EDGEWATER

_____DESCRIPTION_____  CLASS FRM TO   _____VENDOR_____ STYLE __
                         SEA SEA        NAME      NUM   NUM NU

5PC.PLSETTING              50         NORITAKE CO I  142  6207
DINNER |||                 50         NORITAKE CO I  142  6201
CUP |||                    50         NORITAKE CO I  142  6202
SAUCER |                   50         NORITAKE CO I  142  6203
SALAD ||                   50         NORITAKE CO I  142  6204
B&B ||                     50         NORITAKE CO I  142  6205
14IN PLATTER               50         NORITAKE CO I  142  6212
SUGAR                      50         NORITAKE CO I  142  6213
CREAMER                    50         NORITAKE CO I  142  6214
20 PC.SET                  50         NORITAKE CO I  142  6208
5PC.COMPLETER              50         NORITAKE CO I  142  6209


  W O R K S H E E T           DATE 08/23/88 PAGE     1 OF  1
                                 STORE:  19
PC TYPE     EFFECTIVE DATE    CANCEL DATE      REPRINT
               08/30/88                          N

   MARKED/APPROVED BY_____DATE __/__/__


__COLOR____  SIZE UOM    OLD        NEW  C __QUANTITY___

M  DESCRPTN  DESC      RETAIL     RETAIL R SEAS TKT MRKD LN IT
              EA        77.50      81.50  ___ ____  0  01
              EA        24.50      25.50  ___ ____  3  02
              EA        23.50      24.50  ___ ____  3  03
              EA        11.50      12.00  ___ ____  1  04
              EA        16.00      17.00  ___ ____  2  05
              EA        12.00      12.50  ___ ____  2  06
              EA        95.00     100.00  ___ ____  0  07
              EA        54.00      55.00  ___ ____  (  08
              EA        35.50      37.00  ___ ____  )  09
              EA       310.00     326.00  ___ ____     10
              EA       249.50     257.00  ___ ____  (  11

                 QUANTITY TOTALS:              11
```

FIGURE H10   A price change worksheet.

PART TWO

DISCUSSION

Thalhimers is in no way responsible for the following views or presentation. They remain solely the responsibility of the author.

THE FLOW OF THE PROCESS AND OF INFORMATION

During the course of a day the sales associates at Thalhimers do many things. The two chief processes in which they get involved are stocking, which occurs in the morning, and the sales transaction itself, which occurs all day long. Figure H11 diagrams both of those key processes. For the most part, these processes are fairly well structured. While there is interaction between customer and sales associate during the transaction, particularly in the presentation of the merchandise and the closing of the sale, for the most part the process is a fairly

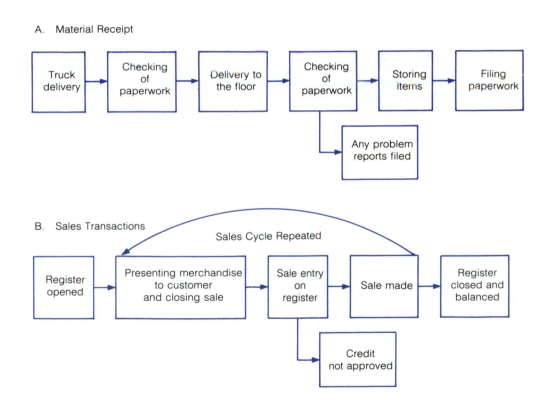

FIGURE H11   Rough process flows for material receipt and for sales transactions at the Cloverleaf Mall Store of Thalhimers.

A. Information flowing down from management
Merchandising shifts
- sales and promotions
- changes to merchandise presentations
- layout and space allocation changes

B. Information flowing up from the sales associates and support staff
Irregularities
- in transactions
- in inventory status
- in paperwork
- in cash balances
- other errors

Inventory counts

FIGURE H12   Examples of information flows at the Cloverleaf
Mall Store of Thalhimers.

standard one, well known to both parties. Similarly the stocking process is something that is a routine occurrence, day in and day out.

Information flows are also fairly standardized, although there are many different kinds of information that can be passed down from management to the sales associate and up from the sales associate to management. Figure H12 indicates some of the key information flows that go both down and up in the organization. For the most part, the information flowing downward relates to merchandising shifts, whereas the information flowing upward deals with irregularities in the routine of the inventory and sales transactions.

DEMANDS OF THE PROCESS ON THE WORKFORCE

Because Thalhimers is open so many hours, one key demand on the workforce relates to the scheduling of everybody's work hours. Evening and weekend work is sometimes required.

Another key demand relates to coping with the change that goes on around the employees all the time. Layouts are changing, inventory stock levels are changing, the character of the inventory is changing, and the workforce must cope with all of these changes. To help cope

with these changes, there is considerable training. Workers are not only trained initially, but spend appreciable amounts of time week in and week out in training, both to improve their skills and to educate them about the merchandising changes in the store.

Another demand on the workforce is accuracy, both with respect to the transactions made and the inventory overseen. Such accuracy is crucial to the success of the store, and it is a routine to which they must apply themselves all the time.

DEMANDS OF THE PROCESS ON MANAGEMENT

Like the sales associates and others of the workforce, management must also adhere to and foster the importance of controls on transactions and inventory. The customer service department and the journal room are the store's nerve centers for control, and they must be supported.

Personnel functions are also extremely important, as indicated by the personnel department's status within the organization chart. Hiring—perhaps better termed "casting"—training, and advancement all consume a lot of management time and care. Sales associates' skills are critical to the sales volume that the store generates, and thus hiring and training are key functions of management.

The other key elements to sales increases are the layout itself and the presentation of merchandise. Good layout and good merchandise presentation depended not only on the analysis of sales per square foot figures, but also on intuition and feel for the psychology of the sales situation. The worth of powerful fashion "statements" throughout the store is a clear indication of the art involved in a store's design and operation.

Because fashion can be so fickle and people's tastes can change so abruptly, it is important for the store to monitor sales and costs on a continual basis. Thus not only are there controls on the transaction and inventory levels themselves, but also controls on sales and selling costs. Store management has to be ready to examine these costs and use them to make decisions about staffing levels and the responsibilities of sales associates and support personnel within the store.

QUESTIONS

1. Visit a department store near you. In what ways is its layout similar to that of Thalhimers? In what ways is it different?

2. One might suppose that theoretically, at least, departments ought to be given space in such proportions that the sales per square foot would be roughly the same anywhere in the store. Why is the point of view impractical?

3. Express some of the fashion statements you have observed at department stores. How does the store's layout, fixtures, and decor contribute to those statements?

4. What kinds of questions would you ask of a prospective employee to work in the Liz Claiborne shop? What characteristics would you look for? How would those characteristics and questions differ for someone working in the jewelry department?

5. What is the argument for paying sales associates on a commission basis? What is the argument for paying them on an hourly basis?

6. Outline the various controls that exist on the store's inventory. Why are there so many controls?

SITUATION FOR STUDY H-1

ADAMS CONVENIENCE STORE

The Adams Convenience Store, located on a major state artery, was open from 6 A.M. until midnight every day and sold grocery items, fast food, and gasoline. Gasoline, dispensed through self-service pumps, represented about 50 percent of total sales. Customers paid first and pumped afterward. All of the gasoline pumps were multiproduct dispensers that handled both leaded and unleaded gasoline.

The store was laid out as in Figure H13. Seventy percent of the people who entered the store to do something other than to pay for gasoline, bought some sort of beverage. The fast food available at the rear of the store was

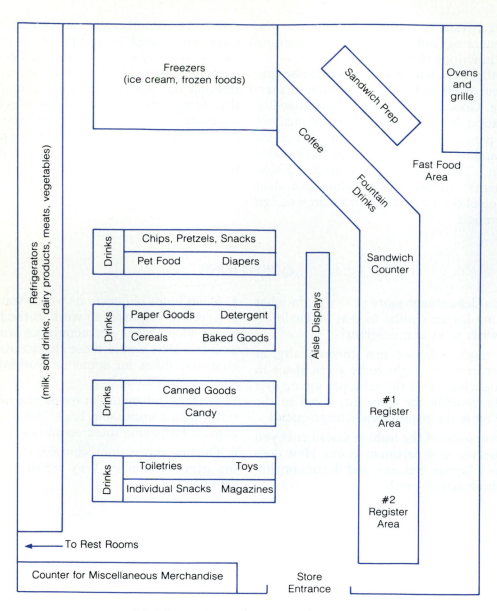

FIGURE H13   Adams Convenience Store layout.

the highest-margin item for sale; gasoline was the lowest-margin item. The Adams Convenience Store offered hot dogs, sandwiches, pastries, coffee, and other food items. There were two peaks for food item demand: breakfast and the four to six, after-work period. Sales slowed in mid-morning, early afternoon, and late at night.

The store operated two shifts. The store manager was there to open the store in the morning and stayed during the first shift; the store's assistant manager handled the second shift. During the busiest period of each shift there were a total of four working in the store: two to handle the fast food area and two to tend the registers. In addition to the manager and the assistant manager there were two full-time workers and the remaining were part-time employees. Pay averaged just above the minimum wage, and turnover rates often ran as high as 300 percent a year.

The manager and the assistant manager trained the others and determined which items met the company's definition for being short in supply and needed to be reordered. Orders were placed to a central Adams warehouse. The managers were also responsible for controls on cash and inventory so that everything was accounted for.

1. Why is the layout designed as it is? Give a rationale and explain what you think the typical traffic flow would be.
2. Why would the Adams Convenience Store invest in multiproduct gasoline pumps?
3. What do you see as the biggest challenges for the store's owners?

A PROFESSIONAL SERVICE
Arthur Andersen & Company
Accounting and Auditing Services
Charlotte, North Carolina

Arthur Andersen & Company was one of the world's leading accounting firms, with headquarters offices in Geneva, Switzerland, and in Chicago. As of the end of 1990, the firm employed 56,800 people worldwide and operated 299 offices in 66 countries. In the United States, Arthur Andersen was the largest of the so-called "Big Six" accounting firms. The firm was operated as a partnership; that is, the firm was owned and managed by a group of partners (in 1990 the partners totaled 2292) who had been elected by other partners in the firm.

The Charlotte office was the administrative office that coordinated the practice for the two Carolinas, including Columbia, South Carolina; Greensboro, North Carolina; and Raleigh, North Carolina. These four offices combined employed about 400 people, of whom 325 were professionals and 75 were support personnel. Of the 400, 290 worked in the Charlotte office. The office served over 450 clients in a variety of industries.

PART ONE

PROCESS DESCRIPTION

LINES OF BUSINESS

Although Arthur Andersen & Co. was known as a Big Six accounting firm, it did more than provide classic auditing attest services (described in detail later). Its practice also included tax, consulting, other accounting services, and a series of specialized services for smaller businesses. In fact, the auditing and entreprise groups accounted for a little less than half of revenues; the tax and consulting divisions accounted for the remainder. While this tour examines mainly the accounting and auditing practice of Arthur Andersen & Co.—

Charlotte, the tax, consulting, and enterprise practices merit brief separate attention.

The Tax Practice

The tax practice in Charlotte employed 29 professionals. It was an important component of the office's business, and was well established in the firm at large. There were two major components to the tax practice. One dealt with client compliance issues—preparing returns for clients, representing clients at Internal Revenue Service reviews, and providing defense for any clients in legal matters involving taxation. The other component involved advice on business transactions that had significant tax implications, such as mergers, acquisitions, and syndications. This kind of work typically complemented that of investment bankers and lawyers. Tax work consisted of many short-term jobs that often had to be done quickly but that typically did not involve substantial resources over extended periods of time; consulting or auditing services were longer term. Although there was considerable demand for tax services all through the year, the peak demands naturally occurred each February through April.

The Consulting Practice

The management information consulting practice was now formally known as Andersen Consulting. For the Carolinas, it was concentrated in Charlotte; of the 169 professionals employed in the consulting practice, 145 worked in the Charlotte office. The mission of the practice was to help clients obtain the information they needed to plan and manage their organizations. The practice was broad-ranging and included the development and installation of computer systems for financial and operations control of both manufacturing and service businesses. Much of this work was customized for the client and thus called for skill in drawing together the disparate elements of each project.

Consulting revenues were evenly divided among banking, manufacturing/textiles, health care, government work, and some miscellaneous businesses. The projects performed by the consulting division varied widely in their scope and duration, although many were accomplished for previous clients. Typical projects included the development of a new deposit system for bank customers and the installation of management information systems for a group of trucking companies.

Enterprise Practice

The enterprise practice was technically a part of the audit practice of the firm; however, attest services accounted for only about half of the entreprise practice's revenues. The remaining revenues were generated by a diverse array of services: assistance in preparing financial forecasts and cash-flow projections, development of long- and short-range business plans, consultation in approaches to minimize taxes and maximize after-tax cash flow, evaluation of lease vs. buy decisions, break-even analysis, capital budgeting, assistance with purchases, merger, or divestiture plans, and an assortment of other services. The enterprise practice had been created nearly 45 years ago and had grown in Charlotte since the office opened in 1958. By 1990 there were 70 clients, and some of the original clients' businesses had grown so much that they were no longer served out of the entreprise practice.

For the professionals in the entreprise practice, the demands were quite diverse. While each professional had fine auditing skills, they also had to foster other skills so that the wide-ranging demands of entreprises could be met: use of the microcomputer, preparation of busi-

The reception area, with Arthur Andersen's trademark doors.
(*Courtesy of Arthur Andersen & Co, Charlotte, North Carolina*)

ness plans, private placements for company stock, and so on. For these reasons, the enterprise practice was often difficult to schedule and manage—but it had been a significant source of growth over the years for all of Arthur Andersen & Co.

AUDITING ATTEST SERVICES

All publicly held companies (those regulated by the Securities and Exchange Commission) are required to have independent certified public accountants report whether their financial statements fairly represented their financial positions and results of operations. These reports were usually found in company annual reports (see Figure I1). Usually, these opinions were unqualified, but occasionally, accounting firms had to qualify them to alert shareholders to particular risks or problems in the financial statements of the company. Thus reports were important safeguards to company shareholders. The audit was the procedure by which Arthur Andersen & Co. developed and substantiated reports about the financial records of the client companies.

Arthur Andersen & Co.'s auditing methodology was designed to identify and concentrate on the risk areas of greatest exposure to a company's financial statements and the related key control elements put in place by management to mitigate the risks. It was a "top down" approach that addressed risks in two stages.

Report of Independent Public Accountants

▮▮▮▮▮▮▮▮▮▮▮▮▮▮▮▮▮▮▮▮▮▮▮▮▮

To the Shareholders of International Paper Company:

We have audited the accompanying consolidated balance sheets of International Paper Company (a New York corporation) and subsidiaries as of December 31, 1990 and 1989, and the related consolidated statements of earnings, common shareholders' equity and cash flows for each of the three years in the period ended December 31, 1990. These financial statements are the responsibility of the Company's management. Our responsibility is to express an opinion on these financial statements based on our audits.

We conducted our audits in accordance with generally accepted auditing standards. Those standards require that we plan and perform the audit to obtain reasonable assurance about whether the financial statements are free of material misstatement. An audit includes examining, on a test basis, evidence supporting the amounts and disclosures in the financial statements. An audit also includes assessing the accounting principles used and significant estimates made by management, as well as evaluating the overall financial statement presentation. We believe that our audits provide a reasonable basis for our opinion.

In our opinion, the financial statements referred to above present fairly, in all material respects, the financial position of International Paper Company and subsidiaries as of December 31, 1990 and 1989, and the results of their operations and their cash flows for each of the three years in the period ended December 31, 1990 in conformity with generally accepted accounting principles.

Arthur Andersen & Co.

New York, N.Y.
February 8, 1991

FIGURE I1   An opinion.

The first stage focused on the principal external and internal factors influencing the client company's operations, such as the industry in which it operated, the nature and complexity of its businesses and products, the operational and financial planning and control activities of management, and management's attitudes toward risk control and financial reporting.

The second stage provided a systematic basis for evaluating individual internal controls over an organization's significant cycles of activity.

Much as a manufacturer is concerned about the integrity and capabilities of the manufacturing process, so, too, an Arthur Andersen & Co. auditor is concerned about the integrity and capability of the internal controls in the client company for capturing accurately all of the accounts billed and collected, the invoices to be paid, the expenses authorized, the inventory, and other data.

The audit attest process could be broken into three major phases: the planning phase, the specific risk analysis phase, and the final phase. These different phases required different amounts of time and attention from the various members of the "engagement team" working on the audit.

The planning phase, for example, accounted for about 10 percent of the typical audit's effort, although much more than 10 percent of the time of the senior members of the engagement team. (The senior members of the team were the partner or partners in charge and the manager—the professional with rank just below partner, who had anywhere from 5 to 12 years of service. The manager was actually in charge of the day-to-day progress of the audit.) The planning phase was concerned with defining the scope of work to be done and its timing. It identified those risks that seemed to be important and those changes in the character of the business that ought to be scrutinized. Once the planning phase was over, the partners returned periodically to review progress, to help work through particular problems, and to monitor the quality of the work in progress.

The specific risk analysis phase was concerned with fact finding. In this phase, the junior members of the engagement team (i.e., the staff accountants, and the "seniors") investigated the paperwork and controls of the client company. They checked to see that the controls actually worked. Sometimes, they identified substantive problems that had to be dealt with by the senior members of the engagement team. This phase could range from perhaps 30 percent of the time and effort in a small audit to 60 percent of the time and effort in a large audit.

The final audit phase led to the generation of the report. This phase updated the results of the specific risk analysis phase, involved substantiation of year-end balances, and included a review of the client company's financial statements and their accompanying footnotes. For large audits, this final phase might account for only 30 percent of Arthur Andersen & Co. time and effort, whereas for smaller audits it could account for fully 60 percent.

The total professional time devoted to an audit by the Charlotte office could vary dramatically, from 80 hours each year for the smallest audits to over 4000 hours each year for the largest. The median amount of time consumed for an audit was estimated at about 500 hours each year. The calendar for an audit varied somewhat. For the largest audits, Arthur Andersen & Co. professionals were in and out of the client company all the time; they were often part of quarterly reviews of the client's business. At more typical audits, however, the planning phase was done about 6 months into each fiscal year, the specific risk

A partner's office. (*Courtesy of Arthur Andersen & Co, Charlotte, North Carolina*)

analysis work was accomplished in months 7 through 11, and the final work was done in the first 3 months after the close of the fiscal year.

Although two-thirds of the Charlotte office's audit work was repeat business, the office was always eager to attract new business. It secured this business by drafting a proposal to the target client company. This proposal discussed the key aspects of Arthur Andersen & Co. and what differentiated it from other accounting firms. It introduced the firm's other major clients in Charlotte and the services the entire office could provide a client company. It discussed in some detail the audit approach that Arthur Andersen & Co. planned to take. It profiled the personnel making up the engagement team, and it discussed how that personnel would be managed over time to provide the kind of continuity that was valued by clients. The proposal also established the fee schedule.

MANAGING THE PRACTICE

The Pyramid

Arthur Andersen & Co.'s auditing practice in the Carolinas totaled 140 professionals. Of this number, there were 10 partners, 22 managers, 38 seniors, and 70 staff accountants. These different classifications could be depicted as a pyramid, with the staff as the base and the partners at the top. Each engagement had its own pyramid, as well, with one or two partners at the top, one or more managers, and various seniors and staff accountants.

Staff, seniors, and managers were all salaried positions. The partners were not salaried; their compensation depended on the profitability of the firm. If the partners were to make money, they had to use the salaried personnel of the firm effectively. This placed several demands on the partners: (1) effective scheduling of personnel so that everybody was working as much

as possible but without excessive overtime (scheduling will be discussed later), and (2) fostering as much responsibility in the lower ranks as possible. Firm profitability was enhanced to the degree that effort and responsibility were thrust to lower levels of the pyramid (this was termed as having a broad-based pyramid). If the higher-paid senior members of an engagement team were free to perform other duties, such as bringing in additional business, everyone benefited and the firm as a whole would be profitable.

If a job could be run with a broad-based pyramid, it would likely be profitable. If, however, the pyramid were slender, it was likely not to be as profitable because the time of the senior members of the engagement team would not be as effectively levered as when the pyramid was broad.

Accounting firms like Arthur Andersen & Co. primarily operated on an "up-or-out" basis; professionals within the firm were either promoted to the next-higher rank or they left the firm. In a sense, the partners were the survivors, those who had proven themselves over the years by their performance and had demonstrated the qualities that made for effective partners.

Arthur Andersen & Co. was continually looking for innovative ways to keep highly trained individuals within the firm, even though they could not rise to partnership rank. In 1991, the firm created the "national partner," a senior level position just short of the equity partner rank, as a way to hang on to talented people.

There was considerable turnover at the lower levels of the pyramid. If one followed the path of new recruits in the Charlotte office, for example, one typically found that only 50 percent of each new cohort of college recruits were still with the firm after 3 years, and that only 25 percent of them were still with the firm after 6 years. After 10 years, the figure was 14 percent. Only about 9 percent reached partnership.

About three-quarters of the time, professionals left the firm of their own volition. Some left for more attractive, higher-paying positions in other companies. Others left because of personal incompatibilities with the degree of overtime or travel, or because of other elements of life on the job. Some returned to school. The remainder left at the suggestion of the firm itself; those were split between professionals who lacked the technical ability to do the job and those who lacked some other personal qualities that were required. While an up-or-out system might be viewed as harsh, Arthur Andersen & Co. was famed for its concern for the placement of its accountants in other jobs, often with valued clients.

The pyramid concept raised several issues. Broad pyramids meant lower costs, but because they encouraged the delegation of responsibilities to lower levels of the pyramid, there were quality control and cost-effectiveness issues associated with doing business that way. For example, staff might not have the capability or the training to do all that may be requested of them, or they might take a longer time than some more senior professionals in doing some job, thus raising the specter of cost-effectiveness.

Another issue involved client perceptions. The accountants at client companies were knowledgeable about firm operations, since they were frequently former Big Six professionals themselves. Often they would prefer experienced personnel and staff continuity. The Charlotte office was careful about making its assignments and had to consider each client's perceptions on staffing.

The Cycle of Business Planning

Given the labor intensity of auditing, business planning for the Charlotte office was very much an exercise in determining labor needs.

The business plan for the audit practice started with a forecast of chargeable hours for the Carolinas' four offices. The area had been growing and Arthur Andersen & Co. had been developing additional clients, so chargeable hours were growing at a healthy clip.

Once a defensible forecast of chargeable hours was developed, "productivity factors" (i.e., average chargeable hours per person per year) were applied to determine how many people would be needed. In general, the Charlotte office averaged between 1400 and 1500 chargeable hours per person per year. That number differed by level within the pyramid: Partners generally had the fewest and seniors the most. Nonchargeable hours were taken up in a variety of activities, including training, marketing and sales promotion, community service, administrative tasks, recruiting, and, of course, inefficiently used time. Given the application of the productivity factors to the forecasted chargeable hours, the office figured out how many people it needed. This, in turn, determined how many new recruits were needed to make up for any personnel shortfalls.

The business plan's forecast of profitability depended on the fee structure as well as the manpower plan. With the growth of business in the Carolinas, there was heavy competition for new business. The Charlotte office took this into account in developing its fee structure to match its chargeable hour plan and its staffing plan.

Scheduling

As mentioned previously, personnel scheduling was a key determinant of the office's profitability. The Charlotte office had responsibility for scheduling not only its own operation but coordinating those of Greensboro, Raleigh, and Columbia as well. One manager-level professional was assigned full time to the scheduling and other division administration functions.

The scheduling process began with the managers in charge of the office's engagements. For each engagement, the manager indicated how many hours of which levels of professional were required (see Figure I2). The manager was also permitted to request specific people to be part of the engagement. This planning could often be done far in advance, due to the recurring nature of much of the office's business. Each December and January, a rough-cut schedule for the following May through April period was devised.

The allocation of people to jobs was depicted on two different schedules: (1) the long-range schedule that planned people's assignments for an entire year, by week, and (2) the 20-week schedule, which showed the assignments of all staff persons, by week, for the next 20 weeks. The 20-week schedule used the same data base as the long-range schedule. Both of these schedules were known in the office as the "railroad," largely because the schedule itself looked like tracks on paper (see Figures I3 and I4).

The scheduling manager was responsible for pulling together the original schedule and then for adjusting it as time wore on and inevitable changes occurred. Naturally, the Charlotte office's profitability was greatest when everybody was fully scheduled and working effectively. The inexperienced staff accountants tended to have the most "open holes" in their schedules, and thus they were a constant concern for the scheduling manager. He reviewed the schedule continually to see whether all of the jobs being worked on were properly staffed. Was the pyramid broad enough for them? Were some projects in need of help? Were the Greensboro, Raleigh, and Columbia offices busy? Could professionals be swapped between offices? The Charlotte scheduler acted as the central clearing-

CLIENT NAME AND # Company 555
DESCRIPTION OF WORK Audit
CLIENT YEAR-END 12-31
LOCATION OF WORK CCT
DATE SUBMITTED

MASTER JOB SCHEDULING SHEET

SUGGESTED STAFF PERSONS-
IN CHARGE AA
STAFF BB

THIS JOB CAN BE STAFF WITH ONE-
MID-YEAR OR INTERN ☐ TAX ASSISTANT ☐ MIC ASSISTANT ☐

	Staff Class	Total Hours	P	F	April					May				June				July					Aug.				Sept.				
					1	8	15	22	29	6	13	20	27	3	10	17	24	1	8	15	22	29	5	12	19	26	2	9	16	23	30
PARTNER		20																													
MANAGER		68																													
STAFF	SR	102																													
	EA	156																													
	IA	168																													
TOTAL STAFF HRS.		426																													

	Oct.				Nov.				Dec.					Jan.				Feb.				March				
	7	14	21	28	4	11	18	25	2	9	16	23	30	6	13	20	27	3	10	17	24	3	10	17	24	30
PARTNER	2					2																			16	
MANAGER	4					8																	8	8	40	
STAFF	16	-	X		X	16	X																16	16	40	
	16	-	X		X	40	X																40	10	40	
					X	40	X						8										40	40	40	
TOTAL STAFF HRS.																										

Instructions:

1. Total hours shown should agree with the hours shown in the fee estimate or reconciled thereto. "Cushion" should not be scheduled.
2. Requirements for recurring type audits of Foundations, Pension Plans, etc., may be submitted separately or included as part of the time submitted for recurring audit of the Company financial statements. However, such time should be clearly identified as part of "Description of Work." If submitted on a combined basis, there should *still* be a clear indication of the pension, quarterly review, recurring registration statement, etc., time requirements.
3. Requirements for nonrecurring work including registration statements and major one-time projects should be submitted on a separate job scheduling sheet.
4. Requirements should be shown based on the ideal scheduling, however, where the time is absolutely not flexible, please so indicate in the "Comments" section. *This is extremely important.*
5. Known overtime requirements should be shown by merely showing Total Hours as 45 or 55 or whatever. In general, we would prefer to minimize overtime in all cases. When required, we should, in particular, attempt to avoid overtime at preliminary dates and for extended periods of time. Our office policy on overtime requires Partner/Manager approval of overtime.
6. In determining time requirements for each job, please consider the office need to achieve a balance of various experience levels on each job. Approximately 50% of our staff has one year or less experience.
7. Minimize situations where staff are assigned to a job for periods of time exceeding 400–500 hours or for extended periods on one job, particularly if out of town. Consider expanding the number of staff assigned.

Comments

Final is not flexible

FIGURE I2 A sample master job scheduling sheet, as used at Arthur Andersen & Co.—Charlotte.

```
CAROLINAS                           PERSONNEL SCHEDULING          PAGE-  2        AS OF-20 NOV 85
DIVISION: CHARLOTTE AUDIT            PERSONNEL ASSIGNMENTS                         PREP-20 NOV 85
TO: 450561                                                                        RPT-EC01000
FOR: CC, ASR, 206440
HOME ADDRESS:
                            LOC        SCHED
                            OFF  PTR MGR  ** HRS TO 11   12              01        02          03           04          05
JOB NUMBER/JOB YEAR END     DIV  SENIOR   STAT DATE  18 25 02 09 16 23 30 06 13 20 27 03 10 17 24 03 10 17 24 31 07 14 21 28 05 12
 JOB NAME/DESCRIPTION
----------------------------------   ----  -------   -- -- -- -- -- -- -- -- -- -- -- -- -- -- -- -- -- -- -- -- -- -- -- -- -- --
                            FGJ DLW   I   140:40   :  36:40 24:32   :      :50 50:50 50:50 50:   :    :    :    :    :    :    :
COMPANY 234                 CC                     :     :    :    :   :      :     :     :    :    :    :    :    :    :    :    :
                            PAG RDG   I   240:      :     :    :    :   :      :     :     :    :    :10  :24  8:   :    :40 40>
COMPANY 345                 CC                     :     :    :    :   :      :     :     :    :    :    :    :    :    :    :    :
                            FGJ JCH   I   260:      :     :    :    :   :      :     :     :    :    :    :    :    :40  :    :
COMPANY 456                 CC                     :     :    :    :   :      :     :     :    :    :    :    :    :    :    :    :
4310000                     BRG                    :     :   3:    :    :   :      :     :     :    :    :    :    :    :    :    :
AUDIT DIV MEETING                                  :     :    :    :   :      :     :     :    :    :    :    :    :    :    :    :
555                                               : 2    :    :    :   :      :     :     :    :    :    :    :    :    :    :    :
P/M MEETING OR OFFICE MTG                          :     :    :    :   :      :     :     :    :    :    :    :    :    :    :    :
815                                               : 16:   :   16: 8   :      :     :     :    :    :    :    :    :    :    :    :
HOLIDAY
                                                  -- --  -- -- -- -- --     -- -- -- -- -- -- --  -- --       --    -- -- --
•TOTAL ASSIGNED HOURS                             42 16  39 40 40 40        50 50 50 50 50 50 10  24  8       40    40 40
•TOTAL CHARGEABLE HOURS                           40     36 40 24 32        50 50 50 50 50 50 10  24  8       40    40 40
•TOTAL NON-CHARGEABLE HOURS                        2 16   3    16  8
•TOTAL OVERTIME HOURS                              2                        50 50 50 50 50 50 10  24  8       40    40 40
 TOTAL TENTATIVE HOURS
 TOTAL UNASSIGNED HOURS                              24 40  1       40 40 40              30 40 16 32 40 40    40

SCHEDULING CONSIDERATIONS:
-----------------------------------------------------------------------------   ••••••••••••••••••••••••••••
•   INCLUDES TENTATIVE HOURS                                                     •                          •
•• STATUS - I = INCHARGE PERSON, R = REVISED SCHEDULE, T, X = TENTATIVE ASSIGNMENT, • CC                    •
          C = STAFFING CONTINUITY, O = OUT OF TOWN                               • 18 NOV 85   WEEKS 1 - 26 •
                                                                                ••••••••••••••••••••••••••••
```

FIGURE I3   A long-range schedule, as used at Arthur Andersen & Co.—Charlotte.

```
CLT.RR RR-PGM                    AUDIT DIVISION PERSONNEL ASSIGNMENTS        DATE: 10/06/88
RAILROAD-PRINT                        TWENTY WEEKS REPORT                    PAGE: 1

DIVISION: CHARLOTTE AUDIT

EMPLOYEE │ CLASS │ 10/10/88 │ 10/17/88 │ 10/24/88 │ 10/31/88 │ 11/07/88 │ 11/14/88 │ 11/21/88 │ 11/28/88 │ 12/05/88 │ 12/12/88

AA    AESR   CO1  -40                                              HOLI  -16        CO1  -40  CO1  -40
                                                                                              CO37 -16
BB    AESR   CO2   -8  CO2  -40  CO2  -40  CO2  -40  CO14 -40  CO14 -40  VACAT -24  CO2  -40  CO14 -30  CO14 -50
             CO3   -8                                    CO2  -20       HOLI  -16  CO5  -40
CC    AESR   VACAT -40  VACAT -40                         CO20 -40       CO23 -24  CO23 -40             CO38 -24
                                                                        HOLI  -16
DD    AESR   CO4  -40  CO4  -40  CO4  -40  CO17  -8           CO17 -40  HOLI  -16  CO27 -40  CO27 -40  CO32 -40
EE    AESR                                                              CO24 -24
                                                                        HOLI  -16
FF    AESR                                CO4  -40  CO4  -40  CO4  -40  HOLI  -16             CO31 -40
GG    ASR    CO5  -16  CO12 -50  CO12 -50  CO12 -50  CO6  -16  CO6  -40  CO6  -24  CO6  -40  CO6  -40
             VACAT -8                                CO12 -24            HOLI  -16
             CO6  -16
HH    ASR    CO7  -40  CO7  -40  CO15 -40           CO21 -24  CO21 -40  CO21 -24             CO1  -40  CO1  -40
                                                                        HOLI  -16
II    ASR    VACAT -16           CO18 -24  CO19 -40  CO19 -24  HOLI  -16                      CO33 -40
             CO8  -24            CO19 -16
JJ    ASR                                                    HOLI  -16  CO36 -32             CO34 -40
KK    ASR    CO9  -16  CO10 -32  CO10 -50  CO10 -50  CO10 -24           HOLI  -16             CO11 -16
             CO10  -8
             CO11 -16
LL    ASR              CO13 -40  CO13 -40  CO13 -24  CO14 -40  CO14 -40  CO25 -24  CO29 -40  CO14 -30  CO14 -50
                                                                        HOLI  -16
MM    ASSR   VACAT -16  CO14 -40  CO14  -8  VACAT -16 CO18 -24           CO22 -24  CO25 -24  CO30 -24  CO14 -30  CO14 -50
                                                                        HOLI  -16
NN    ASSR                        CO16 -40 CO16 -40  CO16 -40           HOLI  -16  CO5  -40
OO    AEST   CO1  -40  CO1   -8  CO10 -50  CO10 -50  CO19 -40  CO19 -16  CO23 -24  CO23 -40  CO1  -40  CO1  -40
             CO10 -24                                                   HOLI  -16
PP    AEST   CO35  -8  CO13 -40  CO13 -40  CO18 -16  CO14 -40  CO14 -40  CO26  -8             CO14 -30  CO14 -50
             TEACH -16                                                  HOLI  -16
             VACAT -16
```

FIGURE I4   A 20-week schedule, as used at Arthur Andersen & Co.—Charlotte.

```
CAROLINAS                              PERSONNEL SCHEDULING                    PAGE-  1      AS OF-20 NOV 85
DIVISION: CHARLOTTE AUDIT                                                                   PREP-20 NOV 85
TO: 450561                      PERSONNEL AVAILABILITY (CONDENSED)                          RPT-EC03100

                PERS              AVAIL 11    12            01          02          03              04              05
CLASS GROUP     NUMBER CLAS  PERSONNEL NAME  TENT  18 25 02 09 16 23 30 06 13 20 27 03 10 17 24 03 10 17 24 31 07 14 21 28 05 12
------------   ------ ----  --------------  ---- -- -- -- -- -- -- -- -- -- -- -- -- -- -- -- -- -- -- -- -- -- -- -- -- -- --
SENIORS         115312 ASR   AA             AVAIL:    24:      :      :      :  40:      :  40:40 40:40 40:40 16:   --  :40 40:40 40:
                206440 ASR   CC             AVAIL:    24:40 1:      :  40:40 40:    :      :    :30 40:16 32:40 40:   40:   :
                295868 ASR   DD             AVAIL:38 24:40 37:40 24:32 40:24   :  :40 40:40 40:    :24 32:40 40:40 40:40 40:
                374725 ASR   EE             AVAIL:38 24:     1:    24:32 40:    :  :40   :    :40 40:40 40:40 40:40 40:40 40:
                595772 ASR   HH             AVAIL:    :     1:    24:32 40:    :    :    :  40:40 40:    :40 40:40 40:40 40:40 40:
                657221 ASR   JJ             AVAIL:    24:40 37:40   :16 40:    :    :    :    :  40:   40:40 40:40 40:40 40:40 40:
                692981 ASR   KK             AVAIL:    :    37:40 24:32   :    :    :    :    :  40:40 40:40 40:40 40:40 40:
                743739 ASR   LL             AVAIL:    :    37:40 24:24   :    :    :    :    :40 40:40 40:40 40:40 40:40 32:
                778001 ASR   MM             AVAIL: 6  :     1:40   :    :    :    :    :  40:40 40:    40:40 40:40 40:40 40:
                941701 ASR   NN             AVAIL:38 24:40 37:40 24:32 40:40 40:40 40:40 40:40 40:40 40:40 40:40 40:40 40:40 40:

                STANDARD WORKLOAD  40 24 40 40 40 24 32 40 40 40 40 40 40 40 40 40 40 40 40 40 40 40 40 40 40 40

                                                                            •••••••••••••••••••••••••
                                                                            •  20 NOV 85   WEEKS 1-26  •
                                                                            •••••••••••••••••••••••••
```

FIGURE I5   A personnel availability chart, as used at Arthur Andersen & Co.—Charlotte.

house for the staffing needs of all of the Carolinas' offices. In trying to fill the open holes in the schedule, he relied most on the 20-week schedule, which was the one that was painstakingly updated with all the changes, and which was the office's controlling document for the next 20 weeks. That document was organized by individual and displayed the open holes for each professional in the Carolinas, as well as which engagements everyone was scheduled for. An associated document was generated automatically to show the converse—which professionals in the office were available in each week. This document identified candidate professionals who could help with a new or existing engagement (see Figure I5).

As might be expected, there were continual inconsistencies in the schedule that had to be resolved. These inconsistencies resulted from many things—new work, changes in personnel, or changes in the scope of an existing engagement. Consider the following situation. The Greensboro office not only had been very busy but was short of seniors as well. The existing seniors there were working too much overtime and needed to be relieved in some way. As it happened, a new senior had been transferred to the Charlotte office from Dallas and was due

soon. The scheduler's dilemma was how to help the Greensboro office without giving them the new senior for too long a time, because, after all, he had been transferred into Charlotte and not Greensboro. The scheduler knew that it would be toughest to adjust the schedule during the peak time of the year (February and March), so he concentrated on helping Greensboro during those two months. By consulting the schedule, he knew that the new senior could be used on the Company 792 engagement for the weeks of February 17, February 24, and March 3. That engagement also called for a senior's time during the week of December 2, which was fine, as the new senior was going to be in Charlotte at that time. In addition, Greensboro needed help on the Company 409 engagement for the weeks of February 3 and February 10. That also called for work in the week of December 23. This, too, fitted well into the new senior's schedule.

However, there were two other engagements, Company 642 and Company 322, that would need additional manpower and could not be assigned to the new senior. The scheduler looked to the 20-week schedule for holes and then also looked at the personnel availability sheet. No one senior had time available in all of the time periods required for the jobs under consideration: Thus there were some incompatibilities that the scheduler had to check into and resolve. Upon investigation, two seniors' schedules showed some promise. In evaluating the jobs assigned to each of them during this period, it looked like one would be the more available. However, he would not be free for all of the 3 weeks slated for those jobs. He was available during the weeks of December 9 and December 16, but he was not available during the week of January 20, as he was scheduled on another job in the Greensboro office. The scheduler then had to do some horse trading. In order for him to assign the senior to the Greensboro office for the 642 and 322 jobs, he needed to take the senior off his other assignment at Greensboro. By calling around to the interested parties, it was agreed that the senior could be taken off the Greensboro job for the week of January 20 and thus assigned to the 642 and 322 jobs that were coming up immediately. The hope was that the Greensboro office could delay work on that job until sometime later, or if that was not possible, perhaps use another senior for the week of January 20.

The scheduler continually made trade-offs in this way, for the most part following a hierarchy of preferences: (1) substitute a similar-level professional for the professional initially planned for (possible for preliminary audit work but not for final audit work), (2) substitute an available professional from a different level, if he or she were qualified, and (3) let the schedule slip so that the desired personnel could perform the job at a more convenient date. During the course of any week, there might be two or three problems that consumed up to 4 hours each to resolve. In addition, there were a host of smaller problems that might be resolved much more easily, say within an hour's time each.

The scheduler was in repeated contact with the managers who oversaw the office's engagements. Those managers informed the scheduler of changes in their needs, and they occasionally lobbied him for particular professionals to work their engagements. Sometimes, particularly with new clients, staff and seniors would also lobby the scheduler for those jobs. An important task for the scheduler was balancing the training and development needs of staff accountants and seniors against the practice's need for their services and the client company's desires for continuity in the personnel assigned to them.

Personnel Policy

While profitability was tied to the pyramid and to the efficient scheduling of people, the Charlotte office's managing partner believed that in the long run, Arthur Andersen & Co. would be successful only if it served its clients well. His major concern was with the development of the professionals in the office and the culture in which they worked. Culture, of course, is an elusive concept, but a variety of indications of the Arthur Andersen & Co. culture could be cited:

- The Arthur Andersen motto: think straight, talk straight.
- No secrets were kept around the office.
- Team work and attention to detail were highly valued.
- Also valued was client service and the urgency in that. There was a "code red" program that called for some member of the engagement team to return a client's phone call within 60 minutes, no matter where the engagement's lead partner or manager might find themselves.
- The pride that Arthur Andersen had in having the widely acknowledged premier training program of any of the Big Six.

- The desire that Charlotte's managing partner had for everyone to have fun doing their jobs. In support of this, the goal was to have every professional assigned to the industry of his preference at least half the time.

- All the office's professionals were reviewed each year. They filled out a form (Figure 16) and then met with a partner to discuss their performance and career development.

Training

One of the partners was responsible for monitoring the training of all the audit professionals. Training at Arthur Andersen & Co. was done either within the local office or at the firm-wide level. To this end, the firm operated a training center in St. Charles, Illinois, and brought professionals from all over the world to it on a regular basis. Continuing professional education was required of all CPAs by the national and various states' accounting societies (120 hours over 3 years with no less than 20 hours per year in any year), so even partners and managers spent time in training. For the most part, partners and managers averaged at least 40 hours per year, for example, most staff accumulated between 150 and 200 hours of training, with 100 to 120 hours spent locally, and the balance at either the regional level or at the St. Charles campus. Seniors did not spend quite that much time, but they too spent a considerable amount of time in training.

Training was considered as important as a client assignment for the professionals in the office. The training plan was usually developed well in advance, and put out "on the railroad." Good attendance was viewed as important for career development. The slow period from April through August was the best time of the year for training, although training programs were scheduled throughout the year.

The training program was considerably varied. Some of the programs were basic for the first- and second-year professionals, but other programs were termed either intermediate or advanced. Some of these courses dealt with interpersonal relationships, marketing, effective presentations, and other aspects of management development. Still other courses had industry orientations, so that professionals could be better steeped in the nature and problems of the specific industries they served.

Recruiting

Recruiting was another important function that was overseen by a partner and that demanded considerable time from the senior professionals in the office. Recruiting concentrated on local schools, although all the recruiting was done for the firm at large and not just for the Charlotte office. Recruiting was done at both the undergraduate and MBA levels.

The Manager's Life

Managers oversaw the day-to-day operations of each audit engagement. It was their job to see that the audit was done well and on time and that the fee level was appropriate. They were responsible for developing the staffing plan and budget for the engagement and for billing and collecting from the client. Managers wanted to avoid any surprises on an engagement, and thus it was in their interest to have seniors and staff that could be trusted and whose competence and capabilities would be unquestioned.

Managers spent a good deal of their time outside the office — perhaps 55 to 65 percent. Most of that time was spent at existing clients, attending to their needs and solving any problems that had cropped up. About 20 to 25 percent of the manager's time was spent with

TO: _____
                    Reviewing Partner

FROM: _____

DATE: _____

SUBJECT: ANNUAL REVIEW _____

1. Am I satisfied with the assignments I have had? What particular assignments (industry, technical specialization, research, compliance, etc.) would I like in the next year?

    _____

    _____

    _____

2. What, if any, particular training would I like to receive during the next year?

    _____

    _____

    _____

3. Are the demands made upon me as a professional (overtime, travel, work pace, etc.) acceptable?

    _____

    _____

    _____

4. Am I progressing at a pace satisfactory to me?

    _____

    _____

    _____

5. Is my performance being evaluated currently and adequately?

    _____

    _____

    _____

6. My interests in professional and civic activities are:

    _____

    _____

    _____

7. Ideas or suggestions for the Firm:

    _____

    _____

    _____

8. Any concerns:

    _____

    _____

    _____

FIGURE I6   An annual review form used at Arthur Andersen & Co.—Charlotte.

administrative chores. And because managers were partners-in-training to a large degree, they spent some of their time developing new clients and new business for the firm.

In the main, managers were pleased with their lives at Arthur Andersen & Co. Their jobs were viewed as challenging, motivating, and educational, and the partners gave them a good deal of autonomy. The partners were also regarded as fair, and the up-or-out nature of the business was not unduly confining because the managers felt that they could easily leave the firm for attractive jobs elsewhere.

The Partner's Life

Many duties fell on the partners in the Charlotte office. The typical partner incurred about 1000 hours each year that were charged to the engagements on which he was the senior member. Of that time, perhaps two-thirds was spent at the client, discussing the client's business and supervising the audit there. Other chargeable hours were spent in the planning phase and final report phases of the actual audit. As an accountant and auditor, the partner was responsible for solving problems, for analyzing risk, and for monitoring quality.

The other 1400 to 1500 hours of a partner's year were spent in many different activities: recruiting; training, including teaching the junior people; practice development, which was the marketing and sales function for generating new business; personnel matters, such as counseling junior professionals and reviewing their progress; and providing counsel to other Arthur Andersen & Co. professionals on matters of the partner's own expertise (e.g., the Charlotte office was viewed as the firm's source of expertise for textiles).

The partners, as a group, are responsible for a multitude of functions. Not only did they have line responsibility for the audits being done, but they had to divide up the overhead functions necessary to further the practice: training, recruiting, expertise in special industry specializations (such as health care, manufacturing, closely held businesses), practice development, administration, and quality control.

Partners were well compensated. Their high incomes were a motivation for many of the junior people. Arthur Andersen & Co. partners were paid by splitting the worldwide profits of the company according to how many "units" they held. Each unit was stated in U.S. dollars. All partners started with a fixed quantity of units, and over time, these units were added onto, based upon seniority and merit. In keeping with the no-secrets aspect of the firm, files were kept on all partners, and all partners knew what everyone else in the firm made.

A manager was not promoted to partner based solely on the evaluations of the office in which he or she served. There were uniform standards for the promotion to partner that were overseen by a firm-wide committee. Upon becoming partners, managers were required to contribute some capital to the firm, but this was a rather modest amount relative to the income the partner made.

PART TWO

DISCUSSION

Arthur Andersen & Co. is in no way responsible for the following views and presentation. They remain solely the responsibility of the author.

The Flow of the Process and of Information

Figure I7 is a very rough process flow diagram of the audit process. It shows the degree of involvement by the various levels comprising the engagement team. This process flow also includes the proposal stage and the selection and scheduling stage for the development of the audit team. The diagram depicts only the major phases of the audit process. One could, of course, develop a more detailed schedule of the work for any engagement, but that detailed schedule would be very particular to the client, and complex as well. It would include many of the work elements of Arthur Andersen & Co.'s audit procedures; as such it is beyond the scope of this service tour.

Similarly, constructing a diagram of the information flow in a process like auditing is well-nigh impossible. The flows of information are so numerous and so idiosyncratic to the client and to the needs of the engagement team itself that it is not worth characterizing here. A Big Six accounting organization simply lives on information, and that information flow must be swift, it must involve all layers of the company, and it must carry differing quantities of information.

There are no real distinctions in either the process or information flows during peak versus nonpeak times. During peak times, of course, lots of overtime is put in, and people are perhaps more charged-up about work, but all of the steps must still be done, and all of the information must still flow as it does at nonpeak times.

	Proposal Writing	Engagement Team Selection and Scheduling	Planning Phase	Specific Risk Analysis Phase	Final Audit Phase
Senior Members of the Engagement Team (Partners and Managers)	High involvement with client and among themselves	High involvement, especially by managers. Scheduler is very much involved	High involvement	Little involvement by partner(s), except for periodic troubleshooting and for quality reviews Day-to-day involvement by manager	High involvement
Junior Members of the Engagement Team (Seniors and Staff)	Little or no involvement	Some preferences for work are stated	Little involvement by staff High involvement by seniors	High involvement. Much work done at client	High involvement

FIGURE I7   Rough process flow diagram for auditing services at Arthur Andersen & Co.—Charlotte.

DEMANDS OF THE PROCESS ON THE WORKFORCE

The workforce at Arthur Andersen & Co. is composed of the salaried professionals (managers, seniors, staff) and the less numerous support people. The process demands on them are essentially a high degree of training and a tremendous amount of flexibility. The firm's accountants have to move from job to job, to put in long hours when called upon, and to be out in the field for, perhaps, long stretches of time. Some of the work is tedious, although much of the work can be challenging and rewarding.

The firm's accountants, at all levels, must make decisions all the time. Even in the preliminary phase of an audit, there are decisions to be made and many issues that need to be resolved. Thus the members of an engagement team have to be constantly on the lookout for problems or opportunities (e.g., add-on business) that they should surface to the senior members of the engagement team for resolution. This is one reason why so much continual training is required to keep everyone fresh and knowledgeable as times and clients change.

Accountants must also face the up-or-out nature of advancement within the firm. Happily, the alternatives to employment are generally good. Arthur Andersen & Co. accountants are typically in high demand by client companies and others. The care with which advancement decisions are made and the counsel given accountants through this process helps to ease any stress that this kind of advancement places on the accountants proceeding through the ranks.

THE PYRAMID[1]

Let us now depart from the organizational structure of Arthur Andersen & Co. to examine in more general terms the nature of the professional pyramid (Figure 18) as it applies in all sorts of professional service firms. For the sake of simplification, consider only three ranks of professionals: partner, manager, and junior. The partner is chiefly responsible for client relations and the running of the firm itself. The manager is the one responsible for the day-to-day activities of the project (engagement)—what there is to do, how it is to be done, and whether it is being done properly, on schedule, and within budget. The junior professional is assigned many of the tasks that constitute the "legwork" of the project. This dispersion of responsibility among partners, managers, and juniors has led some wags to term this hierarchy one of "finders, minders, and grinders."

What determines the shape of the pyramid—whether it is broad or narrow? There are two major determinants of the pyramid's shape:

1. The composition of work on the projects undertaken.
2. The utilization rates of the firm's professionals.

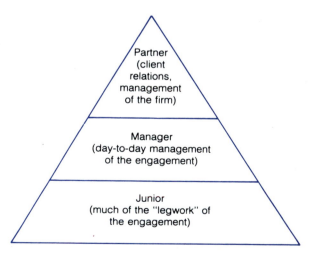

FIGURE 18   The professional pyramid.

Let's explore these factors in greater detail. Assume, for example, that a typical project of 600 hours required certain time commitments from the firm's professionals and that the professionals were used on projects to the degree shown in the following table.

PROFESSIONALS	TYPICAL TIME COMMITMENT[a] (hours)	UTILIZATION RATE[a] (percent)
Partner	60	50
Manager	140	70
Junior	400	85
	600	

[a]These are meant to be illustrative and do not reflect figures for Arthur Andersen & Co.

The utilization rates reflect the facts that other things take up professionals' time (training, recruiting, scheduling, counseling, soliciting new business) and that these other activities are relatively more demanding of the senior professionals in the firm. Given a planning figure of 2000 hours of work per person per year, one partner could sustain supervision of 16.67 average projects (calculated as 1000 hours of partner time available for projects divided by 60 hours of time per typical project, or 16.67 projects per year). A manager, on the other hand, could sustain direction of 10 projects (1400 hours available and 140 hours average for each project). Thus for the firm to be balanced, for every partner there have to be 1.67 managers. Similarly, a junior can work on 4.25 projects each year (1700 hours available and 400 hours for each project). In this way, the pyramid's shape is determined: Every partner supports 1.67 managers and 3.92 juniors.

The shape of the pyramid, in turn, largely determines the rate at which professionals can be advanced and how selective that advancement has to be. Assume, for example, that juniors are 6 years in rank and that managers are 4 years in rank. If there are 10 partners, the pyramid would look as in Figure I9. Suppose that every year a partner has to be added, either to sustain the growth of the firm or to replace a partner who retires or otherwise leaves the firm. Every year, then, a manager has to be

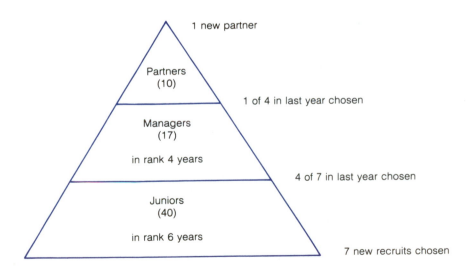

FIGURE I9   The pyramid and advancement.

made a partner in order for the pyramid to be in balance. If, on average each year, 4 of the 17 managers are in the fourth year, the advancement rate has to be 1 in 4. Similarly, if during any year, four new managers are required and there are seven juniors in their final sixth year, then roughly half of the juniors can be promoted to manager status. In this way, the pyramid sustains itself. Of the 11 juniors and managers in their last year, 1 becomes a partner, 4 become managers, and 6 leave the firm. Of the 7 new recruits, only 1 can expect to make partner after 10 years.

Demands of the Process on Management

The partners in a professional firm like Arthur Andersen & Co. have a number of responsibilities that they alone must shoulder. Many of these responsibilities relate to the workforce. Managing the pyramid well, as discussed earlier, is an absolute requirement. Partners must take the lead in recruiting new accountants, in training them, and in managing their advancement through the firm. These activities take up tremendous chunks of partner time.

Partners must also use that pyramid well in order to make money. Thus the assignment of accountants to engagements and the scheduling of those accountants are important areas of partner concern. Even though the actual schedule itself is accomplished by a manager level professional, the partners must develop the strategy underlying the schedules developed.

Partners are also primarily responsible for procuring new business. They are the ones who must contact the clients, develop proposals, and establish fees. And, of course, the partners must be, in themselves, excellent accountants. They must be intimately involved in the general risk analysis and the final phase of each audit. They are called on to troubleshoot and to review the quality of any audit through all its phases, particularly the preliminary phase. In addition to this line management authority, they are also called on by others in the firm as experts in certain areas (e.g., industries and certain accounting practices). Thus partners must wear many hats and are pulled in many different directions.

QUESTIONS

1. What advantages do you see to the hierarchical structure of an accounting firm like Arthur Andersen & Co.? Do you see any disadvantages?

2. What kinds of factors do you suppose would affect the fee schedule advanced in a proposal to a client?

3. In a partner-manager-junior type of organization, as described in the discussion to this service tour, suppose the time commitments were changed from a 60-140-400 breakdown to one of 100-200-300. Assuming the same utilization rates, how would the pyramid change, if it were to remain perfectly balanced?

4. What benefits and costs are involved in increasing the average chargeable hours in the office from its current level of 1400 to 1500 hours per year?

5. What kinds of trade-offs are the hardest for the office's scheduler? What kinds are the easiest?

6. What are the characteristics you would look for in the typical recruit for a staff accountant position?

JOHNSON AND TROTTER ADVERTISING

The Johnson and Trotter Advertising Agency was a large, New York based agency with a wide variety of clients. J&T was organized into six functional departments: account management, research, creativity, production, media, and legal. The account management department was charged with client relations and was responsible for coordinating the efforts of all of the other departments on the client's behalf. The research department was responsible for uncovering useful facts about the products and services advertised that may be of use in developing new ads, and they were also charged with substantiating the advertising claims made by the ads. The creativity department was the lead department in developing new ideas and the means by which the client's products and services could be advertised. The production department was responsible for making the print, radio, and TV ads that had been conceived by the creativity department. The media department interacted with all of the media in which the agency's clients advertised, while the legal department handled any legal issues raised by the agency's efforts.

The agency accomplished its work in teams. Members of the various departments generally worked on more than one account at the same time except for the account executive, who was charged with keeping all members of the product team informed about such things as the client's situation, research findings, creative ideas, production status, and any other information on the account. The account executive was also charged with keeping the entire effort on the established timetable and within budget. Members of each product team could expect to spend several years working for a client, before being rotated off. Any differences of opinion were generally dealt with by compromise within the product team. Account executives did not have the authority to overrule a specialist on a question involving that specialist's realm of expertise. If a difference of opinion persisted, the team could take its dispute to its superiors within each department, where it was usually resolved.

There were three quality control mechanisms at J&T: a strategy review board, a creative review board, and a media review board. At key stages of the firm's process, these quality control mechanisms could be invoked.

Promotion for the professionals of the agency came from within each department. For example, within account management, one typically started as an assistant account executive, moved up to account executive, then, on to account supervisor, management supervisor, and then executive vice-president for client services. A general manager oversaw all of the six functional groups.

1. What strengths do you see in the organization of Johnson and Trotter?

2. What potential disadvantages exist with this operation?

REFERENCE NOTE

1.   Much of the discussion is based upon David Maister, "Balancing the Professional Service Firm," *Sloan Management Review* (Fall 1982): 15–29.

A PROJECT
Geupel DeMars, Inc.
Indianapolis, Indiana

Geupel DeMars, a subsidiary of the DeMars Corporation, was an Indianapolis-based construction management firm for industrial and commercial properties. The company was formed in 1927 and first worked on department store construction. Since that time it had gradually broadened its client base. In 1991 the firm employed 250 people. See the organization charts in Figure J1.

Geupel DeMars (GDI), as a construction management firm, acted as the agent for an Owner. It did no actual construction itself, but rather planned for the construction of the building, arranged for Owner-held contracts for the different aspects of the job, oversaw the work, and reported to the Owner. For its effort, the Owner paid Geupel DeMars a fee to cover the salaries and benefits of its employees on the job, plus a percentage markup.

As a construction management firm, Geupel DeMars was distinct from (1) an architectural/engineering (A/E) firm that designs buildings and does the engineering required to translate the design into documents that serve as the basis of contractor bids on the project, (2) a design/build firm that combines the design, engineering, and construction of a building into a single one-stop package for the Owner, or (3) a contractor that only builds.

The workings of a construction management firm are best seen by examining a particular project. For this tour, the project is the construction of Research Building No. 48 for Eli Lilly & Company. This building (see Figure J2) of 550,000 square feet, was slated for completion in early 1993 at a total cost of $135.2 million. The design envisioned a building of four stories, laid out in three wings (Biology, Chemistry, and Animal) tied to a central administrative section. The nature of the building called for features that were distinct from typical office buildings: special ventilation for the 162 labs, greater capacity HVAC (heating, ventilation, and air conditioning), purified water sources, and special drainage systems, among others.

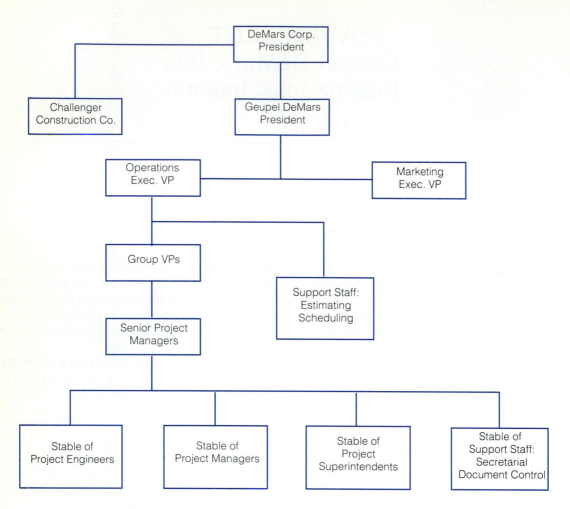

FIGURE J1   A. Simplified Organization Chart for Geupel DeMars

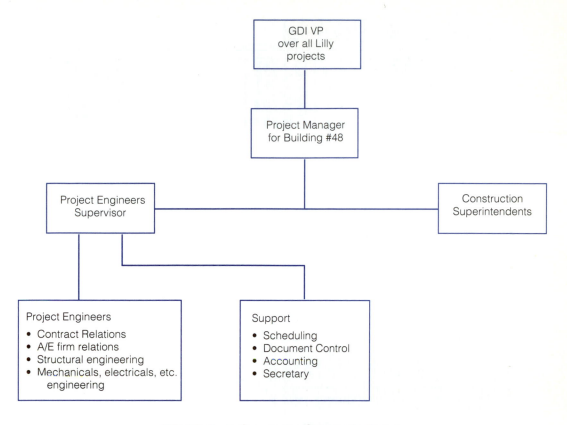

FIGURE J1   B. Organization Chart for the Project

PART ONE

PROCESS DESCRIPTION

PROJECT ORGANIZATION AND TIMETABLE

Research Building No. 48 was scheduled to take roughly three years to build from ground-breaking to completion. Prior to that time, considerable planning had to be done. Geupel De-Mars became involved with the project about two years before construction began. The na-ture of its involvement in the project changed over time. To wit:

Year One, 1988

- *Personnel.* 0 in field; 2 in the office.
- *Tasks.* Gather data; estimate costs and con-structability; develop overall schedule.

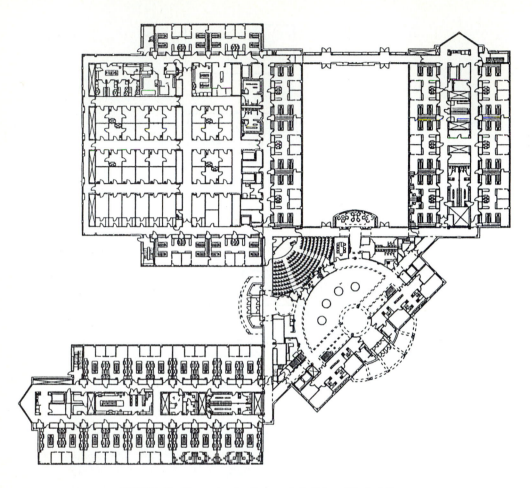

FIGURE J2   Floor plan for Research Building 48's first floor.

When Geupel DeMars entered the project, Lilly had already engaged the architectural/engineering firm (CUH2A of Princeton, New Jersey) and the building's design was proceeding. Although the location for the building was set, there was still considerable fluidity to the design: how many stories, how stretched out, how the various research wings would be connected. Geupel DeMars's role in this conceptual phase of the project involved the following tasks:

- Reviewing the plans and estimating the costs of various alternative designs.
- Reviewing the constructability of the building and identifying the cost-feature tradeoffs that the Owner (Lilly) would have to make, tasks known as value engineering.

- Identifying schedule parameters and inputs. Preparing and presenting alternative schedule strategies for delivery of the project.

In this conceptual phase, GDI provided cost estimates to the Owner with an accuracy of within 10–20 percent of actual costs for each of the building configurations seriously considered. These estimates were based on costs per square foot of interior space, costs per square foot for the face of the building (stone, glass, concrete), and costs for HVAC, plumbing, electrical requirements, and so on.

Year Two, 1989

- *Personnel.*   0 in the field; 2 in the office.
- *Tasks.*   Continue Year One tasks but with further refinement, establish budget.

The second year of the project continued the tasks begun in Year One. Over time, decisions were made that refined the project more and more. The project moved from "conceptual drawings" of the building to "schematic drawings," which were half to two-thirds complete in nature and on which materials began to be specified. These schematic drawings required renewed cost estimation and value engineering.

The character of Geupel DeMars's value engineering could include the following types of suggestions:

- *Selection of materials.*   The choice of materials can often affect the cost of the project by five to ten percent. Such choices can include the use of concrete versus structural steel, the type of roofing materials to use, and the choice of pipe materials for purified water systems, etc.
- *Design modifications.*   The construction manager can suggest which design details are likely to be expensive to render and which

are not. Examples include roofing details such as flashings and eaves, structural concrete details on how the concrete is to be formed, and finish details such as door jambs and lighting fixture wiring systems.

The schematic drawings were followed by the even more specific "design development drawings" and the first of the "construction documents" upon which supplier and contractor bids could be made.

Year Three, 1990

- *Personnel.*   1 in the field; 4 in the office.
- *Tasks.*   Supervise excavation and initial construction; begin bidding procedures for contractors; establish procedures for scheduling, quality control and cost control.

Excavation of the site began in February 1990. During the rest of that year, the foundations were poured and the erection of the steel superstructure for the building was started. One superintendent in the field oversaw this work.

This year also saw the start of what was termed the "fast-track construction design." This process called for the identification of sequential "bid packages" of work, the scheduling of the design process to support this sequence, bidding, the award of initial contracts, and the subsequent commencement of work. The development and awarding of bid packages was done on a rolling basis and would continue for a year and a half. Major parts of the construction were treated in this way: excavation, foundation, concrete, structural steel, enclosure packages (skin, roof, doors); elevators, mechanicals (HVAC, fire protection, plumbing); electricals; and various interior finishes (painting, flooring, carpeting, ceiling, partitions, wall coverings, etc.). Likewise, as part of this process, came the purchase of equipment for the building by the owner.

This was also the year to establish the balance of the procedures that would govern the construction of the project as time went on. These procedures were essential to the control of the project (schedule, cost, quality) so that the Owner actually received what was designed and specified. These controls are explained in more detail later.

Year Four, 1991

- *Personnel.*   5 in the field; 9 in the office.
- *Tasks.*   Continue construction supervision; continue the development and award of bid packages; monitor project schedule, costs, and quality and report to the Owner.

In this year, the building took shape and the need for field monitoring and supervision escalated dramatically. Bid packages continued to be assembled, bid, and awarded.

The year also saw more "administration and management" of the project as opposed to the earlier planning and estimating work that dominated the efforts of Geupel DeMars. This "management" essentially consisted of pursuing the control procedures that had been laid out in the previous year, reporting on progress to the Owner, and dealing with the myriad field questions and problems, both large and small, that inevitably accompanied the work of the contractors as they actually built the research building and as procurement proceeded.

Year Five, 1992

- *Personnel.*   8 in the field; 10 in the office.
- *Tasks.*   Continue supervision of the construction; continue management of the project; start-up of major systems.

Year Six, 1993

- *Personnel.*   8 in the field; 10 in the office.
- *Tasks.*   Close out the project for the Owner.

In this final year the construction was completed and the project closed out. Closing out a project meant several things:

- All contractors had completed their work, including their "punch lists" (i.e., the lists of repairs and inadvertent omissions that are identified prior to an Owner taking possession).
- The contractors had supplied the "as built" drawings, warranties, and guarantees.
- All paperwork had been finished, including all letters answered and items resolved.
- The project file was completed, including all copies of correspondence, all shop drawings (the detailed drawings from the contractors of which there were thousands of pages), all as-built drawings, all quality control documents, and all test reports.

THE MANUAL AND ITS CONTROLS

Prior to the start of construction, Geupel DeMars and Lilly prepared a procedures manual that outlined the specific duties of each party. This manual was the basic control document for the project. In it, procedures were described for such items as bidding, materials substitutions, submittals, quality control, requests for information (usually from a contractor to the A/E firm and the Owner), design changes, and progress payments to the contractors.

Submittals were a particularly important point of control. A submittal was the contractor's submission for review of any of several types of documents:

- Catalog and performance data for the materials to be used
- "Shop drawings"—the contractor's drawings detailing what actually was to be done
- Guarantees
- Job standards, consisting of certifications or approved samples that could be of use as field examples
- Maintenance data and operating instructions
- Installation instructions
- Physical samples of material and assemblies

Once the submittals were reviewed and approved by the A/E, the contractor could procure and install the material and equipment. Submittals thus led to the approval of materials, and materials were seen as driving both the schedule and the budget. Geupel DeMars was insistent that contractors follow a submittal schedule so that they performed on time with the proper information.

Substitutions were another important item to control. Substitutions of materials were usually made to reduce cost, to provide a better product (perhaps as a result of value engineering), or to speed up an aspect of the project that had fallen behind schedule. Substitutions required the approval of the Owner and the A/E firm.

There were a host of other controls on quality. For example, the superintendents completed reports each day on the work accomplished. There were reports on nonconforming work, which typically would necessitate repair. Independent tests were made to verify that items such as the soil, concrete, structural work, roofing, HVAC, and high voltage met design requirements. A periodic report was distributed to the Owner on (1) the construction progress, (2) the topics at morning meetings, and, (3) most importantly, a listing of "open issues."

CONTRACTOR BIDDING

Geupel DeMars did no real construction on Building No. 48, but acted only as the agent for the Owner. A major task for the company was to assist the Owner in assembling the best available stable of contractors who (1) had the expertise, (2) met the Owner's financial requirement, and (3) were aware of the Owner's high quality standards. Working from the plans and engineering specifications of the architectural/engineering firm, Geupel DeMars assembled a total of about 30 bid packages for the diverse pieces of the construction. A bid package contained (1) a summary of work that spelled out all that the contractor was to do; (2) a manual detailing procedures used by Geupel DeMars and Lilly; and (3) the applicable construction drawings and technical specifications by the A/E firm. A bid package took about a week to assemble. Once assembled, the bid package was sent to between 4 and 6 contractors that had been invited to bid on the job. These contractors were generally well known to Geupel DeMars either from past work or from the formal approval process and were all considered to be capable of performing quality work of the scale required. (Often, the Owner would have its own preferred list). New or previously unknown contractors would be visited and their operations and financial conditions reviewed.

After the bidding documents were out for 2 weeks, Geupel DeMars held a pre-bid meeting with the invited contractors. At this meeting, GDI reviewed the manual and the summary of work and then, with the assistance of the A/E and the Owner, answered questions. Following the meeting, an addendum with meeting minutes and revisions was issued.

It took the contractors between 3 and 6 weeks to prepare their bids. After review of the bids, the apparent low bidder was invited to attend a pre-contract meeting between the contractor, the Owner, the A/E (if necessary), and GDI. This was a technical meeting that sometimes involved Geupel DeMars's field superintendents. If all issues were resolved, the Owner then signed a contract with the successful bidder. Construction then began on that contracted work.

Dealing with Contractors on the Job

Once construction began, Geupel DeMars passed on information between contractors and the Owner and the A/E firm. (The Lilly team for this project numbered about 10 people. The A/E firm, at one point, had as many as 300 on the design effort, with about 10 on the continuing supervisory team.) Sometimes, the contractor would have questions about particular details of the construction, questions that were generally overlooked by all parties before then. These questions triggered a formal RFI (Request for Information). Geupel DeMars's role was (1) to follow up on the RFI, (2) to distribute the information to the appropriate people, and (3) to communicate any decided changes to the scope of work.

Geupel DeMars felt that it had failed if any major problems occurred. The company was always looking ahead to such things as the schedule of submittals, approved materials, and test results. Fortunately for all concerned, there were few problems of any magnitude with the Research Building #48 project.

Field Supervision

As the building progressed, more and more attention was devoted to planning and supervising the daily construction activities. This was the responsibility of the field's construction superintendents and field engineers. Initially, there was only one to supervise the excavation and foundation work. As the superstructure was set in place, more superintendents joined the project and a greater division of labor ensued. For Research Building #48, there were different superintendents for each of the major trades represented by the contractors (e.g., structure, mechanicals, electricals).

The superintendents oversaw the quality of work being done, any testing of materials, and the coordination of the contractors' work. They reviewed requests for information. They also determined the percentage completion of various aspects of the job, which affected scheduling and the progress payments Lilly made each month.

Scheduling

After quality control in importance came the control of the project's schedule. The schedule affected both quality and cost. Although keeping on schedule was everybody's concern, there was a scheduler whose task was to document the schedule and the project's conformance to that schedule.

The schedule was derived in backwards fashion. The Owner typically had a completion date in mind. From that date the scheduler worked back to the relevant start dates.

The raw data for the schedule were the numerous activities that had to be accomplished. Some of these activities had to precede others, although many activities could be accomplished simultaneously. For the most part, the activities comprising the schedule followed the various bid packages. The contractors responsible for each bid package detailed their construction schedules, floor-by-floor, and this information served as a basis for the schedule. Geupel DeMars then applied its own experience and its knowledge of supplier timetables, especially for long leadtime items.

A computer program, Primavera Primavision, was used to document, monitor, and make changes to the schedule. Primavera kept track of all the designated activities, their expected times-to-completion, and their relationships to each other. The input to Primavera was built up bit-by-bit as GDI developed each bid package. The program's input began shortly after excavation was begun. By the end of the project, it was expected that 2000 activities would be captured. Of those 2000 activities, for example, 640 would pertain to mechanical and electrical installations. Only about 20 percent of the activities pertained directly to actual installation; the other 80 percent referred to various reviews, approvals, bidding, comments, documentation, and so on.

Primavera Primavision generated several documents that helped with the project's scheduling:

1. *Gantt Charts.*   Figure J3 is a Gantt chart generated by Primavera. Gantt charts, discussed in Chapter 9, are simply bar charts that plot time for each activity. They show expected start and stop times, and the bars could be filled in to show the extent of actual progress at each point in time. Note that in Figure J3 the project is up-to-date. All of the activities that should have been accomplished by the time of the chart's date, April 1991, have indeed been accomplished.

The solid color bars represent activities that are designated as "critical." More formally, these are activities that lie on the so-called "critical path" (see Chapter 9). Activities on the critical path are those which, if delayed, will delay the entire project. For them, there is no spare catch-up time avail-

able to be used. They must be accomplished on time or the project gets delayed.

Sometimes, of course, an activity might be hurried up in order to put a delayed project back on track. Often, such hustling cost money in addition to the extra effort of those involved. This hustling was typically termed "crashing" an activity. For example, for Research Building 48, the erection of the curtain wall began 7 weeks late. The contractor, however, captured back some of the lost time by putting more people on the task than was originally planned.

2. *Time Logic Diagrams.*   Another output of Primavera is a time logic diagram, such as depicted in Figure J4. This diagram depicts the precedence relationships among activities. In the diagram and the accompanying sheet of detail, the ties between activities become clear. For example, Figure J4 relates to Bid Group VII-2 (Fire protection), and among its first set of activities, there are these ties to other activities:

Activity 1110 – The bidding was due on December 12, 1990. Prior to the bidding, two things had to happen: Lilly had to sign the Phase II construction documents (Activity 1024), and GDI had to prepare the bidding documents themselves (Activity 1026). These two things were completed in October. The successor activity (1112) was the awarding of the contract. The expected duration of the activity, 37 days, is given in the ORIG DUR (Original Duration) column.

Activity 1112 – This is the successor activity to Activity 1110. It was accomplished by 1/11/91. We

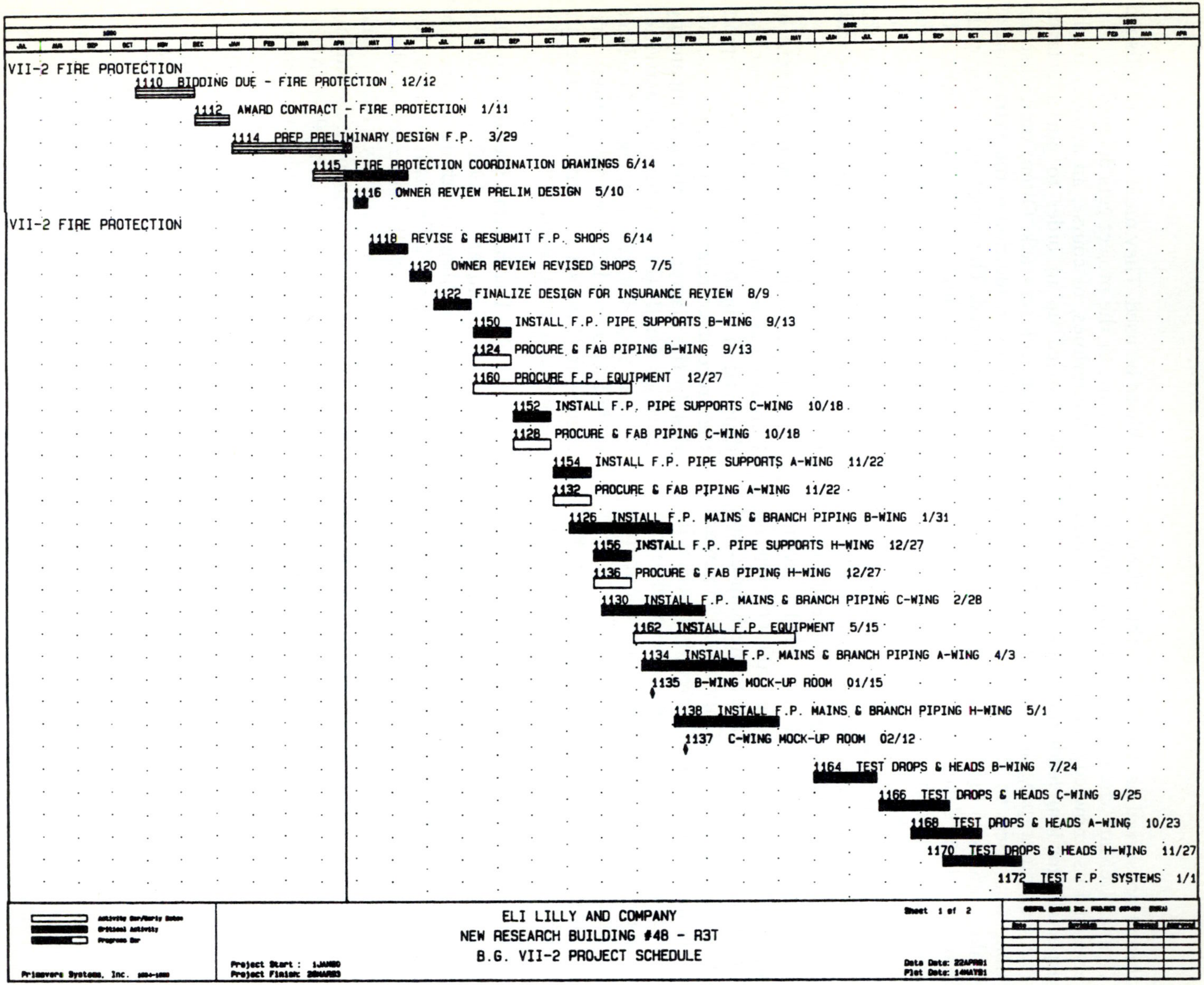

FIGURE J3   A portion of the project schedule for Research Building 48—Gantt chart.

ACTIVITY ID	ORIG DUR	REM DUR	PCT	CODE	ACTIVITY DESCRIPTION	EARLY START	EARLY FINISH	LATE START	LATE FINISH	TOTAL FLOAT
VII-2 FIRE PROTECTION										
1110	37	0	100	44	BIDDING DUE - FIRE PROTECTION 12/12	23OCT90A	12DEC90A			
1112	22	0	100	44	AWARD CONTRACT - FIRE PROTECTION 1/11	13DEC90A	11JAN91A			
1114	75	5	93	44B	PREP PRELIMINARY DESIGN F.P. 3/29	14JAN91A	26APR91		26APR91	0
1115	60	40	33	44B	FIRE PROTECTION COORDINATION DRAWINGS 6/14	25MAR91A	14JUN91		14JUN91*	0
1116	10	10	0	44B	OWNER REVIEW PRELIM DESIGN 5/10	29APR91	10MAY91	29APR91	10MAY91	0
1118	25	25	0	44B	REVISE & RESUBMIT F.P. SHOPS 6/14	13MAY91	14JUN91	13MAY91	14JUN91	0
1120	15	15	0	44B	OWNER REVIEW REVISED SHOPS 7/5	17JUN91	5JUL91	17JUN91	5JUL91	0
1122	25	25	0	44B	FINALIZE DESIGN FOR INSURANCE REVIEW 8/9	8JUL91	9AUG91	8JUL91	9AUG91*	0
1150	25	25	0	44 BB	INSTALL F.P. PIPE SUPPORTS B-WING 9/13	12AUG91	13SEP91	12AUG91	13SEP91	0
1124	25	25	0	44B	PROCURE & FAB PIPING B-WING 9/13	12AUG91	13SEP91	16SEP91	18OCT91*	25
1160	100	100	0	44	PROCURE F.P. EQUIPMENT 12/27	12AUG91	27DEC91	26AUG91	10JAN92	10
1152	25	25	0	44 CC	INSTALL F.P. PIPE SUPPORTS C-WING 10/18	16SEP91	18OCT91	16SEP91	18OCT91	0
1128	25	25	0	44C	PROCURE & FAB PIPING C-WING 10/18	16SEP91	18OCT91	21OCT91	22NOV91*	25
1154	25	25	0	44 AA	INSTALL F.P. PIPE SUPPORTS A-WING 11/22	21OCT91	22NOV91	21OCT91	22NOV91	0
1132	25	25	0	44A	PROCURE & FAB PIPING A-WING 11/22	21OCT91	22NOV91	25NOV91	27DEC91*	25
1126	65	65	0	44 BB	INSTALL F.P. MAINS & BRANCH PIPING B-WING 1/31	4NOV91*	31JAN92	4NOV91	31JAN92*	0
1156	25	25	0	44 HH	INSTALL F.P. PIPE SUPPORTS H-WING 12/27	25NOV91	27DEC91	25NOV91	27DEC91*	0
1136	65	65	0	44H	PROCURE & FAB PIPING H-WING 12/27	25NOV91	27DEC91	30DEC91	31JAN92*	25
1130	65	65	0	44 CC	INSTALL F.P. MAINS & BRANCH PIPING C-WING 2/28	2DEC91*	28FEB92	2DEC91	28FEB92*	0
1162	100	100	0	44	INSTALL F.P. EQUIPMENT 5/15	30DEC91	15MAY92	13JAN92	29MAY92*	10
1134	65	65	0	44 AA	INSTALL F.P. MAINS & BRANCH PIPING A-WING 4/3	6JAN92*	3APR92	6JAN92	3APR92*	0
1135	0	0	0	44 BB	B-WING MOCK-UP ROOM 01/15	15JAN92*	15JAN92	15JAN92	15JAN92*	0
1138	65	65	0	44 HH	INSTALL F.P. MAINS & BRANCH PIPING H-WING 5/1	3FEB92*	1MAY92	3FEB92	1MAY92*	0
1137	0	0	0	44 CC	C-WING MOCK-UP ROOM 02/12	12FEB92*	12FEB92	12FEB92	12FEB92*	0
1164	40	40	0	44 BB	TEST DROPS & HEADS B-WING 7/24	1JUN92*	24JUL92	1JUN92	24JUL92	0
1166	45	45	0	44 CC	TEST DROPS & HEADS C-WING 9/25	27JUL92	25SEP92	27JUL92	25SEP92*	0
1168	45	45	0	44 AA	TEST DROPS & HEADS A-WING 10/23	24AUG92	23OCT92	24AUG92	23OCT92*	0
1170	50	50	0	44 HH	TEST DROPS & HEADS H-WING 11/27	21SEP92	27NOV92	21SEP92	27NOV92*	0
1172	25	25	0	44	TEST F.P. SYSTEMS 1/1	30NOV92	1JAN93	30NOV92	1JAN93*	0

FIGURE J3   (continued)

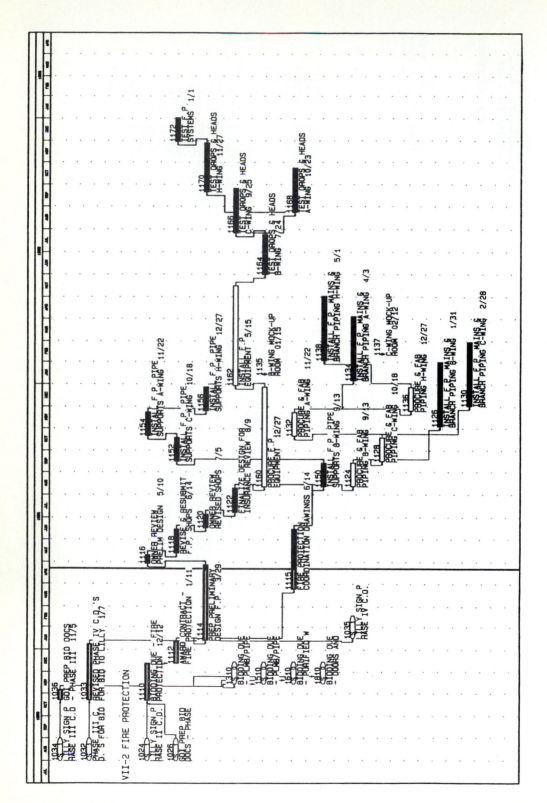

FIGURE J4 A portion of the project schedule for Research Building 48—Activity descriptions

VII-2 FIRE PROTECTION

ACTIVITY ID	ORIG DUR	REM DUR	PCT	CODE	ACTIVITY DESCRIPTION	EARLY START	EARLY FINISH	LATE START	LATE FINISH	TOTAL FLOAT
1110	37	0	100	44	BIDDING DUE - FIRE PROTECTION 12/12	23OCT90A	12DEC90A			
1024	4	0	100	C 6 PRED	LILLY SIGN PHASE II C.D.'S 10/12	9OCT90A	12OCT90A			
1026*	10	0	100	PRED	GDI PREP BID DOCS - PHASE II 10/22	9OCT90A	22OCT90A			
1112*	22	0	100	SUCC	AWARD CONTRACT - FIRE PROTECTION 1/11	13DEC90A	11JAN91A			
1112	22	0	100	44	AWARD CONTRACT - FIRE PROTECTION 1/11	13DEC90A	11JAN91A			
1110*	37	0	100	PRED	BIDDING DUE - FIRE PROTECTION 12/12	23OCT90A	12DEC90A			
1114*	75	5	93	SUCC	PREP PRELIMINARY DESIGN F.P. 3/29	14JAN91A	26APR91		26APR91	0
1115	60	40	33	C 50 SUCC	FIRE PROTECTION COORDINATION DRAWINGS 6/14	25MAR91A	14JUN91		14JUN91*	0
1114	75	5	93	44	PREP PRELIMINARY DESIGN F.P. 3/29	14JAN91A	26APR91		26APR91	0
1112*	22	0	100	PRED	AWARD CONTRACT - FIRE PROTECTION 1/11	13DEC90A	11JAN91A			
1116*	10	10	0	SUCC	OWNER REVIEW PRELIM DESIGN 5/10	29APR91	10MAY91	29APR91	10MAY91	0
1115	60	40	33	44	FIRE PROTECTION COORDINATION DRAWINGS 6/14	25MAR91A	14JUN91		14JUN91*	0
1112	22	0	100	C 50 PRED	AWARD CONTRACT - FIRE PROTECTION 1/11	13DEC90A	11JAN91A			
1150*	25	25	0	C 40 SUCC	INSTALL F.P. PIPE SUPPORTS B-WING 9/13	12AUG91	13SEP91	12AUG91	13SEP91	0
1116	10	10	0	44	OWNER REVIEW PRELIM DESIGN 5/10	29APR91	10MAY91	29APR91	10MAY91	0
1114*	75	5	93	PRED	PREP PRELIMINARY DESIGN F.P. 3/29	14JAN91A	26APR91		26APR91	0
1118*	25	25	0	SUCC	REVISE & RESUBMIT F.P. SHOPS 6/14	13MAY91	14JUN91	13MAY91	14JUN91	0
1118	25	25	0	44	REVISE & RESUBMIT F.P. SHOPS 6/14	13MAY91	14JUN91	13MAY91	14JUN91	0
1116*	10	10	0	PRED	OWNER REVIEW PRELIM DESIGN 5/10	29APR91	10MAY91	29APR91	10MAY91	0
1120*	15	15	0	SUCC	OWNER REVIEW REVISED SHOPS 7/5	17JUN91	5JUL91	17JUN91	5JUL91	0
1120	15	15	0	44	OWNER REVIEW REVISED SHOPS 7/5	17JUN91	5JUL91	17JUN91	5JUL91	0
1118*	25	25	0	PRED	REVISE & RESUBMIT F.P. SHOPS 6/14	13MAY91	14JUN91	13MAY91	14JUN91	0
1122*	25	25	0	SUCC	FINALIZE DESIGN FOR INSURANCE REVIEW 8/9	8JUL91	9AUG91	8JUL91	9AUG91*	0
1122	25	25	0	44	FINALIZE DESIGN FOR INSURANCE REVIEW 8/9	8JUL91	9AUG91	8JUL91	9AUG91*	0
1120*	15	15	0	PRED	OWNER REVIEW REVISED SHOPS 7/5	17JUN91	5JUL91	17JUN91	5JUL91	0
1124*	25	25	0	SUCC	PROCURE & FAB PIPING B-WING 9/13	12AUG91	13SEP91	16SEP91	18OCT91*	25
1150*	25	25	0	SUCC	INSTALL F.P. PIPE SUPPORTS B-WING 9/13	12AUG91	13SEP91	12AUG91	13SEP91	0
1160*	100	100	0	SUCC	PROCURE F.P. EQUIPMENT 12/27	12AUG91	27DEC91	26AUG91	10JAN92	10
1150	25	25	0	44	INSTALL F.P. PIPE SUPPORTS B-WING 9/13	12AUG91	13SEP91	12AUG91	13SEP91	0
1115*	60	40	33	C 40 PRED	FIRE PROTECTION COORDINATION DRAWINGS 6/14	25MAR91A	14JUN91		14JUN91*	0
1122*	25	25	0	PRED	FINALIZE DESIGN FOR INSURANCE REVIEW 8/9	8JUL91	9AUG91	8JUL91	9AUG91*	0
1152*	25	25	0	SUCC	INSTALL F.P. PIPE SUPPORTS C-WING 10/18	16SEP91	18OCT91	16SEP91	18OCT91	0
1124	25	25	0	44	PROCURE & FAB PIPING B-WING 9/13	12AUG91	13SEP91	16SEP91	18OCT91*	25
1122*	25	25	0	PRED	FINALIZE DESIGN FOR INSURANCE REVIEW 8/9	8JUL91	9AUG91	8JUL91	9AUG91*	0
1126	65	65	0	SUCC	INSTALL F.P. MAINS & BRANCH PIPING B-WING 8/9 1/31	4NOV91*	31JAN92	4NOV91	31JAN92*	0
1128*	25	25	0	SUCC	PROCURE & FAB PIPING C-WING 10/18	16SEP91	18OCT91	21OCT91	22NOV91*	25
1160	100	100	0	44	PROCURE F.P. EQUIPMENT 12/27	12AUG91	27DEC91	26AUG91	10JAN92	10
1122*	25	25	0	PRED	FINALIZE DESIGN FOR INSURANCE REVIEW 8/9	8JUL91	9AUG91	8JUL91	9AUG91*	0
1162*	100	100	0	SUCC	INSTALL F.P. EQUIPMENT 5/15	30DEC91	15MAY92	13JAN92	29MAY92*	10
1152	25	25	0	44	INSTALL F.P. PIPE SUPPORTS C-WING 10/18	16SEP91	18OCT91	16SEP91	18OCT91	0
1150*	25	25	0	PRED	INSTALL F.P. PIPE SUPPORTS B-WING 9/13	12AUG91	13SEP91	12AUG91	13SEP91	0
1154*	25	25	0	SUCC	INSTALL F.P. PIPE SUPPORTS A-WING 11/22	21OCT91	22NOV91	21OCT91	22NOV91	0

FIGURE J4 (continued)

know this is so for several reasons: the progress shown on the chart's bar, the 100% completion shown in the PCT column, the "0" in the REM DUR (Remaining Duration) column, and the "A", for Actual, after the early start and early finish dates. Activity 1112 has two successor activities: preparing a preliminary design and working up fire protection coordination drawings.

Activity 1115 — This activity is one of the successors to Activity 1112. It is on schedule, but it has not yet reached its conclusion. The percentage completion is 33 percent, with 40 of the expected 60 days' worth of work remaining. It has one successor activity, installing the fire protection pipe supports in the B-wing, slated to begin on August 12, 1991.

Observe that not all of the activities noted in the diagram have bars shown. The complexity of the relationships has forced Primavera to show them on a different sheet. This diagram is useful for depicting the relationships. Also observe that some of the forthcoming activities are colored solid, indicating that they are critical activities, but that some are not filled in. These latter activities are not on the critical path and thus have some slack associated with them.

Primavera was capable of generating other useful documents and to sort the activities in different ways. Sorts could be done by contractor or by floor of the building. By examining the scheduled activities in these ways, conflicts among contractors, and any subsequent rescheduling, could be held to a minimum.

Weekly meetings were held to track the schedule. Data for the schedule monitoring was provided by the superintendents on the site and by the GDI project engineers who followed the submittals and other paperwork. Every two weeks a rolling 4-week summary schedule was generated.

BUDGETING

The budget's major components included the following:

	Original Budget (in millions)
Construction (building materials, labor, Owner furnished equipment...)	$ 99.4
Engineering (Owner's cost, A/E firm, GDI)	23.8
Contingency	12.0
Total	$135.2

Geupel DeMars monitored these costs and managed the construction costs, the contingency costs, and its own costs closely.

PART TWO

DISCUSSION

Geupel DeMars, Inc. is in no way responsible for the following views and presentation. They remain solely the responsibility of the author.

THE FLOW OF THE PROCESS AND OF INFORMATION

Figures J5 and J6 display process flow and information flow diagrams for a project like the construction of Building #48. The diagrams are fairly general and include numerous entries that can be repeated. For example, there is a whole sequence of the process flow diagram that repeats for each of the 30 bid packages. With the information flow diagram, the interactions of RFIs, submittals, and approvals are repeated almost a countless number of times as the project continues. Geupel DeMars is really the "man in the middle" for a

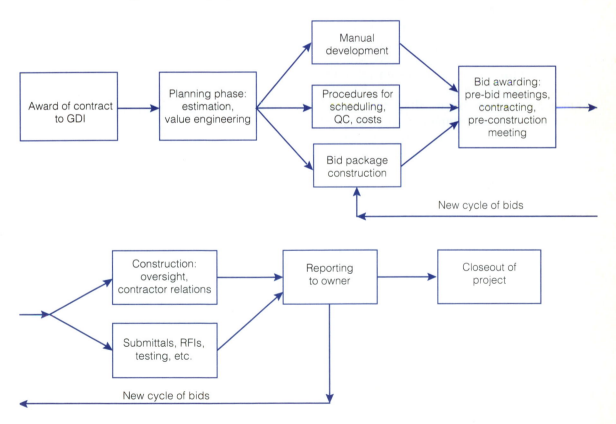

FIGURE J5   Process Flow Diagram for Geupel DeMars Project.

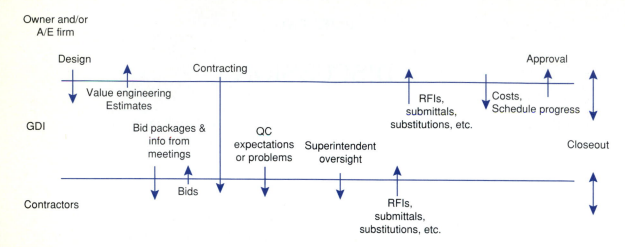

FIGURE J6   Information Flow Diagram for Geupel DeMars Project.

lot of these transactions, filtering information, monitoring progress and costs, making recommendations to the various parties, etc. Coordination and communication are essential to the smooth running of the project.

This is an example of so-called "fast track" construction that is completely analogous to concurrent engineering in a factory setting dealing with new product development. Fast track construction overlaps the phases of construction so that detailed design aspects of the building proceed even as the initial construction is being done. Such management of the process calls for a great deal of coordination among the players, and that is just what Geupel DeMars provides.

FEATURES OF THE PROCESS

There are a number of interesting and distinctive aspects of the project that differentiate it from the other processes investigated in this book:

1. The manning of this process is constantly changing. Not only are there varying num-

bers of construction workers on site, but there are variable numbers of architect/engineers, and importantly, varying numbers of GDI workers as well. There is a constant ebb and flow of people of various skills and in various quantities as the project progresses.

2. A variety of specialized talents are called for. Again, these are talents not only in the architecture/engineering and construction trades themselves, but also with GDI itself as it manages the project. GDI, for example, needs specialists for construction superintendents, cost estimation, scheduling, and contractor and A/E firm relations.

3. Projects take a significant degree of up-front planning in order to be carried out well. The budget for Building #48 reveals this degree of planning. So much of the project involves one-of-a-kind design that such planning cannot be avoided.

4. Constant coordination is required among the parties.

5. The means by which resources can be husbanded is generally by stretching out the timetable. The contractors do this to save costs, and only when forced by slips in the

schedule do they apply more labor or equipment to "crash" an activity. It is the same for GDI. Not all of the bid packages are prepared at the same time or bid on at the same time. Rather, because the job does not require it, and because it would tax GDI too much at the same time, the schedule of bid packages is spaced out. Such smoothing out of the work helps all concerned.

DEMANDS OF THE PROCESS ON MANAGEMENT AND THE WORKFORCE

At GDI the workforce and management are almost indistinguishable, as one could look at the project engineers as lower levels of management. For everybody, the demands of the process are quite similar. The process demands extraordinary planning skills of its workforce as well as tremendous attention to detail. The days of nearly everyone at GDI are filled with the follow-up and communication of details and, when called upon, the development of future needs and controls of all sorts. People live by their calendars, beepers, voice mail, faxes, and their ilk.

PLANNING AND CONTROL

The project lives and dies by the excellence of the planning that has gone into it and by the efficiency and thoroughness of the controls it has in place. All of these controls have to be planned for in advance so that all parties are clear about what is required of them. The chief controls are on:

- Quality (knowing in advance what the requirements of the job are and what constitutes satisfactory accomplishment)
- Schedule (knowing when the job has to be started and stopped)
- Cost (knowing what the flow of expenses is)
- Documentation (keeping a trail of paper about everything decided, everything accomplished, and everything changed).

The planning and control for the project outdoes that of the job shop and other kinds of processes. Similar kinds of documentation are kept: drawings, process and routing sheets (or their equivalents), tests of materials, change documents, quality documents, and schedules, but the quantities of them and care with which they are handled are typically extraordinary. After all, with a project you only get one shot at it, and it better be right the first time or other projects are not likely to be coming your way.

Drawings are an example. Consider the different kinds that are maintained: conceptual drawings, schematics, shop drawings, and as-built drawings. Everything about the building's construction is captured in these drawings. Everybody pours over them, communicates through them, and formally approves or changes them. And, they come by the hundreds and thousands.

QUESTIONS

1. Why does Geupel DeMars Inc.'s organization chart look the way its does?

2. What are the differences between Figures J3 and J4?

3. Describe your concept of a typical working day for a GDI project engineer.

4. Describe the relationship between project quality, cost, and schedule.

A COMPARISON OF PRODUCTION AND SERVICE PROCESSES

In the tours, ten different production and service processes were introduced and their key features described. There were threads in common among some of them as well as some sharp differences. The purpose of this chapter is to tie together the common threads and expose the sharp differences by comparing the processes described. Part One of this summary chapter deals with manufacturing and Part Two deals with services.

PART ONE

MANUFACTURING

TRENDS

First we should recognize that there are some clearly discernible trends among the six manufacturing processes introduced. The trends become evident when the processes are arrayed in the following order, a sort of "spectrum" of production processes:

- Job shop (example: Norcen Industries)
- Batch flow (example: Jos. A. Bank)
- Worker-paced line flow (example: Burger King Restaurant, Noblesville, Indiana)[a]
- Machine-paced line flow (example: GM C-P-C Group, Oklahoma City, Oklahoma)
- Continuous flow (example: International Paper Company, Androscoggin Mill)

The hybrid process represented by the Stroh brewery at Winston-Salem, which displayed aspects of both a batch flow and a continuous flow process, is somewhat more difficult to place since it is not as pure an example as the others.

Analyzing the Trends

To provide some cohesion and organization, this discussion of process trends is divided into six parts, each part devoted to a comparison and analysis of some specific features: (1) prod-

---

[a]The Burger King restaurant can be viewed usefully as both a manufacturing process (a burger factory) and a service process (a fast-food restaurant). For this reason, it is included in both parts of this chapter.

802

uct features, (2) general process features, (3) materials-oriented features, (4) information-oriented features, (5) labor-oriented features, and (6) management features. These six features will be addressed in turn.

Product Features

Arrayed as they have been here, the different processes introduced demonstrate some distinct trends involving the types of products manufactured and how those products compete against others. Specifically, the more one goes from a job shop toward a continuous flow process, the more it is generally the case that:

1. The number of different kinds of products made declines.
2. Product volumes increase to the point where the continuous flow process is essentially producing a commodity for the mass market.
3. Product customization declines and product standardization increases.
4. New product introductions become less frequent and are more costly.
5. Competition is more likely to center on price.
6. Competition, at least in the middle ranges of the array, is more likely to emphasize aspects like workmanship, product performance, and product reliability; but as the process becomes more and more a continuous flow, these differences between rival products become narrower and narrower.

General Process Features

There are complementary trends in some general process features as well. For example, as one progesses down the array of processes from job shop to continuous flow, it is generally true that:

1. The pattern of the process becomes more rigid, and the routing of products through their various process steps becomes less individual, less unprescribed, and better defined.
2. Process segments become more tightly linked together.
3. Equipment becomes more specialized.
4. The operation becomes huge, and economies of scale are possible.
5. More, and generally larger, equipment is part of the process.
6. Equipment is less likely to be idle. Pieces of equipment become better balanced in size and speed to one another.
7. Equipment setups are fewer and run lengths are longer.
8. The pace of the process is determined largely by machine capabilities or regulated by machines or conveyors.
9. The pace of production keeps increasing.
10. The notion of capacity becomes less ambiguous and more measurable in physical, rather than dollar, units.
11. Additions to capacity come in large chunks, and incremental additions to capacity become less viable.
12. Bottlenecks become less and less movable and thus better understood.
13. Incremental change to the nature of the process itself becomes relatively more frequent and routine, but the impact of radical change to the process is likely to be more sweeping and thus scarier to contemplate.
14. Process layouts, with like machines grouped together, give way to line flow layouts.

Materials-Oriented Features

Again, keeping the array of different processes ordered from job shop through continuous flow process, some general trends in materials-oriented features can be observed. For example, as the process becomes more and more a continuous flow, it becomes more and more the case that:

1. The span of the process (vertical integration) becomes broader. A plant is more and more likely to start with "very raw" raw materials and transform them into products that may need little or no "finishing" before consumers purchase them.

2. Most processes take uniform materials and make differentiated products from them. Norcen Industries takes uniform blocks, bars, or rods of steel and machines a multitude of different parts from them. Jos. A. Bank uses the same materials to produce suits in different styles and sizes. The parts going into a Buick Century are much the same as go into a Pontiac 6000; moreover, General Motors goes to great lengths to make sure that cars that are supposed to be exactly alike contain exactly the same components. Such is not the case with many continuous flow processes. Often these processes take differentiated raw materials and make uniform products from them. One log of timber may not be the same as the next log, and so pulp and paper mills are constantly monitoring their raw materials and adjusting them so that the finished product, be it publication gloss or newsprint, looks exactly the same at the end of the roll as it does at the start. Similarly, steel plant blast furnaces make constant changes to their raw materials input mix.

3. As the time of actual production draws closer, materials requirements are generally known with more certainty.

4. Raw materials requirements are large, but their purchase and delivery can be made steady.

5. Supplier ties are long term, with frequent deliveries.

6. Because of large production volumes and steady purchases, control over suppliers for price, delivery, quality, design, and the like is great.

7. Control over the delivery time of the finished product becomes greater.

8. Work-in-process inventories, because of process design, become scant. Queues of work cease to exist.

9. Finished goods inventories are larger (relative to other inventories).

10. Finished goods are sold through formal distribution channels and can sometimes be forced down those channels for the sake of keeping production running smoothly.

Importantly, however, the batch/continuous flow hybrid process does not follow all of these trends. There is a definite work-in-process inventory situated between the batch flow portion of the process and the continuous flow portion. At least some of the raw materials are inventoried as a buffer stock to be drawn down in perhaps unexpected or haphazard fashion, while other raw materials are purchased to coincide with the production plan.

Information-Oriented Features

A number of trends are evident as well for many information-oriented features of the various production processes. As the process changes from job shop to continuous flow, generally it is more and more likely that:

1. Production has not been instigated by a bidding procedure.

2. Longer-term sales forecasts are used, and orders are "frozen" long before production is scheduled to start.

3. The corporation outside the plant is an integral part of the plant's scheduling and materials movement tracking.

4. Order scheduling is done on a very sophisticated basis.

5. A finished goods inventory is managed.

6. The flow of information and paperwork between management and workers is less.

7. Quality control measures become formal.

8. Inventory adjustments become important in responding to seasonal or business cycle changes in demand.

9. The process is less flexible in making swift adjustments to demand changes, and so production must be carefully planned in advance.

Labor-Oriented Features

Again, trends are evident across the spectrum of production processes explored, this time concerning labor issues. Progressing from job shop to continuous flow process, it is more likely that:

1. The labor content of the product becomes smaller and smaller, relative to the product's value.

2. Job contents diminish, although "art" is more likely to be found at either end of the process spectrum.

3. Labor is paid by the hour rather than by some incentive system. In fact, the progression of wage payment schemes tends to go from hourly or individual incentive rates for the job shop, through individual and then group incentive schemes, and then on to hourly rates.

4. The importance of setting standards for labor remains high. The mechanization of the continuous flow process, however, means that such standards are usefullness to define the process and its capacity than to assign the workforce to the equipment.

5. As production moves more and more to mechanical or technological pacing, the scramble to complete a lot of production to meet monthly goals or billings becomes less and less prevalent.

6. The path of worker advancement becomes better defined and even formal.

Management Features

Finally, some trends can be identified as well for several aspects of the management of these diverse production processes. Progressing from job shop to continuous flow process, it is more and more the case that:

1. Staff operations concerning such topics as materials movement, scheduling, capacity planning, new technology planning, and quality control become more important relative to line operations.

2. The size of the plant's management (line and staff) is often larger relative to the size of the workforce both because the capital intensity of the operation is greater and because staff operations are more important. Table TS1 uses statistics from the processes we have discussed to show the ratio of management staff to total employees.

3. Given that the plant involved is part of a multiplant company, the involvement of managers situated at the corporate offices (rather than at the plant itself) becomes greater. The corporation's influence may extend to operations as well as to capital planning and spending.

4. The operation is controlled more as a cost center, as opposed to a profit center.

5. The major challenges that management faces are significantly altered, largely shifting from day-to-day operational considerations to very long-term, high-expense items.

TABLE TS1 Ratio of management staff to total employment		
COMPANY	PROCESS	MANAGEMENT STAFF/ TOTAL EMPLOYMENT
Norcen Industries	Job shop	7/41 = 0.171
Jos. A. Bank	Batch flow	43/625 = 0.069
Burger King, Noblesville	Worker-paced line flow	7/67 = 0.104
C-P-C, Oklahoma City	Machine-paced line flow	430/5300 = 0.081
Stroh, Winston-Salem	Batch/continuous flow hybrid	80/670 = 0.119
International Paper, Androscoggin	Continuous flow	175/1200 = 0.146

THE PROCESS SPECTRUM

By comparing these trends it is fairly plain that entire lists of characteristics hang together to describe particular processes. Table TS2 is an effort to compile just such lists for the range of production processes that have been introduced. Note that the batch/continuous flow hybrid, exemplified by the Stroh brewery, has been segregated from the others so that the process spectrum stretching from job shop through to continuous flow process remains as pure as we can reasonably expect.

Table TS2 offers some generalizations about particular types of production processes, drawing mainly from the processes discussed in the plant tours. Not all of the generalizations may ring true for all of the production processes one may conceivably classify in each category from job shop to continuous flow process. Most of the generalizations, however, are representative of a typical production process in each category. For completeness, one could also place the project in the spectrum, to the left of the job shop and sharing a number of its characteristics.

PART TWO

SERVICE OPERATIONS

COMPARISON OF SERVICES

The characterization of services as service factories, service shops, mass service, and professional service can be used as well for comparing service processes in much the same way that Table TS2 compared manufacturing processes. This is accomplished in Table TS3. The various features compared there are placed in to various groups: service, process, customer-orientation, labor, and management. These features undergird the challenges for management that were introduced earlier.

TABLE TS2 Comparing processes of different types

FEATURES	(1) JOB SHOP (Example: Norcen Industries)	(2) BATCH FLOW (Example: Jos. A. Bank)	(3) WORKER-PACED ASSEMBLY LINE (Example: Burger King Restaurant, Noblesville)	(4) MACHINE-PACED ASSEMBLY LINE (Example: C-P-C Group, Oklahoma City)	(5) CONTINUOUS FLOW (Example: International Paper, Androscoggin Mill)	(6) BATCH/CONTINUOUS FLOW HYBRID (Example: Stroh Brewery, Winston-Salem)
Product Features						
Product mix	Generally custom products	Lots of generally own-designed products	Mostly standard products; some opportunities for selected options	Same as for 3	Standard products with little or no customization possible; not produced in discrete units, so has to be measured in tons, barrels, and so on	Same as for 3 and 4
Products compete largely on:	Speed of delivery, product customization, new product introduction	Product performance, product reliability and workmanship, delivery reliability, new product introduction, flexibility to produce either low or high volumes	Product performance, price, product reliability and workmanship, delivery reliability	Product performance, price, product reliability and workmanship, delivery reliability	Price	Product reliability and workmanship, price
Products unlikely to compete on:	Price				New product introduction, product customization	
New product introduction	All the time; easy	Frequent; routine	Sometimes	Sometimes; generally expensive	Hardly at all; very costly	Infrequent; expensive

TABLE TS2 Comparing processes of different types (continued)

FEATURES	(1) JOB SHOP (Example: Norcen Industries)	(2) BATCH FLOW (Example: Jos. A. Bank)	(3) WORKER-PACED ASSEMBLY LINE (Example: Burger King Restaurant, Noblesville)	(4) MACHINE-PACED ASSEMBLY LINE (Example: C-P-C Group, Oklahoma City)	(5) CONTINUOUS FLOW (Example: International Paper, Androscoggin Mill)	(6) BATCH/CONTINUOUS FLOW HYBRID (Example: Stroh Brewery, Winston-Salem)
Process Features						
Process pattern	No rigid pattern; product can be routed anywhere; sometimes a dominant flow	Not all procedures performed on all products; product can be routed many ways; often a dominant flow	Clear pattern, though special treatment of some products sometimes permitted	Clear, rigid pattern, though some off-line work possible	Clear, very rigid pattern	Same as for 5
Linking of process segments	Very loose	Loose, but "cells" can be created	Between tight and loose; tighter if JIT followed	Tight, especially if JIT followed	Very tight	Loose link between different process types
Type of equipment	General purpose	Mostly general purpose	Specialized	Same as for 3	Same as for 3	Same as for 3, 4, and 5
Balance of equipment	Balance of speed and time done in only the grossest, long-run terms. At any one time, an imbalance is likely to exist	Balance likely to be imperfect between segments of process but better coordinated than typical job shop	Machinery speed and size in good balance with peak needs	Speed and size of equipment very well balanced. Capable of being adjusted together over small changes in line speed	Good balance of speed and size. Any excess capacity often placed in latter portion of process to provide insurance against breakdowns or unusual order requests	Same as for 5
Type of layout	Process layout; similar machines grouped together	Process layout; similar tasks and equipment grouped together	Line flow layout; distinct for product produced	Same as for 3	Same as for 3	Two distinct layouts, one for batch and one for the continuous flow
Capital use	Labor intensive; machines frequently idle	Labor intensive, although less machine idleness	Although equipment is specialized, it is fairly cheap; labor still a big item	Capital intensive	Very capital intensive; machines nearly always used	Same as for 5

Typical size of operation	Generally small	Generally medium-sized	Variable	Large	Mammoth	Same as for 4
Economies of scale	None	Few, if any	Same as for 2	Some, perhaps	Yes	Same as for 5
Yields	Often dependent on setup and product complexity	Often dependent on workers; yields usually high	Same as for 2	Often dependent on both workers and equipment function; yields usually high	Often dependent on raw materials; yields can be variable	Can vary depending on type of hybrid
Notion of capacity	Very fuzzy; definable vaguely in dollar terms; useful only in long run	Fuzzy; product mix implies a dollar definition only	Increasingly clear; some physical unit measures possible	Clear; physical unit measures	Same as for 4	Same as for 4 and 5
Additions to capacity	Incremental over full range of possible capacity	Larger increments possible, but over full range of possible capacity	Same as for 2	Changes can be made, but costly to do so because capacity comes in large chunks; otherwise, mild fluctuations possible within relatively narrow limits (such as line rebalance)	Only modest range for incremental change; otherwise, huge chunks of capacity required	Comparatively modest range for incremental change; then huge chunks of capacity required
Speed of process (dollars of output/unit of time or dollars of output/dollars of input/unit of time)	Slow	Reasonably slow	Increasingly fast	Fast	Fast, sometimes astounding	Same as for 5.
Pacing	Worker discretion key	Worker-paced	Worker-paced but set within some bounds by management action (such as lne balance)	Machine-paced but can be a management goal as well (such as line balance)	Determined technologically; built into equipment	Generally determined technologically but some leeway available
Bottlenecks	Movable; frequent	Movable, but often predictable	Occasionally movable, but often predictable	Generally known and stationary	Known, stationary	Same as for 5

TABLE TS2 Comparing Processes of different types (continued)

FEATURES	(1) JOB SHOP (Example: Norcen Industries)	(2) BATCH FLOW (Example: Jos. A. Bank)	(3) WORKER-PACED ASSEMBLY LINE (Example: Burger King Restaurant, Noblesville)	(4) MACHINE-PACED ASSEMBLY LINE (Example: C-P-C Group, Oklahoma City)	(5) CONTINUOUS FLOW (Example: International Paper, Androscoggin Mill)	(6) BATCH/CONTINUOUS FLOW HYBRID (Example: Stroh Brewery, Winston-Salem)
Nature of process change	Incremental	Mostly incremental; some significant radical changes possible	Mostly routine (rebalance); sometimes radical (equipment)	Same as for 3	Most change increment but radical change possible; means big bucks and sweeping conversions	Sometimes radical; sometimes incremental
Place of technological change in process itself	Little impact; unlikely to be revolutionary	Important once in a while; usually incremental	Increasingly important; embodied in equipment	Important; embodied in equipment	Far-reaching surprisingly regular	Same as for 4
Setups	Many; varied expense	Some setups needed, but generally easy to do	Same as for 2	No setup required; line already set up	Few and expensive, if any; process organized to simplify most kinds of setups	Little or no setup required process simplifies need for most kinds of setups
Run lengths	Short	Medium	Some long, some short	Long	Very long	Same as for 5
Materials-Oriented Features						
Materials requirements	Uncertain	Can often be placed statistically within reasonably narrow bounds	Known statistically within fairly close limits	Certain once production plan established	Same as for 4	Some aspects certain, others known statistically
Character of materials	Set bill of materials; uniform quality; can adapt to nonstandard types	Set bill of materials; uniform quality; often many types possible	Set bill of materials; uniform quality; can adapt to nonstandard types	Same as for 3	Sometimes variable bill of materials; variable quality, since often processes are agricultural/mining extracts	Sometimes variable quality, sometimes uniform quality

	1	2	3	4	5	6
Vertical integration	None	Sometimes backward, sometimes forward	Same as for 2	Sometimes backward, often forward	Often backward and forward	Same as for 2 and 3
Inventories: raw materials	Small; most raw materials purchased to coincide with orders	Moderate; some purchased to coincide with orders and some purchased to provide buffer stock	Varies; often steadily purchased since material needs are generally known within reasonably narrow bounds	Varies; often steadily purchased to coincide with production plan	Often large, but can vary and be steadily purchased to coincide with production plan	Varies; some purchased for buffer stock and some purchased to coincide with production plan
Work-in-process	Large	Moderate	Little	Same as for 3	Very little	Inventory placed between batch and continuous flow segments of process; moderate batch WIP, little continuous flow WIP
Finished goods	Low, if any	Varies	Same as for 2;	Can vary; often thrust down distribution channels	Varies; often thrust down distribution channels; safety stocks often required	Same as for 5
Control over suppliers	Low	Moderate	Great	Same as for 3	Same as for 3	Same as for 3, 4, and 5
Control over customers	Little or none	Same as for 1	Same as for 1	Same, as to delivery	Same as for 4	Sames as for 4 and 5
Supplier ties	Informal; spot buys	Some spot buys, some longer term contracts	Contracts increasingly long term	Formal; long term	Formal; generally long term	Same as for 5
Customer ties	Informal; repeat business encouraged, however	Some informal, some formal distribution	Can be informal or formal	Formal distribution channels	Same as for 4	Same as for 4 and 5

TABLE TS2　Comparing Processes of different types (*continued*)

FEATURES	JOB SHOP (Example: Norcen Industries) (1)	BATCH FLOW (Example: Jos. A. Bank) (2)	WORKER-PACED ASSEMBLY LINE (Example: Burger King Restaurant, Noblesville) (3)	MACHINE-PACED ASSEMBLY LINE (Example: C-P-C Group, Oklahoma City) (4)	CONTINUOUS FLOW (Example: International Paper, Androscoggin Mill) (5)	BATCH/CONTINUOUS FLOW HYBRID (Example: Stroh Brewery, Winston-Salem) (6)
Information-Oriented Features						
Order handling and sales	Often bid for; sales are to order	Varies; some to order, and some from stock with lagged adjustments to that stock	Same as for 2	Sales well established ahead of time or from finished goods inventory; lagged adjustments to stock	Sales well established ahead of time or from finished goods inventory	Same as for 5
Degree of information coordination outside factory	Needed only for bids, receipt of any supplied materials, and to initiate supplies	Needed only to monitor sales and to initiate supplies	Needed to monitor sales and to place orders for supplies; sometimes for order scheduling	Elaborate order scheduling, materials tracking; various levels of forecasts; great deal of corporate communication	Elaborate order scheduling, materials tracking, forecasting; great deal of corporate communication	Monitoring sales, placing supply order; little corporate communication on day-to-day operations
Information systems within factory	Elaborate, viewed as central; lots of flow between factory workers and management; lots of paperwork	Less elaborate, but still considerable; less feedback required; considerable paperwork	Little information needed, basically just to communicate order; flow from management to workers. Opposite flow used primarily to signal breakdowns; generally more informal little paperwork	Same as for 3	Little information needed, basically just to communicate product change; flow from management to workers. Opposite flow used primarily to signal breakdowns; generally more informal	Same as for 5

Trigger for production	Order itself; expediting common	Could be order or level of finished goods inventory; some expediting	Level of finished goods inventory, longer term forecasts, or "frozen" orders; little expediting	Level of finished goods inventory, longer term forecasts, or "frozen" orders; no expediting possible	Same as for 4	First portion of process triggered by forecasts; second portion triggered by "frozen" orders
Scheduling	Uncertain, flexible; always subject to change	Flexible, but not as uncertain; less subject to change	Process designed around fixed schedule	Same as for 3	Easy to group similar jobs or orders; fixed schedule often set technologically and well in advance	Similar jobs or orders grouped together; fixed schedule set in advance, at least for second portion of process
Quality control	Informal, by each worker; spot checks; machine capabilities can be determined	Can be more formal, though often is not	Some on-line checks, some postassembly checks; process capability can be monitored	Same as for 3	Essentially done by process control people; designed into process; periodic sampling often done and process capability monitored	Much of quality designed into process; periodic sampling often done and process capability monitored
Response to cyclicality in demand	Overtime, shift work, some subcontract, hire/fire	Overtime; adjustments can be made by building and depleting inventory, shift work, and hire/fire	Overtime work or closing line early; some adjustments can be made by building and depleting inventory, rebalance, and close down shift or plant	Same as for 3	Inventory adjustments; otherwise, shift or plant is shut down	Same as for 5
Worker payment	Hourly; piece rate or other incentive wage	Often piece rate or other incentive wage	Hourly base; sometimes also tied to percentage of standard that crew achieves	Hourly	Same as for 4	Same as for 4 and 5

TABLE TS2 Comparing Processes of different types *(continued)*

FEATURES	(1) JOB SHOP (Example: Norcen Industries)	(2) BATCH FLOW (Example: Jos. A. Bank)	(3) WORKER- PACED ASSEMBLY LINE (Example: Burger King Restaurant, Noblesville)	(4) MACHINE- PACED ASSEMBLY LINE (Example: C-P-C Group, Oklahoma City)	(5) CONTINUOUS FLOW (Example: International Paper, Androscoggin Mill)	(6) BATCH/ CONTINUOUS FLOW HYBRID (Example: Stroh Brewery, Winston-Salem)
End-of-month syndrome (more produced at end of month for billing than at beginning)	Inevitable	Often happens	Close to nonexistent	Same as for 3	Nonexistent	Same as for 5
Labor-Oriented Features						
Advancement for worker	Knowledge of more and more machinery and machine capabilities. With more skills and/or seniority acquired, greater pay and responsibilities given, such as "lead worker." Seniority can lead to change in shift assignment	Knowledge of more and more machinery and machine capabilities. With more skills and/or seniority acquired, greater responsibilities given, such as "lead worker." Seniority can lead to change in shift assignment or department	With more skills and/or seniority acquired, greater responsibilities given, such as "lead worker" or "group leader." Seniority can lead to change in department or shift assignment	Seniority can lead to change in department or shift assignment	Seniority can lead to change in shift or, sometimes, department assignment	Same as for 5
Labor content per $1 of product value	Very high	High	Medium	Low	Very low	Same as for 4
Job content	Large	Medium	Small	Same as for 3	Often small "push button" stuff; but can be "art" as well	Same as for 5

Management Features

Importance and development of standards for labor	Key for scheduling planning growth; big purpose of information system; sometimes key for wage payment	Often key for wage payment; needed also for scheduling product mix	Crucial for process design	Same as for 3	Important largely for assigning workforce to equipment	Same as for 5
Staff-line needs	Small staff (information, quotes, and new product development). Much line supervision needed	Small staff (generally methods related). Line supervision more critical	Large staff for process redesign, methods, forecasting, capacity planning, and scheduling. Line supervision and troubleshooting still critical	Same as for 3	Large staff for technology and capacity planning, and scheduling. Line supervision less crucial; maintenance key	Same as for 5
Degree of corporate influence on operations, if plant within a multiplant company	Modest	Same as for 1	Same as for 1	Great, for both operations and capital expenditures	Same as for 4	Variable; more likely that corporate influence is with capital expenditures than with day-to-day operations
Means of control	Usually a profit center	Can be either a profit center or a cost center	Same as for 2	Usually a cost center	Same as for 4	Same as for 4 and 5
Challenges to management	Scheduling, bidding, information flows, expediting product innovation, shifting bottlenecks	Order processing, labor issues and pay, handling cyclicality	Balance (process design), product design, managing workforce, technological advances, capacity planning, materials management	Same as for 3	Capital needs, maintenance, site, technological change, vertical integration, raw materials sourcing	Same as for 5

TABLE TS3 A service comparison

	SERVICE FACTORY (Example: Burger King Restaurant)	SERVICE SHOP (Example: Ogle-Tucker Buick)	MASS SERVICE (Example: Thalhimers-Cloverleaf)	PROFESSIONAL SERVICE (Example: Arthur Andersen & Co.)
Service Features				
Mix of services	Limited.	Diverse.	Limited.	Diverse.
Products compete largely on:	Price, speed, perceived "warmth" or "excitement."	Wide choice, competence.	Price, choice, perceived "warmth" or "excitement."	Competence, range of expertise.
New or unique services introduced or performed	Infrequent.	Routine.	Limited experimentation.	Routine.
Process Features				
Capital intensity	High.	High.	Low.	Low.
Pattern of process	Rigid.	Adaptable.	Rigid.	Very loose.
Ties to equipment	Integral part of process, little choice applies.	Equipment important to process, but usually several options exist for its use.	Limited ties to equipment, more tied to plant and layout.	No close ties to plant or equipment.
Importance of balance of tasks and any equipment to smooth process functioning	Balance critical.	Balance often not critical.	Balance not critical.	Balance can be critical.
Tolerance for excess capacity	Excess capacity abhorred.	Excess capacity often not a problem.	Excess capacity implies workforce adjustment that is fairly easily made.	Excess capacity abhorred.
Ease of scheduling	Sometimes tough to schedule, peak demand can be difficult.	Scheduling more easily done.	Scheduling easily done.	Sometimes tough to schedule, peak demand can be difficult.
Economies of scale	Some.	Some—permits better equipment use and thus justification.	Few, if any, except those related to any inventories.	Few, if any, although some specialization can occur.
Notion of capacity	Fairly clearcut, sometimes definable in physical terms.	Fuzzy, very dependent on mix of demands. Only definable in dollar terms.	Not as fuzzy as with service shop. Limits are often due to plant, not processing time.	Fuzzy.
Layout	Line flow-like preferred.	Job-shop or fixed position.	Typically fixed position although layout may change frequently, customers move through layout.	Job-shop frequently.

Additions to capacity	Can be in variable increments, requires balance of capital and labor.	Can be in variable increments, aspects of balance more murky.	Often takes big changes to plant to enact. Processing can sometimes be sped up by adding some labor.	Means adding primarily to labor in incremental fashion.
Bottlenecks	Occasionally movable, but often predictable.	Movable, frequent.	Typically well known, predictable.	Can sometimes be forecast, but otherwise are uncertain.
Nature of process change	Sometimes routine (rebalance), sometimes radical (new equipment).	Occasionally radical (new equipment and procedures).	Process change seldom occurs, although it can be radical (such as big change to plant).	Mostly incremental.
Importance of material flow to service provision	Both inventories and flow are important.	Inventories important but not so much the flow.	Inventories are often important and must be controlled.	Incidental to most services.
Customer-Oriented Features				
Importance of attractive physical surroundings to marketing of service	Can be critical.	Often insignificant.	Critical.	Often insignificant.
Interaction of customer with process	Little, brief.	Can be great.	Some.	Typically, very great.
Customization of service	Scant.	Significant.	Scant.	Significant.
Ease of managing demand for peaks and nonpeaks.	Can be done through price.	Some promotion of off-peak times can be done, but often difficult.	Same as for service shop.	Often very difficult to manage demand, may not be responsive to price.
Process quality control	Can be formal, amenable to standard methods (such as control charts).	Can be formal. Checkpoints can easily be established. Training can be critical to quality.	Mainly informal. Training critical to quality.	Mainly informal. Training critical to quality.

	SERVICE FACTORY (Example: Burger King Restaurant)	SERVICE SHOP (Example: Ogle-Tucker Buick)	MASS SERVICE (Example: Thalhimers-Cloverleaf)	PROFESSIONAL SERVICE (Example: Arthur Andersen & Co.)
Labor-Related Features				
Pay	Typically hourly.	Varies, could include individual incentive or commission schemes.	Same as for service shop.	Salary, often with bonus of some type.
Skill levels	Generally lower skills	High skills.	Variable, but most often lower skill.	Very high skills.
Job content	Small.	Large.	Often medium, but variable.	Very large.
Advancement	With more skills and/or seniority acquired, greater responsibility given. Seniority can lead to change in department or shift assignment.	Often, worker is an independent operator of sorts and can exert some control on what he gives and gets from job; limited hierarchical progression.	Often a hierarchy to progress upward through.	Often a pyramid, up or out. Top of pyramid exerts leverage over bottom of pyramid.
Management Features				
Staff-line needs	Large staff for process redesign, methods, forecasting, capacity planning, and scheduling. Line supervision and troubleshooting still critical.	Limited staff, mostly line operation.	Some staff, often focused on personal issues.	Limited staff, many line managers wear multiple hats.
Means of control	Variable. Can be cost or profit	Usually a profit center.	Usually a profit center.	Usually a profit center.

INDEX